Lecture Notes in Mathematics 2039

For further volumes:
http://www.springer.com/series/304

Olaf Post

Spectral Analysis on Graph-Like Spaces

Springer

Olaf Post
Durham University
Department of Mathematical Sciences
Science Laboratories
South Road
Durham DH1 3LE
United Kingdom
olaf.post@durham.ac.uk

ISBN 978-3-642-23839-0 e-ISBN 978-3-642-23840-6
DOI 10.1007/978-3-642-23840-6
Springer Heidelberg Dordrecht London New York

Lecture Notes in Mathematics ISSN print edition: 0075-8434
ISSN electronic edition: 1617-9692

Library of Congress Control Number: 2011940288

Mathematics Subject Classification (2010): 35PXX; 47A10; 35J25; 05C50; 34B45; 47F05; 58J50

Printed on acid-free paper

Springer is part of Springer Science+Business Media (www.springer.com)

meinen Eltern gewidmet

Preface

In this monograph, we analyse thin tubular structures, so-called "graph-like spaces", and their natural limits, when the radius of a graph-like space tends to zero. The limit space is typically a metric graph, i.e. a graph, where each edge is associated a length, and therefore, the space turns into a one-dimensional manifold with singularities at the vertices. On both, the graph-like spaces and the metric graph, we can naturally define Laplace-like differential operators. We are interested in asymptotic properties of such operators. In particular, we show norm resolvent convergence, convergence of the spectra and resonances.

Tubular structures with small radius have attracted a lot of attention in the last years. Tubular structures are frequently used in different areas such as mathematical physics to describe properties of nano-structures, in spectral geometry to provide examples with given spectral properties, or in global analysis to calculate spectral invariants.

Since the underlying spaces in the thin radius limit change, and even become singular in the limit, we develop new tools such as

- Norm convergence of operators acting in different Hilbert spaces.
- An extension of the concept of boundary triples to partial differential operators.
- An abstract definition of resonances via boundary triples.

These tools are formulated in an abstract framework, independent of the original problem of graph-like spaces, and in a way that they may be applied in many other situations, for example when the underlying space is geometrically perturbed.

We briefly outline the content of this work. Chapter 1 is devoted to an exemplary overview of the results proven in this work, their history and a discussion of further research. In Chap. 2, we introduce the necessary concepts for discrete and metric graphs, especially Laplace-type operators. Chapter 3 contains partially new material, in particular, boundary triples associated with a quadratic form with applications to the PDE case. Moreover, we derive an abstract version of complex scaling via boundary triples and introduce resonances, i.e. poles of a meromorphic continuation of the resolvent. Chapter 4 provides new material on (norm-)convergence of operators and quadratic forms in different Hilbert spaces. We extend these concepts to non-self-adjoint operators. Chapter 5 provides perturbation arguments for manifolds under a change of the Riemannian metric. Moreover, we present how a tubular neighbourhood can be reduced to the underlying one-dimensional space using separation of variables.

In Chap. 6, we define abstract graph-like manifolds associated with a star-shaped graph. We allow different scaling behaviours of the vertex neighbourhood and different boundary conditions, leading to different limit operators. Finally, in Chap. 7, we combine the convergence results for star-shaped graphs in order to get convergence results for general metric graphs and graph-like manifolds. We use the language of boundary triples in order to combine the local convergence results for star-graphs to convergence results for general graphs. Combining and extending existing results in the literature, we show norm-resolvent convergence of the corresponding Laplacians, convergence of the spectrum and of resonances.

Berlin *Olaf Post*

Acknowledgements

I am grateful for stimulating meetings and discussions, especially with Jussi Behrndt, Malcolm Brown, Pavel Exner, Martial Hille, Daniel Grieser, Maria Korotyaeva, Vadim Kostrykin, Peter Kuchment, Daniel Lenz, Fernando Lledó, Jörn Müller, Thomas Neukirchner, Norbert Peyerimhoff, Stefan Teufel, Luca Tenuta, Ivan Veselić and many other people not mentioned here. Last but not least, I would like to thank Jochen Brüning for his continuous support. I also thank the SFB 647 for financial support.

Contents

Chapter 1
Introduction

1.1 About This Monograph

The first aim of this monograph is to introduce into the asymptotic analysis of *graph-like spaces* (also called *graph-like manifolds*, *tubular branched manifolds*, *graph neighbourhoods*, *fat graphs*, *network-shaped domains*, *inflated* or *d-dimensional graphs*) in the 0-thickness limit and the convergence of associated operators and related objects. The second aim is to provide necessary tools from functional analysis and operator theory to treat such convergence problems, where operators act in different spaces, and where spaces are coupled according to a graph.

A graph-like space is, roughly speaking, a Riemannian manifold X_ε (or more precisely, a family of manifolds $\{X_\varepsilon\}_\varepsilon$) which converges to a metric graph as $\varepsilon \to 0$, i.e., a topological graph, where each edge is assigned a length and therefore each edge is isometric to an interval (see Fig. 1.1).

The parameter ε describes the lengths scale of the collapsing transversal space of the part shrinking to an edge. We allow different scaling behaviours at the part shrinking to a vertex. A simple example is given by the ε-neighbourhood of a metric graph embedded in $\mathbb{R}^2$. More abstractly, we can think of a graph-like space as a space coupled from building blocks according to a given combinatorial graph.

1.1.1 Convergence of Laplacians on Graph-Like Spaces

The first aim of this monograph is the asymptotic analysis of graph-like spaces in the 0-thickness limit. We provide in particular:

- Norm convergence of *resolvents*, operator functions and *spectra* of the Laplacian
  - With *Neumann* boundary condition on graph-like spaces with boundary ("Neumann case")

O. Post, *Spectral Analysis on Graph-Like Spaces*, Lecture Notes in Mathematics 2039,
DOI 10.1007/978-3-642-23840-6_1, © Springer-Verlag Berlin Heidelberg 2012

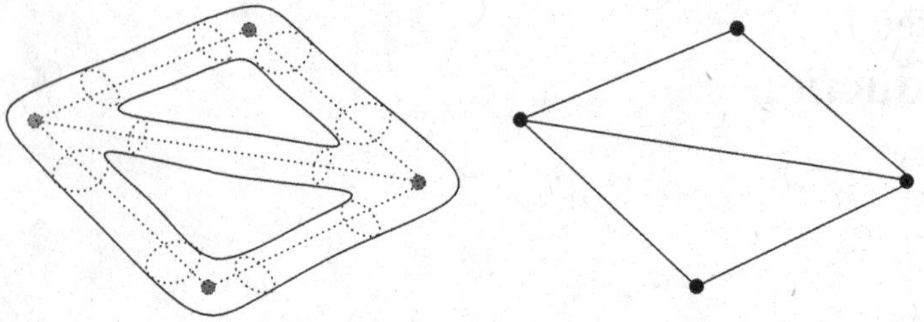

Fig. 1.1 A graph-like space X_ε, here as a boundaryless manifold, the surface of a pipeline network with thin radius of order ε (*left*), together with the underlying metric graph X_0 (*right*)

  - On graph-like spaces without boundary ("boundaryless case") (Theorems 1.3.1 and 1.3.2)
  - With *Dirichlet* boundary conditions on graph-like spaces with boundary ("Dirichlet case") under a certain "smallness condition" on the vertex neighbourhood (Theorems 1.3.1 and 1.3.3)

  to a natural Laplacian on the underlying metric graph. The underlying spaces are allowed to be non-compact under some uniformity assumptions.
- Treatment of different scalings near a vertex for the Neumann and boundaryless case ("fast decaying", "slowly decaying" and "borderline case" determined by the speed of convergence of the ratio "vertex volume" to "transversal volume", see (1.11) and below).
- Convergence of *resonances* on spaces with cylindrical ends shrinking to a half-line (Theorem 1.4.4).
- Treatment of a star-graph and associated graph-like spaces first; the general case follows by decomposing a graph into star graphs (in the middle of each edge) and a coupling argument via boundary maps (star graphs and neighbourhoods in Chap. 6, general graphs in Chap. 7, coupling via boundary maps Sect. 3.9 and Proposition 4.8.3).
- Precise error estimates in terms of geometry and elementary spectral data of the building blocks (like lower bounds on the edge lengths and first eigenvalues of the vertex building block X_v of the unscaled manifold) (for edge neighbourhoods see the theorems in Chap. 5, and for vertex neighbourhoods in Chap. 6).
- Analysis of the limit spaces: *quantum graphs* (i.e., metric graphs together with a natural Laplacian Δ_0 on it); spectral relation between quantum graphs and discrete graphs (Sects. 2.2 and 2.4).
- *Extended quantum graphs* as natural spaces in the limit of the "borderline case" and their spectral analysis (Sect. 2.3).

1.1.2 Tools from Functional Analysis and Operator Theory

The second aim of this monograph is to provide a toolbox of mathematical methods for the convergence of Laplacians on shrinking spaces (see Chaps. 3 and 4). We like to remark, that principally we present the concepts in a very abstract and general way. In order to keep this work at a reasonable size, we only present graph-like spaces as application. Nevertheless, we are sure that the general concepts introduced here have many other applications.

- Norm convergence of non-negative operators and a related notion for quadratic forms acting in different Hilbert spaces (Sects. 4.2 and 4.4), convergence of resolvents, convergence of spectra (discrete, essential) and convergence of eigenfunctions (Sect. 4.3).
- Norm convergence of certain non-self-adjoint operators (typically not of type A nor B in the sense of Kato) acting in different Hilbert spaces: convergence of resolvents, convergence of eigenvalues and resonances (Sects. 4.5–4.9).
- Treatment of boundary value problems for (second order elliptic) partial differential operators via so-called boundary maps and boundary triples starting from a (first order) quadratic form; coupling of such problems (Sects. 3.4–3.6).
- Abstract definition of resonances via boundary triples: typically, resonances and embedded eigenvalues appear when a space is obtained by coupling a compact ("interior") part with a non-compact ("exterior") part which is cylindrical. We provide an abstract framework for the "complex dilation method" Sects. 3.6–3.8).
- A "black box" convergence result for resonances (Sect. 4.9); without concrete knowledge of the interior part; only some estimates on the difference of the operators in the interior parts are needed (Theorem 4.9.12 (1)).

1.1.3 Outline of the Work

Let us briefly outline the content of this work in the order of appearance. Section 1.2 is devoted to a short outline of the history, main results and motivation for the study of graph-like spaces. We provide a more detailed and exemplary exposition of our main results on graph-like spaces and the tools from functional analysis and operator theory in Sects. 1.3 and 1.4. In Sect. 1.5 we present some consequences of the spectral convergence. In Sect. 1.6 we give an outlook on some ideas beyond the scope of this monograph.

In Chap. 2 we introduce the necessary concepts for discrete and metric graphs. Most of the material is standard, and in parts taken from [P09b]. Generalised discrete Laplacians have already been introduced in [P09a, P08]. The concept of extended metric graphs defined via boundary triples is presented in Sect. 2.3.

Chapter 3 contains the material related to a single operator and associated scales of Hilbert spaces and boundary triples. In particular, it contains partially new material: boundary triples associated with a quadratic form with applications to the PDE case. Moreover, we derive an abstract version of complex scaling via boundary triples, and an abstract definition of a resonance.

Chapter 4 provides new material on (norm-)convergence of operators and quadratic forms in different Hilbert spaces. We derive consequences such as spectral convergence and convergence of resonances. The basic ideas were introduced in [P06, EP07]. We extend these concepts to non-self-adjoint operators.

Chapter 5 provides perturbation arguments for manifolds and for tubular neighbourhoods. We present how a tubular neighbourhood can be reduced to the underlying one-dimensional space using separation of variables.

In Chap. 6 we define abstract graph-like manifolds associated with a star-shaped graph. We allow different scaling behaviour of the vertex neighbourhood and different boundary conditions, leading to different limit operators. We use Lipschitz continuous metrics in contrast to the original papers [EP05, KuZ03, KuZ01, RuS01a]. For the quadratic form on a graph-like manifold we find a δ-partial isometric quadratic form on the associated metric graph, which depends on the scaling behaviour of the graph-like manifold. This is a generalisation of results in [EP05, KuZ03, KuZ01, RuS01a], where only the convergence of eigenvalues is shown.

Finally, in Chap. 7, we combine the convergence results for star-shaped graphs provided in Chap. 6 in order to get convergence results for general metric graphs and graph-like manifolds. We use the language of boundary maps and boundary triples in order to combine the local convergence results for star-graphs to convergence results for general graphs. In particular, we show norm-resolvent convergence of the corresponding Laplacians, convergence of the spectrum and of resonances, combining and extending results of [EP07, P05, EP05, KuZ03, KuZ01, RuS01a, Sa00].

1.1.4 Related Topics Not Included in This Work

The aim of this monograph is to show how far the very general and flexible tools of convergence of operators in different Hilbert spaces developed here can be used for the analysis of operators on graph-like spaces. In concrete examples, one can obtain stronger results using different methods such as *matched asymptotic expansions* used for problems involving regions that scale in different ways (see e.g. [GJ09, G08a, JS10]), or *matching of scattering solutions* (see e.g. [MV07, G08a]). For example, Grieser obtained the asymptotics $\lambda_k(\varepsilon) = \lambda_k(0) + \mathrm{O}(\varepsilon)$ instead of $\mathrm{O}(\varepsilon^{1/2})$, which is the convergence speed we obtain by our methods (see Theorem 1.3.2).

To summarise, we have not included the following subjects in this monograph; mainly to keep it at a reasonable size, and since the methods differ from ours:

- The general Dirichlet (or other) boundary condition case.
- Asymptotic expansions of eigenvalues and eigenfunctions.
- Scattering properties.
- Schrödinger operators (i.e., Laplacians with magnetic and electric potential), for such operators see e.g. in [EP07].
- Laplacians on differential forms and other vector bundles.
- Strong or weak convergence of operators acting in different Hilbert spaces.

1.2 History, Results and Motivation

Let us start with a brief history on the convergence of Laplacians on graph-like spaces. We explain further motivations in Sects. 1.2.4–1.2.7, as well as the history of results on the tools from functional analysis and operator theory in Sects. 1.2.8–1.2.9.

Already half a century ago Ruedenberg and Scherr [RSc53] used graph models, elaborating an idea of L. Pauling, to calculate spectra of aromatic carbohydrate molecules. In particular, Ruedenberg and Scherr claimed that an electron in an organic molecule is approximately described by confining it to the bonds of the molecule. In this approximation, the other electrons lead to an effective potential, confining the electron under consideration into a small neighbourhood of the bonds. Therefore, Ruedenberg and Scherr used a one-dimensional graph-model, called in their paper *free-electron network model*, as a simple but powerful tool. They achieved a reasonable accuracy for such a simple model, which nowadays is called *quantum graph model* with Kirchhoff vertex conditions. In the approximation, the electron confinement can be realised by an ε-neighbourhood X_ε of the one-dimensional graph model X_0, and the corresponding Schrödinger operator H_ε is the Laplacian with Dirichlet boundary conditions on ∂X_ε. An heuristic argument was given[1] in order to support the claim that the graph spectrum arises in the thin-neighbourhood limit of a Dirichlet Laplacian acting on the graph neighbourhood.

We provide more examples where graph-like spaces appear in Mathematics and applications in the subsequent sections. Depending on the application, graph-like spaces are also called *tubular neighbourhoods of graphs* [CdV86], *graph-like manifolds [EP05], tubular branched manifolds, fat graphs, graph neighbourhoods, network-shaped domains, inflated* or *d-dimensional graphs*), and they have some common features. First, a family of graph-like spaces $\{X_\varepsilon\}_\varepsilon$ converges to the underlying metric graph in the Gromov-Hausdorff distance, where ε is roughly the transversal length scale of X_ε. Moreover, graph-like spaces are *almost one-dimensional spaces*, i.e. a single length scale (namely the longitudinal one on each edge) dominates all others.

Analytically, a graph-like space can be treated via separation of variables and ODE techniques along the edges, but near vertices, i.e. the singularities of the metric graph, the separation of variables technique breaks down. For a recent survey on thin tubes including graph-like spaces we refer to [G08b].

---

[1] In particular, Ruedenberg and Scherr assumed, that a family of eigenfunctions, such that the associated eigenvalues minus the transversal eigenmode (proportional to ε^{-2}) converge, remains *bounded*. But a simple counterexample shows that there are eigenfunctions concentrating around a vertex neighbourhood, see Sect. 6.11.5. If such eigenfunctions are normalised with respect to an L_2-norm on a shrinking domain, then they cannot be bounded. If one excludes such eigenvalues, then still the limit operator is generically decoupled, and a Kirchhoff operator in the limit, as claimed by Ruedenberg and Scherr, is very unlikely (see the discussion in Sect. 1.2.2).

1.2.1 Convergence of Laplacians on Graph-Like Spaces: The Neumann Case

Colin de Verdière gave the first mathematical treatment of spectral asymptotics of graph-like spaces ("*voisinages tubulaires des graphes*") in [CdV86], where he proved the following:

Theorem 1.2.1. *Let M be a closed manifold of dimension* $\dim M \geq 3$ *and* $\mu \in \mathbb{N}$. *Then there exists a Riemannian metric g_μ on M such that the second (first non-zero) eigenvalue of the Laplacian on (M, g_μ) has multiplicity μ.*

The proof uses the argument that the spectrum of the Neumann Laplacian on a thin tubular neighbourhood of a graph embedded in a manifold converges to the spectrum of the Kirchhoff Laplacian on the graph (see Sect. 1.2.6 for more details). Since this result is used as an intermediate step only and presented in a brief way, the paper seemed to be overlooked in much of the mathematical physics community until recently.

Freidlin and Wentzell considered the problem from a probabilistic point of view in [FW93] (see also [Fre96]). They show in their Theorem 7.3 the pointwise convergence of the resolvent of the Neumann Laplacian on graph-like space to the resolvent of the Kirchhoff Laplacian on the underlying metric graph after a suitable identification. This result is a by-product of the convergence of a diffusion process of a space X_ε on which the process moves fast in certain directions whereas the motion in other directions is slow. The so-called *averaging principle* states that if one identifies the fast directions to a point in a new space X_0, then the original process converges to a process on X_0. The case of a graph-like space in our sense appears as a special example. A similar result, namely the weak convergence of the resolvents for tree graphs, is proven by Saito in [Sa00].

The spectral convergence of the Neumann Laplacian on a thin graph neighbourhood with curved edges towards the Laplacian with Kirchhoff boundary on the metric graph is shown by Rubinstein and Schatzman [RuS01a]. Rubinstein and Schatzman also allow a magnetic field, and have some application in superconductivity in mind (see [RuS01b]).

Independently, Kosugi proved in [Ko00, Ko02] the uniform convergence of the solution of a semilinear elliptic equation with Neumann boundary conditions on a graph-like space to a corresponding solution on the metric graph (with Kirchhoff conditions) after a suitable identification.

Kuchment and Zeng [KuZ01] simplify the arguments of [RuS01a], allowed a variable radius $r_\varepsilon(s) = \varepsilon r(s)$ for the edge neighbourhoods (where s is the coordinate on the edge), and extend the result to a slower length scaling at the vertex neighbourhood (ε^α, $0 < \alpha < 1$) [KuZ03]. The limit operator depends on α, and will lead to different vertex conditions on the corresponding quantum graph. For a 2-dimensional ε-neighbourhood of the graph, there are three cases, the *fast decaying case* $1/2 < \alpha \leq 1$, the *slowly decaying case* $0 < \alpha < 1/2$ and the *borderline case* $\alpha = 1/2$. Heuristically, the cases are distinguished by the ratio of the area of the edge neighbourhood and of the vertex neighbourhood. Roughly speaking, one has a decoupled operator in the slowly decaying case (the vertex neighbourhood acts as an "obstacle" preventing the coupling of different edges at a vertex), and Kirchhoff conditions in the fast decaying case. In the borderline case, a so-called "Wentzell" condition appears on the quantum graph (see [FW93, (3.2)]).

In [EP05] we extend the analysis of [RuS01a, KuZ01, KuZ03] to more general geometries. In particular, we introduced *graph-like manifolds*[2] and stressed the geometric point of view of the analysis.

In all of the above examples (except the result of Saito [Sa00] and [PWZ08]), the underlying spaces are assumed to be compact, and therefore, only the discrete spectrum was considered. The main idea for the spectral convergence in [CdV86, RuS01a, KuZ01, KuZ03, EP05] is to compare the Rayleigh quotients using identification operators for the quadratic form domains (we call them here *first order identification operators*). In [P06] we develop a method how to deal with the convergence of operators acting in two different Hilbert spaces, and show in particular the convergence of the discrete and essential spectrum for non-compact graph-like spaces (in the fast decaying case). Pinchover, Wolansky and Zelig [PWZ08] proved the spectral convergence of graph-like spaces for certain non-compact tree graphs (see Sect. 1.2.5 for more details) in which the spectrum is still discrete, using again first order identification operators and comparison of the Rayleigh quotients.

Let us stress the fact that the geometry of the (unscaled) vertex neighbourhood is not important for the limit of the Neumann Laplacian. Similarly, if X_ε is a neighbourhood of an embedded graph in $\mathbb{R}^2$, then the limit operator does not contain any information on the embedding (e.g. the curvature of the edges or the angles of the edges at the vertex). This robustness of the convergence is maybe the reason why the problem with Neumann boundary conditions was solved earlier than the problem with Dirichlet boundary conditions.

Recently, Joly and Semin considered graph-like spaces in [JS10] converging to a metric graph as models for the propagation of acoustic waves using the techniques of "matched asymptotic expansions" and proposed also an "*improved Kirchhoff condition*" for the asymptotic expansion.

1.2.2 Convergence of Laplacians on Graph-Like Spaces: The Dirichlet Case

The analysis of the Laplacian with Dirichlet boundary conditions on a graph neighbourhood is more complicated. In contrast to the case of Neumann boundary

[2]The notion "graph-like manifold" is sometimes used for a manifold embedded in $\mathbb{R}^n$ which can be written as the graph of a function. In [KZh98], the notion of *G-like manifold* is introduced, where G is a (discrete) graph. A *G-like manifold* is a manifold which is *roughly isometric* with the (discrete) graph G, meaning that the manifold is metrically close to the (metric or discrete) graph. This definition is different from ours.

conditions or the case of manifolds without boundary, the lowest transversal eigenfunction is no longer constant. In order to expect convergence of the spectrum, one either has to rescale the Dirichlet Laplacian, or use an asymptotic expansion of the eigenvalues in dependence of the transversal length scale $\varepsilon > 0$.

Under a certain smallness condition on the vertex neighbourhood excluding the natural ε-neighbourhood, is was shown by the author in [P05], that the Dirichlet Laplacian shifted by the lowest transversal eigenvalue converges to a *decoupled* graph operator, namely the graph Laplacian with Dirichlet vertex conditions. This result was somehow unexpected and "disappointing" from a physical point of view, since a decoupled operator does not carry any information of the network structure.

The asymptotic behaviour of the Dirichlet spectrum for arbitrary vertex neighbourhoods has been solved recently by Molchanov and Vainberg [MV07] and Grieser [G08a]. The problem here is that the asymptotic splitting into eigenvalues of a related graph problem and remaining eigenvalues as $\varepsilon \to 0$ appears only in the second term of the asymptotics, and not in leading order as in the Neumann case. A first step to deal the problem is to turn it into a *scattering* problem. One way to do so is to rescale the problem to a graph neighbourhood with transversal length scale of order 1 and longitudinal length scale of order ε^{-1}. In the limit, such a graph neighbourhood "converges" to the disjoint union of star-graph-like neighbourhoods X_v^∞ with infinite ends. It is now a *scattering problem* which determines the asymptotic behaviour of the eigenvalues. After rescaling, the eigenvalue expansion of the Dirichlet Laplacian on the compact graph-like space X_ε (with transversal length scale ε) is

$$\lambda_k(\varepsilon) = \varepsilon^{-2}\tau_k + \mathrm{O}(\mathrm{e}^{-c/\varepsilon}), \qquad (k = 1, \dots, k_0) \quad \textit{and}$$

$$\lambda_k(\varepsilon) = \varepsilon^{-2}\nu_0 + \mu_{k-k_0} + \mathrm{O}(\varepsilon), \qquad (k = 1, \dots),$$

where k_0 denotes the number of L_2-eigenvalues τ_k of $\bigoplus_v \Delta_{X_v^\infty}$ below or equal to the lowest (unscaled) transversal eigenvalue ν_0, and where μ_k are the eigenvalues associated with a graph Laplacian with vertex condition at the vertex v determined by the scattering matrix $S_v(\nu_0)$ of X_v^∞ at the energy ν_0. It turns out that $S_v(\nu_0)$ is a unitary involution, i.e. $S_v(\nu_0)$ has only eigenvalue 1 and -1. Moreover, the vertex condition is given by

$$\underline{f}(v) = \{f_e(v)\}_{e \in E_v} \in \mathscr{V}_v := \ker(S_v(\nu_0) - 1) \qquad \text{and} \tag{1.1a}$$

$$\underline{f}'(v) = \{f_e'(v)\}_{e \in E_v} \in \mathscr{V}_v^\perp = \ker(S_v(\nu_0) + 1), \tag{1.1b}$$

where E_v denotes the set of edges adjacent with the vertex v, and where $f_e(v)$ and $f_e'(v)$ denote the value of f_e and the inward derivative of f_e at the vertex v. Grieser [G08a] also obtained complete asymptotic expansions for the k-th eigenvalue and the eigenfunctions, uniformly for $k \le C\varepsilon^{-1}$. Other boundary conditions with non-zero transversal eigenmode like Robin boundary conditions can be treated in the same way.

Generically, the vertex condition is decoupled Dirichlet (i.e. $S_v(\nu_0) = -\mathbb{1}$ or $\mathscr{V}_v = 0$), so it is unlikely to find some non-trivial coupling by choosing an arbitrary vertex neighbourhood. The existence of non-trivial vertex conditions (i.e., $\mathscr{V}_v \neq 0$) is related to the existence of *threshold energy resonances* at the energy $\lambda = \nu_0$ of X_v^∞, i.e., eigenfunctions fulfilling the eigenvalue equation, being bounded but not in $\mathsf{L}_2(X_v^\infty)$. Molchanov and Vainberg [MV08, Thm. 13] showed that if there is such an eigenfunction, then the limit operator fulfils Kirchhoff conditions and dim $\mathscr{V}_v = 1$.

Molchanov and Vainberg [MV07] considered the wave equation at energies μ/ε^2 where μ is *above* the threshold ν_0, and gave an asymptotic description of the corresponding wave function and a related function the metric graph. Note that for such μ, the vertex condition ("gluing condition" in [MV07]) on the metric graph depends on μ, and becomes independent only at the threshold $\mu = \nu_0$ (see also Theorem 16 of [G08a] for a quantitative discussion). In [MV08, Thm. 11], Molchanov and Vainberg proved a norm resolvent convergence of the operator $\varepsilon^2 \Delta_{X_\varepsilon}^{\mathrm{D}}$ at energies $\mu = \nu_0 + \mathrm{O}(\varepsilon^2)$. In our language, they proved that

$$\|(\varepsilon^2 \Delta_{X_\varepsilon}^{\mathrm{D}} - \mu)^{-1} - J(\varepsilon^2 \Delta_{(X_0,\mathscr{V})} - \mu)^{-1} J^* \| \leq C\varepsilon,$$

where $(Jf)_e = f_e \otimes \varphi_{\varepsilon,e}$ and $(Jf)_v = 0$ (see Sect. 1.3.4) and where $\Delta_{(X_0,\mathscr{V})}$ is the metric graph Laplacian with vertex condition given by (1.1). Here, $\varphi_{\varepsilon,e}$ is the normalised first Dirichlet eigenfunction on the transversal manifold of the edge e, and $(Jf)_v = 0$ means that the function is set to 0 on the vertex neighbourhood. The Dirichlet case has also been treated by Pavlov [Pav07] using Dirichlet-to-Neumann techniques.

The case of a Dirichlet Laplacian on a graph neighbourhood is closely related to the confinement of a particle to the graph neighbourhood by a strong narrow potential. Again, the lowest transversal eigenvalue is positive and divergent in the confinement limit. In [DeA07, DT06, DT04], Dell'Antonio and Tenuta use the idea, that from a dynamic point of view, the longitudinal scale is much smaller than the transversal scale, and they can find a limit dynamic on the quantum graph for certain geometries. For tubular neighbourhoods (without branching, i.e., singularities), a detailed analysis of a particle confined to a submanifold by a squeezing potential can be found in [TW09]. Note that in this situation, one can use (at least locally) separation of variable techniques. In [SmS06], a product model $G \times \mathbb{R}$ with strong confining potentials has been studied instead of the usual graph neighbourhood (see also [MV06] for a related model with Dirichlet boundary conditions).

1.2.3 Convergence of Resonances

The question of resonances in non-relativistic quantum mechanics is one of the most interesting and challenging mathematical problems, and has been studied by many mathematicians and physicists. A *resonance* is a pole of a meromorphic continuation of the resolvent. For a physicist, a resonance z is an almost stable state with decay rate proportional to $\operatorname{Im} \sqrt{z} > 0$. Resonances of a Laplace-type (self-adjoint) operator occur for example on spaces with cylindrical ends or spaces which look like $\mathbb{R}^d$ at infinity. Such operators have essential spectrum $[\nu_0, \infty)$ and possibly eigenvalues embedded in the essential spectrum. Such an eigenvalue is a resonance without decay (i.e. a stable state), but there might be more resonances in the complex plane. For a nice survey on resonances we refer to [Zw99] (see also [Ba10] and references therein for a more recent article).

To our knowledge, the problem of the convergence of resonances of a graph-like space to the resonances of the corresponding quantum graph was first treated in [EP07]. Typically, the spectrum of a non-compact quantum graph with compact interior part and a finite number of infinite leads attached to the interior part has eigenvalues embedded in the absolutely continuous spectrum $[0, \infty)$. These eigenvalues arise e.g. from compactly supported eigenfunctions on loops in the interior part. In contrast, for an elliptic partial differential operator, the unique continuation principle prohibits compactly supported eigenfunctions. Of course, the quantum graph may also have resonances with non-vanishing imaginary part (e.g. a loop with one lead attached, Sect. 1.4.2 and Fig. 1.4 on page 41). Using the method of *complex scaling* (see below), we show that for each resonance of the quantum graph, there is a resonance of the Laplacian with Neumann boundary conditions on an associated graph-like manifold with cylindrical ends nearby. It is unlikely (but not proven nor disproven), that a resonance on the graph-like manifold arising from an embedded eigenvalue of the graph is itself an embedded eigenvalue of the manifold.

In this monograph, we also extend the class of examples presented in [EP07] by graph-like manifolds with different vertex neighbourhood scaling (the borderline and slowly decaying case) and with Dirichlet boundary conditions. Moreover, we provide a simple way how to calculate the resonances (cf. Theorem 3.8.4). For decoupling limit operators, we obtain the convergence of resonances towards embedded eigenvalues similar to the convergence of resonances of a Helmholtz resonator (see Remark 1.4.6).

The method of *complex scaling* or *complex dilation* developed and applied by many authors (see e.g. [Co69, AC71, BC71, Si72, CT73, Si79, CDKS87, BCD89] or [RS80, Sect. XII.6 and XIII.10]), transforms the operator by a non-unitary operator with the aim to rotate the essential spectrum uncovering a part of the "second sheet" while leaving the poles at place. The advantage of the complex scaling method is, that the poles of the resolvent become discrete eigenvalues of the (non-self-adjoint) transformed operators, Hence, perturbation theory of resonances is reduced to perturbation theory of discrete eigenvalues.

The method developed here is based on an *exterior scaling* introduced by [Si79] (see also [CDKS87]). In particular, the dilation operator acts non-trivially on the cylindrical ends (the so-called "exterior" part) only, and the transformed operator has a domain depending on the scaling parameter. A smooth version leaving the operator domain invariant was developed by Hunziker [Hu86] under the name *distortion analyticity* (see also [HiS89]).

In the following subsections we present different motivations for the study of graph-like spaces and the functional analytic and operator theoretic tools, as well as related results.

1.2.4 Mathematical Physics

Graph-like spaces appear in Mathematical Physics in physical networks, where waves or (quantum mechanical) particles are confined to the vicinity of a one-dimensional branched space.

If the length scale is reduced to some nanometres, the system is too small to be treated using purely classical physics. Such networks are realised for example in semi-conductors, carbon nano-structures or optical fibres, see e.g. the work of Ruedenberg and Scherr mentioned in the beginning of Sect. 1.2. All these models are governed by a Laplace-like Hamiltonian; and spectral properties of this operator give information on the physical system (see e.g. [EŠ89]).

In a semi-conductor with network-like geometry, we consider an electron confined to the network by the geometry. The corresponding Schrödinger operator determines the main physical properties of the material. The spectrum of the Schrödinger operator describes the energy in the one-electron model of solids. In a periodic arrangement, the spectrum consists of *bands*, i.e. a locally finite union of compact intervals. The *gap* or *forbidden zone* between the first and second band is a measurement for the material to be a *conductor*, a *semi-conductor* or an *insulator*, depending on the size of the gap. For example, an electron with an energy in the forbidden zone cannot propagate through the material. Spectral gaps typically occur in periodic media with high contrast (see e.g. [HP03]). Moreover, impurities in the periodic arrangement can lead to additional levels inside the forbidden zone, mathematically described by an eigenvalue in the spectral gap. Such impurity levels are important in the theory of the colour of crystals (see e.g. [ADH89, AADH94, P03a] and references therein).

Similarly, in a *photonic crystal*, i.e. a periodic arrangement of optical fibres, light waves are confined to the fibre due to the difference in the dielectric constant of the fibres and the embedding space. Again, a spectral gap tells us that electromagnetic waves with certain frequencies cannot propagate through the photonic crystal. Photonic crystals are often viewed as optical analogue of semi-conductors. For a survey on the mathematical treatment of photonic crystals we refer to [Ku01] and references therein.

All these systems have in common that the longitudinal length scale along the network is much larger than the transversal length scale. It is therefore tempting to ask the following:

(A) *Does a proper one-dimensional branched space provides a good approximation of the network model with small transversal length scale?*

(B) *Does the one-dimensional model yield information such as conductivity of the original physical system?*

Proper one-dimensional networks, also called *quantum networks* or *quantum graphs*, have a long history in Chemistry and Physics (see e.g., [RSc53, Al83]). As already mentioned in the beginning of Sect. 1.2, Ruedenberg and Scherr [RSc53] believed by their heuristic arguments, that the Dirichlet Laplacian shifted by the lowest transversal eigenvalue converges to the Kirchhoff Laplacian on the graph. Although they obtained a reasonable accuracy for the spectra of aromatic carbohydrate, the conclusion cannot hold, at least not for the ε-neighbourhood of a graph with straight edges and degree ≥ 3 embedded in $\mathbb{R}^2$. Simple examples using variational arguments show, that the Dirichlet Laplacian associated with the ε-neighbourhood has eigenvalues below the lowest transversal eigenvalue (see e.g. [SRW89, ABGM91] or Sect. 6.11.5).

Recently, quantum graph models have successfully be applied to carbon nano-structures, i.e. molecules composed entirely of carbon. For example in graphene, the carbon atoms are situated at the vertices of a hexagonal lattice in the plane, whereas for carbon nano-tubes, the hexagonal lattice is rolled up into a cylinder, i.e. the atoms are situated on a tube. As in the previous example, the electric conductivity is determined by the spectrum of the underlying Schrödinger operator on the one-dimensional network. In particular, the conductivity depends on how the nano-tube is rolled up into a cylinder (see for example [ALM04, BBK07, KL07, KuP07]).

Mathematically, a quantum graph consists of a metric graph, i.e. a graph, where each edge is considered as a one-dimensional space, and a self-adjoint (pseudo-) differential operator acting on the edges. Typically, the operator acts as a system of differential operators on each edge, and the system is coupled via *vertex conditions* on the boundary values of each edge, in order to assure the self-adjointness. For a recent survey on the subject, we refer to [Ku08] in [EKK$^+$08], as well as [GnS06, Ku05, Ku04, KS99] and the references therein. Quantum graphs are also considered as *solvable models* since many properties can be calculated explicitly (see the monograph [AGH$^+$05] and the appendix by Pavel Exner).

Since there are a lot of self-adjoint extensions for symmetric differential operators on metric graphs, we have now the following questions:

(C) *Which vertex conditions are "natural"? Does a given quantum graph occur as a limit of a network, for which the transversal length scale tends to zero?*
(D) *Can we give a recipe how to construct a network converging to the quantum graph?*

The physical motivation of the questions is the following: If we can prove such convergence results, we know that the simpler quantum graph model is a good approximation of the real system. On the other hand, we might find interesting features for certain quantum graphs and ask whether they can be manufactured as nano-structure having the same features. As an example we may think of a *quantum switch* (see e.g. Pavlov [Pav02] and the references therein): It is easy to give an example of a vertex condition depending on a parameter such that a wave is transmitted or reflected, depending on the parameter. Can we find a corresponding model in which for example the parameter has a geometric meaning? We will give partial answers to these questions. In particular, the Kirchhoff Laplacian on the metric graph is natural in the sense of question (C) above, by the results of [RuS01a, KuZ01, EP05, P06] (see Theorem 1.3.2 (1)).

Note that the dependency of the limit operator depends on the scaling behaviour at the vertex neighbourhood (see [KuZ03, EP05]): Roughly speaking, one has a decoupled operator in the slowly decaying case, and Kirchhoff conditions in the fast decaying case. This gives a partial answer to question (D) above. In particular, one can construct a "quantum switch" by changing the geometry near a vertex using the scaling rate α.

Note also, that in [EP07] we extended the convergence results in the fast decaying case to magnetic Schrödinger operators on *non-compact* graph-like manifolds. In particular, using results of [BGP07], we can show that certain magnetic Schrödinger operators on a non-compact graph-like space have asymptotically a fractal spectrum (see also Theorem 1.5.5).

Approximation of Quantum Graph Vertex Conditions by Graph-Like Spaces

Let us make a short digression and give some partial answers to questions (C) and (D) above. The following vertex conditions occur as limits of operators on graph-like spaces:

1. **Kirchhoff (or standard) vertex conditions:** This vertex condition corresponds to the vertex space $\mathscr{V}_v = \mathbb{C}(1,\dots,1)$, i.e. f is continuous at the vertex v and $\sum_{e \in E_v} f_e'(v) = 0$, and occurs as limit of the Neumann Laplacian on a graph-neighbourhood or as limit of a graph-like manifold without boundary (see [RuS01a, KuZ01, EP05, P06], see Theorem 1.3.2 (1)). The scaling at the vertex neighbourhood has to be fast enough. Allowing different radii for the (unscaled) transversal direction, one obtains a weighted version of the Kirchhoff condition (i.e. $\mathscr{V}_v = \mathbb{C}p(v)$ with $p(v) = \{p_e(v)\}$, $p_e(v) > 0$). Similarly, starting with a magnetic Schrödinger operator, one ends up with vertex conditions with *complex* weights (see [RuS01a, KuZ01, EP07]).
2. **Other energy-independent vertex conditions:** A vertex condition is called *energy-independent* if it is of the form $\underline{f}(v) \in \mathscr{V}_v$, $\underline{f}'(v) \in \mathscr{V}_v^\perp$ as above. The name "energy-independent" reflects the fact that the corresponding scattering matrix is independent of the spectral parameter (the "energy"). In principle, any linear space $\mathscr{V}_v \subset \mathbb{C}^{E_v}$ could be obtained by a suitable choice of the rescaled vertex neighbourhood X_v^∞. If in addition, X_v^∞ could be chosen such that there are no L_2-eigenvalues below the threshold ν_0, then the spectrum of the shifted Dirichlet Laplacian $\Delta_{X_\varepsilon}^{\mathrm{D}} - \nu_0/\varepsilon^2$ on the corresponding scaled manifold X_ε converges to a quantum graph with the prescribed vertex conditions $\mathscr{V}_v$.

   Unfortunately, it is unlikely to find manifolds X_v^∞ with non-trivial vertex conditions $\mathscr{V}_v \neq 0$. Generically, one obtains the physically less interesting

decoupled Dirichlet vertex conditions as in [P05]. Note that up to now, there is no formal proof how to construct X_v^∞ from a given space $\mathscr{V}_v$. The very sensitive dependence on the geometry of the vertex neighbourhood is in contrast to case (1) above.

Another case, in which the *decoupled Dirichlet vertex condition* occurs as limit, is the Neumann Laplacian on a graph-like manifold with *slowly decaying* vertex neighbourhoods (see [KuZ03, EP05]). Here, the vertex neighbourhood is large compared to the transversal length scale, and has the effect of an obstacle; an extra state remains from the lowest eigenmode on the vertex neighbourhood (see Theorem 1.3.2 (3)).

3. **A coupled vertex condition:** The Neumann Laplacian on a graph neighbourhood with scaling $\varepsilon^{1/2}$ at the vertex neighbourhood X_v leads to an interesting vertex condition, coupling the values of the function f at a vertex with an extra state. The corresponding eigenvalue equation leads to the vertex condition that f is continuous at v with value $f(v)$ and that

$$\frac{1}{\deg v} \sum_{e \in E_v} f_e'(v) = \lambda \cdot (\operatorname{vol} X_v) f(v). \qquad (1.2)$$

This condition looks like a delta-interaction with energy-depending strength (see [KuZ03, EP05], Sect. 1.3.2 and Theorem 1.3.2 (2)), i.e. the spectral parameter λ enters also in the vertex condition itself, and is sometimes called "Wentzell" condition, see [FW93, (3.2)]. Later on we present a nice general framework for such vertex conditions. Kant, Klauss, Voigt and Weber [KKVW09] showed that the heat operator associated with the Laplacian with Wentzell condition is submarkovian; they interpret the extra states as been implemented via extra *point masses* at the vertices.

4. **Delta- and delta'-interactions:** In [EP09] we introduce a method how to obtain delta- and delta'-interactions at a vertex (and possibly more general energy-dependent vertex conditions) by a Schrödinger operator. Roughly speaking, one has to add a potential at the vertex neighbourhood of strength ε^{-1}, in order to approximate an delta-interaction. A delta'-interaction can be approximated by delta-interactions using ideas of [CE04]. In the same way, one can approximate arbitrary vertex conditions by properly scaled (magnetic) Schrödinger operators using ideas of [CET10]. Another idea of approximating delta-conditions is to use Robin type conditions with a properly scaled factor on the boundary, such that the lowest transversal eigenmode converges to 0 as $\varepsilon \to 0$ (see [MNP10]).

For related results on neighbourhoods of a graph having two infinitely long edges joint at one vertex leading to a non-trivial coupling, we refer to [ACF07, CE07]. We do not focus on such results in this work.

Joly and Semin provided a model for acoustic waves in [JS10]: They consider the wave equation on a star-shaped graph-like space (with infinite edges) and Neumann boundary conditions. Using the method of *matched asymptotic expansions* they provide an asymptotic solution for the wave equation on the graph-like manifold constructed from the wave equation on the underlying metric graph with Kirchhoff conditions. Moreover, they present an asymptotic solution constructed from an "*improved Kirchhoff condition*" on the metric graph; an ε-depending solution where the function and its derivatives on the metric graph are evaluated at distance ε from the vertex and related with the use of a Dirichlet-to-Neumann operator on the vertex neighbourhood. In the limit $\varepsilon \to 0$, the solution with "improved Kirchhoff conditions" converge to the solution of the wave equation on the metric graph with (ordinary) Kirchhoff conditions.

1.2.5 Models from Mathematical Biology

Quantum graphs and corresponding graph-like spaces have also been used as models in Mathematical Biology. One example is to consider tree-like structures with no positive lower bound on the edge length (so-called "*fractal*" metric graphs). Pinchover, Wolansky and Zelig considered in [PWZ08] non-compact tree-like spaces with discrete spectrum with Neumann conditions on the boundary (except at the root and at infinity). Such models can be used as models for a human lung, vascular trees or neural networks (see e.g. [Ni87, Ze05, PWZ08]).

Moreover, Penner et al used in [PKWA10] graph-like spaces with a large number of vertex and edge neighbourhoods in order to encode the structure of a protein. They are interested in the geometry of such spaces and how to encode them for computer access. It is for example allowed that the edge neighbourhoods are "twisted".

1.2.6 Spectral Geometry and Spectral Invariants

1.2.6.1 Spectral Geometry

In Spectral Geometry, one investigates relations of the spectrum of the Laplacian (or related operators) on a Riemannian manifold to its geometry. Graph-like manifolds may serve as toy models in order to show certain properties, or to disprove a conjecture. In particular, such spaces can be used in *spectral engineering*, i.e., in constructing spaces with given spectral properties. Colin de Verdière [CdV86] used graph-like spaces in order to prove Theorem 1.2.1 about the multiplicity of the first non-zero eigenvalue of the Laplacian on a manifold M.

In dimension 2, Cheng [Che76] proved that the multiplicity of the k-th eigenvalue of a connected compact surface M of genus g is bounded by $(2g+k)(2g+k+1)/2$ where $\lambda_1(M) = 0 < \lambda_2(M) \le \dots$. This result follows from a study of the nodal lines, i.e., the set $\varphi_k^{-1}\{0\}$ where φ_k is the k-th eigenfunction.

In contrast, in dimension 3 or higher, there is no restriction on the multiplicity by Theorem 1.2.1. In the proof of this theorem, Colin de Verdière embeds a complete metric graph G with n vertices in (M, g). Such an embedding is possible, since $\dim M \geq 3$. Then he defines a family of metrics $\{g_\varepsilon\}$ on M to be of thin tube type on an ε-neighbourhood G_ε of G and small outside. In a first step, Colin de Verdière shows that the eigenvalues of M are close to the Neumann eigenvalues of G_ε. Since g_ε is a conformal perturbation of g, the dimension assumption $\dim M \geq 3$ also enters in this step (see [P03b, Sect. 4] for a similar argument).

In a second step, it can be seen that the Neumann eigenvalues of G_ε converge to the eigenvalues of the graph Laplacian, using methods discussed below. A topological perturbation argument allows to show that one may change the metric such that the multiplicity is preserved.

A similar construction is used in [CdV87] in order to show the following more general result: Let $\dim M \geq 3$, and let $\lambda_1 = 0 < \lambda_2 \leq \cdots \leq \lambda_n$ be a sequence of n numbers. Then there exists a metric g such that the corresponding Laplacian has $\lambda_1, \ldots, \lambda_n$ as its first n eigenvalues.

1.2.6.2 Convergence of Manifolds and Spaces and Convergence of Spectra

Another interest of research is devoted to the study of spaces of manifolds and convergence of manifolds in such spaces and its implications, like convergence of spectra.

The convergence of eigenvalues of a sequence of compact manifolds M_ε converging to a (possibly degenerated) manifold M_0 has been studied since a long time. Rauch and Taylor [RT75] considered open subsets Ω_n in $\mathbb{R}^m$ converging "metrically" to a subset Ω, and consider the convergence of the associated Dirichlet Laplacians. In particular, they obtain strong convergence of spectral projectors and the solution of the wave equation and applied this convergence to wildly perturbed sets Ω_n, e.g., a domain with many tiny obstacles removed. In particular, they consider the so-called "*crushed ice problem*" where Ω_n is an open set Ω with n balls of radius r_n removed, where the n small balls are evenly spaced in some subregion $\Omega' \subset \Omega$. This situation models the physical problem of the heat flow in Ω_n where the balls are little coolers maintained at temperature zero. The critical parameter is nr_n: if $nr_n \to 0$, then the balls are negligible in the limit, and if $nr_n \to \infty$, then the limit operator is the Dirichlet Laplacian on $\Omega \setminus \Omega'$ (see [RT75, Sect. 4]). The case when nr_n is bounded is delicate and needs more assumptions on the placements of the balls.

Chavel and Feldman [CF78,CF81] showed that one can remove small balls or add small handles of a manifold, i.e., change the topological type, while the convergence of eigenvalues $\lambda_k(M_\varepsilon) \to \lambda_k(M_0)$ still holds. One motivation was the question of Kac [Ka66] "Can one hear the shape of a drum". Chavel and Feldman [CF81] showed that if one adds small handles to a manifold (which changes the Euler characteristic), the eigenvalues still converge as the handles shrink to a point. In particular, an approximate knowledge of the eigenvalues is not enough. One would have to know a priori all the eigenvalues with accuracy uniform in k, but this cannot be true, see Remark 1.5.2.

Fissmer and Hamenstädt [FH05] showed spectral convergence below the essential spectrum, if the manifolds converge in the Gromov-Hausdorff sense (more precisely, in the Lipschitz topology). They use the result in order to construct manifolds with non-trivial essential spectrum and arbitrary high multiplicities for an arbitrarily large number of eigenvalues below the essential spectrum. In dimension 2, the metrics can be chosen to have constant curvature -1.

There are a lot of more results on convergence of spectra of families of manifolds of specific type (like removing small balls, adding handles), for example [Oz81, Oz82, A87, A90, AC93, A94, AC95, Ju01, H06] or more generally, on elliptic theory on varying domains, cf. [Da08]. Moreover there is a vast literature on homogenisation problems (see e.g. [Me08, Zh02, Me03, CLPZ02, Me01, KOZ94, OSY92, SP80] and references therein), i.e., on partial differential equations, where the coefficients have rapidly oscillating coeffients, or where the domain itself is "rapidly oscillating".

1.2.6.3 Spectral Invariants

A spectral invariant of a compact manifold is typically a function on the eigenvalues of its Laplacian. If a spectral invariant is also a topological invariant, one may change the metric on the manifold (and therefore also the eigenvalues) without changing the invariant. Assume e.g. that $X = X^+ \cup X^-$ where $Y = X^+ \cap X^-$ is a hypersurface separating X^+ and X^-. Then we can introduce a long cylindrical part $[-N, N] \times Y$ at Y by changing the metric. Since the topology does not change, the invariant remains the same also in the limit $N \to \infty$, and may be calculated using the building blocks X^+ and X^-. Note that after rescaling of the manifold, the limit $N \to \infty$ corresponds to a thin radius limit. This idea has been used e.g. in [MM06] in order to calculate spectral invariants like the determinant. A similar degeneration of the metrics has been used in [Ju01] in order to show the convergence of eigenfunctions.

Spectral invariants like the determinant usually contain all eigenvalues *simultaneously*. Our spectral convergence gives only control over a *finite* number of the eigenvalues, see the discussion in Remark 1.5.2); the *large* eigenvalues usually have to be controlled by other techniques. For more details and references, we refer to the survey of Grieser [G08b].

1.2.7 Global Analysis

In global analysis, graph-like spaces can be used in order to construct "exotic" spaces, i.e., spaces with unusual properties like heat kernel estimates with untypical exponents.

The heat kernel of a Riemannian manifold X is the smallest positive fundamental solution to the heat equation $-\partial_t u = \Delta_X u$ where $\Delta_X := d^*d \geq 0$ is non-negative. For $X = \mathbb{R}^d$, the heat kernel is given by the classical formula

$$p_t(x, y) = \frac{1}{(4\pi t)^{d/2}} \exp\Big(-\frac{|x-y|^2}{4t}\Big)$$

for $t > 0$ and $x, y \in \mathbb{R}^d$. On a general manifold X, upper and lower estimates on the heat kernel are only known under additional assumptions. A remarkable result was shown by Li and Yau [LY86]. If X is a complete Riemannian manifold with non-negative Ricci curvature, then

$$C_- \frac{1}{\operatorname{vol} B_x(\sqrt{t})} \exp\Big(-\frac{d(x,y)^2}{c_- t}\Big) \leq p_t(x, y) \leq C_+ \frac{1}{\operatorname{vol} B_x(\sqrt{t})} \exp\Big(-\frac{d(x,y)^2}{c_+ t}\Big)$$

for $x, y \in X$ and $t > 0$. Here, $d(x, y)$ denotes the geodesic distance between the points $x, y \in X$, $\operatorname{vol} B_x(r)$ denotes the Riemannian volume of a geodesic ball $B_x(r)$ and $c_\pm, C_\pm$ are positive constants independent of the manifold. If the Riemannian manifold can be decomposed into finitely many building blocks on which upper and lower bounds on the heat kernel are known, one obtains similar bounds on the global heat kernel (see [GSC99]).

The notion of a heat kernel can be extended to metric measure spaces with a local regular Dirichlet form, for example to a fractal space with its Hausdorff measure (for details we refer to [Gr10, Gr03] and references therein). Typically, on a fractal, the heat kernel fulfils upper and lower estimates as below, but now with Gaussians of the form

$$\frac{1}{\mu(B_x(t^{1/\beta}))} \exp\Big(-\Big(\frac{d(x,y)^\beta}{ct}\Big)^{\frac{1}{\beta-1}}\Big)$$

for some $\beta > 2$, e.g., $\beta = \log 5/\log 2$ for the Sierpiński gasket fractal, a compact subset of $\mathbb{R}^2$. Note that $\beta = 2$ corresponds to the classical manifold case discussed above.

Up to recent time it was believed that Gaussian estimates with $\beta > 2$ are typical only for fractals, but unlikely for smooth spaces. Surprisingly, one can construct a so-called *fractal-like*[3] Riemannian manifold X having roughly the Gaussian estimate with $\beta = \log 5/\log 2 > 2$ for *large* times t, and the classical Gaussian estimate $\beta = 2$ for *short* times (see [BBK06] and references therein). The complete Riemannian manifold X is a graph-like manifold, constructed from a finite number of building blocks according to a self-similar graph (see Fig. 1.5 for the Sierpiński graph). From a probabilistic point of view, this behaviour can be understand as

[3]The name "*fractal-like manifold*" is used e.g. in [KZh98, BBK06]. More precisely we should speak of a "graph-like manifold according to a self-similar graph", since the notion "fractal (metric) graph" is already reserved for a non-compact metric graph such that there is no positive lower bound on the edge lengths, see Sects. 1.2.5 and 1.6.7.

follows: $p_t(x, y)$ is the probability density that a particle starting at the point x is at the point y in time t. A particle moving on a fractal-like manifold sees the smooth structure for short times, but for large times, the fractal nature becomes apparent. Fractal-like manifolds were also used in [KZh98] in order to show some untypical behaviour of the wave equation, namely that the average speed of effective energy propagation of the waves satisfying a certain "long wave length condition" is asymptotically zero.

1.2.8 Convergence of Operators Acting in Different Spaces

Many asymptotic problems deal with a converging family of manifolds (or more generally, metric spaces) in a suitable topology. For example, in our main example, we have a family of graph-like spaces X_ε converging to the underlying metric graph X_0 in the so-called *Gromov-Hausdorff topology* (see below). One is then interested in the topology itself or in convergence properties of operator like the Laplacians on these spaces.

In this monograph, we go a step further and develop an asymptotic theory for operators acting in different Hilbert spaces. One motivation was to remove the restriction of compactness of the underlying spaces (like in [RuS01a,KuZ01,EP05]), and to prove norm resolvent convergence of the Laplacians in a suitable sense as in [P06]. In particular, we introduce here the more general notion of "δ-partial isometric equivalence" of the quadratic forms on the graph-like space and the metric graph.

Let us give a brief history of related convergence results.

1.2.8.1 Convergence of Manifolds

We already mentioned in Sect. 1.2.6 the convergence of manifolds and of their Laplacians and spectra in some particular cases. There is also a vast literature on the study of spaces of manifolds and their topology, and the implications of convergence of manifolds.

Let us mention here some works with a functional analytic viewpoint: Rauch and Taylor [RT75] showed strong convergence of spectral projectors and the solution of the wave equation for the Dirichlet Laplacian on convergent subsets in $\mathbb{R}^m$ (after some natural identifications).

Gromov [Gro81] (see also [Gro99]) defined the – now called – *Gromov-Hausdorff distance* of two metric spaces X and Y as the infimum of the numbers $d_Z(f(X), g(Y))$ for all isometric embeddings $f: X \longrightarrow Z$ and $g: Y \longrightarrow Z$ and all metric spaces (Z, d_Z), where $d_Z(A, B)$ denotes the *Hausdorff distance* for subsets A, B in the metric space (Z, d_Z) (see Definition A.1.1). The distance is for example used in order to analyse the topology of certain families of manifolds, and to establish (pre)compactness theorems on classes of compact Riemannian.

In particular, if $\mathsf{Mfd}(n, D, \kappa)$ denotes the set of isometry classes of compact Riemannian manifolds of dimension n with diameter $\operatorname{diam}(M) \leq D$ and Ricci curvature $\geq -(n-1)\kappa^2$, then $\mathsf{Mfd}(n, D, \kappa)$ is precompact (see also [ACh92] for a related result with a lower bound on the injectivity radius instead of the upper bound on the diameter).

Fukaya [Fu87] defined a related notion of convergence, the convergence $M_\alpha \to M$ of manifolds in $\mathsf{Mfd}(n, D, \kappa)$ in the so-called *measured Hausdorff topology*, and showed the convergence of eigenvalues $\lambda_k(M_\alpha) \to \lambda_k(M)$ under a uniform lower bound on the Ricci curvature and an upper bound on the diameter. Kasue and Kumura [KK94, KK96] defined a new distance between two compact Riemannian manifolds, the *spectral distance* (see also [Ka02, Ka06]). A similar notion was introduced by Bérard, Besson and Gallot [BBG94]. The topology induced by the spectral distance is finer than the Gromov-Hausdorff topology on $\mathsf{Mfd}(n, D, \kappa)$. Both results [KK94, BBG94] show the convergence of eigenvalues as $M_\alpha \to M$ in the spectral distance ([KK94] also showed convergence of eigenfunctions and weak convergence of the resolvents).

Most of the above convergence results, e.g. on spectra, need *curvature bounds* and are formulated for *compact manifolds*. Results without compactness and curvature assumptions are given by Colbois and Courtois [CC91]: They provide an equivalent characterisation of the convergence of eigenvalues below the essential spectrum in terms a lower bound on the Dirichlet spectrum around the ends of the manifolds. Moreover, they do not need curvature assumptions. Fissmer and Hamenstädt [FH05] showed spectral convergence for certain non-compact manifolds below the essential spectrum. For a more complete review on related convergence results we refer to [KSh03].

Let us remark that all these results do not directly apply to our situation of a graph-like manifold X_ε converging to a metric graph X_0 for the following reasons:

- The limit space X_0 is not a manifold since it has singularities at the vertices;
- The curvature of the family $\{X_\varepsilon\}_\varepsilon$ is not bounded from below, since at least as the vertex neighbourhoods shrink to points, the curvature tends to $-\infty$;
- The spaces X_ε and X_0 are allowed to be non-compact (with certain uniformity assumptions).

1.2.8.2 Convergence of Hilbert Spaces and Operators

Some of the previous work already contain sections on a convergence theory of operators or quadratic forms acting on different Hilbert spaces. We mention here exemplarily Fukaya [Fu87, Sect. 4], Mosco [Mo94], Kuwae and Shioya [KSh03, Ch. 2] (see also [Ka02, Ka06] for the convergence of Dirichlet forms and references therein) as well as the vast literature on homogenisation problems [SP80, OSY92, KOZ94, Me01, Me03].

Fukaya [Fu87] consider the convergence of eigenvalues: Typically, one has to transplant eigenfunctions from the limit space onto the approximating space and vice versa. The next step is to compare the resulting Rayleigh quotients. This eigenvalue comparison is the core of the spectral convergence for graph-like spaces in [RuS01a, KuZ01, KuZ03, EP05, P06], where the limit space is a metric graph and the approximating space consists of graph-like manifolds.

Mosco, Kuwae and Shioya [Mo94, KSh03] developed a theory of *strong* convergence of resolvents, spectral projections and other related objects of operators A_α and A in Hilbert spaces $\mathscr{H}_\alpha$ and $\mathscr{H}$. This convergence is enough for the spectral convergence *from inside*, i.e., $\sigma(A_\alpha) \nearrow \sigma(A)$ (see Sect. A.1) which means that for each $\lambda \in \sigma(A)$ there is a sequence $\lambda_\alpha \in \sigma(A_\alpha)$ such that $\lambda_\alpha \to \lambda$ (see [KSh03, Prp. 2.5]). If the spectrum is discrete, the convergence of the eigenvalues follows.

Kuwae and Shioya [KSh03] also introduce the notion of a *spectral structure*

$$\Sigma = \bigl(\Delta, \mathfrak{d}, \mathbb{1}_{(\cdot)}(\Delta), \{e^{-t\Delta}\}, \{(\Delta - \zeta)^{-1}\bigr),$$

where Δ is a non-negative operator with associated quadratic form $\mathfrak{d}$. They show that if the manifolds $M_\alpha \to M$ converge in the so-called *compact Lipschitz topology*, then the associated spectral structures converge strongly (meaning that the resolvents of the Laplacians converge strongly in an appropriate sense). This convergence of manifolds usually preserves dimensions, and cannot be applied to our case, since we have a *collapsing* of the graph-like manifolds X_ε as they converge to the metric graph X_0, and *singularities* at the vertices of the limit space X_0.

When dealing with homogenisation problems (see the end of Sect. 1.2.6.2), an abstract notion of strong (or weak) convergence of operators acting in different Hilbert spaces has been defined (see e.g. [ZP07, Pas04] and references therein). The Hilbert spaces there are for example L_2-spaces of parameter-depending measures on a given domain, and weak or strong convergence of resolvents is shown. In other problems of homogenisations, abstract schemes for strong resolvent convergence are established e.g. in [Me03, Me01, KOZ94, OSY92]. In most of these works, the approximating resolvents are assumed to be compact, and as a result of the abstract theory, the convergence of eigenvalues is established.

In this monograph, we use a similar approach and reduce the concrete problem of convergence of operators on graph-like spaces to a problem of operators acting in different Hilbert spaces. In contrast to the previous works, we develop a theory of *norm* resolvent convergence, and we do not pose any compactness conditions on the resolvents, i.e., on the spectrum of the operators. The convergence of the spectra (in particular, discrete and essential) and the norm convergence of the resolvents (and other functions of the operators) are consequences of our general concept. To the best of our knowledge, this approach is new, and surprisingly simple: The estimates needed to ensure the convergence are easily shown in concrete examples, and seem to be very natural. There is no doubt that the concept of strong convergence developed e.g. in [KSh03] can also be applied in our situation, but would lead to less strong results.

More precisely we generalise the concept of norm resolvent convergence and unitary equivalence, where δ measures the deviation of being unitary equivalent and the distance of the two operators. The so-called δ-*quasi-unitary* resp. δ-*partial isometric equivalence* of the corresponding operators (or quadratic forms) reduces to unitary equivalence if $\delta = 0$.

Interestingly, the necessary estimates needed for the proof of the δ-partial isometry of the quadratic forms are very similar to the ones needed for the eigenvalue convergence only (see [EP05, P06]).

1.2.8.3 Convergence of Non-self-adjoint Operators in Different Hilbert Spaces and Convergence of Resonances

We are not aware of a systematic treatment of norm convergence results for *non-self-adjoint* operators acting in different Hilbert spaces. We develop a theory for such operators in order to treat the exterior scaling method described below in Sect. 1.2.3. This treatment is somehow technical since the exterior scaling leads to a holomorphic family of operators $\{H_\varepsilon^\theta\}_\theta$ which is *not* of type A nor B, where H_ε^0 is the original operator, in our case the Laplacian on X_ε. The reason is that functions in the domain of the form resp. operator domain have a jump in the function value resp. also in its derivative (see Sect. 1.4.2). Moreover, in order to keep track of the ε-dependence, we need to define some classes of operators with given *resolvent norm profile*, i.e., $\|(H_\varepsilon^\theta - z)^{-1}\| \le \gamma(z)$ for a given function γ on an open set $U \subset \mathbb{C} \setminus \sigma(H_\varepsilon^\theta)$. This is developed in Sect. 3.3, where we also introduce an adequate substitute for a first order space (on which the sesquilinear form is defined). Another difficulty appears here from the fact that the sesquilinear form $\mathfrak{h}_\varepsilon^\theta$ associated with H_ε^θ has an *asymmetric* domain $\mathscr{H}^{1,\overline{\theta}} \times \mathscr{H}^{1,\theta}$ if θ is complex. We develop a convergence theory for such operators and associated sesquilinear forms (see Sects. 4.5–4.7). Let us mention that Mugnolo, Nittka and the present author developed in [MNP10] a convergence theory for sectorial operators acting in different Hilbert spaces. Note that such operators arise from sesquilinear forms with *symmetric* domain.

The merits for the abstract convergence results for non-self-adjoint operators is a very general "black box" criterion for the convergence of resonances in Sect. 4.9. We only need some abstract information of the interior part (expressed as δ-quasi unitarity of the associated quadratic forms) and some natural conditions on the exterior part.

1.2.9 *Boundary Triples*

1.2.9.1 Boundary Triples Associated with Quadratic Forms and Abstract Complex Scaling

The concept of boundary triples was originally introduced for the study of extensions of symmetric operators in a Hilbert space. The study of extensions goes back to von Neumann [vN30], Friedrichs [Fri34], Krein [Kr47], Vishik [Vi52] and

Birman [Bi53]. For a general treatment of boundary triples we refer exemplarily to [BGP08, DHMS06, DM95, DM91, GG91].

Since boundary triples are central objects in our analysis, we start with a brief introduction. An *(ordinary) boundary triple* $(\Gamma_0, \Gamma_1, \mathscr{G})$ associated with a closed symmetric operator A in a Hilbert space $\mathscr{H}$ is given by a Hilbert space $\mathscr{G}$ and linear operators $\Gamma_0, \Gamma_1 \colon \operatorname{dom} A^* \longrightarrow \mathscr{G}$, such that $(\Gamma_0, \Gamma_1) \colon \operatorname{dom} A^* \longrightarrow \mathscr{G} \oplus \mathscr{G}$ is surjective and that the abstract Green's identity

$$\langle A^* f, g\rangle_{\mathscr{H}} - \langle f, A^* g\rangle_{\mathscr{H}} = \langle \Gamma_0 f, \Gamma_1 g\rangle_{\mathscr{G}} - \langle \Gamma_1 f, \Gamma_0 g\rangle_{\mathscr{G}}. \tag{1.3}$$

holds for all $f, g \in \operatorname{dom} A^*$. A very simple example is given by $\mathscr{H} := \mathsf{L}_2(0,1)$,

$$Af := -f'', \quad \operatorname{dom} A := \{\, f \in \mathsf{H}^2(0,1) \mid f(0) = 0,\, f(1) = 0,\, f'(1) = 0 \,\} \tag{1.4}$$

$$\Gamma_0 f := f(1), \quad \Gamma_1 f := f'(1), \qquad \mathscr{G} := \mathbb{C}. \tag{1.5}$$

Note that we can describe all self-adjoint extensions of A in the form A_L where

$$\operatorname{dom} A_L := \{\, f \in \operatorname{dom} A^* \mid \Gamma_1 f = L\Gamma_0 f \,\},$$

and where $L \in \mathbb{R}$, and $A^{\mathrm{D}} := A^* \restriction_{\ker \Gamma_0}$ (formally, we have $A^{\mathrm{D}} = A_\infty$). Here, $\operatorname{dom} A^* = \{\, f \in \mathsf{H}^2(0,1) \mid f(0) = 0 \,\}$. A *Weyl function* $\Lambda(z)$ can be associated with a boundary triple by mapping φ to $\Lambda(z)\varphi = \Gamma_1 h_{\varphi,z}$, where $h_{\varphi,z}$ is the (unique) solution of the *Dirichlet problem*

$$(A^* - z)h = 0 \qquad \text{and} \qquad \Gamma_0 h = \varphi,$$

and where $S(z)\varphi := h_{\varphi,z}$ defines the so-called *Dirichlet solution operator* or *Krein Γ-field* $S(z) \colon \mathscr{G} \longrightarrow \mathscr{H}$. This works for z not in the Dirichlet spectrum $\sigma(A^{\mathrm{D}})$. In the simple example, we have

$$\sigma(A^{\mathrm{D}}) = \{\, k^2\pi^2 \mid k = 1, 2, \dots \}, \quad h_{\varphi,z}(s) = \frac{\sin(s\sqrt{z})}{\sin\sqrt{z}} \quad \text{and} \quad \Lambda(z) = \sqrt{z}\cot\sqrt{z}.$$

An important tool of the theory of boundary triples is *Krein's resolvent formula*

$$(A_L - z)^{-1} - (A^{\mathrm{D}} - z)^{-1} = S(z)(\Lambda(z) + L)^{-1} S(\overline{z})^*$$

for z not in the spectrum of A_L and A^{D}, comparing the two resolvents of A_L and the Dirichlet operator A^{D}. In particular, we can characterise the spectrum of A_L by

$$\lambda \in \sigma(A_L) \qquad \Longleftrightarrow \qquad \Lambda(\lambda) + L = 0$$

for $\lambda \notin \sigma(A^{\mathrm{D}})$. In the simple example, we have $\lambda \in \sigma(A_L)$ if and only if $\cos\sqrt{\lambda} + L\sqrt{\lambda}\sin\sqrt{z} = 0$. The above example easily extends to higher dimensional boundary spaces $\mathscr{G}$: If we use the point 0 also as boundary and set $\mathscr{G} = \mathbb{C}^2$, $\operatorname{dom} A = \mathring{\mathsf{H}}^2(0,1)$, $\Gamma_0 f = (f(0), f(1))$ and $\Gamma_1 f = (-f'(0), f'(1))$, then the Dirichlet-to-Neumann operator is a (2×2)-matrix, and in the above spectral characterisation, we have $\ker(\Lambda(\lambda) + L) \neq 0$. More complicated examples are provided by quantum graphs. These examples show that boundary triples are a useful tool for the extension theory of closed operators and for Sturm-Liouville problems.

Vishik [Vi52] was the first who applied this concept to elliptic boundary value problem. Several subsequent works treated elliptic boundary value problems in an abstract framework, we only mention here [Gru68, GG91, BMNW08, Pc08, BGW09, Ma10, Gru10b] and references therein. Let us present an example showing the difficulties in applying the concept of boundary triples to elliptic boundary value problems:

Let X be a bounded subset of $\mathbb{R}^d$ with smooth boundary $\partial X \neq \emptyset$. We set

$$\mathscr{H} := \mathsf{L}_2(X); \quad Au := \Delta u, \quad \operatorname{dom} A := \mathring{\mathsf{H}}^2(X)$$

$$\Gamma_0 u := u \restriction_{\partial X}, \quad \Gamma_1 u := \partial_{\mathrm{n}} u \restriction_{\partial X}, \quad \mathscr{G} := \mathsf{L}_2(\partial X)$$

where $\Delta u = -\sum_i \partial_{ii} u$, where $\mathring{\mathsf{H}}^2(X)$ is the completion of $\mathsf{C}_{\mathrm{c}}^{\infty}(X)$ under a Sobolev norm of order 2, where $\partial_{\mathrm{n}} u$ denotes the normal outwards derivative of u near ∂X, and where ∂X carries the natural Riemannian measure of dimension $d-1$.

The above assumptions of a boundary triple cannot be fulfilled with $\operatorname{dom} A^*$ since

$$\operatorname{dom} A^* = \{\, f \in \mathsf{L}_2(X) \mid \Delta f \in \mathsf{L}_2(X) \,\}$$

is *not* subset of the Sobolev space $\mathsf{H}^2(X)$ (and not even of $\mathsf{H}^1(X)$). Moreover, on $\operatorname{dom} A^*$, the boundary maps Γ_0 and Γ_1 do *not* map into $\mathsf{L}_2(\partial X)$ (actually, one can extend the boundary operators such that $\Gamma_0(\operatorname{dom} A^*) = \mathsf{H}^{-1/2}(\partial X)$ and $\Gamma_1(\operatorname{dom} A^*) = \mathsf{H}^{-3/2}(\partial X)$, see e.g. [LM68]). Moreover, the (joint) surjectivity of the boundary maps $(\Gamma_0, \Gamma_1)\colon \operatorname{dom} A^* \longrightarrow \mathscr{G} \oplus \mathscr{G}$, $(\Gamma_0, \Gamma_1) f := \Gamma_0 f \oplus \Gamma_1 f$ is not fulfilled in the above example.

Elliptic boundary value problems in combination with extension theory and Krein's resolvent formula have attracted a lot of attention in recent years, see [BL10, Gru10b, Gru10a, Ma10, BGW09, BHM+09, PcR09, Gru08, BMNW08, Pc08, Ry07, P07, BL07] and references therein. There are successful attempts to use *ordinary* boundary triples also for elliptic PDE operators in [Gru10a, Ma10, BGW09, BHM+09, BMNW08], where usually, a *regularised* version of the normal derivative is used. In particular, (minus) the Weyl-Titchmarsh function is *not* the usual Dirichlet-to-Neumann map. In another attempt, Behrndt and Langer (cf. [BL10, BL07] restrict the operator A^* to a subspace like $\mathsf{H}^2(X)$ or $\mathsf{H}^{3/2}(X)$. Then the boundary operators map into the right space, but the restricted operator is no longer closed.

In this work, we develop a theory of boundary triples $(\Gamma, \Gamma', \mathscr{G})$ associated with a non-negative *quadratic form* $\mathfrak{h} \geq 0$ instead of an *operator* as for ordinary boundary triples above. To our knowledge, this extension of the usual theory of boundary triples is new (apart from the work of Arlinskii [Ar00] mentioned below). We drop the condition of *surjectivity* on the boundary operators Γ_0 and Γ_1 (they are called Γ and Γ' in our approach, where usually, Γu is the restriction of a function to the boundary, and $\Gamma' u$ the restriction of the normal derivative, for a simple example, see Sect. 1.4.1).

In particular, our approach is very useful for our purposes because of the following reasons:

- It has a direct application to elliptic PDE operators on manifolds.
- We can control the norms of several (first) order spaces (this is important since our manifolds later on depend on a parameter ε).
- The Weyl function associated with our boundary triple $(\Gamma, \Gamma', \mathscr{G})$ is the Dirichlet-to-Neumann operator of the corresponding manifold problem.

In the literature, the focus often lies on the extension theory of symmetric operators, this was not our aim here.

We apply our concept of boundary triples in order to define an abstract version of the complex scaling method on a purely functional-analytic level. In particular, we simultaneously define the operator transformed via complex scaling on the manifold and on the metric graph. In this approach, resonances can be calculated in an easy way using the Dirichlet-to-Neumann operator (see the end of Sect. 1.4.2).

We also use boundary triple (more precisely only the boundary map Γ) in order to decompose a graph-like space into its star-graph components at each vertex. To recover the global space, we define boundary triples coupled via graphs.

A related approach using sectorial forms has been given by Arlinskii [Ar00] in order to characterise all maximal sectorial extensions of a given closed densely defined sectorial operator. There, it is assumed that the boundary maps are *surjective* and they are defined on the domain of the *maximal* operator.

We introduce the Dirichlet-to-Neumann operator via a naturally defined quadratic form. Arendt and ter Elst [AtE10] used a similar approach in order to define the Dirichlet-to-Neumann operator on domains in $\mathbb{R}^d$ with "rough" boundary.

There is also some work on *first* order operators like a Dirac operator. For example in [BBC08] (see also [BCh05]), so-called *Dirac systems* are introduced, a similar approach as our half-line boundary triples in Sect. 3.5. It would be interesting to compare this approach with the one in [P07].

1.3 Convergence of Operators and Spectra: A Brief Overview

Let us present exemplary results following from this work in a slightly informal way. Precise definitions and generalisations are given in the subsequent chapters.

1.3.1 Graph-Like Spaces

We can think of the construction of a graph-like space or manifold as a plumber's shop in which a pipeline network is constructed from a collection of building blocks (fulfilling some uniformity assumptions) according to the given metric graph G. Moreover, we introduce a shrinking parameter ε describing roughly the radius or thickness of each pipe in the pipeline network X_ε. In a certain sense, the spaces X_ε converge to the given metric graph G, and the space X_ε is also called a *thick graph*. Here and in the sequel, we will often speak of "the" space X_ε, meaning implicitly the *family* of spaces $\{X_\varepsilon\}_{0<\varepsilon\leq\varepsilon_0}$ for some ε_0 small enough, and similarly for operators on X_ε etc.

We will now describe the construction of a graph-like space with different scaling at the vertex neighbourhoods and different boundary conditions in more details.

1.3.1.1 Setting for Neumann Boundary Conditions

Let G be a metric graph, i.e. a discrete graph with vertices V, and edges E and a length function $\ell: E \longrightarrow (0,\infty)$ associating a length ℓ_e to each edge $e \in E$. In particular, each edge can be identified with an interval $I_e = [0,\ell_e]$. For ease of exposition, we assume that the metric graph G is embedded in $\mathbb{R}^2$ and has straight edges, i.e. V is a discrete set of points in $\mathbb{R}^2$ joined by non-intersecting straight line segments. We denote by $\psi_e: I_e \longrightarrow \mathbb{R}^2$ the arc-length parametrisation of the line segment corresponding to the edge e with tangent vector $\mathrm{t}_e \in \mathbb{R}^2$ and normal vector $\mathrm{n}_e \in \mathbb{R}^2$.

We assume the uniform lower length bound

$$\ell_e \geq \ell_- \qquad \forall\, e \in E, \tag{1.6}$$

for a certain constant $\ell_- \in (0,1]$, say, $\ell_- = 1$. Let e, e' be two edges meeting in a common vertex v. We denote by $\measuredangle_v(e,e')$ the angle between the tangent vectors t_e, $\mathrm{t}_{e'}$ at the vertex v, and we assume that there is a lower bound $\kappa_0 \in (0,\pi/2]$ on this angle, i.e.,

$$\measuredangle_v(e,e') \geq \kappa_0 \qquad \forall\, e,e' \in E_v, \quad \forall\, v \in V. \tag{1.7}$$

Let $0 < \varepsilon \leq 1$ and $0 < \alpha \leq 1$. For each edge $e \in E$ and vertex $v \in V$ we associate the closed sets

$$\begin{aligned} X_{\varepsilon,e} &:= \bigl\{\, \psi_e(s) + \varepsilon y \mathrm{n}_e \in \mathbb{R}^2 \,\big|\, (\varepsilon^\alpha/2)\ell_e \leq s \leq (1-\varepsilon^\alpha/2)\ell_e,\ y \in [-1/2,1/2] \bigr\},\\ X_{\varepsilon,v} &:= \bigcup_{e\in E_v} \bigl\{ v + s\mathrm{t}_e + \varepsilon^\alpha y \mathrm{n}_e \in \mathbb{R}^2 \,\big|\, 0 \leq s \leq (\varepsilon^\alpha/2)\ell_e,\ y \in [-1/2,1/2] \bigr\}, \end{aligned} \tag{1.8}$$

called the *edge neighbourhood* and *vertex neighbourhood*, respectively. We call $Y_e = [-1/2,1/2]$ the *transversal manifold* (for ease of exposition, we assume here

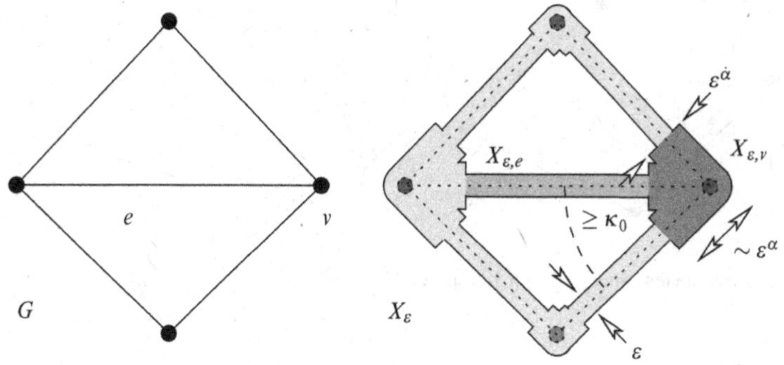

Fig. 1.2 A simple example of a graph-like space with boundary and scaling rate α at the vertex neighbourhood $X_{v,e}$

that the transversal manifolds are the same for all edges). Moreover, E_v denotes the set of edges adjacent to v. Here, $0 < \alpha \leq 1$ denotes the *scaling rate* of the vertex neighbourhood $X_{\varepsilon,v}$, i.e., the vertex neighbourhood scales with ε^α in each direction. In particular, if $\alpha < 1$ then $X_{\varepsilon,v}$ has a larger volume compared with the edge neighbourhood $X_{\varepsilon,e}$ as $\varepsilon \to 0$. The global assumptions (1.6) and (1.7) with $\ell_- = 1$ assure that $X_{\varepsilon,v}$ has a shape as in Fig. 1.2. The total space is then

$$X_\varepsilon := \bigcup_{e \in E} X_{\varepsilon,e} \cup \bigcup_{v \in V} X_{\varepsilon,v}, \tag{1.9}$$

a simple example for a *graph-like space*. For $\alpha = 1$, the closed set X_ε is just the $\varepsilon/2$-neighbourhood of G in $\mathbb{R}^2$ (note that there are other choices for the unscaled vertex neighbourhood $X_{1,v} = X_v$). We consider the Laplacian $\Delta_\varepsilon = \Delta_{X_\varepsilon}$ on X_ε with Neumann boundary conditions on ∂X_ε.

1.3.1.2 Graph-Like Spaces Without Boundary

Assume here that the metric graph G is embedded in $\mathbb{R}^3$ and that $X_\varepsilon \subset \mathbb{R}^3$ consists of the points with distance ε from G for $0 < \varepsilon \leq \varepsilon_0$, i.e. the surface of a pipeline network around G. In order to avoid dealing with non-smooth surfaces, we slightly modify the neighbourhood near a vertex such that the surface is smooth. We decompose X_ε into closed components $X_{\varepsilon,v}$ and $X_{\varepsilon,e}$ such that (1.9) holds and such that any two members of the family $\{X_{\varepsilon,v}\}_{v \in V}$ and $\{X_{\varepsilon,e}\}_{e \in E}$ intersect in subsets of (2-dimensional) measure 0. Moreover, we assume that $X_{\varepsilon,v}$ is *ε-homothetic* with a fixed Riemannian manifold (X_v, g_v), i.e. $X_{\varepsilon,v}$ is isometric to the Riemannian manifold $(X_v, \varepsilon^2 g_v)$; for short $X_{\varepsilon,v} = \varepsilon X_v$. We also write X_v for the Riemannian manifold (X_v, g_v). We do not specify the precise global conditions on X_v here (see e.g. Sect. 7.1), but assume for simplicity, that each Riemannian manifold of the family $\{X_v\}_{v \in V}$ is isometric to one member in a *finite* family $\mathscr{X}_V$ of Riemannian

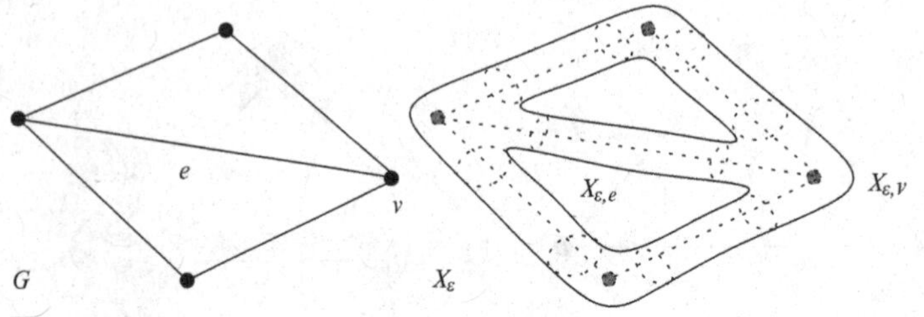

Fig. 1.3 A boundaryless scaled graph-like manifold

manifolds. Roughly speaking, we can think of X_ε as being constructed from a finite number of (ε-scaled) building blocks in $\mathscr{X}_V$ and cylinders $X_{\varepsilon,e}$ of length $\ell_e - \mathrm{O}(\varepsilon)$ and transversal manifold $\varepsilon\mathbb{S}^1$, the 1-dimensional sphere of radius ε (see Fig. 1.3).

The error term $\mathrm{O}(\varepsilon)$ in the longitudinal direction comes from the fact that we need a certain part of the edge for the vertex neighbourhood in order to embed the total space in $\mathbb{R}^3$. In Chaps. 6 and 7 we will use an abstract definition of a scaled graph-like manifold without such a longitudinal error correction. These abstractly defined scaled graph-like manifolds need not to be embedded in an ambient space any more; however they are easier to treat in calculations. The embedded situation is then treated as a perturbation of the abstract spaces (see Chap. 5)

On the graph-like manifold, we consider the Laplace operator $\Delta_\varepsilon = \Delta_{X_\varepsilon}$ on the Riemannian manifold X_ε.

1.3.1.3 Setting for Dirichlet Boundary Conditions

Again, we assume that G is embedded in $\mathbb{R}^2$, and that $X_\varepsilon = \overline{X}_\varepsilon \subset \mathbb{R}^2$ is a graph-like space as in the Neumann case. Here, we allow slightly more general vertex neighbourhoods $X_{\varepsilon,v}$ not necessarily bounded by straight lines. In particular, we assume that $X_{\varepsilon,v}$ is ε-homothetic with a closed subset X_v in $\mathbb{R}^2$. Moreover, we assume that the boundary of $X_v \subset \mathbb{R}^2$ has $\deg v$-many subsets $\partial_e X_v$ isometric to $[0,1]$ ($\deg v$ denotes the *degree* of the vertex v). Denote by $\partial_0 X_v = \overline{\partial X_v \setminus \bigcup_{e \in E_v} \partial_e X_v}$ the *transversal boundary*(i.e. the boundary induced by the boundary of the transversal manifold $[-1/2, 1/2]$), and by $\mathring{\partial} X_v = \bigcup_{e \in E_v} \partial_e X_v$ the *internal* boundary, i.e. the topological boundary of $X_v \subset X_1$ w.r.t. the relative topology. We assume that the vertex neighbourhoods X_v are *small* in the following spectral sense, namely that the *spectral vertex neighbourhood condition*

$$\inf_{v \in V} \lambda_1^{\partial_0 X_v}(X_v) > \lambda_1^{\mathrm{D}}([-1/2, 1/2]) = \pi^2 \tag{1.10}$$

is fulfilled (see Fig. 6.12 on page 355 for an example). Here, $\lambda_1^{\partial_0 X_v}(X_v)$ denotes the first eigenvalue of the Laplacian on X_v with Dirichlet boundary conditions on $\partial_0 X_v$ (and Neumann boundary conditions on the remaining boundary components $\partial_e X_v$).

The condition is a sort of *spectral smallness condition* since roughly speaking, a "small" manifold has large eigenvalues (e.g., if $\hat{X}_v \subset X_v$ is a closed subset having the same internal boundary, i.e. $\mathring{\partial}\hat{X}_v = \mathring{\partial}X_v$, then $\lambda_1^{\partial_0 X_v}(\hat{X}_v) \geq \lambda_1^{\partial_0 X_v}(X_v)$). In particular, the ε-neighbourhood as constructed in the Neumann case (with $\alpha = 1$) does *not* fulfil the condition (1.10) (see Sect. 6.11.5).

Since the lowest transversal eigenvalue is no longer 0 (as it is in the case of Neumann boundary conditions or the boundaryless case), we have to rescale the operator in order to expect a convergence result. Note that the lowest Dirichlet eigenvalue of the scaled transversal manifold $\varepsilon Y_e \cong [0,\varepsilon]$ is $\lambda_1^{\mathrm{D}}(\varepsilon Y_e) = \varepsilon^{-2}\lambda_1^{\mathrm{D}}([0,1]) = \pi^2/\varepsilon^2$ and diverges as $\varepsilon \to 0$. We therefore set

$$\Delta_\varepsilon := \Delta_{X_\varepsilon}^{\mathrm{D}} - \frac{\pi^2}{\varepsilon^2}$$

as operator on the closed set X_ε, where $\Delta_{X_\varepsilon}^{\mathrm{D}}$ denotes the Laplacian with Dirichlet boundary conditions on ∂X_ε. Note that the spectral vertex neighbourhood condition (1.10) implies that $\Delta_\varepsilon \geq 0$ (see Proposition 6.11.3).

1.3.2 The Limit Hilbert Spaces Associated with the Graph Models

1.3.2.1 Neumann Boundary Conditions and Boundaryless Case

Let us now describe the spaces and operators expected in the limit $\varepsilon \to 0$. The limit behaviour is determined by the ratio

$$\mathrm{vol}(\varepsilon, v) := \frac{\mathrm{vol}_2 X_{\varepsilon,v}}{\mathrm{vol}_1 \mathring{\partial} X_{\varepsilon,v}} = \varepsilon^{2\alpha-1}\frac{\mathrm{vol}_2 X_v}{\mathrm{vol}_1 \mathring{\partial} X_v}, \tag{1.11}$$

where $\mathring{\partial}X_{\varepsilon,v}$ is the internal boundary of $X_{\varepsilon,v}$, i.e. the topological boundary of $X_{\varepsilon,v}$ as subset of the topological space X_ε (and not of the ambient space $\mathbb{R}^2$ or $\mathbb{R}^3$), and similarly for $\mathring{\partial}X_v = \mathring{\partial}X_{1,v}$. In particular, we distinguish three cases, depending on the value of the limit $\mathrm{vol}(0,v) := \lim_{\varepsilon\to 0}\mathrm{vol}(\varepsilon, v)$, namely

- **Fast decaying vertex volume:** $\frac{1}{2} < \alpha \leq 1$ $(\mathrm{vol}(0,v) = 0)$,
- **The borderline case:** $\alpha = \frac{1}{2}$ $(\mathrm{vol}(0,v) \in (0,\infty))$,
- **Slowly decaying vertex volume:** $0 < \alpha < \frac{1}{2}$ $(\mathrm{vol}(0,v) = \infty)$.

For simplicity, we assume that the scaling rate α does not depend on the vertex. Roughly speaking, in the fast decaying case, the vertex neighbourhood is not seen in the limit, and the limit operator on the graph is the usual standard Laplacian; whereas in the slowly decaying case, the vertex neighbourhood serves as huge obstacle and leads to a decoupled limit operator on the extended graph (see below). In the borderline case, we obtain a coupled operator on the extended graph.

If X_ε is a manifold without boundary, we can also introduce other scaling rates than $\alpha = 1$. However, for ease of exposition we restrict ourselves here to the simple situation $\alpha = 1$..

1.3.2.2 Fast Decaying Vertex Volume

If $1/2 < \alpha \leq 1$ or if we are in the boundaryless case, the limit operator Δ_0 is the *standard Laplacian* or *Kirchhoff Laplacian* acting in the Hilbert space

$$\mathscr{H}_0 = \mathsf{L}_2(G) = \bigoplus_{e \in E} \mathsf{L}_2(I_e).$$

That is, the operator Δ_0 is defined by $(\Delta_0 f)_e = -f_e''$ with

$$f \in \operatorname{dom} \Delta_0 = \Big\{ f \in \mathsf{H}^2_{\max}(G) := \bigoplus_{e \in E} \mathsf{H}^2(I_e) \,\Big|\, f \text{ continuous}, \sum_{e \in E_v} \overset{\curvearrowright}{\underline{f}}{}'_e(v) = 0 \Big\},$$

where the last condition is valid for all vertices $v \in V$. Here, the continuity of f is a condition on the function at the vertex only. Namely, that $f(v) := \underline{f}_e(v)$ is *independent* of $e \in E_v$, where for a function $f \in \mathsf{H}^1_{\max}(G)$, we denote by

$$\underline{f}_e(v) := \begin{cases} f_e(0), & \text{if } v = \partial_- e, \\ f_e(\ell_e), & \text{if } v = \partial_+ e, \end{cases} \qquad \text{and} \qquad \overset{\curvearrowright}{\underline{f}}_e(v) := \begin{cases} -f_e(0), & \text{if } v = \partial_- e, \\ f_e(\ell_e), & \text{if } v = \partial_+ e, \end{cases}$$

the *unoriented evaluation* and *oriented evaluation*, respectively. It can be seen that under the assumption (1.6), where it was assumed that the length is uniformly bounded from below, the operator Δ_0 is self-adjoint and non-negative (see Proposition 2.2.10).

1.3.2.3 The Borderline Case

If $\alpha = 1/2$, then the limit operator Δ_0 acts in the *extended* Hilbert space

$$\mathscr{H}_0 = \mathsf{L}_2(G) \oplus \ell_2(V) = \bigoplus_{e \in E} \mathsf{L}_2(I_e) \oplus \ell_2(V).$$

The space $\ell_2(V) = \ell_2(V, \deg)$ consists of elements $F = \{F(v)\}_{v \in V} \in \ell_2(V)$ with finite weighted ℓ_2-norm

$$\|F\|^2_{\ell_2(V)} := \sum_{v \in V} |F(v)|^2 \deg v.$$

The domain of Δ_0 couples the metric and the discrete components, namely,

$$\operatorname{dom} \Delta_0 = \{\, \hat{f} = (f, F) \in \mathsf{H}^2_{\max}(G) \oplus \ell_2(V) \mid f \text{ continuous}, \quad f(v) = L(v) F(v) \,\},$$

where the last condition is valid for all vertices and where $L(v) := (\operatorname{vol}_2 X_v)^{-1/2}$ denotes the coupling constant. The operator Δ_0 acts as

$$(\Delta_0 \hat{f})_e = -f_e'' \qquad \text{and} \qquad (\Delta_0 \hat{f})(v) = \frac{L(v)}{\deg v} \sum_{e \in E_v} \underset{\sim}{f}_e'(v) \tag{1.12}$$

on the components $\bigoplus_e \mathsf{L}_2(I_e)$ and $\ell_2(V)$, respectively. This vertex coupling is sometimes also called *Wentzell condition.* We refer to Sect. 2.3 for a systematic treatment of extended spaces and operators generalising this Wentzell condition.

1.3.2.4 Slowly Decaying Vertex Volume

If $0 < \alpha < 1/2$, then the limit operator Δ_0 acts again in the *extended* Hilbert space $\mathscr{H}_0 = \mathsf{L}_2(G) \oplus \ell_2(V)$. The operator Δ_0 is the decoupled Dirichlet operator, namely,

$$\Delta_0 = \bigoplus_{e \in E} \Delta^{\mathrm{D}}_{I_e} \oplus 0$$

with respect to the above decomposition of $\mathscr{H}_0$. Here, $\Delta^{\mathrm{D}}_{I_e}$ is the Laplacian on the interval $I_e = [0, \ell_e]$ with Dirichlet boundary conditions at 0 and ℓ_e.

Note that the slowly decaying case corresponds to $L(v) = 0$. The fast decaying case can be obtained formally as a limit $L(v) \to \infty$ in the sense that $f(v) = L(v)F(v)$ enforces $F(v) = 0$ in the limit $L(v) \to \infty$, i.e. the extra space $\ell_2(V)$ is no longer present. In particular, the $\ell_2(V)$-component of $\Delta_0 \hat{f}$ in (1.12) has to vanish, and we obtain the so-called Kirchhoff condition on the vertex derivatives as in the fast decaying case.

1.3.2.5 The Decoupled Dirichlet Case

If we have Dirichlet boundary conditions, the limit operator for $\Delta_\varepsilon = \Delta^{\mathrm{D}}_{X_\varepsilon} - \pi^2 \varepsilon^{-2}$ is the decoupled operator Δ_0 with Dirichlet boundary conditions at the vertices, i.e.

$$\Delta_0 = \bigoplus_{e \in E} \Delta^{\mathrm{D}}_{I_e}$$

acting in the usual space $\mathscr{H}_0 = \mathsf{L}_2(G)$. In particular, it is not necessary to introduce an extra space as in the slowly decaying case.

1.3.3 Convergence Results for Operators in Different Hilbert Spaces

In all of the above examples, the operators on the graph-like spaces and the operators on the graph act in different Hilbert spaces $\mathscr{H}_\varepsilon = \mathsf{L}_2(X_\varepsilon)$ and $\mathscr{H}_0$. We therefore have to develop a concept of norm convergence of (non-negative) operators Δ_0, Δ_ε in different Hilbert spaces. Basically, we introduce identification operators $J_\varepsilon \colon \mathscr{H}_0 \longrightarrow \mathscr{H}_\varepsilon$ and $J'_\varepsilon \colon \mathscr{H}_\varepsilon \longrightarrow \mathscr{H}_0$ which are not unitary, but unitary up to an error: Let $\delta \geq 0$ (typically, $\delta = \delta_\varepsilon$, but this dependence is not of importance for the δ-quasi-unitarity). We say that J_ε and J'_ε are *δ-quasi-unitary* if the following four inequalities hold:

$$\|J_\varepsilon\| \leq 2, \qquad \|J_\varepsilon^* - J'_\varepsilon\| \leq \delta, \tag{1.13a}$$

$$\|(\mathrm{id}_{\mathscr{H}_0} - J'_\varepsilon J_\varepsilon)(\Delta_0 + 1)^{-1}\| \leq \delta \quad \text{and} \quad \|(\mathrm{id}_{\mathscr{H}_\varepsilon} - J_\varepsilon J'_\varepsilon)(\Delta_\varepsilon + 1)^{-1}\| \leq \delta. \tag{1.13b}$$

Note that if $\delta = 0$, then J_ε is unitary.

We say that the operators Δ_0 and Δ_ε are *δ-quasi-unitarily equivalent* if there is a δ-quasi unitary operator J_ε such that

$$\|J_\varepsilon(\Delta_0 + 1)^{-1} - (\Delta_\varepsilon + 1)^{-1} J_\varepsilon\| \leq \delta. \tag{1.14}$$

Note that if $\delta = 0$, then the operators Δ_0 and Δ_ε are unitarily equivalent. We say that Δ_ε *converges* to Δ_0 *in the generalised sense* ($\Delta_\varepsilon \to \Delta_0$), if there is a family $\{\delta_\varepsilon\}_{\varepsilon>0}$ such that $\lim_{\varepsilon\to 0} \delta_\varepsilon = 0$ and Δ_ε, Δ_0 are δ_ε-quasi-unitarily equivalent. The function $\varepsilon \mapsto \delta_\varepsilon$ is called the *convergence speed* or *error*.

In Sect. 4.1 we also introduce the slightly stricter concept of δ-partial isometry (namely, we assume $J'_\varepsilon = J_\varepsilon^*$ and $J_\varepsilon^* J_\varepsilon = \mathrm{id}_{\mathscr{H}_0}$). In addition, we introduce identification operators on the first order spaces (i.e. on the quadratic form domains) and define δ-partial isometric equivalence for the corresponding quadratic forms. For more details and precise definitions, we refer to Chap. 4.

Various convergence results follow from results in Chap. 4, in particular in Theorems 4.2.10, 4.2.14, 4.2.16, and 4.3.3–4.3.5.

Theorem 1.3.1. *Assume that $\Delta_\varepsilon \to \Delta_0$ (in the generalised sense) with convergence speed δ_ε, where Δ_0 and Δ_ε are non-negative operators in $\mathscr{H}_\varepsilon$ and $\mathscr{H}_0$, respectively. Then the following assertions are true:*

1. *The discrete spectrum of Δ_ε converges to the discrete spectrum of Δ_0 respecting multiplicity.*
2. *If λ_0 is a discrete eigenvalue of multiplicity 1 with corresponding normalised eigenvector φ_0, then there exists a family of normalised eigenvectors $\{\varphi_\varepsilon\}_\varepsilon$ such that $\|J_\varepsilon\varphi_0 - \varphi_\varepsilon\| \leq C_1\delta_\varepsilon$, where C_1 only depends on λ_0 and on the distance between λ_0 and $\sigma(\Delta_0) \setminus \{\lambda_0\}$.*

3. *The spectrum of Δ_ε converges uniformly on any compact interval to the spectrum of Δ_0. The same statement is true for the essential spectrum.*
4. *Let ψ be a bounded, measurable function on $\mathbb{R}_+$ such that ψ is continuous in a neighbourhood of $\sigma(\Delta_0)$ and such that $\lim_{\lambda\to\infty}\psi(\lambda)$ exists. Then there exists a constant $C_\psi > 0$ such that*

$$\|\psi(\Delta_\varepsilon)J_\varepsilon - J_\varepsilon\psi(\Delta_0)\| \le C_\psi\delta_\varepsilon \qquad \text{and} \qquad \|\psi(\Delta_\varepsilon) - J_\varepsilon\psi(\Delta_0)J'_\varepsilon\| \le C_\psi\delta_\varepsilon.$$

   In particular, we can choose $\psi = \mathbb{1}_I$ where $I \subset \mathbb{R}_+$ is an interval such that $\partial I \cap \sigma(\Delta_0) = \emptyset$. Moreover, we can choose $\psi(\lambda) = \mathrm{e}^{-t\lambda}$ and obtain the corresponding convergence of the heat operators $\mathrm{e}^{-t\Delta_0}$ *and* $\mathrm{e}^{-t\Delta_\varepsilon}$.
5. *Let $\lambda_0 \in \mathbb{R}_+ \setminus \sigma(\Delta_0)$. Then there exist constants $C_0, C_1 > 0$ depending only on λ_0 such that*

$$\|(\mathrm{e}^{\mathrm{i}t\Delta_\varepsilon}J_\varepsilon - J_\varepsilon\mathrm{e}^{\mathrm{i}t\Delta_0})P_0\| \le (C_1 t + C_0)\delta_\varepsilon \quad \text{and}$$
$$\|(\mathrm{e}^{\mathrm{i}t\Delta_\varepsilon} - J_\varepsilon\mathrm{e}^{\mathrm{i}t\Delta_0}J'_\varepsilon)P_\varepsilon\| \le (C_1 t + C_0)\delta_\varepsilon,$$

   where $P_0 = \mathbb{1}_{[0,\lambda_0]}(\Delta_0)$ and $P_\varepsilon = \mathbb{1}_{[0,\lambda_0]}(\Delta_\varepsilon)$. In particular, if $t = \mathrm{O}(\delta_\varepsilon^{-\gamma})$ for some $0 < \gamma < 1$, then the above operator norms are of order $\delta_\varepsilon^{1-\gamma}$.

Let us stress that the above convergence results are formulated in a way that the error terms can be expressed completely in terms of $\delta = \delta_\varepsilon$, of the operator function ψ and of the neighbourhood U, on which ψ is continuous, e.g. $\psi = \mathbb{1}_I$ and U open such that $\partial I \cap U = \emptyset$. No other information on the operators or the spaces enter into the error, even if the Hilbert spaces, their norms or the operators depend on additional parameters.

The results in Chap. 4 can be applied to many other situations, in which the Hilbert spaces vary, especially when the Hilbert space is an L_2-space with perturbed underlying manifold. For example we may remove discs, add handles etc. Such results are not included in this work and will be treated elsewhere.

1.3.4 Convergence Results for Graph-Like Spaces

The above concept of generalised convergence applies to the graph-like spaces and their limits, once the uniformity assumptions (1.6) and (1.7) are fulfilled. Roughly speaking, the identification operator J_ε maps functions on the metric graph to the graph-like space by associating to a function f_e on each edge the product $f_e \otimes \varphi_{\varepsilon,e}$, where $\varphi_{\varepsilon,e}$ is the normalised lowest eigenfunction on the transversal manifold $Y_{\varepsilon,e} = \varepsilon Y_e$. In our example we have $Y_e = [-1/2, 1/2]$ or $Y_e = \mathbb{S}^1$, independently of the edge. In the case of Neumann boundary conditions or in the boundaryless case, $\varphi_{\varepsilon,e}$ is constant; in the Dirichlet case, $\varphi_{\varepsilon,e}$ is the first Dirichlet eigenfunction.

Let us describe the identification operators for Neumann boundary conditions (or boundaryless transversal manifolds) in the fast decaying case (say, $\alpha = 1$). In this case we set

$$(J_\varepsilon f)_e(s, y) := \varepsilon^{-1/2} f(\tau_\varepsilon(s)) \qquad \text{and} \qquad (J_\varepsilon f)_v := 0$$

with respect to the decomposition

$$\mathsf{L}_2(X_\varepsilon) = \bigoplus_{e \in E} \mathsf{L}_2(X_{\varepsilon,e}) \oplus \bigoplus_{v \in V} \mathsf{L}_2(X_{\varepsilon,v}).$$

Here, $\tau_\varepsilon \colon [\varepsilon\ell_e/2, (1-\varepsilon/2)\ell_e] \longrightarrow [0, \ell_e]$ is the affine linear (orientation preserving) bijection correcting the longitudinal error of the embedding. Moreover, (s, y) are the coordinates of the edge neighbourhood $X_{\varepsilon,e}$ (see (1.8)). For the identification operator in the opposite direction, we set

$$(J'_\varepsilon u)_e(s) := \varepsilon^{1/2} \int_{-1/2}^{1/2} u(\tau_\varepsilon^{-1}(s), y) \mathrm{d}y.$$

In order to verify (1.13) and (1.14) we use Sobolev trace estimates and Poincaré-type inequalities like

$$\|u - f_v u\|^2_{\mathsf{L}_2(X_{\varepsilon,v})} \le \frac{\varepsilon^2}{\lambda_2(X_v)} \|\nabla u\|^2_{\mathsf{L}_2(X_{\varepsilon,e})},$$

where $f_v u = (\operatorname{vol} X_v)^{-1} \int_{X_v}$ is the normalised average of u, and $\lambda_2(X_v)$ is the second (first non-vanishing) Neumann eigenvalue of the (unscaled) set X_v. For example, for the proof of the $\mathrm{O}(\varepsilon^{1/2})$-quasi-unitary equivalence we will show the estimate

$$\begin{aligned} \|u - J_\varepsilon J'_\varepsilon u\|^2_{\mathsf{L}_2(X_\varepsilon)} &= \sum_{v \in V} \|u\|^2_{\mathsf{L}_2(X_{\varepsilon,v})} + \sum_{e \in E} \|u - P_0 u\|^2_{\mathsf{L}_2(X_{\varepsilon,e})} \\ &\le \mathrm{O}(\varepsilon)\big(\|u\|^2_{\mathsf{L}_2(X_\varepsilon)} + \|\nabla u\|^2_{\mathsf{L}_2(X_\varepsilon)}\big), \end{aligned} \tag{1.15}$$

where $(P_0 u)(s) = \int_{-1/2}^{1/2} u(s, y) \mathrm{d}y$ is the projection onto the first transversal (constant) eigenfunction. Here, the error $\mathrm{O}(\varepsilon)$ depends only on the constants ℓ_- and κ_0 of (1.6) and (1.7), or more generally, of the class of building blocks $\mathscr{X}_V$ out of which the (unscaled) vertex neighbourhoods are chosen. In particular, we have $\|(\mathrm{id}_{\mathsf{L}_2(X_\varepsilon)} - J_\varepsilon J'_\varepsilon)(\Delta_\varepsilon + 1)^{-1/2}\| \le \mathrm{O}(\varepsilon^{1/2})$ since $\Delta_\varepsilon = \nabla^* \nabla$. Roughly speaking, estimate (1.15) means that a spectrally bounded function u (i.e. a function in the range of the spectral projection $\mathbb{1}_{[0,\lambda_0]}(\Delta_\varepsilon)$ for some $\lambda_0 > 0$) cannot concentrate on $X_{\varepsilon,v}$ and its transversally non-constant parts $u - P_0 u$ are small as well.

We will prove these facts in detail in Chap. 6 for the various graph-like spaces associated with star graphs, and in Chap. 7 for general graphs. Since we define an abstract graph-like space in Chap. 6 without the longitudinal length correction (i.e. the edge neighbourhood is defined as $X_\varepsilon = I_e \times \varepsilon Y_e$), we can even show the $\mathrm{O}(\varepsilon^{1/2})$-partial isometric equivalence for the corresponding quadratic forms.

We end the section by stating the main result for the Laplacian on a graph-like space with Neumann boundary conditions (or the boundaryless case):

Theorem 1.3.2. *Similar as above, let Δ_ε and Δ_0 denote the operators on the graph-like spaces and on the corresponding limit, respectively. We then have $\Delta_\varepsilon \to \Delta_0$ in the generalised sense with the following convergence speed δ_ε:*

1. *In the fast decaying case $1/2 < \alpha \leq 1$, we have $\delta_\varepsilon = \mathrm{O}(\varepsilon^{\alpha-1/2})$.*
2. *In the borderline case $\alpha = 1/2$, we have $\delta_\varepsilon = \mathrm{O}(\varepsilon^{1/2-})$ (i.e. $\delta_\varepsilon \varepsilon^{-(1/2-\gamma)} \to 0$ for all $0 < \gamma < 1/2$).*
3. *In the slowly decaying case $0 < \alpha < 1/2$, we have $\delta_\varepsilon = \mathrm{O}(\varepsilon^{\min\{\alpha,1/2-\alpha\}})$.*

In all cases, the error depends only on the lower length bound ℓ_- and on the isometry classes of building blocks $\mathscr{X}_v$ out of which the vertex neighbourhoods X_v are chosen.

A similar statement holds in the decoupled Dirichlet case:

Theorem 1.3.3. *Let $\Delta_\varepsilon = \Delta^{\mathrm{D}}_{X_\varepsilon} - \pi^2\varepsilon^{-2}$ be the rescaled Dirichlet operator on the graph-like space, and let $\Delta_0 = \bigoplus_e \Delta^{\mathrm{D}}_{I_e}$ be the decoupled Dirichlet operator on the graph. Then $\Delta_\varepsilon \to \Delta_0$ in the generalised sense with convergence speed $\delta_\varepsilon = \mathrm{O}(\varepsilon^{1/2})$, provided $\inf_{v\in V} \lambda_2^{\partial_0 X_v}(X_v) > \pi^2$ and $\inf_e \ell_e > 0$. Moreover, the error depends only on these two infima.*

1.4 Boundary Triples and Convergence of Resonances: A Brief Overview

1.4.1 Boundary Triples Associated with Quadratic Forms

In order to explain our concept of boundary triple associated with a quadratic form, we express the examples of Sect. 1.2.9 in terms of our approach here. Let $\mathscr{H} := \mathsf{L}_2(0,1)$,

$$\mathfrak{h}(f) := \int_0^1 |f'(s)|^2 \mathrm{d}s, \quad \mathscr{H}^1 := \operatorname{dom}\mathfrak{h} := \{\, f \in \mathsf{H}^1(0,1) \mid f(0) = 0 \,\}.$$

We define the boundary map by $\Gamma f := f(1)$. It is now easily seen that

$$\Gamma \colon \mathscr{H}^1 \longrightarrow \mathscr{G} := \mathbb{C} \tag{1.16}$$

is bounded and surjective. Moreover, endowing $\mathscr{H}^1$ with its natural norm defined by $\|f\|^2_{\mathscr{H}^1} := \mathfrak{h}(f) + \|f\|^2$, then we have the orthogonal decomposition

$$\mathscr{H}^1 = \mathscr{H}^{1,\mathrm{D}} \oplus \mathscr{N}^1, \qquad \text{where} \qquad \mathscr{H}^{1,\mathrm{D}} := \ker \Gamma$$

and where $\mathscr{N}^1$ consists of the space of solutions $-h'' + h = 0$, so that in the example, $\mathscr{N}^1 = \mathbb{C}h$ with $h(s) = \sinh s/\sinh 1$. We can now invert Γ on the orthogonal complement $\mathscr{N}^1$ of $\ker \Gamma$, and obtain the Dirichlet solution operator $S: \mathscr{G} \longrightarrow \mathscr{N}^1 \subset \mathscr{H}^1$. We define now the Dirichlet-to-Neumann operator Λ via the quadratic form $\mathfrak{l}$ given by

$$\mathfrak{l}(\varphi) := \|S\varphi\|^2_{\mathscr{H}^1} = \mathfrak{h}(S\varphi) + \|S\varphi\|^2_{\mathscr{H}}.$$

It is easily seen that $\mathfrak{l}(\varphi) = |\varphi|^2 \coth 1$ in the simple example here, agreeing with the value $\Lambda(-1) = \mathrm{i}\cot \mathrm{i} = \coth 1$ in Sect. 1.2.9. Note that we did not use the second boundary map Γ' up to now. It enters in the theory as additional information: Let $\mathscr{W}^2 := \mathsf{H}^2(0,1) \cap \mathscr{H}^1$, then $\Gamma': \mathscr{W}^2 \longrightarrow \mathscr{G}$ is bounded and we have the following version of Green's formula

$$\langle f, \check{H} g\rangle_{\mathscr{H}} = \mathfrak{h}(f,g) - \langle \Gamma f, \Gamma' g\rangle_{\mathscr{G}} \quad \forall\, f \in \mathscr{H}^1, g \in \mathscr{W}^2, \tag{1.17}$$

where $\check{H} g = -g''$ for $g \in \mathscr{W}^2 = \operatorname{dom} \check{H}$.

Up to now, the concept of boundary triples associated with a quadratic form is not much different from the ordinary one. Its power can be seen in the example of a Laplacian on a subset X of $\mathbb{R}^d$ with smooth boundary ∂X. Here, we set

$$\mathscr{H} := \mathsf{L}_2(X), \quad \mathfrak{h}(u) := \|du\|^2, \quad \mathscr{H}^1 := \operatorname{dom} \mathfrak{h} := \mathsf{H}^1(X)$$

We define the boundary map here by $\Gamma u := u \restriction_{\partial X}$. Again, it is not difficult to see that Γ as operator (1.16) is bounded. Note that here, Γ is no longer *surjective*, but only has *dense range* which we call $\mathscr{G}^{1/2} := \operatorname{ran} \Gamma$. It follows from Sobolev space theory, that $\mathscr{G}^{1/2} = \mathsf{H}^{1/2}(\partial X)$ in the example here, explaining the notation. Note that the operator H^{N} associated with $\mathfrak{h}$ is the usual *Neumann* Laplacian, whereas the operator associated with $\mathfrak{h}^{\mathrm{D}} := \mathfrak{h} \restriction_{\ker \Gamma}$ is the *Dirichlet* Laplacian on X.

We define the associated Dirichlet solution operator as before, noting that is is now only defined on the *range* of Γ, namely, we have $S: \mathscr{G}^{1/2} \longrightarrow \mathscr{H}^1$, and S is unbounded as operator in $\mathscr{G} \to \mathscr{H}^1$. We define the Dirichlet-to-Neumann operator again via its quadratic form given by $\mathfrak{l}(\varphi) := \|S\varphi\|^2_{\mathscr{H}^1}$, i.e., $\mathfrak{l}(\varphi)$ is the *energy* of its 1-harmonic extension $h = S\varphi$, defined only for $\varphi \in \mathscr{G}^{1/2}$. The associated operator Λ is now *unbounded*, and defines a scale of Hilbert spaces on $\mathscr{G}$, namely $\mathscr{G}^m = \operatorname{dom} \Lambda^m$, which agrees with $\mathsf{H}^m(\partial X)$ in our example.

We define the second boundary map Γ' on $\mathscr{W}^2 := \mathsf{H}^2(X)$ as $\Gamma' u := \partial_{\mathrm{n}} u \restriction_{\partial X}$, which leads to a bounded operator $\Gamma': \mathscr{W}^2 \longrightarrow \mathscr{G}$. Moreover, setting $\check{H} := \Delta$ on $\operatorname{dom} \check{H} := \mathscr{W}^2$, we have again Green's formula (1.17).

After these two examples, let us now briefly outline the concept of boundary triples associated with a non-negative quadratic form and of abstract elliptic theory (see Sect. 3.4) in an abstract way. Note that the quadratic form setting fits well to our applications: On parameter-depending manifolds, it is easy to control first order objects like the norm of the operator (1.16), while second order objects like Sobolev spaces of order 2 or estimates in terms of the graph norm of the Laplacian are difficult to control quantitatively (in dependence on the parameter).

To our knowledge, the concept of boundary triples associated with quadratic forms is new, although there are strong links to existing concepts (see below). We stress that our main purpose here is not to characterise all self-adjoint extensions of a given symmetric operator, but to show that the concept of boundary triples can successfully be applied to the PDE case, namely to Laplacians on parameter-depending manifolds with boundary.

Let $\mathscr{H}, \mathscr{G}$ be two Hilbert spaces, $\mathfrak{h}$ a non-negative closed quadratic form on $\mathscr{H}$ with domain $\mathscr{H}^1 := \operatorname{dom} \mathfrak{h}$, and let $\Gamma: \mathscr{H}^1 \longrightarrow \mathscr{G}$ be a bounded operator such that

$$\mathscr{H}^{1,\mathrm{D}} := \ker \Gamma \subset \mathscr{H} \qquad \text{and} \qquad \mathscr{G}^{1/2} := \operatorname{ran} \Gamma \subset \mathscr{G}$$

are dense subspaces. We denote by H^{N} the *Neumann operator* associated with $\mathfrak{h}$, and by H^{D} the *Dirichlet operator* associated with the closed form $\mathfrak{h} \restriction_{\mathscr{H}^{1,\mathrm{D}}}$.

Let

$$S := (\Gamma \restriction_{\mathscr{N}^1})^{-1} : \mathscr{G}^{1/2} \longrightarrow \mathscr{N}^1 \tag{1.18}$$

be the inverse of Γ restricted to $\mathscr{N}^1 := \mathscr{H}^1 \ominus \ker \Gamma$, where the orthogonal complement is taken with respect to the Hilbert space $\mathscr{H}^1$ with norm $\|u\|^2_{\mathscr{H}^1} = \mathfrak{h}(u) + \|u\|^2$. It is easy to see that the quadratic form $\mathfrak{l}$ defined by

$$\mathfrak{l}(\varphi) := \|S\varphi\|^2_{\mathscr{H}^1}, \qquad \operatorname{dom} \mathfrak{l} = \mathscr{G}^{1/2}$$

is *closed*, and that $\mathfrak{l}(\varphi) \geq \|\varphi\|^2_{\mathscr{G}} / \|\Gamma\|^2_{1\to 0}$, where $\|\Gamma\|_{1\to 0}$ denotes the norm of the operator $\Gamma: \mathscr{H}^1 \longrightarrow \mathscr{G}$. We define the *Dirichlet-to-Neumann map* as the operator associated with $\mathfrak{l}$.

It can be seen that Λ is bounded iff $\operatorname{ran} \Gamma = \mathscr{G}$ (see Proposition 3.4.11). In the unbounded case, we define a scale of Hilbert spaces on $\mathscr{G}$ via

$$\mathscr{G}^m := \operatorname{dom} \Lambda^m \qquad \text{and} \qquad \|\varphi\|_m := \|\Lambda^m \varphi\|.$$

Let us now define the concept of a *boundary triple* in our sense: A triple $(\Gamma, \Gamma', \mathscr{G})$ is called a *boundary triple associated with the quadratic form* $\mathfrak{h}$ if the following hold:

1. The boundary operator $\Gamma: \mathscr{H}^1 \longrightarrow \mathscr{G}$ is bounded, and $\ker \Gamma$, $\operatorname{ran} \Gamma$ are dense in $\mathscr{H}, \mathscr{G}$, respectively.
2. There is a Hilbert space $\mathscr{W}^2$ such that $\mathscr{W}^2 \subset \mathscr{H}^1$ and $\|u\|_{\mathscr{H}^1} \leq \|u\|_{\mathscr{W}^2}$.
3. The boundary operator $\Gamma': \mathscr{W}^2 \longrightarrow \mathscr{G}$ is bounded.
4. There is an operator $\check{H}$ in $\mathscr{H}$ with $\operatorname{dom} \check{H} = \mathscr{W}^2$, bounded as operator $\check{H}: \mathscr{W}^2 \longrightarrow \mathscr{H}$ such that the abstract Green's formula (1.17) holds.
5. The space $\mathring{\mathscr{W}}^2 := \ker \Gamma \cap \ker \Gamma'$ is dense in $\mathscr{H}$.

We say that $(\Gamma, \Gamma', \mathscr{G})$ is *bounded* iff $\operatorname{ran} \Gamma = \mathscr{G}$ (i.e. if the Dirichlet-to-Neumann map is bounded).

For $z \in \mathbb{C} \setminus \sigma(H^{\mathrm{D}})$, we can define an operator $S(z): \mathscr{G}^{1/2} \longrightarrow \mathscr{H}^1$, the *Dirichlet solution map*, by associating to a boundary value φ the corresponding unique *weak* solution of the Dirichlet problem, namely $h = S(z)\varphi$ if

$$\mathfrak{h}(f,h) = z\langle f,h\rangle \quad \forall\, f \in \mathscr{H}^{1,\mathrm{D}} \qquad \text{and} \qquad \Gamma h = \varphi.$$

Note that $S(-1) = S$ where S is defined in (1.18).

We now make some assumptions in order to assure that $S(z)\varphi$ is also a *strong solution*. These assumptions are an abstract version of facts fulfilled in the PDE case. Namely, we say that $(\Gamma, \Gamma', \mathscr{G})$ is *elliptic*iff

$$\operatorname{dom} H^{\mathrm{D}} \subset \mathscr{W}^2 \qquad \text{and} \qquad \operatorname{dom} H^{\mathrm{N}} \subset \mathscr{W}^2$$

and if the operators

$$\Gamma\colon \mathscr{W}^2 \longrightarrow \mathscr{G}^{3/2} \qquad \text{and} \qquad \Gamma'\colon \mathscr{W}^2 \longrightarrow \mathscr{G}^{3/2}$$

are bounded and surjective (see Definition 3.4.21).

It follows now from Green's formula that for an elliptic boundary triple and for $\varphi \in \mathscr{G}^{3/2}$, the solution of the Dirichlet problem $h = S(z)\varphi$ is a *strong* solution, i.e.

$$(\check{H} - z)h = 0 \qquad \text{and} \qquad \Gamma h = \varphi.$$

For an elliptic boundary triple, we can define the Dirichlet-to-Neumann map for arbitrary $z \in \mathbb{C} \setminus \sigma(H^{\mathrm{D}})$ by

$$\Lambda(z)\colon \mathscr{G}^{3/2} \longrightarrow \mathscr{G}^{1/2}, \qquad \Lambda(z)\varphi := \Gamma' S(z)\varphi,$$

i.e. the Dirichlet-to-Neumann map takes a boundary value φ and associates the "normal derivative" of the corresponding Dirichlet solution. It can be seen that for $z = -1$, the Dirichlet-to-Neumann map actually agrees with Λ defined above as the operator associated with the form $\mathfrak{l}$ (more precisely, $\langle \varphi, \Gamma' S\varphi\rangle = \|S\varphi\|^2_{\mathscr{H}^1}$ for $\varphi \in \mathscr{G}^{3/2}$). Moreover, it can be seen (see Lemma 3.4.29) that $-(\Gamma'(H^{\mathrm{D}} - \overline{z})^{-1})^*$ extends the operator $S(z)$ uniquely to a bounded operator $\mathscr{G} \to \mathscr{H}$ (denoted by the same symbol).

We show Krein-type formulas like

$$(H^{\mathrm{N}} - z)^{-1} - (H^{\mathrm{D}} - z)^{-1} = S(z)\Lambda(z)^{-1} S(\overline{z})^*\colon \mathscr{H} \to \mathscr{H}$$

for $z \in \mathbb{C} \setminus (\sigma(H^{\mathrm{N}}) \cup \sigma(H^{\mathrm{D}}))$. Moreover, we obtain a similar expression for the resolvent difference as operator $\mathscr{H} \to \mathscr{W}^2$ using the (abstract) ellipticity.

Let us mention some relation to other concepts of boundary triples:

- Assume that $(\Gamma, \Gamma', \mathscr{G})$ is an elliptic boundary triple and that $\Gamma'(\operatorname{dom} H^{\mathrm{D}}) = \mathscr{G}^{1/2}$, then $(\Gamma \restriction_{\mathscr{W}^2}, \Gamma', \mathscr{G})$ is a *quasi-boundary triple* associated with $\mathring{H}^*$, where $\mathring{H} = \check{H} \restriction_{\ker \Gamma \cap \ker \Gamma'}$. The concept of quasi-boundary triples was introduced in [BL07] (see Definition 3.4.35).

- If $(\Gamma, \Gamma', \mathscr{G})$ is an elliptic and bounded boundary triple and if $\Gamma'(\operatorname{dom} H^{\mathrm{D}}) = \mathscr{G}$, then $(\Gamma \restriction_{\mathscr{W}^2}, \Gamma', \mathscr{G})$ is an *ordinary boundary triple* associated with $\mathring{H}^* = \check{H}$, i.e. Γ, Γ' are linear (not necessarily bounded) operators from $\mathscr{W}^2$ into $\mathscr{G}$, the operator $\mathscr{W}^2 \to \mathscr{G} \oplus \mathscr{G}$, $f \mapsto (\Gamma f, \Gamma' f)$ is *surjective* and the following abstract version of Green's formula holds, namely

$$\langle \check{H} f, g\rangle_{\mathscr{H}} - \langle f, \check{H} g\rangle_{\mathscr{H}} = \langle \Gamma f, \Gamma' g\rangle_{\mathscr{G}} - \langle \Gamma' f, \Gamma g\rangle_{\mathscr{G}} \qquad \forall\, f, g \in \mathscr{W}^2.$$

The concept of (ordinary) boundary triples has successfully been applied to the theory of self-adjoint extensions of symmetric operators (see Sect. 1.2.9 for further details and references).

Let us now describe the main examples we have in mind, the first one have been mentioned already before:

Example 1.4.1. Assume that X is a compact Riemannian manifold with smooth boundary $Y = \partial X$ (or a smooth subset $Y \subset \partial X$, see Theorems 3.4.40 and 3.4.41 for details) and define

$$\mathscr{H} := \mathsf{L}_2(X), \quad \mathscr{G} := \mathsf{L}_2(\partial X), \quad \mathfrak{h}(u) := \|du\|^2, \quad \mathscr{H}^1 := \mathsf{H}^1(X), \quad \mathscr{W}^2 := \mathsf{H}^2(X).$$

Moreover, $\check{H}$ is the usual Laplacian with domain $\mathsf{H}^2(X)$. The boundary operators are given by

$$\Gamma u := u \restriction_Y \qquad \text{and} \qquad \Gamma' u := \partial_{\mathrm{n}} u \restriction_Y,$$

where $\partial_{\mathrm{n}} u$ is the outwards normal derivative near Y. It follows then from Green's formula and from the elliptic theory for PDEs on manifolds, that $(\Gamma, \Gamma', \mathscr{G})$ is an elliptic boundary triple associated with $\mathfrak{h}$ (see e.g. [LM68, Gru68, Grv85, Ta96] resp. Theorems 3.4.39–3.4.41). Moreover, $\Lambda(z)$ is the usual Dirichlet-to-Neumann operator associating to a function φ on the boundary the normal derivative of the corresponding Dirichlet solution. Furthermore, $\mathscr{G}^m = \mathsf{H}^m(\partial X)$.

Example 1.4.2. Another example is given by a metric graph G with boundary $\partial G \subset V$ together with standard (Kirchhoff) vertex conditions. We assume that the degree (number of adjacent edges) of the vertices in ∂G is globally bounded by a number $d_0 \in \mathbb{N}$ (this is of course fulfilled if ∂G is finite). We set[4]

[4]We use here an *unweighted* boundary space $\ell_2(\partial G)$, in order to have a simple coupling formula for the interior and exterior part in the next section. In Sect. 2.2 (especially the example with the weighted standard vertex space, see (2.15)), we use a weighted version $\ell_2(\partial G, \deg)$ with $\|F\|^2_{\ell_2(\partial G\, \deg)} = \sum_{v \in \partial G} |F(v)|^2 \deg v$. In this way, one can avoid the global bound on the degree $\deg v \le d_0$. Roughly speaking, the difference of the unweighted and weighted space is to consider the combinatorial and normalised discrete Laplacian given by

$$\mathscr{H} := \mathsf{L}_2(G), \quad \mathscr{G} := \ell_2(\partial G), \quad \mathfrak{h}(f) := \|f'\|^2 = \sum_e \|f_e'\|^2_{I_e}, \quad \mathscr{H}^1 := \mathsf{H}^1_{\mathrm{cont}}(G),$$

$$\mathscr{W}^2 := \{\, f \in \mathsf{H}^2_{\mathrm{cont}}(X) \mid \sum_{e \in E_v} \underrightarrow{f}'_e(v) = 0 \ \forall v \in V \setminus \partial G \,\} \quad \text{and} \quad (\check{H} f)_e = -f_e'',$$

where $\mathsf{H}^k_{\mathrm{cont}}(G) := \mathsf{H}^k_{\max}(G) \cap \mathsf{C}(G)$. The boundary operators are given by

$$\Gamma f := \{f(v)\}_{v \in \partial V} \qquad \text{and} \qquad \Gamma' f := \Big\{ \sum_{e \in E_v} \underrightarrow{f}'_e(v) \Big\}_{v \in \partial V}. \tag{1.19}$$

Then $(\Gamma, \Gamma', \mathscr{G})$ is a bounded, elliptic boundary triple associated with $\mathfrak{h}$, i.e. the triple $(\Gamma \restriction_{\mathscr{W}^2}, \Gamma, \mathscr{G})$ is an ordinary boundary triple. The Dirichlet-to-Neumann map of this boundary triple has a nice interpretation: Assume that $F \in \ell_2(\partial G)$ describes a voltage applied to the boundary vertices of a graph. The Dirichlet-to-Neumann map gives the current at each boundary vertex (associated with the "energy" z). Actually, such measurements are commonly used in practise; and it is an interesting question what information on the (hidden) entire graph can be deduced from $\Lambda(z)$.

1.4.2 Resonances and Complex Dilation

Let us briefly explain the concept of complex dilation in order to describe resonances (i.e. poles of a meromorphic continuation of the resolvent) as eigenvalues of a non-unitarily transformed operator. We assume that the space consists of an interior and exterior part,that the interior part is compact and that the exterior part is a cylindrical end (not necessarily connected). As simple example, let G be the metric graph embedded in $\mathbb{R}^2$ consisting of a half-line $G_{\mathrm{ext}} \cong \mathbb{R}_+ = [0, \infty)$ (the *exterior edge*) attached to a loop $G_{\mathrm{int}} \cong (2\pi)^{-1}\mathbb{S}^1$ of perimeter 1 at the vertex v.[5] The spectrum of the corresponding standard Laplacian $H^0 = \Delta_G$ is purely essential and given by $[0, \infty)$. Moreover, we have eigenvalues embedded in the essential spectrum arising from compactly supported eigenfunctions $f_{\mathrm{int}}(s) = \sin(2\pi k s)$ for $k \in \mathbb{N}$ and $f_{\mathrm{ext}} = 0$, since

$$f_{\mathrm{int}}(v) := f_{\mathrm{int}}(0) = f_{\mathrm{int}}(1) = 0 \qquad \text{and} \qquad f_{\mathrm{int}}'(v) := f_{\mathrm{int}}'(1) - f_{\mathrm{int}}'(0) = 0$$

$$(\ddot{\Delta}_G^{\mathrm{comb}} F)(v) = \sum_{w \sim v} (F(v) - F(w)) \qquad \text{and} \qquad (\ddot{\Delta}_G^{\mathrm{norm}} F)(v) = \frac{1}{\deg v} \sum_{w \sim v} (F(v) - F(w))$$

on $\ell_2(V)$ and $\ell_2(V, \deg)$, respectively.

[5] In contrast to Sect. 1.3.1, we allow here curved edges. It is shown in Sect. 5.4 that the deviation of the tubular neighbourhood of a curved edge from the tubular neighbourhood of a straight edge is small, and that the corresponding quadratic forms are $\mathrm{O}(\varepsilon)$-quasi-unitarily equivalent.

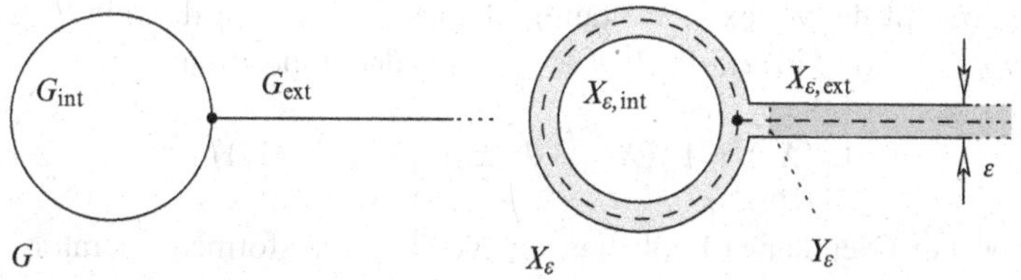

Fig. 1.4 A simple example of a graph and the corresponding graph-like space with one exterior edge and a loop attached. Here, the (scaled) transversal space Y_ε is isometric to the interval $[-\varepsilon/2, \varepsilon/2]$

on the loop. Such eigenvalues induce a pole in the resolvent. If the resolvent is continued meromorphically onto the second sheet of $\sqrt{z}$ (cut along $[0, \infty)$), the continuation might have more poles, called *resonances*. For simplicity, we call any pole of the resolvent a resonance, i.e. an embedded real eigenvalue is also called a resonance.

The complex dilation (or scaling) method provides a simple method how to obtain these poles as eigenvalues of a complexly dilated operator H^θ: On the half-line G_{ext}, we have an unitary action of the group $\mathbb{R}$ by

$$(\hat{U}^\theta f)(s) := e^{\theta/2} f(e^\theta s)$$

for *real* $\theta \in \mathbb{R}$. In particular, we define the *exterior dilation operator* U_0^θ by $U_0^\theta f := f_{\text{int}} \oplus \hat{U}^\theta f_{\text{ext}}$ on $\mathsf{L}_2(G) = \mathsf{L}_2(G_{\text{int}}) \oplus \mathsf{L}_2(G_{\text{ext}})$, which in fact is unitary. We set

$$H_0^\theta := U_0^\theta \Delta_G U_0^{-\theta}.$$

For real $\theta \in \mathbb{R}$, the domain of H_0^θ is given by

$$\begin{aligned} \operatorname{dom} H_0^\theta = \bigl\{\, f \in \mathsf{H}^2_{\text{cont}}([0,1]) \oplus \mathsf{H}^2(\mathbb{R}_+) \bigm| f_{\text{ext}}(0) = e^{\theta/2} f_{\text{int}}(v), \\ f'_{\text{ext}}(0) = e^{3\theta/2} f'_{\text{int}}(v) \,\bigr\}, \end{aligned} \tag{1.20a}$$

where $f_{\text{int}} \in \mathsf{H}^2_{\text{cont}}([0,1])$ iff $f_{\text{int}} \in \mathsf{H}^2([0,1])$ and $f(0) = f(1)$. Moreover,

$$H_0^\theta f = -f''_{\text{int}} \oplus (-e^{-2\theta} f''_{\text{ext}}). \tag{1.20b}$$

The idea of exterior complex dilation is to use (1.20) as definition for *complex* $\theta \in S_\vartheta$, where $S_\vartheta := \{\, \theta \in \mathbb{C} \mid |\operatorname{Im} \theta| < \vartheta/2 \,\}$ denotes the horizontal strip around $\mathbb{R}$ of thickness $\vartheta \in (0, \pi/2)$ (Fig. 1.4).

We consider a similar situation on the associated graph-like space $X_\varepsilon = X_{\varepsilon,\text{int}} \cup X_{\varepsilon,\text{ext}}$, where $X_{\varepsilon,\text{ext}} = \mathbb{R}_+ \times Y_\varepsilon$ and where $X_{\varepsilon,\text{int}}$ is a graph-like manifold (with vertex neighbourhood scaling rate $\alpha = 1$) associated with G_{int} with an additional boundary component $X_{\varepsilon,\text{int}} \cap X_{\varepsilon,\text{ext}} \cong Y_\varepsilon$ (identified with Y_ε in the sequel) in the vertex neighbourhood of the vertex v. We define the unitary exterior dilation operator by $U_\varepsilon^\theta u := u_{\mathrm{int}} \oplus (\hat{U}_{\mathrm{ext}} \otimes \mathrm{id}) u_{\mathrm{ext}}$ for $\theta \in \mathbb{R}$ w.r.t. the decomposition

$$\mathsf{L}_2(X_\varepsilon) = \mathsf{L}_2(X_{\varepsilon,\mathrm{int}}) \oplus (\mathsf{L}_2(\mathbb{R}_+) \otimes \mathsf{L}_2(Y_\varepsilon)).$$

Let Δ_{X_ε} be the (Neumann) Laplacian on X_ε. The transformed operator $H_\varepsilon^\theta := U_\varepsilon^\theta \Delta_{X_\varepsilon} U_\varepsilon^{-\theta}$ then has domain

$$\begin{aligned} \operatorname{dom} H_\varepsilon^\theta = \{ u \in \mathsf{H}^2(X_{\varepsilon,\mathrm{int}}) \oplus \mathsf{H}^2(X_{\varepsilon,\mathrm{ext}}) \mid u_{\mathrm{ext}} = \mathrm{e}^{\theta/2} u_{\mathrm{int}}, \\ u'_{\mathrm{ext}} = \mathrm{e}^{3\theta/2} u'_{\mathrm{int}} \text{ on } Y_\varepsilon \}, \end{aligned} \tag{1.21a}$$

where u'_{int} denotes the outward normal derivative on $X_{\varepsilon,\mathrm{int}}$ and u'_{ext} denotes the inward normal derivative on $X_{\varepsilon,\mathrm{ext}}$ (i.e. the derivative in the longitudinal direction).

$$H_\varepsilon^\theta u = \Delta_{X_{\varepsilon,\mathrm{int}}} u_{\mathrm{int}} \oplus \Big(-\mathrm{e}^{-2\theta} u''_{\mathrm{ext}} + \frac{1}{\varepsilon^2} (\mathrm{id} \otimes \Delta_Y) u_{\mathrm{ext}} \Big). \tag{1.21b}$$

Note that $Y_\varepsilon \cong \varepsilon Y$ where $Y := Y_1$. Again, we use (1.21) as definition for *complex* $\theta \in \mathsf{S}_\vartheta$.

Using the language of boundary triples, we can abstractly define the complexly dilated operator H^θ from two boundary triples $(\Gamma_{\mathrm{int}}, \Gamma'_{\mathrm{int}}, \mathscr{G})$ and $(\Gamma_{\mathrm{ext}}, \Gamma'_{\mathrm{ext}}, \mathscr{G})$ on the interior and exterior parts. Note that above, we associated an elliptic boundary triple to a *compact* Riemannian manifold only. Since in the situation here, X_ε has cylindrical ends, the theory can be carried over to such "nice" non-compact manifolds.

The complexly dilated operator H^θ has domain

$$\operatorname{dom} H^\theta = \{ u \in \mathscr{W}^2_{\mathrm{int}} \oplus \mathscr{W}^2_{\mathrm{ext}} \mid \Gamma_{\mathrm{ext}} u = \mathrm{e}^{\theta/2} \Gamma_{\mathrm{int}} u, \quad \Gamma'_{\mathrm{ext}} u = \mathrm{e}^{3\theta/2} \Gamma'_{\mathrm{int}} u \},$$

generalising the above two examples. The coupled operator can be understood as the Neumann operator of a coupled, complexly dilated boundary triple. Moreover, we show the Krein-type formula

$$(H^\theta - z)^{-1} - (H^{\theta,\mathrm{D}} - z)^{-1} = S^\theta(z) \Lambda(z)^{-1} S^{\overline{\theta}}(\overline{z})^* \tag{1.22}$$

for *real* $\theta \in \mathbb{R}$, where $H^{\theta,\mathrm{D}} = H^{\mathrm{D}}_{\mathrm{int}} \oplus H^{\theta,\mathrm{D}}_{\mathrm{ext}}$ is decoupled, and $H^{\mathrm{D}}_{\mathrm{int}}$, $H^{\theta,\mathrm{D}}_{\mathrm{ext}}$ are the Dirichlet operators associated with the interior and exterior boundary triple. One of the main observation is here, that there is no parameter θ entering in the coupled Dirichlet-to-Neumann map $\Lambda(z) = \Lambda_{\mathrm{int}}(z) + \Lambda_{\mathrm{ext}}(z)$ (see Sect. 3.6.2). The coupled operator H^θ for *complex* values $\theta \in \mathsf{S}_\vartheta$ is then defined via its resolvent in (1.22). The resolvent formula indeed defines a closed operator H^θ for complex θ, such that $(H^\theta - z)^{-1}$ depends holomorphically on θ. This takes some effort and the details are given in Sects. 3.5–3.7, based on a concrete PDE example developed in [CDKS87]. The advantage of the abstract formulation in terms of coupled boundary triples is a common treatment of both the PDE and metric graph case, and the possibility to apply the results to other models in an easy way.

Let us now state an important result on the operators H_ε^θ ($\varepsilon \geq 0$). Note that the result is valid for any coupled complexly dilated operator defined via boundary triples satisfying some natural conditions (see Proposition 3.8.1, Theorems 3.8.2 and 3.8.4).

Theorem 1.4.3.

1. *The operator H_ε^θ has spectrum in the sector $\Sigma_\vartheta = \{ z \in \mathbb{C} \mid |\arg z| \leq \vartheta \}$, and the family $\{H_\varepsilon^\theta\}_{\theta \in S_\vartheta}$ is self-adjoint, i.e. $(H_\varepsilon^\theta)^* = H_\varepsilon^{\overline{\theta}}$. Moreover, for $z \in \mathbb{C} \setminus \Sigma_\vartheta$, the resolvent $(H_\varepsilon^\theta - z)^{-1}$ depends holomorphically on $\theta \in S_\vartheta$.*
2. *The essential spectrum of H_ε^θ is given by*[6]
$$\sigma_{\mathrm{ess}}(H_0^\theta) = \mathrm{e}^{-2\mathrm{i}\,\mathrm{Im}\,\theta}[0,\infty) \quad \textit{resp.}$$
$$\sigma_{\mathrm{ess}}(H_\varepsilon^\theta) = \mathrm{e}^{-2\mathrm{i}\,\mathrm{Im}\,\theta}[0,\infty) \cup \bigcup_{k \geq 2} \Big\{ \frac{\lambda_k(Y)}{\varepsilon^2} + \mathrm{e}^{-2\mathrm{i}\,\mathrm{Im}\,\theta}\kappa \,\Big|\, \kappa \in [0,\infty) \Big\}$$
for $\varepsilon = 0$ resp. $\varepsilon > 0$. The singular continuous spectrum of H_0 is empty.
3. *The discrete spectrum $\sigma_{\mathrm{disc}}(H_\varepsilon^\theta)$ is locally constant in θ. Moreover, if $0 < \mathrm{Im}\,\theta_1 \leq \mathrm{Im}\,\theta_2 < \vartheta/2$ then $\sigma_{\mathrm{disc}}(H_\varepsilon^{\theta_1}) \subset \sigma_{\mathrm{disc}}(H_\varepsilon^{\theta_2})$.*
4. *For $\theta \notin \mathbb{R}$ we have $\sigma(H_\varepsilon^\theta) \cap [0,\infty) = \sigma_{\mathrm{p}}(H_\varepsilon^0)$, where $\sigma_{\mathrm{p}}(H_\varepsilon^0)$ denotes the set of eigenvalues of the undilated operator $H_0^0 = \Delta_G$ resp. $H_\varepsilon^0 = \Delta_{X_\varepsilon}$, i.e. the eigenvalues of H_ε^0 embedded in the continuous spectrum are unveiled (see Fig. 3.2 on page 176).*
5. *The set of discrete eigenvalues of H_ε^θ coincides with the poles of a meromorphic continuation of the resolvent of H_ε^0. More precisely, $\lambda \in \Sigma_\vartheta$ is a discrete eigenvalue of H_ε^θ iff there exists $\varphi \in \mathscr{H}_\varepsilon$ such that the meromorphic continuation of $\langle \varphi, (H_\varepsilon - z)^{-1}\varphi \rangle$ has a pole in λ.*

The above theorem allows us to define a *resonance* of Δ_G resp. Δ_{X_ε} as a discrete eigenvalue of the coupled complexly dilated operator H_ε^θ for $\mathrm{Im}\,\theta$ large enough. The *multiplicity of a resonance* is the multiplicity of the corresponding eigenvalue. This definition does not depend on θ and agrees with the definition of a resonance as a pole (see the preceding theorem).

Let us calculate the resonances z on the loop graph G. We use the ansatz

$$f_{\mathrm{int}}(s) = F \frac{1}{\sin\sqrt{z}} \Big(\sin((1-s)\sqrt{z}) + \sin(s\sqrt{z}) \Big) \quad \text{and}$$
$$f_{\mathrm{ext}}(s) = \mathrm{e}^{\theta/2} F \exp(s \cdot \mathrm{e}^{\theta}\mathrm{i}\sqrt{z})$$

[6] In our example, we have $Y = [-1/2, 1/2]$, so that $\lambda_k(Y) = \pi^2(k-1)^2$.

for some coefficient $F \in \mathbb{C}$. Here, $\sqrt{z}$ is the complex root cut along $[0,\infty)$. Note that $f_{\mathrm{ext}} \in \mathsf{L}_2(\mathbb{R}_+)$, since $\operatorname{Im}(e^{\theta}\sqrt{z}) > 0$ for $z \in \mathbb{C} \setminus \Sigma_\vartheta$ and since $\theta \in S_\vartheta$. In particular, f solves the equation $H_0^\theta f = zf$ and the first vertex condition $\Gamma_{\mathrm{ext}} f = e^{\theta/2}\Gamma_{\mathrm{int}} f$. Moreover, the condition on the derivatives reads as

$$e^{3\theta/2} i\sqrt{z} F = f'_{\mathrm{ext}}(0) = e^{3\theta/2} f'_{\mathrm{int}}(v) = e^{3\theta/2}\Lambda_{\mathrm{int}}(z) F,$$

where $\Lambda_{\mathrm{int}}(z) := \Gamma'_{\mathrm{int}} S_{\mathrm{int}}(z)$ denotes the Dirichlet-to-Neumann map of the interior graph (here, the Dirichlet-to-Neumann map is just multiplication with the complex number $\Lambda_{\mathrm{int}}(z)$). In our example, we have

$$\Lambda_{\mathrm{int}}(z) = -\frac{2\sqrt{z}}{\sin\sqrt{z}}(1-\cos\sqrt{z}) = -2\sqrt{z}\tan(\sqrt{z}/2).$$

In particular, for $\sqrt{z} \notin \pi\mathbb{N}_0$, the function f is a non-trivial solution of $H_0^\theta f = zf$ iff $i\sqrt{z} = \Lambda_{\mathrm{int}}(z)$, or equivalently,

$$i\sin\sqrt{z} = -2 + 2\cos\sqrt{z}, \quad \text{or} \quad (w-3)(w-1) = w^2 - 4w + 3 = 0,$$

where $w = e^{i\sqrt{z}}$. In particular, $w = 1$ or $w = 3$ are (formal) solutions, i.e. $z = (2\pi k)^2$ and $z = (2\pi k - i\log 3)^2$ for $k \in \mathbb{Z}$. Note that the solutions $z = (2\pi k)^2$ arise from the embedded eigenvalues of the loop states. The Dirichlet-to-Neumann map is not defined for these values since $(2\pi k)^2$ is part of the Dirichlet spectrum on G_{int}, but can be extended holomorphically there. The solutions $z = (2\pi k - i\log 3)^2$ are non-real resonances.

Krein's resolvent formula (1.22) allows us to generalise the above considerations and to give a simple criterion for resonances also for more complicated spaces: let $z \in \mathbb{C} \setminus [0,\infty)$. Then

$$z \text{ is a resonance} \quad \Leftrightarrow \quad \ker \Lambda(z) \neq 0,$$

where $\Lambda(z)$ is the (maximal holomorphic extension of the) Dirichlet-to-Neumann map of the (dilated or undilated) coupled boundary triple (see Theorem 3.8.4). In particular, we can calculate $\Lambda(z)$ as

$$\Lambda(z) = \Lambda_{\mathrm{int}}(z) + \Lambda_{\mathrm{ext}}(z),$$

where $\Lambda_{\mathrm{ext}}(z) = -i\sqrt{z}$ resp. $\Lambda_{\mathrm{ext}}(z) = -i\sqrt{z - \Delta_Y/\varepsilon^2}$ for an exterior edge resp. a cylindrical end $X_{\varepsilon,\mathrm{ext}}$.

1.4.3 Convergence of Resonances on Graph-Like Spaces

Let us now formulate the convergence result for resonances in a slightly more general way, taking the different vertex neighbourhood scalings and boundary conditions of Sects. 1.3.1 and 1.3.2 into account.

Assume that $G = G_{\mathrm{int}} \cup G_{\mathrm{ext}}$ is a graph with a compact interior part G_{int} and an exterior part G_{ext} isometric to a finite number of copies of half-lines. Let X_ε be an associated graph-like space with operator Δ_ε (the Neumann Laplacian or $\Delta^{\mathrm{D}}_{X_\varepsilon} - \pi^2/\varepsilon^2$ in the Dirichlet case).

Theorem 1.4.4. *Let $\lambda_0 \in \mathbb{C}$ be a resonance with multiplicity $\mu > 0$ of the limit operator Δ_0 on the metric graph G. Then there exist μ resonances $\lambda_{j,\varepsilon}$ of the operator Δ_ε (not all necessarily different) such that $\lambda_{j,\varepsilon} \to \lambda_0$ for $j = 1, \dots, \mu$ (with the same convergence speed as in Theorems 1.3.2 and 1.3.3).*

In particular, an embedded eigenvalue of Δ_0 is approached by a resonance of the operator Δ_ε.

The above theorem also gives interesting results for the *decoupled* limit operators $\Delta_0 = \Delta_{0,\mathrm{int}} \oplus \Delta_{0,\mathrm{ext}}$ in the slowly decaying (Neumann) case and the decoupled Dirichlet case: Recall that the spectrum of the interior graph G_{int} is given by $\sigma(\Delta_{0,\mathrm{int}}) = \{0\} \cup \Sigma^{\mathrm{D}}$ resp. $\sigma(\Delta_{0,\mathrm{int}}) = \Sigma^{\mathrm{D}}$ in the slowly decaying resp. Dirichlet case, where

$$\Sigma^{\mathrm{D}} := \Big\{ \frac{\pi^2 k^2}{\ell_e^2} \Bigm| e \in E_{\mathrm{int}}, k = 1, 2, \dots \Big\}, \tag{1.23}$$

since the operator $\Delta_{0,\mathrm{int}}$ is an orthogonal sum of Dirichlet operators. In particular, we have:

Corollary 1.4.5. *Let $\lambda_0 \in \sigma(\Delta_{0,\mathrm{int}})$ be a discrete eigenvalue of multiplicity $\mu > 0$ of the interior operator $\Delta_{0,\mathrm{int}}$ in the slowly decaying (Neumann) case resp. the Dirichlet case. Then there exist μ resonances $\lambda_{j,\varepsilon}$ of the (Neumann) Laplacian Δ_{X_ε} resp. the rescaled Dirichlet Laplacian $\Delta^{\mathrm{D}}_{X_\varepsilon} - \pi^2/\varepsilon^2$ converging to λ_0.*

Remark 1.4.6. The effect is similar to a Helmholtz resonator: A *Helmholtz resonator* consists of a compact cavity joined by a thin connection with a non-compact exterior part. Although the eigenvalues of the interior problem (without the thin connection) are no longer eigenvalues of the entire problem, they give rise to *resonances*. A simple example is obtained from the disconnected set $X_0 := \{ x \in \mathbb{R}^3 \mid 0 < |x| < r_1 \text{ or } r_2 < |x| \}$ for $0 < r_1 < r_2$ by joining the two components with a thin cylinder of radius ε. The spectrum of the (Dirichlet or Neumann) Laplacian on the resulting space X_ε is purely essential. Moreover, the eigenvalues of the interior ball are limits of *resonances* of the Laplacian on X_ε as $\varepsilon \to 0$. Such problems have been treated e.g. in [Be73, HiM91, Hi92, BHM94, HiL01].

The effect on a graph-like space is similar: Assume that we put a slowly decaying vertex neighbourhood at the vertex v_0 joining the interior and exterior part, but use fast decaying vertex neighbourhoods on the remaining vertices of the interior part. Then the limit operator decouples at v_0, and the spectrum of the interior graph appears as embedded eigenvalues in $[0, \infty)$ (the spectrum of the exterior part). The embedded eigenvalues on the interior graph are again limits of resonances of the Laplacian on the graph-like space.

1.5 Consequences of the Convergence Results

Let us state some exemplary consequences of the convergence results on graph-like spaces. In particular, we give partial answers to questions (A) and (B) of Sect. 1.2.4 on conductivity (spectral gaps) and impurity levels (eigenvalues in gaps). More generally, we are interested in *spectral engineering*, i.e., in constructing spaces with given spectral properties similar as in Colin de Verdière's work (cf. Theorem 1.2.1 and Sect. 1.2.6). Note that we already gave partial answers to questions (C) and (D) on which vertex conditions are obtained by graph-like spaces and how to construct given vertex conditions on a metric graph by graph-like spaces in Sect. 1.2.4.

Recall that the graph-like spaces associated with a graph are assumed to fulfil some uniformity conditions on the vertex neighbourhoods and the edge length (e.g., that X_v is isometric to one member of a *finite* family of Riemannian manifolds $\mathscr{X}_V$ and that $\inf_e \ell_e > 0$). We call such graphs and graph-like spaces *uniform*. The operators Δ_ε acting on graph-like spaces X_ε are given by the (Neumann) Laplacian $\Delta_\varepsilon = \Delta_{X_\varepsilon}$ resp. the rescaled Dirichlet Laplacian $\Delta_\varepsilon = \Delta^{\mathrm{D}}_{X_\varepsilon} - \pi^2 \varepsilon^{-2}$, see Sect. 1.3.1. The operators Δ_0 on the limit space are explained in Sect. 1.3.2.

1.5.1 Spectral Gaps and Covering Manifolds

A *spectral gap* of a (non-negative) operator Δ is an interval $I \subset [0, \infty)$ such that

$$\sigma(\Delta) \cap I = \emptyset.$$

Of course, if Δ is a Laplace-type operator on a compact space, Δ has purely discrete spectrum and spectral gaps always exists. Therefore, we are interested in spectral gaps of the *essential* spectrum of non-compact spaces. A simple consequence of the spectral convergence of Theorems 1.3.1 and 1.3.2–1.3.3 for operators on graph-like spaces is the following existence result for spectral gaps:

Theorem 1.5.1. *Suppose that G is a (non-compact) graph with associated uniform graph-like manifold X_ε. If the limit operator Δ_0 has a spectral gap (a, b), then the operator Δ_ε on the graph-like space has a spectral gap close to (a, b) provided ε is small enough.*

An important class of *uniform* graph-like spaces are covering spaces with a compact quotient: Clearly, a covering metric graph $\widetilde{G} \to G$ with compact quotient G fulfils the global lower length bound $\inf_e \ell_e > 0$. Similarly, an associated family of graph-like covering manifold $\widetilde{X}_\varepsilon \to X_\varepsilon$ with compact quotient X_ε has only finitely many isometry classes of vertex neighbourhoods. For examples of quantum graph with spectral gaps we refer to [LP08a, EP05] and references therein.

Let $\widetilde{X} \to X$ be a covering with abelian covering group $\mathbb{Z}^r$. Here, X is either a graph-like manifold or a metric graph. The corresponding Laplacian $\widetilde{\Delta}$ on $\widetilde{X}$

commutes with the covering group. Moreover, the spectrum of $\widetilde{\Delta}$ can be analysed using *Floquet theory*. Floquet theory uses basically a partial Fourier transformation in order to transform the periodic operator $\widetilde{\Delta}$ on the non-compact space into a family of so-called *θ-periodic* operators $\{\Delta^\theta\}_\theta$ on the compact quotient. Here, θ belongs to the dual $\hat{\mathbb{Z}}^r = \mathbb{T}^r$, and we identify the torus $\mathbb{T}^r$ (as measure space, not as topological space!) with $W := [0, 2\pi]^r$. As main result we obtain for the spectrum of the periodic operator that

$$\sigma(\widetilde{\Delta}) = \bigcup_{k \in \mathbb{N}} B_k, \qquad \text{where} \qquad B_k = \{\, \lambda_k(\Delta^\theta) \,|\, \theta \in W = [0, 2\pi]^r \,\}$$

denotes the k-th *band*. Each band is a compact interval, and the union of the bands is locally finite. Moreover, $\lambda_k(\Delta^\theta)$ is the k-th eigenvalue of Δ^θ (ordered w.r.t. multiplicity). The operator Δ^θ can be defined on θ-periodic functions

$$u(\gamma x) = \mathrm{e}^{\mathrm{i}\theta \cdot \gamma} u(x) \qquad \text{for} \qquad x \in X \quad \text{and} \quad \gamma \in \mathbb{Z}^r \tag{1.24}$$

(see e.g. [LP08b, LP07, P03b] and references therein). Note that a band may reduce to a point (i.e. an L_2-eigenvalue of $\widetilde{\Delta}$). In general, the intervals may overlap or touch, so that in the extreme case, $\sigma(\widetilde{\Delta})$ has no gap at all.

The existence of spectral gaps for the operator on a graph-like space is in particular ensured, if we are in the slowly decaying case. The spectrum of the decoupled limit operator is $\sigma(\widetilde{\Delta}_0) = \{0\} \cup \Sigma^{\mathrm{D}}$ (see (1.23)). Since the quotient is compact, the set of different length ℓ_e is finite, and $\sigma(\widetilde{\Delta}_0)$ is a discrete set in $\mathbb{R}_+$. In particular, any interval $[a, b] \subset \mathbb{R}_+ \setminus \sigma(\widetilde{\Delta}_0)$ is a spectral gap for $\widetilde{\Delta}_\varepsilon$ provided ε is small enough.

Nevertheless, we cannot conclude in general that the Laplacian on the graph-like manifold X_ε has infinitely many spectral gaps:

Remark 1.5.2. The spectrum of a periodic operator typically has band-gap structure, but in higher dimension, the bands tend to overlap. The *Bethe-Sommerfeld conjecture* states that a periodic Laplace-type operator in dimension 2 or more has only finitely many gaps in its spectrum. This conjecture is confirmed e.g. for periodic Schrödinger operators (see [Par08, Sk85, DT82, Sk79]). On singular spaces like metric graphs or an array of attached manifolds it is in general easy to construct examples with *infinitely* many gaps (see e.g. [BGL05, BEG03, EG96]). On an equilateral metric graph, the Laplacian has infinitely many spectral gaps iff the corresponding discrete Laplacian $\dddot{\Delta}$ has spectrum different from $[0, 2]$ (see Theorem 2.4.1). Note that $\sigma(\dddot{\Delta}) \subset [0, 2]$.

Our analysis here in general does not prove nor disprove the Bethe-Sommerfeld conjecture for periodic graph-like manifolds: The spectral convergence implies the convergence $\lambda_k(\Delta_\varepsilon^\theta) \to \lambda_k(\Delta_0^\theta)$ as $\varepsilon \to 0$ for each $k \in \mathbb{N}$ uniformly in $\theta \in [0, 2\pi]$. If this convergence would be *uniform* in $k \in \mathbb{N}$, then we could construct a counterexample to the Bethe-Sommerfeld conjecture for the Laplacian on a manifold.

But the convergence cannot be uniform in $k \in \mathbb{N}$: Otherwise, the *theta-function*

$$\Theta_\varepsilon^\theta(t) := \operatorname{tr} e^{-t\Delta_{X_\varepsilon}^\theta} = \sum_k e^{-t\lambda_k(\Delta_\varepsilon^\theta)}$$

would converge to $\Theta_0^\theta(t)$. More precisely, if $|\lambda_k(\Delta_\varepsilon^\theta) - \lambda_k(\Delta_0^\theta)| \leq \eta$ for ε small enough *uniformly* in k, then

$$\frac{|\Theta_\varepsilon^\theta(t) - \Theta_0^\theta(t)|}{\Theta_0^\theta(t)} \leq \eta e^{t\eta},$$

so that the fraction converges to 0 uniformly in $t \in (0, 1]$ as $\eta \to 0$. But the Weyl asymptotics are different in the two cases,

$$\Theta_\varepsilon^\theta(t) \sim \frac{\operatorname{vol}_d X_\varepsilon}{(4\pi t)^{d/2}}, \qquad \text{whereas} \qquad \Theta_0^\theta(t) \sim \frac{\operatorname{vol}_1 G}{(4\pi t)^{1/2}}$$

as $t \to 0$ (cf. [Ch84, Sect. VI.4] and Theorem 2.5.5 resp. [Rot84, Thm. 1], [KPS07, Thm. 4.1]). Recall that $d \geq 2$ and $\operatorname{vol}_1 G := \sum_e \ell_e$, i.e. the total length of the graph.

A different procedure of creating gaps in a metric graph is provided by the so-called *graph-decoration*. Roughly speaking, the new graph $\widetilde{G}$ is obtained from a given infinite graph G by attaching a fixed (compact) graph G_0 to each vertex v of G. The Laplacian on $\widetilde{G}$ now has spectral gaps. This result has been established for a discrete graph in [AS00]. The general case for quantum graphs is announced in [Ku05], and proved in the case when G is a compact graph (in the sense that there is no spectrum of the decorated graph near (certain parts) of the spectrum of the decorating graph). For quantum graphs, there are related examples leading to gaps (cf. [AEY94, Ex95]). A similar effect by attaching a single loop to each vertex of a periodic graph has been used in [EP05, Sect. 9]. The case of periodically arranged manifolds connected by line segments or through points has been analysed in [BEG03, BGL05].

1.5.2 Eigenvalues in Gaps

Suppose that G is a uniform metric graph such that its standard Laplacian has a spectral gap, namely, $\sigma(\Delta_G) \cap (a, b) = \emptyset$. By the previous example, the (Neumann) Laplacian on a corresponding uniform graph-like manifold X_ε (with fast decaying vertex neighbourhoods) has also a spectral gap close to (a, b), provided ε is small enough.

Let us now change the metric graph locally, e.g. by attaching a loop of length ℓ_1 at a fixed vertex $v_1 \in V$, and call the decorated graph $\hat{G}$. The corresponding standard Laplacian on $\hat{G}$ has additional eigenvalues $\lambda_k = (2\pi k/\ell_1)^2$ with eigenfunctions located on the loop and vanishing at v_1 and on the rest of the graph (there might be additional eigenvalues). For example, if $2\pi/\sqrt{b} < \ell_1 < 2\pi/\sqrt{a}$ then the ground state of the loop lies in (a, b). Note that the essential spectrum of G and the decorated metric graph $\hat{G}$ are the same since the perturbation is compact; in particular, only *discrete* eigenvalues occur in the spectral gap. Due to the spectral convergence, a corresponding (uniform) graph-like manifold now must have (at least) one eigenvalue in (a, b). Furthermore, if the corresponding eigenvalue of the quantum graph is simple (i.e. there are no other eigenvalues of the decorated graph $\hat{G}$ at $\lambda_1 = (2\pi/\ell_0)^2$), then there is a unique eigenvalue close to λ_1 of multiplicity 1. The corresponding normalised eigenfunction φ_0 of the graph-like manifold is close to the normalised eigenfunction φ_ε of the metric graph, i.e.

$$\|J\varphi - \varphi_\varepsilon\| \to 0$$

(see Theorem 1.3.1). Roughly speaking, the convergence says, that an eigenfunction on the graph-like manifold is approximately given by the corresponding eigenfunction on the graph, extended constantly in the transversal direction (and set to 0 in the vertex neighbourhood).

Eigenvalues in gaps have been discussed e.g. in [AADH94, P03a] (see also the references therein). One can interpret such a local perturbation as an *impurity of a periodic structure*, say, a crystal or a semi-conductor. The additional eigenvalue in the gap of the unperturbed operator and the corresponding eigenfunction corresponds to an additional energy level and a bound state, respectively.

1.5.3 Equilateral Graphs

Let us consider now a special class of metric graphs closely related to the corresponding discrete graph. Assume that the metric graph is *equilateral*, i.e. that all edges have the same length, say $\ell_e = 1$. We assume here, that we are in the fast decaying or borderline case, hence the limit operator on the graph is the standard Laplacian or the coupled operator Δ_0 on the extended metric graph defined in (1.12).

For equilateral graphs, we have a simple spectral relation of the corresponding metric and discrete graph Laplacians. The discrete Laplacian $\dddot{\Delta}$ is defined in (2.4) and has spectrum in $[0, 2]$. The spectral relation is given in Theorem 2.4.1 (see also [vB85, Ni87, Ct97, Ex97, Pan06, P08, BGP08]).

Theorem 1.5.3 (Fast decaying vertex volume). *Suppose that G is an equilateral graph with associated graph-like manifold X_ε and vertex neighbourhoods $X_{\varepsilon,v} = \varepsilon X_v$ with scaling rate $\alpha = 1$. If the corresponding* discrete *Laplacian $\dddot{\Delta}$ has a spectral gap $(a, b) \subset (0, 2)$, then the metric graph Laplacian Δ_0 has spectral gaps I_k where I_k are the connected components of*

$$\{\lambda \geq 0 \,|\, 1 - \cos\sqrt{\lambda} \in (a, b)\}.$$

Moreover, the Laplacian Δ_ε on the graph-like manifold has spectral gaps close to the gaps I_k provided ε is small enough.

Note that if the discrete Laplacian has a spectral gap (a,b), then the metric graph Laplacian has infinitely many spectral gaps I_k since the function $1-\cos$ is periodic. Nevertheless, we do not know whether the number of gaps of $\sigma(\Delta_{X_\varepsilon})$ is infinite or not, see Remark 1.5.2.

A similar result can be proven for the borderline case (see the spectral relation in this case in Theorem 2.4.3):

Theorem 1.5.4 (The borderline case). *Suppose that $\widetilde{G}$ is an abelian covering of a compact equilateral graph G with associated 2-dimensional graph-like manifolds $\widetilde{X}_\varepsilon \to X_\varepsilon$ such that the vertex neighbourhoods $X_{\varepsilon,v} = \varepsilon^\alpha X_v$ have scaling rate $\alpha = 1/2$ and that the volume*

$$\operatorname{vol}_2 X_v = c$$

is independent of $v \in V$. Then the Laplacian $\Delta_{\widetilde{X}_\varepsilon}$ on the graph-like covering manifold has spectral gaps close to

$$\{\, \lambda \geq 0 \,|\, m_c(\lambda) := c\sqrt{\lambda}\sin\sqrt{\lambda} + 1 - \cos\sqrt{\lambda} \notin [0,2] \,\}$$

provided ε is small enough. In particular, given $n \in \mathbb{N}$, there exists $\varepsilon_n > 0$ such that $\sigma(\Delta_{\widetilde{X}_{\varepsilon n}})$ has at least N gaps.

The existence of an infinite sequence of spectral gaps for the extended metric graph Laplacian $\Delta_{0,\widetilde{G}}$ defined in (1.12) is ensured by the fact that $m_c(\lambda)$ oscillates between $\pm c\sqrt{\lambda}$ as $\lambda \to \infty$, but the discrete operator $\ddot{\Delta}$ on $\widetilde{G}$ only has spectrum in $[0,2]$. The restriction to abelian covering manifolds is necessary since Theorem 2.4.3 states the spectral relation only for *eigenvalues*. We are confident that the result extends to the entire spectrum, so that Theorem 1.5.4 should be valid for any equilateral graph and associated graph-like manifolds X_ε with scaling rate $\alpha = 1/2$.

Another class of examples is given by *self-similar metric graphs*,i.e. graphs arranged in a self-similar manner (see Fig. 1.5). Note that the Sierpiński metric graph is non-compact. We assume that each edge has length 1. The main point here is that the metric graph spectrum has a fractal structure, so if ε is small enough then the corresponding Laplacian on a graph-like manifold has an arbitrary (but a priori finite) number of spectral gaps in a *compact* spectral interval $[0,\lambda_0]$.

Let G be the equilateral metric Sierpiński graph given in Fig. 1.5. Note that G has one vertex of degree 2, all other vertices have degree 4. The nature of the spectrum of the discrete Laplacian $\ddot{\Delta}$ was calculated by [Tep98, Thm. 2]: Namely, we have

$$\sigma(\ddot{\Delta}) = J \cup D, \qquad D := \Big\{\frac{3}{2}\Big\} \cup \bigcup_{n=0}^{\infty} p^{-n}\Big\{\frac{3}{4}\Big\}, \qquad J := \overline{D}, \tag{1.25}$$

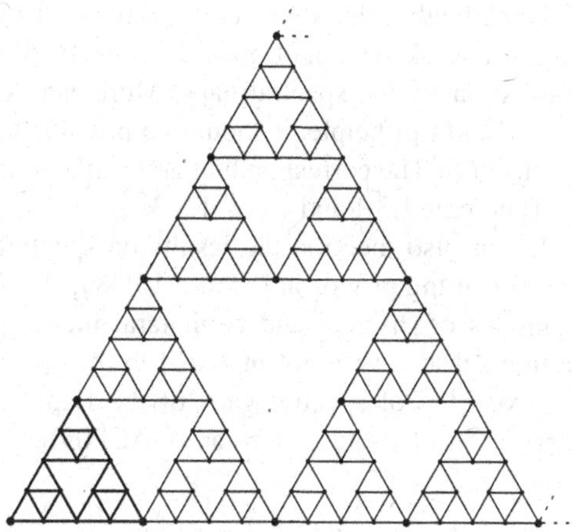

Fig. 1.5 The Sierpiński graph G. Note that G is a non-compact equilateral metric graph

where $p(z) := z(5-4z)$ and p^{-n} is the n-th pre-image, i.e. $p^{-n}\{3/4\} = \{z \in \mathbb{R} \mid p^{\circ n}(z) = 3/4\}$. The spectrum of $\ddot{\Delta}$ is pure point and each eigenvalue has multiplicity ∞, therefore the spectrum is also purely essential. The set D consists of the isolated eigenvalues and the set J is a Cantor set of Lebesgue measure 0 (the Julia set of the polynomial p). Due to Theorem 2.4.1 the spectrum of the equilateral metric graph Laplacian Δ_G is pure point and purely essential. It is given by

$$\sigma(\Delta_G) \setminus \Sigma^{\mathrm{D}} = m^{-1}(D) \cup m^{-1}(J \setminus \{0\}).$$

Since $m(\lambda) = 1 - \cos\sqrt{\lambda}$ is a local homeomorphism mapping measure 0 set into measure 0 sets and vice versa, $m^{-1}(D)$ consists of isolated eigenvalues of Δ_G of infinite multiplicity and $m^{-1}(J\setminus\{0\})$ is a Cantor set of measure 0 (away from the Dirichlet spectrum $\Sigma^{\mathrm{D}} = \{(\pi k)^2 \mid k = 1, 2, \dots\}$). A similar statement holds for m_c restricted to λ such that $m_c(\lambda) \in [0, 2]$.

Now our spectral convergence result on graph-like manifolds leads to an example of a family of smooth manifolds approaching a fractal spectrum:

Theorem 1.5.5. *Suppose X_ε is a family of graph-like manifolds associated with the equilateral metric Sierpiński graph. Then the essential spectrum of the Laplacian on X_ε approaches the fractal spectrum of Δ_G in any fixed interval $[0, \lambda_0]$. The discrete spectrum of Δ_{X_ε} is either empty or merges into the essential spectrum as $\varepsilon \to 0$. In particular, $\sigma(\Delta_{X_\varepsilon})$ has an arbitrary large number of spectral gaps in the* compact *interval $[0, \lambda_0]$ provided ε is small enough.*

Again, we cannot conclude that the number of gaps in the interval $[0, \lambda_0]$ is infinite, since the uniform convergence of the essential spectrum allows us to control the sets up to an error only (as $\varepsilon \to 0$). In order to conclude that X_ε has fractal spectrum, one has to use different methods.

In particular, the above constructions of manifolds with spectral gaps may be seen as a weak version of *spectral engineering* in Sect. 1.2.6: We are able to localise approximately the spectral gaps. Moreover, let S be a closed and unbounded set $S \subset \mathbb{R}_+$. In principle, it should be possible to construct a quantum graph G with spectrum S. The corresponding graph-like manifold then has a spectrum close to S by Theorems 1.3.1 and 1.3.2–1.3.3.

Let us also mention the results on spectral gaps for coverings with *residually finite* covering groups in [LP08b, LP08a]. In particular, we constructed in [LP08a] examples of discrete and equilateral metric graphs where we could assure that certain subsets were not included in the spectrum of the covering Laplacian. For the existence of spectral gaps of the Laplacian on differential forms on certain $\mathbb{Z}$-periodic manifolds, we refer to [ACP09].

1.5.4 Spectral Band Edges

Let us mention another interesting example on the *spectral band edges*:It was commonly believed by Physicists that the k-th band of a $\mathbb{Z}^r$-periodic operator (i.e., an operator on a $\mathbb{Z}^r$-covering space commuting with the group action) is actually given as the range of

$$\hat{B}_k := \{ \lambda_k(\Delta^\theta) \mid \theta \in \partial W \},$$

where W is the so-called *Brillouin zone*, a fundamental domain of the dual lattice which is identified with $\mathbb{Z}^r$ in our case here. More precisely, if the covering space has more symmetries, on considers the *reduced* Brillouin zone, a subset of the full Brillouin zone, taking the additional (discrete) symmetries into account.

Harrison, Kuchment, Sobolev and Winn [HKSW07] gave interesting examples of periodic operators on discrete graphs where $\hat{B}_k \subsetneq B_k$, i.e., where the band edges are attained by *interior* points $\theta \in (0, 2\pi)^r$, disproving the commonly believed argument above. Exner, Kuchment and Winn [EKW10] even gave an example of a $\mathbb{Z}$-periodic metric graph with $\hat{B}_k \subsetneq B_k$. The periodic metric graph graph in their example has a (suitable) fundamental domain, which is connected with its forward neighbour by more than one edge.

Using the spectral relation between discrete and equilateral metric graphs of Sect. 1.5.3 we can construct periodic graphs with the same properties, i.e., where $\hat{B}_k \subsetneq B_k$. Since the eigenvalues of a graph-like manifold converge to the ones of an appropriate quantum graph, i.e., $\lambda_k(\Delta_\varepsilon^\theta) \to \lambda_k(\Delta_0^\theta)$ for $\theta \in \partial W$, and since $\hat{B}_k(\Delta_0) = \{ \lambda_k^\theta(\Delta_0) \mid \theta \in \partial W \} \subsetneq B_k(\Delta_0)$, we also have an example of a periodic manifold on which the band edges are not achieved by the boundary values $\theta \in \partial W$.

1.5.5 The Decoupled Case

Let us consider again the slowly decaying case (the decoupled Dirichlet case can be treated similarly). Recall that the (Neumann) Laplacian on X_ε converges to the *decoupled* limit operator $\Delta_0 := \bigoplus_e \Delta_{I_e}^{\mathrm{D}} \oplus 0$ acting in the extended Hilbert space $\mathscr{H}_0 := \mathsf{L}_2(G) \oplus \ell_2(V)$, where $\Delta_{I_e}^{\mathrm{D}}$ is the Dirichlet operator on the interval $I_e = [0, \ell_e]$ and 0 is the trivial operator on $\ell_2(V)$. In particular, the spectrum of the limit operator Δ_0 is pure point and given by (1.23). Moreover the multiplicity of λ is $|V|$ if $\lambda = 0$ and $|\{ e \in E \mid \lambda = \lambda_k(e) \}|$ if $\lambda > 0$.

Since the graph structure is no more visible in the limit due to the decoupling, we just give a simple example of a graph consisting of a half-line with one vertex of degree 1 at 0 and the others of degree 2. We label the edges by $n \in \mathbb{N}$. By an appropriate choice of the length ℓ_n we can construct a manifold with certain spectral properties. If we consider e.g. $\ell_n := \sqrt{n}/\pi$ then $\lambda_k(n) = k^2/n$. Since every rational number r can be written in the form $r = k^2/n$ ($r = p/q = p^2/(pq)$), the operator Δ_0 has dense point spectrum consisting of all non-negative rational numbers. Therefore the spectrum of Δ_0 equals $[0, \infty)$ and is purely essential. Other choices of the length are possible, e.g., a set of rationally independent length ℓ_e, $e \in E$. Once one is able to determine $\sigma(\Delta_0)$ by number theoretical arguments one has a Laplacian Δ_{X_ε} with a spectrum close to $\sigma(\Delta_0)$.

The spectral convergence result here is too weak to say anything else about the nature of the Laplacian on the graph-like manifold X_ε with slowly decaying vertex neighbourhoods than that it is purely essential in any bounded spectral interval. Moreover, we cannot apply the results on resolvent convergence since the eigenvalues $\lambda \in \sigma_{\mathrm{p}}(\Delta_0)$ are *not* isolated. We expect that Δ_{X_ε} also has pure point spectrum, but one needs further arguments to prove this.

1.6 Outlook and Remarks

Finally, we outline several topics of further research which in principle can be treated by similar methods as developed in this work.

1.6.1 Geometric Perturbations

We are confident that the convergence methods developed here can be applied in many other situations: For example, one can show convergence results for the Laplacians on topologically perturbed manifolds (removing discs, adding handles etc.), or on conformally perturbed metrics. Such problems have been treated e.g. in [CF78, CF81, Oz81, Oz82, A87, A90, AC93, A94, AC95, H06, Da08, ACP09], but mostly only the eigenvalues on a compact space have been considered.

Formally, adding long thin handles to a family of compact manifolds labelled by the vertices of a graph corresponds to the case when the shrinking rate at the vertex neighbourhoods is $\alpha = 0$. In particular, for such graph-like manifolds, the vertex neighbourhood do not shrink to a point.

We are confident that we can show the convergence of resonances also for such geometrically or conformally perturbed spaces, and obtain results on Helmholtz-resonator-type systems. A *Helmholtz resonator* is a system with a cavity joint by a thin connection with a non-compact exterior part. Although the eigenvalues of the interior problem (without the thin connection) are no longer eigenvalues of the entire problem, they give rise to *resonances*, i.e. poles of a meromorphic continuation of the resolvent. Such problems have been treated e.g. in [Be73,HiM91,Hi92,BHM94, HiL01]. The basic property for all these results is a *decoupling* of the limit problem, leading to embedded eigenvalues in the continuous spectrum. For example, if one approximates hybrid spaces as hedgehog-shaped manifolds presented in [BG03] by smooth manifolds, the limit problem decouples in many situations. Nevertheless, we still obtain the non-trivial result that the embedded eigenvalues of the decoupled limit problem are approached by resonances. Such approximations are interesting in understanding manifolds with cusps (see e.g. [BG05]).

One should also analyse the convergence speed of resonances for our graph-like manifolds. If a resonance approaches e.g. an embedded eigenvalue, then the convergence speed gives the decay rate of the corresponding quasi-stable state.

In addition, one can use the abstract convergence arguments for wildly perturbed domains as in [RT75] (see also [HSS91,Si96]). In particular, it should be possible to construct manifolds with given spectral properties (*spectral engineering*). For such results, one needs scattering theory adopted to our situation, see below.

1.6.2 Scattering Properties

For graph-like spaces, the convergence of scattering data like the scattering matrix and the coefficients of reflection and transmission is of interest (see e.g. [Pav07, Pav02]). More generally, one may develop a scattering theory for operators in two Hilbert spaces as in [K67] making use of our zeroth and first order identification operators. Using similar methods, it should also be possible to show the convergence of the heat kernels. Note that in Theorem 1.3.1 we showed the convergence of the heat operators in the operator norm only, not in the Hilbert-Schmidt norm.

1.6.3 Convergence of Differential Forms and First Order Operators

On graph-like spaces, the convergence of differential forms is also of interest. Here, the topology of the graph-like manifold is of importance. For graph-like spaces homotopy-equivalent with the original metric graph, 1-forms should converge to "1-forms" on a graph. Here, an abstract setting using first order boundary triples as in [P07] combined with the two Hilbert space formalism and identification operators might be helpful. One should obtain stability results on the index as in Sect. 2.4.2. Stability results on the index can be useful in order to decide whether a certain vertex condition can be approximated by purely geometric operators (like the Laplacian on forms) or not. If the indices of the graph operator and the manifold operator are different, there cannot hold such a convergence result.

In [ACP09], we show that for $n \in \mathbb{N}$ and a $\mathbb{Z}$-covering $X \to M$ of a d-dimensional compact manifold M, there is a metric g_n on $X \to M$ such that the spectrum of the Laplacian on p-forms on (X, g_n) has at least n spectral gaps provided the $(d-1)/2$-Betti number of Σ vanishes. The main idea is to deform the compact quotient manifold M in a neighbourhood $U \cong (-1,1) \times \Sigma$ using a warped product on U, where Σ is a non-disconnecting hypersurface in M. In [ACP09], only the convergence of the discrete spectrum is shown (making use of Floquet theory). In principle, it should be possible to carry over the convergence results to general non-compact spaces, and to show for example the convergence of the resolvents.

1.6.4 Convergence of Boundary Triples Coupled via Graphs

We define boundary triples coupled via graphs. In the spirit of our generalised convergence of operators in different Hilbert spaces, one can define the "convergence" of such boundary triples and show e.g. the convergence of the coupled operators. In particular, all our results on graph-like spaces can be understood in this way; and maybe some proofs simplify. Note e.g. that in all our examples we use the natural idea to associate to a function f_e on the metric graph edge e the function $f_e \otimes \varphi_e$ of the associated edge neighbourhood of the graph-like space, where φ_e is an eigenfunction of the transversal manifold. In addition, one might find other results by choosing φ_e to be the second or a higher eigenfunction etc.

A convergence criteria for boundary triples should also yield the convergence of the associated Dirichlet-to-Neumann maps under suitable assumptions. Such results are of interest e.g. for graph-like spaces and their limits (see e.g. [Pav02]).

1.6.5 Metric Structure on the Space of Operators

Given two non-negative and self-adjoint operators Δ and $\widetilde{\Delta}$ on the Hilbert spaces $\mathscr{H}$ and $\widetilde{\mathscr{H}}$, respectively, we can define a sort of "distance"

$$d(\Delta, \widetilde{\Delta}) := \inf\{\, \delta \geq 0 \,|\, \Delta, \widetilde{\Delta} \text{ are } \delta\text{-quasi-unitarily equivalent} \,\}.$$

In this way, we have shown in Theorem 1.3.1 that e.g. the spectrum depends continuously on Δ. Moreover, for the Laplacian Δ_ε on a graph-like manifold and its corresponding graph operator Δ_0 we have shown that $d(\Delta_\varepsilon, \Delta_0) \le O(\varepsilon^\beta)$ for some $\beta > 0$ depending on the geometry.

Several questions arise: What is the metric structure of this space of non-negative operators? A systematic approach of the associated topology as it was e.g. done for the strong convergence of such operators by Kuwae and Shioya in [KSh03] and a comparison of the different concepts deserves a systematic treatment.

Another question is whether our (upper) error estimates on the distance $d(\Delta_\varepsilon, \Delta_0)$ are optimal in the sense, that we can find a lower bound on $d(\Delta_\varepsilon, \Delta_0)$ of order $O(\varepsilon^\beta)$.

1.6.6 Dirichlet-to-Neumann Map and Inverse Problems

The treatment of boundary triples associated with quadratic forms developed in this work provides an abstract way to deal with the Dirichlet-to-Neumann map on a manifold. There are interesting inverse problems concerning the Dirichlet-to-Neumann map: Can one reconstruct the inner domain (or the inner metric graph) from the data of the Dirichlet-to-Neumann map? Such problems occur very often in applications like electrical resistivity tomography. For example, a typical measurement is the following: One applies a voltage to the boundary, and measures the current at the boundary. This measurement is precisely, what the Dirichlet-to-Neumann map gives as a result. We refer to [Na88, LU01] and [CM91, CMM94, CIM98, CM00] for solutions of the inverse problem and related results on manifolds and discrete graphs, respectively.

1.6.7 Fractal Metric Graphs

If the global lower length bound (1.6) is violated, we obtain *fractal* metric graphs. On such graphs, the minimal Laplacian (defined as closure of functions in $\mathsf{H}^2_{\max}(G)$ with *compact* support) is (in general) no longer essentially self-adjoint. In particular, one has to define boundary conditions at "infinity" (see e.g. [Ca00, So04], where radial trees are considered). The corresponding graph-like spaces can be used e.g. in modelling blood-flow in a lung (see [PWZ08]).

Chapter 2
Graphs and Associated Laplacians

In this chapter, we provide the necessary notation and basic results on spectral graph theory needed in the subsequent chapters. Results on spectral theory of combinatorial Laplacians can be found e.g. in [Do84, MW89, CdV98, CGY96, Chu97, HS99, Sh00, HS04, Su08]. For metric graph Laplacians we mention the works [Rot84, vB85, Ni87, KS99, Ha00, KS03, Ku04, FT04a, Ku05, KN05, BF06, KS06, Pan06, BR07, Ku08, P08, HiP09, P09b]. Discrete and metric graphs often serve as a toy model in spectral geometry (see e.g. [CGY96, CGY00]).

Let us briefly motivate the generalisation of the usual discrete (or combinatorial) Laplacian on a graph presented in Sect. 2.1. Generalised discrete Laplacians as defined below in Definition 2.1.10 occur naturally in the Dirichlet-to-Neumann operator of a boundary triple associated with the corresponding (equilateral) metric graph together with a self-adjoint Laplacian acting on it, see Sect. 2.2.3 (the notion of boundary triples will be introduced in Sect. 3.4). The self-adjoint (energy-independent) vertex conditions of a metric graph Laplacian can be encoded in a certain *vertex space* $\mathscr{V} = \bigoplus_v \mathscr{V}_v$. Here, $\mathscr{V}_v$ is a subspace of the $\deg v$-dimensional space $\mathbb{C}^{E_v}$, where E_v is the set of edges adjacent to v, and $\deg v = |E_v|$ is the degree of the vertex. The generalised discrete Laplacian will be an operator acting on $\mathscr{V}$, generalising the usual discrete Laplacian defined on $\ell_2(V)$.

In Sect. 2.1 we introduce discrete Laplacians and their generalisations. Section 2.2 contains material on metric graphs and associated operators, as well as quantum graphs (a metric graph with vertex space). Moreover, we associate boundary triples to a quantum graph, declaring a subset of the vertices as boundary. In Sect. 2.3 we define so-called *extended quantum graphs*, which appear naturally as limit spaces for certain vertex neighbourhood scalings (the "borderline case" in Chap. 6). In all cases (discrete graph, quantum graph and extended quantum graph), we introduce the Laplacians Δ via an *exterior derivative* d by setting $\Delta := \mathrm{d}^*\mathrm{d}$.

The theory of boundary triples gives us a spectral relation between the discrete and metric graph Laplacian, at least for equilateral graphs, see Sect. 2.4. The interpretation of the corresponding discrete operator as a new type of combinatorial Laplacian might be of its own interest (see also [Sm07] for a related generalisation of combinatorial Laplacians via a scattering approach). Moreover, we show in

O. Post, *Spectral Analysis on Graph-Like Spaces*, Lecture Notes in Mathematics 2039,
DOI 10.1007/978-3-642-23840-6_2, © Springer-Verlag Berlin Heidelberg 2012

Theorems 2.4.5 and 2.4.7 the equality of a naturally defined *index* for discrete and metric graph Laplacians, defined as the Fredholm index of d (cf. also [FKW07]).

We end this chapter with some trace formulas for discrete and quantum graph Laplacians (Sect. 2.5). In particular, we can interpret the constant term in the trace formula of Kostrykin, Potthoff and Schrader [KPS07] in Theorem 2.5.5 as index of the associated quantum graph.

2.1 Discrete Graphs and Generalised Laplacians

The aim of the present section is to define the spaces and operators associated with a discrete graph and to conclude some simple consequences needed later on.

2.1.1 Discrete Graphs and Vertex Spaces

Let us first fix the notation for graphs.

Definition 2.1.1.

1. A *discrete graph* G is given by (V, E, ∂), where V, E are countable. Here, $V = V(G)$ denotes the set of vertices, $E = E(G)$ denotes the set of edges, and $\partial\colon E \longrightarrow V \times V$ is a map associating to each edge e the pair $(\partial_- e, \partial_+ e)$ of its *initial vertex* and *terminal vertex*, the *connection map*. [1] In particular, ∂e fixes an orientation of the edge e.[2] Abusing the notation, we also denote by ∂e the *set* $\{\partial_- e, \partial_+ e\}$.
2. For each vertex $v \in V$ we define the (*set of outgoing* (−) resp. *incoming* (+)) *edges adjacent to* v by

$$E_v^{\pm} := \{\, e \in E \mid \partial_{\pm} e = v \,\} \qquad \text{and} \qquad E_v := E_v^{+} \mathbin{\dot\cup} E_v^{-},$$

   i.e. $E_v^{\pm}$ consists of all edges starting (−) resp. ending (+) at v and E_v their *disjoint* union.[3]

---

[1]The use of a connection map allows us to consider multiple edges. Such graphs are sometimes called *multi-graphs*, in contrast to a *simple* graph, where ∂ is assumed to be *injective*. For a simple graph, we may define the set of edges as a subset of $V \times V$.

[2]Some authors (see e.g. [Su08]) prefer to use $E^{\mathrm{unor}} := E \mathbin{\dot\cup} \bar{E}$ for the "doubled" edge set, i.e. each edge $e \in E$ has a counterpart $\bar{e}$ with the opposite orientation. Starting with a set E^{unor}, an *orientation of a graph* can be defined as a subset $E \subset E^{\mathrm{unor}}$ such that either e or $\bar{e}$ belongs to E. In our setting, we automatically fix an orientation. Note that the orientation is important for the exterior derivative $\overset{\dots}{\mathrm{d}}$ in (2.5), but not for the Laplacian $\overset{\dots}{\Delta}$ in (2.4); the former is of "first order", while the second is of "second order".

[3]Note that the *disjoint* union is necessary in order to generate two formally different labels for a *self-loop* e, i.e. an edge with $\partial_- e = \partial_+ e$. Moreover, a loop is counted *twice* in the degree of

3. The *degree of* $v \in V$ is defined by

$$\deg v := |E_v| = |E_v^+| + |E_v^-|,$$

   i.e. the number of adjacent edges at v. In order to avoid trivial cases, we assume that $\deg v \geq 1$, i.e. no vertex is isolated. We also assume that $\deg v$ is finite for each vertex.
4. A discrete graph is *(edge-)*weighted , if there is a function $\ell \colon E \longrightarrow (0, \infty)$ associating to each edge $e \in E$ a *length* $\ell_e > 0$. Alternatively, we may think of $1/\ell_e$ as a *weight* associated with the edge e.[4] If $\ell_e = 1$ for all edges, we call the graph *equilateral.*

Let us slightly generalise the concept of a graph by allowing edges with *infinite* length, i.e. having its terminal vertex at "infinity":

Definition 2.1.2.

1. We call $G = (V, E, \partial)$ a *discrete exterior graph* if we have a decomposition of E into *interior* and *exterior* edges, i.e. $E = E_{\mathrm{int}} \mathbin{\dot\cup} E_{\mathrm{ext}}$. In addition, we assume that $\partial \colon E \longrightarrow (V \times V) \mathbin{\dot\cup} V$ maps an interior edge $e \in E_{\mathrm{int}}$ onto the pair $\partial e = (\partial_- e, \partial_+ e)$ of its initial and terminal vertex, and an exterior edge $e \in E_{\mathrm{ext}}$ onto its initial vertex $\partial e = \partial_- e$ only. The set E_v of edges adjacent to v and the degree $\deg v$ is defined as above. Note that an exterior edge $e \in E_{\mathrm{ext}}$ belongs only to the set E_v^- of outgoing edges since $v = \partial_- e$.
2. We call $G = (V, E, \partial, \ell)$ an *(edge-)weighted discrete exterior graph* if (V, E, ∂) is a discrete exterior graph, and $\ell \colon E \longrightarrow (0, \infty]$ is a length function such that

$$E_{\mathrm{int}} = \{\, e \in E \mid \ell_e < \infty \,\} \qquad \text{and} \qquad E_{\mathrm{ext}} = \{\, e \in E \mid \ell_e = \infty \,\}.$$

Note that for a weighted discrete exterior graph, we may interpret interior edges as edges of finite length, and exterior edges as edges of infinite length with terminal vertex at "infinity".

We will use frequently the following elementary fact about reordering a sum over edges and vertices, namely

$$\sum_{e \in E_{\mathrm{int}}} F(\partial_+ e, e) = \sum_{v \in V} \sum_{e \in E_v^+} F(v, e) \quad \text{and} \quad \sum_{e \in E} F(\partial_- e, e) = \sum_{v \in V} \sum_{e \in E_v^-} F(v, e) \tag{2.1}$$

for a function $(v, e) \mapsto F(v, e)$ depending on v and $e \in E_v$ with the convention that a sum over the empty set is 0. Note that this equation is also valid for self-loops and

---

a vertex. This convention is useful when comparing discrete and equilateral metric graphs (see Theorem 2.4.1).

[4]The interpretation "length" will become clear when defining *metric graphs* in Definition 2.2.1, as well as the interpretation of $1/\ell_e$ as a *weight* or *conductivity*, see the definition of the norm on $\ell_2(E)$ in (2.6).

multiple edges. The reordering is a bijection since the unions $E_{\rm int} = \dot\bigcup_{v\in V} E_v^+$ and $E = \dot\bigcup_{v\in V} E_v^-$ are *disjoint*. For a graph with finite edge set, the relation

$$2|E_{\rm int}| + |E_{\rm ext}| = \sum_{v\in V} \deg v \tag{2.2}$$

follows by setting $F(v,e) = 1$. For simplicity, we assume for the rest of *this* section that the graph has no exterior edges, i.e. that $E = E_{\rm int}$.

Let us make the following assumption on the lower bound of the edge lengths:

Assumption 2.1.1. Throughout this work we assume that there is a constant $\ell_- > 0$ such that

$$\ell_e \geq \ell_- \qquad \forall\, e \in E, \tag{2.3}$$

i.e. that the weight function ℓ^{-1} is bounded. Without loss of generality and for convenience, we assume that $\ell_- \leq 1$.

We want to introduce a vertex space allowing us to define Laplace-like combinatorial operators motivated by general vertex boundary conditions on metric graphs.

The usual discrete (weighted) Laplacian is defined on *scalar* functions $F: V \longrightarrow \mathbb{C}$ on the vertices V, namely

$$\dddot{\Delta} F(v) = -\frac{1}{\deg v} \sum_{e\in E_v} \frac{1}{\ell_e}\big(F(v_e) - F(v)\big), \tag{2.4}$$

where v_e denotes the vertex on e opposite to v. Note that $\dddot{\Delta}$ can be written as $\dddot{\Delta} = \dddot{\rm d}^*\dddot{\rm d}$ with

$$\dddot{\rm d}: \ell_2(V) \longrightarrow \ell_2(E), \qquad (\dddot{\rm d}F)_e = F(\partial_+ e) - F(\partial_- e). \tag{2.5}$$

Here, $\ell_2(V) = \ell_2(V, \deg)$ and $\ell_2(E) = \ell_2(E, \ell^{-1})$ carry the *weighted* norms defined by

$$\|F\|^2_{\ell_2(V)} := \sum_{v\in V} |F(v)|^2 \deg v \quad \text{and} \quad \|\eta\|^2_{\ell_2(E)} := \sum_{e\in E} |\eta_e|^2 \frac{1}{\ell_e}, \tag{2.6}$$

and $\dddot{\rm d}^*$ denotes the adjoint with respect to the corresponding inner products. We sometimes refer to functions in $\ell_2(V)$ and $\ell_2(E)$ as 0*-forms* and 1*-forms*, respectively.

We would like to carry over the above concept for the "vertex space" $\ell_2(V)$ to more general vertex spaces $\mathscr{V} = \bigoplus_v \mathscr{V}_v$. The main motivation to do so are metric graph Laplacians with general vertex boundary conditions as defined in Sect. 2.2.2 and their relations with discrete graphs (cf. Sect. 2.4).

Definition 2.1.3.

1. Denote by $\mathscr{V}_v^{\max} := \mathbb{C}^{E_v}$ the *maximal vertex space* at the vertex $v \in V$, i.e. a value $F(v) = \{F_e(v)\}_{e \in E_v} \in \mathscr{V}_v^{\max}$ has $\deg v$ components, one for each adjacent edge. A *vertex space* at the vertex v is a linear subspace $\mathscr{V}_v$ of $\mathscr{V}_v^{\max}$.
2. The corresponding (total) vertex spaces associated with the graph (V, E, ∂) are

$$\mathscr{V}^{\max} := \bigoplus_{v \in V} \mathscr{V}_v^{\max} \quad \text{and} \quad \mathscr{V} := \bigoplus_{v \in V} \mathscr{V}_v,$$

respectively. Elements of $\mathscr{V}$ are also called *0-forms*. The space $\mathscr{V}$ carries its natural Hilbert norm, namely

$$\|F\|^2_{\mathscr{V}} := \sum_{v \in V} |F(v)|^2 = \sum_{v \in V} \sum_{e \in E_v} |F_e(v)|^2.$$

Associated with a vertex space is an orthogonal projection $P = \bigoplus_{v \in V} P_v$ in $\mathscr{V}^{\max}$, where P_v is the orthogonal projection in $\mathscr{V}_v^{\max}$ onto $\mathscr{V}_v$.
3. We call a general subspace $\mathscr{V}$ of $\mathscr{V}^{\max}$ *local* if it decomposes with respect to $\mathscr{V}^{\max} = \bigoplus_v \mathscr{V}_v^{\max}$, i.e. if $\mathscr{V} = \bigoplus_v \mathscr{V}_v$ and $\mathscr{V}_v \subset \mathscr{V}_v^{\max}$. Similarly, an operator A on $\mathscr{V}$ is called *local* if it is decomposable with respect to the above direct sum.
4. The *dual vertex space* associated with $\mathscr{V}$ is defined by $\mathscr{V}^{\perp} := \mathscr{V}^{\max} \ominus \mathscr{V}$ and has projection $P^{\perp} = \mathbb{1} - P$.

Note that a local subspace $\mathscr{V}$ is closed since $\mathscr{V}_v \leq \mathscr{V}_v^{\max}$ is finite dimensional. Alternatively, a vertex space is characterised by fixing an orthogonal projection P in $\mathscr{V}$ which is local. In view of the corresponding notation on a metric graph (see Definition 2.2.5), one may call the pair $(G, \mathscr{V})$ a *discrete quantum graph*.

Example 2.1.4. The names of the vertex spaces in the examples are borrowed from the corresponding examples in the metric graph case, see below. For more general cases defined via vertex spaces, e.g. the discrete magnetic Laplacian, we refer to [P09a].

1. Choosing $\mathscr{V}_v = \mathbb{C}\mathbb{1}(v) = \mathbb{C}(1, \ldots, 1)$, we obtain the *standard vertex space* denoted by $\mathscr{V}_v^{\text{std}}$, also called *continuous vertex space* or *Kirchhoff vertex space*. The associated projection is

$$P_v = \frac{1}{\deg v} \mathbb{E}$$

where $\mathbb{E}$ denotes the square matrix of rank $\deg v$ where all entries equal 1. This case corresponds to the standard discrete case mentioned before. Namely, the natural identification

$$\widetilde{\bullet} \colon \mathscr{V}^{\text{std}} := \bigoplus_v \mathscr{V}_v^{\text{std}} \longrightarrow \ell_2(V), \qquad F \mapsto \widetilde{F}, \qquad \widetilde{F}(v) := F_e(v),$$

(the latter value is independent of $e \in E_v$) is isometric, since the weighted norm in $\ell_2(V)$ and the norm in $\mathscr{G}^{\rm std}$ agree, i.e.

$$\|F\|^2_{\mathscr{V}^{\rm std}} = \sum_{v\in V}\sum_{e\in E_v} |F_e(v)|^2 = \sum_{v\in V} |\widetilde{F}(v)|^2 \deg v = \|\widetilde{F}\|^2_{\ell_2(V)}.$$

2. More generally, we can fix a vector $p(v) = \{p_e(v)\}_{e\in E_v}$ with non-negative entries $p_e(v) \neq 0$ and define the *weighted standard* vertex space by $\mathscr{V}_v^p := \mathbb{C}p(v)$. The corresponding projection is given by

$$P_v F(v) = \frac{1}{|p(v)|^2}\langle p(v), F(v)\rangle\, p(v).$$

As in the previous example, we have an isometry

$$\widetilde{\bullet}\colon \mathscr{V}^p := \bigoplus_v \mathscr{V}_v^{p(v)} \longrightarrow \ell_2(V, |p|^2), \qquad F \mapsto \widetilde{F}, \qquad \widetilde{F}(v) := \frac{F_e(v)}{p_e(v)}$$

(the latter value is independent of $e \in E_v$), since

$$\|F\|^2_{\mathscr{V}^p} = \sum_{v\in V}\sum_{e\in E_v} |F_e(v)|^2 = \sum_{v\in V} |\widetilde{F}(v)|^2 |p(v)|^2 =: \|\widetilde{F}\|^2_{\ell_2(V,|p|^2)}.$$

3. A special case of a weighted standard vertex space is given by vectors $p(v)$ such that $|p_e(v)| = 1$, we call such a vertex space a *magnetic perturbation* of the standard one. An example is giving in the following way: Let $\alpha \in \mathbb{R}^E$ be a function associating to each edge e the *magnetic vector potential* $\alpha_e \in \mathbb{R}$ and set

$$p_e(v) = \mathrm{e}^{-\mathrm{i}\overset{\curvearrowleft}{\alpha}_e(v)/2}$$

where $\overset{\curvearrowleft}{\alpha}_e(v) := \pm\alpha_e$ if $v = \partial_\pm e$. We call the associated vertex space $\mathscr{V}_v^{\rm mag,\alpha}$ *magnetic*.
4. We call $\mathscr{V}_v^{\min} := 0$ the *minimal vertex space* or *Dirichlet vertex space*. Similarly, $\mathscr{V}^{\max}$ is called the *maximal vertex space* or *Neumann vertex space*. The corresponding projections are $P = 0$ and $P = \mathbb{1}$.
5. Assume that $\deg v = 4$ and define a vertex space of dimension 2 by

$$\mathscr{V}_v = \mathbb{C}(1,1,1,1) \oplus \mathbb{C}(1,\mathrm{i},-1,-\mathrm{i}).$$

The corresponding orthogonal projection is

$$P_v = \frac{1}{4}\begin{pmatrix} 2 & 1+\mathrm{i} & 0 & 1-\mathrm{i} \\ 1-\mathrm{i} & 2 & 1+\mathrm{i} & 0 \\ 0 & 1-\mathrm{i} & 2 & 1+\mathrm{i} \\ 1+\mathrm{i} & 0 & 1-\mathrm{i} & 2 \end{pmatrix}.$$

In Example 2.1.7 (2), we will show some invariance properties of this vertex space.

In contrast to the standard vertex space, the vertex space may "decouple" some or all of the adjacent edges $e \in E_v$ at a vertex v, e.g., if the vertex space is $\mathscr{V}^{\max}$. "Decoupling" here means, that we may split the graph at a vertex space $\mathscr{V}_v$ such that the corresponding projection P_v has block structure w.r.t. a non-trivial decomposition $\mathscr{V}_v = \mathscr{V}_{1,v} \oplus \mathscr{V}_{2,v}$. We call a vertex space $\mathscr{V}_v$ without such a decomposition *irreducible*. Similarly, we say that $\mathscr{V} = \bigoplus_v \mathscr{V}_v$ is *irreducible*, if all its local subspaces $\mathscr{V}_v$ are irreducible. For more details, we refer to [P09b].

In [P09a, Lem. 2.13] we showed the following result on symmetry of a vertex space:

Proposition 2.1.5. *Assume that the vertex space $\mathscr{V}_v$ of a vertex v with degree $d = \deg v$ is invariant under permutations of the coordinates $e \in E_v$, then $\mathscr{V}_v$ is one of the spaces $\mathscr{V}_v^{\min} = 0$, $\mathscr{V}_v^{\max} = \mathbb{C}^{E_v}$, $\mathscr{V}_v^{\mathrm{std}} = \mathbb{C}(1,\dots,1)$ or $(\mathscr{V}^{\mathrm{std}})^{\perp}$, i.e. only the minimal, maximal, standard and dual standard vertex spaces are invariant.*

If we only require invariance under the cyclic group of order d, we have the following result:

Proposition 2.1.6. *Assume that the vertex space $\mathscr{V}_v$ of a vertex v with degree $d = \deg v$ is invariant under cyclic permutation of the coordinates $e \in E_v = \{e_1,\dots,e_d\}$, i.e. $e_i \mapsto e_{i+1}$ and $e_d \mapsto e_1$, then $\mathscr{V}_v$ is an orthogonal sum of spaces of the form $\mathscr{V}_v^k = \mathbb{C}(1,\theta^k,\theta^{2k},\dots,\theta^{(d-1)k})$ for $k = 0,\dots,d-1$, where $\theta = \mathrm{e}^{2\pi \mathrm{i}/d}$.*

Proof. The (representation-theoretic) irreducible vector spaces invariant under the cyclic group are one-dimensional (since the cyclic group is Abelian) and have the form $\mathscr{V}_v^k$ as given above. □

We call $\mathscr{V}_v^k$ a *magnetic perturbation* of $\mathscr{V}_v^{\mathrm{std}}$, i.e. the components of the generating vector $(1,\dots,1)$ are multiplied with a phase factor (see e.g. [P09a, Ex. 2.10 (vii)]).

Example 2.1.7.

1. If we require that the vertex space $\mathscr{V}_v$ is cyclic invariant with *real* coefficients in the corresponding projections, then $\mathscr{V}_v$ is spanned by either $(1,\dots,1)$ or $(1,-1,\dots,1,-1)$ (if d even) or both. But the sum is not irreducible since

$$\begin{aligned}\mathscr{V}_v &= \mathbb{C}(1,\dots,1) \oplus \mathbb{C}(1,-1,\dots,1,-1)\\ &= \mathbb{C}(1,0,1,0,\dots,1,0) \oplus \mathbb{C}(0,1,0,1,\dots,0,1)\end{aligned}$$

   and the latter two spaces are standard with degree $d/2$. In other words, the irreducible graph at v associated with the boundary space $\mathscr{V}_v$ splits the vertex v into two vertices v_1 and v_2 adjacent with the edges with even and odd labels, respectively. The corresponding vertex spaces are standard.
2. The sum of two cyclic invariant spaces is not always reducible: Take the cyclic invariant vertex space $\mathscr{V}_v = \mathscr{V}_v^0 \oplus \mathscr{V}_v^1 \subset \mathbb{C}^4$ of dimension 2 given in

Example 2.1.4 (5). Note that $\mathscr{V}_v$ is irreducible, since the associated projection P_v does not have block structure. This vertex space is maybe the simplest example of a (cyclic invariant) irreducible vertex space which is not standard or dual standard. Note that if $\deg v = 3$, then an irreducible vertex space is either standard or dual standard (or the corresponding version with weights and magnetic perturbations, i.e. $(1, \ldots, 1)$ replaced by a vector $p(v)$ with non-zero entries).

2.1.2 Operators Associated with Vertex Spaces

Let us now define a generalised *coboundary operator* or *exterior derivative associated with a vertex space*. We use this exterior derivative for the definition of an associated Laplace operator below:

Definition 2.1.8. Let $\mathscr{V}$ be a vertex space of the graph G. The *exterior derivative* on $\mathscr{V}$ is defined via

$$\dddot{\mathrm{d}}_{\mathscr{V}}: \mathscr{V} \longrightarrow \ell_2(E), \qquad (\dddot{\mathrm{d}}_{\mathscr{V}} F)_e := F_e(\partial_+ e) - F_e(\partial_- e),$$

mapping 0-forms onto 1-forms.

We often drop the subscript $\mathscr{V}$ for the vertex space or write $\dddot{\mathrm{d}}_{\mathrm{std}}$ instead of $\dddot{\mathrm{d}}_{\mathscr{V}\mathrm{std}}$ etc. The proof of the next lemma is straightforward using (2.1) (see e.g. [P09a, Lem. 3.3]):

Lemma 2.1.9. *Assume the lower lengths bound* (2.3), *then* $\dddot{\mathrm{d}}$ *is norm-bounded by* $\sqrt{2/\ell_-}$. *The adjoint*

$$\dddot{\mathrm{d}}^*_{\mathscr{V}}: \ell_2(E) \longrightarrow \mathscr{V}$$

fulfils the same norm bound and is given by

$$(\dddot{\mathrm{d}}^* \eta)(v) = P_v\Bigl(\Bigl\{\frac{1}{\ell_e} \overset{\curvearrowright}{\eta}_e(v)\Bigr\}_{e \in E_v}\Bigr) \in \mathscr{V}_v,$$

where $\overset{\curvearrowright}{\eta}_e(v) := \pm \eta_e$ *if* $v = \partial_\pm e$ *denotes the* oriented evaluation *of* η_e *at the vertex* v.

Definition 2.1.10. The *discrete generalised Laplacian* associated with a vertex space $\mathscr{V}$ is defined by $\dddot{\Delta}_{\mathscr{V}} := \dddot{\mathrm{d}}^*_{\mathscr{V}} \dddot{\mathrm{d}}_{\mathscr{V}}$, i.e.

$$(\dddot{\Delta}_{\mathscr{V}} F)(v) = P_v\Bigl(\Bigl\{\frac{1}{\ell_e}(F_e(v) - F_e(v_e))\Bigr\}_{e \in E_v}\Bigr)$$

for $F \in \mathscr{V}$, where v_e denotes the vertex on $e \in E_v$ opposite to v.

Remark 2.1.11.

1. From Lemma 2.1.9 it follows that $\ddot{\Delta}_{\mathscr{V}}$ is a bounded, non-negative operator on $\mathscr{V}$ with norm estimated from above by $2/\ell_-$. In particular, $\sigma(\ddot{\Delta}_{\mathscr{V}}) \subset [0, 2/\ell_-]$.
2. We can also define a Laplacian $\ddot{\Delta}^1_{\mathscr{V}} := \ddot{d}_{\mathscr{V}}\ddot{d}^*_{\mathscr{V}}$ acting on the space of "1-forms" $\ell_2(E)$ (and $\ddot{\Delta}^0_{\mathscr{V}} := \ddot{\Delta}_{\mathscr{V}} = \ddot{d}^*_{\mathscr{V}}\ddot{d}_{\mathscr{V}}$). For more details and the related supersymmetric setting, we refer to [P09a]. In particular, we have

$$\sigma(\ddot{\Delta}^1_{\mathscr{V}}) \setminus \{0\} = \sigma(\ddot{\Delta}^0_{\mathscr{V}}) \setminus \{0\}.$$

   Moreover, in [P09a, Ex. 3.16–3.17] we discussed how these generalised Laplacians can be used in order to analyse the (standard) Laplacian on the *line graph* and *subdivision graph* associated with G (see also [Sh00]).

The next example shows that we have indeed a generalisation of the standard discrete Laplacian:

Example 2.1.12.

1. For the standard vertex space $\mathscr{V}^{\mathrm{std}}$, it is convenient to use the unitary transformation from $\mathscr{V}^{\mathrm{std}}$ onto $\ell_2(V)$ associating to $F \in \mathscr{V}$ the (common value) $\widetilde{F}(v) := F_e(v)$ as in Example 2.1.4 (1). Then the exterior derivative $\ddot{d}_{\mathrm{std}}$ and its adjoint $\ddot{d}^*_{\mathrm{std}}$ are unitarily equivalent with

$$\ddot{d}\colon \ell_2(V) \longrightarrow \ell_2(E), \qquad (\ddot{d}\widetilde{F})_e = \widetilde{F}(\partial_+e) - \widetilde{F}(\partial_-e)$$

   and

$$(\ddot{d}^*\eta)(v) = \frac{1}{\deg v} \sum_{e \in E_v} \frac{1}{\ell_e} \overset{\curvearrowright}{\eta}_e(v),$$

   i.e. $\ddot{d}$ is the classical *coboundary operator* already defined in (2.5) and $\ddot{d}^*$ its adjoint.

   Moreover, the corresponding discrete Laplacian $\ddot{\Delta}_{\mathrm{std}} := \ddot{\Delta}_{\mathscr{V}^{\mathrm{std}}}$ is unitarily equivalent with the usual discrete Laplacian $\ddot{\Delta} = \ddot{d}^*\ddot{d}$ defined in (2.4) as one can easily check.

   Similarly, for the weighted standard vertex space $\mathscr{V}^p$ with $p_e(\partial_-e) = p_e(\partial_+e)$ $(=: p_e)$, the generalised discrete Laplacian expressed on the space $\ell_2(V, |p|^2)$ is given by

$$\ddot{\Delta}_p \widetilde{F}(v) = -\frac{1}{|p(v)|^2} \sum_{e \in E_v} \frac{|p_e|^2}{\ell_e} \big(\widetilde{F}(v_e) - \widetilde{F}(v)\big), \tag{2.7}$$

   where $|p(v)|^2 = \sum_{e \in E_v} |p_e(v)|^2$.

For the magnetically perturbed standard vertex space we have

$$(\ddot{\Delta}_{\mathrm{mag},\alpha}\widetilde{F})(v) = -\frac{1}{\deg v}\sum_{e\in E_v}\frac{1}{\ell_e}(\mathrm{e}^{+\mathrm{i}\widehat{\alpha}_e(v)}\widetilde{F}(v_e)) - \widetilde{F}(v),$$

i.e. the *discrete magnetic Laplacian* associated with the magnetic potential $\alpha = \{\alpha_e\}_e$. Note that the θ-periodic boundary condition of Floquet theory as in (1.24) lead to a magnetic perturbation of a vertex space.

2. For the minimal vertex space $\mathscr{V}^{\min} = 0$, we have $\ddot{\mathrm{d}} = 0$, $\ddot{\mathrm{d}}^* = 0$ and $\ddot{\Delta}_{\mathscr{V}^{\min}} = 0$. For the maximal vertex space, we have

$$(\ddot{\Delta}_{\mathscr{V}^{\max}}F)_e(v) = \Big\{\frac{1}{\ell_e}(F_e(v) - F_e(v_e))\Big\}_{e\in E_v}$$

and

$$\ddot{\Delta}_{\mathscr{V}^{\max}} \cong \bigoplus_{e\in E}\ddot{\Delta}_e, \qquad \text{where} \qquad \ddot{\Delta}_e \cong \frac{1}{\ell_e}\begin{pmatrix} 1 & -1 \\ -1 & 1\end{pmatrix}.$$

In particular, in both cases, the Laplacians are decoupled and any connection information of the graph is lost.

Let us now assume that the graph is equilateral (i.e. $\ell_e = 1$) and the graph has no multiple edges (i.e. ∂ is injective). Then we can write the Laplacian in the form

$$\ddot{\Delta}_{\mathscr{V}} = \mathbb{1} - M_{\mathscr{V}}, \qquad M_{\mathscr{V}} := PA^{\max},$$

where $M_{\mathscr{V}}: \mathscr{V} \longrightarrow \mathscr{V}$ is called the *principle part* of the generalised discrete Laplacian, and $A^{\max}: \mathscr{V}^{\max} \longrightarrow \mathscr{V}^{\max}$ the *generalised adjacency matrix*, defined by

$$A^{\max}\{F(w)\}_w = \{A^{\max}(v,w)F(w)\}_v, \quad A^{\max}(v,w): \mathbb{C}^{E_w} \longrightarrow \mathbb{C}^{E_v}$$

for $F \in \mathscr{V}^{\max}$. Furthermore, $A^{\max}(v,w) = 0$ if v, w are not joined by an edge and

$$A^{\max}(v,w)_{e,e'} = \delta_{e,e'}, \quad e \in E_v,\ e' \in E_w$$

otherwise. In particular, written as a matrix, $A^{\max}(v,w)$ has only one entry 1 and all others equal to 0. The principle part of the Laplacian then has the form

$$(M_{\mathscr{V}}F)(v) = \sum_{e\in E_v} A_{\mathscr{V}}(v, v_e)F(v_e),$$

for $F \in \mathscr{V}$ similar to the form of the principle part of the standard Laplacian defined for $\mathscr{V}^{\mathrm{std}} \cong \ell_2(V)$, where

$$A_{\mathscr{V}}(v,w) := P_v A^{\max}(v,w)P_w: \mathscr{V}_w \longrightarrow \mathscr{V}_v.$$

Equivalently,

$$M_{\mathscr{V}} = \bigoplus_{v \in V} \sum_{w \in V} A_{\mathscr{V}}(v, w) \tag{2.8}$$

where the sum is actually only over those vertices w, which are connected with v. In particular, in the standard case $\mathscr{V} = \mathscr{V}^{\mathrm{std}}$, the matrix $A_{\mathscr{V}^{\mathrm{std}}}(v, w)$ consists of one entry only since $\mathscr{V}_v^{\mathrm{std}} \cong \mathbb{C}(\deg v)$ isometrically. Namely, we have $A_{\mathscr{V}^{\mathrm{std}}}(v, w) = 1$ if v and w are connected and 0 otherwise, i.e. $A_{\mathscr{V}^{\mathrm{std}}}$ is (unitarily equivalent with) the standard adjacency operator in $\ell_2(V)$.

Let us return to the general situation (i.e. general lengths ℓ_e and possibly multiple edges). Let G be a discrete graph. We define a *Hilbert chain associated with a vertex space* $\mathscr{V}$ *on* G by

$$\mathscr{C}_{(G,\mathscr{V})}: 0 \longrightarrow \mathscr{V} \xrightarrow{\dddot{\mathrm{d}}_{\mathscr{V}}} \ell_2(E) \longrightarrow 0.$$

Obviously, the chain condition is trivially satisfied since only one operator is non-zero. In this situation and since we deal with Hilbert spaces, the associated cohomology spaces (with coefficients in $\mathbb{C}$) can be defined by

$$H^0(G, \mathscr{V}) := \ker \dddot{\mathrm{d}}_{\mathscr{V}} \cong \ker \dddot{\mathrm{d}}_{\mathscr{V}} / \operatorname{ran} 0,$$
$$H^1(G, \mathscr{V}) := \ker \dddot{\mathrm{d}}_{\mathscr{V}}^* = (\operatorname{ran} \dddot{\mathrm{d}}_{\mathscr{V}})^{\perp} \cong \ker 0 / \operatorname{ran} \dddot{\mathrm{d}}_{\mathscr{V}}$$

where $\operatorname{ran} A := A(\mathscr{H}_1)$ denotes the range ("image") of the operator $A: \mathscr{H}_1 \longrightarrow \mathscr{H}_2$. The *index* or *Euler characteristic* of the Hilbert chain $\mathscr{C}_{(G,\mathscr{V})}$ is defined by

$$\operatorname{ind}(G, \mathscr{V}) := \dim \ker \dddot{\mathrm{d}}_{\mathscr{V}} - \dim \ker \dddot{\mathrm{d}}_{\mathscr{V}}^*,$$

i.e. the *Fredholm index* of $\dddot{\mathrm{d}}_{\mathscr{V}}$, provided at least one of the dimensions is finite. Note that for the standard vertex space $\mathscr{V}^{\mathrm{std}} \cong \ell_2(V)$, the exterior derivative is just (unitarily equivalent with) the classical coboundary operator defined in (2.5). In particular, the corresponding homology spaces are the classical ones, and $\dim H^p(G, \mathscr{V}^{\mathrm{std}})$ counts the number of components ($p = 0$) and edges not in a spanning tree ($p = 1$).

Using the stability of the index under continuous perturbations, we can calculate the index via simple (decoupled) model spaces and obtain the following result (see [P09a, Sect. 4]); alternatively, we can apply standard arguments from linear algebra:

Proposition 2.1.13. *Let* $\mathscr{V}$ *be a vertex space associated with the finite discrete graph* $G = (V, E, \partial)$, *then*

$$\operatorname{ind}(G, \mathscr{V}) = \dim \mathscr{V} - |E|.$$

Note that in particular, if $\mathscr{V} = \mathscr{V}^{\mathrm{std}}$, i.e. if $\mathscr{V} \cong \ell_2(V)$ is the standard vertex space, we recover the well-known formula for (standard) discrete graphs, namely

$$\operatorname{ind}(G, \mathscr{V}^{\mathrm{std}}) = |V| - |E|,$$

i.e. the index equals the Euler characteristic $\chi(G) := |V| - |E|$ of the graph G. On the other hand, in the "extreme" cases, we have

$$\operatorname{ind}(G, \mathscr{V}^{\max}) = |E| \quad \text{and} \quad \operatorname{ind}(G, \mathscr{V}^{\min}) = -|E|.$$

since $\dim \mathscr{V}^{\max} = \sum_{v \in V} \deg v = 2|E|$ and $\dim \mathscr{G}^{\min} = 0$. Note that the first index equals the Euler characteristic of the "decoupled" graph G^{dec} consisting of the disjoint union of the graphs $G_e = (\partial e, e)$ having only two vertices and one edge, i.e. we have

$$\operatorname{ind}(G, \mathscr{V}^{\max}) = \chi(G^{\mathrm{dec}}) = \sum_e \chi(G_e) = |E|,$$

since $\chi(G_e) = 2 - 1 = 1$. Similarly, the second index equals the *relative Euler characteristic*, i.e.

$$\operatorname{ind}(G, \mathscr{V}^{\min}) = \chi(G^{\mathrm{dec}}, \partial G^{\mathrm{dec}}) := \chi(G^{\mathrm{dec}}) - \chi(\partial G^{\mathrm{dec}}) = -|E|,$$

where $\partial G^{\mathrm{dec}} = \dot{\bigcup}_e \partial G_e$ and $\partial G_e = \partial e$.

In [P09a, Lem. 4.4] we established a general result on the cohomology of the dual $\mathscr{V}^{\perp}$ of a vertex space $\mathscr{V}$. It shows that actually, $\mathscr{V}^{\perp}$ and the *oriented* version of $\mathscr{V}$, i.e. $\overset{\curvearrowright}{\mathscr{V}} = \{\, F \in \mathscr{V}^{\max} \mid \overset{\curvearrowright}{F} \in \mathscr{G}\,\}$, are related:

Proposition 2.1.14. *Assume that the global length bound*

$$\ell_- \leq \ell_e \leq \ell_+ \qquad \textit{for all } \mathrm{e} \in E \tag{2.9}$$

holds for some constants $0 < \ell_- \leq \ell_+ < \infty$. *Then* $H^0(G, \mathscr{V}^{\perp})$ *and* $H^1(G, \overset{\curvearrowright}{\mathscr{V}})$ *are isomorphic. In particular, if G is finite, then* $\operatorname{ind}(G, \mathscr{G}^{\perp}) = -\operatorname{ind}(G, \overset{\curvearrowright}{\mathscr{V}})$.

The orientation also occurs in the metric graph case, see e.g. Lemma 2.2.8.

2.2 Metric Graphs, Quantum Graphs and Associated Operators

In this section, we fix the basic notion for metric and quantum graphs and derive some general assertion needed later on. Most of the material is standard, and we refer to the literature for further results and references, see e.g. [Ku08, Ku05, Ku04, KS99, KSm99].

2.2.1 Metric Graphs

Definition 2.2.1. Let $G = (V, E, \partial)$ be a discrete (exterior) graph. A *topological graph* G^{top} associated with G is a CW complex containing only 0-cells and 1-cells, such that the 0-cells are the vertices V and the 1-cells are labelled by the edge set E, respecting the graph structure in the obvious way.

A *metric graph* G^{met} associated with a weighted discrete graph (V, E, ∂, ℓ) is a topological graph associated with (V, E, ∂) such that for every edge $e \in E$ there is a continuous map $\Phi_e \colon I_e \longrightarrow G^{\mathrm{met}}$, $I_e := [0, \ell_e]$ (resp. $I_e := [0, \infty)$ if $e \in E_{\mathrm{ext}}$), such that $\Phi_e(\mathring{I}_e)$ is the 1-cell corresponding to e, and the restriction $\Phi_e \colon \mathring{I}_e \longrightarrow \Phi(\mathring{I}_e) \subset G^{\mathrm{met}}$ is a homeomorphism. The maps Φ_e induce a metric on G^{met}. In this way, G^{met} becomes a metric space.

By abuse of notation, we omit the labels $(\cdot)^{\mathrm{top}}$ and $(\cdot)^{\mathrm{met}}$ for the topological and metric graph associated with the discrete weighted graph, and simply write G or (V, E, ∂, ℓ).

Given a weighted discrete graph, we can abstractly construct the associated metric graph as the disjoint union of the intervals I_e for all $e \in E$ and appropriate identifications of the end-points of these intervals (according to the combinatorial structure of the graph), namely

$$G^{\mathrm{met}} = \dot{\bigcup_{e \in E}} I_e / \sim. \tag{2.10}$$

We denote the union of the 0-cells and the union of the (open) 1-cells (edges) by G^0 and G^1, i.e.

$$G^0 = V \hookrightarrow G^{\mathrm{met}}, \qquad G^1 = \dot{\bigcup_{e \in E}} I_e \hookrightarrow G^{\mathrm{met}},$$

and both subspaces are canonically embedded in G^{met}.

Remark 2.2.2.

1. The metric graph G^{met} becomes canonically a *metric measure space* by defining the distance of two points to be the length of the shortest path in G^{met}, joining these points. We can think of the maps $\Phi_e \colon I_e \longrightarrow G^{\mathrm{met}}$ as coordinate maps and the Lebesgue measures $\mathrm{d}s_e$ on the intervals I_e induce a (Lebesgue) measure on the space G^{met}. We will often omit the coordinate map Φ_e, e.g., for functions f on G^{met} we simply write $f_e := f \circ \Phi_e$ for the corresponding function on I_e. If the edge e is clear from the context, we also omit the label $(\cdot)_e$.
2. Note that two metric graphs G_1^{met} and G_2^{met} can be isometric as metric spaces, such that the underlying discrete graphs G_1 and G_2 are not isomorphic: The metric on a metric graph G^{met} cannot distinguish between a single edge e of length ℓ_e in G_1 and two edges e', e'' of length $\ell_{e'}$, $\ell_{e''}$ with $\ell_e = \ell_{e'} + \ell_{e''}$ joined by a vertex of degree 2 in G_2: The underlying graphs are not (necessarily) isomorphic. For a discussion on this point, see for example [BR07, Sect. 2].

2.2.2 Operators on Metric Graphs

Since a metric graph is a topological space, and isometric to intervals outside the vertices, we can introduce the notion of measurability and differentiate function on the edges. We start with the basic Hilbert space

$$\mathsf{L}_2(G) := \bigoplus_{e \in E} \mathsf{L}_2(I_e) \qquad \text{and} \qquad \|f\|^2 = \|f\|^2_{\mathsf{L}_2(G)} := \sum_{e \in E} \int_{I_e} |f_e(s)|^2 \mathrm{d}s,$$

where $f = \{f_e\}_e$ with $f_e \in \mathsf{L}_2(I_e)$.

In order to define Laplacian-like differential operators in the Hilbert space $\mathsf{L}_2(G)$ we introduce the *maximal Sobolev space* or *decoupled Sobolev space* of order k as

$$\mathsf{H}^k_{\max}(G) := \bigoplus_{e \in E} \mathsf{H}^k(I_e),$$

$$\|f\|^2_{\mathsf{H}^k_{\max}(G)} := \sum_{e \in E} \|f_e\|^2_{\mathsf{H}^k(I_e)},$$

where $\mathsf{H}^k(I_e)$ is the classical Sobolev space on the interval I_e, i.e. the space of functions with (weak) derivatives in $\mathsf{L}_2(I_e)$ up to order k. We define the *unoriented evaluation at a vertex* and *oriented evaluation at a vertex* of f on the edge e at the vertex v by

$$\underline{f}_e(v) := \begin{cases} f_e(0), & \text{if } v = \partial_- e, \\ f_e(\ell(e)), & \text{if } v = \partial_+ e \end{cases} \quad \text{and} \quad \underline{\overset{\curvearrowright}{f}}_e(v) := \begin{cases} -f_e(0), & \text{if } v = \partial_- e, \\ f_e(\ell(e)), & \text{if } v = \partial_+ e. \end{cases}$$

Note that $\underline{f}_e(v)$ and $\underline{\overset{\curvearrowright}{f}}_e(v)$ are defined for $f \in \mathsf{H}^1_{\max}(G)$. Recall that $\mathscr{V}^{\max} = \bigoplus_v \mathscr{V}^{\max}_v = \bigoplus_v \mathbb{C}^{E_v}$.

Lemma 2.2.3. *Assume the lower lengths bound* (2.3)*, then the evaluation operators*

$$\underline{\bullet}\colon \mathsf{H}^1_{\max}(G) \longrightarrow \mathscr{V}^{\max}, \qquad \underline{\overset{\curvearrowright}{\bullet}}\colon \mathsf{H}^1_{\max}(G) \longrightarrow \mathscr{V}^{\max},$$

$$f \mapsto \underline{f} = \{\{\underline{f}_e(v)\}_{e \in E_v}\}_v \in \mathscr{V}^{\max} \quad \textit{and} \quad f \mapsto \underline{\overset{\curvearrowright}{f}} = \{\{\underline{\overset{\curvearrowright}{f}}_e(v)\}_{e \in E_v}\}_v \in \mathscr{V}^{\max}$$

are bounded by $(2/\ell_-)^{1/2}$.

Proof. By Corollary A.2.18, the boundedness of the individual evaluation operator $\mathsf{H}^1(I_e) \to \mathbb{C}$, $f_e \mapsto \pm f_e(v)$ is bounded by $(2/\min\{1, \ell_e\})^{1/2}$. The boundedness of the global evaluation operators follow immediately. □

These two evaluation maps allow a very simple formula of a partial integration formula on the metric graph, namely

$$\langle f', g\rangle_{\mathsf{L}_2(G)} = \langle f, -g'\rangle_{\mathsf{L}_2(G)} + \langle \underline{f}, \underset{\frown}{g}\rangle_{\mathscr{V}^{\max}}, \tag{2.11}$$

where $f' = \{f'_e\}_e$ and similarly for g. Basically, the formula follows from partial integration on each interval I_e and a reordering of the sum as in (2.1).

Remark 2.2.4. If we distinguish between functions (0-forms) and vector fields (1-forms), we can say that 0-forms are evaluated *unoriented,* whereas 1-forms are evaluated *oriented.* In this way, we should interpret f' and g as 1-forms and f, g' as 0-forms.

Let us now introduce another data in order to define operators on the metric graph:

Definition 2.2.5. A *quantum graph* $(G, \mathscr{V})$ is given by a metric graph G together with a vertex space $\mathscr{V}$ associated with G (i.e. a local subspace of $\mathscr{V}^{\max}$, see Definition 2.1.3). In particular, a quantum graph is fixed by the data $(V, E, \partial, \ell, \mathscr{V})$.

Note that in the literature (see e.g. [Ku08]), a *quantum graph* is sometimes defined as a metric graph together with a self-adjoint (pseudo-)differential operator acting on it. This definition is more general, since we only associate the Laplacian $\Delta_{\mathscr{V}}$ defined below with a quantum graph $(G, \mathscr{V})$.

Associated with a quantum graph $(G, \mathscr{V})$, we define the Sobolev spaces

$$\mathsf{H}^k_{\mathscr{V}}(G) := \{\, f \in \mathsf{H}^k_{\max}(G) \mid \underline{f} \in \mathscr{V}\,\} \quad \text{and} \quad \mathsf{H}^k_{\overset{\frown}{\mathscr{V}}}(G) := \{\, f \in \mathsf{H}^k_{\max}(G) \mid \underset{\frown}{f} \in \mathscr{V}\,\}.$$

By Lemma 2.2.3, these spaces are closed in $\mathsf{H}^k_{\max}(G)$ as pre-image of the closed subspace $\mathscr{V}$ and the bounded operators $\underline{\bullet}$ and $\underset{\frown}{\bullet}$, respectively; and therefore themselves Hilbert spaces.

On the Sobolev space $\mathsf{H}^1_{\mathscr{V}}(G)$, we can rewrite the vertex term in the partial integration formula (2.11) and obtain

$$\langle f', g\rangle_{\mathsf{L}_2(G)} = \langle f, -g'\rangle_{\mathsf{L}_2(G)} + \langle \underline{f}, P\, \underset{\frown}{g}\rangle_{\mathscr{V}} \tag{2.12}$$

for $f \in \mathscr{V}$ and $g \in \mathscr{V}^{\max}$, where P denotes the orthogonal projection of $\mathscr{V}$ in $\mathscr{V}^{\max}$.

Let us now mimic the concept of exterior derivative:

Definition 2.2.6. The *exterior derivative associated with a quantum graph* $(G, \mathscr{V})$ is the unbounded operator $\mathrm{d}_{\mathscr{V}}$ in $\mathsf{L}_2(G)$ defined by $\mathrm{d}_{\mathscr{V}} f := f'$ for $f \in \operatorname{dom} \mathrm{d}_{\mathscr{V}} := \mathsf{H}^1_{\mathscr{V}}(G)$.

Remark 2.2.7.

1. Note that $\mathrm{d}_{\mathscr{V}}$ is a closed operator (i.e. its graph is closed in $\mathsf{L}_2(G) \oplus \mathsf{L}_2(G)$), since $\mathsf{H}^1_{\mathscr{V}}(G)$ is a Hilbert space and the graph norm of $\mathrm{d} = \mathrm{d}_{\mathscr{V}}$, given by $\|f\|^2_{\mathrm{d}} := \|\mathrm{d}f\|^2 + \|f\|^2$, is the Sobolev norm, i.e, $\|f\|_{\mathrm{d}} = \|f\|_{\mathsf{H}^1_{\max}(G)}$.

2. We can think of d as an operator mapping 0-forms into 1-forms. Obviously, on a one-dimensional *smooth* space, there is no need for this distinction, but the distinction between 0- and 1-forms makes sense through the vertex conditions $\underline{f} \in \mathscr{V}$, see also the next lemma.

The adjoint of $\mathrm{d}_{\mathscr{V}}$ can easily be calculated from the partial integration formula (2.12), namely the vertex term $P \underset{\frown}{\underline{g}}$ has to vanish for functions g in the domain of $\mathrm{d}_{\mathscr{V}}^*$:

Lemma 2.2.8. *The adjoint of* $\mathrm{d}_{\mathscr{V}}$ *is given by* $\mathrm{d}_{\mathscr{V}}^* g = -g'$ *with domain* $\operatorname{dom} \mathrm{d}_{\mathscr{V}}^* = \mathsf{H}^1_{\widetilde{\mathscr{V}}^\perp}(G)$.

As for the discrete operators, we define the Laplacian via the exterior derivative:

Definition 2.2.9. The *Laplacian associated with a quantum graph* $(G, \mathscr{V})$ is defined by

$$\Delta_{\mathscr{V}} = \Delta_{(G,\mathscr{V})} := \mathrm{d}_{\mathscr{V}}^* \mathrm{d}_{\mathscr{V}}$$

with domain $\operatorname{dom} \Delta_{\mathscr{V}} := \{\, f \in \operatorname{dom} \mathrm{d}_{\mathscr{V}} \mid \mathrm{d}_{\mathscr{V}} f \in \operatorname{dom} \mathrm{d}_{\mathscr{V}}^* \,\}$.

Let us collect some simple facts about the Laplacian:

Proposition 2.2.10. *Let* $(G, \mathscr{V})$ *be a quantum graph with lower lengths bound* $\inf_e \ell_e \geq \ell_-$, $\ell_- \in (0, 1]$.

1. *The Laplacian* $\Delta_{\mathscr{V}} = \mathrm{d}_{\mathscr{V}}^* \mathrm{d}_{\mathscr{V}}$ *is self-adjoint and non-negative. Moreover, the Laplacian is the operator associated with the closed quadratic form* $\mathfrak{d}_{\mathscr{V}}(f) := \|\mathrm{d}_{\mathscr{V}} f\|_G^2$ *and* $\operatorname{dom} \mathfrak{d}_{\mathscr{V}} = \mathsf{H}^1_{\mathscr{V}}(G)$.
2. *The domain of the Laplacian* $\Delta_{\mathscr{V}} = \mathrm{d}_{\mathscr{V}}^* \mathrm{d}_{\mathscr{V}}$ *is given by*

$$\operatorname{dom} \Delta_{\mathscr{V}} = \{\, f \in \mathsf{H}^2_{\max}(G) \mid \underline{f} \in \mathscr{V}, \quad \underset{\frown}{\underline{f}}' \in \mathscr{V}^\perp \,\}.$$

Proof. The self-adjointness follows immediately from the definition of the Laplacian. Moreover,

$$\langle f, \Delta_{\mathscr{V}} g \rangle = \langle \mathrm{d}_{\mathscr{V}} f, \mathrm{d}_{\mathscr{V}} g \rangle = \mathfrak{d}_{\mathscr{V}}(f, g)$$

for all $f \in \operatorname{dom} \mathrm{d}_{\mathscr{V}}$ and $g \in \operatorname{dom} \Delta_{\mathscr{V}}$. Hence, $\Delta_{\mathscr{V}}$ is the operator associated with $\mathscr{V}$ (see Theorem 3.1.1). Finally, the domain characterisation is easily seen using Lemma 2.2.8. □

The condition $\underline{f} \in \mathscr{V}$, $\underset{\frown}{\underline{f}}' \in \mathscr{V}^\perp$ will be called *vertex condition*, and similarly, $\underline{f}(v) \in \mathscr{V}_v$, $\underset{\frown}{\underline{f}}'(v) \in \mathscr{V}_v^\perp$ *vertex condition at the vertex* v.

An important example is the quantum graph with *standard* vertex space

$$\mathscr{V}^{\mathrm{std}} = \bigoplus_v \mathscr{V}_v^{\mathrm{std}}, \qquad \mathscr{V}_v^{\mathrm{std}} = \mathbb{C}(1, \ldots, 1) \subset \mathbb{C}^{E_v}, \tag{2.13a}$$

respectively, its weighted version

$$\mathscr{V}^p = \bigoplus_v \mathscr{V}_v^{p(v)}, \qquad \mathscr{V}_v^{p(v)} = \mathbb{C}p(v) \subset \mathbb{C}^{E_v}, \tag{2.13b}$$

where $p(v) = \{p_e(v)\}_{e \in E_v}$ and $p_e(v) \neq 0$.

Note that $\mathsf{H}^1_{\mathscr{V}^{\rm std}}(G)$ consists of *continuous* functions on the topological graph G: On each edge, we have the embedding $\mathsf{H}^1(I_e) \subset \mathsf{C}(I_e)$, i.e. f is already continuous inside each edge. Moreover, the evaluation $\underline{f}_e(v)$ is *independent* of $e \in E_v$. This is the reason why we call $\mathscr{V}^{\rm std}$ also the *continuous* vertex space.

Corollary 2.2.11. *Let $(G, \mathscr{V}^p)$ be a quantum graph with weighted standard vertex space $\mathscr{V}^p$ and lower lengths bound* $\inf_e \ell_e \geq \ell_-$, $\ell_- \in (0, 1]$.

1. *The domain of the adjoint $\mathrm{d}_p^* := \mathrm{d}_{\mathscr{V}^p}^*$ is given by*

$$\operatorname{dom} \mathrm{d}_p^* = \Big\{ g \in \mathsf{H}^1_{\max}(G) \,\Big|\, \sum_{e \in E_v} p_e(v) \, \underset{\sim}{\overset{\curvearrowright}{g}}_e(v) = 0 \quad \forall v \in V \Big\}.$$

2. *The domain of the Laplacian $\Delta_p := \Delta_{\mathscr{V}^p}$ is given by*

$$\operatorname{dom} \Delta_p = \Big\{ f \in \mathsf{H}^2_{\max}(G) \,\Big|\, \underline{f}(v) \in \mathbb{C}p(v), \ \sum_{e \in E_v} p_e(v) \, \overset{\curvearrowright}{\underline{f}}{}'_e(v) = 0 \ \forall v \in V \Big\}.$$

In particular, a function f is in the domain of the standard Laplacian *or* Kirchhoff Laplacian $\Delta_{\rm std} = \Delta_{\mathscr{V}^{\rm std}}$ *iff $f \in \mathsf{H}^2_{\max}(G)$, f is continuous and if the* flux condition on the derivatives

$$\sum_{e \in E_v} f_e'(v) = 0 \qquad \forall v \in V.$$

is fulfilled.

Remark 2.2.12.

1. Let us make a short remark on the extremal vertex spaces $\mathscr{V}_v^{\max} = \mathbb{C}^{E_v}$ and $\mathscr{V}^{\min} = 0$: Assume that a quantum graph has $\mathscr{V}_v^{\max/\min}$ as vertex space at the vertex v. Then the corresponding Laplacian fulfils the vertex conditions

$$f_e(v) = 0 \quad \forall e \in E_v \qquad \text{resp.} \qquad f_e'(v) = 0 \quad \forall e \in E_v,$$

i.e. the function f fulfils *individual* Dirichlet resp. Neumann vertex conditions at the vertex v. This is the reason for the name *Dirichlet vertex space* resp. *Neumann vertex space* in Example 2.1.4 (4).

In particular, if $\mathscr{V} = 0$ and $\mathscr{V} = \mathscr{V}^{\max}$ are the minimal and maximal vertex spaces, then

$$\Delta_0 = \bigoplus_e \Delta_{I_e}^{\partial I_e} \qquad \text{and} \qquad \Delta_{\max} = \bigoplus_e \Delta_{I_e},$$

respectively, i.e. the operators are *decoupled*. Here $\Delta_{I_e}^{\partial I_e}$ is the Laplacian on I_e with Dirichlet boundary conditions on ∂I_e, and similarly, Δ_{I_e} is the Laplacian on I_e with Neumann boundary conditions on ∂I_e.

2. If $\deg v = 2$ and $\mathscr{V}_v = \mathbb{C}(1,1)$ then functions f in the domain of the associated Laplacian "do not see" the vertex condition at v: Namely, f *and* f' are continuous along the vertex in an appropriate orientation of the two edges $e_1, e_2 \in E_v$. In particular, one may join the two edges to one edge of length $\ell_e = \ell_{e_1} + \ell_{e_2}$ and omit the vertex v (see also Remark 2.2.2 (2).
3. There are other possibilities how to define self-adjoint extensions of a Laplacian, see e.g. [Ha00, Ku04, FKW07] and (4) below. In particular, for a self-adjoint (bounded) operator L on $\mathscr{V}$, we can define a self-adjoint Laplacian $\Delta_{(\mathscr{V},L)}$ with domain

$$\operatorname{dom} \Delta_{(\mathscr{V},L)} := \{\, f \in \mathsf{H}^2_{\mathscr{V}}(G) \,|\, P \underline{\overset{\curvearrowright}{f}}' = L\underline{f} \,\},$$

where P is the projection in $\mathscr{V}^{\max}$ onto the space $\mathscr{V}$. The vertex conditions $\underline{f} \in \mathscr{V}$ and $P \underline{\overset{\curvearrowright}{f}}' = L\underline{f}$ at the vertex v split into three different parts, namely the *Dirichlet part* $\underline{f} \in \mathscr{V}$, the *Neumann part* $P \underline{\overset{\curvearrowright}{f}}' \in (\ker L)^\perp \subset \mathscr{V}$ and the *Robin part* $P \underline{\overset{\curvearrowright}{f}}' = L\underline{f}$ on $(\ker L)^\perp$ (see e.g. [FKW07] for details). If $L = 0$, then the Robin part is not present, as it is the case in Proposition 2.2.10.
4. One can encode the vertex boundary conditions also in a (unitary) operator S on $\mathscr{V}^{\max}$, the *scattering operator* (see e.g. [KSm97, KS99, KSm03, KPS07]). In general, $S = S(\lambda)$ depends on the eigenvalue ("energy") parameter λ, namely, $S(\lambda)$ is (roughly) defined by looking how incoming and outgoing waves (of the form $x \mapsto \mathrm{e}^{\pm \mathrm{i}\sqrt{\lambda}x}$) propagate through a vertex. In our case (i.e. if $L = 0$ in $\Delta_{(\mathscr{V},L)}$ described above), one can show that S is independent of the *energy*, namely,

$$S = \begin{pmatrix} \mathbb{1} & 0 \\ 0 & -\mathbb{1} \end{pmatrix} = 2P - \mathbb{1} \tag{2.14}$$

with respect to the decomposition $\mathscr{V}^{\max} = \mathscr{V} \oplus \mathscr{V}^\perp$, and where P is the orthogonal projection of $\mathscr{V}$ in $\mathscr{V}^{\max}$. In particular, the so-called *energy-independent* vertex conditions are precisely the ones without Robin part (i.e. $L = 0$), see (3) above.
5. As in the discrete case, we can consider $\Delta^0_{\mathscr{V}} := \Delta_{\mathscr{V}}$ as the Laplacian on 0-forms, and $\Delta^1_{\mathscr{V}} := \mathrm{d}_{\mathscr{V}} \mathrm{d}^*_{\mathscr{V}}$ as the Laplacian on 1-forms. By supersymmetry, we have the spectral relation

$$\sigma(\Delta^1_{\mathscr{V}}) \setminus \{0\} = \sigma(\Delta^0_{\mathscr{V}}) \setminus \{0\}.$$

For more details we refer to [P09a, Sect. 5].

We have the following simple, but useful observation (recall the definition $A \le B$ for operators in Definition 3.1.3):

Proposition 2.2.13. *Assume that $\mathscr{V}_1$ and $\mathscr{V}_2$ are two vertex spaces with $\mathscr{V}_1 \subset \mathscr{V}_2$, then the corresponding Laplacians fulfil $\Delta_{\mathscr{V}_2} \le \Delta_{\mathscr{V}_1}$.*

Proof. The assertion follows directly from the inclusion $\mathsf{H}^1_{\mathscr{V}_1}(G) \subset \mathsf{H}^1_{\mathscr{V}_2}(G)$ and the fact that the quadratic forms agree on $\operatorname{dom} \mathfrak{d}_{\mathscr{V}_1}$. Recall that the forms are given by $\mathfrak{d}_{\mathscr{V}_i}(f) = \|\mathrm{d} f\|^2_{\mathsf{L}_2(G)}$ and $\operatorname{dom} \mathfrak{d}_{\mathscr{V}_i} = \mathsf{H}^1_{\mathscr{V}_i}(G)$. □

We say that a quantum graph $(G, \mathscr{V})$ is *compact* if the underlying metric graph G is compact as topological space. In particular, G is compact iff $|E|$ is finite and all edges have finite length.

Proposition 2.2.14. *Assume that $(G, \mathscr{V})$ is a compact quantum graph, then the resolvent $(\Delta_{\mathscr{V}} + 1)^{-1}$ of the associated Laplacian is a compact operator. In particular, $\Delta_{\mathscr{V}}$ has purely discrete spectrum, i.e. there is an infinite sequence $\{\lambda_k\}_k$ of eigenvalues, where $\lambda_k = \lambda_k(\Delta_{\mathscr{V}})$ denotes the k-th eigenvalue (repeated according to its multiplicity) and $\lambda_k \to \infty$ as $k \to \infty$.*

Proof. The compactness of the resolvent follows easily from the estimate $\Delta_{\mathscr{V}} \ge \Delta_{\mathscr{V}\max} = \bigoplus_e \Delta_{I_e}$ of Proposition 2.2.13. In particular, the operator inequality implies the opposite inequality for the resolvents in -1, i.e.

$$0 \le (\Delta_{\mathscr{V}} + 1)^{-1} \le (\Delta_{\mathscr{V}\max} + 1)^{-1} = \bigoplus_{e \in E} (\Delta_{I_e} + 1)^{-1}.$$

Since E is finite and since $\ell_e < \infty$ for all $e \in E$, the latter operator is compact as direct sum of the compact resolvents of the Neumann Laplacians Δ_{I_e}. □

Combining the last two results together with the variational characterisation of the eigenvalues (see Theorem 3.2.3), we obtain the following eigenvalue inequalities:

Corollary 2.2.15. *Assume that G is a compact metric graph and that $\mathscr{V}_1$, $\mathscr{V}_2$ are two vertex spaces with $\mathscr{V}_1 \subset \mathscr{V}_2$. Then*

$$\lambda_k(\Delta_{\mathscr{V}_2}) \le \lambda_k(\Delta_{\mathscr{V}_1})$$

for all $k \in \mathbb{N}$. For a single vertex space $\mathscr{V}$, we have

$$\lambda_k^{\mathrm{N}}\Big(\dot{\bigcup_e} I_e\Big) = \lambda_k(\Delta_{\max}) \le \lambda_k(\Delta_{\mathscr{V}}) \le \lambda_k(\Delta_{\min}) = \lambda_k^{\mathrm{D}}\Big(\dot{\bigcup_e} I_e\Big)$$

where $\lambda_k^{\mathrm{D/N}}(\dot{\bigcup}_e I_e)$ is the k-th eigenvalue (respecting multiplicity) of the decoupled Dirichlet/Neumann Laplacian $\Delta_{\min/\max} = \bigoplus_e \Delta_{I_e}^{\mathrm{D/N}}$.

In particular, if G is equilateral (i.e, $\ell_e = 1$), then

$$(m-1)^2\pi^2 \le \lambda_k(\Delta_{\mathscr{V}}) \le m^2\pi^2$$

for $k = (m-1)|E|+1,\ldots,m|E|$ *and* $m = 1,2,\ldots$, *and the eigenvalues of* $\Delta_{\mathscr{V}}$ *group into sets of cardinality* $|E|$ *(respecting multiplicity) lying inside the intervals* $K_m := [(m-1)^2\pi^2, m^2\pi^2]$.

Note that $\lambda_k^{\mathrm{N}}(\dot{\bigcup}_e I_e) = 0$ for $k = 1,\ldots,|E|$, and $\lambda_{k+|E|}^{\mathrm{N}}(\dot{\bigcup}_e I_e) = \lambda_k^{\mathrm{D}}(\dot{\bigcup}_e I_e)$, and the latter sequence is a reordering of the individual Dirichlet eigenvalues $\lambda_k^{\mathrm{D}}(I_e) = k^2\pi^2/\ell_e^2$ repeated according to multiplicity.

2.2.3 Boundary Triples Associated with Quantum Graphs

We need the notion of a *boundary* of a graph:

Definition 2.2.16. A *boundary of a metric graph* G is a discrete subset ∂G of G. We call vertices in $\mathring{V} := V \setminus \partial G$ *interior vertices.*

Remark 2.2.17. The boundary of a metric graph need not necessarily consist of vertices. In Chap. 7, we decompose a metric graph G into its star-graph components $G = \overline{\dot{\bigcup}_v} G_v$ by cutting the metric space G in the middle of each interior edge. Then G_v consists of the interior half-edges and the exterior edges adjacent to v. In particular, G_v is a star-shaped graph. It is convenient in Chap. 6 that the points in ∂G_v are not considered as vertices. Otherwise, if we glued the star-shaped graphs G_v together, we would have an additional vertex of degree 2 in the middle of each edge on the entire metric graph. The graph obtained in this way from G is called the *subdivision graph.*

However, in *this* section, it is convenient to assume that $\partial G \subset V$: If $w \in \partial G \setminus V$, then w lies in the interior of an edge e. Splitting this edge into two edges e_1, e_2 joined by the point w, we obtain a new underlying graph with an additional vertex w, and two new edges e_1, e_2 replacing e. As vertex space we use $\mathscr{V}_w = \mathbb{C}(1,1)$. We have seen in Remark 2.2.12 (2) that such vertex spaces are not seen by the corresponding Laplacian.

In this section, we assume for simplicity that $\partial G \subset V$. In particular, we do not exclude the extreme case $\partial G = V$.

Let us now define a boundary triple associated with a quantum graph $(G, \mathscr{V})$ (see also [Pan06, BGP07, BGP08, P08]). The concept of a boundary triple is explained in detail in Sect. 3.4. Note that a boundary triple in our sense is associated with a non-negative quadratic form $\mathfrak{d}$ (see Definition 3.4.3). The definition slightly differs from the usual one, since we assume that the boundary map $\Gamma\colon \mathscr{H}^1 \longrightarrow \mathscr{G}$ is defined on the "first order space" $\mathscr{H}^1 = \operatorname{dom} \mathfrak{d}$, while the boundary map $\Gamma'\colon \mathscr{W}^2 \longrightarrow \mathscr{G}$ is defined on the "second order space" $\mathscr{W}^2$. The (abstract) Green's formula in this context reads as

$$\langle f, \check{\Delta} g\rangle_{\mathscr{H}} = \mathfrak{d}(f,g) - \langle \Gamma f, \Gamma' g\rangle_{\mathscr{G}}$$

for all $f \in \mathscr{H}^1$ and $g \in \mathscr{W}^2$. Here, $\check{\Delta}$ is the "maximal" operator with $\operatorname{dom} \check{\Delta} = \mathscr{W}^2$. Nevertheless, in the context here, our definition of a boundary triple embeds naturally in the usual one (see Corollary 3.4.37). In particular, for a quantum graph $(G, \mathscr{V})$ we set

$$\mathscr{H} := \mathsf{L}_2(G), \qquad \mathscr{H}^1 := \mathsf{H}^1_{\mathscr{V}}(G), \qquad \mathfrak{d}(f) := \|\mathrm{d}_{\mathscr{V}} f\|^2_G,$$

$$\mathscr{G} := \bigoplus_{v \in \partial G} \mathscr{V}_v, \qquad \Gamma f := \{\underline{f}(v)\}_{v \in \partial G}.$$

Moreover, we need

$$\mathscr{W}^2 := \{\, f \in \mathsf{H}^2_{\mathscr{V}}(G) \mid P_v \overset{\curvearrowright}{\underline{f}}{}'(v) = 0 \quad \forall\, v \in \mathring{V} \,\},$$

$$\Gamma' f := \{P_v \overset{\curvearrowright}{\underline{f}}{}'(v)\}_{v \in \partial G}, \qquad (\check{\Delta} f)_e = -f_e'',$$

where $P = \bigoplus_v P_v$ denotes the orthogonal projection of $\mathscr{V}$ in $\mathscr{V}^{\max}$. In particular, functions in $\mathscr{W}^2$ fulfil the vertex conditions $\underline{f}(v) \in \mathscr{V}_v$ and $\overset{\curvearrowright}{\underline{f}}{}'(v) \in \mathscr{V}_v^{\perp}$ for all *inner* vertices, whereas for the boundary vertices $v \in \partial G$, only $\underline{f}(v) \in \mathscr{V}_v$ is assumed. Note that $\mathscr{W}^2$ is a Hilbert space as closed subspace of $\mathsf{H}^2_{\max}(G)$ by Lemma 2.2.3.

If $\mathscr{V} = \mathscr{V}^p$ is a weighted standard vertex space, we can use the equivalent space $\ell_2(V, |p|^2)$ (see Example 2.1.4 (2)). Moreover, for $\underline{f} \in \mathscr{V}^p = \mathbb{C}p$ we use the notation

$$\underline{f}(v) = f(v)p(v), \qquad \text{where} \qquad f(v) := \frac{f_e(v)}{p_e(v)} \in \mathbb{C}, \tag{2.15}$$

since the quotient is independent of $e \in E_v$. In this case, the boundary operators are given by

$$(\Gamma f)(v) = f(v) \qquad \text{and} \qquad (\Gamma' f)(v) = \frac{1}{|p(v)|^2} \sum_{e \in E_v} \overline{p}_e(v) \overset{\curvearrowright}{\underline{f}}{}'_e(v).$$

Proposition 2.2.18. *Assume that $(G, \mathscr{V})$ is a quantum graph with boundary $\partial G \subset V$ and lower length bound $\ell_e \geq \ell_-$ for some $\ell_- \in (0, 1]$. Then we have:*

1. *The operator $\Gamma \colon \mathscr{H}^1 \longrightarrow \mathscr{G}$ is a non-proper boundary map associated with the quadratic form $\mathfrak{d}$. Moreover, $\|\Gamma\|_{1 \to 0} \leq (2/\ell_-)^{1/2}$.*
2. *The triple $(\Gamma, \Gamma', \mathscr{G})$ is a bounded boundary triple associated with $\mathfrak{d}$. The Neumann operator of this boundary triple (i.e., the operator associated with $\mathfrak{d}$) is the Laplacian $\Delta_{\mathscr{V}}$ defined in Definition 2.2.9.*
3. *The boundary triple $(\Gamma, \Gamma', \mathscr{G})$ is elliptic (see Definition 3.4.21).*
4. *The boundary triple $(\Gamma, \Gamma', \mathscr{G})$ is associated with an ordinary boundary triple (see Corollary 3.4.37).*

Proof. Let us start with a general observation: Let $\chi \in \mathsf{C}^\infty_{\mathrm{c}}(\mathbb{R})$ be a function such that $0 \leq \chi \leq 1$, $\chi(s) = 1$ near 0 and $\chi(s) = 0$ if $s \geq \ell_-/2$. For $F \in \mathscr{G}$ and $e \in E$ set

$$h_e(s) := F_e(\partial_- e)\chi(s) + F_e(\partial_+ e)\chi(\ell_e - s), \tag{2.16}$$

where we extend F by $F(v) = 0$ for $v \in \mathring{V}$. Then $h_e \in \mathsf{C}^\infty(I_e)$, and

$$\begin{aligned}\|h\|^2_{\mathsf{H}^k_{\mathscr{V}}(G)} &= \sum_{e\in E}(|F_e(\partial_- e)|^2 + |F_e(\partial_+ e)|^2)\|\chi\|^2_{\mathsf{H}^k(\mathbb{R})}\\ &= \sum_{v\in V}\sum_{e\in E_v}|F_e(v)|^2\|\chi\|^2_{\mathsf{H}^k(\mathbb{R})} = \|F\|^2_{\mathscr{G}}\|\chi\|^2_{\mathsf{H}^k(\mathbb{R})}.\end{aligned}$$

(1) We have to show that Γ is bounded, that $\mathscr{H}^{1,\mathrm{D}} := \ker\Gamma$ is dense in $\mathscr{H}$ and that $\operatorname{ran}\Gamma = \mathscr{G}$. The boundedness of Γ together with the norm-bound follows as in Lemma 2.2.3. The density of $\ker\Gamma = \bigoplus_e \mathring{\mathsf{H}}^1(I_e)$ follows from the standard fact that $\mathring{\mathsf{H}}^1(I_e) = \{ f \in \mathsf{H}^1(I_e) \,|\, f\restriction_{I_e} = 0\}$ is dense in $\mathsf{L}_2(I_e)$. The surjectivity of Γ (i.e., that Γ is non-proper) follows from the consideration above for $k = 1$: For $F \in \mathscr{G}$, let h be as in (2.16). We have $h \in \mathscr{H}^1 = \mathsf{H}^1_{\mathscr{V}}(G)$ and clearly, $\Gamma h = F$.

(2) Recall the notion of a boundary triple Definition 3.4.3. The embedding $\mathscr{W}^2 \subset \mathscr{H}^1$ is norm-bounded by 1, since

$$\|f\|^2_{\mathscr{H}^1} := \|f\|^2 + \|\mathrm{d}_{\mathscr{V}} f\|^2 \le \|f\|^2 + \|f'\|^2 + \|f''\|^2 =: \|f\|^2_{\mathscr{W}^2}.$$

The boundedness of Γ' follows as before, namely we have

$$\begin{aligned}\|\Gamma' f\|^2_{\mathscr{G}} &\le \sum_v \|\underset{\sim}{\overset{\curvearrowright}{f}}{}'(v)\|^2_{\mathscr{V}^{\max}} = \sum_e(|f_e'(\partial_- e)|^2 + |f_e'(\partial_+ e)|^2)\\ &\le \frac{4}{\ell_-}\sum_e(\|f_e'\|^2_{I_e} + \|f_e''\|^2_{I_e}) \le \frac{4}{\ell_-}\|f\|^2_{\mathscr{W}^2}\end{aligned}$$

as in Lemma 2.2.3. For the density of $\operatorname{ran}\Gamma'$ we actually show that $\Gamma'(\mathscr{W}^2) = \mathscr{G}$: Let $\psi \in \mathsf{C}^\infty_{\mathrm{c}}(\mathbb{R})$ such that $\psi(s) = s$ near 0 and $\psi(s) = 0$ for $s \ge \ell_-/2$. For $F \in \mathscr{G}$ we define h as in (2.16) with χ replaced by $-\psi$. It is easily seen that $\underset{\sim}{\overset{\curvearrowright}{h}}{}' = F$ and that $\underline{h} = 0$. Moreover, $\|h\|^2_{\mathscr{W}^2} = \|F\|^2_{\mathscr{G}}\|\psi\|^2_{\mathsf{H}^2(\mathbb{R})} < \infty$. Clearly, $\check{\Delta}$ is bounded (by 1) as operator $\mathscr{W}^2 \to \mathscr{H}$. Green's formula follows from partial integration (2.12), namely

$$\langle f, \check{\Delta} g\rangle = \langle f, -g''\rangle = \langle f', g'\rangle - \langle \underline{f}, P\,\underset{\sim}{\overset{\curvearrowright}{g}}{}'\rangle_{\mathscr{V}} = \mathfrak{d}(f,g) - \langle \Gamma f, \Gamma' g\rangle_{\mathscr{G}}$$

for $f \in \mathscr{H}^1$ and $g \in \mathscr{W}^2$ since $P_v \underset{\sim}{\overset{\curvearrowright}{g}}{}'(v) = 0$ for $v \in \mathring{V}$ by definition of $\mathscr{W}^2$. The density of $\mathring{\mathscr{W}}^2 = \ker\Gamma \cap \ker\Gamma'$ follows from the density of $\mathring{\mathsf{H}}^2(I_e)$ in $\mathsf{L}_2(I_e)$ and the fact that $\bigoplus_e \mathring{\mathsf{H}}^2(I_e) \subset \mathring{\mathscr{W}}^2$. The assertion $\Delta^{\mathrm{N}} = \Delta_{\mathscr{V}}$ follows from Proposition 2.2.10.

(3) We have already seen that $\Gamma(\mathscr{H}^1) = \mathscr{G}$ and $\Gamma'(\mathscr{W}^2) = \mathscr{G}$. It remains to show that the domain of the Dirichlet and Neumann operators are subsets of $\mathscr{W}^2$, i.e., that $\operatorname{dom} \Delta^{\mathrm{D}} \subset \mathscr{W}^2$ and $\operatorname{dom} \Delta^{\mathrm{N}} \subset \mathscr{W}^2$. We recall form Sect. 3.4 that the Dirichlet operator is the operator associated with the form $\mathfrak{d}\restriction_{\ker \Gamma}$, while the Neumann operator is the operator associated with $\mathfrak{d}$ itself. We use Proposition 3.4.24 in order to show the remaining assertion of ellipticity: In particular, we have

$$\|f\|^2_{\mathscr{W}^2} = \|f\|^2 + \|f'\|^2 + \|f''\|^2 \overset{\text{CY}}{\leq} \frac{3}{2}(\|f\|^2 + \|f''\|^2) = \frac{3}{2}(\|f\|^2 + \|\Delta^{\mathrm{D}} f\|^2)$$

for $f \in \mathscr{W}^{2,\mathrm{D}} := \ker \Gamma \cap \mathscr{W}^2$ using the partial integration (2.12) for the term $\|f'\|^2 = \langle f', f'\rangle$ and the fact that the boundary terms vanish. The same argument is true for the Neumann operator and $f \in \mathscr{W}^{2,\mathrm{N}} := \ker \Gamma' \subset \mathscr{W}^2$.

(4) It remains to show that $\Gamma'(\mathscr{H}^{2,\mathrm{D}}) = \mathscr{G}$, where $\mathscr{H}^{2,\mathrm{D}} = \operatorname{dom} \Delta^{\mathrm{D}}$ is the domain of the Dirichlet operator, see above. Actually, we have shown this already in (2). The result follows now from Corollary 3.4.37. □

For the next proposition, we need some more notation. Let $\sqrt{z}$ be the square root cut along the positive axis $\mathbb{R}_+ = [0, \infty)$, so that in particular, $\operatorname{Im} \sqrt{z} > 0$. We denote by

$$\sin_{z,e,+}(s) := \frac{\sin_z s}{\sin_z \ell_e} \qquad \text{and} \qquad \sin_{z,e,-}(s) := \frac{\sin_z(\ell_e - s)}{\sin_z \ell_e}, \tag{2.17a}$$

where $\sin_z(s) := \sin(\sqrt{z}\, s)$ for $z \in \mathbb{C} \setminus \mathbb{R}_+$, the two *fundamental solutions* of $-f_e'' = z f_e$ on I_e with $\sin_{z,e,+}(0) = 1$, $\sin_{z,e,+}(\ell_e) = 0$ and vice versa for $\sin_{z,e,-}$.[5] Moreover, we set

$$\tan_z(s) = \tan(\sqrt{z}\, s) \qquad \text{and} \qquad \cot_z(s) := \cot(\sqrt{z}\, s). \tag{2.17b}$$

If the boundary consists of *all* vertices, i.e. $\partial G = V$, then we can give explicit formulas for the Dirichlet solution operator and the Dirichlet-to-Neumann map (for the notation, we refer to Sect. 3.4, similar results can be found in [P08, BGP08]). Note that in this case, the boundary space of the boundary triple agrees with the entire vertex space, i.e. $\mathscr{G} = \mathscr{V}$:

Proposition 2.2.19. *Assume that $(G, \mathscr{V})$ is a quantum graph with boundary $\partial G = V$ and lower length bound $\ell_e \geq \ell_-$ for some $\ell_- \in (0, 1]$. Then for $z \in \mathbb{C} \setminus \mathbb{R}_+$, we have:*

1. *The Dirichlet operator Δ^{D} associated with the boundary triple $(\Gamma, \Gamma', \mathscr{V})$ is the Laplacian associated with the minimal vertex space $\mathscr{V}^{\min} = 0$, i.e.*

---

[5] We also need the analytic continuation in $z = 0$, i.e. we set $\sin_{0,e,+}(s) := \dfrac{s}{\ell_e}$ and $\sin_{0,e,-}(s) := 1 - \dfrac{s}{\ell_e}$.

$$\Delta^{\mathrm{D}} = \Delta_0 = \bigoplus_e \Delta_{I_e}^{\partial I_e}.$$

In particular, Δ^{D} is decoupled. The spectrum of Δ^{D} is

$$\sigma(\Delta^{\mathrm{D}}) = \Big\{ \frac{k^2\pi^2}{\ell_e^2} \,\Big|\, e \in E,\ k = 1, 2, \dots \Big\}.$$

Moreover, if the graph is equilateral, then

$$\sigma(\Delta^{\mathrm{D}}) = \big\{ k^2\pi^2 \,|\, k = 1, 2, \dots \big\} =: \Sigma^{\mathrm{D}}.$$

2. *The Neumann operator associated with the boundary triple $(\Gamma, \Gamma', \mathscr{V})$ is the Laplacian associated with the quantum graph $(G, \mathscr{V})$, i.e.*

$$\Delta^{\mathrm{N}} = \Delta_{\mathscr{V}}.$$

3. *The Dirichlet solution operator $S(z)\colon \mathscr{V} \longrightarrow \mathscr{W}^2$, defined by $f = S(z)F$, where f is the (unique) solution of the Dirichlet problem $(\check{\Delta} - z)f = 0$, $\Gamma f = F$, is given by*

$$f_e(s) = \begin{cases} F_e(\partial_- e) \sin_{z,e,-}(s) + F_e(\partial_+ e) \sin_{z,e,+}(s), & \text{if } e \in E_{\mathrm{int}} \\ F_e(\partial_- e) \mathrm{e}^{\mathrm{i}\sqrt{z}s}, & \text{if } e \in E_{\mathrm{ext}}, \end{cases}$$

where $F = \{F(v)\}_{v \in V} \in \bigoplus_v \mathscr{V}_v = \mathscr{V}$.

4. *The Dirichlet-to-Neumann map $\Lambda\colon \mathscr{V} \longrightarrow \mathscr{V}$ is given by*

$$(\Lambda(z)F)(v) = P_v H(v), \qquad H(v) = \{H_e(v)\}_{e \in E_v}$$

where

$$H_e(v) = \begin{cases} \dfrac{\sqrt{z}}{\sin_z \ell_e}\Big((\cos_z \ell_e) F_e(v) - F_e(v_e) \Big), & \text{if } e \in E_{\mathrm{int}} \\ \mathrm{i}\sqrt{z} F_e(\partial_- e), & \text{if } e \in E_{\mathrm{ext}}, \end{cases}$$

and where v_e denotes the vertex adjacent with e opposite of v. In particular, if the graph is equilateral ($\ell_e = 1$), there are no exterior edges and we have

$$\Lambda(z) = \frac{\sqrt{z}}{\sin \sqrt{z}} \Big(\ddot{\Delta}_{\mathscr{V}} - (1 - \cos \sqrt{z}) \Big),$$

i.e. the Dirichlet-to-Neumann map is an affine-linear (z-depending) function of the discrete Laplacian $\ddot{\Delta}_{\mathscr{V}}$.

For star graphs, we calculate the Dirichlet solution and Dirichlet-to-Neumann operators in Sect. 6.1.

Proof. The first assertion is obvious. The second assertion follows from the characterisation of the domain of $\Delta_{\mathscr{V}}$ in Proposition 2.2.10. Note that if $\partial G = V$, then $\mathscr{W}^2 = \mathsf{H}^2_{\mathscr{V}}(G)$. For the third assertion, note that $f = S(z)F$ solves the differential equation on each edge, that $\Gamma f = F$ by the ansatz for f and that $f \in \mathsf{L}_2(G)$. From the inequality

$$\|f\|^2_{\mathscr{W}^2} = (1+|z|^2)\|f\|^2 + \|f'\|^2 \le \Big((1+|z|^2) + \frac{1025}{\ell_-^2} + 2|z|^2\Big)\|f\|^2 =: C(z)^2\|f\|^2$$

using Proposition A.2.19 on each interval I_e individually, it follows that $f \in \mathscr{W}^2$. The last assertion is easily seen by a straightforward calculation. □

2.3 Extended Quantum Graphs

Let us provide Laplacians defined on an extended space associated with a quantum graph $(G, \mathscr{V})$. This Laplacian appeared first in an article of Freidlin and Wentzell [FW93] in a probabilistic context. In [KKVW09], Kant, Klauss, Voigt and Weber interpreted the extra states on the vertices as extra *point masses* at the vertices, and they showed that the heat operator associated with the Laplacian with Wentzell condition is submarkovian. Both articles are restricted to the *standard* vertex space, i.e., to continuous functions on the metric graph. Here, we allow more general vertex spaces $\mathscr{V}$, and we give a factorisation of the corresponding Laplacian in the form $\widehat{\Delta}_{(\mathscr{V},L)} = (\widehat{\mathrm{d}}_{(\mathscr{V},L)})^* \widehat{\mathrm{d}}_{(\mathscr{V},L)}$, where L is a non-negative local operator on $\mathscr{V}$ (see below and Proposition 2.3.3).

We always assume that the lower length bound $\ell_e \ge \ell_- > 0$ is fulfilled. The operators on the extended space described below occur naturally in the limit of graph-like manifolds, in particular, when the shrinking rate of the vertex neighbourhood volume is of the same order as the transversal volume, see Chap. 6.

Definition 2.3.1. An *extended quantum graph* is a triple $(G, \mathscr{V}, L)$, where $(G, \mathscr{V})$ is a quantum graph (i.e. G is a metric graph and $\mathscr{V} = \bigoplus_v \mathscr{V}_v$ a vertex space), and L a self-adjoint, bounded operator on $\mathscr{V}$. Moreover, we assume that L is local, i.e. $L = \bigoplus_v L(v)$, where $L(v)$ acts on $\mathscr{V}_v$.

We sometimes refer to a quantum graph $(G, \mathscr{V})$ and the related operators and spaces (see Sect. 2.2.2) as *simple*.

We now associate the Hilbert spaces and operators to an extended metric graph. Basically, we define the extended spaces and operators as the extended corresponding objects associated with the boundary triple $(\Gamma, \Gamma', \mathscr{V})$, see Definition 3.4.47. In particular, the *extended Hilbert space* is given by

$$\mathscr{H} := \mathsf{L}_2(G) \oplus \mathscr{V},$$

(see Proposition 3.4.48).

The *extended exterior derivative* $\hat{\mathrm{d}}_{(\mathscr{V},L)}$ is defined via

$$\hat{\mathrm{d}}_{(\mathscr{V},L)} \colon \hat{\mathscr{H}}_L^1 \longrightarrow \mathsf{L}_2(G), \qquad \hat{\mathrm{d}}_{(\mathscr{V},L)} \hat{f} := f' \qquad \text{for} \qquad \hat{f} = (f, F) \in \hat{\mathscr{H}}_L^1,$$

where

$$\hat{\mathscr{H}}_L^1 := \{ (f, F) \in \mathsf{H}^1_{\mathscr{V}}(G) \oplus \mathscr{V} \mid \underline{f} = LF \} \tag{2.18}$$

is the *extended space of order* 1 associated with $(G, \mathscr{V}, L)$. The following result follows straightforward:

Lemma 2.3.2. *The adjoint of* $\hat{\mathrm{d}}_{(\mathscr{V},L)}$ *is given by*

$$(\hat{\mathrm{d}}_{(\mathscr{V},L)})^* g = (-g', LP \, \underset{\sim}{\overset{\curvearrowright}{g}}) \qquad \textit{and} \qquad \mathrm{dom}(\hat{\mathrm{d}}_{(\mathscr{V},L)})^* = \mathsf{H}^1_{\max}(G),$$

where P *denotes the projection onto* $\mathscr{V}$ *in* $\mathscr{V}^{\max}$.

From Proposition 3.4.48 we conclude the following assertions:

Proposition 2.3.3. *Assume that* $(G, \mathscr{V})$ *is a quantum graph and* L *a bounded operator on* $\mathscr{G}$.

1. *The extended quadratic form* $\hat{\mathfrak{d}}_L$ *associated with* $(G, \mathscr{V})$ *and* L *is given by*

$$\hat{\mathfrak{d}}_L(\hat{f}) = \|\hat{\mathrm{d}}_{(\mathscr{V},L)} \hat{f}\|_G^2$$

with domain $\mathrm{dom}\, \hat{\mathfrak{d}}_L = \hat{\mathscr{H}}_L^1$ *(see* (2.18)*). In particular,* $\hat{\mathfrak{d}}_L$ *is closed and non-negative.*

2. *The extended Laplacian* $\hat{\Delta}_{(\mathscr{V},L)} = (\hat{\mathrm{d}}_{(\mathscr{V},L)})^* \hat{\mathrm{d}}_{(\mathscr{V},L)}$ *is non-negative and acts as*

$$\hat{\Delta}_{(\mathscr{V},L)}(f, F) = (-f'', LP \, \underset{\sim}{\overset{\curvearrowright}{f}}{}')$$

with domain given by

$$\mathrm{dom}\, \hat{\Delta}_{(\mathscr{V},L)} = \{ \hat{f} = (f, F) \in \mathsf{H}^2_{\max}(G) \oplus \mathscr{G} \mid \underline{f} \in \mathscr{V}, \quad \underline{f} = LF \}.$$

In particular, for $L = 0$ we obtain (see Corollary 3.4.49 and Proposition 2.2.19):

Corollary 2.3.4. *Assume that* $(G, \mathscr{V})$ *is a quantum graph. Then the extended quadratic form* $\hat{\mathfrak{d}}_0$ *and the corresponding operator* $\hat{\Delta}_{(\mathscr{V},0)}$ *associated with* $(G, \mathscr{V})$ *and the trivial operator* $L = 0$ *are decoupled, i.e.*

$$\hat{\mathfrak{d}}_0 = \bigoplus_{e \in E} \mathfrak{d}_{I_e}^{\partial I_e} \oplus \mathfrak{o} \qquad \textit{and} \qquad \hat{\Delta}_{(\mathscr{V},0)} = \bigoplus_{e \in E} \Delta_{I_e}^{\partial I_e} \oplus 0$$

with respect to the decomposition $\mathscr{H} = \bigoplus_e \mathsf{L}_2(I_e) \oplus \mathscr{G}$, *where* $\mathfrak{d}_{I_e}^{\partial I_e}$ *and* $\Delta_{I_e}^{\partial I_e}$ *denote the Dirichlet quadratic form and the Dirichlet Laplacian on* I_e, *respectively.*

If $\mathscr{V} = \mathscr{V}^p$ is a weighted standard vertex space, we can use the equivalent space $\ell_2(V, |p|^2)$ (see Example 2.1.4 (2)). Now, the local, bounded operator L viewed as operator on $\ell_2(V, |p|^2)$ can be identified with a bounded, real-valued sequence $\{L(v)\}_{v \in V}$:

Corollary 2.3.5. *Let G be a metric graph with weighted standard vertex space* $\mathscr{V}^p$ *and* $\{L(v)\}_{v \in V}$ *a bounded, real-valued sequence. Then the following assertions are true:*

1. *The adjoint* $(\mathrm{d}_{(p,L)})^*$ *of* $\mathrm{d}_{(p,L)} := \mathrm{d}_{(\mathscr{V}^p,L)}$ *acts as*

$$(\hat{\mathrm{d}}_{(p,L)})^* g = \Big(-g', \Big\{\frac{L(v)}{|p(v)|^2} \sum_{e \in E_v} \overline{p}_e(v) \, \underset{\sim}{\overset{\curvearrowright}{g}}_e(v)\Big\}_{v \in V}\Big)$$

with domain $\operatorname{dom}(\hat{\mathrm{d}}_{(p,L)})^* = \mathsf{H}^1_{\max}(G)$.

2. *The extended Laplacian* $\hat{\Delta}_{(p,L)} := \hat{\Delta}_{(\mathscr{V}^p,L)}$ *acts as*

$$\hat{\Delta}_{(p,L)}(f, F) = \Big(-f'', \Big\{\frac{L(v)}{|p(v)|^2} \sum_{e \in E_v} \overline{p}_e(v) \, \underset{\sim}{\overset{\curvearrowright}{f}}{}'_e(v)\Big\}_{v \in V}\Big)$$

with domain given by

$$\operatorname{dom} \hat{\Delta}_{(p,L)} = \Big\{ f \in \mathsf{H}^2_{\max}(G) \oplus \ell_2(V, |p|^2) \,\Big|\, \underline{f}(v) = f(v) p(v),$$
$$f(v) = L(v) F(v) \;\; \forall v \in V \Big\}.$$

Remark 2.3.6. We can define an extended Laplacian acting on "1-forms" by

$$\hat{\Delta}^1_{(\mathscr{V},L)} = \hat{\mathrm{d}}_{(\mathscr{V},L)} (\hat{\mathrm{d}}_{(\mathscr{V},L)})^*$$

acting as $\hat{\Delta}^1_{(\mathscr{V},L)} g = -g''$ with domain given by

$$\operatorname{dom} \hat{\Delta}^1_{(\mathscr{V},L)} = \{ g \in \mathsf{H}^2_{\max}(G) \,|\, \underset{\sim}{\overset{\curvearrowright}{g}}{}' \in \mathscr{V}, \;\; \underline{g}' + L^2 P \, \underset{\sim}{\overset{\curvearrowright}{g}} = 0 \}.$$

In particular, $\hat{\Delta}^1_{(\mathscr{V},L)}$ represents a Laplacian on a (simple) quantum graph, i.e. the Laplacian acts in the (simple) Hilbert space $\mathsf{L}_2(G)$. For details on this point of view, we refer to [P09a]. If $\mathscr{V}^p$ is a weighted standard vertex space, then the Laplacian $\hat{\Delta}^1_{(\mathscr{V},L)}$ can be interpreted as a delta'-interaction, see e.g. [EP09].

2.4 Spectral Relations Between Discrete and Metric Graphs

In this section, we provide two results of a spectral relation between the discrete and quantum graph Laplacian. The first one is true for the entire spectrum, but only for *equilateral* graphs, the second is valid for general metric graphs, but only at the *bottom* of the spectrum.

2.4.1 Spectral Relation for Equilateral Graphs

The spectral relation between the metric and combinatorial operator for the standard vertex space is well-known, see for example [vB85, Ni87] for the compact case and [Ct97] for the general case (see also [Ku04, Pan06, P08, BGP08] and the references therein). Moreover, in [Ex97], delta- and delta'-vertex conditions are considered. Dekoninck and Nicaise [DN00] proved spectral relations for fourth order operators, and Cartwright and Woess [CW07] used integral operators on the edge.

Let us combine the concrete information on the boundary triple $(\Gamma, \Gamma', \mathscr{V})$ with Theorems 3.4.44 and 3.4.45, in order to obtain a spectral relation between the quantum and discrete graph spectrum:

Theorem 2.4.1. *Assume that $(G, \mathscr{V})$ is a quantum graph with lower length bound $\ell_e \geq \ell_-$ for some $\ell_- \in (0, 1]$. Then the following assertions are true:*

1. *For $z \in \mathbb{C} \setminus \sigma(\Delta^{\mathrm{D}})$ we have the explicit formula for the eigenspaces*

$$\ker(\Delta_{\mathscr{V}} - z) = S(z) \ker \Lambda(z).$$

In particular, if the graph is equilateral (i.e. $\ell_e = 1$) and $z \notin \Sigma^{\mathrm{D}} = \{ (\pi k)^2 \mid k = 1, 2, \dots \}$, then

$$\ker(\Delta_{\mathscr{V}} - z) = \frac{\sqrt{z}}{\sin\sqrt{z}} S(z) \ker\big(\dddot{\Delta}_{\mathscr{V}} - (1 - \cos\sqrt{z})\big).$$

Here, $\dddot{\Delta}_{\mathscr{V}}$ is the discrete Laplacian associated with the vertex space $\mathscr{V}$ (see Definition 2.1.10).

2. *For $z \notin \sigma(\Delta_{\mathscr{V}}) \cup \sigma(\Delta_0)$ we have $0 \notin \sigma(\Lambda(z))$ and Krein's resolvent formula*

$$(\Delta_{\mathscr{V}} - z)^{-1} = (\Delta_0 - z)^{-1} + S(z)\Lambda(z)^{-1}S(\overline{z})^*$$

holds, where $S(\overline{z})^*$ is the adjoint of $S(\overline{z}) \colon \mathscr{V} \longrightarrow \mathsf{L}_2(G)$.

3. *We have the spectral relation*

$$\sigma_\bullet(\Delta_{\mathscr{V}}) \setminus \sigma(\Delta_0) = \{\lambda \in \mathbb{C} \setminus \sigma(\Delta_0) \,|\, 0 \in \sigma_\bullet(\Lambda(\lambda))\},$$

where $\bullet \in \{\emptyset, \mathrm{pp}, \mathrm{disc}, \mathrm{ess}\}$, *i.e. the spectral relation holds for the entire, the pure point (set of all eigenvalues), discrete and essential spectrum.*

Assume in addition that the graph is equilateral and that $\lambda \notin \Sigma^{\mathrm{D}}$. *Then we have the spectral relation*

$$\lambda \in \sigma_\bullet(\Delta_{\mathscr{V}}) \quad \Leftrightarrow \quad (1 - \cos\sqrt{\lambda}) \in \sigma_\bullet(\dddot{\Delta}_{\mathscr{V}})$$

for all spectral types, namely, $\bullet \in \{\emptyset, \mathrm{pp}, \mathrm{disc}, \mathrm{ess}, \mathrm{ac}, \mathrm{sc}, \mathrm{p}\}$, *i.e. the spectral relation holds for the entire, pure point, discrete, essential, absolutely continuous, singular continuous and point spectrum* ($\sigma_{\mathrm{p}}(A) = \overline{\sigma_{\mathrm{pp}}(A)}$). *Finally, the multiplicity of an eigenspace is preserved.*

Proof. The first assertion is immediate from Proposition 2.2.19 (2) and Theorem 3.4.45 (1). The second assertion follows from (3.45a) in Theorem 3.4.44. For the last assertion, we use Theorem 3.4.45 (3) and (4) with $m(z) = 1 - \cos\sqrt{z}$ and $n(z) = (\sin\sqrt{z})/\sqrt{z}$ (with the analytic continuation $n(0) := 1$). Note that the zeros of n agree with the Dirichlet spectrum Σ^{D}. □

For the weighted standard vertex space $\mathscr{V} = \bigoplus_v \mathbb{C}p(v)$, identified with $\ell_2(V, |p|^2)$ as in Example 2.1.4 (2), we immediately obtain the above result with the vertex space $\mathscr{V}$ replaced by $\ell_2(V, |p|^2)$. Moreover, $\dddot{\Delta}_{\mathscr{V}}$ is replaced by $\dddot{\Delta}_p$, the discrete weighted standard Laplacian given in (2.7).

Remark 2.4.2.

1. We do not make an assertion about the Dirichlet spectrum. For weighted standard vertex spaces and equilateral graphs, we have given a topological interpretation of the corresponding eigenspaces associated with eigenvalues $\lambda_k = k^2\pi^2 \in \Sigma^{\mathrm{D}}$ in terms of the homology of the graph, see [LP08a] and references therein.
2. One can use the above spectral relation of discrete and metric graphs for an eigenvalue bracketing argument in the *discrete* case. Note that one has the eigenvalue monotony Proposition 2.2.13 in the metric case. Such estimates can be used for example to ensure *spectral gaps* for the discrete Laplacian on an infinite covering of a finite graph with *residually finite* covering group, see [LP08a] for details.

Let us now compare the extended Laplacian $\hat{\Delta}_{(\mathscr{V},L)}$ with the corresponding discrete operators as in Theorem 3.4.50. The Dirichlet solution and the Dirichlet-to-Neumann operator for the boundary triple $(\Gamma, \Gamma', \mathscr{V})$ are given in Proposition 2.2.19.

Theorem 2.4.3. *Assume that $(G, \mathscr{V})$ is a quantum graph with lower length bound $\ell_e \geq \ell_-$ for some $\ell_- \in (0, 1]$, and that L is a local, bounded operator on $\mathscr{V}$. Then the following assertions are true:*

1. *For $z \notin \sigma(\Delta^{\mathrm{D}})$, we have*
$$\ker(\hat{\Delta}_{(\mathscr{V},L)} - z) = \hat{S}(z)\ker\bigl(L\Lambda(z)L - z\bigr), \tag{2.19}$$
where $\hat{S}(z): \mathscr{V} \longrightarrow \hat{\mathscr{H}}^2_L \subset \mathsf{H}^2_{\mathscr{V}}(G) \oplus \mathscr{V}$ and $\hat{S}(z)F := (S(z)LF, F)$. Moreover, $\hat{S}(z)$ is an isomorphism between the above spaces.
2. *Assume that $\lambda \in \mathbb{R}\setminus\sigma(\Delta^{\mathrm{D}})$, then λ is an eigenvalue of $\hat{\Delta}_{(\mathscr{V},L)}$ iff $\ker(L\Lambda(z)L - z)$ is non-trivial. Moreover, the multiplicity of the (eigen)spaces is preserved.*
3. *Assume in addition that the graph is equilateral (i.e. $\ell_e = 1$) and that $\lambda \notin \Sigma^{\mathrm{D}} = \{\,(\pi k)^2 \,|\, k = 1, 2, \ldots\}$. Moreover, assume that $L = L_0\,\mathrm{id}_{\mathscr{V}}$ for some $L_0 \in \mathbb{R}\setminus\{0\}$. Then we have the spectral relation*
$$\lambda \text{ is an eigenvalue of } \hat{\Delta}_{(\mathscr{V},L)} \quad\Leftrightarrow\quad \hat{m}_{L_0}(\lambda) \text{ is an eigenvalue of } \dddot{\Delta}_{\mathscr{V}},$$
where
$$\hat{m}_{L_0}(\lambda) = L_0^{-2}\sqrt{\lambda}\sin\sqrt{\lambda} + (1 - \cos\sqrt{\lambda}).$$
Finally, the multiplicity is preserved.

Proof. We have shown in Proposition 2.2.18 that $(\Gamma, \Gamma', \mathscr{V})$ is a bounded, elliptic boundary triple. The assertions follow now from Theorem 3.4.50 and the concrete expressions for the boundary triple objects in Proposition 2.2.19. □

Again, the above theorem immediately applies to the case $\mathscr{V} = \bigoplus_v \mathbb{C}p(v)$; and we obtain a similar assertion with the vertex space $\mathscr{V}$ replaced by $\ell_2(V, |p|^2)$. Moreover, $\dddot{\Delta}_{\mathscr{V}}$ is replaced by the discrete weighted standard Laplacian $\dddot{\Delta}_p$; and the local bounded self-adjoint operator L on $\mathscr{V}$ is replaced by multiplication with the bounded sequence $\{L(v)\}_{v\in V}$ of real numbers.

Remark 2.4.4.

1. Note that for a quantum graph $(G, \mathscr{V})$ with invertible operator L, the eigenvalue equation $\hat{\Delta}_{(\mathscr{V},L)}\hat{f} = \lambda\hat{f}$ for $\hat{f} = (f, F) \in \hat{\mathscr{H}}^2_L$ is equivalent with
$$-f_e'' = \lambda f_e, \qquad L^2 P\,\underline{\widetilde{f}}' = \lambda\underline{f}$$
since $F = L^{-1}\underline{f}$. For example, for a weighted standard vertex space, we have
$$-f_e'' = \lambda f_e, \qquad \frac{L(v)^2}{|p(v)|^2}\sum_{e\in E_v}\overline{p}_e(v)\,\underline{\widetilde{f}}_e'(v) = \lambda f(v). \tag{2.20}$$

In particular (for invertible operators L), the eigenvalue equation can be expressed completely in terms of the function f, without reference to the auxiliary vector $F \in \mathscr{V}$. Nevertheless, the vertex condition now depends on the spectral parameter λ.

2. If we consider the Laplacian $\Delta_{(\mathscr{V},L)}$ acting in $\mathsf{L}_2(G)$ (see Remark 2.2.12 (3)), the eigenvalue equation $\Delta_{(\mathscr{V},L)} f = \lambda f$ reads as $-f_e'' = \lambda f_e$ and

$$P \underline{\overset{\curvearrowright}{f}}{}' = L \underline{f} \qquad \text{and} \qquad \frac{1}{|p(v)|^2} \sum_{e \in E_v} \overline{p}_e(v) \underline{\overset{\curvearrowright}{f}}{}'_e(v) = L(v) f(v)$$

for a general vertex space $\mathscr{V}$ and a weighted standard one, respectively. The vertex condition in the latter case is sometimes also referred to as *delta-interaction* with vertex potential (proportional to) $L(v)$. In particular, in the weighted standard case, the eigenvalue equation (2.20) can be interpreted as a delta-interaction with *energy-dependent* vertex potential $\lambda L(v)^{-2}$.

2.4.2 Spectral Relation at the Bottom of the Spectrum

Let us analyse the spectrum at the bottom of $\Delta_{\mathscr{V}}$ in more detail. For simplicity, we assume that there are no exterior edges, i.e. $\ell_e < \infty$ for all edges $e \in E$. As in the discrete case, we define the Hilbert chain associated with the exterior derivative $\mathrm{d}_{\mathscr{V}}$ by

$$\mathscr{C}_{(G^{\mathrm{met}},\mathscr{V})} \colon 0 \longrightarrow \mathsf{H}^1_{\mathscr{V}}(G^{\mathrm{met}}) \xrightarrow{\mathrm{d}_{\mathscr{V}}} \mathsf{L}_2(G^{\mathrm{met}}) \longrightarrow 0$$

and call elements of the first non-trivial space *0-forms*, and of the second space *1-forms*. The associated cohomology spaces (with coefficients in $\mathbb{C}$) are defined by

$$H^0(G^{\mathrm{met}}, \mathscr{V}) := \ker \mathrm{d}_{\mathscr{V}} \cong \ker \mathrm{d}_{\mathscr{V}} / \operatorname{ran} 0,$$
$$H^1(G^{\mathrm{met}}, \mathscr{V}) := \ker \mathrm{d}^*_{\mathscr{V}} = (\operatorname{ran} \mathrm{d}_{\mathscr{V}})^{\perp} \cong \ker 0 / \operatorname{ran} \mathrm{d}_{\mathscr{V}}$$

The *index* or *Euler characteristic* of the Hilbert chain $\mathscr{C}_{(G^{\mathrm{met}},\mathscr{V})}$ associated with the quantum graph $(G^{\mathrm{met}}, \mathscr{V})$ is then defined by

$$\operatorname{ind}(G^{\mathrm{met}}, \mathscr{V}) := \dim \ker \mathrm{d}_{\mathscr{V}} - \dim \ker \mathrm{d}^*_{\mathscr{V}},$$

i.e. the *Fredholm index* of $\mathrm{d}_{\mathscr{V}}$, provided at least one of the dimensions is finite.

We have the following result (for more general cases cf. [P09a], and for related results, see e.g. [FKW07, Ks08, KPS07, GO91]):

Theorem 2.4.5. *Assume that G is a weighted discrete graph with lower lengths bound* (2.3)*, and that $(G^{\mathrm{met}}, \mathscr{V})$ is a quantum graph, where G^{met} denotes the metric graph associated with G. Then there is an isomorphism $\Phi^* = \Phi^*_0 \oplus \Phi^*_1$ with*

$$\Phi^*_p \colon H^p(G^{\mathrm{met}}, \mathscr{V}) \longrightarrow H^p(G, \mathscr{V}).$$

More precisely, Φ^* is induced by a Hilbert chain morphism Φ , i.e.

$$\begin{array}{lcccccc}
\mathscr{C}_{(G^{\mathrm{met}},\mathscr{V})}: 0 & \longrightarrow & \mathsf{H}^1_{\mathscr{V}}(G^{\mathrm{met}}) & \xrightarrow{\mathrm{d}_{\mathscr{V}}} & \mathsf{L}_2(G^{\mathrm{met}}) & \longrightarrow & 0 \\
 & & \Big\downarrow \Phi_0 & & \Big\downarrow \Phi_1 & & \\
\mathscr{C}_{(G,\mathscr{V})}: 0 & \longrightarrow & \mathscr{V} & \xrightarrow{\dddot{\mathrm{d}}_{\mathscr{V}}} & \ell_2(E) & \longrightarrow & 0
\end{array}$$

is commutative, where

$$\Phi_0 f := \underline{f}, \qquad \Phi_1 g := \Big\{\int_{I_e} g_e(s)\mathrm{d}s\Big\}_e.$$

In particular, if G is finite (and therefore G^{met} compact), then

$$\operatorname{ind}(G^{\mathrm{met}}, \mathscr{V}) = \operatorname{ind}(G, \mathscr{V}) = \dim \mathscr{V} - |E|.$$

For general results on Hilbert chains and their morphisms we refer to [Lü02, Ch. 1] or [BL92].

Proof. The operators Φ_p are bounded. Moreover, that Φ is a chain morphism follows from

$$(\Phi_1 \mathrm{d}_{\mathscr{V}} f)_e = \int_{I_e} f_e'(s)\mathrm{d}s = f_e(\ell_e) - f_e(0) = (\dddot{\mathrm{d}}_{\mathscr{V}} \underline{f})_e = (\dddot{\mathrm{d}}_{\mathscr{V}} \Phi_0 f)_e.$$

Furthermore, there is a Hilbert chain morphism Ψ, i.e.

$$\begin{array}{lcccccc}
\mathscr{C}_{(G^{\mathrm{met}},\mathscr{V})}: 0 & \longrightarrow & \mathsf{H}^1_{\mathscr{V}}(G^{\mathrm{met}}) & \xrightarrow{\mathrm{d}_{\mathscr{V}}} & \mathsf{L}_2(G^{\mathrm{met}}) & \longrightarrow & 0 \\
 & & \Big\uparrow \Psi_0 & & \Big\uparrow \Psi_1 & & \\
\mathscr{C}_{(G,\mathscr{V})}: 0 & \longrightarrow & \mathscr{V} & \xrightarrow{\dddot{\mathrm{d}}_{\mathscr{V}}} & \ell_2(E) & \longrightarrow & 0
\end{array}$$

given by

$$\Psi_0 F := S(0)F = \{F_e(\partial_- e)\sin_{0,e,-} + F_e(\partial_+ e)\sin_{0,e,-}\}_e, \qquad \Psi_1 \eta := \{\eta_e \mathbb{1}_{I_e}/\ell_e\}_e$$

(see (2.17a)), i.e. we let $\Phi_0 F$ be the edge-wise affine linear (harmonic) function

$$(S(0)F)_e(s) = F_e(\partial_- e) \cdot \frac{\ell_e - s}{\ell_e} + F_e(\partial_+ e) \cdot \frac{s}{\ell_e}, \qquad s \in I_e; \tag{2.21}$$

and $\Phi_1\eta$ be an (edgewise) constant function. Again, the chain morphism property $\Psi_1\ddot{\mathrm{d}}_{\mathscr{V}} = \mathrm{d}_{\mathscr{V}}\Psi_0$ can easily be seen. Furthermore, $\Phi\Psi$ is the identity on the second (discrete) Hilbert chain $\mathscr{C}_{(G,\mathscr{V})}$. It follows now from abstract arguments (see e.g. [BL92, Lem. 2.9]) that the corresponding induced maps Φ_p^* are isomorphisms on the cohomology spaces. □

Remark 2.4.6. The subcomplex $\Psi(\mathscr{C}_{(G,\mathscr{V})})$ of $\mathscr{C}_{(G^{\mathrm{met}},\mathscr{V})}$ consists of the subspace of edge-wise affine linear functions (0-forms) and of edge-wise constant functions (1-forms). In this way, we can naturally embed the discrete setting into the metric graph one. In particular, assume that $0 < \ell_- \le \ell_e \le \ell_+ < \infty$ for all $e \in E$, then

$$\|\Psi_0 F\|^2 = \sum_e \frac{1}{\ell_e^2}\int_0^{\ell_e} |F_e(\partial_- e)(\ell_e - s) + F_e(\partial_+ e)s|^2 \mathrm{d}s$$
$$= \sum_e \frac{\ell_e}{3}\Big(|F_e(\partial_- e)|^2 + \mathrm{Re}\big(\overline{F_e(\partial_- e)}F_e(\partial_+ e)\big) + |F_e(\partial_+ e)|^2\Big),$$

so that

$$\frac{\ell_-}{6}\|F\|^2_{\mathscr{V}} \overset{\mathrm{CY}}{\le} \|\Psi_0 F\|^2 \overset{\mathrm{CY}}{\le} \frac{\ell_+}{2}\|F\|^2_{\mathscr{V}},$$

i.e. redefining the norm on $\mathscr{V}$ by $\|F\|_{\mathscr{V},1} := \|\Psi_0 F\|$ gives an equivalent norm turning Ψ_0 into an isometry. Moreover, $\|\Psi_1\eta\| = \|\eta\|_{\ell_2(E)}$. For more details on this point of view (as well as "mixed" types of discrete and metric graphs), we refer to [FT04b] and references therein.

Finally, we analyse the spectrum at the bottom of the *extended* model. Let $(G,\mathscr{V})$ be a quantum graph and $L = L^*$ be a local, bounded operator on $\mathscr{V}$. We define the Hilbert chain associated with the exterior derivative $\mathrm{d}_{(\mathscr{V},L)}$ by

$$\mathscr{C}_{(G^{\mathrm{met}},\mathscr{V},L)}\colon 0 \longrightarrow \hat{\mathscr{H}}_L^1 \overset{\mathrm{d}_{(\mathscr{V},L)}}{\longrightarrow} \mathsf{L}_2(G^{\mathrm{met}}) \longrightarrow 0$$

and call elements of the first non-trivial space *extended* 0*-forms*,and of the second space 1*-forms*. Recall that $\mathrm{d}_{(\mathscr{V},L)}(f,F) = f'$. The associated cohomology spaces (with coefficients in $\mathbb{C}$) are defined by

$$H^0(G^{\mathrm{met}},\mathscr{V},L) := \ker \mathrm{d}_{(\mathscr{V},L)} \cong \ker \mathrm{d}_{(\mathscr{V},L)}/\operatorname{ran} 0,$$
$$H^1(G^{\mathrm{met}},\mathscr{V},L) := \ker \mathrm{d}^*_{(\mathscr{V},L)} = (\operatorname{ran} \mathrm{d}_{(\mathscr{V},L)})^\perp \cong \ker 0/\operatorname{ran} \mathrm{d}_{\mathscr{V}}$$

The *index* or *Euler characteristic* of the Hilbert chain $\mathscr{C}_{(G^{\mathrm{met}},\mathscr{V},L)}$ associated with the extended quantum graph $(G^{\mathrm{met}},\mathscr{V},L)$ is then defined by

$$\mathrm{ind}(G^{\mathrm{met}},\mathscr{V},L) := \dim\ker \mathrm{d}_{(\mathscr{V},L)} - \dim\ker \mathrm{d}^*_{(\mathscr{V},L)},$$

i.e. the *Fredholm index* of $\mathrm{d}_{(\mathscr{V},L)}$, provided at least one of the dimensions is finite.

We define the *extended discrete Hilbert chain* associated with the graph G with vertex space $\mathscr{V}$ and operator L by

$$\mathscr{C}_{(G,\mathscr{V},L)}\colon 0 \longrightarrow \mathscr{V} \xrightarrow{\dddot{\mathrm{d}}_{\mathscr{V}} L} \ell_2(E) \longrightarrow 0.$$

Similarly, we denote by $H^p(G,\mathscr{V},L)$ the corresponding cohomology spaces for $p = 0, 1$. The index of the extended discrete Hilbert chain is given by the Fredholm index of $\dddot{\mathrm{d}}_{\mathscr{V}} L$, namely

$$\begin{aligned} \operatorname{ind}(G,\mathscr{V},L) &:= \dim\ker \dddot{\mathrm{d}}_{\mathscr{V}} L - \dim\ker L\, \dddot{\mathrm{d}}^*_{\mathscr{V}} \\ &= \operatorname{ind}(G, \ker L^\perp) + \dim\ker L \\ &= \dim\ker L^\perp - |E| + \dim\ker L = \dim \mathscr{V} - |E|. \end{aligned}$$

We have the following result:

Theorem 2.4.7. *Assume that G is a weighted discrete graph with lower lengths bound (2.3). Assume in addition, that $(G^{\mathrm{met}},\mathscr{V},L)$ is an extended quantum graph, where G^{met} denotes the metric graph associated with G. Then there is an isomorphism $\hat{\Phi}^* = \hat{\Phi}^*_0 \oplus \Phi^*_1$ with*

$$\hat{\Phi}^*_0, \Phi^*_1\colon H^p(G^{\mathrm{met}},\mathscr{V},L) \longrightarrow H^p(G,\mathscr{V},L).$$

More precisely, $\hat{\Phi}^*$ is induced by a Hilbert chain morphism $\hat{\Phi}$, i.e.

$$\begin{array}{ccccccc} \mathscr{C}_{(G^{\mathrm{met}},\mathscr{V},L)}\colon 0 & \longrightarrow & \mathscr{H}^1_L & \xrightarrow{\mathrm{d}_{(\mathscr{V},L)}} & \mathsf{L}_2(G^{\mathrm{met}}) & \longrightarrow & 0 \\ & & \Big\downarrow \hat{\Phi}_0 & & \Big\downarrow \Phi_1 & & \\ \mathscr{C}_{(G,\mathscr{V},L)}\colon 0 & \longrightarrow & \mathscr{V} & \xrightarrow{\dddot{\mathrm{d}}_{\mathscr{V}} L} & \ell_2(E) & \longrightarrow & 0 \end{array}$$

is commutative, where

$$\hat{\Phi}_0(f,F) := F, \qquad \Phi_1 g := \Big\{\int_{I_e} g_e(s)\,\mathrm{d}s\Big\}_e.$$

In particular, if G is finite (and therefore G^{met} compact), then

$$\operatorname{ind}(G^{\mathrm{met}},\mathscr{V},L) = \operatorname{ind}(G,\mathscr{V},L) = \dim\mathscr{V} - |E|.$$

Proof. The operators $\hat{\Phi}_0$ and Φ_1 are bounded. Moreover, that $\hat{\Phi}$ is a chain morphism follows from

$$(\Phi_1 \mathrm{d}_{(\mathscr{V},L)}(f,F))_e = \int_{I_e} f'_e(s)\mathrm{d}s = f_e(\ell_e) - f_e(0)$$
$$= (\dddot{\mathrm{d}}_{\mathscr{V}} \underline{f})_e = (\dddot{\mathrm{d}}_{\mathscr{V}} LF)_e = (\dddot{\mathrm{d}}_{\mathscr{V}} L \hat{\Phi}_0(f,F))_e.$$

Furthermore, there is a Hilbert chain morphism $\hat{\Psi}$, i.e.

$$\begin{array}{ccccccc}
\mathscr{C}_{(G^{\mathrm{met}},\mathscr{V},L)}: 0 & \longrightarrow & \hat{\mathscr{H}}^1_L & \xrightarrow{\mathrm{d}_{(\mathscr{V},L)}} & \mathsf{L}_2(G^{\mathrm{met}}) & \longrightarrow & 0 \\
 & & \uparrow \hat{\Psi}_0 & & \uparrow \Psi_1 & & \\
\mathscr{C}_{(G,\mathscr{V},L)}: 0 & \longrightarrow & \mathscr{V} & \xrightarrow{\dddot{\mathrm{d}}_{\mathscr{V}} L} & \ell_2(E) & \longrightarrow & 0
\end{array}$$

given by

$$\hat{\Psi}_0 F := \hat{S}(0)F = (S(0)LF, F), \qquad \Psi_1 \eta := \{\eta_e \mathbb{1}_{I_e}/\ell_e\}_e,$$

where $S(0)F$ is the affine (harmonic) function as defined in (2.21). It is shown in Theorem 3.4.50 that $\hat{S}(0)$ maps into $\hat{\mathscr{H}}^1_L$. Moreover, the chain morphism property $\Psi_1 \dddot{\mathrm{d}}_{\mathscr{V}} L = \mathrm{d}_{(\mathscr{V},L)} \hat{\Psi}_0$ is easy to see. In addition, $\hat{\Phi}\hat{\Psi}$ is the identity on the second (discrete) Hilbert chain $\mathscr{C}_{(G,\mathscr{V},L)}$. It follows again from abstract arguments (see e.g. [BL92, Lem. 2.9]) that the corresponding induced maps $\hat{\Phi}^*_0$ and Φ^*_1 are isomorphisms on the cohomology spaces. □

2.5 Some Trace Formulas on Metric and Discrete Graphs

Let us finish this chapter with some results concerning the trace of the heat operator. Trace formulas for metric graph Laplacians appeared first in an article of Roth [Rot84], where standard (Kirchhoff) boundary conditions are used; more general self-adjoint vertex conditions (energy-independent, see Remark 2.2.12 (4)) are treated in [KS06, KPS07]. Trace formulas are useful for inverse problems, see [KN05, KN06, No07, Ks08] and references therein.

We first need some (technical) notation; which is inevitable in order to properly write down the trace formula. For an illustration of the next definition see Fig. 2.1.

Definition 2.5.1. Let G be a discrete graph. A *combinatorial path* c in G is a sequence

$$c = (e_0, v_0, e_1, v_1, \dots, e_n, v_n, e_{n+1}),$$

where $v_i \in \partial e_i \cap \partial e_{i+1}$ for $i = 0, \dots, n$. We call $|c| := n + 1$ the *combinatorial length* of the path c (the number of vertices passed *inside* the path), and $e_-(c) := e_0$ resp. $e_+(c) := e_{n+1}$ the *initial edge* resp. *terminal edge* of c. Similarly, we denote

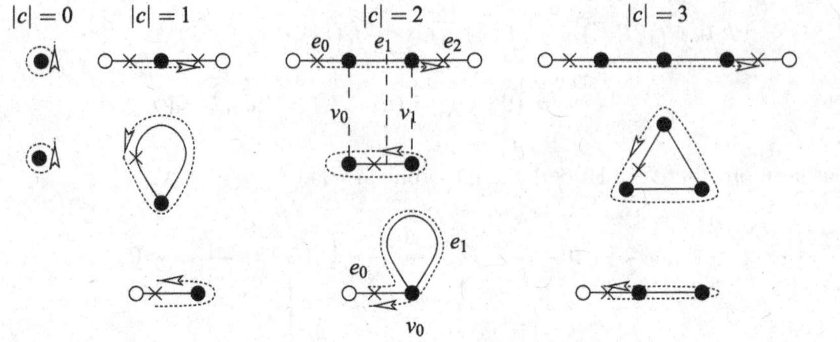

Fig. 2.1 Different types of paths: In the first row, we have general paths, in the second properly closed and in the last, non-properly closed paths. The cross indicates the start and endpoint of each path. The combinatorial paths of length $|c| = 2$ are given by $c = (e_0, v_0, e_1, v_1, e_2)$ with $e_-(c) = e_0$, $e_+(c) = e_2$, $\partial_- c = v_0$, $\partial_+ c = v_1$ by $c = (e_1, v_0, e_1, v_1, e_1)$ with $e_-(c) = e_1 = e_+(c)$, $\partial_- c = v_0 \neq \partial_+ c = v_1$ and by $c = (e_0, v_0, e_1, v_0, e_0)$ with $e_-(c) = e_0 = e_+(c) = e_0$, $\partial_- c = v_0 = \partial_+ c = v_0$ (*top down*)

by $\partial_- c := v_0$ and $\partial_+ c := v_n$ the *initial vertex* resp. *terminal vertex* of c, i.e. the first resp. last vertex in the sequence c. We call a path with $e_-(c) = e_+(c)$ *closed*. A closed path is *properly closed* if c is closed and $\partial_- c \neq \partial_+ c$.[6] Denote by C_m the set of all properly closed paths of combinatorial length m, and by C the set of all properly closed paths.

If the graph does not have multiple edges, a properly closed combinatorial path can equivalently be described by the sequence $c = (v_0, \ldots, v_n)$ of vertices passed by. In particular, if the graph has no self-loops then $|C_0| = |V|$, $|C_1| = 0$ and $|C_2| = 2|E|$. Moreover, $C_{2k+1} = \emptyset$ for all $k \geq 1$ is equivalent with the fact that G is bipartite. A graph G is called *bipartite*, if $V = V_+ \mathbin{\dot\cup} V_-$ and all edges join exactly one vertex in V_- with exactly one vertex in V_+.

Definition 2.5.2. Two properly closed paths c, c' are called *equivalent* if they can be obtained from each other by successive application of the cyclic transformation

$$(e_0, v_0, e_1, v_1, \ldots, e_n, v_n, e_0) \to (e_1, v_1, \ldots, e_n, v_n, e_0, v_0, e_1.)$$

The corresponding equivalence class is called *cycle* and is denoted by $\widetilde{c}$. The set of all cycles is denoted by $\widetilde{C}$. Given $p \in \mathbb{N}$ and a cycle $\widetilde{c}$, denote by $p\widetilde{c}$ the cycle obtained from $\widetilde{c}$ by repeating it p-times. A cycle $\widetilde{c}$ is called *prime*, if $\widetilde{c} = p\widetilde{c}'$ for any other cycle $\widetilde{c}'$ implies $p = 1$. The set of all prime cycles is denoted by $\widetilde{C}_{\text{prim}}$.

[6] A closed path of (combinatorial) length 0 consists by definition of a single vertex and is by definition properly closed. A closed path of length 1 is never properly closed. For a self-loop e (i.e. an edge e with $\partial_- e = \partial_+$) we make the following convention: the closed path $c = (e, v, e)$ has length 1 and is properly closed (by definition).

Definition 2.5.3. Let $\gamma\colon [0,1] \longrightarrow G^{\mathrm{met}}$ be a *metric path* in the metric graph G^{met}, i.e. a continuous function which is of class C^1 on each edge and $\gamma'(t) \neq 0$ for all $t \in [0,1]$ such that $\gamma(t) \in G^1 = G^{\mathrm{met}} \setminus V$, i.e. $\gamma'(t)$ never vanishes inside an edge. In particular, a path in G^{met} cannot turn its direction inside an edge. We denote the set of all paths from x to y by $\Gamma(x,y)$.

Associated with a metric path $\gamma \in \Gamma(x,y)$ there is a unique combinatorial path c_γ determined by the sequence of edges and vertices passed along $\gamma(t)$ for $0 < t < 1$. It is not excluded that the initial point $\gamma(0)$ or the terminal point $\gamma(1)$ of the path are vertices, but these vertices are not encoded in the sequence c.

On the other hand, a combinatorial path c and two points x, y being on the initial resp. terminal edge, i.e. $x \in I_{e_-(c)}$, $y \in I_{e_+(c)}$, but different from the initial resp. terminal vertex, i.e. $x \neq \partial_-(c)$ and $y \neq \partial_+(c)$, uniquely determine a metric path $\gamma = \gamma_c \in \Gamma(x,y)$ (up to a change of velocity). Denote the set of such combinatorial paths from x to y by $C(x,y)$.

Definition 2.5.4. The *length* of the metric path $\gamma \in \Gamma(x,y)$ is defined as $\ell(\gamma) := \int_0^1 |\gamma'(s)| \, \mathrm{d}s$. In particular, if $c = c_\gamma = (e_0, v_0, \dots, e_n, v_n, e_{n+1})$ is the combinatorial path associated with γ, then

$$d_c(x,y) := \ell(\gamma) = |x - \partial_- c_\gamma| + \sum_{i=1}^n \ell_{e_i} + |y - \partial_+ c_\gamma|,$$

where $|x-y| := |x_e - y_e|$ denotes the distance of x, y being inside the same edge e, and x_e, $y_e \in I_e$ are the corresponding coordinates (cf. Remark 2.2.2 (1)). Note that there might be a shorter path between x and y *outside* the edge e. For a properly closed path c we define the *metric length* of c as $\ell(c) = \ell(\gamma_c)$ and similarly, $\ell(\widetilde{c}) := \ell(c)$ for a cycle. Note that the latter definition is well-defined.

Finally, we define the *scattering amplitude associated with a vertex space* $\mathscr{V}$ and a combinatorial path $c = (e_0, v_0, \dots, e_n, v_n, e_{n+1})$. Denote by $P = \bigoplus_v P_v$ the orthogonal projection in $\mathscr{V}^{\max}$ onto $\mathscr{V}$. Denote by $S := 2P - \mathbb{1}$ the corresponding scattering matrix defined in (2.14). In particular, S is local, i.e. $S = \bigoplus_v S_v$. We define

$$S_{\mathscr{V}}(c) := \prod_{i=0}^n S_{e_i, e_{i+1}}(v_i),$$

where $S_{e,e'}(v) = 2P_{e,e'}(v) - \delta_{e,e'}$ for $e, e' \in E_v$. For a cycle, we set $S(\widetilde{c}) := S(c)$, and this definition is obviously well-defined, since multiplication of complex numbers is commutative.

For example, the standard vertex space $\mathscr{V}^{\mathrm{std}}$ has projection $P = (\deg v)^{-1}\mathbb{E}$, where $\mathbb{E}$ denotes the $(\deg v \times \deg v)$-matrix such that all entries are 1. In particular,

$$S^{\mathrm{std}}_{e,e'}(v) = \frac{2}{\deg v}, \quad \text{if} \quad e \neq e', \qquad \text{and} \qquad S^{\mathrm{std}}_{e,e}(v) = \frac{2}{\deg v} - 1.$$

If, in addition, the graph is regular, i.e, $\deg v = r$ for all $v \in V$, then one can simplify the scattering amplitude of a combinatorial path c to

$$S^{\mathrm{std}}(c) = \Big(\frac{2}{r}\Big)^a \Big(\frac{2}{r} - 1\Big)^b$$

where b is the number of reflections in c ($e_i = e_{i+1}$) and a the number of transmissions $e_i \neq e_{i+1}$ in c.

We can now formulate the trace formula for a compact quantum graph with Laplacian $\Delta_{\mathscr{V}}$ (cf. [Rot84, Thm. 1], [KPS07, Thm. 4.1]):

Theorem 2.5.5. *Assume that* $(G^{\mathrm{met}}, \mathscr{V})$ *is a compact quantum graph (without self-loops) and* $\Delta_{\mathscr{V}}$ *the associated self-adjoint Laplacian (cf. Proposition 2.2.10). Then we have*

$$\begin{aligned}\operatorname{tr} e^{-t\Delta_{\mathscr{V}}} &= \frac{\operatorname{vol}_1 G^{\mathrm{met}}}{(4\pi t)^{1/2}} + \frac{1}{2}\big(\dim \mathscr{V} - |E|\big) \\ &\quad + \frac{1}{(4\pi t)^{1/2}} \sum_{\widetilde{c} \in \widetilde{C}_{\mathrm{prim}}} \sum_{p \in \mathbb{N}} S_{\mathscr{V}}(\widetilde{c})^p \ell(\widetilde{c}) \exp\Big(-\frac{p^2 \ell(\widetilde{c})^2}{4t}\Big)\end{aligned}$$

for $t > 0$*, where* $\operatorname{vol}_1 G^{\mathrm{met}} = \sum_e \ell_e$ *is the total length of the metric graph* G^{met}.

Remark 2.5.6.

1. The first term on the RHS is the term expected from the Weyl asymptotics. The second term is precisely $1/2$ of the index $\operatorname{ind}(G^{\mathrm{met}}, \mathscr{V})$ of the metric graph G^{met} with vertex space $\mathscr{V}$, i.e. the Fredholm index of $d_{\mathscr{V}}$. In Theorem 2.4.5 we showed that the index is the same as the discrete index $\operatorname{ind}(G, \mathscr{V})$ (the Fredholm index of $\dddot{d}_{\mathscr{V}}$). In [KPS07], the authors calculated the second term as $(\operatorname{tr} S)/4$, but since $S = 2P - \mathbb{1}$, we have $\operatorname{tr} S = 2\dim \mathscr{V} - \dim \mathscr{V}^{\max} = 2(\dim \mathscr{V} - |E|)$. The last term in the trace formula comes from a combinatorial expansion. Nicaise [Ni87] proved a similar formula using the spectral relation for equilateral graphs Theorem 2.4.1.
2. The sum over prime cycles of the metric graph G^{met} is an analogue of the sum over primitive periodic geodesics on a manifold in the celebrated Selberg trace formula, as well as an analogue of a similar formula for (standard) discrete graphs, see Theorem 2.5.7.
3. Trace formulas can be used to solve the inverse problem: For example, Gutkin, Smilansky and Kurasov, Nowaczyk [GSm01, Ks08, KN06, KN05] showed that if G^{met} does not have self-loops, multiple edges, and if all its lengths are rationally independent, then the metric structure of the graph is uniquely determined. Further extensions are given e.g. in [KPS07]. Some results can be extended to the case of (trivially or weakly) rationally dependent edge lengths (see [No07]), but counterexamples in [Rot84, GSm01, BSS06] show that one

needs some conditions on the edge lengths. In particular, there are isospectral, non-homeomorphic graphs.

The proof of Theorem 2.5.5 uses the expansion of the heat kernel, namely one can show that

$$p_t(x,y) = \frac{1}{2(\pi t)^{1/2}}\Big(\delta_{x,y}\exp\Big(-\frac{|x-y|^2}{4t}\Big) + \sum_{c\in C(x,y)} S(c)\exp\Big(-\frac{d_c(x,y)^2}{4t}\Big)\Big),$$

where $\delta_{x,y} = 1$ if x, y are inside the same edge (and not both on opposite sides of ∂e) and 0 otherwise. The trace of $e^{-t\Delta_{\mathscr{V}}}$ can now be calculated as the integral over $p_t(x,x)$. The first term in the heat kernel expansion gives the volume term, the second splits into *properly* closed paths leading to the third term (the sum over prime cycles), and the index term in the trace formula is the contribution of non-properly closed paths. More precisely, a non-properly closed path runs through its initial and terminal edge (which are the same by definition of a closed path) in opposite directions. For more details, we refer to [Rot84] or [KPS07].

Let us finish with some trace formulas for discrete graphs. Assume for simplicity, that G is a simple discrete graph, i.e. G has no self-loops and no multiple edges. Moreover, we assume that G is equilateral, i.e. $\ell_e = 1$. For simplicity, we write $v \sim w$ if v, w are connected by an edge. Let $\mathscr{V}$ be an associated vertex space. Since $\Delta_{\mathscr{V}} = \mathbb{1} - M_{\mathscr{V}}$ and $M_{\mathscr{V}}$ (see (2.8)) are bounded operators on $\mathscr{V}$, we have

$$\operatorname{tr} e^{-t\Delta_{\mathscr{V}}} = e^{-t}\operatorname{tr} e^{tM_{\mathscr{V}}} = e^{-t}\sum_{n=0}^{\infty}\frac{t^n}{n!}\operatorname{tr} M_{\mathscr{V}}^n.$$

Furthermore, using (2.8) n-times, we obtain

$$M_{\mathscr{V}}^n = \bigoplus_{v_0}\sum_{v_1\sim v}\cdots\sum_{v_n\sim v_{n-1}} A_{\mathscr{V}}(v_0,v_1)A_{\mathscr{V}}(v_1,v_2)\cdot\ldots\cdot A_{\mathscr{V}}(v_{n-1},v_n),$$

and

$$\operatorname{tr} M_{\mathscr{V}}^n = \sum_{v_0}\sum_{v_1\sim v_0}\cdots\sum_{v_{n-1}\sim v_{n-2}} \operatorname{tr} A_{\mathscr{V}}(v_0,v_1)A_{\mathscr{V}}(v_1,v_2)\cdot\ldots\cdot A_{\mathscr{V}}(v_{n-1},v_0).$$

Note that the sum is precisely over all combinatorial (properly) closed paths $c = (v_0,\ldots,v_{n-1}) \in C_n$. Denoting by

$$W_{\mathscr{V}}(c) := \operatorname{tr} A_{\mathscr{V}}(v_0,v_1)A_{\mathscr{V}}(v_1,v_2)\cdot\ldots\cdot A_{\mathscr{V}}(v_{n-1},v_0)$$

the *weight* associated with the path c and to the vertex space $\mathscr{V}$, we obtain the following general trace formula. In particular, we can write the trace as a (discrete) "path integral":

Theorem 2.5.7. *Assume that G is a discrete, finite graph with weights* $\ell_e = 1$ *without self-loops and multiple edges. Then*

$$\operatorname{tr} e^{-t\ddot{\Delta}_{\mathscr{V}}} = e^{-t} \sum_{n=0}^{\infty} \sum_{c \in C_n} \frac{t^n}{n!} W_{\mathscr{V}}(c) = e^{-t} \sum_{c \in C} \frac{t^{|c|}}{|c|!} W_{\mathscr{V}}(c). \tag{2.22}$$

Let us interpret the weight in the standard case $\mathscr{V} = \mathscr{V}^{\mathrm{std}}$. Here, $A_{\mathscr{V}^{\mathrm{std}}}(v, w)$ can be interpreted as operator from $\mathbb{C}(\deg w)$ to $\mathbb{C}(\deg v)$ (the degree indicating the corresponding ℓ_2-weight) with $A_{\mathscr{V}^{\mathrm{std}}}(v, w) = 1$ if v, w are connected and 0 otherwise. Viewed as multiplication in $\mathbb{C}$ (without weight), $A_{\mathscr{V}^{\mathrm{std}}}(v, w)$ is unitarily equivalent to the multiplication with $(\deg v \deg w)^{-1/2}$ if $v \sim w$ resp. $A_{\mathscr{V}^{\mathrm{std}}}(v, w) = 0$ otherwise. In particular, if $c = (v_0, \ldots, v_{n-1})$ is of length n, then the weight is

$$W^{\mathrm{std}}(c) = \frac{1}{\deg v_0} \cdot \frac{1}{\deg v_1} \cdot \ldots \cdot \frac{1}{\deg v_{n-1}}.$$

If, in addition, G is a regular graph, i.e. $\deg v = r$ for all $v \in V$, then $W^{\mathrm{std}}(c) = r^{-n}$. Then the trace formula (2.22) reads as

$$\operatorname{tr} e^{-t\ddot{\Delta}} = e^{-t} \sum_{n=0}^{\infty} \frac{t^n}{r^n n!} |C_n| = e^{-t} \Big(|V| + \frac{|E|}{r^2} t^2 + \frac{|C_3|}{6r^3} t^3 + \ldots \Big),$$

since $|C_0| = |V|$, $|C_1| = 0$ (no self-loops) and $|C_2| = 2|E|$. In particular, one can determine the coefficients $|C_n|$ form the trace formula expansion.

The weight $W^{\mathrm{std}}(c)$ for the standard vertex space is a sort of probability of a particle choosing the path c (with equal probability to go in any adjacent edge at each vertex). It would be interesting to give a similar meaning to the "weights" $W_{\mathscr{V}}(c)$ for general vertex spaces.

Chapter 3
The Functional Analytic Part: Scales of Hilbert Spaces and Boundary Triples

In this chapter, we present the necessary *quantitative* analysis for a single operator in a Hilbert space. Basically, we introduce the natural scale of Hilbert spaces associated with a non-negative operator (Sect. 3.2) and to general closed operators (Sect. 3.3). We need quantitative norm estimates on the resolvent and other functions of such closed operator as well as norm bounds on related embeddings especially when dealing with *families* of operators in Chap. 4. The closed operators we are dealing here with, are more general than sectorial operators. Note that for such operators H, there is usually no easily expressed upper estimate on the norm of the resolvent $\|(H-z)^{-1}\|$ like $\|(\Delta - z)^{-1}\| = \operatorname{dist}(z, \sigma(\Delta))^{-1}$ for a self-adjoint operator Δ. Therefore we use an estimate $\|(H-z)^{-1}\| \le \gamma(z)$ as assumption, and call γ the *resolvent norm profile* of H.

The remaining part of this chapter (Sects. 3.4–3.9) is devoted to the theory of boundary triples associated with non-negative quadratic forms, e.g. arising from Laplacians on manifolds or metric graphs. Boundary triples are introduced for several purposes: First, to introduce abstractly the notion of *resonances* or eigenvalues of a complexly dilated operator (Sects. 3.6–3.8). Second, via boundary triples (or, more precisely, boundary maps) it is easy to carry over the analysis of objects associated with a star graph to general graphs (see Sect. 3.9).

The treatment of what we call *elliptic boundary triple* is stated on a purely functional analytic level. Moreover, quadratic form techniques are used in order to obtain "natural" estimates. These estimates can be controlled also when the underlying space depends on a parameter. We give a more detailed introduction to boundary triples at the beginning of Sect. 3.4 (see also Sects. 1.2.9 and 1.4.1).

3.1 Sesquilinear Forms, Associated Operators and Dense Subspaces

Let $\mathscr{H}$ be a Hilbert space with inner product $\langle \cdot, \cdot \rangle = \langle \cdot, \cdot \rangle_{\mathscr{H}}$ and norm $\|\cdot\| = \|\cdot\|_{\mathscr{H}}$. Throughout the paper we use the convention that $\langle \cdot, \cdot \rangle$ and other sesquilinear forms are *anti-linear* in the first and linear in the second argument. In what follows,

O. Post, *Spectral Analysis on Graph-Like Spaces*, Lecture Notes in Mathematics 2039,
DOI 10.1007/978-3-642-23840-6_3, © Springer-Verlag Berlin Heidelberg 2012

we skip subscripts whenever it is clear from the context which Hilbert space is meant. Moreover, for a bounded operator A in $\mathscr{H}$, we denote its usual operator norm by $\|A\|$. Let $\sigma(A)$ and $\rho(A) = \mathbb{C} \setminus \sigma(A)$ denote the *spectrum* and *resolvent set* of A, respectively.

Let us give a short overview on sesquilinear forms and the associated operators. For more details, we refer to the book of Kato [K66, Chap. VI]. For the moment, we only consider *sesquilinear forms* with *symmetric domains*, i.e. we consider sesquilinear forms

$$\mathfrak{h}\colon \mathscr{D} \times \mathscr{D} \longrightarrow \mathbb{C} \tag{3.1}$$

defined on the product $\mathscr{D} \times \mathscr{D}$ of the *same* space $\mathscr{D}$; the non-symmetric case will appear in Definition 3.3.16. We will always assume that $\mathfrak{h}$ is densely defined, i.e. that $\mathscr{D}$ is dense in $\mathscr{H}$. We call

$$\mathfrak{h}(u) := \mathfrak{h}(u, u) \tag{3.2}$$

the *quadratic form* associated with $\mathfrak{h}$. We set $\operatorname{dom} \mathfrak{h} := \mathscr{D}$. Note that any sesquilinear form is uniquely determined by its quadratic form due to the *polarisation identity*

$$\mathfrak{h}(u, v) = \sum_{k=0}^{3} \mathrm{i}^k \mathfrak{h}(u + \mathrm{i}^k v) = \mathfrak{h}(u+v) - \mathfrak{h}(u-v) + \mathrm{i}\mathfrak{h}(u+\mathrm{i}v) - \mathrm{i}\mathfrak{h}(u-\mathrm{i}v). \tag{3.3}$$

Therefore, we use the same symbol for a sesquilinear form and its associated quadratic form.

We denote by $\mathfrak{o}$ and $\mathfrak{1}$ the quadratic form with domain $\mathscr{H}$ given by

$$\mathfrak{o}(u) = 0 \qquad \text{and} \qquad \mathfrak{1}(u) = \|u\|^2,$$

respectively. For $0 \le \vartheta < \pi/2$ let

$$\Sigma_\vartheta := \{ z \in \mathbb{C} \mid |\arg z| \le \vartheta \}. \tag{3.4}$$

be the *sector* with angle ϑ at the origin. We say that $\mathfrak{h}$ is ϑ-*sectorial*, IF

$$\mathfrak{h}(u) \in \Sigma_\vartheta$$

for all $u \in \operatorname{dom} \mathfrak{h}$. We say that a ϑ-sectorial form $\mathfrak{h}$ is *closed* , if $\mathfrak{h}$ is densely defined and if

$$\mathscr{H}^1 := \mathscr{H}^1(\mathfrak{h}) := \operatorname{dom} \mathfrak{h} \tag{3.5}$$

is complete with respect to the norm defined by

$$\|u\|_1^2 := \|u\|_{\mathfrak{h}}^2 := \operatorname{Re} \mathfrak{h}(u) + \|u\|^2. \tag{3.6}$$

Here,

$$\operatorname{Re}\mathfrak{h}(u,v) := \frac{1}{2}\big(\mathfrak{h}(u,v)+\overline{\mathfrak{h}(v,u)}\big) \qquad \text{and} \qquad \operatorname{Im}\mathfrak{h}(u,v) := \frac{1}{2\mathrm{i}}\big(\mathfrak{h}(u,v)-\overline{\mathfrak{h}(v,u)}\big).$$

Since

$$\operatorname{Re}^2 z \le |z|^2 = \operatorname{Re}^2 z + \operatorname{Im}^2 z \le \big(1+\tan^2\vartheta\big)\operatorname{Re}^2 z = \frac{1}{\cos^2\vartheta}\operatorname{Re}^2 z$$

by elementary geometry, we can use the equivalent norm $|\mathfrak{h}(u)| + \|u\|^2$ instead of $\|u\|_{\mathfrak{h}}^2$. Note that $\operatorname{Re}\mathfrak{h}$, $\operatorname{Im}\mathfrak{h}$ and $\mathfrak{h} = \operatorname{Re}\mathfrak{h} + \mathrm{i}\operatorname{Im}\mathfrak{h}$ are bounded with respect to the norm $\|\cdot\|_{\mathfrak{h}}$.

Let H be a densely defined operator in $\mathscr{H}$. The operator H is *closed* if the graph $\operatorname{gr} H = \{(u, Hu) \mid u \in H\}$ of H is a closed subspace of $\mathscr{H}\times\mathscr{H}$. Equivalently, H is closed if

$$\mathscr{H}^2 := \mathscr{H}^2(H) := \operatorname{dom} H \tag{3.7}$$

endowed with the norm

$$\|u\|_2 := \|u\|_H := \|(H - z_0)u\| \tag{3.8}$$

is complete for some (any) z_0 in the resolvent set $\rho(H)$.

We say that H is *ϑ-sectorial* if the *numerical range* of H is in the sector Σ_ϑ, i.e. if

$$\langle u, Hu\rangle \in \Sigma_\vartheta$$

for all $u \in \operatorname{dom} H$. We say that H is *maximally ϑ-sectorial* if H is ϑ-sectorial, closed and has no ϑ-sectorial extension (following the terminology of Kato [K66, Sec. V.3.10, p. 280]).

Let H be a ϑ-sectorial operator. Then the sesquilinear form $\langle u, Hv\rangle$ with $u, v \in \operatorname{dom} H$ has a closed extension $\mathfrak{h} = \mathfrak{h}_H$, the *sesquilinear form associated with the operator H* (see [K66, Thm. VI.1.27]).

From [K66, Thm. VI.2.1] we obtain:

Theorem 3.1.1. *Let $\mathfrak{h}$ be a closed, ϑ-sectorial sesquilinear form ($0 \le \vartheta < \pi/2$). Then there exists a maximal-ϑ-sectorial operator H such that:*

1. $\operatorname{dom} H \subset \operatorname{dom}\mathfrak{h}$ *and* $\mathfrak{h}(u,v) = \langle u, Hv\rangle$ *for all* $u \in \operatorname{dom}\mathfrak{h}$ *and* $v \in \operatorname{dom} H$,
2. $\operatorname{dom} H$ *is dense in* $(\mathscr{H}^1, \|\cdot\|_{\mathfrak{h}})$,
3. *if* $v \in \operatorname{dom}\mathfrak{h}$, $w \in \mathscr{H}$ *and*
$$\mathfrak{h}(u,v) = \langle u, w\rangle$$
for all $u \in \operatorname{dom}\mathfrak{h} = \mathscr{H}^1$ *(or for all u in a dense subspace $\mathscr{D} \subset \mathscr{H}^1$), then* $v \in \operatorname{dom} H$ *and* $Hv = w$.

The operator is uniquely determined by (1).

Remark 3.1.2.

1. By the spectral calculus for non-negative operators (see Theorem 3.2.2), one can show that $\operatorname{dom}\mathfrak{h} = \operatorname{dom} H^{1/2}$ and $\mathfrak{h}(u) = \|H^{1/2}u\|^2$.
2. The above theorem and other assertions mentioned above extend easily to sesquilinear forms with numerical range in the sector $\Sigma_\vartheta - \lambda_0$ with vertex at $\lambda_0 \in \mathbb{R}$, by replacing $\mathfrak{h}$ with $\mathfrak{h} - \lambda_0 \mathbb{1}$. In particular, the theorem applies to sesquilinear forms with range in $[\lambda_0, \infty)$ for some $\lambda_0 \in \mathbb{R}$, i.e. $\mathfrak{h}(u) \geq \lambda_0 \|u\|^2$ for $u \in \operatorname{dom}\mathfrak{h}$.

Quadratic forms allow us to define an order relation on the set of self-adjoint operators:

Definition 3.1.3. Let $\mathfrak{a}$ and $\mathfrak{b}$ be two closed quadratic forms with range in $[\lambda_0, \infty)$ for some $\lambda_0 \in \mathbb{R}$. We define $\mathfrak{a} \leq \mathfrak{b}$ if $\operatorname{dom}\mathfrak{a} \supset \operatorname{dom}\mathfrak{b}$ and $\mathfrak{a}(u) \leq \mathfrak{b}(u)$ for all $u \in \operatorname{dom}\mathfrak{b}$.

Similarly, for two self-adjoint operators A and B, bounded from below, we define $A \leq B$ if $\mathfrak{a}_A \leq \mathfrak{b}_B$, where $\mathfrak{a}_A$ and $\mathfrak{b}_B$ denote the corresponding forms, respectively.

There is an intuitive "justification" of the – at first sight – unintuitive order relation for the domains: Let $\mathfrak{a}$ be a quadratic form, bounded from below. We extend $\mathfrak{a}$ onto $\mathscr{H}$ by setting $\widetilde{\mathfrak{a}}(u) = \infty$ if $u \in \mathscr{H} \setminus \operatorname{dom}\mathfrak{a}$. Then we have $\mathfrak{a} \leq \mathfrak{b}$ (in the sense defined above) iff $\widetilde{\mathfrak{a}}(u) \leq \widetilde{\mathfrak{b}}(u)$ for all $u \in \mathscr{H}$ (see e.g. [D95] for details).

Let us give a useful quantitative definition of a dense embedding:

Definition 3.1.4. Let $C > 0$. We say that $\mathscr{D}$ is *C-densely embedded* in $\mathscr{H}$ if $\mathscr{D}$ is dense in $\mathscr{H}$, if $\mathscr{D}$ carries a norm $\|\cdot\|_{\mathscr{D}}$ such that $\mathscr{D}$ itself is a Hilbert space, and if the embedding $\mathscr{D} \hookrightarrow \mathscr{H}$ is bounded by C, i.e. $\|u\|_{\mathscr{H}} \leq C\|u\|_{\mathscr{D}}$.

We will often omit C if $0 < C \leq 1$. Note that the precise value of C does not matter, when dealing with a single subspace $\mathscr{D} \subset \mathscr{H}$. Later on we will consider *families* of C-densely embedded subspaces $\mathscr{D}_\varepsilon$ in $\mathscr{H}_\varepsilon$, where not only $\mathscr{D}_\varepsilon$ but also $\mathscr{H}_\varepsilon$ or their norms are allowed to depend on ε.

Let us give some typical examples which are discussed in more detail in the next sections. The value of C is of particular interest for *families* of operators H_ε as in Example 3.1.5 (2):

Example 3.1.5.

1. Let Δ be a non-negative operator and set $\mathscr{D} := \mathscr{H}^k = \operatorname{dom}\Delta^{k/2}$ with norm $\|u\|_{\mathscr{D}} := \|(\Delta + 1)^{k/2}u\|$, then $\mathscr{D}$ is 1-densely embedded.
2. More generally, if H is a closed operator with bounded inverse (i.e. $0 \in \rho(H)$), then $\mathscr{D} := \operatorname{dom} H$ with $\|u\|_{\mathscr{D}} := \|Hu\|$ is C-densely embedded with $C = \|H^{-1}\|$.

In the following, we state that the above Example 3.1.5 (1) (with $A = (\Delta + 1)^k$) covers already the general case:

Lemma 3.1.6. *Assume that $\mathscr{D}$ is C-densely embedded in $\mathscr{H}$, then there exists a strictly positive operator $A \geq 1/C$ with associated quadratic form $\mathfrak{a}$ such that*

$$\|u\|^2_{\mathscr{D}} = \|A^{1/2}u\|^2 = \mathfrak{a}(u).$$

Proof. Define $\mathfrak{a}(u) := \|u\|_{\mathscr{D}}^2$ with $\operatorname{dom}\mathfrak{a} = \mathscr{D}$, then it is easily seen that $\mathfrak{a}$ is a densely defined, non-negative quadratic form. Furthermore, from the C-boundedness of the embedding, we obtain the estimate $\mathfrak{a}(u) \geq C^{-1}\|u\|^2$. Now, A is the operator associated with $\mathfrak{a}$ by Theorem 3.1.1, and the equality on the norm of $\mathscr{D}$ follows. □

We call A the *generating operator*of the C-densely embedded space $\mathscr{D}$.

Remark 3.1.7. Note that the above constructed non-negative operator A associated with a C-densely embedded space is not always the best choice to work with: For example, if H is closed with $0 \in \rho(H)$, then

$$\|u\|_{\mathscr{D}}^2 = \|Hu\|^2 = \langle u, H^*H\rangle u = \|(H^*H)^{1/2}u\|^2,$$

i.e. the space $\mathscr{D}$ is generated by the positive operator $A = H^*H$. Nevertheless, it is sometimes useful to keep the original operator: If e.g. H is a differential operator, $A^{1/2}$ could be a pseudo-differential operator, and more difficult to treat.

We define the *dual* $\mathscr{D}^*$ of $\mathscr{D}$ as subspace of $\mathscr{H}$ as follows:

Definition 3.1.8. Let $\mathscr{D}$ be C-densely embedded in $\mathscr{H}$. The *dual* $\mathscr{D}^*$ is defined by

$$\mathscr{D}^* := \{\varphi\colon \mathscr{D} \longrightarrow \mathbb{C} \mid \varphi \text{ anti-linear and bounded}\}, \tag{3.9a}$$

with norm

$$\|\varphi\|_{\mathscr{D}^*} := \sup_{u\in\mathscr{D}} \frac{|\varphi(u)|}{\|u\|_{\mathscr{D}}}. \tag{3.9b}$$

Proposition 3.1.9. *Let $\mathscr{D}$ be C-densely embedded and $A \geq 1/C$ be the corresponding generating operator. Then $\mathscr{D}^*$ can be defined by the (abstract) completion of $\mathscr{H}$ with respect to the norm $\|v\|_{\mathscr{D}^*} = \|A^{-1/2}v\|$, $v \in \mathscr{H}$. In particular, the dual space $\mathscr{D}^*$ is a Hilbert space.*

Proof. Let $\iota v := \langle\cdot, v\rangle$, then it is easy to see that $\iota v \in \mathscr{D}^*$. Moreover, we have

$$\|\iota v\|_{\mathscr{D}^*} = \sup_{u\in\mathscr{D}} \frac{|\langle u, v\rangle|}{\|u\|_{\mathscr{D}}} = \sup_{u\in\mathscr{D}} \frac{|\langle u, v\rangle|}{\|A^{1/2}u\|} = \sup_{w\in\mathscr{H}} \frac{|\langle A^{-1/2}w, v\rangle|}{\|w\|} = \|A^{-1/2}v\|$$

showing the equality of the norms. In particular, the completion of $\iota\mathscr{H}$ is a subspace of $\mathscr{D}^*$.

Let us now show that $\iota\mathscr{H}$ is dense in $\mathscr{D}^*$ (hence the completion of $\iota\mathscr{H}$ *equals* $\mathscr{D}^*$): Assume that $\varphi \in \mathscr{D}^*$ and $\eta > 0$. By the Riesz-Fischer theorem (in the Hilbert space $(\mathscr{D}, \|\cdot\|_{\mathscr{D}})$) and Lemma 3.1.6, there exists $v \in \mathscr{D}$ such that

$$\varphi(u) = \langle u, v\rangle_{\mathscr{D}} = \mathfrak{a}(u, v)$$

for all $u \in \mathscr{D}$. By Theorem 3.1.1 (2), there exists $w \in \operatorname{dom} A$ such that $\|v - w\|_{\mathfrak{a}} \le \eta$. Now,

$$\big|\varphi(u) - \langle u, Aw\rangle\big| = \big|\mathfrak{a}(u, v - w)\big| \overset{\text{CS}}{\le} |\mathfrak{a}(u)||\mathfrak{a}(v - w)| \le \|u\|_{\mathscr{D}}\|v - w\|_{\mathfrak{a}} \le \eta\|u\|_{\mathscr{D}}$$

for all $u \in \mathscr{D}$, since A is the operator associated with $\mathfrak{a}$. In particular, we have shown $\|\varphi - \iota(Aw)\|_{\mathscr{D}^*} \le \eta$, i.e. the density.

Finally, $\mathscr{D}^*$ is a Hilbert space, since its norm fulfils the parallelogram equality on a dense subset. □

We will often identify $\mathscr{H}$ with its canonical image $\iota\mathscr{H}$ in $\mathscr{D}^*$.

Lemma 3.1.10. *Let $\mathscr{D}$ be C-densely embedded in $\mathscr{H}$, then $\mathscr{H}$ is C-densely embedded in $\mathscr{D}^*$ via $\iota\colon v \mapsto \langle\cdot, v\rangle$. In other words, the embeddings*

$$\mathscr{D} \hookrightarrow \mathscr{H} \hookrightarrow \mathscr{D}^*$$

are bounded by C and have dense ranges.

Proof. We have

$$\|\iota v\|_{\mathscr{D}^*} = \sup_{u\in\mathscr{D}} \frac{|\langle u, v\rangle|}{\|u\|_{\mathscr{D}}} \overset{\text{CS}}{\le} \sup_{u\in\mathscr{D}} \frac{\|u\|\,\|v\|}{\|u\|_{\mathscr{D}}} \le C\,\|v\|$$

showing the boundedness of the embedding. The density of $\mathscr{H}$ in $\mathscr{D}^*$ was already shown in the previous proposition. □

3.2 Scale of Hilbert Spaces Associated with a Non-negative Operator

Let us define in this section the scale of Hilbert spaces associated with a non-negative operator operator Δ in a Hilbert space $\mathscr{H}$. We will always assume that Δ is a closed operator; in particular, Δ is self-adjoint. More details on scales of Hilbert spaces can be found e.g. in [RS80], [GG91] or [AK00].

The scale of Hilbert spaces associated with Δ is defined as follows: For $k \ge 0$ we set

$$\mathscr{H}^k := \mathscr{H}^k(\Delta) := \operatorname{dom}(\Delta + 1)^{k/2}, \qquad \|u\|_k := \|(\Delta + 1)^{k/2}u\|, \quad k \ge 0. \tag{3.10}$$

For negative powers, we set $\mathscr{H}^{-k} := (\mathscr{H}^k)^*$ (see Definition 3.1.8). For the natural pairing $\mathscr{H}^k \times \mathscr{H}^{-k}$ we also write

$$\langle\cdot,\cdot\rangle = \langle\cdot,\cdot\rangle_{k,-k}\colon \mathscr{H}^k \times \mathscr{H}^{-k} \longrightarrow \mathbb{C}.$$

Note that $\mathscr{H} = \mathscr{H}^0$ embeds naturally into $\mathscr{H}^{-k}$ via $\iota: v \mapsto \langle \cdot, v \rangle$ since

$$\|\iota v\|_{-k} = \sup_{u \in \mathscr{H}^k} \frac{|\langle u, v \rangle|}{\|u\|_k} = \sup_{w \in \mathscr{H}^0} \frac{|\langle R^{k/2} w, v \rangle|}{\|w\|_0} = \|R^{k/2} v\|_0,$$

where

$$R := (\Delta + 1)^{-1}, \tag{3.11}$$

i.e. $\mathscr{H}^{-k}$ can be viewed as the completion of $\mathscr{H}$ with respect to the norm

$$\|v\|_{-k} := \|(\Delta + 1)^{-k/2} v\|. \tag{3.12}$$

We call the family $\{\mathscr{H}^k\}_k = \{\mathscr{H}^k(\Delta)\}_k$ the *scale of Hilbert spaces associated with* Δ.

If Δ is defined via the closed non-negative quadratic form

$$\mathfrak{d}(u) = \|\Delta^{1/2} u\|^2, \qquad \operatorname{dom} \mathfrak{d} = \mathscr{H}^1 = \operatorname{dom} \Delta^{1/2}, \tag{3.13}$$

(see Theorem 3.1.1) then

$$\|u\|_1^2 = \|u\|^2 + \mathfrak{d}(u). \tag{3.14}$$

Moreover, if $v \in \mathscr{H} \subset \mathscr{H}^{-1}$, then $\|v\|_{-1} \leq \delta$ is equivalent with the fact that

$$\big|\langle u, v \rangle\big| \leq \delta \|u\|_1 \tag{3.15}$$

for all $u \in \mathscr{H}^1$. In addition, we have the following estimate:

Lemma 3.2.1. *The norm* $\|\cdot\|_k$ *and the graph norm of* $\Delta^{k/2}$ *are equivalent, i.e.*

$$\|f\|^2 + \|\Delta^{k/2} f\|^2 \leq \|f\|_k^2 \leq 2^k \big(\|f\|^2 + \|\Delta^{k/2} f\|^2\big)$$

for all $f \in \mathscr{H}^k$.

Proof. We have

$$\|f\|^2 + \|\Delta^{k/2} f\|^2 = \langle f, (1 + \Delta^k) f \rangle \qquad \text{and} \qquad \|f\|_k^2 = \langle f, (1 + \Delta)^k f \rangle$$

for $f \in \mathscr{H}^{2k} = \operatorname{dom} \Delta^k \subset \mathscr{H}^k$. By the spectral calculus, the estimates follow from

$$1 + \lambda^k \leq (1 + \lambda)^k = \sum_{i=0}^{k} \binom{k}{i} \lambda^i \leq \sum_{i=0}^{k} \binom{k}{i} (1 + \lambda^k) = 2^k (1 + \lambda^k)$$

for $\lambda \geq 0$. □

For non-negative operators (or more generally, self-adjoint or normal operators), we have the L_∞-functional calculus:

Theorem 3.2.2. *Let* $\varphi \in \mathsf{L}_\infty(\mathbb{R}_+)$, *then* $\|\varphi(\Delta)\| = \|\varphi\|_{\mathsf{L}_\infty(\mathbb{R}_+)}$.

Let Δ be a non-negative operator with *purely discrete* spectrum, i.e.

$$\sigma(\Delta) = \{\, \lambda_k \mid k = 1, 2, \dots \},$$

where λ_k denotes the k-th eigenvalue written in ascending order and repeated according to multiplicity such that $\lambda_k \to \infty$ as $k \to \infty$. In this case, we can employ the *min-max principle* (see e.g. [D97] for the formulation used here):

Theorem 3.2.3. *Let* $\mathfrak{d}$ *be a non-negative closed quadratic form such that the associated operator has purely discrete spectrum. Then the* k*-th eigenvalue of* Δ *is given by*

$$\lambda_k = \inf_{D_k} \sup_{f \in D_k \setminus \{0\}} \frac{\mathfrak{d}(f)}{\|f\|^2} \tag{3.16}$$

where the infimum is taken over all k*-dimensional subspaces* D_k *of* $\mathscr{H}^1 = \operatorname{dom} \mathfrak{d}$.

Clearly, the previous theorem holds not only for non-negative quadratic forms but also for forms bounded from below.

3.3 Scale of Hilbert Spaces Associated with a Closed Operator

Although we mainly use non-negative operators like the Laplacian on a manifold in our applications, we also need to treat non-self-adjoint operators. Such operators arise naturally when dealing with resonances. A *resonance* will be defined as the (discrete) eigenvalues of a "complexly dilated" non-negative operator. The complexly dilated operator is no longer self-adjoint, but usually has spectrum inside a sector in the right half-plane. We will explain the setting in detail in Sects. 3.5–3.8. In what follows, we present a scale of Hilbert spaces associated with a closed operator, as well as a quantitative analysis of norm-bounds of various associated embeddings and operators, starting from a norm estimate on the resolvent.

Since we mostly deal with second order operators, we refer to the operator domain as *second order* and to the corresponding sesquilinear form domain as *first order*.

3.3.1 Scale of Hilbert Spaces of Second Order Associated with a Closed Operator

Let H be a closed, densely defined operator in the Hilbert space $\mathscr{H}$. We fix a point z_0 in the resolvent set $\rho(H) = \mathbb{C} \setminus \sigma(H)$ of H throughout this section. In most of our applications we choose $z_0 = -1$. Denote

$$R(z) := (H - z)^{-1}, \qquad R := R(z_0) \tag{3.17}$$

the resolvent of H in $z \in \rho(H)$ resp. z_0. Let

$$\mathscr{H}^2(H) := \operatorname{dom} H, \qquad \|u\|_{2,H} := \|(H - z_0)u\| \tag{3.18}$$

denote the *scale of order* 2 *associated with* H.[1] Note that $\mathscr{H}^2(H)$ with the given norm is a Hilbert space, since H is closed. For the adjoint H^*, we use the norm $\|u\|_{2,H^*} := \|(H^* - \bar{z}_0)u\|$. Denote by $\mathscr{H}^{-2}(H) := (\mathscr{H}^2(H^*))^*$ the dual of $\mathscr{H}^2(H^*)$ with respect to $\mathscr{H}$, see Definition 3.1.8. We call $\mathscr{H}^{-2}(H)$ the *scale of order* -2 *associated with* H. It follows from Proposition 3.1.9 (with $A = (H - z_0)(H^* - \bar{z}_0)$) that we can embed $\mathscr{H}$ into $\mathscr{H}^{-2}(H)$ via $\iota: u \mapsto \langle \cdot, u\rangle$ and define $\mathscr{H}^{-2}(H)$ as the completion of $\mathscr{H}$ in the norm

$$\|u\|_{-2,H} := \|(H - z_0)^{-1}u\| = \|Ru\| = \|(R^*R)^{1/2}u\| = \|A^{-1/2}u\|. \tag{3.19}$$

The proof of the following lemma is obvious:

Lemma 3.3.1. *The maps*

$$(H - z_0): \mathscr{H}^2(H) \longrightarrow \mathscr{H} \quad \text{and} \quad R: \mathscr{H} \longrightarrow \mathscr{H}^2(H) \tag{3.20a}$$

are isometries and inverse to each other. Similarly,

$$(H - z_0): \mathscr{H} \longrightarrow \mathscr{H}^{-2}(H) \quad \text{and} \quad R: \mathscr{H}^{-2}(H) \longrightarrow \mathscr{H} \tag{3.20b}$$

are isometries and inverse to each other. Here, the operator $H - z_0$ in (3.20b) *is defined by $(H - z_0)g := \langle (H^* - z_0)(\cdot), g\rangle$, and $R(\iota g) := Rg$ extends to an isometry on $\mathscr{H}^{-2}(H)$.*[2]

Lemma 3.3.2. *The function $\rho(H) \to (0, \infty)$, $z \mapsto \|R(z)\|$ depends continuously on z.*

Proof. The assertion follows immediately from the holomorphy of $z \mapsto \langle f, R(z)g\rangle$. □

In order to formulate *quantitative* assertions on *families* of closed operators, we need the following definition:

---

[1] The notion "order" refers to the fact that H will be a second order elliptic differential operator in our applications.

[2] In order not to overwhelm the notation, we use the same symbol $H - z_0$ resp. R acting on different scales of Hilbert spaces and similarly for other operators. We indicate the precise meaning by the map notation, e.g., we write $(H - z_0): \mathscr{H} \longrightarrow \mathscr{H}^{-2}(H)$.

Definition 3.3.3. Let $U \subset \mathbb{C}$, $z_0 \in U$ and $\gamma\colon U \longrightarrow (0,\infty)$ be a continuous function.

1. We say that H has *resolvent norm profile γ on U (of order* 0*)*if $U \subset \rho(H)$ and

$$\|R(z)\| \le \gamma(z) \tag{3.21}$$

for all $z \in \rho(H)$. We denote by $\mathscr{R}(\gamma, U) = \mathscr{R}_0(\gamma, U)$ the set of all closed operators with resolvent norm profile γ on U of order 0.
2. If U contains only one point z_0 and the map has the single value $C > 0$ we set $\mathscr{R}(C, z_0) := \mathscr{R}(z_0 \mapsto C, \{z_0\})$.

Lemma 3.3.4. *Let $H \in \mathscr{R}(C, z_0)$, then $\mathscr{H}^2(H)$ is C-bounded in $\mathscr{H}$. In particular, the natural embeddings*

$$\mathscr{H}^2(H) \hookrightarrow \mathscr{H} \overset{\iota}{\hookrightarrow} \mathscr{H}^{-2}(H)$$

are bounded by C and the ranges of the inclusions are dense.

Proof. The continuity of the embedding $\mathscr{H}^2(H) \hookrightarrow \mathscr{H}$ follows from

$$\|u\| = \|R(H - z_0)u\| \le \|R\|\,\|(H - z_0)u\| = \|R\|\,\|u\|_{2,H} \le C\,\|u\|_{2,H}.$$

The remaining assertion follows from Lemma 3.1.10. □

Moreover, we can naturally extend the resolvents and the operator on the scales of second order:

Lemma 3.3.5. *Assume that $H \in \mathscr{R}(\gamma, U)$ and $k \in \{-2, 0, 2\}$, then the resolvent as operator*

$$R(z)\colon \mathscr{H}^k(H) \longrightarrow \mathscr{H}^k(H) \tag{3.22a}$$

is bounded by $\gamma(z)$. Moreover, the resolvents as operators

$$\mathscr{H}^{-2}(H) \overset{R(z)}{\longrightarrow} \mathscr{H} \overset{R(z)}{\longrightarrow} \mathscr{H}^2(H) \tag{3.22b}$$

are invertible and bounded by

$$\gamma_0(z) := 1 + |z - z_0|\gamma(z) \tag{3.22c}$$

for $z \in U$. Its inverses are given by

$$\mathscr{H}^2(H) \overset{H-z}{\longrightarrow} \mathscr{H} \overset{H-z}{\longrightarrow} \mathscr{H}^{-2}(H) \tag{3.22d}$$

and bounded by $1 + |z_0 - z|\gamma(z_0)$. In particular, the operators

$$\mathscr{H}^2(H) \overset{H}{\longrightarrow} \mathscr{H} \overset{H}{\longrightarrow} \mathscr{H}^{-2}(H) \tag{3.22e}$$

have norms bounded by $1 + |z_0|\gamma(z_0)$.

Proof. The first claim follows from the fact that $R(z)$ and H commute, and therefore $\|R(z)\|_{k,H\to k,H} = \|R(z)\| \le \gamma(z)$. For the second norm estimate, note that

$$\begin{aligned}\|R(z)f\|_{2,H} &= \|(H-z_0)R(z)f\| = \big\|\big(\mathrm{id}+(z-z_0)R(z)\big)f\big\| \\ &\le \big(1+|z-z_0|\gamma(z)\big)\|f\|,\end{aligned}$$

i.e. $\|R(z)\|_{0\to 2,H} \le 1+|z-z_0|\gamma(z)$, and the boundedness of $R(z)$ from $\mathscr{H}^{-2}(H)$ into $\mathscr{H}$ follows by duality. The remaining assertions can be seen similarly. □

We will also need the following holomorphic functional calculus. Let $D \subset \mathbb{C}$ be an open set with piecewise smooth boundary ∂D. We choose an orientation of ∂D, such that D "lies on the left side" of ∂D. We do not exclude the case that ∂D "passes" infinity, i.e. that D is unbounded.

Theorem 3.3.6. *Assume that H is a closed operator with resolvent norm profile γ on $U \subset \rho(H)$, i.e. $H \in \mathscr{R}(\gamma, U)$. Assume in addition that D is an open set in $\mathbb{C}$ such that $\partial D \subset U$ (and therefore $\partial D \subset \rho(H)$). Let $\varphi \in \mathsf{L}_1(\partial D, \gamma(z)\mathrm{d}|z|)$ such that φ is holomorphic in a neighbourhood of $\overline{D}$. Then the operator*

$$\varphi_D(H) = -\frac{1}{2\pi\mathrm{i}}\oint_{\partial D}\varphi(z)R(z)\mathrm{d}z = \frac{1}{2\pi\mathrm{i}}\oint_{\partial D}\frac{\varphi(z)}{z-H}\mathrm{d}z$$

is well-defined and bounded. Moreover, the integral exists in the operator norm topology. Finally, the norm can be estimated by

$$\|\varphi_D(H)\| \le \frac{1}{2\pi}\oint_{\partial D}|\varphi(z)|\gamma(z)\mathrm{d}|z| =: C_D(\varphi,\gamma).$$

For the definition and some results on vector-valued integrals we refer to Sect. A.2.1.

Remark 3.3.7. It can be seen that $\varphi_D(H)$ depends only on the homotopy type of ∂D in the resolvent set $\rho(H) = \mathbb{C}\setminus\sigma(H)$, i.e. on the connected components of $\sigma(H)$ contained in D. If $\sigma(H) \subset D$, then we simply write $\varphi(H)$ instead of $\varphi_D(H)$.

3.3.2 Scale of Hilbert Spaces of First Order Associated with a Closed Operator

The aim of the following considerations is to give a "good" first order Hilbert space scale $\mathscr{H}^1(H)$ associated with the closed operator H (for the notion "order" see

Footnote 1 on page 105). Basically, we want to assure that $\mathscr{H}^1(H)$ and its dual behave like the "natural" scale of Hilbert spaces. Note that for sectorial operators, one could define the square root $(H+1)^{1/2}$ as natural scale, but it is in general a difficult task to determine the domain of $(H+1)^{1/2}$. In particular, if H is a second order elliptic partial differential operator, then it is difficult to decide whether $\operatorname{dom}(H+1)^{1/2}$ is a Sobolev space of order 1. This question is also known under the name "Kato's square root problem", see e.g. [McI72, AHL$^+$01]. Moreover, the expression for $\|(H+1)^{1/2}f\|^2$ is (in general) no longer a differential operator, and therefore difficult to handle with. At first sight, our approach looks more technical, but we will see that the necessary conditions are naturally fulfilled in many cases.

Let us give a motivation why we are interested in such a scale of order 1 associated with H: In our applications, we easily obtain estimates given in the "free" scale $\mathscr{H}^k = \mathscr{H}^k(\Delta)$ associated with Δ (or the corresponding quadratic form $\mathfrak{d}$). Let e.g. $A: \mathscr{H}^1 \longrightarrow \mathscr{H}$ be an operator on the free scale of order 1 such that

$$\|Av\| \leq \delta\|v\|_1 = \delta\|(\Delta+1)^{1/2}v\| = \delta\big(\|v\|^2 + \mathfrak{d}(v)\big) \tag{3.23}$$

holds for all $v \in \mathscr{H}^1$. But when dealing with the scale of Hilbert spaces associated with a closed operator H, we finally need an estimate in terms of the norm $\|\cdot\|_{2,H}$.

We will use the "free" operator Δ in order to define a scale of order 1 associated with H:

Definition 3.3.8. Let H be a closed operator and Δ a non-negative operator in the Hilbert space $\mathscr{H}$.

1. We say that H is *compatible with Δ (with compatibility operator T)* if $T: \mathscr{H} \longrightarrow \mathscr{H}$ is a bounded and invertible operator such that

$$T(\operatorname{dom} H) \subset \mathscr{H}^1 \qquad \text{and} \qquad T^*(\operatorname{dom} H^*) \subset \mathscr{H}^1, \tag{3.24a}$$

   where $\mathscr{H}^1 = \operatorname{dom}\mathfrak{d}$ is the scale of order 1 associated with Δ (the domain of the associated quadratic form $\mathfrak{d}$). We sometimes refer to $\mathfrak{d}$ (resp. Δ) as the *free* or *reference* form (resp. operator) associated with H.
2. We set $\mathscr{H}^1(H) := T^{-1}(\mathscr{H}^1)$ and $\mathscr{H}^1(H^*) := (T^*)^{-1}(\mathscr{H}^1)$ endowed with the norms

$$\|u\|_{1,H} := \|Tu\|_1 = \|(\Delta+1)^{1/2}Tu\|, \tag{3.24b}$$

$$\|u\|_{1,H^*} := \|T^*u\|_1 = \|(\Delta+1)^{1/2}T^*u\|. \tag{3.24c}$$

3. The dual of $\mathscr{H}^1(H)$ is defined by $\mathscr{H}^{-1}(H) := (\mathscr{H}^1(H^*))^*$ with its natural norm (see Definition 3.1.8).

Remark 3.3.9.

1. The operator T restricted to $\mathscr{H}^1(H)$, i.e. $T^1 \colon \mathscr{H}^1(H) \longrightarrow \mathscr{H}^1$, $u \mapsto Tu$, is unitary.
2. The natural embedding $\iota \colon \mathscr{H} \hookrightarrow \mathscr{H}^{-1}(H)$, $\iota v = \langle \cdot, v \rangle$, gives

$$\begin{aligned}\|\iota v\|_{-1,H} &= \sup_{u \in \mathscr{H}^1(H^*)} \frac{|\langle u, v\rangle|}{\|u\|_{1,H^*}} \\ &= \sup_{w \in \mathscr{H}} \frac{|\langle (T^{-1})^*(\Delta+1)^{-1/2} w, v\rangle|}{\|w\|} \\ &= \sup_{w \in \mathscr{H}} \frac{|\langle w, (\Delta+1)^{-1/2} T^{-1} v\rangle|}{\|w\|} = \|(\Delta+1)^{-1/2} T^{-1} v\|.\end{aligned}$$

   In particular, we can alternatively define $\mathscr{H}^{-1}(H)$ as the completion of $\mathscr{H}$ in the norm $\|v\|_{-1,H} := \|(\Delta+1)^{-1/2} T^{-1} v\|$. Moreover, T^{-1} extends naturally to a unitary operator

$$T^{-1} \colon \mathscr{H}^{-1}(H) \longrightarrow \mathscr{H}^{-1},$$

   since $\|T^{-1} v\|_{-1} = \|(\Delta+1)^{-1/2} T^{-1} v\| = \|v\|_{-1,H}$.
3. Note that $\|\cdot\|_{1,H}$ is the norm associated with the non-negative quadratic form $\mathfrak{d} \circ T$ defined by $f \mapsto \mathfrak{d}(Tf)$ on $\mathscr{H}^1(H)$ in the Hilbert space $(\mathscr{H}, \|T\cdot\|)$ with norm given by $f \mapsto \|Tf\|$, since

$$\|u\|_{1,H}^2 = \mathfrak{d}(Tu) + \|Tu\|^2.$$

The scale of order 1 should behave like a scale of Hilbert spaces, e.g., the embedding $\mathscr{H}^2(H) \hookrightarrow \mathscr{H}^1(H)$ should be bounded. This will be ensured by the next definition and the following lemmata. Note that for operators $A \colon \mathscr{H}^{-1}(H) \longrightarrow \mathscr{H}^1(H)$ and $B \colon \mathscr{H} \longrightarrow \mathscr{H}^1(H)$, the corresponding operator norms are given by

$$\|A\|_{-1,H\to 1,H} = \|(\Delta+1)^{1/2} TAT (\Delta+1)^{1/2}\| \quad \text{and} \tag{3.25a}$$

$$\|B\|_{0\to 1,H} = \|(\Delta+1)^{1/2} TB\|. \tag{3.25b}$$

Definition 3.3.10. Let $\gamma \colon U \longrightarrow (0,\infty)$ be a continuous function on $U \subset \mathbb{C}$.

1. We say that the closed operator H has *resolvent norm profile* γ *on* U *of order* 1 (with respect to Δ), if $U \subset \rho(H)$ and if H is compatible with Δ with compatibility operator T such that

$$\|T\| \le \gamma(z_0), \qquad \|T^{-1}\| \le \gamma(z_0) \quad \text{and} \tag{3.26a}$$

$$\|R(z)\|_{-1,H\to 1,H} \le \gamma(z) \tag{3.26b}$$

for all $z \in U$. We denote the set of all closed operators H having resolvent norm profile γ on U of order 1 w.r.t. Δ by $\mathscr{R}_1(\gamma, U)$ (omitting the dependence on Δ in the notation).
2. If $U = \{z_0\}$ contains only one point and if $C = \gamma(z_0) \geq 1$, then we also say that H is *C-compatible of order* 1 w.r.t. Δ, for short $H \in \mathscr{R}_1(C, z_0)$.

Remark 3.3.11. Note that this setting is a generalisation of the scale of Hilbert spaces associated with a non-negative operator: Assume that $H = \Delta$ is non-negative and set $z_0 = -1$. Then we can choose $T = \mathrm{id}$ and $\gamma(z) = \mathrm{dist}(z, \mathbb{R}_+)^{-1}$ for $z \in U = \mathbb{C} \setminus \mathbb{R}_+$. Moreover, $\gamma(z_0) = 1 = \|T\|$.

Lemma 3.3.12. *Assume that H has resolvent norm profile γ on U of order* 1 *and set $C := \gamma(z_0)$. Then*

$$\|R(z)\|_{0\to 1,H} \leq C\gamma(z) \qquad \text{and} \qquad \|R(z)\|_{-1,H\to 0} \leq C\gamma(z), \tag{3.27a}$$

$$\|R(z)\|_{0\to 0} \leq C^2\gamma(z) \tag{3.27b}$$

for all $z \in U$. In particular, H has resolvent norm profile $C^2\gamma$ on U of order 0, *i.e.* $\mathscr{R}_1(\gamma, U) \subset \mathscr{R}_0(C^2\gamma, U)$.

Proof. The norm estimates of the resolvent can be seen as follows

$$\begin{aligned}\|R(z)\|_{0\to 1,H} &= \|(\Delta+1)^{1/2}TR(z)T(\Delta+1)^{1/2}(\Delta+1)^{-1/2}T^{-1}\| \\ &\leq \|R(z)\|_{-1,H\to 1,H}\|(\Delta+1)^{-1/2}\|\,\|T^{-1}\| \leq \gamma(z)C,\end{aligned}$$

and similarly for the estimate on $\|R(z)\|_{-1,H\to 0}$. Moreover,

$$\begin{aligned}\|R(z)\|_{0\to 0} &= \|T^{-1}(\Delta+1)^{-1/2}(\Delta+1)^{1/2}TR(z)\| \\ &\leq \|T^{-1}\|\,\|(\Delta+1)^{-1/2}\|\,\|R(z)\|_{0\to 1,H} \leq C^2\gamma(z),\end{aligned}$$

for all $z \in U$. □

Note that (3.27a) is just the desired estimate on $Av = ATu$ in (3.23) in terms of the scale associated with H, namely

$$\|ATu\| \leq \delta\|(\Delta+1)^{1/2}Tu\| = \delta\|u\|_{1,H} \leq C^2\delta\|u\|_{2,H}.$$

The next lemma shows that the embeddings are bounded in terms of universal constants depending only on the resolvent norm profile γ and z_0.

Lemma 3.3.13. *Let $H \in \mathscr{R}_1(\gamma, U)$ and set $C := \gamma(z_0)$. Then the natural embeddings*

$$\mathscr{H}^2(H) \hookrightarrow \mathscr{H}^1(H) \qquad \text{and} \qquad \mathscr{H}^{-1}(H) \hookrightarrow \mathscr{H}^{-2}(H)$$

are bounded by C^2. *Moreover,*

$$\mathscr{H}^1(H) \hookrightarrow \mathscr{H} \hookrightarrow \mathscr{H}^{-1}(H)$$

are C*-densely embedded (i.e. they are bounded by* C *and have dense range).*

Proof. The boundedness of the first embedding follows from

$$\|u\|_{1,H} = \|(\Delta+1)^{1/2}Tu\| \le \|R\|_{0\to 1,H}\|(H-z_0)u\| \le C^2\|u\|_{2,H}$$

and (3.27a). Moreover, for the embedding $\mathscr{H}^1(H) \hookrightarrow \mathscr{H}$, we estimate

$$\|u\| = \|T^{-1}(\Delta+1)^{-1/2}(\Delta+1)^{1/2}Tu\| \le C\|(\Delta+1)^{1/2}Tu\| = C\|u\|_{1,H}.$$

The space $\mathscr{H}^1(H) = T^{-1}(\mathscr{H}^1)$ is dense in $\mathscr{H}$ since $\mathscr{H}^1$ is dense in $\mathscr{H}$ and T^{-1} is bounded. □

Remark 3.3.14. A priori, it is not clear whether the density of $\mathscr{H}^2(H)$ in $\mathscr{H}^1(H)$ already follows from our assumptions, i.e. whether

$$\check{\mathscr{H}}^1(H) := \overline{\operatorname{dom} H}^{\|\cdot\|_{1,H}} \subseteq \mathscr{H}^1(H) \tag{3.28}$$

already *equals* $\mathscr{H}^1(H)$. We do not try to prove (or disprove) this fact here. A similar remark holds for H^*.

Lemma 3.3.15. *Let* $H \in \mathscr{R}_1(\gamma, U)$ *and set* $C := \gamma(z_0)$. *Then the resolvent as operator*

$$R(z) \colon \mathscr{H}^k(H) \longrightarrow \mathscr{H}^k(H)$$

is bounded by $C^2\gamma(z)$ *for* $k = 0, \pm 1, \pm 2$. *Moreover, the resolvent as operator*

$$\mathscr{H}^{-2}(H) \xrightarrow{R(z)} \mathscr{H}^{-1}(H) \qquad \textit{resp.} \qquad \mathscr{H}^1(H) \xrightarrow{R(z)} \mathscr{H}^2(H)$$

is bounded by

$$(1 + |z - z_0| C^2\gamma(z))C.$$

Finally, the resolvents as operators

$$\mathscr{H}^{-1}(H) \xrightarrow{R(z)} \mathscr{H} \xrightarrow{R(z)} \mathscr{H}^1(H)$$

are bounded by $C\gamma(z)$.

Proof. For $k = 1$, the boundedness of the resolvent as operator $R(z)$ in $\mathscr{H}^k(H)$ is a consequence of

$$\begin{aligned}\|R(z)\|_{1,H\to 1,H} &= \|(\Delta+1)^{1/2}TR(z)T^{-1}(\Delta+1)^{-1/2}\| \\ &\le \|R(z)\|_{0\to 1,H}\|T^{-1}\|\|(\Delta+1)^{-1/2}\| \le \gamma(z)C^2\end{aligned}$$

using (3.27a). For $k = -1$ we obtain the result by duality, and for $k = 0, \pm 2$, the estimate follows from $\|R(z)\|_{k,H\to k,H} = \|R(z)\|_{0\to 0}$ and (3.27b). Here, $\|A\|_{k,H\to k,H}$ denotes the operator norm of an operator $A\colon \mathscr{H}^k(H) \longrightarrow \mathscr{H}^k(H)$, and we omit H if $k = 0$. Finally, we have

$$\|R(z)u\|_{2,H} \le \bigl(1 + |z - z_0|C^2\gamma(z)\bigr)\|u\| \le \bigl(1 + |z - z_0|C^2\gamma(z)\bigr)C\|u\|_{1,H}$$

by (3.22c), (3.27b) and Lemma 3.3.13. Moreover,

$$\|R(z)u\|_{1,H} = \|(\Delta+1)^{1/2}TR(z)u\| \le C\gamma(z)\|u\|$$

by (3.27a). The norm estimates for the negative scales follow by duality. □

Definition 3.3.16.

1. Let H be C-compatible of order 1 w.r.t. Δ. We say that $\mathfrak{h}$ is a *sesquilinear form associated with* H if
   a. The form $\mathfrak{h}\colon \mathscr{H}^1(H^*) \times \mathscr{H}^1(H) \longrightarrow \mathbb{C}$ is sesquilinear and bounded, i.e.
   $$|\mathfrak{h}(u,v)| \le \|\mathfrak{h}\|_{1\times 1}\|u\|_{1,H^*}\|v\|_{1,H}$$
   for all $u \in \mathscr{H}^1(H^*)$ and $v \in \mathscr{H}^1(H)$;
   b. We have
   $$\mathfrak{h}(u,v) = \langle u, Hv\rangle$$
   for all $u \in \mathscr{H}^1(H^*)$ and $v \in \mathscr{H}^2(H)$.
2. Let $\mathfrak{h}\colon \mathscr{H}^1(H^*) \times \mathscr{H}^1(H) \longrightarrow \mathbb{C}$ be sesquilinear and bounded. The *operator* $H_{\mathfrak{h}}\colon \mathscr{H}^1(H) \longrightarrow \mathscr{H}^{-1}(H)$ *induced by* $\mathfrak{h}$ is defined by
   $$(H_{\mathfrak{h}}v)u := \mathfrak{h}(u,v), \qquad u \in \mathscr{H}^1(H^*),\ v \in \mathscr{H}^1(H).$$

The operator $H_{\mathfrak{h}}$ induced by $\mathfrak{h}$ fits naturally in the scale of Hilbert spaces:

Lemma 3.3.17. *Let $\mathfrak{h}$ be a sesquilinear form associated with H and let $H_{\mathfrak{h}}$ be the operator induced by $\mathfrak{h}$, then*

$$H_{\mathfrak{h}}\colon \mathscr{H}^1(H) \longrightarrow \mathscr{H}^{-1}(H)$$

is bounded by $\|\mathfrak{h}\|_{1\times 1}$. Moreover, the induced operator extends the original operator H, i.e. $H_{\mathfrak{h}}v = \langle\cdot, Hv\rangle$ or

$$(H_{\mathfrak{h}}v)(u) = \langle u, Hv\rangle$$

for $u \in \mathscr{H}^1(H^*)$ and $v \in \mathscr{H}^2(H)$.

Proof. The proof follows immediately from Definition 3.3.16. □

The associated sesquilinear form is unique under the following condition:

Lemma 3.3.18. *If $\mathscr{H}^2(H)$ is dense in $\mathscr{H}^1(H)$, then the sesquilinear form $\mathfrak{h}$ associated with H is uniquely determined.*

Proof. If $\mathfrak{h}_1$ and $\mathfrak{h}_2$ are both associated with H, then

$$\mathfrak{h}_1(u, v) = \langle u, Hv \rangle = \mathfrak{h}_2(u, v)$$

for all $u \in \mathscr{H}^1(H^*)$ and $v \in \mathscr{H}^2(H)$. Since both forms are bounded on $\mathscr{H}^1(H^*) \times \mathscr{H}^1(H)$, and $\mathscr{H}^2(H)$ is dense in $\mathscr{H}^1(H)$, we have $\mathfrak{h}_1 = \mathfrak{h}_2$. □

Remark 3.3.19.

1. In the sequel we will not always distinguish between H and $H_\mathfrak{h}$, although $H_\mathfrak{h}$ may contain information not included in H, e.g. if $\mathscr{H}^2(H)$ is not dense in $\mathscr{H}^1(H)$.
2. *Existence of associated sesquilinear forms:* A priori, it is not clear what is the "right" sesquilinear form associated with H: Let $\mathfrak{h}$ be a bounded sesquilinear form defined on $\mathscr{D}' \times \mathscr{D}$, where $\mathscr{D}'$ and $\mathscr{D}$ are C-densely embedded subspaces of $\mathscr{H}$. We say that $\mathfrak{h}$ is a *sesquilinear form associated with H (in the generalised sense)* if $\operatorname{dom} H \subset \mathscr{D}$, $\operatorname{dom} H^* \subset \mathscr{D}'$ and

   $$\mathfrak{h}(u, v) = \langle u, Hv \rangle$$

   for all $u \in \mathscr{D}'$ and $v \in \operatorname{dom} H$ With this definition we can always associate sesquilinear forms to H, namely

   $$\mathfrak{h}_1 \colon \mathscr{H} \times \mathscr{H}^2(H) \longrightarrow \mathbb{C}, \qquad \mathfrak{h}_1(u, v) := \langle u, Hv \rangle \qquad \text{or}$$
   $$\mathfrak{h}_2 \colon \mathscr{H}^2(H^*) \times \mathscr{H} \longrightarrow \mathbb{C}, \qquad \mathfrak{h}_2(u, v) := \langle H^* u, v \rangle .$$

   Of course, one wants a more "symmetric" domain. It is not clear to us whether one can always find a domain as in Definition 3.3.16. A possibility would be to use interpolation spaces in order to interpolate between the two extreme cases $\mathfrak{h}_1$ and $\mathfrak{h}_2$.
3. Note that we do not always need the boundedness of $\mathfrak{h}$ or of the induced operator $H_\mathfrak{h} \colon \mathscr{H}^1(H) \longrightarrow \mathscr{H}^{-1}(H)$. For example one application is an estimate on the operator

   $$J^{-1} H_\mathfrak{h} - \widetilde{H}_{\widetilde{\mathfrak{h}}} J^1 \colon \mathscr{H}^1(H) \longrightarrow \mathscr{H}^{-1}(\widetilde{H})$$

   for two operators H and $\widetilde{H}$ and identification operators J^1 and J^{-1} (see Definition 4.7.1), which can be expressed in terms of the associated sesquilinear forms (see (4.46c')). In this estimate, the individual boundedness of the operators

   $$\mathscr{H}^1(H) \xrightarrow{H_\mathfrak{h}} \mathscr{H}^{-1}(H) \xrightarrow{J^{-1}} \mathscr{H}^{-1}(\widetilde{H})$$

   and a similar expression for $\widetilde{H}_{\widetilde{\mathfrak{h}}} J^1$ is not needed.

As in Theorem 3.3.6, we have the following holomorphic functional calculus:

Theorem 3.3.20. *Assume that H is a closed operator with resolvent norm profile γ of order 1 on $U \subset \rho(H)$, i.e. $\mathscr{R}_1(\gamma, U)$. Assume in addition that D is an open set in $\mathbb{C}$ such that $\partial D \subset U$ (and therefore $\partial D \subset \rho(H)$). Let $\varphi \in \mathsf{L}_1(\partial D, \gamma(z)\mathrm{d}|z|)$ such that φ is holomorphic in a neighbourhood of $\overline{D}$. Then the operator*

$$\varphi_D(H) = -\frac{1}{2\pi \mathrm{i}} \oint_{\partial D} \varphi(z) R(z) \mathrm{d}z = \frac{1}{2\pi \mathrm{i}} \oint_{\partial D} \frac{\varphi(z)}{z - H} \mathrm{d}z \colon \mathscr{H}^{-1}(H) \longrightarrow \mathscr{H}^{1}(H)$$

is well-defined and bounded. Moreover, the integral exists in the corresponding operator norm topology. Finally, the norm can be estimated by

$$\|\varphi_D(H)\|_{-1,H\to 1,H} \le \frac{1}{2\pi} \oint_{\partial D} |\varphi(z)| \gamma(z) \mathrm{d}|z| =: C_D(\varphi, \gamma).$$

3.4 Boundary Triples and Abstract Elliptic Theory

As already mentioned in the introduction, the concept of boundary triples was originally introduced for the study of extensions of symmetric operators in a Hilbert space, going back to von Neumann [vN30], Friedrichs [Fri34], Krein [Kr47], Vishik [Vi52] and Birman [Bi53]. Vishik [Vi52] was the first who applied this concept to elliptic boundary value problem. Several works later on treated elliptic boundary value problems in an abstract framework, we only mention here [Gru68, GG91, BMNW08, Pc08, BGW09, Ma10, Gru10b] and references therein. For a general treatment of boundary triples we refer exemplarily to [BGP08, DHMS06, DM95, DM91].

We briefly recall the definition of a boundary triple, named here *ordinary boundary triple*, in order to distinguish the standard concept from ours introduced below (see also Sect. 1.2.9). Let A be a closed, symmetric operator in a Hilbert space $\mathscr{H}$. Then $(\Gamma_0, \Gamma_1, \mathscr{G})$ is said to be an *(ordinary) boundary triple for A^** if $\mathscr{G}$ is a Hilbert space and $\Gamma_0, \Gamma_1 \colon \operatorname{dom} A^* \longrightarrow \mathscr{G}$ are linear operators, such that $(\Gamma_0, \Gamma_1) \colon \operatorname{dom} A^* \longrightarrow \mathscr{G} \oplus \mathscr{G}$ is surjective and that the abstract Green's formula[3]

[3] We use the sign convention for Γ_0 and Γ_1 such that the abstract Green's formula agrees with the partial integration formula

$$\int_a^b (-\overline{f}'') g \mathrm{d}s - \int_a^b \overline{f} (-g'') \mathrm{d}s = [\overline{f} g']_a^b - [\overline{f}' g]_a^b,$$

where $\mathscr{H} = \mathsf{L}_2(a,b)$, $Af = -f''$ for $f \in \operatorname{dom} A = \mathring{\mathsf{H}}^2(a,b)$, choosing $\mathscr{G} = \mathbb{C}^2$ and $\Gamma f = (f(a), f(b))$ and $\Gamma' f = (-f'(a), f'(b))$.

$$\langle A^* f, g \rangle_{\mathscr{H}} - \langle f, A^* g \rangle_{\mathscr{H}} = \langle \Gamma_0 f, \Gamma_1 g \rangle_{\mathscr{G}} - \langle \Gamma_1 f, \Gamma_0 g \rangle_{\mathscr{G}} \tag{3.29}$$

holds for all $f, g \in \operatorname{dom} A^*$. Note that a boundary triple exists for A^* if the defect indices $\dim \ker(A \pm \mathrm{i})^*$ are equal (cf. [BGP08, Prp. 1.10]).

Besides operator extension theory, the concept was introduced in order to treat boundary value problems of elliptic partial differential operators like the Laplacian. For example if $X \subset \mathbb{R}^d$ is an open set with smooth (non-trivial) boundary, then $\mathscr{H} := \mathsf{L}_2(X)$, $\mathscr{G} := \mathsf{L}_2(\partial X)$ and $A = \Delta \geq 0$ is the closure of the Laplacian acting on $\mathsf{C}_c^\infty(X)$. Although one has Green's formula in this situation with boundary maps $\Gamma_0 f := f \restriction_{\partial X}$ and $\Gamma_1 f := \partial_n f \restriction_{\partial X}$ (the outwards normal derivative) for $f \in \mathsf{H}^2(X)$, the above concept does not apply, since $\operatorname{dom} A^*$ is *strictly* larger than $\mathsf{H}^2(X)$, and the boundary operators Γ_0 and Γ_1 do not extend to $\operatorname{dom} A^*$ without leaving the space $\mathscr{G} = \mathsf{L}_2(\partial X)$ (see e.g. [Gru68, BMNW08] and Sect. 1.2.9). Either one has to work on a dense subspace $\mathscr{W}^2$ of $\operatorname{dom} A^*$ (see e.g. the concept of *quasi-boundary triples* introduced in [BL07]), or one has to modify the boundary operators (see e.g. [Gru68, BMNW08, Pc08, BGW09, Ma10]).

Boundary triples as defined here are associated with a non-negative *quadratic form*, a concept which seems to be new, to our knowledge. As in [BL07], we work on a restriction of the operator A^*, but we allow different domains for the boundary maps. In the case of the Laplacian on a manifold with boundary, the Sobolev trace operator $\Gamma u := u \restriction_{\partial X}$ is naturally defined on the quadratic form domain $\mathsf{H}^1(X)$ of the Laplacian, and its norm can be estimated in geometric terms in an easy way. We define the Dirichlet-to-Neumann operator in a natural way via its quadratic form (see also [AtE10] for a similar approach to the Dirichlet-to-Neumann operator on domains in $\mathbb{R}^d$ with "rough" boundary). The advantage of our approach is that the Dirichlet-to-Neumann operator is the usual Dirichlet-to-Neumann operator on a manifold with boundary, associating to a suitable function φ on the boundary the normal derivative $\partial_n h$ of the solution to the Dirichlet problem $(\Delta - z)h = 0$ with $h = \varphi$ on the boundary.

We used a similar approach to boundary triples using quadratic forms instead of operators for *first order operators* in [P07].

Our main purpose here is not to characterise all self-adjoint extensions of a given symmetric operator, but to show that the concept of boundary triples can successfully be applied to the PDE case, namely to Laplacians on a manifold with boundary. In particular, we are interested in *quantitative* norm estimates on the operators involved. Moreover, the concept of boundary triple as presented here leads to an abstract and unified approach to resonances as eigenvalues of non-self-adjoint "complexly dilated" Laplacians in Sects. 3.5 and 3.8. The quantitative control of the norm estimates will be needed especially in Sect. 4.9 where we show the convergence of resonances. The concrete PDE case with the direct application to the resonances of certain Schrödinger operators in $\mathbb{R}^d$ was already considered in [CDKS87].

3.4.1 Boundary Triples Associated with Quadratic Forms

Definition 3.4.1. Let $\mathfrak{h}$ be a closed non-negative quadratic form in the Hilbert space $\mathscr{H}$ with domain $\mathscr{H}^1 := \operatorname{dom} \mathfrak{h}$. We endow $\mathscr{H}^1$ with its natural norm given by the quadratic form,

$$\|u\|_1^2 := \|u\|_{\mathfrak{h}}^2 := \mathfrak{h}(u) + \|u\|^2. \tag{3.30}$$

Moreover, let

$$\Gamma : \mathscr{H}^1 \longrightarrow \mathscr{G}$$

be a bounded map, where $\mathscr{G}$ is another Hilbert space. We denote the norm of the operator Γ by $\|\Gamma\|_{1\to 0}$.

1. We say that Γ is a *boundary map associated with the quadratic form* $\mathfrak{h}$ if the following conditions are fulfilled:

   a. $\mathscr{H}^{1,\mathrm{D}} := \ker \Gamma$ is dense in $\mathscr{H}$.
   b. $\mathscr{G}^{1/2} := \operatorname{ran} \Gamma$ is dense in $\mathscr{G}$.

2. We denote by $\mathfrak{h}^{\mathrm{D}} := \mathfrak{h} \restriction_{\mathscr{H}^{1,\mathrm{D}}}$ the quadratic form[4] restricted to $\mathscr{H}^{1,\mathrm{D}}$ and denote the inclusion $\mathscr{H}^{1,\mathrm{D}} \hookrightarrow \mathscr{H}^1$ by ι_{DN}.
3. If $\mathscr{G}^{1/2} \neq \mathscr{G}$, then we call the boundary map Γ *proper*, otherwise *non-proper*.

Note that the natural embeddings

$$\mathscr{H}^{1,\mathrm{D}} \overset{\iota_{\mathrm{DN}}}{\hookrightarrow} \mathscr{H}^1 \longrightarrow \mathscr{H} \longrightarrow \mathscr{H}^{-1} \overset{\iota_{\mathrm{DN}}^*}{\twoheadrightarrow} \mathscr{H}^{-1,\mathrm{D}} \tag{3.31}$$

are all bounded by 1. We now define the operators corresponding to the quadratic forms $\mathfrak{h}^{\mathrm{D}}$ and $\mathfrak{h}$ (see Theorem 3.1.1), the names are borrowed from the concrete PDE example (see Theorem 3.4.39):

Definition 3.4.2. Let Γ be a boundary map for $\mathfrak{h}$.

1. The self-adjoint and non-negative operator H^{D} associated with the quadratic form $\mathfrak{h}^{\mathrm{D}}$ is called the *Dirichlet operator*.
2. Similarly, the self-adjoint and non-negative operator H^{N} associated with the quadratic form $\mathfrak{h}$ is called the *Neumann operator*.
3. We denote by $\mathscr{H}^{k,\mathrm{D}}$ resp. $\mathscr{H}^{k,\mathrm{N}}$ the natural scales of Hilbert spaces associated with the self-adjoint operators H^{D} resp. H^{N} as defined in Sect. 3.2.

Let us now define a boundary triple:

Definition 3.4.3. Let Γ be a boundary map for $\mathfrak{h}$. We say that $(\Gamma, \Gamma', \mathscr{G})$ is a *boundary triple associated with the quadratic form* $\mathfrak{h}$ if the following conditions are fulfilled:

---

[4] Note that $\mathfrak{h}^{\mathrm{D}}$ is a closed form since Γ is bounded on $\mathscr{H}^1$.

1. There is a Hilbert space $\mathscr{W}^2 \hookrightarrow \mathscr{H}^1$ with $\|u\|_{\mathscr{H}^1} \le \|u\|_{\mathscr{W}^2}$ such that
$$\Gamma' : \mathscr{W}^2 \longrightarrow \mathscr{G} \tag{3.32a}$$
is a bounded operator.
2. There is an (unbounded) operator $\check{H}$ in $\mathscr{H}$ with $\operatorname{dom} \check{H} = \mathscr{W}^2$, bounded as operator $\check{H} : \mathscr{W}^2 \longrightarrow \mathscr{H}$, such that the *(abstract) Green's formula*
$$\langle f, \check{H} g \rangle_{\mathscr{H}} = \mathfrak{h}(f, g) - \langle \Gamma f, \Gamma' g \rangle_{\mathscr{G}} \tag{3.32b}$$
holds for all $f \in \mathscr{H}^1$ and $g \in \mathscr{W}^2$.
3. The space
$$\mathring{\mathscr{W}}^2 := \ker \Gamma \cap \ker \Gamma' \subset \mathscr{W}^2 \tag{3.32c}$$
is dense in $\mathscr{H}$.

We say that the boundary triple $(\Gamma, \Gamma', \mathscr{G})$ is *unbounded* if Γ is a proper boundary map, i.e. if $\mathscr{G}^{1/2} = \operatorname{ran} \Gamma \neq \mathscr{G}$, otherwise the boundary triple is called *bounded*.

Remark 3.4.4.

1. The attributes "bounded" and "unbounded" become clear in Proposition 3.4.11.
2. We will show in Corollary 3.4.37 that certain bounded boundary triples have a canonical associated *ordinary* boundary triple. For simplicity, we mostly omit the attribute "associated with a quadratic form" and simply speak of "boundary triples" here.
3. The reader should have the following example in mind: The above abstract setting mimics the situation where $\check{H}$ is the Laplacian on a compact manifold X or on an open domain $X \subsetneq \mathbb{R}^d$ with smooth boundary $\partial X \neq \emptyset$: We set
$$\mathscr{H} := \mathsf{L}_2(X), \qquad \mathscr{H}^1 := \mathsf{H}^1(X), \qquad \mathfrak{h}(u) := \|du\|^2,$$
$$\mathscr{G} := \mathsf{L}_2(\partial X), \qquad \Gamma u := u \restriction_{\partial X}, \qquad \Gamma' u := \partial_n u \restriction_{\partial X},$$
where $\partial_n u$ is the outwards normal derivative on ∂X. It is easily seen that $(\Gamma, \Gamma', \mathscr{G})$ is indeed an (unbounded) boundary triple with the choices $\mathscr{W}^2 := \mathsf{H}^2(X)$ (the usual second order Sobolev space without boundary conditions) and $\check{H} u = \Delta_X$ being the usual Laplacian (see also Theorem 3.4.39 for further properties). In particular, the Dirichlet and Neumann operators are the usual Dirichlet and Neumann Laplacians, respectively, i.e., $H^{\mathrm{D}} = \Delta_X^{\mathrm{D}}$ and $H^{\mathrm{N}} = \Delta_X^{\mathrm{N}}$.
   In particular, in this situation, $\check{H}$ with domain $\mathscr{W}^2 = \mathsf{H}^2(X)$ is *not* a closed operator in $\mathscr{H} = \mathsf{L}_2(X)$. Its operator closure is the *maximal* operator defined on the domain of functions such that f and $\check{H} f$ are in $\mathscr{H}$. In the PDE case, the maximal operator domain is *strictly* larger than the second order Sobolev space $\mathscr{W}^2$ (provided $\partial X \neq \emptyset$). In particular, $\check{H}$ is *not* the maximal operator used in *ordinary* boundary triples (see Corollary 3.4.37).

4. We do not assume a *maximality* condition on the domain of $\check{H}$. If we set $\mathscr{W}^2 = \mathsf{H}^k(X)$ for some $k > 2$ in the previous example, then we still obtain a boundary triple. A sufficiently large domain for $\check{H}$ will be required in Sect. 3.4.2, namely we want that $\operatorname{dom} H^{\mathrm{D}}$ and $\operatorname{dom} H^{\mathrm{N}}$ are subsets of $\mathscr{W}^2$. This is actually the core of elliptic regularity theory.

Let us state a simple consequence of Definition 3.4.3, justifying the labels $(\cdot)^{\mathrm{D}}$ and $(\cdot)^{\mathrm{N}}$. The equality with the domains of the Dirichlet and Neumann operator will be an extra assumption ("ellipticity") in Sect. 3.4.2. The inclusion may be strict, see Remark 3.4.4 (4) above.

Proposition 3.4.5. *We have*

$$\mathscr{W}^{2,\mathrm{D}} := \{ u \in \mathscr{W}^2 \mid \Gamma u = 0 \} \subset \operatorname{dom} H^{\mathrm{D}}, \tag{3.33a}$$

$$\mathscr{W}^{2,\mathrm{N}} := \{ u \in \mathscr{W}^2 \mid \Gamma' u = 0 \} \subset \operatorname{dom} H^{\mathrm{N}}, \tag{3.33b}$$

and the operators act as $H^{\mathrm{D}} f = \check{H} f$ and $H^{\mathrm{N}} f = \check{H} f$ for $f \in \mathscr{W}^{2,\mathrm{D}}$ and $f \in \mathscr{W}^{2,\mathrm{N}}$, respectively.

Proof. The proof follows easily from the characterisation of the operator associated with a quadratic form (see Theorem 3.1.1 (3)) and Green's formula (3.32b). □

We will now define so-called "solution operators" $S(z) = S^{\mathrm{D}}(z)$ resp. $S^{\mathrm{N}}(z)$ associating to given boundary data $\varphi \in \mathscr{G}$ the corresponding solution $h \in \mathscr{W}^2$ of the Dirichlet resp. Neumann problem, i.e. a solution of

$$(\check{H} - z)h = 0, \qquad \Gamma h = \varphi \quad \text{resp.} \quad \Gamma' h = \varphi$$

for some $z \in \mathbb{C} \setminus [0, \infty)$. For an unbounded boundary triple (i.e. Γ is not surjective, or, equivalently, $\mathscr{G}^{1/2} \subsetneq \mathscr{G}$), we have to define a natural scale of Hilbert spaces on $\mathscr{G}$ and take care of the order of the boundary elements φ.

We start with the Dirichlet solution operator and $z = -1$:

Definition 3.4.6. Let $\mathscr{N}^1$ be the orthogonal complement of $\ker \Gamma$ in $\mathscr{H}^1$. We call the inverse

$$S := (\Gamma \restriction_{\mathscr{N}^1})^{-1} \colon \mathscr{G}^{1/2} \longrightarrow \mathscr{N}^1$$

of the bijective map $\Gamma \restriction_{\mathscr{N}^1} \colon \mathscr{N}^1 \longrightarrow \mathscr{G}^{1/2} := \operatorname{ran} \Gamma$ the *(weak) Dirichlet solution operator* (at the point $z = -1$).

The name "Dirichlet solution operator" is justified by the following obvious fact (following again from Green's formula (3.32b)):

Lemma 3.4.7. *We have*

$$\mathscr{N}^2 := \{ u \in \mathscr{W}^2 \mid (\check{H} + 1)u = 0 \} \subset \mathscr{N}^1. \tag{3.34}$$

Moreover, if $\varphi \in \mathscr{G}^{1/2}$ is such that the Dirichlet solution $h = S\varphi$ is in $\mathscr{W}^2$, then $(\check{H} + 1)h = 0$ and $\Gamma h = \varphi$.

The solution operator is defined also for elements $h = S\varphi \in \mathcal{N}^1 \setminus \mathcal{N}^2$. In this case, $(\check{H} + 1)h = 0$ has the meaning

$$\big(\iota_{\mathrm{DN}}^*(H^{\mathrm{N}} + 1)\big)h = 0, \qquad \text{where} \qquad \iota_{\mathrm{DN}}^*(H^{\mathrm{N}} + 1)\colon \mathcal{H}^1 \longrightarrow \mathcal{H}^{-1,\mathrm{D}},$$

in the natural scales of Hilbert spaces. Note that we use the symbol H^{N} also for extensions onto the scale of Hilbert spaces.[5] In particular, $H^{\mathrm{N}}h$ is here defined by $H^{\mathrm{N}}h(f) := \mathfrak{h}(f, h)$; and the equation $(\iota_{\mathrm{DN}}^*(H^{\mathrm{N}} + 1))h = 0 \in \mathcal{H}^{-1,\mathrm{D}}$ is equivalent with

$$\mathfrak{h}(f, h) + \langle f, h\rangle_{\mathcal{H}} = 0 \qquad \forall\, f \in \mathcal{H}^{1,\mathrm{D}}.$$

But the latter condition is just an orthogonal condition w.r.t. the inner product on $\mathcal{H}^1$, i.e. saying that $h \in \mathcal{H}^1 \ominus \mathcal{H}^{1,\mathrm{D}}$.

The Dirichlet solution operator S allows us to define a natural norm on the range $\mathcal{G}^{1/2}$ of Γ:

Definition 3.4.8. We set

$$\|\varphi\|_{\mathcal{G}^{1/2}} := \|S\varphi\|_{\mathcal{H}^1},$$

i.e. the norm of the boundary element φ is given by the $\mathcal{H}^1$-norm of its Dirichlet solution.

Clearly, the operator $S\colon \mathcal{G}^{1/2} \longrightarrow \mathcal{H}^1$ is isometric and its left inverse $\Gamma\colon \mathcal{N}^1 \longrightarrow \mathcal{G}^{1/2}$ is unitary. In particular, $\mathcal{G}^{1/2}$ is itself a Hilbert space (with its inner product induced by $\|\cdot\|_{\mathcal{G}^{1/2}}$). Moreover, the natural inclusion $\mathcal{G}^{1/2} \hookrightarrow \mathcal{G}$ is bounded. In particular,

$$\|\varphi\|_{\mathcal{G}} = \|\Gamma S\varphi\|_{\mathcal{G}} \le \|\Gamma\|_{1\to 0}\|S\varphi\|_{\mathcal{H}^1} = \|\Gamma\|_{1\to 0}\|\varphi\|_{\mathcal{G}^{1/2}}, \tag{3.35}$$

where $\|\Gamma\|_{1\to 0}$ is the norm of Γ as operator $\Gamma\colon \mathcal{H}^1 \longrightarrow \mathcal{G}$.

Proposition 3.4.9. *Let Γ be a boundary map associated with $\mathfrak{h}$, then we have:*

1. *The Dirichlet solution operator S is closed and densely defined as operator in $\mathcal{G} \to \mathcal{H}^1$. Its domain is given by* $\operatorname{dom} S = \mathcal{G}^{1/2}$.
2. *The quadratic form $\mathfrak{l}$ defined by $\mathfrak{l}(\varphi) := \|S\varphi\|_1^2$ with* $\operatorname{dom}\mathfrak{l} = \mathcal{G}^{1/2}$ *is a closed quadratic form in $\mathcal{G}$. Moreover,*

$$\mathfrak{l}(\varphi) \ge \frac{1}{\|\Gamma\|_{1\to 0}^2}\|\varphi\|_{\mathcal{G}}^2,$$

and the constant is optimal. Moreover, we have $\|\Gamma\|_{1\to 0} = \|\Lambda^{-1/2}\|$.

---

[5] More precisely, we should use a notation like $H^{k\to k-2,\mathrm{N}}$ for the operator from the scale $\mathcal{H}^{k,\mathrm{N}}$ into $\mathcal{H}^{k-2,\mathrm{N}}$. In particular, we have used the short hand $H^{\mathrm{N}} = H^{1\to -1,\mathrm{N}}$ here. In order not to overwhelm the notation, we refrain from doing so. Instead, we use the notation $H^{\mathrm{N}}\colon \mathcal{H}^{k,\mathrm{N}} \longrightarrow \mathcal{H}^{k-2,\mathrm{N}}$ to indicate the precise meaning if necessary. Moreover, we sometimes also omit the embedding ι_{DN} in the notation.

Proof. (1) Let $\varphi_n \to \varphi$ in $\mathscr{G}$ with $\varphi_n \in \operatorname{dom} S = \mathscr{G}^{1/2}$ and $S\varphi_n \to h$ in $\mathscr{H}^1$. Since $\{S\varphi_n\}_n$ is a Cauchy sequence, $\{\varphi_n\}_n$ is a Cauchy sequence in $\mathscr{G}^{1/2}$ by definition of the norm on $\mathscr{G}^{1/2}$ and therefore, $\varphi_n \to \widetilde{\varphi} \in \mathscr{G}^{1/2}$ and $S\varphi_n \to S\widetilde{\varphi}$. By (3.35), the convergence $\varphi_n \to \widetilde{\varphi}$ also holds in $\mathscr{G}$ and therefore $\varphi = \widetilde{\varphi}$ and $S\varphi = h$. In particular, we have shown that S is closed.

(2) The lower bound on $\mathfrak{l}$, the optimality and the norm equality for $\|\Gamma\|_{1\to 0}$ follow immediately from (3.35). Let $(\varphi_n)_n$ be a Cauchy sequence in $\mathscr{G}^{1/2}$ with respect to $\mathfrak{l}$, then $(\varphi_n)_n$ is also a Cauchy sequence in $\mathscr{G}$, hence converges in $\mathscr{G}$ to an element $\varphi \in \mathscr{G}$. Moreover, $(S\varphi_n)_n$ is a Cauchy sequence in $\mathscr{H}^1$, hence also convergent to $h \in \mathscr{H}^1$. Since S is closed it follows that $\varphi \in \operatorname{dom} S = \mathscr{G}^{1/2}$ and $S\varphi = h$, i.e., $\varphi \in \operatorname{dom} \mathfrak{l}$ and $\mathfrak{l}(\varphi_n - \varphi) \to 0$. □

Let us now associate a natural operator Λ to a boundary map Γ. It will turn out later on (cf. Proposition 3.4.13) that Λ is the Dirichlet-to-Neumann operator,[6] i.e., $\Lambda\varphi$ associates to a boundary value φ the "normal derivative" of the associated solution of the Dirichlet problem $h = S\varphi$.

Definition 3.4.10. Let Λ be the operator associated with the quadratic form $\mathfrak{l}$. Then Λ is called the *Dirichlet-to-Neumann operator* (at the point $z = -1$) associated with the boundary map Γ and the quadratic form $\mathfrak{h}$. We denote by $\mathscr{G}^k$ the natural scale of Hilbert spaces associated with the self-adjoint operator Λ, i.e. we set

$$\mathscr{G}^k := \operatorname{dom} \Lambda^k, \qquad \|\varphi\|_k := \|\Lambda^k \varphi\|_{\mathscr{G}}.$$

Note that $\|\varphi\|^2_{1/2} = \|\Lambda^{1/2}\varphi\|^2 = \mathfrak{l}(\varphi) = \|\varphi\|^2_{\mathscr{G}^{1/2}}$, i.e. the setting is compatible with our previously defined norm in Definition 3.4.8. The exponents in the scale of Hilbert spaces $\mathscr{H}^k$ and $\mathscr{G}^k$ will be consistent with the regularity order of Sobolev spaces in our main example in Theorem 3.4.39, a boundary triple associated with a Laplacian on a manifold with boundary.

In the following proposition, we denote the adjoints[7] of $\Gamma\colon \mathscr{H}^1 \longrightarrow \mathscr{G}$ w.r.t. the inner products in $\mathscr{H}^1$ and $\mathscr{G}$ by Γ^{1*}. Similarly, the adjoint of the operator S viewed as (possibly unbounded) operator from $\mathscr{G}$ into $\mathscr{H}^1$ with domain $\mathscr{G}^{1/2}$ is denoted by S^{*1}.

Proposition 3.4.11. *Let Γ be a boundary map associated with $\mathfrak{h}$.*

1. We have ($\mathscr{G}^1 =$) $\operatorname{dom} \Lambda = \operatorname{dom} S^{*1}S$ and

$$\Lambda = S^{*1}S \geq \frac{1}{\|\Gamma\|^2_{1\to 0}}. \tag{3.36}$$

---

[6] Actually, Λ will be *minus* the usual Weyl function due to our sign convention.

[7] It is easy to see that $\Gamma^{1*} = R\Gamma^*\colon \mathscr{G} \longrightarrow \mathscr{H}^1$ and $S^{*1} = S^*(H^{\mathrm{N}} + 1)\colon \mathscr{H}^1 \longrightarrow \mathscr{G}$, where $\Gamma^*\colon \mathscr{H}^{-1} \longrightarrow \mathscr{G}$ and $S^*\colon \mathscr{G} \longrightarrow \mathscr{H}^{-1}$ are the duals with respect to the pairing $\langle\cdot,\cdot\rangle_{-1,1}\colon \mathscr{H}^{-1} \times \mathscr{H}^1 \longrightarrow \mathbb{C}$.

In particular, $\Lambda^{-1} = \Gamma\Gamma^{1*}$ exists and is a bounded operator in $\mathscr{G}$ with norm bounded by $\|\Gamma\|^2_{1\to 0}$.

2. *The operators Λ and Λ^{-1} extend to bounded operators*

$$\Lambda = S^*(H^{\mathrm{N}}+1)S\colon \mathscr{G}^{1/2} \xrightarrow{S} \mathscr{H}^1 \xrightarrow{H^{\mathrm{N}}+1} \mathscr{H}^{-1} \xrightarrow{S^*} \mathscr{G}^{-1/2},$$

$$\Lambda^{-1} = \Gamma(H^{\mathrm{N}}+1)^{-1}\Gamma^*\colon \mathscr{G}^{-1/2} \xrightarrow{\Gamma^*} \mathscr{H}^{-1} \xrightarrow{R^{\mathrm{N}}} \mathscr{H}^1 \xrightarrow{\Gamma} \mathscr{G}^{1/2}.$$

respectively.

3. *The boundary map Γ is proper (i.e., the boundary triple $(\Gamma, \Gamma', \mathscr{G})$ is unbounded) iff Λ is unbounded.*

Proof. (1) The lower bound on Λ follows from (3.35). Moreover, by definition of the associated operator (see e.g. [K66, Thm. VI.2.1]) $\varphi \in \operatorname{dom}\Lambda$ iff

$$\mathfrak{l}(\cdot,\varphi) = \langle \cdot,\varphi\rangle_{\mathscr{G}^{1/2}} = \langle S\cdot, S\varphi\rangle_{\mathscr{H}^1}$$

extends to a bounded functional $\mathscr{G} \to \mathbb{C}$, i.e., iff $S\varphi \in \operatorname{dom} S^{*1}$. Moreover,

$$\langle \varphi, \Lambda\varphi\rangle_{\mathscr{G}} = \langle \varphi,\varphi\rangle_{\mathscr{G}^{1/2}} = \langle S\varphi, S\varphi\rangle_{\mathscr{H}^1} = \langle \varphi, S^{*1}S\varphi\rangle_{\mathscr{G}}$$

for $\varphi \in \operatorname{dom}\Lambda$. Since S is closed, densely defined and $S^{-1} = \Gamma\colon \mathscr{N}^1 \longrightarrow \mathscr{G}$ is bounded, it follows that the operator S^{*1} is invertible and $(S^{*1})^{-1} = \Gamma^{1*}$ (cf. [K66, Thm. III.5.30]), hence $\Lambda^{-1} = \Gamma\Gamma^{1*}$.

(2) follows immediately from (1) and Footnote 7.

(3) Assume that $\operatorname{ran}\Gamma = \mathscr{G}$. Since $\Gamma\restriction_{\mathscr{N}^1}\colon \mathscr{N}^1 \longrightarrow \mathscr{G}$ is bounded and bijective, its inverse S is bounded as well by the open mapping theorem. Hence $\Lambda = S^{*1}S$ is bounded. On the other hand, if Λ is bounded, then $\mathfrak{l}(\varphi) = \|S\varphi\|^2_{\mathscr{H}^1}$ is a bounded and everywhere defined quadratic form. In particular, $S\colon \mathscr{G} \longrightarrow \mathscr{H}^1$ is everywhere defined and bounded. For $\varphi \in \mathscr{G}$ we then have $\varphi = \Gamma S\varphi \in \operatorname{ran}\Gamma$, i.e., $\operatorname{ran}\Gamma = \mathscr{G}$. □

Remark 3.4.12.

1. If the boundary triple is bounded, then Λ is bounded and $\mathscr{G}^k = \mathscr{G}$ for all $k \geq 0$. In particular, we have

$$\|\Gamma\|^{-2k}_{1\to 0}\|\varphi\|_{\mathscr{G}} \leq \|\varphi\|_{\mathscr{G}^k} \leq \|\Lambda\|^k\|\varphi\|_{\mathscr{G}}.$$

2. We will show in Corollary 3.4.37 that certain bounded boundary triples naturally induce *ordinary boundary triples*. For more results and references on ordinary boundary triples, we refer to the recent review [BGP08], where also new results on the absolute and singular continuous spectrum of the Dirichlet-to-Neumann map and of self-adjoint restrictions of $\check{H}$ are proven.

In Definition 3.4.31, we define the Dirichlet-to-Neumann operator $\Lambda(z)$ for values $z \notin \sigma(H^{\mathrm{D}})$ other than $z = -1$. In this case, (minus) the Dirichlet-to-Neumann operator $\Lambda(z)$ is also called *Krein's Q-function*, (operator-valued) *Weyl-Titchmarsh*, *Herglotz* or *Nevanlinna function* in the context of ordinary boundary triples.

3. Note that

$$\|\varphi\|_{1/2}^2 = \|\Lambda^{1/2}\varphi\|_{\mathscr{G}}^2 = \langle \varphi, S^{*1} S\varphi \rangle_{\mathscr{H}^1} = \|\varphi\|_{\mathscr{G}^{1/2}}^2,$$

i.e. the setting is compatible with our previously defined norm on the scale of order $1/2$ in Definition 3.4.8. Moreover, since the Dirichlet-to-Neumann operator Λ is a pseudo-differential operator of order 1 in the PDE case, we do not use $k/2$ as exponent in the norm definition, in order to be consistent with the regularity orders in the PDE case (see also Corollary 3.5.5).

The name "Dirichlet-to-Neumann operator" will be clear from the following proposition:

Proposition 3.4.13. *Assume that for $\varphi \in \mathscr{G}^{1/2}$, we have additionally $h = S\varphi \in \mathscr{W}^2$, then $\Lambda\varphi = \Gamma' S\varphi$, i.e. Λ maps the Dirichlet data onto the Neumann data of the solution h.*

Proof. By assumption, $h \in \mathscr{W}^2$, so we can apply Green's formula (3.32b). In particular,

$$\mathfrak{l}(\psi, \varphi) = \langle S\psi, h \rangle_{\mathscr{H}^1} = \mathfrak{h}(S\psi, h) + \langle S\psi, h \rangle_{\mathscr{H}} = \langle \psi, \Gamma' h \rangle_{\mathscr{G}}$$

for all $\psi \in \mathscr{G}^{1/2}$, using also Lemma 3.4.7 and the fact that $(\check{H} + 1)h = 0$ in the last equality. Since Λ is the operator associated with $\mathfrak{l}$, it follows that $\Lambda\varphi = \Gamma' h = \Gamma' S\varphi$. □

Let us now define the Dirichlet solution operator for general values $z \notin \sigma(H^{\mathrm{D}})$. To simplify matters, we assume in the sequel that $z \in \mathbb{C} \setminus [0, \infty)$ if not stated otherwise. We start with the definition of *weak solutions*:

Definition 3.4.14. Set

$$\mathscr{N}^1(z) := \{ g \in \mathscr{H}^1 \mid \mathfrak{h}(f, g) = z\langle f, g \rangle \; \forall f \in \mathscr{H}^{1,\mathrm{D}} \}.$$

We say that h is a *(weak) solution of the Dirichlet problem* with boundary data $\varphi \in \mathscr{G}^{1/2}$ if $\Gamma h = \varphi$ and $h \in \mathscr{N}^1(z)$, or, equivalently, $\iota_{\mathrm{DN}}^*(H^{\mathrm{N}} - z)h = 0 \in \mathscr{H}^{-1,\mathrm{D}}$.

The following lemma follows easily from the fact that $\mathfrak{h}^{\mathrm{D}}$ is non-degenerative for $z \in C \setminus \sigma(H^{\mathrm{D}})$:

Lemma 3.4.15. *The space $\mathscr{N}^1(z)$ is closed in $\mathscr{H}^1$. Moreover, if $z \in \mathbb{C} \setminus \sigma(H^{\mathrm{D}})$, then*[8]

[8]Here, $\mathscr{H} = \mathscr{H}_1 \dotplus \mathscr{H}_2$ means that $\mathscr{H}$ is the *topological sum* of $\mathscr{H}_1$ and $\mathscr{H}_2$, i.e. the sum is direct (but not necessarily orthogonal), and $\mathscr{H}_1$, $\mathscr{H}_2$ are closed in $\mathscr{H}$.

$$\mathscr{H}^1 = \mathscr{H}^{1,\mathrm{D}} \dotplus \mathscr{N}^1(z).$$

If $z < 0$, then the sum is orthogonal.

Denote by[9]

$$R^{\mathrm{D}}(z) := (H^{\mathrm{D}} - z)^{-1} \quad \text{and} \quad R^{\mathrm{D}} := R^{\mathrm{D}}(-1)$$

the resolvent of the Dirichlet operator in $z \notin \sigma(H^{\mathrm{D}})$ and in -1. We use a similar notion for the resolvent of the Neumann operator. Note that we use the symbols H^{N} and $R^{\mathrm{D}}(z)$ also for the corresponding extensions as operators on the associated scales of Hilbert spaces.

We will often need the following constant when dealing with norm resolvent estimates:

Lemma 3.4.16. *Let $z, w \in \mathbb{C} \setminus [0, \infty)$, then*

$$1 \le C(z, w) := \sup_{\lambda \ge 0} \frac{|\lambda - w|}{|\lambda - z|} < \infty. \tag{3.37}$$

Moreover, if $H \ge 0$, then

$$\|(H - w)(H - z)^{-1}\| \le C(z, w)$$

for $z, w \in \mathbb{C} \setminus \mathbb{R}_+$.

Proof. The boundedness follows easily from the fact that the limit of $|\lambda - w|/|\lambda - z|$ is 1 as $\lambda \to \infty$, i.e. the fraction extends to a continuous function on $[0, \infty]$ and attains therefore its maximum (depending of course on z and w). The norm estimate follows from the continuous functional calculus. □

The operator $S(z) \colon \mathscr{G}^{1/2} \longrightarrow \mathscr{H}^1$ appearing in the next proposition is named *(weak) Dirichlet solution operator* (in z). Recall that $S \colon \mathscr{G}^{1/2} \longrightarrow \mathscr{H}^1$ is the (weak) Dirichlet solution operator of Definition 3.4.6 (in $z = -1$). Moreover, ι_{DN} is given in (3.31).

Proposition 3.4.17. *Let $z \in \mathbb{C} \setminus \sigma(H^{\mathrm{D}})$ and $\varphi \in \mathscr{G}^{1/2}$, then the weak Dirichlet solution h with boundary data φ is unique. Moreover, $h = S(z)\varphi$ with*

$$S(z) := S^{\mathrm{D}}(z) := \big(\mathrm{id}_{\mathscr{H}^1} - \iota_{\mathrm{DN}} R^{\mathrm{D}}(z) \iota_{\mathrm{DN}}^*(H^{\mathrm{N}} - z)\big) S \colon \mathscr{G}^{1/2} \longrightarrow \mathscr{N}^1(z) \subset \mathscr{H}^1$$

is the weak solution of the Dirichlet problem with boundary data φ. Finally, the operator $S(z) \colon \mathscr{G}^{1/2} \longrightarrow \mathscr{H}^1$ has norm bounded by

$$\|S(z)\|_{1/2 \to 1} \le 1 + C(z, -1)C(-1, z).$$

[9] We often omit the argument z if $z = -1$, i.e. we write $R^{\mathrm{D}} = R^{\mathrm{D}}(-1)$, $S = S(-1)$ etc.

Proof. The uniqueness of the weak Dirichlet problem follows from the fact that $\mathfrak{h}^{\mathrm{D}} - z\mathbb{1}$ is a non-degenerated sesquilinear form on $\mathscr{H}^{1,\mathrm{D}}$ if $z \notin \sigma(H^{\mathrm{D}})$. Moreover, the operator $S(z)$ is properly defined on the corresponding scales of Hilbert spaces. It is an easy calculation that $h = S(z)\varphi$ is the weak solution with boundary data φ. The norm bound follows from the fact that S is an isometry by definition and that

$$\begin{aligned}\|S(z)\|_{1/2\to 1} &\le 1 + \big\|(H^{\mathrm{N}}+1)^{1/2}\iota_{\mathrm{DN}}R^{\mathrm{D}}(z)\iota_{\mathrm{DN}}^{*}(H^{\mathrm{N}}+1)(R^{\mathrm{N}})^{1/2}\big\|_{0\to 0}\\ &\le 1 + \|(H^{\mathrm{D}}+1)R^{\mathrm{D}}(z)\|\,\|(R^{\mathrm{D}})^{1/2}\iota_{\mathrm{DN}}^{*}(H^{\mathrm{N}}+1)^{1/2}\|\,\|R^{\mathrm{N}}(H^{\mathrm{N}}-z)\|\\ &\le 1 + C(z,-1)C(-1,z).\end{aligned}$$

Note that we used $(H^{\mathrm{N}}+1)^{1/2}\iota_{\mathrm{DN}} = (H^{\mathrm{D}}+1)^{1/2}$ and $\|(R^{\mathrm{D}})^{1/2}\iota_{\mathrm{DN}}^{*}(H^{\mathrm{N}}+1)^{1/2}\| = 1$. □

We call a weak solution $h = S\varphi$ with $h \in \mathscr{W}^2$ a *strong solution of the Dirichlet problem*. This is justified, since

$$\mathscr{N}^2(z) := \ker(\check{H} - z) \subset \mathscr{N}^1(z),$$

as the next proposition shows:

Proposition 3.4.18. *Let $z \in \mathbb{C} \setminus \sigma(H^{\mathrm{D}})$.*

1. *If $\varphi = \Gamma g$ with $g \in \mathscr{W}^2$, then $\widetilde{S}(z)\varphi := g - R^{\mathrm{D}}(z)(\check{H} - z)g$ is well defined.*
2. *$h := \widetilde{S}(z)\varphi \in \mathscr{N}^1(z)$, i.e., h is the weak solution of the Dirichlet problem (cf. Definition 3.4.14), and $\widetilde{S}(z)\varphi = S(z)\varphi$. Moreover, h is unique.*
3. *If[10] $h = S(z)\varphi \in \mathscr{W}^2$, then h is the strong solution of the Dirichlet problem with boundary value $\varphi = \Gamma h$, i.e.,*

$$(\check{H} - z)h = 0, \qquad \Gamma h = \varphi. \tag{3.38}$$

In particular, $\Gamma S(z)\varphi = \varphi$.

Proof. (1) Assume that $\Gamma g_1 = \Gamma g_2$ and $g_1, g_2 \in \mathscr{W}^2$, then $g_1 - g_2 \in \mathscr{W}^{2,\mathrm{D}} \subset \operatorname{dom} H^{\mathrm{D}}$ and $R^{\mathrm{D}}(z)(\check{H} - z)(g_1 - g_2) = g_1 - g_2$, hence $\widetilde{S}(z)\varphi$ is well-defined.

(2) Let $g \in \mathscr{W}^2$, then $R^{\mathrm{D}}(z)\iota^*(H^{\mathrm{N}} - z)g = R^{\mathrm{D}}(z)(\check{H} - z)g$: Namely we have

$$\begin{aligned}\langle R^{\mathrm{D}}(z)\iota^*(H^{\mathrm{N}}-z)g, f\rangle &= \langle (H^{\mathrm{N}}-z)g, \iota R^{\mathrm{D}}(\overline{z})f\rangle_{-1,1} = (\mathfrak{h} - z)(g, R^{\mathrm{D}}(\overline{z})f)\\ &= \langle (\check{H}-z)g, R^{\mathrm{D}}(\overline{z})f\rangle = \langle R^{\mathrm{D}}(z)(\check{H}-z)g, f\rangle\end{aligned}$$

[10]Note that $h \in \mathscr{W}^2$ iff $g \in \mathscr{W}^2$ is only true if $\operatorname{ran} R^{\mathrm{D}}(z) = \mathscr{H}^{2,\mathrm{D}} \subset \mathscr{W}^2$. This is actually an additional information, provided in the concept of *elliptic* boundary triples, see Definition 3.4.21.

for $f \in \mathscr{H}$, where we used Green's the third equality. The equality $\widetilde{S}(z)\varphi = S(z)\varphi \in \mathscr{N}^1(z)$ now follows from Proposition 3.4.17. (3) follows similarly as in Lemma 3.4.7. □

We can now address the solution of the *Neumann problem*:

Definition 3.4.19. We say that $h \in \mathscr{W}^2$ is a *(strong) solution of the Neumann problem* with boundary data $\psi \in \operatorname{ran} \Gamma'$ if $h \in \mathscr{N}^2(z)$ and $\Gamma' h = \psi$, or, equivalently,

$$(\check{H} - z)h = 0 \qquad \text{and} \qquad \Gamma' h = \psi.$$

As for the (strong) Dirichlet problem, we can show:

Proposition 3.4.20. *Let* $z \in \mathbb{C} \setminus \sigma(H^{\mathrm{N}})$.

1. *The solution of the Neumann problem is unique. If* $\psi = \Gamma' g$ *with* $g \in \mathscr{W}^2$*, then* $S^{\mathrm{N}}(z)\psi := g - R^{\mathrm{N}}(z)(\check{H} - z)g$ *is well defined.*
2. *If* $h = S^{\mathrm{N}}(z)\psi \in \mathscr{W}^2$ *then* h *is the solution of the Neumann problem with boundary value* $\psi = \Gamma' h$*. In particular,* $\Gamma' S^{\mathrm{N}}(z)\psi = \psi$*. Moreover,* h *is unique.*

We call $S^{\mathrm{N}}(z) \colon \operatorname{ran} \Gamma' \longrightarrow \mathscr{W}^2$ the *Neumann solution operator*. We will specify $\operatorname{ran} \Gamma'$ in the next section.

3.4.2 Elliptic Boundary Triples

We now require more assumptions on the boundary triple in order to ensure that the space $\mathscr{W}^2$ is chosen large enough. In particular, we assume that the domains of the Dirichlet and Neumann operator are subsets of $\mathscr{W}^2$, which is actually the core of elliptic regularity theory for partial differential operators on manifolds with boundary. Guided by the example of the Laplacian on a manifold with boundary (see Theorem 3.4.39), we also assume the following conditions on the regularity order of the boundary maps.

Definition 3.4.21. We call the boundary triple $(\Gamma, \Gamma', \mathscr{G})$ associated with the quadratic form $\mathfrak{h}$ *elliptic* if the following conditions hold:

1. $\mathscr{H}^{2,\mathrm{D}} = \operatorname{dom} H^{\mathrm{D}} \subset \mathscr{W}^2$ and $\mathscr{H}^{2,\mathrm{N}} = \operatorname{dom} H^{\mathrm{N}} \subset \mathscr{W}^2$,
2. $\Gamma \colon \mathscr{W}^2 \longrightarrow \mathscr{G}^{3/2}$ is bounded and surjective, and
3. $\Gamma' \colon \mathscr{W}^2 \longrightarrow \mathscr{G}^{1/2}$ is bounded and surjective.

We call the last two items the *regularity assumptions* of the boundary maps.

For elliptic boundary triples, we actually have the natural interpretation of $\ker \Gamma \cap \mathscr{W}^2 =: \mathscr{W}^{2,\mathrm{D}}$ and $\ker \Gamma' =: \mathscr{W}^{2,\mathrm{N}}$:

Proposition 3.4.22. *Assume that the boundary triple is elliptic (actually, condition* (1) *of Definition 3.4.21 is enough). Then the domains of the Dirichlet and Neumann operators are given by*

$$\mathscr{H}^{2,\mathrm{D}} = \operatorname{dom} H^{\mathrm{D}} = \{u \in \mathscr{W}^2 \mid \Gamma u = 0\}, \tag{3.40a}$$

$$\mathscr{H}^{2,\mathrm{N}} = \operatorname{dom} H^{\mathrm{N}} = \{u \in \mathscr{W}^2 \mid \Gamma' u = 0\} \tag{3.40b}$$

and $\mathring{\mathscr{W}}^2 = \mathscr{H}^{2,\mathrm{D}} \cap \mathscr{H}^{2,\mathrm{N}}$. *Moreover,* $\mathscr{W}^2$ *is dense in* $\mathscr{H}^1$.

Proof. The assertions on the domains of the Dirichlet and Neumann operators follow immediately from Proposition 3.4.5 and Definition 3.4.21 (1). For the last assertion, note that $\mathscr{H}^{2,\mathrm{N}} \subset \mathscr{W}^2 \subset \mathscr{H}^1$. Since the operator domain $\mathscr{H}^{2,\mathrm{N}} = \operatorname{dom} H^{\mathrm{N}}$ is a form core for the corresponding quadratic form domain $\mathscr{H}^1 = \operatorname{dom} \mathfrak{h}$, cf. [K66, Thm. VI.2.1], the assertion follows. □

For elliptic boundary triples, we have the following bounded embeddings (recall that we endow $\operatorname{dom} H^{\mathrm{D}} = \mathscr{H}^{2,\mathrm{D}}$ with its graph norm given by the natural scale of Hilbert spaces, and similarly for the Neumann operator):

Proposition 3.4.23. *Let* $(\Gamma, \Gamma', \mathscr{G})$ *be an elliptic boundary triple then*

$$\iota_{\mathrm{ell,D}} \colon \mathscr{H}^{2,\mathrm{D}} \hookrightarrow \mathscr{W}^2 \qquad \text{and} \qquad \iota_{\mathrm{ell,N}} \colon \mathscr{H}^{2,\mathrm{N}} \hookrightarrow \mathscr{W}^2$$

are bounded operators.

Proof. The embedding $\iota_{\mathrm{ell,D}} \colon \mathscr{H}^{2,\mathrm{D}} \hookrightarrow \mathscr{W}^{2,\mathrm{D}}$ is actually a bijection with bounded inverse, since

$$\|f\|_{2,\mathrm{D}} = \|(H^{\mathrm{D}} + 1) f\| = \|(\check{H} + 1) f\| \le (\|\check{H}\|_{\mathscr{W}^2 \to \mathscr{H}} + \|\iota_{\mathscr{W}^2 \hookrightarrow \mathscr{H}}\|) \|f\|_{\mathscr{W}^2}$$

for $f \in \mathscr{H}^{2,\mathrm{D}} \subset \mathscr{W}^2$. By the closed graph theorem, it follows that $\iota_{\mathrm{ell,D}}$ is also bounded. The argument in the Neumann case is similar. □

Let us give a condition assuring the ellipticity; the converse of Proposition 3.4.23. Actually, we could use (3.41) as definition of ellipticity instead of the first condition in Definition 3.4.21.

Proposition 3.4.24. *Let* $(\Gamma, \Gamma', \mathscr{G})$ *be a boundary triple such that there exists a constant* $C > 0$ *with*

$$\|f\|^2_{\mathscr{W}^2} \le C (\|\check{H} f\|^2 + \|f\|^2) \tag{3.41}$$

for all $f \in \mathscr{W}^{2,\mathrm{D}} \cup \mathscr{W}^{2,\mathrm{N}}$. *Then the boundary triple fulfils the first condition of Definition 3.4.21, i.e.,* $\operatorname{dom} H^{\mathrm{D}} \subset \mathscr{W}^2$ *and* $\operatorname{dom} H^{\mathrm{N}} \subset \mathscr{W}^2$.

Proof. Note first, that for $f \in \mathscr{W}^{2,\mathrm{D}} := \mathscr{W}^2 \cap \operatorname{dom} H^{\mathrm{D}}$, we have $\check{H} f = H^{\mathrm{D}} f$ (Proposition 3.4.5). Let $\check{H}^{\mathrm{D}}$ be the restriction of $\check{H}$ to $\mathscr{W}^{2,\mathrm{D}}$. If $f_n \in \mathscr{W}^{2,\mathrm{D}}$ and $f, g \in \mathscr{H}$ with $f_n \to f$ and $\check{H}^{\mathrm{D}} f_n \to g$, then (f_n) is a Cauchy sequence in $\mathscr{W}^{2,\mathrm{D}}$ by (3.41). Moreover, $\mathscr{W}^{2,\mathrm{D}}$ is closed in $\mathscr{W}^2$, and therefore itself complete. Thus, $f_n \to \widetilde{f}$ in $\mathscr{W}^{2,\mathrm{D}} \hookrightarrow \mathscr{H}$. Since $f_n \to f$ in $\mathscr{H}$, we have $f = \widetilde{f}$. It is now easy to see that $\check{H}^{\mathrm{D}} f = g$, and therefore, $\check{H}^{\mathrm{D}}$ is closed. Moreover, $\check{H}^{\mathrm{D}}$ is non-negative.

Since $\operatorname{dom} \check{H}^{\mathrm{D}} \subset \operatorname{dom} H^{\mathrm{D}}$ and $\langle \check{H}^{\mathrm{D}} f, g\rangle = \mathfrak{h}^{\mathrm{D}}(f, g)$ for all $f \in \operatorname{dom} \check{H}^{\mathrm{D}}$ and $g \in \operatorname{dom} \mathfrak{h}^{\mathrm{D}}$ using Green's formula, one sees that $\check{H}^{\mathrm{D}} = H^{\mathrm{D}}$: Indeed, the equality follows by the uniqueness of the operator H^{D} associated with the non-negative form $\mathfrak{h}^{\mathrm{D}}$ (cf. [K66, Thm. IV.2.1]).

The argument for the Neumann operator is similar. □

Remark 3.4.25. The boundedness of the embeddings $\iota_{\mathrm{ell,D}}\colon \mathscr{H}^{2,\mathrm{D}} \hookrightarrow \mathscr{W}^2$ for the Dirichlet domain and $\iota_{\mathrm{ell,N}}\colon \mathscr{H}^{2,\mathrm{N}} \hookrightarrow \mathscr{W}^2$ for the Neumann domain is basically the core of elliptic (L_2-)regularity theory in the case of PDEs: The boundedness means that we can estimate the Sobolev norm on $\mathscr{W}^2$ by the graph norm of H^{D}, i.e. that

$$\|u\|^2_{\mathscr{W}^2} \le C\,\|u\|^2_{2,\mathrm{D}} = C\,\|(H^{\mathrm{D}}+1)u\|^2$$

for $u \in \mathscr{H}^{2,\mathrm{D}}$, and similarly for the Neumann operator (as in the previous proposition). Note that the constant C will depend on the concrete manifold or on the open set in a complicated way, and it seems to be hopeless to control the constant if the manifold depends on some shrinking parameter as in our case. We therefore have to avoid this embedding in *quantitative* estimates. Nevertheless, the boundedness is needed in *qualitative* statements, e.g., for the holomorphy of certain families of resolvents in Sect. 3.6. We will sometimes refer to norms, which can be expressed in terms of $\|\Gamma\|_{1\to 0}$ and of the resolvent parameter z only, as *naturally bounded*. Note that $\|\Gamma\|_{1\to 0}$ can be controlled explicitly in our examples (see e.g. Propositions 6.1.4 and 6.2.9).

Let us now analyse when a boundary triple with *finite*-dimensional boundary space is elliptic. Note that such a boundary triple is bounded since Λ is bounded, cf. Proposition 3.4.11).

Proposition 3.4.26. *If the boundary space $\mathscr{G}$ of a boundary triple $(\Gamma, \Gamma', \mathscr{G})$ is finite-dimensional, if $\Gamma'(\mathscr{W}^2) = \mathscr{G}$ and if* (3.41) *holds for $f \in \mathscr{W}^{2,\mathrm{D}} \cup \mathscr{W}^{2,\mathrm{N}}$, then the boundary triple is elliptic.*

Proof. We have $\Gamma'(\mathscr{W}^2) = \mathscr{G} = \mathscr{G}^{1/2}$ by assumption. Moreover, let $\varphi \in \mathscr{G}$. Since $\operatorname{ran}\Gamma = \mathscr{G}$ (again by density of the range), there exists $f \in \mathscr{H}^1$ such that $\Gamma f = \varphi$. Moreover, by Proposition 3.4.24, the first condition of Definition 3.4.21 is fulfilled, hence we can apply Proposition 3.4.22, and therefore, $\mathscr{W}^2$ is dense in $\mathscr{H}^1$. In particular, there exist $f_n \in \mathscr{W}^2$ such that $f_n \to f$ in $\mathscr{H}^1$, and hence $\Gamma f_n \to \Gamma f = \varphi$ in $\mathscr{G}$. We therefore have shown that $\Gamma(\mathscr{W}^2)$ is dense in $\mathscr{G}$, hence $\Gamma(\mathscr{W}^2) = \mathscr{G}$. □

Note that we cannot drop the assumption (3.41). An example is given by a half-line model on $\mathscr{H} = \mathsf{L}_2(\mathbb{R}_+)$ with $\mathscr{G} = \mathbb{C}$ and $K = 0$ (see Sect. 3.5.1 and Remark 3.4.4 (4)). If we choose $\mathscr{W}^2$ too small, for example $\mathscr{W}^2 = \mathsf{H}^3(\mathbb{R}_+)$, then $\operatorname{dom} H^{\mathrm{D}} = \{\, f \in \mathsf{H}^2(\mathbb{R}_+) \mid f(0) = 0\,\}$ is *not* a subset of $\mathscr{W}^2$.

Let us now specify how the regularity assumptions of Definition 3.4.21 affect the solution operators. For the rest of the section, we assume that $z \in \mathbb{C} \setminus [0, \infty)$.

Proposition 3.4.27. *Let $(\Gamma, \Gamma', \mathscr{G})$ be an elliptic boundary triple. Then we have:*

1. *Let $z \notin \sigma(H^{\mathrm{D}})$, then the restriction of the weak Dirichlet solution operator (cf. Proposition 3.4.17) to $\mathscr{G}^{3/2}$ maps into $\mathscr{W}^2$ and is bounded as operator*

$$S(z): \mathscr{G}^{3/2} \longrightarrow \mathscr{W}^2 \tag{3.42a}$$

with right inverse $\Gamma: \mathscr{W}^2 \longrightarrow \mathscr{G}^{3/2}$. Moreover, $S(z): \mathscr{G}^{3/2} \longrightarrow \mathscr{N}^2(z)$ is a topological isomorphism, and $P^{2,\mathrm{D}}(z) := S(z)\Gamma: \mathscr{W}^2 \longrightarrow \mathscr{W}^2$ is the projection onto $\mathscr{N}^2(z)$ along $\mathscr{H}^{2,\mathrm{D}}$.

2. *Let $z \notin \sigma(H^{\mathrm{N}})$, then the Neumann solution operator (cf. Proposition 3.4.20) is bounded as operator*

$$S^{\mathrm{N}}(z): \mathscr{G}^{1/2} \longrightarrow \mathscr{W}^2 \tag{3.42b}$$

with right inverse $\Gamma': \mathscr{W}^2 \longrightarrow \mathscr{G}^{1/2}$. Moreover, $S^{\mathrm{N}}(z): \mathscr{G}^{1/2} \longrightarrow \mathscr{N}^2(z)$ is a topological isomorphism, and $P^{2,\mathrm{N}}(z) := S^{\mathrm{N}}(z)\Gamma': \mathscr{W}^2 \longrightarrow \mathscr{W}^2$ is the projection onto $\mathscr{N}^2(z)$ along $\mathscr{H}^{2,\mathrm{N}}$.

Proof. (1) Let $\varphi \in \mathscr{G}^{3/2}$, then by Definition 3.4.21 (2) and Proposition 3.4.18, there exist $g \in \mathscr{W}^2$ such that $\Gamma\varphi = g$ and $S(z)\varphi = g - R^{\mathrm{D}}(z)(\check{H} - z)g \in \mathscr{W}^2$, and finally, $\Gamma S(z)\varphi = \varphi$. Moreover, since $\ker(\Gamma: \mathscr{W}^2 \longrightarrow \mathscr{G}) = \mathscr{H}^{2,\mathrm{D}}$ by Proposition 3.4.22, the operator $\Gamma: \mathscr{N}^2(z) \longrightarrow \mathscr{G}^{3/2}$ is injective, and bounded and surjective by Definition 3.4.21, hence, by the closed graph theorem, its inverse $S(z)$ is also bounded. (2) The same arguments holds for the Neumann case. □

In contrast to the case of ordinary boundary triples (where one can choose the graph norm of $\check{H}$ on $\mathscr{W}^2$, see [BGP08, Prp. 1.9]), it is not clear to us whether the norm on $\mathscr{W}^2(z)$ is in general equivalent to the norm on $\mathscr{H}$.

The following is a useful way how to prove that a boundary is elliptic:

Proposition 3.4.28. *Assume that $(\Gamma, \Gamma', \mathscr{G})$ is a boundary triple such that:*

1. *The elliptic estimate* (3.41) *holds (or, alternatively, we have* $\operatorname{dom} H^{\mathrm{D}} \subset \mathscr{W}^2$ *and* $\operatorname{dom} H^{\mathrm{N}} \subset \mathscr{W}^2$),
2. *The boundary maps $\Gamma: \mathscr{W}^2 \longrightarrow \mathscr{G}^{3/2}$ and $\Gamma': \mathscr{W}^2 \longrightarrow \mathscr{G}^{1/2}$ are bounded, and that*
3. *The Dirichlet solution operator (in $z = -1$) fulfils $S\varphi \in \mathscr{W}^2$ for all $\varphi \in \mathscr{G}^{3/2}$.*

Then the boundary triple is elliptic.

Proof. The only point to check is the surjectivity of the boundary maps: For Γ this is clear, since the Dirichlet solution operator is a right inverse for Γ, hence $\varphi = \Gamma S\varphi$ for $\varphi \in \mathscr{G}^{3/2}$. For Γ', we show that

$$\widetilde{S}^{\mathrm{N}} := S\Lambda^{-1}: \mathscr{G}^{1/2} \longrightarrow \mathscr{W}^2$$

is indeed the Neumann solution operator: Let $h := \widetilde{S}^{\mathrm{N}}\psi$ for $\psi \in \mathscr{G}^{1/2}$. Then

$$\langle \Gamma f, \Gamma' h\rangle = (\mathfrak{h} + 1)(f, h) = \langle \Gamma f, \Lambda \Gamma h\rangle_{-1/2,1/2}$$

for $f \in \mathscr{N}^1$ using Green's formula (3.32b), since $h \in \mathscr{W}^2$ and $(\check{H} - z_0)h = 0$. But since $\Lambda \Gamma h = \psi$, we have $\Gamma' h = \psi$. By the uniqueness of the Neumann solution, we obtain $\widetilde{S}^{\mathrm{N}}\varphi = S^{\mathrm{N}}\varphi$ (see Proposition 3.4.20). Hence, we have $\psi = \Gamma' \widetilde{S}^{\mathrm{N}}\psi$ for $\psi \in \mathscr{G}^{1/2}$, and the surjectivity of Γ' follows. □

Let us now describe the solution operator via its adjoint. The following operator $B(z)$ will be useful in Krein's resolvent formula of Theorem 3.4.44:

Lemma 3.4.29. *Assume that the boundary triple $(\Gamma, \Gamma', \mathscr{G})$ is elliptic and that $z \in \mathbb{C} \setminus \sigma(H^{\mathrm{D}})$. Then the operator*

$$B(z) := -\Gamma' \iota_{\mathrm{ell,D}} R^{\mathrm{D}}(z) \colon \mathscr{H} \overset{R^{\mathrm{D}}(z)}{\longrightarrow} \mathscr{H}^{2,\mathrm{D}} \hookrightarrow \mathscr{W}^2 \overset{\Gamma'}{\longrightarrow} \mathscr{G}$$

is bounded. Moreover, for $\varphi \in \mathscr{G}^{1/2}$, we have

$$B(\overline{z})^* \varphi = S(z)\varphi$$

and $S(z)$ extends to a bounded operator $\mathscr{G} \to \mathscr{H}$.

Proof. The boundedness of the operator $B(z)$ follows from Proposition 3.4.23 and Definition 3.4.21 (3). Let $\varphi \in \mathscr{G}^{1/2}$, then $h := S(z)\varphi \in \mathscr{H}^1$. Moreover, for $g \in \mathscr{H}^{2,\mathrm{D}}$ we have

$$\langle (H^{\mathrm{D}} - \overline{z})g, h \rangle = \mathfrak{h}(g, h) - z\langle g, h \rangle - \langle \Gamma' g, \Gamma h \rangle_{\mathscr{G}} = -\langle \Gamma' g, \varphi \rangle_{\mathscr{G}}$$

using Green's formula and the fact that h is a weak solution. In particular, by the definition of the adjoint map, $(H^{\mathrm{D}} - \overline{z})g \in \operatorname{dom} S(z)^*$ and $S(z)^*(H^{\mathrm{D}} - \overline{z})g = -\Gamma' g$, i.e.

$$S(z)^* = -\Gamma' R^{\mathrm{D}}(\overline{z}) = B(\overline{z}) \colon \mathscr{H} \longrightarrow \mathscr{G}.$$

Finally, $S(z)$ extends to a bounded operator $S(z)^{**} = B(\overline{z})^*$ from $\mathscr{G}$ into $\mathscr{H}$ since $B(\overline{z})$ is bounded. □

The Neumann solution operator $S^{\mathrm{N}}(z)$ is related to the adjoint of another naturally defined operator:

Lemma 3.4.30. *Assume that $z \in \mathbb{C} \setminus \sigma(H^{\mathrm{N}})$, then the operator*

$$A(z) := \Gamma R^{\mathrm{N}}(z) \colon \mathscr{H} \overset{R^{\mathrm{N}}(z)}{\longrightarrow} \mathscr{H}^{2,\mathrm{N}} \hookrightarrow \mathscr{H}^1 \overset{\Gamma}{\longrightarrow} \mathscr{G}$$

is bounded. Moreover, for $\varphi \in \mathscr{G}^{1/2}$, we have

$$A(\overline{z})^* \varphi = S^{\mathrm{N}}(z)\varphi$$

and the Neumann solution operator $S^{\mathrm{N}}(z)$ extends to a bounded operator $\mathscr{G} \to \mathscr{H}$.

Proof. Let $\varphi \in \mathscr{G}^{1/2}$, then $h := S^{\mathrm{N}}(z)\varphi \in \mathscr{W}^2$ by Proposition 3.4.27 (2). Moreover, for $g \in \mathscr{H}^{2,\mathrm{N}}$ we have

$$\langle (H^{\mathrm{N}} - \overline{z})g, h\rangle = \langle g, (H^{\mathrm{N}} - z)h\rangle + \langle \Gamma g, \Gamma' h\rangle_{\mathscr{G}} = \langle \Gamma g, \varphi\rangle_{\mathscr{G}}.$$

Here, we have used Green's formula twice and the fact that h is a solution. In particular, $(H^{\mathrm{N}} - \overline{z})g \in \operatorname{dom} S^{\mathrm{N}}(z)^*$ and $S^{\mathrm{N}}(z)^*(H^{\mathrm{N}} - \overline{z})g = \Gamma g$, i.e.

$$S^{\mathrm{N}}(z)^* = \Gamma R^{\mathrm{N}}(\overline{z}) = A(\overline{z}) \colon \mathscr{H} \longrightarrow \mathscr{G}.$$

Finally, $S^{\mathrm{N}}(z)$ extends to a bounded operator $S^{\mathrm{N}}(z)^{**} = A(\overline{z})^*$ from $\mathscr{G}$ into $\mathscr{H}$ and the result follows. □

We can now define the Dirichlet-to-Neumann map for general $z \notin \sigma(H^{\mathrm{D}})$:

Definition 3.4.31. Let $(\Gamma, \Gamma', \mathscr{G})$ be an elliptic boundary triple. We call the operator

$$\Lambda(z) := \Gamma' S(z) \colon \mathscr{G}^{3/2} \xrightarrow{S(z)} \mathscr{W}^2 \xrightarrow{\Gamma'} \mathscr{G}^{1/2}$$

the *Dirichlet-to-Neumann operator* in $z \in \mathbb{C} \setminus \sigma(H^{\mathrm{D}})$.

Note that $\Lambda(-1)\varphi = \Lambda\varphi$ for $\varphi \in \mathscr{G}^{3/2}$ (see Proposition 3.4.13), where Λ was defined in (3.36)). We need the following relation of $\Lambda(z)$ with Λ:

Proposition 3.4.32. *Assume that $(\Gamma, \Gamma', \mathscr{G})$ is an elliptic boundary triple, and that $z \in \mathbb{C} \setminus \sigma(H^{\mathrm{D}})$, then $\Lambda(z) - \Lambda$ extends to a bounded operator (denoted by the same symbol)*

$$\Lambda(z) - \Lambda = (z+1)BB(\overline{z})^* \colon \mathscr{G} \longrightarrow \mathscr{G}.$$

Moreover, $\Lambda(z) \colon \mathscr{G}^1 \longrightarrow \mathscr{G}$ is bounded.

Proof. Let $\varphi, \psi \in \mathscr{G}^{3/2}$, then by ellipticity, $f = S\varphi$ and $g = S(z)\psi$ are both in $\mathscr{W}^2$. Now, we have

$$\begin{aligned} 0 &= \langle f, (\check{H} - z)g\rangle - \langle (\check{H} + 1)f, g\rangle \\ &= \langle \Gamma' f, \Gamma g\rangle_{\mathscr{G}} - \langle \Gamma f, \Gamma' g\rangle_{\mathscr{G}} - (z+1)\langle f, g\rangle \\ &= \langle \Lambda\varphi, \psi\rangle_{\mathscr{G}} - \langle \varphi, \Lambda(z)\psi\rangle_{\mathscr{G}} - (z+1)\langle S\varphi, S(z)\psi\rangle \end{aligned}$$

using Green's formula (3.32b) for the second equality, and the definition of Λ and $\Lambda(z)$ in the third equality. By Lemma 3.4.29, $S(z)$ extends to a bounded operator $B(\overline{z})^* \colon \mathscr{G} \longrightarrow \mathscr{H}$, and similarly for $S = S(-1)$, so that the boundedness of $\Lambda(z) - \Lambda$ follows. Finally,

$$\|\Lambda(z)\varphi\|_{\mathscr{G}} \le \|\Lambda\varphi\|_{\mathscr{G}} + |z+1|\,\|BB(\overline{z})^*\varphi\|_{\mathscr{G}} \le \|\varphi\|_{\mathscr{G}^1} + |z+1|\,\|B\|\,\|B(\overline{z})\|\,\|\varphi\|_{\mathscr{G}}$$

and we are done. □

The boundedness of $\Lambda(z): \mathscr{G}^1 \longrightarrow \mathscr{G}$ can be interpreted by saying that $\Lambda(z)$ is an *operator of order* 1.

Remark 3.4.33. Note that the difference $\Lambda(z) - \Lambda$ is not *naturally bounded* (see Remark 3.4.25), since $\|S\| = \|B\| \le \|\Gamma'\|_{2\to 0}\|\iota_{\mathrm{ell,D}}\|$ by Lemma 3.4.29 and similarly for $\|S(z)\| = \|B(z)\|$, i.e. the norm contains estimates on Γ' and the elliptic embedding.

Proposition 3.4.34. *Assume that $z \in \mathbb{C} \setminus \mathbb{R}_+$, then the Dirichlet-to-Neumann map $\Lambda(z): \mathscr{G}^{3/2} \longrightarrow \mathscr{G}^{1/2}$ is bijective. Moreover, its inverse $\Lambda(z)^{-1}: \mathscr{G}^{1/2} \longrightarrow \mathscr{G}^{3/2}$ extends to a bounded operator (denoted by the same symbol) $\Lambda(z)^{-1}: \mathscr{G}^{-1/2} \longrightarrow \mathscr{G}^{1/2}$ given by*

$$\Lambda(z)^{-1} = \Gamma R^{\mathrm{N}}(z)\Gamma^* : \mathscr{G}^{-1/2} \xrightarrow{\Gamma^*} \mathscr{H}^{-1} \xrightarrow{R^{\mathrm{N}}(z)} \mathscr{H}^1 \xrightarrow{\Gamma} \mathscr{G}^{1/2}$$

and this operator is naturally bounded *by*

$$\|\Lambda(z)^{-1}\|_{-1/2\to 1/2} = \|A(\overline{z})\|_{-1\to 1/2} \le C(z,-1). \tag{3.43}$$

Proof. Consider $A(\overline{z})$ as operator

$$A(\overline{z}) = \Gamma R^{\mathrm{N}}(\overline{z}) : \mathscr{H}^{-1} \xrightarrow{R^{\mathrm{N}}(\overline{z})} \mathscr{H}^1 \xrightarrow{\Gamma} \mathscr{G}^{1/2}.$$

Let $\varphi \in \mathscr{G}^{3/2}$. Since $\Lambda(z)\varphi \in \mathscr{G}^{1/2}$, we have $h = A(\overline{z})^*\Lambda(z)\varphi = S^{\mathrm{N}}(z)\Lambda(z)\varphi$ by Lemma 3.4.30. Moreover, we have $\Gamma h = \varphi$, so that $\varphi = \Gamma A(\overline{z})^*\Lambda(z)\varphi$. In particular, we have shown that $\Lambda(z)$ is injective as operator $\mathscr{G}^{3/2} \to \mathscr{G}^{1/2}$. By definition and the regularity assumptions, $\Lambda(z)$ is also surjective, hence bijective. Clearly, $\Gamma R^{\mathrm{N}}(z)\Gamma^*$ as operator $\mathscr{G}^{-1/2} \to \mathscr{G}^{1/2}$ extends the inverse as operator $\mathscr{G}^{1/2} \to \mathscr{G}^{3/2}$. The norm estimate follows from

$$\begin{aligned}\|\Lambda(z)^{-1}\|_{-1/2\to 1/2} &= \|\Gamma A(\overline{z})^*\|_{-1/2\to 1/2} = \|A(\overline{z})^*\|_{-1/2\to 1} = \|A(\overline{z})\|_{-1\to 1/2}\\ &= \|\Gamma R^{\mathrm{N}}(\overline{z})\|_{-1\to 1/2} = \|R^{\mathrm{N}}(\overline{z})\|_{-1\to 1} \le C(z,-1),\end{aligned}$$

where $C(z,-1)$ is defined in (3.37) □

Note that we have just shown the following relation between the Dirichlet and Neumann solution operator

$$S(z) = S(z)^{\mathrm{N}}\Lambda(z): \mathscr{G}^{3/2} \longrightarrow \mathscr{W}^2 \quad \text{and} \quad S^{\mathrm{N}}(z) = S(z)\Lambda(z)^{-1}: \mathscr{G}^{1/2} \longrightarrow \mathscr{W}^2. \tag{3.44}$$

3.4.3 Relation with Other Concepts of Boundary Triples and Examples

In this subsection, we compare our notion of boundary triple with the one introduced in [BL07, Def. 2.1] and with ordinary boundary triples. Moreover, we present our two main examples, namely boundary triples associated with a manifold with boundary and to a metric graph.

Definition 3.4.35. Let A be a closed operator in a Hilbert space $\mathscr{H}$. We call $(\Gamma_0, \Gamma_1, \mathscr{G})$ a *quasi-boundary triple* for A^* if there is a dense subspace $\mathscr{W}^2$ of $\operatorname{dom} A^*$ such that $\Gamma_0, \Gamma_1 : \mathscr{W}^2 \longrightarrow \mathscr{G}$ are linear maps with the following properties:

1. The joint boundary operator $(\Gamma_0, \Gamma_1) : \mathscr{W}^2 \longrightarrow \mathscr{G}$ defined by $(\Gamma_0, \Gamma_1) f = \Gamma_0 f \oplus \Gamma_1 f$ has dense range;
2. $A^{\mathrm{D}} := A \restriction_{\ker \Gamma_0}$ is self-adjoint;
3. Green's formula (3.29) holds for $f, g \in \mathscr{W}^2$.

Proposition 3.4.36. *Assume that $(\Gamma, \Gamma', \mathscr{G})$ is an elliptic boundary triple associated with the quadratic form $\mathfrak{h}$ such that $\Gamma'(\mathscr{H}^{2,\mathrm{D}}) = \mathscr{G}^{1/2}$. Then $(\Gamma \restriction_{\mathscr{W}^2}, \Gamma', \mathscr{G})$ is a quasi-boundary triple associated with $\mathring{H}^*$, where $\mathring{H} := \check{H} \restriction_{\mathring{\mathscr{W}}^2}$.*

Proof. We set $A := \mathring{H}$. Note first that A is a closed operator as the intersection of the two self-adjoint operators H^{D} and H^{N} and that $A \restriction_{\ker \Gamma_0} = H^{\mathrm{D}}$ is self-adjoint (see Proposition 3.4.22), using the first assumption of ellipticity only. It is easy to see from Green's formula (3.32b) for the operator and quadratic form that Green's formula (3.29) for the operator holds. Moreover, $\mathscr{W}^2$ is dense by assumption and $\mathscr{W}^2 \subset \operatorname{dom} A^*$ follows easily by (3.29), since the boundary terms vanish.

For the density condition on the joint boundary operator, let $\varphi, \psi \in \mathscr{G}$. Since $\mathscr{G}^{3/2}$ and $\mathscr{G}^{1/2}$ are dense in $\mathscr{G}$, we have sequences $\{\varphi_n\}_n \subset \mathscr{G}^{3/2}$ and $\{\psi_n\}_n \subset \mathscr{G}^{1/2}$ such that $\varphi_n \to \varphi$ and $\psi_n \to \psi$ in $\mathscr{G}$. Moreover, $\Lambda \varphi_n \in \mathscr{G}^{1/2}$ by Definition 3.4.10. Now, by assumption, there is $f_n \in \mathscr{H}^{2,\mathrm{D}}$ such that $\Gamma' f_n = \psi_n - \Lambda \varphi_n$. Finally, we set $h_n := S\varphi_n + f_n$, then $\Gamma h_n = \varphi_n \to \varphi$ and $\Gamma' h_n = \Gamma' S \varphi_n + \psi_n - \Lambda \varphi_n = \psi_n \to \psi$, which shows the density. □

In particular, we can use a characterisation of quasi-boundary triples for being ordinary boundary triples (namely, we only have to replace "has dense range" by "is surjective" in Definition 3.4.35 (1)), given in [BL07, Cor. 3.2]:

Corollary 3.4.37. *Assume that $(\Gamma, \Gamma', \mathscr{G})$ is a bounded, elliptic boundary triple associated with the quadratic form $\mathfrak{h}$ such that $\Gamma'(\mathscr{H}^{2,\mathrm{D}}) = \mathscr{G}$. Then $\check{H} = \mathring{H}^*$ and $(\Gamma \restriction_{\mathscr{W}^2}, \Gamma', \mathscr{G})$ is an ordinary boundary triple associated with $\mathring{H}^*$, where $\mathring{H} := \check{H} \restriction_{\mathring{\mathscr{W}}^2}$.*

Proof. Since $(\Gamma \restriction_{\mathscr{W}^2}, \Gamma', \mathscr{G})$ is a quasi-boundary triple by the previous proposition, we only have to show that the joint boundary operator $(\Gamma, \Gamma') : \mathscr{W}^2 \longrightarrow \mathscr{G} \oplus \mathscr{G}$

is surjective (see [BL07, Cor. 3.2]): Let $\varphi, \psi \in \mathscr{G}$, then by assumption, there is $f \in \mathscr{H}^{1,\mathrm{D}}$ such that $\Gamma' f = \psi - \Lambda\varphi$. Set $h := S\varphi + f$, then $h \in \mathscr{W}^2$ since the boundary triple is elliptic. Finally, we have $\Gamma h = \varphi$ and $\Gamma' h = \psi$, and the surjectivity follows. □

Remark 3.4.38. The condition $\Gamma'(\mathscr{H}^{2,\mathrm{D}}) = \mathscr{G}^{1/2}$ (resp. $\Gamma'(\mathscr{H}^{2,\mathrm{D}}) = \mathscr{G}$) in the previous two results is an extra condition, which does not follow from our abstract setting. For example, the condition follows from the existence of an "extension operator", a right inverse $E: \mathscr{G}^{3/2} \longrightarrow \mathscr{W}^2$ of Γ (i.e., $\Gamma E\varphi = \varphi$) with the additional property that $u = E\varphi$ fulfils $\Gamma' u = 0$ (see the proof of Theorem 3.4.39 (4) below). In our main examples below, the existence of such a right inverse is assured.

The following theorem presents the main example from PDE theory, explaining the notion "elliptic". We call the boundary triple constructed below the *boundary triple associated with a manifold*. For the notion of Sobolev spaces of order 1 and other notation related to manifolds see Sect. 5.1.3. Sobolev spaces of higher order on compact manifolds can be introduced via a covariant derivative (see e.g. [He96]) or locally via a partition of unity (see e.g. [Ros97, LPPV08]).

Theorem 3.4.39. *Assume that X is a compact d-dimensional manifold ($d \geq 2$) with smooth boundary ∂X. We set*

$$\mathscr{H} := \mathsf{L}_2(X), \qquad \mathscr{H}^1 := \mathsf{H}^1(X) \qquad \mathfrak{h}(u) := \|\mathrm{d}u\|^2,$$
$$\mathscr{G} := \mathsf{L}_2(\partial X), \qquad \Gamma u := u \restriction_{\partial X}, \qquad \Gamma' u := \partial_{\mathrm{n}} u \restriction_{\partial X},$$

where $\partial_{\mathrm{n}} u$ is the outwards normal derivative on ∂X. Then the following assertions hold:

1. *$(\Gamma, \Gamma', \mathscr{G})$ is an elliptic boundary triple with $\mathscr{W}^2 = \mathsf{H}^2(X)$ and $\check{H}$ is the Laplacian on $\mathsf{H}^2(X)$ fulfilling no boundary condition.*
2. *H^{D} and H^{N} are the Laplacians on X with Dirichlet and Neumann boundary conditions on ∂X, respectively. Moreover, $\Lambda(z)$ is the usual Dirichlet-to-Neumann operator.*
3. *The scale of Hilbert spaces on the boundary space is given by $\mathscr{G}^k = \mathsf{H}^k(\partial X)$.*
4. *We have $\Gamma'(\mathscr{H}^{2,\mathrm{D}}) = \mathscr{G}^{1/2}$ so that $(\Gamma \restriction_{\mathscr{W}^2}, \Gamma', \mathscr{G})$ is a quasi-boundary triple associated with $\mathring{H}^*$, where $\mathring{H}$ is the minimal operator, the closure of the Laplacian on smooth functions with support away from the boundary.*

Proof. (1) Most of the assertions of a boundary triple are easily seen. For the ellipticity, we recall that the a priori estimate (3.41) is fulfilled by standard elliptic regularity theory (see e.g. [Ta96, Thm. 5.1.3]). The regularity assumptions on the boundary maps, i.e., $\Gamma(\mathsf{H}^2(X)) = \mathsf{H}^{3/2}(\partial X)$ and $\Gamma'(\mathsf{H}^2(X)) = \mathsf{H}^{1/2}(\partial X)$ are also standard facts, see e.g. [LM68, Gru68]. Therefore, the boundary triple is elliptic.

(2) follows from Proposition 3.4.22 and Definition 3.4.31. Moreover, (3) follows from the fact that the Dirichlet-to-Neumann operator Λ is an elliptic pseudo-differential operator of order 1, and hence $\|\varphi\|_k = \|\Lambda^k \varphi\|$ defines a norm equivalent to any norm on $\mathsf{H}^k(\partial X)$.

For the final assertion (4), consider an extension operator $E: \mathsf{H}^{3/2}(\partial X) \longrightarrow \mathsf{H}^2(X)$, i.e., a right inverse of Γ ($\Gamma E\varphi = \varphi$) with the additional property that $\Gamma' E\varphi = 0$ (see Remark 3.4.38). Note that such an extension exists: in a collar neighbourhood of ∂X choose $u = E\varphi$ to be constant in the normal direction. Then $\mathscr{G}^{3/2} \subset \Gamma(\mathscr{H}^{2,\mathrm{D}})$ follows: Let $\psi \in \mathscr{G}^{3/2}$ and set $h := S^{\mathrm{N}}\psi$. By elliptic regularity, we have $h \in \mathsf{H}^2(X)$. Set $u = h - E\Gamma h$. Then $\Gamma u = \Gamma h - \Gamma h = 0$ and $\Gamma' u = \Gamma' h - 0 = \psi$, i.e., $\psi \in \Gamma'(\mathscr{H}^{2,\mathrm{D}})$. The assertion on the quasi-boundary triple follows from Proposition 3.4.36. □

Let us now consider the case when we only consider a part Y of the entire boundary ∂X as "boundary" in the sense of boundary triples. On the remaining part $Z := \overline{\partial X \setminus Y}$ we assume a Neumann condition:

Theorem 3.4.40. *Assume that X is a compact d-dimensional manifold ($d \geq 2$) with piecewise smooth boundary ∂X such that $Y \subsetneq \partial X$ is a smooth manifold with boundary of dimension $d-1$ (having possibly several components). We assume that near Y, the manifold has a collar neighbourhood U isometric to a product $[0, \ell_-] \times Y$.*[11] *We set*

$$\mathscr{H} := \mathsf{L}_2(X), \qquad \mathscr{H}^1 := \mathsf{H}^1(X) \qquad \mathfrak{h}(u) := \|du\|^2,$$

$$\mathscr{G} := \mathsf{L}_2(Y), \qquad \Gamma u := u\restriction_Y, \qquad \Gamma' u := \partial_{\mathrm{n}} u \restriction_Y,$$

where $\partial_{\mathrm{n}} u$ is the outwards normal derivative on Y (in the collar coordinates $(s, y) \in U$, $\partial_{\mathrm{n}} u = -\partial_s u$). Then the following assertions hold:

1. *$(\Gamma, \Gamma', \mathscr{G})$ is an elliptic boundary triple with*
$$\mathscr{W}^2 = \{ u \in \mathsf{H}^2(X) \mid \partial_{\mathrm{n}} u \restriction_Z = 0 \},$$
and $\check{H}$ is the Laplacian on $\mathscr{W}^2$ fulfilling a Neumann condition on Z.
2. *H^{D} is the Laplacian on X with Dirichlet condition on Y and Neumann condition on Z, and H^{N} is the Laplacian on X with Neumann condition on ∂X. Moreover, $\Lambda(z)$ is the usual Dirichlet-to-Neumann operator associating to the boundary data φ on Y the normal derivative on Y of the Dirichlet solution $h \in \mathscr{W}^2$ with Neumann condition on Z.*
3. *The scale of Hilbert spaces $\mathscr{G}^k$ (defined as $\operatorname{dom} \Lambda^k$ with norm $\|\varphi\|_{\mathscr{G}^k} = \|\Lambda^k \varphi\|_{\mathscr{G}}$) is equivalent with the scale of Hilbert spaces $\mathscr{G}^k(\Delta_Y^{\mathrm{N}})$ associated with the Neumann Laplacian on Y (with norm $\|\varphi\|_{\mathscr{G}^k(\Delta_Y^{\mathrm{N}})} = \|(\Delta_Y^{\mathrm{N}} + 1)^{k/2}\varphi\|_{\mathscr{G}}$).*
4. *We have $\Gamma'(\mathscr{H}^{2,\mathrm{D}}) = \mathscr{G}^{1/2}$ so that $(\Gamma\restriction_{\mathscr{W}^2}, \Gamma', \mathscr{G})$ is a quasi-boundary triple associated with $\mathring{H}^*$. Here, $\mathring{H}$ is the minimal operator, i.e., the Laplacian with Neumann condition on ∂X and Dirichlet condition on Y.*

[11] We assume the product structure near Y for simplicity only. Smooth deformations of this situations still lead to the same result.

Proof. The proof is very much the same as the one of Theorem 3.4.39. (1) For the ellipticity, we recall that the a priori estimate (3.41) depends only on the local structure of the boundary. We have assumed here the product structure of the collar neighbourhood U, so that we can also use arguments as in Sect. 3.5, especially as in Theorem 3.5.6 for the proof of ellipticity (including the regularity assumptions $\Gamma(\mathscr{W}^2) = \mathscr{G}^{3/2}$ and $\Gamma'(\mathscr{W}^2) = \mathscr{G}^{1/2}$). For a discussion of a priori estimates for domains with non-smooth boundary, we refer to [Grv85], e.g., Thm. 4.3.1.4 for polygons, or to the recent survey [Da08]. (2) is obvious.

For (3), we again refer to the discussion in Sect. 3.5, especially to Corollary 3.5.5. Note that the behaviour of the boundary maps depends only on the behaviour on U. Moreover, Λ is an elliptic pseudo-differential operator of order 1.

The final assertion (4) can be seen as in Theorem 3.4.39. □

Similarly, we consider the case where we assume a Dirichlet condition on the on the remaining part $Z = \overline{\partial X \setminus Y}$. The proof is analogous to the one of Theorem 3.4.40.

Theorem 3.4.41. *Let X, $Y \subset \partial X$ as in Theorem 3.4.40. We set*

$$\mathscr{H} := \mathsf{L}_2(X), \quad \mathscr{H}^1 := \mathsf{H}^1(X, Z) = \{\, u \in \mathsf{H}^1(X) \mid u\restriction_Z = 0 \,\} \quad \mathfrak{h}(u) := \|du\|^2,$$
$$\mathscr{G} := \mathsf{L}_2(Y), \quad \Gamma u := u\restriction_Y, \quad \Gamma' u := \partial_n u\restriction_Y.$$

Then the following assertions hold:

1. *$(\Gamma, \Gamma', \mathscr{G})$ is an elliptic boundary triple with*
$$\mathscr{W}^2 = \{\, u \in \mathsf{H}^2(X) \mid u\restriction_Z = 0 \,\},$$
and $\check{H}$ is the Laplacian on $\mathscr{W}^2$ fulfilling a Dirichlet condition on Z.
2. *H^{D} is the Laplacian on X with Dirichlet condition on ∂X, and H^{N} is the Laplacian on X with Neumann condition on Y and Dirichlet condition on Z. Moreover, $\Lambda(z)$ is the usual Dirichlet-to-Neumann operator associating to the boundary data φ on Y the normal derivative on Y of the Dirichlet solution $h \in \mathscr{W}^2$ with Dirichlet condition on Z.*
3. *The scale of Hilbert spaces $\mathscr{G}^k$ (defined as $\operatorname{dom} \Lambda^k$ with norm $\|\varphi\|_{\mathscr{G}^k} = \|\Lambda^k \varphi\|_{\mathscr{G}}$) is equivalent with the scale of Hilbert spaces $\mathscr{G}^k(\Delta_Y^{\mathrm{D}})$ associated with the Dirichlet Laplacian on Y (with norm $\|\varphi\|_{\mathscr{G}^k(\Delta_Y^{\mathrm{N}})} = \|(\Delta_Y^{\mathrm{D}} + 1)^{k/2} \varphi\|_{\mathscr{G}}$).*
4. *We have $\Gamma'(\mathscr{H}^{2,\mathrm{D}}) = \mathscr{G}^{1/2}$ so that $(\Gamma\restriction_{\mathscr{W}^2}, \Gamma', \mathscr{G})$ is a quasi-boundary triple associated with $\mathring{H}^*$. Here, $\mathring{H}$ is the minimal operator, i.e., the Laplacian with Dirichlet condition on ∂X and Neumann condition on Y.*

Remark 3.4.42.

1. If ∂X is smooth, then the norm bound on $\Gamma \colon \mathsf{H}^1(X) \longrightarrow \mathsf{L}_2(\partial X)$ will depend on the geometry near $\partial X \hookrightarrow X$. Basically, we have the behaviour $\|\Gamma\|^2_{1\to 0} \le C\ell_-^{-1}$, where ℓ_- is the width of a collar neighbourhood $\partial X \times [0, \ell_-]$ of ∂X. Moreover,

the metric on the collar neighbourhood is of the form $ds^2 + h_s$, and C depends on the behaviour of the metric h_s on ∂X compared with h_0. We will calculate $\|\Gamma\|_{1\to 0}$ in an example in Proposition 6.2.9, an abstract version can be found in Corollary A.2.7.

2. One can extend the above example of a boundary triple associated with a manifold X also to certain examples of *non-compact* manifolds X with *compact* boundary ∂X. Actually, the boundary map $\Gamma\colon \mathsf{H}^1(X) \longrightarrow \mathsf{L}_2(\partial X)$ and its derived objects such as the Neumann and Dirichlet operator, the (weak) Dirichlet solution operator and the Dirichlet-to-Neumann map are still defined for a general non-compact manifold with compact boundary ∂X, but for further objects such as the second order space $\mathscr{W}^2$, it is not a priori clear what space to use.

   For example, if we define the norm on $\mathsf{H}^2(X)$ locally via an atlas $\mathscr{A}$, the space $\mathsf{H}^2(X) = \mathsf{H}^2(X, \mathscr{A})$ depends on the choice of $\mathscr{A}$, and may even differ as vector space for different atlases (see e.g. [LPPV08, App.]). In Sects. 3.5–3.8 we present a concept of boundary triples which allows non-compact manifolds, decomposing into a compact part and cylindrical ends. For such spaces, we can control the H^2-norm on the infinite ends (see Proposition 3.5.4).

Example 3.4.43. We constructed a boundary triple associated with a quantum graph $(G, \mathscr{V})$ with boundary $\partial G \subset G$ in Sect. 2.2.3. It is shown in Proposition 2.2.18 that this boundary triple is bounded, elliptic and associated with an ordinary boundary triple in the sense of Corollary 3.4.37.

3.4.4 Krein's Resolvent Formulas and Spectral Relations

Krein's resolvent formula relate the resolvent of different self-adjoint extensions of a symmetric operator. Here, we only compare the Dirichlet and Neumann operator. Recall the definition of $B(z)$ and $A(z)$ in Lemmata 3.4.29 and 3.4.30. The main result in this context is the following:

Theorem 3.4.44 ("Krein's resolvent formula"). *Assume that $(\Gamma, \Gamma', \mathscr{G})$ is an elliptic boundary triple. Let $z \in \mathbb{C} \setminus (\sigma(H^{\mathrm{D}}) \cup \sigma(H^{\mathrm{N}}))$, then the resolvent difference can be expressed by the chain of bounded operators*

$$R^{\mathrm{N}}(z) - R^{\mathrm{D}}(z) = A(\overline{z})^* B(z)\colon \mathscr{H} \xrightarrow{B(z)} \mathscr{G} \xrightarrow{A(\overline{z})^*} \mathscr{H}$$

$$= B(\overline{z})^* \Lambda(z)^{-1} B(z)\colon \mathscr{H} \xrightarrow{B(z)} \mathscr{G} \xrightarrow{\Lambda(z)^{-1}} \mathscr{G} \xrightarrow{B(\overline{z})^*} \mathscr{H} \quad \text{and} \tag{3.45a}$$

$$R^{\mathrm{N}}(z) - R^{\mathrm{D}}(z) = S(z)\Lambda(z)^{-1} B(z)\colon \mathscr{H} \xrightarrow{B(z)} \mathscr{G}^{1/2} \xrightarrow{\Lambda(z)^{-1}} \mathscr{G}^{3/2} \xrightarrow{S(z)} \mathscr{W}^2. \tag{3.45b}$$

Moreover, the resolvent difference extends to a bounded operator

$$R^{\mathrm{N}}(z) - \iota_{\mathrm{DN}} R^{\mathrm{D}}(z) \iota_{\mathrm{DN}}^* = S(z)\Lambda(z)^{-1} S(\overline{z})^*\colon \mathscr{H}^{-1} \xrightarrow{S(\overline{z})^*} \mathscr{G}^{-1/2} \xrightarrow{\Lambda(z)^{-1}} \mathscr{G}^{1/2} \xrightarrow{S(z)} \mathscr{H}^1. \tag{3.45c}$$

Proof. Set $u = R^{\mathrm{N}}(\overline{z})\widetilde{u}$ and $v = R^{\mathrm{D}}(z)\widetilde{v}$. By the ellipticity, namely its first condition, $u, v \in \mathscr{W}^2$. Therefore, we can apply Green's formula (3.32b) twice and obtain

$$\begin{aligned}\langle \widetilde{u}, (R^{\mathrm{N}}(z) - R^{\mathrm{D}}(z))\widetilde{v}\rangle &= \langle u, H^{\mathrm{D}}v\rangle - \langle H^{\mathrm{N}}u, v\rangle = \langle \Gamma u, -\Gamma' v\rangle_{\mathscr{G}} \\ &= \langle \widetilde{u}, A(\overline{z})^* B(z)\widetilde{u}\rangle .\end{aligned}$$

In particular,

$$\begin{aligned}R^{\mathrm{N}}(z) - R^{\mathrm{D}}(z) &= A(\overline{z})^* B(z) = \big[\Gamma R^{\mathrm{N}}(\overline{z})\big]^* B(z) \\ &= \big[\Gamma\big(R^{\mathrm{D}}(z) + A(z)^* B(\overline{z})\big)\big]^* B(z) = B(\overline{z})^*\big[\Gamma A(\overline{z})^*\big]^* B(z)\end{aligned}$$

since $\Gamma R^{\mathrm{D}}(z) = 0$. Moreover,

$$\Gamma A(\overline{z})^* = \Gamma S^{\mathrm{N}}(z) = \Lambda(z)^{-1} : \mathscr{G} \longrightarrow \mathscr{G}$$

due to Lemma 3.4.30 and Proposition 3.4.34. In particular, (3.45a) follows. Equation (3.45b) can be seen from the ellipticity of the boundary triple and Proposition 3.4.34. Equation (3.45c) is a consequence of the fact that $S(\overline{z})^* : \mathscr{H}^{-1} \longrightarrow \mathscr{G}^{-1/2}$ extends $B(z)$ and that $B(\overline{z})^*$ restricts to $S(z)$ by Lemma 3.4.29. □

Let us deduce some consequences of Krein's resolvent formula (based on results in [BGP08]):

Theorem 3.4.45. *Assume that $(\Gamma, \Gamma', \mathscr{G})$ is an elliptic boundary triple. Then the following assertions are true:*

1. *Let $z \notin \sigma(H^{\mathrm{D}})$, then*

$$\ker(H^{\mathrm{N}} - z) = S(z) \ker \Lambda(z), \tag{3.46a}$$

   where $\Lambda(z) = \Gamma' S(z) : \mathscr{G}^{3/2} \longrightarrow \mathscr{G}^{1/2}$, and $S(z)$ is bijective as map between the kernels.

2. *For $\lambda \notin \sigma(H^{\mathrm{D}})$, we have the spectral relation*

$$\lambda \in \sigma(H^{\mathrm{N}}) \quad \Leftrightarrow \quad 0 \in \sigma(\Lambda(\lambda)). \tag{3.46b}$$

   Moreover, the multiplicity of an eigenspace is preserved.

3. *Assume in addition that the boundary triple is bounded (i.e. that Λ is bounded) and that $\Gamma'(\mathscr{H}^{2,\mathrm{D}}) = \mathscr{G}$, then the spectral relation* (3.46b) *remains true for the spectral types* pp, disc, ess, *i.e. for the pure point (set of all eigenvalues), discrete and essential spectrum.*
4. *Assume in addition to* (3) *that $(a, b) \cap \sigma(H^{\mathrm{D}}) = \emptyset$, i.e. (a, b) is a spectral gap for H^{D}. Moreover, assume that the Dirichlet-to-Neumann map has the special form*

$$\Lambda(z) = \frac{\dddot{\Delta} - m(z)}{n(z)},$$

where $\dddot{\Delta}$ *is a* bounded, *self-adjoint operator on* $\mathscr{G}$ *and where* m, n *are functions holomorphic on* $(\mathbb{C} \setminus \mathbb{R}) \cup (a,b)$ *and* $n(\lambda) \neq 0$ *on* (a,b). *Then for* $\lambda \in (a,b)$ *we have*

$$\lambda \in \sigma_\bullet(H^{\mathrm{N}}) \quad \Leftrightarrow \quad m(\lambda) \in \sigma_\bullet(\dddot{\Delta})$$

for all spectral types, namely, $\bullet \in \{\emptyset, \mathrm{pp}, \mathrm{disc}, \mathrm{ess}, \mathrm{ac}, \mathrm{sc}, \mathrm{p}\}$, *the entire, pure point, discrete, essential, absolutely continuous, singular continuous and point spectrum* ($\sigma_\mathrm{p}(A) = \overline{\sigma_{\mathrm{pp}}(A)}$). *Finally, the multiplicity of an eigenspace is preserved.*

Proof. (1) Assume that $h \in \ker(H^{\mathrm{N}} - z)$. By ellipticity, we have $h \in \mathscr{W}^2$ and $\varphi := \Gamma h \in \mathscr{G}^{3/2}$. By definition of the Dirichlet solution operator, we have $h = S(z)\varphi$, so that

$$\Lambda(z)\varphi = \Gamma' S(z)\varphi = \Gamma' h = 0 \tag{3.47}$$

using Proposition 3.4.22 for the characterisation of the Neumann operator domain. For the opposite inclusion, let $\varphi \in \ker \Lambda(z)$. By definition, $\varphi \in \mathscr{G}^{3/2}$, so that the regularity assumptions of the ellipticity imply $h = S(z)\varphi \in \mathscr{W}^2$. It follows from (3.47) and Proposition 3.4.5, that $h \in \operatorname{dom} H^{\mathrm{N}}$ and therefore $h \in \ker(H^{\mathrm{N}} - z)$. In particular, (3.46a) is shown, as well as the surjectivity of $S(z)$ as map between the kernels. The injectivity is clear, since $S(z)$ has a left inverse, namely Γ.

(2) Let $0 \notin \sigma(\Lambda(z)^{-1})$. By (3.45a), $R^{\mathrm{N}}(z) = R^{\mathrm{D}}(z) + B(\overline{z})^* \Lambda(z)^{-1} B(z)$ is bounded, and therefore $z \in \mathbb{C} \setminus \sigma(H^{\mathrm{N}})$. For the opposite inclusion, let $z \in \mathbb{C} \setminus \sigma(H^{\mathrm{N}})$. By Proposition 3.4.34, the operator $\Lambda(z)^{-1} = \Gamma R^{\mathrm{N}}(z) \Gamma^*$ is bounded as operator $\mathscr{G}^{-1/2} \to \mathscr{G}^{1/2}$, hence also as operator on $\mathscr{G}$. In particular, $0 \in \sigma(\Lambda(z)^{-1})$.

(3) follows from the fact that a bounded elliptic boundary triple is naturally associated with an ordinary boundary triple (Corollary 3.4.37) and [BGP08, Thm. 3.3]. Finally, (4) follows from [BGP08, Thm. 3.16]. Note that the special form of $\Lambda(z)$ already implies that $\Lambda = \Lambda(-1)$ is bounded, and therefore the boundary triple is bounded. □

Remark 3.4.46. We used the condition $\Gamma'(\mathscr{H}^{2,\mathrm{D}}) = \mathscr{G}$ only to ensure that we have an associated *ordinary* boundary triple, in order to apply the results of [BGP08]. It is not clear to us whether this condition is really needed in this context. Moreover, we are confident that one may prove a similar result for unbounded boundary triples as in Theorem 3.4.45 (3) and (4) under suitable conditions. We want to treat these questions in a future publication.

We end this section with the construction of an *extended* self-adjoint operator associated with a boundary triple (for simplicity bounded), motivated by a similar construction on quantum graphs (see Sect. 2.3):

Definition 3.4.47. Let $(\Gamma, \Gamma', \mathscr{G})$ be a bounded, elliptic boundary triple, and let L be a bounded operator on $\mathscr{G}$. The associated *extended Hilbert space* is given by $\hat{\mathscr{H}} := \mathscr{H} \oplus \mathscr{G}$, and similarly, the *extended quadratic form* $\hat{\mathfrak{h}}_L$ is defined by

$$\hat{\mathfrak{h}}_L(\hat{f}) := \mathfrak{h}(f), \qquad \hat{f} = (f, F) \in \operatorname{dom} \hat{\mathfrak{h}}_L := \hat{\mathscr{H}}^1_L,$$

where

$$\hat{\mathscr{H}}^1_L := \{ (f, F) \in \mathscr{H}^1 \oplus \mathscr{G} \mid \Gamma f = LF \}. \tag{3.48}$$

Proposition 3.4.48. *Assume that $(\Gamma, \Gamma', \mathscr{G})$ is a bounded, elliptic boundary triple such that $\Gamma'(\mathscr{H}^{2,\mathrm{D}}) = \mathscr{G}$, and L a self-adjoint bounded operator on $\mathscr{G}$. Then the extended quadratic form $\hat{\mathfrak{h}}_L$ is closed and non-negative. Moreover, the associated self-adjoint operator is given by*

$$\hat{H}_L \hat{f} = (\check{H} f, L\Gamma' f) \qquad \textit{for} \qquad \hat{f} = (f, F) \in \operatorname{dom} \hat{H}_L = \hat{\mathscr{H}}^2_L,$$

where

$$\hat{\mathscr{H}}^2_L := \{ (f, F) \in \mathscr{W}^2 \oplus \mathscr{G} \mid \Gamma f = LF \}.$$

Proof. The closeness of $\hat{\mathfrak{h}}_L$ follows from the fact that $\hat{\mathscr{H}}^1_L$ is a closed subspace of $\mathscr{H}^1 \oplus \mathscr{G}$, since Γ and L are bounded. The statement on the associated operator follows by an obvious calculation, using Green's formula (3.32b). Moreover, from Corollary 3.4.37 it follows that $\operatorname{dom} \overset{\circ}{H}{}^* = \mathscr{W}^2$, where $\overset{\circ}{H} = \check{H} \restriction_{\overset{\circ}{\mathscr{W}}{}^2}$, so that in particular, $\operatorname{dom} \hat{H}_L \subset \mathscr{W}^2 \oplus \mathscr{G}$. □

We mention the special case $L = 0$:

Corollary 3.4.49. *Assume that $(\Gamma, \Gamma', \mathscr{G})$ is a bounded, elliptic boundary triple such that $\Gamma'(\mathscr{H}^{2,\mathrm{D}}) = \mathscr{G}$. Then the extended quadratic form $\hat{\mathfrak{h}}_0$ and the corresponding extended operator $\hat{H}_0$ associated with $L = 0$ are given by*

$$\hat{\mathfrak{h}}_0 = \mathfrak{h}^{\mathrm{D}} \oplus \mathfrak{o} \qquad \textit{and} \qquad \hat{H}_0 = H^{\mathrm{D}} \oplus 0,$$

where $\mathfrak{o}$ and 0 are the trivial form and operator on $\mathscr{G}$, respectively.

Let us now relate the spectrum of $\hat{H}_L$ with the one of $\Lambda(z)$:

Theorem 3.4.50. *Assume that $(\Gamma, \Gamma', \mathscr{G})$ is a bounded, elliptic boundary triple such that $\Gamma'(\mathscr{H}^{2,\mathrm{D}}) = \mathscr{G}$. Then the following assertions are true:*

1. *For $z \notin \sigma(H^{\mathrm{D}})$, we have*

$$\ker(\hat{H}_L - z) = \hat{S}(z) \ker\bigl(L\Lambda(z)L - z\bigr) \tag{3.49}$$

   where $\hat{S}(z) \colon \mathscr{G} \longrightarrow \hat{\mathscr{H}}^2_L \subset \mathscr{W}^2 \oplus \mathscr{G}$ and $\hat{S}(z)F := (S(z)LF, F)$. Moreover, $\hat{S}(z)$ is an isomorphism between the above spaces.
2. *Assume that $\lambda \in \mathbb{R} \setminus \sigma(H^{\mathrm{D}})$, then λ is an eigenvalue of $\hat{H}_L$ iff $\ker(L\Lambda(z)L - z)$ is non-trivial. Moreover, the multiplicity of the (eigen)spaces is preserved.*
3. *Assume that $(a, b) \cap \sigma(H^{\mathrm{D}}) = \emptyset$, i.e. (a, b) is a spectral gap for H^{D}. Moreover, assume that $L = L_0 \operatorname{id}_{\mathscr{G}}$ for some $L_0 \in \mathbb{R} \setminus \{0\}$. Assume finally, that the Dirichlet-to-Neumann map has the special form*

$$\Lambda(z) = \frac{\dddot{\Delta} - m(z)}{n(z)},$$

where $\dddot{\Delta}$ *is a* bounded, *self-adjoint operator on* $\mathscr{G}$ *and where* m, n *are functions on* $(\mathbb{C} \setminus \mathbb{R}) \cup (a,b)$ *such that* $n(\lambda) \neq 0$ *for* $\lambda \in (a,b)$. *Then for* $\lambda \in (a,b)$,

$$\lambda \text{ is an eigenvalue of } \hat{H}_L \quad \Leftrightarrow \quad \hat{m}_{L_0}(\lambda) \text{ is an eigenvalue of } \dddot{\Delta},$$

where

$$\hat{m}_{L_0}(\lambda) = L_0^{-2} z n(\lambda) + m(\lambda).$$

Proof. (1) Note first, that $\hat{S}(z)F = (S(z)LF, F)$ fulfils the coupling condition, since $LF = \Gamma S(z)LF$. Let $(f, F) \in \hat{\mathscr{H}}_L^2$ such that $(\hat{H} - z)(f, F) = 0$, then $(\check{H} - z)f = 0$, $L\Gamma' f = zF$ and $\Gamma f = LF$. In particular, $f = S(z)\Gamma f = S(z)LF$, so that $\hat{S}(z)F = (f, F)$. Moreover, $L\Lambda(z)LF = L\Gamma' S(z)LF = L\Gamma' f = zF$ by the definition of $\Lambda(z)$ (see Definition 3.4.31). In particular, the inclusion "$\subset$" follows. The other inclusion can be seen similarly. In addition, $\hat{S}(z)$ is bijective on the given spaces. (2) follows immediately from (1), as well as (3). □

Remark 3.4.51.

1. For brevity, we do not provide a resolvent formula similar to the one given in Theorem 3.4.44. Moreover, as in [BGP08] (see Theorem 3.4.45), one may prove similar results for other spectral types.
2. Formally, the above extended operator converges to the Neumann operator H^{N} in $\mathscr{H}$ if $L_0 \to \infty$, and to the decoupled case of Corollary 3.4.49 if $L_0 \to 0$. The limits can be given a precise meaning. For example, the function $\hat{m}_{L_0}(\lambda)$ tends to $m(\lambda)$ if $L_0 \to \infty$.

3.5 Half-Line Boundary Triples and Complex Dilation

In this subsection we present a simple example of a boundary triple associated with a half-line. Roughly, the idea is to consider elliptic partial differential operators like the Laplacian $\Delta_X = -\partial_{ss} + \Delta_Y$ on a cylinder $X = \mathbb{R}_+ \times Y$ as an operator-valued ordinary differential operator on $\mathbb{R}_+ = [0, \infty)$, and express the conditions on the transversal manifold Y and transversal operator $K = \Delta_Y$ on $\mathscr{G} = \mathsf{L}_2(Y)$ only in functional analytic terms. The construction given in Sect. 3.5.1 is closely related to the setting of abstract fibred spaces in Sect. A.2, see also the book [GG91] for related results and [BCh05] or [BBÇ08] for similar concepts for *first* order operators. The half-line example shows in a simple and abstract way why we loose "half" of the regularity when restricting a function on a manifold to its boundary. Moreover, we obtain explicit formulas for the Dirichlet solution and the Dirichlet-to-Neumann operator. Finally, the abstract setting allows us to treat the metric graph and manifold (and other) models at the same time.

Next, in Sect. 3.5.2, we introduce the so-called *complex dilation*, starting with the usual dilation $\Phi^\theta : \mathbb{R}_+ \longrightarrow \mathbb{R}_+$, $\Phi^\theta(s) := e^\theta s$ for $\theta \in \mathbb{R}$ acting only on the longitudinal half-line $\mathbb{R}_+ = [0, \infty)$. The corresponding unitary action U^θ acts in the obvious way on the Hilbert space $\mathscr{H} := \mathsf{L}_2(\mathbb{R}_+) \otimes \mathscr{G}$. The explicit formulas for the unitarily equivalent operators like $H^\theta := U^\theta H U^{-\theta}$, a priori only defined for *real* $\theta \in \mathbb{R}$, will serve as a definition for *complex* θ in the strip $\mathrm{S}_\vartheta := \{\theta \in \mathbb{C} \mid |\mathrm{Im}\,\theta| < \vartheta/2\}$ for $\vartheta \in (0, \pi/2)$. In the above example of the Laplacian on a cylinder, the (Neumann-)Laplacian $H = -e^{-2\theta}\partial_{ss}^{\mathrm{N}} + \Delta_Y$ becomes $H^\theta = -e^{-2\theta}\partial_{ss}^{\mathrm{N}} + \Delta_Y$. In particular, using the latter expression as a definition for *complex* θ, the essential spectrum $\sigma_{\mathrm{ess}}(H) = [0, \infty)$ of the undilated operator turns into the complex plane as $\sigma_{\mathrm{ess}}(H^\theta) = e^{-2\,\mathrm{Im}\,\theta}[0, \infty)$. Moreover, we can define a *dilated* boundary triple $(\Gamma^\theta, -\Gamma'^\theta, \mathscr{G})$. It is an important fact that the associated Dirichlet-to-Neumann map is *independent* of θ, i.e., $\Lambda^\theta(z) = \Lambda(z)$.

We will see that in more complicated examples (given by *coupled* boundary triples in Sects. 3.6–3.7), eigenvalues (of finite multiplicity) embedded in $[0, \infty)$ for the undilated operator H remain unchanged and therefore become *discrete* eigenvalues of H^θ.

We introduce the complex dilation in this simple half-line model first, in order to illustrate the simple idea of complex dilation. The more complicated discussion involving the domains for coupled boundary triples will be given in Sects. 3.6–3.7.

3.5.1 Half-Line Boundary Triples

We start with a Hilbert space $\mathscr{G}$ and a non-negative operator K with associated quadratic form $\mathfrak{k}$. We assume that $\mathrm{dom}\,\mathfrak{k}$ is dense in $\mathscr{G}$. We associate a boundary triple to these data as follows: The Hilbert space is given by

$$\mathscr{H} = \mathsf{L}_2(\mathbb{R}_+) \otimes \mathscr{G} \cong \mathsf{L}_2(\mathbb{R}_+, \mathscr{G}), \tag{3.50a}$$

where $\mathbb{R}_+ = [0, \infty)$. Moreover, we define a non-negative quadratic form $\mathfrak{h}$ by

$$\mathfrak{h}(f \otimes \varphi) = \|f'\|^2_{\mathsf{L}_2(\mathbb{R}_+)} \|\varphi\|^2_{\mathscr{G}} + \|f\|^2_{\mathsf{L}_2(\mathbb{R}_+)} \mathfrak{k}(\varphi), \tag{3.50b}$$

with $\mathrm{dom}\,\mathfrak{h} =: \mathscr{H}^1$ generated by all $f \otimes \varphi$ with $f \in \mathsf{H}^1(\mathbb{R}_+)$ and $\varphi \in \mathrm{dom}\,\mathfrak{k}$, i.e. the closure of all linear combinations of such elementary tensors with respect to the norm

$$\|f \otimes \varphi\|^2_{\mathscr{H}^1} := \|f \otimes \varphi\|^2_{\mathfrak{h}} = \|f'\|^2\|\varphi\|^2 + \|f\|^2\mathfrak{k}(\varphi) + \|f\|^2\|\varphi\|^2. \tag{3.50c}$$

Next, we denote by $\mathscr{W}^2$ the subspace of $\mathscr{H}^1$ defined by the closure of all linear combinations of elementary tensors $f \otimes \varphi$ with respect to the norm

$$\|f \otimes \varphi\|^2_{\mathscr{W}^2} := \|f\|^2_{\mathsf{H}^2(\mathbb{R}_+)}\|\varphi\|^2_{\mathscr{G}} + \|f'\|^2_{\mathsf{L}_2(\mathbb{R}_+)}\mathfrak{k}(\varphi) + \|f\|^2_{\mathsf{L}_2(\mathbb{R}_+)}\|K\varphi\|^2_{\mathscr{G}}, \tag{3.50d}$$

where $f \in \mathsf{H}^2(\mathbb{R}_+)$ and $\varphi \in \operatorname{dom} K$.

Proposition 3.5.1. *Define*

$$\Gamma : \mathscr{H}^1 \longrightarrow \mathscr{G}, \qquad \Gamma u := u(0) \in \mathscr{G}, \tag{3.51a}$$

$$\Gamma' : \mathscr{W}^2 \longrightarrow \mathscr{G}, \qquad \Gamma' u := u'(0) \in \mathscr{G}, \tag{3.51b}$$

then the operators Γ and Γ' are bounded by $\sqrt{2}$. Moreover, $(\Gamma, -\Gamma', \mathscr{G})$ is a boundary triple with operator $\check{H}$ given by

$$\check{H}u = -u'' + H^{\perp}u, \qquad H^{\perp} = \mathrm{id} \otimes K. \tag{3.51c}$$

where $u'' = (\partial_{ss} \otimes \mathrm{id})u$ is the second derivative with respect to the variable $s \in \mathbb{R}_+$.

For the choice of the sign in Γ' we refer to Remark 3.5.3 (1).

Proof. First, we note that $\mathscr{H}^1 \subset \mathsf{H}^1(\mathbb{R}_+) \otimes \mathscr{G}$ and $\mathscr{W}^2 \subset \mathsf{H}^2(\mathbb{R}_+) \otimes \mathscr{G}$. Moreover, the norm of both inclusions is bounded by 1, since

$$\|f \otimes \varphi\|^2_{\mathsf{H}^1(\mathbb{R}_+)\otimes\mathscr{G}} = (\|f\|^2 + \|f'\|^2)\|\varphi\|^2 \le \|f \otimes \varphi\|^2_{\mathscr{H}^1}.$$

A similar assertion holds for $\mathscr{W}^2$. In particular, we can apply the results of Sect. A.2 and Corollary A.2.8, and the boundedness of Γ and Γ' follows (choosing $s_1 = 0$, $s_2 = 1$, $\rho_\infty = 1$). The density of $\ker \Gamma$ and $\ker \Gamma'$ in $\mathscr{H}$ is a known fact from Sobolev space theory. In order to proof the density of $\operatorname{ran} \Gamma$, assume that $u = f \otimes \varphi$ with $f \in \mathsf{H}^1(\mathbb{R}_+)$ and $\varphi \in \operatorname{dom} \mathfrak{k}$, then $\Gamma u = f(0)\varphi$. Since $\operatorname{dom} \mathfrak{k}$ is dense in $\mathscr{G}$ by assumption, the density of $\operatorname{ran} \Gamma$ in $\mathscr{G}$ follows. Green's formula and the form of $\check{H}$ is easily seen by partial integration and the fact that

$$u'(s) \in \mathscr{G} \qquad \text{and} \qquad u(s) \in \operatorname{dom} K \tag{3.52}$$

for almost all $s \in \mathbb{R}_+$ and all $u \in \mathscr{W}^2$, using the definition of the norm on $\mathscr{W}^2$. Here, we consider u as element in $\mathsf{L}_2(\mathbb{R}_+, \mathscr{G})$. Moreover, the boundedness of $\check{H} : \mathscr{W}^2 \longrightarrow \mathscr{H}$ is easily seen. □

Definition 3.5.2. We call the boundary triple $(\Gamma, -\Gamma', \mathscr{G})$ associated with the quadratic form $\mathfrak{k}$ on $\mathscr{G}$ as constructed above in (3.50)–(3.51), a *half-line boundary triple*. We call $\mathfrak{k}$ resp. the associated operator K in $\mathscr{G}$ the *transversal form* resp. *transversal operator* of the half-line boundary triple.

Remark 3.5.3.

1. We prefer the minus sign inside the definition of Γ' since then, continuity of the "derivative" along the passage from the interior to the exterior part can be expressed as $\Gamma'_{\mathrm{int}}u = \Gamma'_{\mathrm{ext}}u$, see Proposition 3.6.2.
2. The basic example we have in mind is the PDE case where $\mathscr{G} = \mathsf{L}_2(Y)$ and $K = \Delta_Y$. Then $\mathscr{H} = \mathsf{L}_2(X)$ with $X = \mathbb{R}_+ \times Y$, $\mathscr{H}^1 = \mathsf{H}^1(X)$ and $\mathscr{W}^2 = \mathsf{H}^2(X)$. Since this example is pretty simple, we use it as a toy model in order to define the complex scaling, and *separate* the arguments of complex scaling and coupled boundary triples. Moreover, in this example, the Dirichlet-to-Neumann map can be calculated explicitly (see Corollary 3.5.5). The more interesting *coupled* model will be introduced in Sect. 3.6.

Let us now give explicit formulas for the solution operator and the Dirichlet-to-Neumann map. Denote by $\mathscr{G}^m(K) = \operatorname{dom} K^{m/2} \subset \mathscr{G}$ the Hilbert scale associated with K. We assume that K has discrete spectrum $\mu_k = \lambda_k(K)$ (note that the case $\dim \mathscr{G} < \infty$ is included):

Proposition 3.5.4. *Assume that K has the spectral decomposition $K\varphi_k = \mu_k\varphi_k$, where $\{\varphi_k\}$ is an orthonormal basis of $\mathscr{G}$, and where the eigenvalues μ_k form a discrete set. Moreover, we assume that $z \notin \Sigma_\vartheta = \{z \in \mathbb{C} \mid |\arg z| \le \vartheta\}$, where $\vartheta \in (0, \pi/2)$.*

Then the Dirichlet solution operator $S(z): \mathscr{G} \longrightarrow \mathscr{H}$ is given by

$$S(z)\varphi = \sum_k \langle \varphi_k, \varphi \rangle f_k \otimes \varphi_k,$$

where $f_k(s) = \mathrm{e}^{\mathrm{i}\sqrt{z-\mu_k}s}$ and $\sqrt{w}$ is the complex root cut along the positive half-axis $\mathbb{R}_+$. The convergence of the sum in $\mathscr{H} = \mathsf{L}_2(\mathbb{R}_+) \otimes \mathscr{G}$ and $S(z)$ as operator from $\mathscr{G}$ into $\mathscr{H}$ is bounded by $2^{-1/2}(\sin(\vartheta/2))^{-1/2} d_\vartheta(z)^{-1/4}$, where $d_\vartheta(z) = \operatorname{dist}(z, \mathbb{C} \setminus \Sigma_\vartheta)$ (see Fig. 3.1).

Moreover, $S(z)\varphi \in \mathsf{H}^m(\mathbb{R}_+) \otimes \mathscr{G}$ iff $\varphi \in \mathscr{G}^{m-1/2}(K)$ for integer $m \ge 0$, and the sum converges in $\mathsf{H}^m(\mathbb{R}_+) \otimes \mathscr{G}$. In particular, $S(z)\varphi \in \mathscr{H}^1$ iff $\varphi \in \mathscr{G}^{1/2}(K)$ and $S(z)\varphi \in \mathscr{W}^2$ iff $\varphi \in \mathscr{G}^{3/2}(K)$, and the sum converges in $\mathscr{H}^1$ resp. $\mathscr{W}^2$. Finally, $S(z): \mathscr{G}^{1/2}(K) \longrightarrow \mathscr{H}^1$ and $S(z): \mathscr{G}^{3/2}(K) \longrightarrow \mathscr{W}^2$ are bounded by a constant depending only on ϑ and $d_\vartheta(z)$.

Proof. Note first that $\operatorname{Im}\sqrt{z-\mu_k} > 0$ and $\vartheta < \arg(z-\mu_k) < 2\pi - \vartheta$. Moreover,

$$\begin{aligned} \|f_k\|^2 &= \int_0^\infty \mathrm{e}^{-2(\operatorname{Im}\sqrt{z-\mu_k})s} \mathrm{d}s = \frac{1}{2\operatorname{Im}\sqrt{z-\mu_k}} \\ &\le \frac{1}{2\sin(\vartheta/2)|z-\mu_k|^{1/2}} \le \frac{1}{2\sin(\vartheta/2) d_\vartheta(z)^{1/2}} \end{aligned}$$

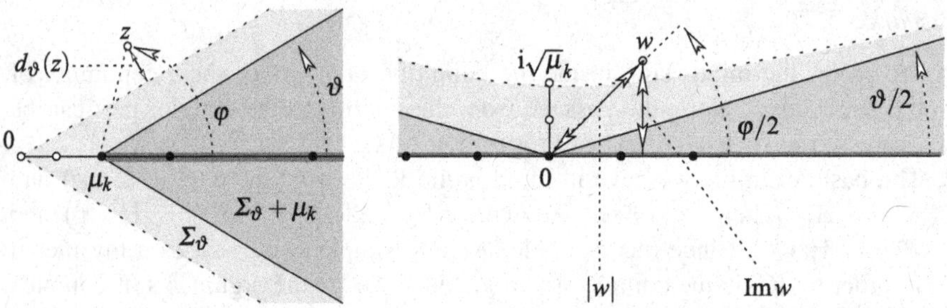

Fig. 3.1 The complex plane z with cut along $[\mu_k, \infty)$ and the complex half-plane of $w = \sqrt{z - \mu_k}$. Here, $\varphi = \arg(z - \mu_k)$

for all k (see Fig. 3.1). Similarly, the m-th derivative is $f_k^{(m)} = (\mathrm{i}\sqrt{z - \mu_k})^m f_k$, and therefore we have

$$\|f_k^{(m)}\|^2 \leq \frac{|z - \mu_k|^{m-1/2}}{2\sin(\vartheta/2)}.$$

Set $u := S(z)\varphi$. Since $\{f_k \otimes \varphi_k\}_k$ is an orthogonal system, we have

$$\|u\|^2 = \sum_k \bigl|\langle\varphi_k, \varphi\rangle\bigr|^2 \|f_k\|^2 \leq \frac{1}{2\sin(\vartheta/2) d_\vartheta(z)} \|\varphi\|^2.$$

This shows the convergence in $\mathscr{H}$ of the sum and the norm bound. Similarly,

$$\begin{aligned}
\|u\|^2_{\mathsf{H}^m(\mathbb{R}_+)\otimes\mathscr{G}} &= \sum_k \bigl|\langle\varphi_k, \varphi\rangle\bigr|^2 \sum_{i=0}^m \|f_k^{(i)}\|^2 \|\varphi\|^2 \\
&= \sum_k \bigl|\langle\varphi_k, \varphi\rangle\bigr|^2 \sum_{i=0}^m \frac{|z - \mu_k|^i}{2|\operatorname{Im}\sqrt{z - \mu_k}|} \\
&\quad \times \begin{Bmatrix} \leq (\sin(\vartheta/2))^{-1} \\ \geq 1 \end{Bmatrix} \cdot \frac{1}{2} \sum_k \bigl|\langle\varphi_k, \varphi\rangle\bigr|^2 \sum_{i=0}^m |z - \mu_k|^{i-1/2},
\end{aligned}$$

and the latter expression is finite iff $\varphi \in \mathscr{G}^{m-1/2}(K)$. Since the coefficients of

$$(\mathbb{1} \otimes \mathfrak{k})(u) = \sum_k \bigl|\langle\varphi_k, \varphi\rangle\bigr|^2 \|f_k\|^2 \mathfrak{k}(\varphi) = \sum_k \bigl|\langle\varphi_k, \varphi\rangle\bigr|^2 \frac{\mu_k}{2|\operatorname{Im}\sqrt{z - \mu_k}|}$$

are of order $\mu_k^{1/2}$ as $k \to \infty$ and similarly, the coefficients of $\|(\mathrm{id} \otimes K)u\|^2$ are of order $\mu_k^{3/2}$, we have $u \in \operatorname{dom} \mathbb{1} \otimes \mathfrak{k}$ and $u \in \operatorname{dom} \mathrm{id} \otimes K$, respectively. In particular, $u \in \mathscr{H}^1$ resp. $u \in \mathscr{W}^2$. The convergence of the sums in the appropriate norms follow from the above estimates, as well as the boundedness of $S(z)$. □

Corollary 3.5.5. *Assume that K has purely discrete spectrum $K\varphi_k = \mu_k\varphi_k$. Then the Dirichlet-to-Neumann operator $\Lambda(z) := -\Gamma' S(z)$ is given by*

$$\Lambda(z)\varphi = -\mathrm{i}\sum_k \sqrt{z-\mu_k}\langle\varphi_k,\varphi\rangle\varphi_k =: -\mathrm{i}\sqrt{z-K}\varphi$$

with domain

$$\operatorname{dom}\Lambda(z) = \Big\{\varphi\in\mathscr{G}\;\Big|\;\sum_k(1+\mu_k)\big|\langle\varphi_k,\varphi\rangle\big| < \infty\Big\} = \operatorname{dom}K^{1/2}.$$

Moreover, $\Lambda(z)^2 = K - z$, and $\Lambda(z)^{-1}$ is compact. In particular, for $z = -1$, we have

$$\Lambda = \Lambda(-1) = \sum_k \sqrt{1+\mu_k}\langle\varphi_k,\cdot\rangle\varphi_k = (K+1)^{1/2}.$$

Finally, the scales of Hilbert spaces $\mathscr{G}^m(K)$ and $\mathscr{G}^m$ (see Definition 3.4.10) agree and have equal norms

$$\|\varphi\|_{\mathscr{G}^m(K)} := \|(K+1)^{m/2}\| = \|\Lambda^m\|\varphi =: \|\varphi\|_{\mathscr{G}^m}.$$

Proof. Since for $\varphi\in\mathscr{G}^{3/2}(K)$, the expansion of $u = S(z)\varphi$ converges in $\mathscr{W}^2$ by the previous lemma, we can interchange the sum and Γ' in the expansion of $\Lambda(z)\varphi = \Gamma' u$, and obtain the desired expression for $\Lambda(z)\varphi$. The other assertions follow easily. □

Theorem 3.5.6. *Let $(\Gamma, -\Gamma', \mathscr{G})$ be the half-line boundary triple associated with $\mathfrak{k}$ on $\mathscr{G}$. If the associated operator K has purely discrete spectrum, then the boundary triple is elliptic.*

Proof. We check the assumptions of Proposition 3.4.28. The Dirichlet and Neumann operator H^{D} and H^{N} associated with the boundary triple are given by

$$H^{\mathrm{D/N}} = -\partial_{ss}^{\mathrm{D/N}} + \mathrm{id}\otimes K = L + H^{\perp},$$

where $L := -\partial_{ss}^{\mathrm{D/N}}$ denotes the Laplacian on $\mathbb{R}_+$ with Dirichlet resp. Neumann boundary condition at 0. By Proposition 3.4.24, we have to show that the $\mathscr{W}^2$-norm can be estimated by the graph norm of the Dirichlet resp. Neumann operator. In particular, for the Dirichlet operator we have

$$\begin{aligned}\|u\|^2_{\mathscr{W}^2} &= \|u\|^2 + \|u'\|^2 + \|u''\|^2 + \|(H^{\perp})^{1/2}u'\|^2 + \|H^{\perp}u\|^2\\ &= \|u\|^2 + \langle u, Lu\rangle + \|Lu\|^2 + \langle Lu, H^{\perp}u\rangle + \|H^{\perp}u\|^2\\ &\overset{\mathrm{CY}}{\le} 2\big(\|u\|^2 + \|Lu\|^2 + \|H^{\perp}u\|^2\big)\end{aligned}$$

for $u \in \mathscr{W}^2$ with $u(0) = 0$ resp. $u'(0) = 0$, since L and $H^\perp$ commute. By the same reason, the spectral calculus applies and we have

$$\|Lu\|^2 + \|H^\perp u\|^2 = \langle u, (L^2 + (H^\perp)^2)u\rangle \le \langle u, (L + H^\perp)^2)u\rangle = \|H^{\mathrm{D/N}} u\|^2,$$

hence we have shown (3.41).

Next, we show that $\Gamma : \mathscr{W}^2 \longrightarrow \mathscr{G}^{3/2}$ is bounded: For $u \in \mathscr{W}^2$, let $\Phi(s) := \|K^{3/4}u(s)\|^2$. It can be seen that $\Phi(s)$ exists and is derivable for almost all $s \in \mathbb{R}_+$, that $\Phi(s) \to 0$ as $s \to \infty$ and that $\Phi'(s) = 2\operatorname{Re}\langle Ku(s), K^{1/2}u'(s)\rangle$. Then we have

$$\|\Gamma u\|^2_{\mathscr{G}^{3/2}} = \Phi(0) = -\int_0^\infty \Phi(s)\mathrm{d}s \overset{\mathrm{CY}}{\le} \int_0^\infty (\|Ku(s)\|^2 + \|K^{1/2}u'(s)\|^2)\mathrm{d}s \le \|u\|^2_{\mathscr{W}^2}$$

using the fact that $\mathscr{G}^m$ is the scale of order m associated with K (see Corollary 3.5.5). Therefore, the boundedness with norm bound 1 is shown.

For $\Gamma' : \mathscr{W}^2 \longrightarrow \mathscr{G}^{1/2}$ we argue similarly, now with $\Psi(s) := \|K^{1/4}u'(s)\|^2$. Here, we have $\Psi'(s) = 2\operatorname{Re}\langle -u''(s), K^{1/2}u(s)\rangle$ and $\|\Gamma' u\|^2_{\mathscr{G}^{1/2}} = \Psi(0) \le \|u\|^2_{\mathscr{W}^2}$ similarly as before.

It remains to show that $S(\mathscr{G}^{3/2}) \subset \mathscr{W}^2$, but this was shown in Proposition 3.5.4. □

3.5.2 *Complex Dilation*

Let us now define the one-parameter unitary group of dilations. Such unitary groups will be used later on in order to define resonances. For a self-adjoint restriction H of $\check{H}$, we define the dilated operator $H^\theta = U^\theta H U^{-\theta}$ for real parameters θ, and use the explicit form of H^θ later on as definition for complex values of θ. A discrete eigenvalue of H^θ will be called a *resonance*; and it can be shown that the definition does not depend on θ provided $\operatorname{Im}\theta$ is large enough. Note that such eigenvalues only occur for coupled systems. In the half-line model here, there will be no resonance (e.g., the spectrum of $H^{\theta,\mathrm{D}}$ – the dilated Dirichlet operator H^{D}) – is purely essential, see Proposition 3.5.14). Nevertheless, we think that it is helpful to treat first the complex dilation in this section, and then coupled systems in Sects. 3.6 and 3.7.

Definition 3.5.7. For $\theta \in \mathbb{R}$ we define $U^\theta : \mathscr{H} \longrightarrow \mathscr{H}$ by

$$U^\theta(f \otimes \varphi) := (\hat{U}^\theta f) \otimes \varphi,$$

where $\hat{U}^\theta$ is the unitary dilation operator in $\mathsf{L}_2(\mathbb{R}_+)$ given by

$$(\hat{U}^\theta f)(s) := \mathrm{e}^{\theta/2} f(\mathrm{e}^\theta s).$$

A simple calculation shows that U^θ is unitary. Let us now define the unitarily transformed quadratic form $\mathfrak{h}^\theta$:

Definition 3.5.8. The *dilated quadratic form* associated with $\mathfrak{h}$ is defined as

$$\mathfrak{h}^\theta(u) := h(U^{-\theta}u), \qquad \operatorname{dom}\mathfrak{h}^\theta := U^\theta(\operatorname{dom}\mathfrak{h}).$$

Note that $\mathfrak{h}^0 = \mathfrak{h}$ is the original form. If H is the operator associated with $\mathfrak{h}$ (the Neumann operator), then

$$H^\theta = U^\theta H U^{-\theta}.$$

Denote by $\mathfrak{h}^\perp$ the form defined by $\mathfrak{h}^\perp(f \otimes \varphi) = \|f\|^2\mathfrak{k}(\varphi)$, i.e. the form associated with $H^\perp = \mathrm{id} \otimes K$.

Lemma 3.5.9. *Let $\theta \in \mathbb{R}$, then the dilated quadratic form $\mathfrak{h}^\theta$ is given by*

$$\mathfrak{h}^\theta(u) = \mathrm{e}^{-2\theta}\|u'\|^2 + \mathfrak{h}^\perp(u) \tag{3.53}$$

for $u \in \operatorname{dom}\mathfrak{h}^\theta = \operatorname{dom}\mathfrak{h} = \mathscr{H}^1$.

Proof. By the definition of U^θ, we have $\mathfrak{h}^\theta(u) = \mathfrak{h}(U^{-\theta}u)$. Moreover,

$$\mathfrak{h}(U^{-\theta}u) = \int_0^\infty \|(U^{-\theta}u)'(s)\|^2_{\mathscr{G}}\mathrm{d}s + \mathfrak{h}^\perp(u)$$

since $\mathfrak{h}^\perp(U^{-\theta}(f \otimes \varphi)) = \mathfrak{h}^\perp(\hat{U}^{-\theta}f \otimes \varphi) = \|\hat{U}^{-\theta}f\|^2\mathfrak{k}(\varphi) = \|f\|^2\mathfrak{k}(\varphi) = \mathfrak{h}^\perp(f \otimes \varphi)$ by (3.50b). The integral equals

$$\int_0^\infty \|(U^{-\theta}u)'(s)\|^2_{\mathscr{G}}\mathrm{d}s = \mathrm{e}^{-2\theta}\int_0^\infty \|u'(e^{-\theta}s)\|^2_{\mathscr{G}}\mathrm{e}^{-\theta}\mathrm{d}s$$

and the formula for $\mathfrak{h}^\theta$ follows by substituting $\widetilde{s} = \mathrm{e}^{-\theta}s$. □

We can now define a *dilated boundary triple* $(\Gamma^\theta, -\Gamma'^\theta, \mathscr{G})$ with

$$\Gamma^\theta := \mathrm{e}^{-\theta/2}\Gamma \qquad \text{and} \qquad \Gamma'^\theta := \mathrm{e}^{-3\theta/2}\Gamma'. \tag{3.54}$$

The following result is now a simple consequence. The ellipticity follows as in Theorem 3.5.6:

Proposition 3.5.10. *If $\theta \in \mathbb{R}$, then $(\Gamma^\theta, -\Gamma'^\theta, \mathscr{G})$ is a boundary triple for $\mathfrak{h}^\theta$. The operator $\check{H}^\theta$ is defined on $\mathscr{W}^2$ and acts as*

$$\check{H}^\theta = -\mathrm{e}^{-2\theta}\partial_{ss} \otimes \mathrm{id} + \mathrm{id} \otimes K \tag{3.55a}$$

where ∂_{ss} denotes the second derivative operator in $\mathsf{L}_2(\mathbb{R}_+)$, and where K is the non-negative operator in $\mathscr{G}$ associated with $\mathfrak{k}$.

Assume in addition, that K has purely discrete spectrum, then the dilated Dirichlet solution operator *is given by*

$$S^\theta(z)\varphi = \sum_k \langle \varphi_k, \varphi \rangle f_k^\theta \otimes \varphi_k, \qquad \textit{where} \qquad f_k^\theta(s) = \mathrm{e}^{\theta/2 + \mathrm{i}\mathrm{e}^\theta \sqrt{z-\mu_k} s}. \tag{3.55b}$$

Moreover, the same convergence assertions as in Proposition 3.5.4 hold.

The dilated Dirichlet-to-Neumann operator, defined by $\Lambda^\theta(z) := -\Gamma'^\theta S^\theta(z)$, is independent *of θ, i.e., $\Lambda^\theta(z) = \Lambda(z)$, where $\Lambda(z) = -\Gamma' S(z)$ is the Dirichlet-to-Neumann operator associated with the undilated boundary triple $(\Gamma, -\Gamma', \mathscr{G})$. In particular, the Hilbert spaces $\mathscr{G}^k$ associated with the Dirichlet-to-Neumann are the same for all θ.*

Finally, the boundary triple is elliptic.

We will now use the above formulas (3.53)–(3.55a) as definition for *complex* values of θ. Let $\vartheta \in [0, \pi/2)$[12] and denote by

$$\mathrm{S}_\vartheta := \{\theta \in \mathbb{C} \mid |\mathrm{Im}\,\theta| \le \vartheta/2\} \tag{3.56}$$

the horizontal strip of thickness ϑ centred around the real axis.

Definition 3.5.11. Let $\theta \in \mathrm{S}_\vartheta$. We say that the boundary triple $(\Gamma^\theta, -\Gamma'^\theta, \mathscr{G})$ associated with $\mathfrak{h}^\theta$ with operator $\check{H}^\theta$ as given in (3.53)–(3.55a) is obtained from the half-line boundary triple $(\Gamma, -\Gamma', \mathscr{G})$ associated with $\mathfrak{h}$ by *complex dilation*. We also call $(\Gamma^\theta, -\Gamma'^\theta, \mathscr{G})$ the *complexly dilated boundary triple*.

Formally, $(\Gamma^\theta, -\Gamma'^\theta, \mathscr{G})$ is a boundary triple associated with the *sectorial* quadratic form $\mathfrak{h}^\theta$ in the sense of Definition 3.4.3, with the corresponding dilated operator $\check{H}^\theta$ given in (3.55a) for $\theta \in \mathrm{S}_\vartheta$. For example, Green's formula in the complexly dilated case reads as

$$\langle u, \check{H}^\theta v \rangle_{\mathscr{H}} = \mathfrak{h}^\theta(u, v) + \langle \Gamma^{\overline{\theta}} u, \Gamma'^\theta v \rangle_{\mathscr{G}}. \tag{3.57}$$

We also extend the notion of "ellipticity" to the case treated here. In particular, we call the complexly dilated boundary triple $(\Gamma^\theta, -\Gamma'^\theta, \mathscr{G})$ *elliptic*, if the conditions of Definition 3.4.21 are fulfilled.

Since the spaces $\mathscr{W}^2$, $\mathscr{H}^{k,\mathrm{D}} = \mathrm{dom}\, H^{\theta,\mathrm{D}}$ and $\mathscr{H}^{k,\mathrm{N}} = \mathrm{dom}\, H^{\theta,\mathrm{N}}$ are independent of θ, and since the dilated and undilated boundary maps differ only by a factor, the ellipticity of the complexly dilated boundary triple can be seen as in Theorem 3.5.6.

We do not provide a general theory of boundary triples associated with sectorial quadratic forms and operators here, for this purpose see e.g. [Ar00].

Note that the domain of the Neumann operator $H^{\theta,\mathrm{N}}$ for a *coupled dilated* boundary triples as defined in Sect. 3.6.2, which plays the role of the *coupled*

[12] We restrict ϑ to $[0, \pi/2)$ in order to obtain *sectorial* operators like $H^{\theta,\mathrm{D}}$ of Proposition 3.5.14 having its spectrum lying in a sector strictly contained in the right half-plane. Some results remain true also for $\pi/2 \le \vartheta < \pi$.

operator will *depend* on θ, causing some technical trouble when showing the holomorphic dependence on θ in Sect. 3.7.

For the rest of this section we assume that $0 \le \vartheta < \vartheta_0 < \pi/2$,

$$\theta \in \mathrm{S}_\vartheta = \{\theta \in \mathbb{C} \,|\, |\mathrm{Im}\,\theta| \le \vartheta/2\} \quad \text{and} \quad z \in \mathbb{C} \setminus \Sigma_{\vartheta_0} = \{z \in \mathbb{C} \,|\, |\arg z| > \vartheta_0\},$$
$$\text{where} \quad \Sigma_{\vartheta_0} := \{z \in \mathbb{C} \,|\, |\arg z| \le \vartheta_0\}, \tag{3.58}$$

if not otherwise stated. Then the sum in the dilated Dirichlet solution operator given in (3.55b) converges in $\mathscr{W}^2$ (resp. in $\mathsf{H}^m(\mathbb{R}_+) \otimes \mathscr{G}$), as in the proof of Proposition 3.5.4.

Remark 3.5.12. Note that we need to restrict z to $\mathbb{C} \setminus \Sigma_{\vartheta_0} \subset \mathbb{C} \setminus \Sigma_\vartheta$ in order that the dilated Dirichlet solution operator converges as in Proposition 3.5.4 (replacing ϑ by $\vartheta_0 - \vartheta$).

Let us now look at the holomorphic dependence on θ and give some precise bounds on certain operators associated with the complexly dilated boundary triple. We first need the following extension of Lemma 3.4.16 for the dilated case:

Lemma 3.5.13. *For* $z, w \in \mathbb{C} \setminus \Sigma_\vartheta$ *and* $\theta \in \mathrm{S}_\vartheta$ *set*

$$C_\theta(z,w) := \sup_{\lambda,\kappa \ge 0} \frac{|\lambda + \kappa - w|}{|\mathrm{e}^{-2\theta}\lambda + \kappa - z|}, \tag{3.59a}$$

$$C^\theta(z,w) := \sup_{\lambda,\kappa \ge 0} \frac{|\mathrm{e}^{-2\theta}\lambda + \kappa - w|}{|\lambda + \kappa - z|}. \tag{3.59b}$$

Then $C_\theta(z,w)$ *and* $C^\theta(z,w)$ *fulfil*

$$1 \le C_\theta(z,w), C^\theta(z,w) < \infty. \tag{3.59c}$$

Both constants depend only on $\mathrm{Im}\,\theta$ *and the dependence is continuous. In particular, the constants*

$$C_{\le\vartheta}(z,w) := \sup_{\theta \in \mathrm{S}_\vartheta} C_\theta(z,w) \qquad \textit{and} \qquad C^{\le\vartheta}(z,w) := \sup_{\theta \in \mathrm{S}_\vartheta} C^\theta(z,w) \tag{3.59d}$$

are monotonous in ϑ *and fulfil the bounds*

$$1 \le C_{\le\vartheta}(z,w), C^{\le\vartheta}(z,w) < \infty. \tag{3.59e}$$

Proof. In order to see the boundedness note that

$$\mathrm{e}^{-2|\mathrm{Re}\,\theta|} \le \liminf_{\lambda,\kappa\to\infty} \frac{|\lambda + \kappa - w|}{|\mathrm{e}^{-2\theta}\lambda + \kappa - z|} \le 1 \le \limsup_{\lambda,\kappa\to\infty} \frac{|\lambda + \kappa - w|}{|\mathrm{e}^{-2\theta}\lambda + \kappa - z|} \le \mathrm{e}^{2|\mathrm{Re}\,\theta|}.$$

In particular, $C_\theta(z,w)$ fulfils the bounds (3.59c). Moreover, replacing λ by $\lambda' = e^{-2\operatorname{Re}\theta}$ one can see that $C_\theta(z,w) = C_{i\operatorname{Im}\theta}(z,w)$, and the latter depends continuously on $\operatorname{Im}\theta$. In particular, $C_{\leq\vartheta}(z,w)$ fulfils the same bounds and is monotonous. The assertions for $C^\theta(z,w)$ follow similarly. □

We collect some obvious facts about the Dirichlet operator. Similar assertions hold for the Neumann operator $H^{\theta,\mathrm{N}}$:

Proposition 3.5.14.

1. *Let $H^{\theta,\mathrm{D}}$ be the Dirichlet operator associated with the dilated boundary triple $(\Gamma^\theta, -\Gamma'^\theta, \mathscr{G})$, then $H^{\theta,\mathrm{D}}$ is given by*

$$H^{\theta,\mathrm{D}} = -e^{-2\theta}\partial_{ss}^{\mathrm{D}} \otimes \mathrm{id} + \mathrm{id} \otimes K, \tag{3.60a}$$

where $-\partial_{ss}^{\mathrm{D}}$ is the Laplacian on $\mathbb{R}_+$ with Dirichlet condition at 0. Moreover, the domain $\operatorname{dom} H^{\theta,\mathrm{D}} = \operatorname{dom} H^{\mathrm{D}}$ is independent of θ.
2. *The operator $H^{\theta,\mathrm{D}}$ is normal, i.e. commutes with its adjoint.*
3. *The operator $H^{\theta,\mathrm{D}}$ commutes with the undilated operator $H^{\mathrm{D}} = H^{0,\mathrm{D}}$.*
4. *The family $\{H^{\theta,\mathrm{D}}\}_\theta$ is self-adjoint, i.e. $(H^{\theta,\mathrm{D}})^* = H^{\overline{\theta},\mathrm{D}}$.*
5. *The family $\{H^{\theta,\mathrm{D}}\}_\theta$ depends holomorphically on $\theta \in \mathrm{S}_\vartheta$, $0 < \vartheta < \pi/2$, i.e. the family is of type A (see [K66, Sect. VII.2]).*
6. *The spectrum of $H^{\theta,\mathrm{D}}$ is purely essential and given by*

$$\sigma(H^{\theta,\mathrm{D}}) = \sigma(K) + e^{-2\operatorname{Im}\theta}[0,\infty) = \{\kappa + e^{-2\operatorname{Im}\theta}\lambda \mid \kappa \in \sigma(K),\ \lambda \geq 0\},$$

and is contained in the sector Σ_ϑ. Moreover, the resolvent $R^{\theta,\mathrm{D}}(z) = (H^{\theta,\mathrm{D}} - z)^{-1}$ fulfils the norm estimate

$$\|R^{\theta,\mathrm{D}}(z)\| \leq \frac{1}{\operatorname{dist}(z, \mathbb{C} \setminus \Sigma_\vartheta)} =: \gamma_\vartheta^{\mathrm{D}}(z), \quad \text{i.e.} \quad H^{\theta,\mathrm{D}} \in \mathscr{R}(\gamma_\vartheta^{\mathrm{D}}, \mathbb{C} \setminus \Sigma_\vartheta)$$

(see Definition 3.3.3).
7. *The resolvent as operator*

$$R^{\theta,\mathrm{D}}(z) \colon \mathscr{H}^{k,\mathrm{D}} \longrightarrow \mathscr{H}^{k+2,\mathrm{D}}$$

is bounded by $C_{\leq\vartheta}(z,-1)$ for $z \notin \Sigma_\vartheta$ and $\theta \in \mathrm{S}_\vartheta$ (see Lemma 3.5.13), where $\mathscr{H}^{k,\mathrm{D}}$ is the scale of Hilbert spaces associated with $H^{\mathrm{D}} = H^{0,\mathrm{D}}$. In particular,

$$\|R^{\theta,\mathrm{D}}(z)\|_{-1\to 1} \leq C_{\leq\vartheta}(z,-1) =: \gamma_\vartheta^{\mathrm{D},1}(z), \quad \text{i.e.} \quad H^{\theta,\mathrm{D}} \in \mathscr{R}_1(\gamma_\vartheta^{\mathrm{D},1}, \mathbb{C} \setminus \Sigma_\vartheta)$$

(see Definition 3.3.10).

Proof. The assertions are easily seen. For example, the explicit expression of $H^{\theta,\mathrm{D}}$ in (3.60a) guarantees the holomorphy as well as the commuting properties. Moreover,

$$\sigma(H^{\theta,\mathrm{D}}) \subset \{\mathrm{e}^{-2\theta}\lambda + \kappa \mid \lambda, \kappa \geq 0\} \subset \Sigma_\vartheta.$$

The boundedness of the resolvent in (7) follows from

$$\|R^{\theta,\mathrm{D}}(z)\|_{k\to k+2} = \|(H^{\mathrm{D}} + 1)R^{\theta,\mathrm{D}}(z)\| \leq C_\theta(z, -1),$$

where we also used (3). □

Due to (3), we can use the scale of Hilbert spaces associated with H^{D}. In particular, we do not need a compatibility operator as in Sect. 3.3, since the domain of $H^{\theta,\mathrm{D}} = \mathscr{H}^{1,\mathrm{D}}$ is independent of θ.

The dilated Dirichlet solution operator $S^\theta(z)$ maps $\mathscr{G}^{1/2}$ into

$$\mathscr{N}^{1,\theta}(z) := \{v \in \mathscr{H}^1 \mid \mathfrak{h}^\theta(u, v) = z\langle u, v\rangle \ \forall\, u \in \mathscr{H}^{1,\mathrm{D}}\}, \tag{3.61}$$

the *weak solution space*.

We have an expression for the solution operator as in Proposition 3.4.17:

Lemma 3.5.15. *The Dirichlet solution operator $S^\theta(z)\colon \mathscr{G}^{1/2} \longrightarrow \mathscr{H}^1$ can be expressed as*

$$S^\theta(z) = \mathrm{e}^{\theta/2}\big(\mathrm{id}_{\mathscr{H}^1} - \iota_{\mathrm{DN}} R^{\theta,\mathrm{D}}(z)\iota_{\mathrm{DN}}^*(H^{\theta,\mathrm{N}} - z)\big)S, \tag{3.62}$$

where S is the Dirichlet solution operator in $z = -1$ for the undilated boundary triple (see Definition 3.4.6), and where $\iota_{\mathrm{DN}}\colon \mathscr{H}^{1,\mathrm{D}} \longrightarrow \mathscr{H}^1$ is the natural embedding.

Moreover, the above chain of operators is bounded by

$$\|S^\theta(z)\|_{1/2\to 1} \leq \mathrm{e}^{\operatorname{Re}\theta/2}\big(1 + C_{\leq\vartheta}(z, -1)C^{\leq\vartheta}(-1, z)\big).$$

Here, $\mathscr{G}^{1/2}$ carries the norm of the undilated boundary triple (see Lemma 3.5.13 for the definition of the constants).

Let us now summarise the assertions on holomorphic dependency on θ and boundedness of certain dilated operators as follows:

Proposition 3.5.16. *Assume that the transversal operator K has purely discrete spectrum. Then the adjoint of*

$$B^\theta(z) = \mathrm{e}^{-3\theta/2}\Gamma'\iota_{\mathrm{ell,D}} R^{\theta,\mathrm{D}}(z)\colon \mathscr{H} \longrightarrow \mathscr{G}$$

extends the Dirichlet solution operator $S^\theta(z)\colon \mathscr{G}^{1/2} \longrightarrow \mathscr{H}^1$. Moreover, the adjoint of the Dirichlet solution operator $S^{\overline{\theta}}(\overline{z})\colon \mathscr{G}^{1/2} \longrightarrow \mathscr{H}^1$ extends $B^\theta(z)$. In addition, the following operators are bounded and depend holomorphically on $\theta \in \mathrm{S}_\vartheta$ as operators

$$B^\theta(z)\colon \mathscr{H} \longrightarrow \mathscr{G}, \qquad B^{\overline{\theta}}(\overline{z})^*\colon \mathscr{G} \longrightarrow \mathscr{H},$$

$$B^\theta(z)\colon \mathscr{H} \longrightarrow \mathscr{G}^{1/2}, \qquad S^\theta(z)\colon \mathscr{G}^{3/2} \longrightarrow \mathscr{W}^2,$$

$$S^{\overline{\theta}}(\overline{z})^*\colon \mathscr{H}^{-1} \longrightarrow \mathscr{G}^{-1/2}, \qquad S^\theta(z)\colon \mathscr{G}^{1/2} \longrightarrow \mathscr{H}^1.$$

Proof. The boundedness and holomorphic dependence of $B^\theta(z)\colon \mathscr{H} \longrightarrow \mathscr{G}$ follows by ellipticity of the boundary triple. Moreover, that the adjoint of $B^{\overline{\theta}}(\overline{z})$ extends the Dirichlet solution operator follows similarly as in Lemma 3.4.29, namely if $\varphi \in \mathscr{G}^{1/2}$, then

$$B^{\overline{\theta}}(\overline{z})^*\varphi = S^\theta(z)\varphi.$$

In particular, the operators in the first line of the second displayed formula are bounded. The boundedness of the operators in the second line is ensured by the ellipticity. The holomorphic dependence of the operators in the second line follows from the holomorphic dependence of the operators in the first line and the fact that an operator-valued function is holomorphic w.r.t. the operator topology iff it is holomorphic w.r.t. the weak topology (see e.g. [K66, Thm. III.3.12]). Indeed, it suffices to consider the weak holomorphy only on a dense subset in the Hilbert space, since the operators depend continuously on θ, and therefore are locally bounded (cf. the remark after Thm. III.3.12 in [K66]): Denote by A the strictly positive operator associated with the 1-densely embedded space $\mathscr{W}^2$ into $\mathscr{H}$ (see Lemma 3.1.6). Then

$$\langle u, S^\theta(z)\varphi\rangle_{\mathscr{W}^2} = \langle Au, S^\theta(z)\varphi\rangle_{\mathscr{H}} = \langle B^{\overline{\theta}}(\overline{z})Au, \varphi\rangle_{\mathscr{H}}$$

for $u \in \operatorname{dom} A \subset \mathscr{W}^2$ and $\varphi \in \mathscr{G}^{3/2}$. But the latter depends holomorphically on θ by the explicit expression for $B^{\overline{\theta}}(\overline{z})$ and Proposition 3.5.14. Moreover, since A is strictly positive, $\operatorname{ran} A = \mathscr{H}$ and therefore $S^\theta(z)\colon \mathscr{G}^{3/2} \longrightarrow \mathscr{W}^2$ is holomorphic. The holomorphy of $B^\theta(z)\colon \mathscr{H} \longrightarrow \mathscr{G}^{1/2}$ follows similarly.

Finally, the boundedness in the third line follow from Lemma 3.5.15, and the holomorphic dependence can be seen similarly as below. □

We will need similar properties as above for the scale of Hilbert spaces on $\mathscr{G}$ associated with a *coupled* Dirichlet-to-Neumann map as defined in the following section.

3.6 Coupled Boundary Triples and Dilation

In this section we define a new boundary triple by coupling two boundary triples having the same boundary space $\mathscr{G}$. If one of the boundary triples is a half-line model, we can define the associated dilated, coupled boundary triple. The associated Neumann operator H^θ fulfils jump conditions at the passage from one boundary triple to the other depending on a (in this section) real parameter θ.

3.6.1 Coupled Boundary Triples

We first define a coupled boundary triple associated with a coupled quadratic form. Note that the construction for *ordinary* boundary triples is well-known, see for example [DHMS00, Sect. 5] and the references therein. Let $(\Gamma_{\text{int}}, \Gamma'_{\text{int}}, \mathscr{G})$ be a boundary triple associated with a quadratic form $\mathfrak{h}_{\text{int}} \geq 0$ on $\mathscr{H}_{\text{int}}$ with operator $\check{H}_{\text{int}}$ as in Sect. 3.4, the *interior boundary triple*. Moreover, let $(\Gamma_{\text{ext}}, -\Gamma'_{\text{ext}}, \mathscr{G})$ be a half-line boundary triple associated with the transversal quadratic form $\mathfrak{k}$ as in Sect. 3.5 having the same boundary space $\mathscr{G}$ as the interior boundary triple. The underlying Hilbert space is $\mathscr{H}_{\text{ext}} = \mathsf{L}_2(\mathbb{R}_+) \otimes \mathscr{G}$. We call $(\Gamma_{\text{ext}}, -\Gamma'_{\text{ext}}, \mathscr{G})$ the *exterior boundary triple*. We assume that K has purely discrete spectrum. The results of this subsection also hold for more general boundary triples $(\Gamma_{\text{ext}}, -\Gamma'_{\text{ext}}, \mathscr{G})$ (see Example 3.9.6).

Remark 3.6.1. We use the subscripts $(\cdot)_{\text{int}}$ or $(\cdot)_{\text{ext}}$ for the corresponding objects of the interior and exterior boundary triple. If it is clear from the context which boundary triple is meant, we often omit the subscripts. For example, we write $\Gamma'_{\text{int}} u$ instead of $\Gamma'_{\text{int}} u_{\text{int}}$ etc.

The aim of the following section is to define a so-called *coupled* boundary triple $(\Gamma, \Gamma', \mathscr{G})$ associated with a *coupled* quadratic form $\mathfrak{h}$. The Hilbert space of the coupled boundary triple is defined by

$$\mathscr{H} := \mathscr{H}_{\text{int}} \oplus \mathscr{H}_{\text{ext}}. \tag{3.63a}$$

For further references, we also define the decoupled spaces

$$\mathscr{H}^{1,\text{dec}} := \mathscr{H}^1_{\text{int}} \oplus \mathscr{H}^1_{\text{ext}} \qquad \text{and} \qquad \mathscr{W}^{2,\text{dec}} := \mathscr{W}^2_{\text{int}} \oplus \mathscr{W}^2_{\text{ext}}. \tag{3.63b}$$

Moreover, we denote by $H^{\text{D}}_\bullet$ and $H^{\text{N}}_\bullet$ the corresponding parts $\bullet = \text{int}, \text{ext}$ (see Definition 3.4.2) and the associated Hilbert scales by $\mathscr{H}^{k,\text{D}}_\bullet$ and $\mathscr{H}^{k,\text{N}}_\bullet$, respectively. The corresponding spaces

$$\mathscr{H}^{k,\text{D}} := \mathscr{H}^{k,\text{D}}_{\text{int}} \oplus \mathscr{H}^{k,\text{D}}_{\text{ext}} \qquad \text{and} \qquad \mathscr{H}^{k,\text{N}} := \mathscr{H}^{k,\text{N}}_{\text{int}} \oplus \mathscr{H}^{k,\text{N}}_{\text{ext}} \tag{3.63c}$$

are decoupled. Note that $\mathscr{H}^{1,\text{N}} = \mathscr{H}^{1,\text{dec}}$.

The coupled quadratic form $\mathfrak{h}$ is defined by

$$\mathfrak{h}(u) := \mathfrak{h}_{\text{int}}(u) + \mathfrak{h}_{\text{ext}}(u)$$

for $u \in \operatorname{dom} \mathfrak{h} := \mathscr{H}^1$ and

$$\mathscr{H}^1 := \{ u \in \mathscr{H}^{1,\text{dec}} \mid \Gamma_{\text{int}} u = \Gamma_{\text{ext}} u \}. \tag{3.63d}$$

On $\mathscr{H}^1$, we define the boundary map

$$\Gamma \colon \mathscr{H}^1 \longrightarrow \mathscr{G}, \qquad \Gamma u = \Gamma_{\mathrm{int}} u = \Gamma_{\mathrm{ext}} u,$$

since the latter are equal. The operator $\check{H}$ is defined on

$$\mathscr{W}^2 := \{ u \in \mathscr{W}^{2,\mathrm{dec}} \mid \Gamma_{\mathrm{int}} u = \Gamma_{\mathrm{ext}} u \} \tag{3.63e}$$

and given by $\check{H} u = \check{H}_{\mathrm{int}} u \oplus \check{H}_{\mathrm{ext}} u$. Finally,

$$\Gamma' \colon \mathscr{W}^2 \longrightarrow \mathscr{G}, \qquad \Gamma' u := \Gamma'_{\mathrm{int}} u - \Gamma'_{\mathrm{ext}} u. \tag{3.63f}$$

Proposition 3.6.2. *Assume that* $(\Gamma_{\mathrm{int}}, \Gamma'_{\mathrm{int}}, \mathscr{G})$ *and* $(\Gamma_{\mathrm{ext}}, -\Gamma'_{\mathrm{ext}}, \mathscr{G})$ *are the two boundary triples associated with* $\mathfrak{h}_{\mathrm{int}}$ *and* $\mathfrak{h}_{\mathrm{ext}}$, *and that* $(\Gamma, \Gamma', \mathscr{G})$ *with the associated objects are defined as in* (3.63).

1. *The coupled space* $\mathscr{H}^1$ *is closed in* $\mathscr{H}^{1,\mathrm{dec}}$. *In particular, the quadratic form* $\mathfrak{h} = (\mathfrak{h}_{\mathrm{int}} \oplus \mathfrak{h}_{\mathrm{ext}}) \restriction_{\mathscr{H}^1}$ *is closed.*
2. *Assume that*
$$\mathscr{G}^{1/2} := \operatorname{ran} \Gamma_{\mathrm{int}} \cap \operatorname{ran} \Gamma_{\mathrm{ext}} \tag{3.64a}$$
is dense in $\mathscr{G}$, *then* $(\Gamma, \Gamma', \mathscr{G})$ *is a boundary triple associated with* $\mathfrak{h}$, *called* coupled boundary triple.
3. *The associated Dirichlet operator* $H^{\mathrm{D}} = H^{\mathrm{D}}_{\mathrm{int}} \oplus H^{\mathrm{D}}_{\mathrm{ext}}$ *is decoupled. Moreover, the associated Neumann operator* $H = H^{\mathrm{N}}$ *has*
$$\mathscr{W}^{2,\mathrm{N}} := \{ u \in \mathscr{W}^{2,\mathrm{dec}} \mid \Gamma_{\mathrm{ext}} u = \Gamma_{\mathrm{int}} u, \quad \Gamma'_{\mathrm{ext}} u = \Gamma'_{\mathrm{int}} u \}$$
in its domain, i.e., $\mathscr{W}^{2,\mathrm{N}} \subset \mathscr{H}^2 := \mathscr{H}^{2,\mathrm{N}} := \operatorname{dom} H^{\mathrm{N}}$.
4. *The associated Dirichlet solution operator is given by*
$$S(z)\varphi = S_{\mathrm{int}}(z)\varphi \oplus S_{\mathrm{ext}}(z)\varphi \tag{3.64b}$$
for $\varphi \in \mathscr{G}^{1/2}$.
5. *The associated Dirichlet-to-Neumann operator is given by*
$$\Lambda = \Lambda_{\mathrm{int}} + \Lambda_{\mathrm{ext}}, \tag{3.64c}$$
where $S_\bullet(z)$ *and* $\Lambda_\bullet$ *are the Dirichlet solution and Dirichlet-to-Neumann operators associated with the interior and exterior boundary triple, respectively. In particular, if the transversal operator* K *of the exterior boundary triple has purely discrete spectrum, then* $\Lambda(z) = \Lambda_{\mathrm{int}}(z) - \mathrm{i}\sqrt{z - K}$ *by Corollary 3.5.5.*
6. *The coupled boundary triple is bounded iff the interior and exterior boundary triples are bounded.*

Proof. (1) The operators $\Gamma_\bullet \colon \mathscr{H}^1_\bullet \longrightarrow \mathscr{G}$ are bounded so $\mathscr{H}^1$ is a closed subspace of $\mathscr{H}^{1,\mathrm{dec}}$ and therefore $\mathfrak{h}$ a closed form. Recall that $\mathfrak{h}_\bullet$ is assumed to be a closed form on $\mathscr{H}^1_\bullet$, i.e. $\mathscr{H}^1_\bullet$ with its natural quadratic form norm is a Hilbert space.

(2) The boundary operator $\Gamma' \colon \mathscr{W}^2 \longrightarrow \mathscr{G}$ is bounded, since Γ'_{int} and Γ'_{ext} are. Moreover, we have

$$\begin{aligned} \mathfrak{h}(u,v) &= \mathfrak{h}_{\text{int}}(u,v) + \mathfrak{h}_{\text{ext}}(u,v) \\ &= \langle u, \check{H}_{\text{int}} v\rangle_{\text{int}} + \langle u, \check{H}_{\text{ext}} v\rangle_{\text{ext}} + \langle \Gamma u, \Gamma'_{\text{int}} v - \Gamma'_{\text{ext}} v\rangle_{\mathscr{G}} \\ &= \langle u, \check{H}_{\text{int}} v\rangle_{\text{int}} + \langle u, \check{H}_{\text{ext}} v\rangle_{\text{ext}} + \langle \Gamma u, \Gamma' v\rangle_{\mathscr{G}} \end{aligned}$$

for $u \in \operatorname{dom} \mathfrak{h}$ and $v \in \mathscr{W}^2$ using Green's formula (3.32b) for the individual boundary triples, so that Green's formula for the coupled triple holds. Next, $\operatorname{ran} \Gamma = \operatorname{ran} \Gamma_{\text{int}} \cap \operatorname{ran} \Gamma_{\text{ext}}$ is dense in $\mathscr{G}$, as well as

$$\ker \Gamma = \ker \Gamma_{\text{int}} \oplus \ker \Gamma_{\text{ext}} \qquad \text{and} \qquad \ker \Gamma' \supset \mathring{\mathscr{W}}^2_{\text{int}} \oplus \mathring{\mathscr{W}}^2_{\text{ext}}$$

are dense in $\mathscr{H}$, since the corresponding property is true for the interior and exterior triple. Finally, it is easily seen that $\check{H} \colon \mathscr{W}^2 \longrightarrow \mathscr{H}$ is bounded, since $\check{H}_\bullet \colon \mathscr{W}^2_\bullet \longrightarrow \mathscr{H}_\bullet$ are bounded.

(3) The assertion on the Dirichlet operator is clear. The assertion on the Neumann operator domain is obvious from Proposition 3.4.5, since $\mathscr{W}^{2,\mathrm{N}} = \ker \Gamma'$.

(4)–(5) The formula for the Dirichlet solution operator is easily seen. In particular, $S^{*1} u = S^{*1}_{\text{int}} u + S^{*1}_{\text{ext}} u$ for $u \in \mathscr{H}^1$, and therefore (3.64c) follows.

(6) Finally, Λ is bounded iff Λ_{int} and Λ_{ext} are bounded, and by Proposition 3.4.11, the same is true for the boundary triples. □

Denote by $\mathscr{G}^k_\bullet$ and $\mathscr{G}^k$ the scale of Hilbert spaces associated with $\Lambda_\bullet$ and $\Lambda = \Lambda_{\text{int}} + \Lambda_{\text{ext}}$, respectively, where $\bullet = \text{int}, \text{ext}$. Let us relate the different scales of Hilbert spaces on $\mathscr{G}$:

Lemma 3.6.3. *We have*

$$\|\varphi\|^2_{\mathscr{G}^{1/2}} = \|S\varphi\|^2_{\mathscr{H}^1} = \|\varphi\|^2_{\mathscr{G}^{1/2}_{\text{int}}} + \|\varphi\|^2_{\mathscr{G}^{1/2}_{\text{ext}}} \ge \|\varphi\|^2_{\mathscr{G}^{1/2}_\bullet}, \tag{3.65}$$

and in particular, the natural embeddings

$$\begin{array}{ccccc} \mathscr{G}^{1/2} & \hookrightarrow & \mathscr{G} & \twoheadrightarrow & \mathscr{G}^{-1/2} \\ \downarrow & & \| & & \uparrow \\ \mathscr{G}^{1/2}_\bullet & \hookrightarrow & \mathscr{G} & \twoheadrightarrow & \mathscr{G}^{-1/2}_\bullet \end{array}$$

are bounded by 1.

Proof. The result follows from

$$\|\varphi\|^2_{\mathscr{G}^{1/2}} = \|S\varphi\|^2_{\mathscr{H}^1} = \|S_{\text{int}}\varphi\|^2_{\mathscr{H}^1_{\text{int}}} + \|S_{\text{ext}}\varphi\|^2_{\mathscr{H}^1_{\text{ext}}} = \|\varphi\|^2_{\mathscr{G}^{1/2}_{\text{int}}} + \|\varphi\|^2_{\mathscr{G}^{1/2}_{\text{ext}}},$$

using the definition of the norm on $\mathscr{G}^{1/2}$ (Definition 3.4.8). □

Remark 3.6.4.

1. In general, we can not expect to have other relations between the different scales $\mathscr{G}^m$ and $\mathscr{G}^m_\bullet$ than the one for $m = 1/2$ in the above lemma. Note that $(0 \le)\ \Lambda_\bullet \le \Lambda$ does not in general imply that $\Lambda_\bullet^k \le \Lambda^k$ for $k > 1$, i.e.

$$\|\varphi\|^2_{\mathscr{G}_\bullet^{k/2}} = \langle \varphi, \Lambda_\bullet^k \varphi\rangle \le \langle \varphi, \Lambda^k \varphi\rangle = \|\varphi\|^2_{\mathscr{G}^{k/2}},$$

since $(\cdot)^k$ is not operator monotone. On the other hand, the function $(\cdot)^k$ is operator monotone for $0 < k \le 1$, and therefore, we have a similar diagram with $1/2$ replaced by $0 < m = k/2 \le 1/2$.
2. Note that ellipticity for the individual boundary triples does not imply in general the ellipticity for the coupled boundary triple, since the scales of Hilbert spaces on $\mathscr{G}$ associated with Λ_{int}, Λ_{ext} and $\Lambda = \Lambda_{\text{int}} + \Lambda_{\text{ext}}$ may be different, as we have seen above, and there is (in general) no relation between the scales of order $|m| > 1/2$.

   But if we know that a priori that (3.66a) holds for some constants $0 < c \le C$, then we can conclude by operator monotonicity that the scale $\mathscr{G}^m$ associated with the coupled Dirichlet-to-Neumann operator Λ agrees with the scale $\mathscr{G}^m_\bullet$ associated with $\Lambda_\bullet$ for $|m| \le 3/2$, moreover, the m-norms are equivalent. Note that then also condition (3.64a) is fulfilled since $\mathscr{G}^{1/2} = \mathscr{G}^{1/2}_\bullet = \operatorname{ran} \Gamma_\bullet$.

In order to assure the ellipticity for the coupled boundary triple (see Definition 3.4.21), we therefore have to assure that the scales of Hilbert spaces $\mathscr{G}^m_\bullet$ and $\mathscr{G}^m$ agree for $m = 3/2$ and therefore for all $|m| \le 3/2$.

Theorem 3.6.5. *Assume that $(\Gamma_{\text{int}}, \Gamma'_{\text{int}}, \mathscr{G})$ is an elliptic boundary triple and that K has purely discrete spectrum (so that $(\Gamma_{\text{ext}}, -\Gamma'_{\text{ext}}, \mathscr{G})$ is also elliptic by Theorem 3.5.6). If*

$$(c\Lambda)^3 \le \Lambda_\bullet^3 \le (C\Lambda)^3 \tag{3.66a}$$

holds for some constants $0 < c \le C$, and if in addition, the domain of the coupled operator H (the Neumann operator for the coupled triple) fulfils

$$\mathscr{H}^{2,\text{N}} = \operatorname{dom} H \subset \mathscr{W}^2, \tag{3.66b}$$

then the coupled boundary triple is elliptic. Moreover, for $z \in \mathbb{C} \setminus \mathbb{R}_+$, the coupled (strong) Dirichlet solution operator is given by

$$S(z)\colon \mathscr{G}^{3/2} \longrightarrow \mathscr{W}^2, \qquad S(z)\varphi = S_{\text{int}}(z)\varphi \oplus S_{\text{ext}}(z)\varphi \tag{3.66c}$$

for $\varphi \in \mathscr{G}^{3/2}$, the coupled Neumann solution operator reads as

$$S^{\mathrm{N}}(z)\colon \mathscr{G}^{1/2} \longrightarrow \mathscr{W}^2, \quad S^{\mathrm{N}}(z)\psi = S^{\mathrm{N}}_{\mathrm{int}}(z)\Lambda_{\mathrm{int}}\Lambda^{-1}\psi \oplus S^{\mathrm{N}}_{\mathrm{ext}}(z)\Lambda_{\mathrm{ext}}\Lambda^{-1}\psi$$
$$= S_{\mathrm{int}}(z)\Lambda^{-1}\psi \oplus S_{\mathrm{ext}}(z)\Lambda^{-1}\psi \qquad (3.66\mathrm{d})$$

for $\psi \in \mathscr{G}^{1/2}$ and the coupled Dirichlet-to-Neumann operator is

$$\Lambda(z) = \Lambda_{\mathrm{int}}(z) + \Lambda_{\mathrm{ext}}(z)\colon \mathscr{G}^{3/2} \longrightarrow \mathscr{G}^{1/2}. \qquad (3.66\mathrm{e})$$

Moreover, all these operators are bounded.

Proof. For the ellipticity, we check the assumptions of Proposition 3.4.28: First, we have to ensure that the domains of the Dirichlet and Neumann operators are actually subsets of $\mathscr{W}^2$. Since $\operatorname{dom} H^{\mathrm{D}} = \operatorname{dom} H^{\mathrm{D}}_{\mathrm{int}} \oplus \operatorname{dom} H^{\mathrm{D}}_{\mathrm{ext}}$, and since both boundary triples are elliptic, we have

$$\operatorname{dom} H^{\mathrm{D}} = (\mathscr{W}^2_{\mathrm{int}} \cap \ker \Gamma_{\mathrm{int}}) \oplus (\mathscr{W}^2_{\mathrm{ext}} \cap \ker \Gamma_{\mathrm{ext}}) = \mathscr{W}^2 \cap \ker \Gamma \subset \mathscr{W}^2.$$

The corresponding assertion for the Neumann operator is part of our assumption. Moreover,

$$\|\Gamma u\|_{\mathscr{G}^{3/2}} \le c^{-3/2}\|\Gamma_{\mathrm{ext}} u\|_{\mathscr{G}^{3/2}_{\mathrm{ext}}} \le c^{-3/2}\|u\|_{\mathscr{W}^2_{\mathrm{ext}}} \le c^{-3/2}\|u\|_{\mathscr{W}^2}$$

(see the proof of Theorem 3.5.6 for the second inequality), and similarly

$$\|\Gamma' u\|_{\mathscr{G}^{1/2}} \le c^{-1/2}\big(\|\Gamma'_{\mathrm{int}} u\|_{\mathscr{G}^{1/2}_{\mathrm{int}}} + \|\Gamma'_{\mathrm{ext}} u\|_{\mathscr{G}^{1/2}_{\mathrm{ext}}}\big)$$
$$\le c^{-1/2}(\|\Gamma'_{\mathrm{int}}\|_{2\to 1/2} + 1)\big(\|u\|_{\mathscr{W}^2_{\mathrm{int}}} + \|u\|_{\mathscr{W}^2_{\mathrm{ext}}}\big)$$
$$\le c^{-1/2}\sqrt{2}(\|\Gamma'_{\mathrm{int}}\|_{2\to 1/2} + 1)\|u\|_{\mathscr{W}^2}.$$

Finally, let $\varphi \in \mathscr{G}^{3/2}$. It follows from (3.66a) that $\varphi \in \mathscr{G}^{3/2}_{\mathrm{int}} \cap \mathscr{G}^{3/2}_{\mathrm{ext}}$. Moreover, by the ellipticity of the individual boundary triples, $S_{\mathrm{int}}\varphi \in \mathscr{W}^2_{\mathrm{int}}$ and $S_{\mathrm{ext}}\varphi \in \mathscr{W}^2_{\mathrm{ext}}$. In particular, we have $S\varphi = S_{\mathrm{int}}\varphi \oplus S_{\mathrm{ext}}\varphi \in \mathscr{W}^2$. The ellipticity now follows from Proposition 3.4.28. □

Remark 3.6.6. We would like to conclude the *coupled* ellipticity only from the individual ellipticity and the compatibility condition (3.66a), i.e., we would like to drop the condition (3.66b). By a slightly different approach to boundary triples we can prove Krein's resolvent formula $R^{\mathrm{N}} = R^{\mathrm{D}} + S\Lambda^{-1}B\colon \mathscr{H} \longrightarrow \mathscr{H}$ (see (3.45a)) without using ellipticity (namely $\operatorname{dom} H^{\mathrm{N}} \subset \mathscr{W}^2$; note that the proof of (3.45a) uses this assumption in a slightly hidden way: we assumed there that $v = R^{\mathrm{N}}(z)\widetilde{v} \in \mathscr{W}^2$ in order to apply Green's formula). Then we have

$$\operatorname{dom} H^{\mathrm{N}} = \operatorname{ran} R^{\mathrm{N}} \subset \operatorname{ran} R^{\mathrm{D}} + \operatorname{ran} S\Lambda^{-1}S^*$$

for the coupled operators, and the ellipticity of the individual boundary triples ensures that the latter space is included in $\mathscr{W}^2$, since

$$S^* f = -\Gamma'_{\rm int} \iota_{\rm int,ell,D} R^{\rm D}_{\rm int} f_{\rm int} + \Gamma'_{\rm ext} \iota_{\rm ext,ell,D} R^{\rm D}_{\rm ext} f_{\rm ext} \in \mathscr{G}^{1/2} \xrightarrow{\Lambda^{-1}} \mathscr{G}^{3/2}$$

for $f \in \mathscr{H}$ and $S(\mathscr{G}^{3/2}) \subset \mathscr{W}^2$ by (3.66c). We will treat this approach in a future publication. In any case, the domain condition (3.66b) is fulfilled in our main examples, see Propositions 3.6.11 and 3.6.13.

Corollary 3.6.7. *Assume that* $(\Gamma_{\rm int}, \Gamma'_{\rm int}, \mathscr{G})$ *and* $(\Gamma_{\rm ext}, -\Gamma'_{\rm ext}, \mathscr{G})$ *are bounded elliptic boundary triples and that* $\operatorname{dom} H = \mathscr{H}^{2,\rm N} \subset \mathscr{W}^2$, *then the coupled boundary triple is elliptic.*

Proof. We only have to check (3.66a) from Theorem 3.6.5. But this is obvious since all Dirichlet-to-Neumann operators Λ, $\Lambda_{\rm int}$ and $\Lambda_{\rm ext}$ are bounded by Proposition 3.4.11 (3). □

3.6.2 Dilated Coupled Boundary Triples

Let us now define the dilated coupled boundary triple, first for dilations with *real* parameter θ. The main observation for dilated coupled boundary triples is, that the Dirichlet-to-Neumann operators associated with the dilated and undilated boundary triple *agree*.

The one-parameter unitary group of dilations on the coupled system defined by

$$U^\theta : \mathscr{H} \longrightarrow \mathscr{H}, \qquad U^\theta u = \mathrm{id}_{\rm int} \oplus U^\theta_{\rm ext}, \tag{3.67}$$

for $\theta \in \mathbb{R}$, where $U^\theta_{\rm ext} = \hat{U}^\theta \otimes \mathrm{id}_{\mathscr{G}}$ was defined in Definition 3.5.7.

The unitarily transformed coupled or *dilated* form $\mathfrak{h}^\theta$ is given by

$$\mathfrak{h}^\theta(u) := \mathfrak{h}(U^{-\theta} u), \qquad \operatorname{dom} \mathfrak{h}^\theta := U^\theta(\operatorname{dom} \mathfrak{h}). \tag{3.68}$$

Note that $\mathfrak{h}^0 = \mathfrak{h}$ is the original form. The proof of the following proposition is now straightforward:

Proposition 3.6.8. *Let* $\theta \in \mathbb{R}$, *then the dilated quadratic form* $\mathfrak{h}^\theta$ *is closed and given by*

$$\mathfrak{h}^\theta(u) = \mathfrak{h}_{\rm int}(u) + \mathfrak{h}^\theta_{\rm ext}(u), \qquad \mathfrak{h}^\theta_{\rm ext}(u) = \mathrm{e}^{-2\theta} \|u'\|^2 + \mathfrak{h}^\perp_{\rm ext}(u) \tag{3.69a}$$

for $u \in \operatorname{dom} \mathfrak{h}^\theta = \mathscr{H}^{1,\theta}$, *where*

$$\mathscr{H}^{1,\theta} = \{ u \in \mathscr{H}^{1,\rm dec} \mid \Gamma_{\rm ext} u = \mathrm{e}^{\theta/2} \Gamma_{\rm int} u \}. \tag{3.69b}$$

The associated operator is $H^\theta = U^\theta H U^{-\theta} \ge 0$ *and acts as*

$$H^\theta u = \check{H}_{\mathrm{int}} u \oplus \check{H}^\theta_{\mathrm{ext}} u, \quad \text{where} \quad \check{H}^\theta_{\mathrm{ext}} = -\mathrm{e}^{-2\theta} \partial_{ss} + \mathrm{id} \otimes K, \tag{3.69c}$$

and K *is the transversal operator associated with the half-line boundary triple,. Moreover,*

$$\operatorname{dom} H^\theta \supset \mathscr{W}^{2,\theta,\mathrm{N}} := \\ \{u \in \mathscr{W}^{2,\mathrm{dec}} \mid \Gamma_{\mathrm{ext}} u = \mathrm{e}^{\theta/2} \Gamma_{\mathrm{int}} u, \quad \Gamma'_{\mathrm{ext}} u = \mathrm{e}^{3\theta/2} \Gamma'_{\mathrm{int}} u\}. \tag{3.69d}$$

We can now define the associated *dilated coupled boundary triple* as follows:

Definition 3.6.9. Let $\theta \in \mathbb{R}$. The *dilated coupled* (or θ*-coupled*) *boundary triple* $(\Gamma^\theta, \Gamma'^\theta, \mathscr{G})$ associated with the quadratic form $\mathfrak{h}^\theta$ is defined by

$$\Gamma^\theta u := \Gamma_{\mathrm{int}} u = \Gamma^\theta_{\mathrm{ext}} u = \mathrm{e}^{-\theta/2} \Gamma_{\mathrm{ext}} u, \qquad u \in \mathscr{H}^{1,\theta}, \tag{3.70a}$$

$$\Gamma'^\theta u := \Gamma'_{\mathrm{int}} u - \Gamma'^\theta_{\mathrm{ext}} u = \Gamma'_{\mathrm{int}} u - \mathrm{e}^{-3\theta/2} \Gamma'_{\mathrm{ext}} u, \qquad u \in \mathscr{W}^{2,\theta}. \tag{3.70b}$$

The operator $\check{H}^\theta$ is given by

$$\check{H}^\theta u = \check{H}_{\mathrm{int}} u \oplus \check{H}^\theta_{\mathrm{ext}} u = \check{H}_{\mathrm{int}} u_{\mathrm{int}} \oplus (-\mathrm{e}^{-2\theta} u''_{\mathrm{ext}} + (\mathrm{id} \otimes K) u_{\mathrm{ext}}) \tag{3.70c}$$

for $u \in \mathscr{W}^{2,\theta}$, where

$$\mathscr{W}^{2,\theta} := \{u \in \mathscr{W}^{2,\mathrm{dec}} \mid \Gamma_{\mathrm{ext}} u = \mathrm{e}^{\theta/2} \Gamma_{\mathrm{int}} u\}$$

with norm as on $\mathscr{W}^{2,\mathrm{dec}}$.

The dilated coupled boundary triple is indeed a boundary triple:

Proposition 3.6.10. *Let* $\theta \in \mathbb{R}$ *and assume that* $\operatorname{ran} \Gamma_{\mathrm{int}} \cap \operatorname{ran} \Gamma_{\mathrm{ext}} = \operatorname{ran} \Gamma$ *is dense in* $\mathscr{G}$ *(see* (3.64a)*). Then the triple* $(\Gamma^\theta, \Gamma'^\theta, \mathscr{G})$ *associated with the (closed) quadratic form* $\mathfrak{h}^\theta$ *is a boundary triple.*

Moreover, the associated Dirichlet operator $H^{\theta,\mathrm{D}} = H^{\mathrm{D}}_{\mathrm{int}} \oplus H^{\theta,\mathrm{D}}_{\mathrm{ext}}$ *is decoupled. The corresponding (weak) Dirichlet solution operator is given by*

$$S^\theta(z)\varphi = S_{\mathrm{int}}(z)\varphi \oplus S^\theta_{\mathrm{ext}}(z)\varphi \tag{3.71a}$$

for $\varphi \in \mathscr{G}^{1/2}$ *and* $z \in \mathbb{C} \setminus \mathbb{R}_+$. *The associated Dirichlet-to-Neumann operator is given by*

$$\Lambda^\theta(z) = \Lambda(z) = \Lambda_{\mathrm{int}}(z) + \Lambda_{\mathrm{ext}}(z), \tag{3.71b}$$

where $\Lambda_\bullet(z)$ are the Dirichlet-to-Neumann operators associated with the interior and exterior boundary triple. In particular, the Dirichlet-to-Neumann operator is independent of θ.

Finally, the dilated coupled boundary triple is elliptic if the (undilated, i.e., $\theta = 0$) coupled boundary triple is elliptic. In particular, $\operatorname{dom} H^\theta = \mathscr{W}^{2,\theta,\mathrm{N}}$ *(see* (3.69d)*).*

Proof. The assertions follow from the fact that U^θ is unitary and that U^θ maps the undilated spaces onto the dilated ones. This is in particular true for the condition $\operatorname{dom} H^\theta \subset \mathscr{W}^{2,\theta}$ of the ellipticity. Moreover, the associated undilated and dilated Dirichlet-to-Neumann operators Λ and Λ^θ are equal (see Proposition 3.5.10), so that the corresponding scales of Hilbert spaces agree. □

Note that the assumptions of the above theorem are fulfilled in our main examples: the PDE case (for the notion $X = X_{\mathrm{int}} \mathbin{\overline{\cup}} X_{\mathrm{ext}}$ see Sect. 5.1.2) and the metric graph case.

Proposition 3.6.11. *Assume that $X = X_{\mathrm{int}} \mathbin{\overline{\cup}} X_{\mathrm{ext}}$ is a manifold with finitely many cylindrical ends $X_{\mathrm{ext}} = \mathbb{R}_+ \times Y$ and compact interior part X_{int} with piecewise smooth boundary $\partial X_{\mathrm{int}}$ such that the common boundary of the interior and exterior part $Y = X_{\mathrm{int}} \cap X_{\mathrm{ext}}$ is smooth and compact. In addition, we assume that the metric on X_{ext} has product structure $\mathrm{d}s^2 + h$, where h is the metric on Y. Moreover, we set*

$$\mathscr{H}_\bullet := \mathsf{L}_2(X_\bullet), \quad \mathscr{H}^1_\bullet := \mathsf{H}^1(X_\bullet), \quad \mathscr{W}^2_\bullet := \mathsf{H}^2(X_\bullet), \quad \mathfrak{h}_\bullet(u) := \|\mathrm{d}u\|^2_\bullet$$

$$\mathscr{G} := \mathsf{L}_2(Y) \qquad \Gamma_\bullet u := u_\bullet \restriction_Y, \qquad \Gamma'_\bullet u := u'_\bullet \restriction_Y, \qquad \check{H}_\bullet := \Delta_{X_\bullet},$$

where $\bullet = \mathrm{int}, \mathrm{ext}$ and u' is the derivative in outgoing direction on the cylinder. Then the dilated coupled boundary triple $(\Gamma^\theta, \Gamma'^\theta, \mathscr{G})$ on $\mathscr{H} = \mathsf{L}_2(X)$ is given by

$$\mathscr{H}^{1,\theta} = \bigl\{ u \in \mathsf{H}^1(X_{\mathrm{int}}) \oplus \mathsf{H}^1(X_{\mathrm{ext}}) \bigm| u_{\mathrm{int}} \restriction_Y = \mathrm{e}^{-\theta/2} u_{\mathrm{ext}} \restriction_Y \bigr\},$$

$$\mathfrak{h}^\theta(u) = \|\mathrm{d}u\|^2_{\mathrm{int}} + \mathrm{e}^{-2\theta} \|u'\|^2_{\mathrm{ext}} + \|\mathrm{d}_Y u\|^2_{\mathrm{ext}},$$

$$\Gamma^\theta u := u_{\mathrm{int}} \restriction_Y = \mathrm{e}^{-\theta/2} u_{\mathrm{ext}} \restriction_Y, \qquad \Gamma'^\theta u := u'_{\mathrm{int}} \restriction_Y - \mathrm{e}^{-3\theta/2} u'_{\mathrm{ext}} \restriction_Y.$$

Moreover, the dilated coupled boundary triple is bounded and elliptic. The transversal operator is given by the (Neumann) Laplacian on Y, i.e, $K = \Delta_Y$, and the Dirichlet-to-Neumann operator is independent of θ, namely

$$\Lambda^\theta(z) = \Lambda(z) = \Lambda_{\mathrm{int}}(z) - \mathrm{i}\sqrt{z - \Delta_Y}.$$

Proof. The individual (undilated) boundary triples are elliptic by Theorems 3.4.39 and 3.4.40. Note that we can always assume the situation of Theorem 3.4.40, namely the existence of a collar neighbourhood $U \cong [0, \ell_-] \times Y$ near Y in X_{int} by cutting the entire manifold on the cylindrical end. Since $\Lambda_\bullet$ and Λ are elliptic pseudo-differential operators of order 1, the norms on the associated scales of order m are all equivalent as in Theorem 3.4.40 (3). In particular, (3.66a) is fulfilled.

For (3.66b), we use the ellipticity of the *coupled* operator, which is just the Laplacian $H = \Delta_X$ on X. In particular, the ellipticity ensures that $\operatorname{dom} \Delta_X \subset \mathscr{W}^2 = \mathsf{H}^2(X)$. The ellipticity now follows from Theorem 3.6.5. □

We have a similar result for Dirichlet boundary conditions on ∂X as in Theorem 3.4.41. In this case, the transversal operator K is the Dirichlet Laplacian Δ_Y^{D} on Y.

The (dilated or undilated) coupled Dirichlet-to-Neumann operator $\Lambda(z) = \Lambda^\theta(z)$ of the above example is indeed the Dirichlet-to-Neumann operator of the manifold in the sense that given a smooth function $\varphi \in \mathsf{L}_2(Y)$ on the common boundary, then $\Lambda\varphi$ is given by the *jump of the normal derivative* of the Dirichlet solution, i.e. $\Lambda\varphi = (u'_{\mathrm{int}} - u'_{\mathrm{ext}}) \restriction_Y$, where u is the solution of $(\Delta - z)u = 0$ and $u \restriction_Y = \varphi$.

Let us now present the second main example: Let G be a metric graph with a finite number of exterior edges. We decompose the graph into its *interior* and *exterior* part $G = G_{\mathrm{int}} \mathbin{\overline{\cup}} G_{\mathrm{ext}}$. Here, G_{int} consists of all edges of finite lengths, whereas G_{ext} contains the exterior edges. G_{ext} may be disconnected. This happens if exterior edges are attached to different vertices. In particular, $Y := G_{\mathrm{int}} \cap G_{\mathrm{ext}}$ consists of finitely many vertices only.

We assume that $(G_\bullet, \mathscr{V}_\bullet)$ are quantum graphs with standard weighted vertex space $\mathscr{V}_\bullet = \bigoplus_v \mathscr{V}_{\bullet,v}$. For the notion of a quantum graph we refer to Definition 2.2.5. In particular, $\mathscr{V}_{\bullet,v} = \mathbb{C}p_\bullet(v)$ (see (2.13)) with weights $p_{\bullet,e}(v) \neq 0$ for $\bullet = \mathrm{int}, \mathrm{ext}$. Moreover, we write

$$\underline{f}_\bullet(v) = \{f_{\bullet,e}(v)\}_{e \in E_{\bullet,v}} = f_\bullet(v) p_\bullet(v) \in \mathscr{V}_{v,\bullet},$$

where $f_\bullet(v) = f_{\bullet,e}(v)/p_{\bullet,e}(v) \in \mathbb{C}$ is independent of $e \in E_{\bullet,v}$. As vertex space on the entire graph at the boundary, we choose the standard weighted vertex space $\mathscr{V}_v := \mathbb{C}p(v)$ with $p(v) = \{p_e(v)\}_{e \in E_v} = (p_{\mathrm{int}}(v), p_{\mathrm{ext}}(v))$.

We define a boundary triple associated with the quantum graphs $(G_\bullet, \mathscr{V}_\bullet)$ with respect to the boundary Y in a slightly different way than in Sect. 2.2.3. In particular, we set

$$\mathscr{H}_\bullet := \mathsf{L}_2(G_\bullet), \qquad \mathscr{H}_\bullet^1 := \mathsf{H}^1_{\mathscr{V}_\bullet}(G_\bullet), \qquad \mathscr{W}_\bullet^2 := \mathsf{H}^2_{\mathscr{V}_\bullet}(G_\bullet),$$

$$\mathfrak{h}_\bullet(f) := \|f'\|^2_{\mathsf{L}_2(G_\bullet)} \qquad \mathscr{G} := \mathbb{C}^Y,$$

$$(\Gamma_\bullet f)(v) := f_\bullet(v), \qquad (\Gamma'_\bullet f)(v) := \sum_{e \in E_{\bullet,v}} \overline{p_{\bullet,e}(v)}\, \underline{\widetilde{f}}'_e(v), \qquad (\check{H}_\bullet f)_e := -f''_e$$

(for the notation of the Sobolev spaces see Sect. 2.2.2). The difference with the setting in Sect. 2.2.3 is that we use an *unweighted* boundary space here (i.e., $\mathbb{C}^Y = \ell_2(Y)$ instead of $\ell_2(Y, |p|^2)$). It is now an easy exercise to check that $(\Gamma_\bullet, \Gamma'_\bullet, \mathscr{G})$ is a boundary triple associated with $\mathfrak{h}_\bullet$ for $\bullet = \mathrm{int}, \mathrm{ext}$ (provided we have a global positive lower bound on the edge lengths if $G_{\rm int}$ is non-compact). Moreover, this boundary triple is elliptic and associated with an ordinary boundary triple. This can be seen similarly as in Proposition 2.2.18.

Remark 3.6.12. The relation of the boundary triple $(\Gamma_{\rm int}, \Gamma'_{\rm int}, \mathscr{G})$ with the boundary triple with *weighted* space $\bar{\mathscr{G}} = \ell_2(Y, |p_{\rm int}|^2)$ is as follows: Let us denote the objects associated with the weighted space by a bar $\bar{\cdot}$. Then $\Gamma_{\rm int} = \bar{\Gamma}_{\rm int}$, $\Gamma'_{\rm int} = |p_{\rm int}|^2 \bar{\Gamma}'_{\rm int}$ and $\Lambda_{\rm int}(z) = |p_{\rm int}|^2 \bar{\Lambda}_{\rm int}(z)$. Here, $|p_{\rm int}|^2$ denotes the multiplication with the number $|p_{\rm int}(v)|^2$ on each component $F(v)$ of $F = \{F(v)\}_{v \in V}$.

Proposition 3.6.13. *Assume that $G = G_{\rm int} \mathbin{\dot\cup} G_{\rm ext}$ is a metric graph with finitely many exterior edges and a positive lower bound on the edge lengths* $\inf_{e \in E_{\rm int}} \ell_e > 0$, *together with the associated boundary triples $(\Gamma_\bullet, \Gamma'_\bullet, \mathscr{G})$ as explained above. Then the dilated coupled boundary triple $(\Gamma^\theta, \Gamma'^\theta, \mathscr{G})$ on $\mathscr{H} = \mathsf{L}_2(G)$ is given by*

$$\mathscr{H}^{1,\theta} = \{\, f \in \mathsf{H}^1_{\mathscr{V}_{\rm int}}(G_{\rm int}) \oplus \mathsf{H}^1_{\mathscr{V}_{\rm ext}}(G_{\rm ext}) \mid f_{\rm int}(v) = \mathrm{e}^{-\theta/2} f_{\rm ext}(v) \; \forall v \in Y \,\},$$

$$\mathfrak{h}^\theta(f) = \|f'\|^2_{\rm int} + \mathrm{e}^{-2\theta} \|f'\|^2_{\rm ext},$$

$$(\Gamma^\theta f)(v) = f_{\rm int}(v) = \mathrm{e}^{-\theta/2} f_{{\rm ext},v}(0) \qquad \textit{and}$$

$$(\Gamma'^\theta f)(v) = \sum_{e \in E_{{\rm int},v}} \overline{p_{{\rm int},e}(v)}\, \underset{\sim}{f}'_e(v) - \mathrm{e}^{-3\theta/2} \sum_{e \in E_{{\rm ext},v}} \overline{p_{{\rm ext},e}(v)}\, f'_{{\rm ext},e}(0).$$

Moreover, the dilated coupled boundary triple is elliptic. The transversal operator is $K = 0$, and we have

$$\Lambda^\theta(z) = \Lambda(z) = \Lambda_{\rm int}(z) + \Lambda_{\rm ext}(z) = \Lambda_{\rm int}(z) - \mathrm{i}\sqrt{z}|p_{\rm ext}|^2.$$

Proof. The proof follows from an application of Proposition 3.6.10 and the fact that the undilated coupled boundary triple associated with the metric graph G is elliptic: To see this, we use Corollary 3.6.7. It is actually shown in Sect. 2.2 that the individual boundary triples are bounded, elliptic and that $\operatorname{dom} H^{\rm N} \subset \mathscr{W}^2 = \mathsf{H}^2_{\mathscr{V}}(G))$. Moreover, it is easily seen that $(\Lambda_{\rm ext}(z)F)(v) = -\mathrm{i}\sqrt{z}|p_{\rm ext}(v)|^2 F(v)$. □

Let us present a particularly simple example, where we calculate the Dirichlet-to-Neumann map explicitly:

Example 3.6.14. If there is at least one edge in the interior metric graph and if all these edges have length 1, if all weights are equal, say, $p_e(v) = 1$, and if all vertices of $G_{\rm int}$ are connected with an exterior edge, i.e., $Y = V_{\rm int} = V$, then

$$\begin{aligned}\Lambda_{\rm int}(z) &= \frac{\sqrt{z}}{\sin\sqrt{z}} \deg_{\rm int}\bigl(\ddot{\Delta}_{\rm int} - (1 - \cos\sqrt{z})\bigr) \\ &= \frac{\sqrt{z}}{\sin\sqrt{z}}\bigl(\ddot{\Delta}^{\rm comb}_{\rm int} - \deg_{\rm int}(1 - \cos\sqrt{z})\bigr),\end{aligned} \tag{3.72}$$

where $\deg_{\mathrm{int}}$ denotes the multiplication with the degree $\deg_{\mathrm{int}}(v)$ of the vertex $v \in V_{\mathrm{int}}$ w.r.t. the interior graph G_{int} (i.e., the number of adjacent edges). Moreover, $\dddot{\Delta}_{\mathrm{int}}$ is the *normalised discrete Laplacian* (see (2.7)) acting on $\ell_2(V, \deg_{\mathrm{int}})$, namely

$$(\dddot{\Delta}_{\mathrm{int}} F)(v) = \frac{1}{\deg_{\mathrm{int}}(v)} \sum_{e \in E_{\mathrm{int},v}} (F(v) - F(v_e)),$$

where v_e is the vertex on e opposite to v. Moreover, $\dddot{\Delta}_{\mathrm{int}}^{\mathrm{comb}}$ denotes the *combinatorial (discrete) Laplacian* acting on $\ell_2(V)$, defined by

$$(\dddot{\Delta}_{\mathrm{int}}^{\mathrm{comb}} F)(v) = \sum_{e \in E_{\mathrm{int},v}} (F(v) - F(v_e)).$$

These facts follow from Proposition 2.2.19 and Remark 3.6.12.

In particular, the coupled Dirichlet-to-Neumann operator is given by

$$\Lambda(z) = \frac{\sqrt{z}}{\sin\sqrt{z}} \Big(\dddot{\Delta}_{\mathrm{int}}^{\mathrm{comb}} - \deg_{\mathrm{int}}(1 - \cos\sqrt{z}) - \mathrm{i}(\sin\sqrt{z}) \deg_{\mathrm{ext}} \Big). \tag{3.73}$$

We can now state the analogue of Theorem 3.4.44 for the θ-coupled case. Note that the Dirichlet-to-Neumann operator in Krein's resolvent formulas below is the *undilated* Dirichlet-to-Neumann operator. We write $\overline{\theta}$ although θ is real here for consistency with the next section, in which θ is allowed to be also complex-valued.

Theorem 3.6.15. *Assume that $\theta \in \mathbb{R}$ and $z \in \mathbb{C} \setminus [0, \infty)$. Assume in addition that the coupled boundary triple $(\Gamma, \Gamma', \mathscr{G})$ is elliptic. Then the resolvent difference can be expressed by the bounded operators*

$$R^\theta(z) - R^{\theta,\mathrm{D}}(z) = W^\theta(z) \qquad : \mathscr{H} \longrightarrow \mathscr{H} \tag{3.74a}$$

$$R^\theta(z) - \iota_{\mathrm{ell}} R^{\theta,\mathrm{D}}(z) = W^{\theta,0\to 2}(z) \qquad : \mathscr{H} \longrightarrow \mathscr{W}^{2,\theta}, \tag{3.74b}$$

$$R^\theta(z) - \iota_{\mathrm{D},\theta} R^{\theta,\mathrm{D}}(z) \iota^*_{\mathrm{D},\overline{\theta}} = W^{\theta,-1\to 1}(z) \qquad : \mathscr{H}^{-1,\theta} \longrightarrow \mathscr{H}^{1,\theta}, \tag{3.74c}$$

where

$$W^\theta(z) := (B^{\overline{\theta}}(\overline{z}))^* \Lambda(z)^{-1} B^\theta(z) \colon \mathscr{H} \xrightarrow{B^\theta(z)} \mathscr{G} \xrightarrow{\Lambda(z)^{-1}} \mathscr{G} \xrightarrow{(B^{\overline{\theta}}(\overline{z}))^*} \mathscr{H},$$

$$W^{\theta,0\to 2}(z) := S^\theta(z) \Lambda(z)^{-1} B^\theta(z) \colon \mathscr{H} \xrightarrow{B^\theta(z)} \mathscr{G}^{1/2} \xrightarrow{\Lambda(z)^{-1}} \mathscr{G}^{3/2} \xrightarrow{S^\theta(z)} \mathscr{W}^{2,\theta},$$

$$W^{\theta,-1\to 1}(z) := S^\theta(z) \Lambda(z)^{-1} (S^{\overline{\theta}}(\overline{z}))^* \colon \mathscr{H}^{-1,\theta} \xrightarrow{(S^{\overline{\theta}}(\overline{z}))^*} \mathscr{G}^{-1/2} \xrightarrow{\Lambda(z)^{-1}} \mathscr{G}^{1/2} \xrightarrow{S^\theta(z)} \mathscr{H}^{1,\theta}$$

are bounded and the operators $W^\theta(z)$ and $W^{\theta,0\to 2}(z)$ agree up to the target space, and the operator $W^{\theta,-1\to 1}(z)$ extends $W^\theta(z)$.

Proof. The operator equalities (3.74a) and (3.74c) follow from Theorem 3.4.44 for the dilated coupled boundary triple $(\Gamma^\theta, \Gamma'^\theta, \mathscr{G})$, which is elliptic by Proposition 3.6.10. Moreover, the boundedness of the operators follows from Proposition 3.5.16 for the exterior boundary triple, and from Lemma 3.4.29 and the ellipticity for the interior boundary triple. □

3.7 Complexly Dilated Coupled Operators

In this section, we extend the previously defined dilated coupled operator H^θ to *complex* values of $\theta \in S_\vartheta$. Since this section is somewhat technical, we give a short outline first. The basic ideas can already be found in [CDKS87] for a concrete PDE example, but not formulated in the language of boundary triples. This conceptual language allows us to treat the metric graph and manifold case simultaneously. Moreover, our results may be applied to many other similar situations in an easy way, once the problem is formulated in terms of boundary triples.

The proof that H^θ depends holomorphically on θ (in the sense that $(H^\theta - z)^{-1}$ is holomorphic in θ for suitable z) needs some effort: We will first extend the definition of H^θ (a priori only defined for $\theta \in \mathbb{R}$) to a *non-self-adjoint* operator H^θ via Krein's resolvent formula

$$R^\theta(z) = R^{\theta,\mathrm{D}}(z) + W^\theta(z)$$

as in Theorem 3.6.15, but now for *non-real* θ. It is easily seen that the RHS is holomorphic in θ (as we show in Sect. 3.7.1). In Sect. 3.7.2, we show that $R^\theta(z)$ is the resolvent of a closed operator $\widetilde{H}^\theta$ such that

$$\operatorname{dom} \widetilde{H}^\theta = \mathscr{W}^{2,\theta,\mathrm{N}}, \quad \text{where} \tag{3.75a}$$

$$\mathscr{W}^{2,\theta,\mathrm{N}} := \{ u \in \mathscr{W}^{2,\mathrm{dec}} \mid \Gamma_{\mathrm{ext}} u = \mathrm{e}^{\theta/2} \Gamma_{\mathrm{int}} u, \quad \Gamma'_{\mathrm{ext}} u = \mathrm{e}^{3\theta/2} \Gamma'_{\mathrm{int}} u \}, \tag{3.75b}$$

and

$$\widetilde{H}^\theta u = \check{H}_{\mathrm{int}} u \oplus \check{H}^\theta_{\mathrm{ext}} = \check{H}_{\mathrm{int}} u_{\mathrm{int}} \oplus -\mathrm{e}^{-2\theta} u''_{\mathrm{ext}} + (\mathrm{id} \otimes K) u_{\mathrm{ext}} \tag{3.75c}$$

for $u \in \operatorname{dom} \widetilde{H}^\theta$. In particular, $\widetilde{H}^\theta = H^\theta$ for *real* θ (and hence, we also use the notation H^θ for complex values $\theta \in S_\vartheta$). Note that H^θ is formally the Neumann operator of the complexly dilated coupled boundary triple. In Sect. 3.7.3 we extend the operator H^θ onto suitable first order spaces $\mathscr{H}^{1,\theta} \to \mathscr{H}^{-1,\theta}$.

Note that the operator domain of H^θ and even the quadratic form domain depend on θ already for real θ (due to the "jump" condition $\Gamma_{\mathrm{int}} u = \mathrm{e}^{-\theta/2} \Gamma_{\mathrm{ext}} u$), so that $\{H^\theta\}_\theta$ is neither a holomorphic family of type A nor of type B in the sense of Kato (see [K66, Sect. VII.2]).

Throughout this section we assume that the (undilated) coupled boundary triple $(\Gamma, \Gamma', \mathscr{G})$ is elliptic. Note that the dilated boundary triple $(\Gamma^\theta, \Gamma'^\theta, \mathscr{G})$

(see Definition 3.6.9) for real θ is elliptic iff the undilated ($\theta = 0$) coupled boundary triple $(\Gamma, \Gamma', \mathscr{G})$ is. For conditions assuring the ellipticity of the latter, see Theorem 3.6.5.

3.7.1 Holomorphic Dependency

We start with the holomorphy and other properties of the decoupled operator $H^{\theta,\mathrm{D}}$. The proof follows straightforward from Proposition 3.5.14.

Proposition 3.7.1. *Assume that assume that $0 \le \vartheta < \pi/2$, $\theta \in \mathrm{S}_\vartheta$ and $z \in \mathbb{C} \setminus \Sigma_\vartheta$. Then the following assertions hold:*

1. *The operator*
$$H^{\theta,\mathrm{D}} = H^{\mathrm{D}}_{\mathrm{int}} \oplus (-\mathrm{e}^{-2\theta} \partial^{\mathrm{D}}_{ss} + \mathrm{id} \otimes K)$$
is the Dirichlet operator associated with the complexly dilated boundary triple $(\Gamma^\theta, \Gamma'^\theta, \mathscr{G})$ formally defined as in Definition 3.6.9 for complex values of θ (recall that $-\partial^{\mathrm{D}}_{ss}$ denotes the Laplacian on $\mathbb{R}_+$ with Dirichlet condition at 0). Moreover, the domain dom $H^{\theta,\mathrm{D}}$ *is independent of θ.*
2. *The operator $H^{\theta,\mathrm{D}}$ is normal, i.e. commutes with its adjoint.*
3. *The operator $H^{\theta,\mathrm{D}}$ commutes with the undilated operator $H^{\mathrm{D}} = H^{0,\mathrm{D}}$.*
4. *The family $\{H^{\theta,\mathrm{D}}\}_\theta$ is self-adjoint, i.e. $(H^{\theta,\mathrm{D}})^* = H^{\overline{\theta},\mathrm{D}}$.*
5. *The family $\{H^{\theta,\mathrm{D}}\}_\theta$ depends holomorphically on $\theta \in \mathrm{S}_\vartheta$, i.e. the family is of type A (see [K66, Sect. VII.2]).*
6. *The spectrum of $H^{\theta,\mathrm{D}}$ is given by*
$$\begin{aligned}\sigma(H^{\theta,\mathrm{D}}) &= \sigma(H^{\mathrm{D}}_{\mathrm{int}}) \cup \big(\sigma(K) + \mathrm{e}^{-2\,\mathrm{Im}\,\theta}[0,\infty)\big)\\ &= \sigma(H^{\mathrm{D}}_{\mathrm{int}}) \cup \{\kappa + \mathrm{e}^{-2\,\mathrm{Im}\,\theta}\lambda \,|\, \kappa \in \sigma(K),\ \lambda \ge 0\},\end{aligned}$$
and is contained in the sector Σ_ϑ. Moreover, the resolvent
$$R^{\theta,\mathrm{D}}(z) = (H^{\theta,\mathrm{D}} - z)^{-1} = (H^{\mathrm{D}}_{\mathrm{int}} - z)^{-1} \oplus (H^{\theta,\mathrm{D}}_{\mathrm{ext}} - z)^{-1}$$
fulfils the norm estimate
$$\|R^{\theta,\mathrm{D}}(z)\| \le \frac{1}{\mathrm{dist}(z, \mathbb{C} \setminus \Sigma_\vartheta)} =: \gamma^{\mathrm{D}}_\vartheta(z), \quad \textit{i.e.} \quad H^{\theta,\mathrm{D}} \in \mathscr{R}(\gamma^{\mathrm{D}}_\vartheta, \mathbb{C} \setminus \Sigma_\vartheta)$$
(see Definition 3.3.3).
7. *The resolvent as operator*
$$R^{\theta,\mathrm{D}}(z) \colon \mathscr{H}^{k,\mathrm{D}} \longrightarrow \mathscr{H}^{k+2,\mathrm{D}}$$

is bounded by $C_{\leq\vartheta}(z,-1)$ *for* $z \in \mathbb{C} \setminus \Sigma_\vartheta$ *and* $\theta \in S_\vartheta$ *(see Lemma 3.5.13), where* $\mathscr{H}^{k,\mathrm{D}}$ *is the scale of Hilbert spaces associated with* $H^{\mathrm{D}} = H^{0,\mathrm{D}}$*. In particular,*

$$\|R^{\theta,\mathrm{D}}(z)\|_{-1\to 1} \leq C_{\leq\vartheta}(z,-1) =: \gamma_\vartheta^{\mathrm{D},1}(z), \quad i.e. \quad H^{\theta,\mathrm{D}} \in \mathscr{R}_1(\gamma_\vartheta^{\mathrm{D},1}, \mathbb{C} \setminus \Sigma_\vartheta)$$

(see Definition 3.3.10).

Let us now define the operator $W^\theta(z)$ appearing in Krein's resolvent formula in Theorem 3.6.15 on different scales for *complex* values of $\theta \in S_\vartheta = \{\theta \in \mathbb{C} \mid |\mathrm{Im}\,\theta| \leq \vartheta/2\}$. We start with the *dilated coupled Dirichlet solution operator*, given by

$$S^\theta(z) \colon \mathscr{G} \longrightarrow \mathscr{H}, \qquad S^\theta(z)\varphi := S_{\mathrm{int}}(z)\varphi \oplus S^\theta_{\mathrm{ext}}(z)\varphi. \tag{3.77}$$

Let us summarise some facts on the coupled, dilated solution operator $S^\theta(z)$ and its adjoint $B^\theta(z) \colon \mathscr{H} \longrightarrow \mathscr{G}$ defined by

$$\begin{aligned} B^{\overline{\theta}}(\overline{z})u &:= -\Gamma'^{\overline{\theta}} R^{\overline{\theta},\mathrm{D}}(\overline{z})u = B_{\mathrm{int}}(\overline{z})u_{\mathrm{int}} + B^{\overline{\theta}}_{\mathrm{ext}}(\overline{z})u_{\mathrm{ext}} \\ &= -\Gamma'_{\mathrm{int}} R^{\mathrm{D}}_{\mathrm{int}}(\overline{z})u_{\mathrm{int}} + \mathrm{e}^{-3\overline{\theta}/2}\Gamma'_{\mathrm{ext}} R^{\overline{\theta},\mathrm{D}}_{\mathrm{ext}}(\overline{z})u_{\mathrm{ext}}. \end{aligned} \tag{3.78}$$

for $\theta \in S_\vartheta$ and $z \in \mathbb{C} \setminus \Sigma_{\vartheta_0}$) from the previous sections (see Remark 3.5.12 for the need of the two variables $\vartheta < \vartheta_0$). Note that $R^{\overline{\theta},\mathrm{D}}(\overline{z})u \in \mathscr{W}^2$ since the (undilated) boundary triple is elliptic, and in particular, $\mathrm{dom}\, H^{\theta,\mathrm{D}} = \mathrm{dom}\, H^{0,\mathrm{D}} \subset \mathscr{W}^2$.

We show the boundedness and holomorphy of these operators on different scales of Hilbert spaces:

Proposition 3.7.2. *Assume that* $0 \leq \vartheta < \vartheta_0 < \pi/2$, $\theta \in S_\vartheta$ *and* $z \in \mathbb{C} \setminus \Sigma_{\vartheta_0}$. *Then the adjoint of* $B^\theta(z) \colon \mathscr{H} \longrightarrow \mathscr{G}$ *extends the Dirichlet solution operator* $S^\theta(z) \colon \mathscr{G}^{1/2} \longrightarrow \mathscr{H}^{1,\theta}$*. Moreover, the adjoint of* $S^{\overline{\theta}}(\overline{z}) \colon \mathscr{G}^{1/2} \longrightarrow \mathscr{H}^{1,\theta}$ *extends* $B^\theta(z)$*. In addition, the following operators are bounded as maps*

$$B^\theta(z) \colon \mathscr{H} \longrightarrow \mathscr{G}, \qquad B^{\overline{\theta}}(\overline{z})^* \colon \mathscr{G} \longrightarrow \mathscr{H}, \tag{3.79a}$$

$$B^\theta(z) \colon \mathscr{H} \longrightarrow \mathscr{G}^{1/2}, \qquad S^\theta(z) \colon \mathscr{G}^{3/2} \longrightarrow \mathscr{W}^{2,\theta}, \tag{3.79b}$$

where $\mathscr{G}^k$ *is the scale associated with the undilated coupled Dirichlet-to-Neumann operator* $\Lambda = S^{*1}S$*. Finally, for* $z \in \mathbb{C} \setminus \Sigma_{\vartheta_0}$*, the corresponding operator-valued functions*

$$S_\vartheta \to \mathscr{L}(\mathscr{H},\mathscr{G}),\ \mathscr{L}(\mathscr{H},\mathscr{G}^{1/2}), \qquad \theta \mapsto B^\theta(z), B^\theta(z),$$

$$S_\vartheta \to \mathscr{L}(\mathscr{G},\mathscr{H}),\ \mathscr{L}(\mathscr{G}^{3/2},\mathscr{W}^{2,\mathrm{dec}}), \qquad \theta \mapsto B^{\overline{\theta}}(\overline{z})^*, S^\theta(z)$$

are holomorphic.

Proof. The assertions on the extension by adjoints follow as in Lemma 3.4.29, and the boundedness and holomorphic dependence of the operators in (3.79a) as in Proposition 3.5.16. The boundedness of the operators in (3.79b) is a consequence of the ellipticity. For the holomorphy, we argue as in the proof of Proposition 3.5.16; here, we embed the θ-dependent space $\mathscr{W}^{2,\theta}$ into $\mathscr{W}^{2,\mathrm{dec}} = \mathscr{W}^2_{\mathrm{int}} \oplus \mathscr{W}^2_{\mathrm{ext}} \subset \mathscr{H}$. □

Denote by $\Lambda(z)$ the Dirichlet-to-Neumann operator associated with the coupled *undilated* boundary triple; see Proposition 3.6.2 for an explicit expression. Combining the various results on holomorphic dependency, we obtain:

Proposition 3.7.3. *Assume that* $0 \le \vartheta < \vartheta_0 < \pi/2$, $\theta \in \mathrm{S}_\vartheta$ *and* $z \in \mathbb{C} \setminus \Sigma_{\vartheta_0}$. *Then the operators*

$$W^\theta(z) := (B^{\overline{\theta}}(\overline{z}))^* \Lambda(z)^{-1} B^\theta(z) \colon \mathscr{H} \xrightarrow{B^\theta(z)} \mathscr{G} \xrightarrow{\Lambda(z)^{-1}} \mathscr{G} \xrightarrow{(B^{\overline{\theta}}(\overline{z}))^*} \mathscr{H},$$

$$W^{\theta,0\to 2}(z) := S^\theta(z)\Lambda(z)^{-1} B^\theta(z) \colon \mathscr{H} \xrightarrow{B^\theta(z)} \mathscr{G}^{1/2} \xrightarrow{\Lambda(z)^{-1}} \mathscr{G}^{3/2} \xrightarrow{S^\theta(z)} \mathscr{W}^{2,\theta}$$

are bounded and the operators $W^\theta(z)$ *and* $W^{\theta,0\to 2}(z)$ *agree up to the target space. In addition, the operator-valued functions*

$$\mathrm{S}_\vartheta \to \mathscr{L}(\mathscr{H}),\ \theta \mapsto W^\theta(z) \quad \text{and} \quad \mathrm{S}_\vartheta \to \mathscr{L}(\mathscr{H}, \mathscr{W}^{2,\mathrm{dec}}),\ \theta \mapsto W^{\theta,0\to 2}(z)$$

are holomorphic.

Proof. The assertions follow from Proposition 3.7.2 for the maps $S^\theta(z)$ and $B^\theta(z)$, and from Proposition 3.4.34 and Definition 3.4.31 (applied to the undilated coupled boundary triple) for $\Lambda(z)$. □

3.7.2 The Complexly Dilated Coupled Operator

In this subsection, we want to extend H^θ defined a priori only for *real* θ also for *complex* θ in the strip S_ϑ. Actually, we start defining an operator family $\{R^\theta(z)\}_{\theta,z}$ via Krein's resolvent formula, and then show that $R^\theta(z)$ is the resolvent of a closed operator $\widetilde{H}^\theta$ given as in (3.75), which agrees with H^θ for real θ.

We set

$$R^\theta(z) := W^\theta(z) + R^{\theta,\mathrm{D}}(z) \colon \mathscr{H} \longrightarrow \mathscr{H} \tag{3.80}$$

for $\theta \in \mathrm{S}_\vartheta$ and $z \in \mathbb{C} \setminus \Sigma_{\vartheta_0}$ (recall that $0 \le \vartheta < \vartheta_0 < \pi/2$). Let us first check that $R^\theta(z)$ is indeed the resolvent of a closed operator:

Proposition 3.7.4.

1. *For* $z \in \mathbb{C} \setminus \Sigma_{\vartheta_0}$, *the family of* $\{R^\theta(z)\}_\theta$ *is holomorphic in* $\theta \in \mathrm{S}_\vartheta$.
2. *The operators* $R^\theta(z)$ *satisfy the resolvent equation*

$$R^\theta(z) - R^\theta(w) = (z-w)R^\theta(z)R^\theta(w)$$

for $w, z \in \mathbb{C} \setminus \Sigma_{\vartheta_0}$.

3. *The kernel of* $R^\theta(z)$ *is trivial.*

In particular, $R^\theta(z)$ *is the resolvent of a closed operator*

$$\widetilde{H}^\theta u := (R^\theta)^{-1}u - u, \qquad u \in \operatorname{dom} \widetilde{H}^\theta := R^\theta(\mathscr{H}) \tag{3.81}$$

where $R^\theta := R^\theta(-1)$ *and the family* $\{\widetilde{H}^\theta\}_\theta$ *is self-adjoint (i.e.* $(\widetilde{H}^\theta)^* = \widetilde{H}^{\overline{\theta}}$*) with spectrum contained in the sector* Σ_{ϑ_0}.

Proof. (1) The first assertion follows immediately from Propositions 3.7.1 and 3.7.3.

(2) The resolvent equation is fulfilled for real θ, since then, the operator is the resolvent of a (self-adjoint) operator by Theorem 3.6.15. Due to holomorphy, the resolvent equation remains true for all $\theta \in \mathrm{S}_\vartheta$. (3) To prove the third assertion, we claim that

$$\langle (H^{\overline{\theta},\mathrm{D}} - \overline{z})u, W^\theta(z)v \rangle = 0 \tag{3.82}$$

for all $u \in \mathring{\mathscr{W}}^2$, for all $v \in \mathscr{H}$, all $z \in \mathbb{C} \setminus \Sigma_{\vartheta_0}$ and all $\theta \in \mathrm{S}_\vartheta$, where $\mathring{\mathscr{W}}^2 := \mathring{\mathscr{W}}^2_{\mathrm{int}} \oplus \mathring{\mathscr{W}}^2_{\mathrm{ext}}$ and $\mathring{\mathscr{W}}^2_\bullet = \ker \Gamma_\bullet \cap \ker \Gamma'_\bullet$, see (3.32c). This claim can easily be seen from

$$\langle (H^{\overline{\theta},\mathrm{D}} - \overline{z})u, W^\theta(z)v \rangle_{\mathscr{H}} = \langle B^{\overline{\theta}}(\overline{z})(H^{\overline{\theta},\mathrm{D}} - \overline{z})u, \Lambda(z)^{-1}B^\theta(z)v \rangle_{\mathscr{G}}$$

using the definition of $W^\theta(z)$ in Proposition 3.7.3 for complex θ. Now the left-hand side of the last inner product vanishes since $B^{\overline{\theta}}(\overline{z})(H^{\overline{\theta},\mathrm{D}} - \overline{z})u = (-\Gamma'_{\mathrm{int}} + \mathrm{e}^{-3\overline{\theta}/2}\Gamma'_{\mathrm{ext}})u = 0$ for $u \in \mathring{\mathscr{W}}^2$ and therefore, (3.82) is fulfilled.

Suppose now, that $R^\theta(z)v = 0$. Then we have

$$0 = \langle (H^{\overline{\theta},\mathrm{D}} - \overline{z})u, R^\theta(z)v \rangle = \langle (H^{\overline{\theta},\mathrm{D}} - \overline{z})u, R^{\theta,\mathrm{D}}(z)v \rangle = \langle u, v \rangle$$

for $u \in \mathring{\mathscr{W}}^2$ due to (3.82). Since $\mathring{\mathscr{W}}^2_\bullet$ is dense in $\mathscr{H}_\bullet$ by assumption (see Definition 3.4.3), $\mathring{\mathscr{W}}^2$ is dense in $\mathscr{H}$, hence $v = 0$.

To conclude, we observe that (2) and (3) imply that $R^\theta(z)$ is a resolvent of a closed operator $\widetilde{H}^\theta$ (cf. [K66, p. 428]) for all $z \in \mathbb{C} \setminus \Sigma_{\vartheta_0}$, and therefore $\sigma(\widetilde{H}^\theta) \subset \Sigma_{\vartheta_0}$. In addition, the family $\{\widetilde{H}^\theta\}_\theta$ is self-adjoint since $(W^\theta)^* = W^{\overline{\theta}}$ and $(R^{\theta,\mathrm{D}})^* = R^{\overline{\theta},\mathrm{D}}$. □

We finally show that $\widetilde{H}^\theta$ actually acts as the Neumann operator H^θ associated with the dilated coupled boundary triple (see Definition 3.6.9) also for complex values of θ:

Proposition 3.7.5. *The operator $\widetilde{H}^\theta$ defined in the previous proposition fulfils the assertions in* (3.75), *i.e.,* dom $\widetilde{H}^\theta = \mathscr{W}^{2,\theta,\mathrm{N}}$ *and* $\widetilde{H}^\theta$ *acts as in* (3.75c). *In particular, for* $\theta \in \mathbb{R}$, *we have* $\widetilde{H}^\theta = H^\theta$.

Proof. Let $v = R^\theta \widetilde{v} \in \operatorname{dom} \widetilde{H}^\theta = \operatorname{ran} R^\theta$ (i.e., we fix $z = -1$ here). Then we have

$$\begin{aligned}\langle \widetilde{H}^{\overline{\theta}} u, v\rangle &= \langle (\widetilde{H}^{\overline{\theta}} + 1)u, v\rangle - \langle u, v\rangle \\ &= \langle (\widetilde{H}^{\overline{\theta},\mathrm{D}} + 1)u, R^{\theta,\mathrm{D}}\widetilde{v}\rangle - \langle u, v\rangle = \langle u, \widetilde{v} - v\rangle = \langle u, \widetilde{H}^\theta v\rangle\end{aligned}$$

for all $u \in \mathring{\mathscr{W}}^2$ using (3.82) in the second and (3.81) in the last equation. On the other hand, $v \in \operatorname{ran} R^\theta$ implies $v \in \mathscr{W}^{2,\theta}$ by the definition (3.80) of R^θ, using Proposition 3.7.3 and $\operatorname{ran} R^{\theta,\mathrm{D}} = \operatorname{dom} H^{\mathrm{D}}_{\mathrm{int}} \oplus \operatorname{dom} H^0_{\mathrm{ext}} \subset \mathscr{W}^{2,\theta}$ (by ellipticity). In particular, we can apply Green's formula (3.32b) for the interior and exterior boundary triple, so that

$$\langle \widetilde{H}^{\overline{\theta}} u, v\rangle = \langle u, \check{H}_{\mathrm{int}} v\rangle_{\mathrm{int}} + \langle u, -\mathrm{e}^{-2\theta} v'' + (\mathrm{id} \otimes K)v\rangle_{\mathrm{ext}}$$

for $u \in \mathring{\mathscr{W}}^2$ using $\widetilde{H}^{\overline{\theta}} u = H^{\overline{\theta},\mathrm{D}} u = H_{\mathrm{int}} u \oplus (-\mathrm{e}^{-2\overline{\theta}} u''_{\mathrm{ext}} + (\mathrm{id} \otimes K) u_{\mathrm{ext}}$ (Proposition 3.7.1 (1)). Hence, we have shown that $\widetilde{H}^\theta$ acts as in (3.75c).

In order to show that $\operatorname{dom} \widetilde{H}^\theta = \operatorname{ran} R^\theta \subset \mathscr{W}^{2,\theta,\mathrm{N}}$ we first note that $\operatorname{ran} W^{\theta,0\to 2} \subset \mathscr{W}^{2,\theta}$ from Proposition 3.7.3, and the definition (3.80) implies that the first condition $\Gamma_{\mathrm{ext}} v = \mathrm{e}^{\theta/2}\Gamma_{\mathrm{int}} v$ of $\mathscr{W}^{2,\theta,\mathrm{N}}$ is fulfilled for $v \in \operatorname{ran} R^\theta$.

For the second condition of $\mathscr{W}^{2,\theta,\mathrm{N}}$, we recall from Proposition 3.7.3 that $\iota_{\mathrm{ell,D}} R^{\theta,\mathrm{D}} + W^{\theta,0\to 2}$ is bounded as map from $\mathscr{H}$ into $\mathscr{W}^{2,\mathrm{dec}}$ and depends holomorphically on θ. Moreover, $\iota_{\mathrm{ell,D}} R^{\theta,\mathrm{D}} + W^{\theta,0\to 2}$ agrees with R^θ for *real* θ. By holomorphy, the operators agree for *all* $\theta \in \mathrm{S}_\vartheta$. Finally, since $\iota_{\mathrm{ell,D}} R^{\theta,\mathrm{D}} + W^{\theta,0\to 2}$ is bounded (as operator from $\mathscr{H}$ into $\mathscr{W}^{2,\mathrm{dec}}$), the same is true for R^θ. To finish the argument, we note that

$$\Gamma'^\theta R^\theta \colon \mathscr{H} \longrightarrow \mathscr{W}^{2,\mathrm{dec}} \longrightarrow \mathscr{H}$$

is also bounded and analytic in θ, where

$$\Gamma'^\theta u := \Gamma'_{\mathrm{int}} u - \Gamma'^\theta_{\mathrm{ext}} u = \Gamma'_{\mathrm{int}} u - \mathrm{e}^{-3\theta/2}\Gamma'_{\mathrm{ext}} u \tag{3.83}$$

for $u \in \mathscr{W}^{2,\theta}$. In addition, the latter operator vanishes for *real* θ due to $\operatorname{dom} H^\theta = \mathscr{W}^{2,\theta,\mathrm{N}}$ and the ellipticity in this case, and therefore for all $\theta \in \mathrm{S}_\vartheta$ again due to holomorphy. Finally, we have shown that if $u \in \operatorname{dom} \widetilde{H}^\theta = \operatorname{ran} R^\theta$, then $u \in \mathscr{W}^{2,\theta,\mathrm{N}}$.

For the opposite inclusion, we have to check that a function $u \in \mathscr{W}^{2,\theta,\mathrm{N}}$ is of the form $u = R^\theta v$. A straightforward calculation (using similar arguments as in Theorem 3.6.15) shows that

$$\check{v} := \left(\check{H}_{\text{int}} u_{\text{int}} \oplus \check{H}_{\text{ext}} u_{\text{ext}}\right) + u$$

is the right candidate. □

Definition 3.7.6. We call the operator $\widetilde{H}^\theta$ defined above (see (3.75)) the *complexly dilated coupled operator* associated with the coupled boundary triple $(\Gamma, \Gamma', \mathscr{G})$. For simplicity, we also write H^θ for the operator $\widetilde{H}^\theta$ for $\theta \in S_\vartheta$ (i.e., we omit the tilde $\widetilde{\cdot}$). We denote the domain by $\mathscr{H}^{2,\theta} := \operatorname{dom} H^\theta$ $(=\mathscr{W}^{2,\theta,\mathrm{N}}$ by Proposition 3.7.5).

Let us finally proof that the complexly dilated boundary maps and the Dirichlet solution operator fit well together with the Dirichlet-to-Neumann operator:

Proposition 3.7.7. *The Dirichlet-to-Neumann operator is naturally related with the boundary maps and the solution operator of the complexly dilated boundary triple, i.e.,*

$$\Gamma'^\theta S^\theta(z) = \Lambda(z) \colon \mathscr{G}^{3/2} \longrightarrow \mathscr{G}^{1/2},$$

and the operator is bounded.

Proof. We have

$$\Gamma'^\theta S^\theta(z)\varphi = \Gamma'_{\text{int}} S_{\text{int}}(z)\varphi - \Gamma'^\theta_{\text{ext}} S^\theta_{\text{ext}}(z)\varphi = \Lambda_{\text{int}}(z)\varphi + \Lambda_{\text{ext}}(z)\varphi = \Lambda(z)\varphi$$

for $\varphi \in \mathscr{G}^{3/2}$ by (3.77) and (3.70b). □

3.7.3 The Complexly Dilated Coupled Operator on the First Order Spaces

In this section, we want to define a suitable sesquilinear form associated with the H^θ. Let us start with a remark:

Remark 3.7.8. If we proceed in the naive way, we could define a quadratic form (and a corresponding sesquilinear form with *symmetric* domain) associated with H^θ by

$$\mathfrak{q}^\theta(u) := \langle u, H^\theta u\rangle = \mathfrak{h}^\theta(u,u) - (1 - \mathrm{e}^{\mathrm{i}\,\mathrm{Im}\,\theta})\langle \Gamma_{\text{int}} u, \Gamma'_{\text{int}} u\rangle$$

for $u \in \mathscr{H}^{2,\theta}$ using Green's formula individually on the interior and exterior part. Note that the boundary term occurs for $\mathrm{Im}\,\theta \neq 0$, since we have to respect the *sesquilinearity* of the inner products. There has been some confusion on the *quadratic* form $\mathfrak{q}^\theta$ due to the anti-linearity of a sesquilinear form in its first argument, and the boundary term seems to have been overlooked in [GY83, (2.14)] (cf. the Mathematical Reviews entry).

For non-real θ, it is not clear to us, whether this form is closable and that the closure is defined on

$$\mathscr{H}^{1,\theta} = \left\{ u \in \mathscr{H}^{1,\text{dec}} \;\middle|\; \Gamma_{\text{ext}} u = \mathrm{e}^{\theta/2}\Gamma_{\text{int}} u \right\}.$$

Moreover, it is difficult to use such expressions as upper estimates. In addition, this quadratic form is not related to an operator H^θ defined from the Hilbert space $\mathscr{H}^{1,\theta}$ into $\mathscr{H}^{-1,\theta} = (\mathscr{H}^{1,\overline{\theta}})^*$. These are basically the reasons why we introduced the scale of order 1 and sesquilinear forms with non-symmetric domains in Sect. 3.3.2.

The above remark explains why we cannot proceed as in [K66] (see Sect. 3.1): the natural candidate for a sesquilinear form associated with H^θ acts on $\mathfrak{h}^\theta : \mathscr{H}^{1,\overline{\theta}} \times \mathscr{H}^{1,\theta} \longrightarrow \mathbb{C}$ (in order to preserve holomorphy in the first argument). In particular, $\mathfrak{h}^\theta$ has a *non-symmetric* domain.

Let us therefore proceed as in Sect. 3.3.2, and define the relevant data in order to endow the first order space $\mathscr{H}^{1,\theta}$ with a suitable norm, serving later on as first order space $\mathscr{H}^1(H^\theta)$ associated with H^θ:

Definition 3.7.9. For $\theta \in \mathrm{S}_\vartheta$, we define an operator

$$T^\theta : \mathscr{H} \longrightarrow \mathscr{H}, \qquad T^\theta u := u_{\mathrm{int}} \oplus \mathrm{e}^{-\theta/2} u_{\mathrm{ext}}.$$

Moreover, we define a norm on $\mathscr{H}^{1,\theta}$ by

$$\|u\|_{1,\theta}^2 := \|u\|_{1,H^\theta}^2 := \|T^\theta u\|_1^2 = \|u_{\mathrm{int}}\|_1^2 + \mathrm{e}^{-\operatorname{Re}\theta} \|u_{\mathrm{ext}}\|_1^2. \tag{3.84}$$

The dual is defined by $\mathscr{H}^{-1,\theta} = (\mathscr{H}^{1,\overline{\theta}})^*$ with its natural norm (see Definition 3.1.8).

The following facts follow immediately from the definition:

Lemma 3.7.10.

1. *The above defined operator T^θ is bounded and invertible. Moreover,*

$$(T^\theta)^* = T^{\overline{\theta}}, \qquad (T^\theta)^{-1} = T^{-\theta} \qquad \textit{and} \qquad \|T^\theta\|, \|T^{-\theta}\| \le \mathrm{e}^{|\operatorname{Re}\theta|/2}.$$

2. *Let $\Delta := H^0$ and $\mathscr{H}^1 = \operatorname{dom} \Delta^{1/2}$ be the corresponding first order space, then*

$$T^\theta(\operatorname{dom} H^\theta) \subset \mathscr{H}^1.$$

Moreover, denoting by $T^{1,\theta}$ the restriction of T^θ to $\mathscr{H}^{1,\theta}$, then

$$T^{1,\theta} : \mathscr{H}^{1,\theta} \longrightarrow \mathscr{H}^1$$

is unitary.

3. *The closed operator H^θ is compatible with $\Delta = H^0$ with compatibility operator T^θ, see Definition 3.3.8.*
4. *The canonical maps*

$$\mathscr{H}^{1,\mathrm{D}} \overset{\iota_{\mathrm{D},\theta}}{\hookrightarrow} \mathscr{H}^{1,\theta} \hookrightarrow \mathscr{H} \twoheadrightarrow \mathscr{H}^{-1,\theta} \overset{\iota^*_{\mathrm{D},\overline{\theta}}}{\twoheadrightarrow} \mathscr{H}^{-1,\mathrm{D}} \tag{3.85}$$

are bounded by $\mathrm{e}^{|\operatorname{Re}\theta|/2}$.

Proof. We only show the norm bounds of the canonical maps (3.85): The second embedding is bounded since

$$\begin{aligned}\|u\|^2 &\le \|u\|^2_{\mathrm{int}} + \mathrm{e}^{\mathrm{Re}\,\theta}\|u\|^2_{\mathrm{ext}} \le \max\{1, \mathrm{e}^{\mathrm{Re}\,\theta}\}\|T^\theta u\|^2 \\ &\le \max\{1, \mathrm{e}^{\mathrm{Re}\,\theta}\}\|T^\theta u\|^2_1 \le \mathrm{e}^{|\mathrm{Re}\,\theta|}\|u\|^2_{1,\theta}\end{aligned}$$

for $u \in \mathscr{H}^{1,\theta}$. Similarly, we have $\|u\|^2_{1,\theta} \le \max\{1, \mathrm{e}^{-\mathrm{Re}\,\theta}\}\|u\|^2_1$ for $u \in \mathscr{H}^{1,\mathrm{D}}$. □

Let us now extend Krein's resolvent formula also for complex values of θ. In particular, we need the following facts on the Dirichlet solution operator:

Proposition 3.7.11. *Assume that* $0 \le \vartheta < \vartheta_0 < \pi/2$, $\theta \in \mathrm{S}_\vartheta$ *and* $z \in \mathbb{C} \setminus \Sigma_{\vartheta_0}$. *Then the operators*

$$S^{\overline{\theta}}(\overline{z})^*\colon \mathscr{H}^{-1,\theta} \longrightarrow \mathscr{G}^{-1/2}, \qquad S^\theta(z)\colon \mathscr{G}^{1/2} \longrightarrow \mathscr{H}^{1,\theta} \tag{3.86a}$$

have norm bounded by

$$\|S^\theta(z)\|_{1/2\to 1,\theta} \le 1 + C_{\le\vartheta}(z,-1)C^{\le\vartheta}(-1,z) =: C_1^\vartheta(z). \tag{3.86b}$$

Moreover $S^{\overline{\theta}}(\overline{z})^*$ *is an extension of* $B^\theta(z)\colon \mathscr{H} \longrightarrow \mathscr{G}$ *of Proposition 3.7.2. Finally, for* $z \in \mathbb{C} \setminus \Sigma_{\vartheta_0}$, *the operator-valued functions*

$$\mathrm{S}_\vartheta \to \mathscr{L}(\mathscr{H}^{-1,\mathrm{N}}, \mathscr{G}^{-1/2}),\ \theta \mapsto S^{\overline{\theta}}(\overline{z})^* \ \text{and}\ \mathrm{S}_\vartheta \to \mathscr{L}(\mathscr{G}^{1/2}, \mathscr{H}^{1,\mathrm{dec}}),\ \theta \mapsto S^\theta(z)$$

are holomorphic.

Proof. The boundedness of the Dirichlet solution operators in (3.86a) follows from Proposition 3.4.17 and Lemma 3.5.15: Note that a priori, the operator $S_{\mathrm{int}}(z)$ is bounded as operator $\mathscr{G}^{1/2}_{\mathrm{int}} \to \mathscr{H}^1_{\mathrm{int}}$, but the embedding of the coupled space $\mathscr{G}^{1/2}$ into $\mathscr{G}^{1/2}_{\mathrm{int}}$ is bounded by 1, see Lemma 3.6.3. In particular, the norm of $S_{\mathrm{int}}(z)$ as operator $\mathscr{G}^{1/2} \to \mathscr{H}^1$ is the same as the one given in Proposition 3.4.17. A similar remark holds for the exterior part. Therefore, the norm bound (3.86b) follows. Note that the extra factor $\mathrm{e}^{\theta/2}$ in $S^\theta_{\mathrm{ext}}(z)$ (see (3.62)) cancels due to the definition of the norm $\|u\|_{1,\theta} = \|T^\theta u\|_1$.

For the holomorphic dependence, we use Proposition 3.5.16. The embedding $\mathscr{H}^{1,\theta} \hookrightarrow \mathscr{H}^{1,\mathrm{dec}} = \mathscr{H}^1_{\mathrm{int}} \oplus \mathscr{H}^1_{\mathrm{ext}}$ allows us to get rid of the θ-dependence in the space. □

The proof of the following assertion follows from the previous proposition and the fact that the inverse of the Dirichlet-to-Neumann operator $\Lambda(z)^{-1}\colon \mathscr{G}^{-1/2} \longrightarrow \mathscr{G}^{1/2}$ is bounded by Proposition 3.4.34:

Proposition 3.7.12. *Assume that the (undilated) coupled boundary triple* $(\Gamma, \Gamma', \mathscr{G})$ *is elliptic. Assume in addition that* $0 \le \vartheta < \vartheta_0 < \pi/2$, $\theta \in \mathrm{S}_\vartheta$

and $z \in \mathbb{C} \setminus \Sigma_{\vartheta_0}$. *Then the operator*

$$W^{\theta,-1\to 1}(z) := S^\theta(z)\Lambda(z)^{-1}(S^{\overline{\theta}}(\overline{z}))^* \colon \mathscr{H}^{-1,\theta} \xrightarrow{(S^{\overline{\theta}}(\overline{z}))^*} \mathscr{G}^{-1/2} \xrightarrow{\Lambda(z)^{-1}} \mathscr{G}^{1/2} \xrightarrow{S^\theta(z)} \mathscr{H}^{1,\theta}$$

is bounded, and the operator $W^{\theta,-1\to 1}(z)$ *extends* $W^\theta(z)$ *of Proposition 3.7.3. In addition, the operator-valued function*

$$\mathrm{S}_\vartheta \to \mathscr{L}(\mathscr{H}^{-1,\mathrm{dec}}, \mathscr{H}^{1,\mathrm{dec}}), \quad \theta \mapsto W^{\theta,-1\to 1}(z)$$

is holomorphic. Finally, we have the norm bounds

$$\|W^{\theta,-1\to 1}(z)\|_{-1,\theta\to 1,\theta} \le \bigl(C_1^\vartheta(z)\bigr)^2 C(z,-1) \quad \textit{and}$$
$$\|W^\theta(z)\|_{0\to 0} \le \mathrm{e}^{|\mathrm{Re}\,\theta|}\bigl(C_1^\vartheta(z)\bigr)^2 C(z,-1).$$

Definition 3.7.13. Let $\theta \in \mathrm{S}_\vartheta$ for $0 \le \vartheta < \pi/2$. The *complexly dilated coupled sesquilinear form*

$$\mathfrak{h}^\theta \colon \mathscr{H}^{1,\overline{\theta}} \times \mathscr{H}^{1,\theta} \longrightarrow \mathbb{C}$$

associated with $\mathfrak{h} = \mathfrak{h}^0$ is defined by

$$\begin{aligned}\mathfrak{h}^\theta(u,v) &= \mathfrak{h}_{\mathrm{int}}(u,v) + \mathfrak{h}^\theta_{\mathrm{ext}}(u,v)\\ &= \mathfrak{h}_{\mathrm{int}}(u,v) + \mathrm{e}^{-2\theta}\langle u', v'\rangle_{\mathrm{ext}} + \mathfrak{h}^\perp_{\mathrm{ext}}(u,v),\end{aligned}$$

where $\mathscr{H}^{1,\theta}$ is defined in (3.69b), but now also for *complex* $\theta \in \mathrm{S}_\vartheta$.

Recall that $C_1^\vartheta(z)$ defined in (3.86b) depends only on ϑ and z, and similarly for $C_{\le\vartheta}(z,-1)$ and $C(z,-1) = C_0(z,-1)$ (see Lemma 3.5.13 for the definition). Let us now present the main result of this section:

Theorem 3.7.14. *Assume that the coupled boundary triple* $(\Gamma, \Gamma', \mathscr{G})$ *is elliptic (see Theorem 3.6.5 for conditions assuring this). Assume in addition that* $0 \le \vartheta < \vartheta_0 < \pi/2$, $\theta \in \mathrm{S}_\vartheta$ *and* $z \in \mathbb{C} \setminus \Sigma_{\vartheta_0}$. *Then we have the following assertions:*

1. *The complexly dilated coupled operator* H^θ *given in Definition 3.7.6 has spectrum in the sector* $\Sigma_{\vartheta_0} = \{\, z \in \mathbb{C} \mid |\arg z| \le \vartheta \,\}$, *and the family* $\{H^\theta\}_{\theta\in\mathrm{S}_\vartheta}$ *is a self-adjoint, holomorphic family of operators.*
2. *We have the unitary equivalence*

$$H^{\theta_1+\theta_2} = U^{\theta_1} H^{\theta_2} U^{-\theta_1} \tag{3.87}$$

for real θ_1 *and complex* $\theta_2 \in \mathrm{S}_\vartheta$.
3. *The complexly dilated coupled operator* H^θ *has the resolvent norm profile*

$$\|R^\theta(z)\|_{0\to 0} \le \frac{1}{\operatorname{dist}(z, \mathbb{C} \setminus \Sigma_{\vartheta_0})} + e^{|\operatorname{Re}\theta|}\big(C_1^\vartheta(z)\big)^2 C(z,-1) =: \gamma(z),$$

i.e. $H^\theta \in \mathscr{R}(\gamma, \mathbb{C} \setminus \Sigma_{\vartheta_0})$.

4. *We have the resolvent norm profile of order one*

$$\|R^\theta(z)\|_{-1,\theta\to 1,\theta} \le \big(e^{|\operatorname{Re}\theta|} + (C_1^\vartheta(z))^2\big) C_{\le\vartheta}(z,-1) =: \gamma^1(z)$$

and in particular, $H^\theta \in \mathscr{R}_1(\gamma^1, \mathbb{C} \setminus \Sigma_{\vartheta_0})$.

5. *The complexly dilated coupled sesquilinear form* $\mathfrak{h}^\theta$ *is bounded by* $e^{|\operatorname{Re}\theta|}$. *Moreover,* $\mathfrak{h}^\theta$ *is associated with* H^θ.

Proof. (1) follows from Propositions 3.7.4 and 3.7.5. (2) The unitary equivalence holds a priori only for real θ_1 and θ_2. But since both sides are holomorphic in θ_2, equality (3.87) extends therefore to complex $\theta_2 \in S_\vartheta$.

(3) We have $R^\theta(z) = R^{\theta,\mathrm{D}}(z) + W^\theta(z)\colon \mathscr{H} \longrightarrow \mathscr{H}$ by the definition (3.80). The norm estimate follows from Proposition 3.7.1 (6) and Proposition 3.7.12.

(4) The operator

$$R^\theta(z) = \iota_{\mathrm{D},\theta} R^{\theta,\mathrm{D}}(z)\iota^*_{\mathrm{D},\overline{\theta}} + W^\theta(z)\colon \mathscr{H}^{-1,\theta} \longrightarrow \mathscr{H}^{1,\theta}$$

is bounded a priori only for real θ, but the equality holds on the dense subset $\mathscr{H} \subset \mathscr{H}^{-1,\theta}$ also for complex θ by (3.80), and the RHS is bounded. Therefore, the operator is bounded for complex θ. The norm estimate follows from Proposition 3.7.1 (7), (3.85) and Proposition 3.7.12. (5) The estimate on $\mathfrak{h}^\theta$ follows from

$$\begin{aligned}|\mathfrak{h}^\theta(f,g)| &\le e^{-2\operatorname{Re}\theta}|\langle f', g'\rangle_{\mathrm{ext}}| + |\mathfrak{h}^\perp_{\mathrm{ext}}(f,g)| + |\mathfrak{h}_{\mathrm{int}}(f,g)| \\ &\overset{\mathrm{CS}}{\le} e^{|\operatorname{Re}\theta|}\big(\mathfrak{h}(T^{\overline{\theta}}f)\mathfrak{h}(T^\theta g)\big)^{1/2} \le e^{|\operatorname{Re}\theta|}\|f\|_{1,\overline{\theta}}\|g\|_{1,\theta}.\end{aligned}$$

for $f \in \mathscr{H}^{1,\overline{\theta}}$ and $g \in \mathscr{H}^{1,\theta}$. Moreover, that $\mathfrak{h}^\theta$ is associated with H^θ in the sense of Definition 3.3.16 follows easily from Green's formula. Note that we can consider H^θ as the Neumann operator associated with the complexly dilated coupled boundary triple defined formally as in Definition 3.6.9 for complex values of θ, i.e. $\Gamma'^\theta u = 0$ for $u \in \operatorname{dom} H^\theta = \mathscr{H}^{2,\theta}$ (see also (3.83)). □

3.8 Resonances

Let us now define resonances as the eigenvalues of the dilated operators H^θ. Throughout this section, we use the notation of Sects. 3.6 and 3.7. In particular, we assume that the coupled boundary triple is elliptic (see Theorem 3.6.5 for conditions assuring this).

First, we make some general observations on the spectrum of H^θ: We assume that the transversal operator K has discrete spectrum denoted here by $\{\lambda_k(K)\}_k$. We do not exclude the case that the sequence is finite, i.e. that $\mathscr{G}$ is finite-dimensional. This case occurs e.g. on a metric graph with finitely many exterior edges E_{ext}, where $K = 0$ and $\mathscr{G} = \mathbb{C}^{E_{\mathrm{ext}}}$. A typical example of an infinite sequence is the manifold case, where $K = \Delta_Y$ on $\mathscr{G} = \mathsf{L}_2(Y) = \bigoplus_{e \in E_{\mathrm{ext}}} \mathsf{L}_2(Y_e)$, and where each Y_e is a compact manifold.

Proposition 3.8.1. *Assume that K has discrete spectrum. Then the coupled Dirichlet-to-Neumann operator Λ has discrete spectrum. Moreover, the essential spectrum of the coupled dilated operator H^θ is given by*

$$\sigma_{\mathrm{ess}}(H^\theta) = \Big(\bigcup_k \lambda_k(K) + \mathrm{e}^{-2\theta}[0,\infty)\Big) \cup \sigma_{\mathrm{ess}}(H^{\mathrm{D}}_{\mathrm{int}})$$

(see Fig. 3.2). If, in addition, $H^{\mathrm{D}}_{\mathrm{int}}$ in $\mathscr{H}_{\mathrm{int}}$ also has discrete spectrum, then

$$\sigma_{\mathrm{ess}}(H^\theta) = \bigcup_{k \in \mathbb{N}} \lambda_k(K) + \mathrm{e}^{-2\theta}[0,\infty).$$

Proof. By Corollary 3.5.5, we have $\Lambda_{\mathrm{ext}}^2 = K+1$, hence $\Lambda_{\mathrm{ext}}^{-1}$ is compact. Moreover, $\Lambda = \Lambda_{\mathrm{int}} + \Lambda_{\mathrm{ext}} \geq \Lambda_{\mathrm{ext}} \geq 0$ (Proposition 3.6.2), so that $0 \leq \Lambda^{-1} \leq \Lambda_{\mathrm{ext}}^{-1}$, and in particular, Λ^{-1} is also compact.

For the assertion on the essential spectrum, we argue as follows: By definition of the resolvent in (3.80), we have $R^\theta - R^{\theta,\mathrm{D}} = W^\theta$, and the latter operator is compact since Λ^{-1} is compact and $W^\theta = S^\theta \Lambda^{-1} B^\theta$. In particular,

$$\sigma_{\mathrm{ess}}(H^\theta) = \sigma_{\mathrm{ess}}(H^{\theta,\mathrm{D}}) = \sigma_{\mathrm{ess}}(H^{\mathrm{D}}_{\mathrm{int}}) \cup \sigma_{\mathrm{ess}}(H^{\mathrm{D}}_{\mathrm{ext}}).$$

Now the latter spectrum is $\sigma_{\mathrm{ess}}(H^{\mathrm{D}}_{\mathrm{ext}}) = \bigcup_k \lambda_k(K) + \mathrm{e}^{-2\theta}[0,\infty)$ due to Proposition 3.5.14 and the first assertion follows. The second assertion is obvious. □

Theorem 3.8.2. *Assume that the coupled boundary triple $(\Gamma, \Gamma', \mathscr{G})$ is elliptic (see Theorem 3.6.5 for conditions assuring this). Assume in addition that K and $H^{\mathrm{D}}_{\mathrm{int}}$ have purely discrete spectrum:*

1. *The spectrum $\sigma(H^\theta)$ depends only on $\operatorname{Im}\theta$ and $\sigma(H^{\overline{\theta}}) = \overline{\sigma(H^\theta)}$ (complex conjugation).*
2. *The discrete spectrum $\sigma_{\mathrm{disc}}(H^\theta)$ is locally constant in θ. Moreover, if $0 < \operatorname{Im}\theta_1 \leq \operatorname{Im}\theta_2 < \vartheta/2$ then $\sigma_{\mathrm{disc}}(H^{\theta_1}) \subset \sigma_{\mathrm{disc}}(H^{\theta_2})$.*
3. *For $\theta \notin \mathbb{R}$ we have $\sigma(H^\theta) \cap [0,\infty) = \sigma_{\mathrm{p}}(H)$, where $\sigma_{\mathrm{p}}(H)$ denotes the set of eigenvalues of H, i.e. the eigenvalues of H embedded in the continuous spectrum are unveiled (see Fig. 3.2).*
4. *The singular continuous spectrum of H is empty.*

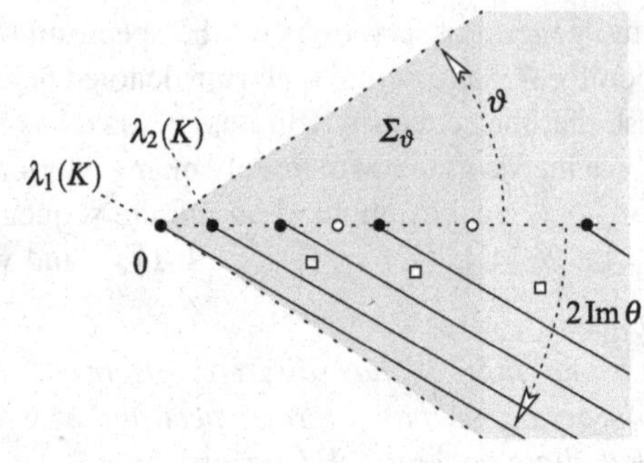

Fig. 3.2 The spectrum of the complexly dilated operator H^θ, the spectrum of the transversal operator K (*black circles*) the unveiled point spectrum (*outlined circles*) and resonances with non-vanishing imaginary part (*outlined squares*)

5. *There is a subspace* $\mathscr{A}$ *satisfying* (3.88) *such that* $\Psi_f(z) := \langle f, (H-z)^{-1} f\rangle$ *has a meromorphic continuation onto the Riemann surface defined by* $w \to \sqrt{w - \lambda_k(K)}$.
6. $\lambda \in \Sigma_{\vartheta_0}$ *is a discrete eigenvalue of* H^θ *iff there exists* $f \in \mathscr{H}$ *such that the meromorphic continuation of* Ψ_f *has a pole in* λ.

Proof. The proof follows closely the proof of [RS80, Thm. XIII.36], so we only comment on the differences. The basic ingredient in the proof is first, the holomorphy of the family $\{H^\theta\}_\theta$ in the sense that the resolvents are holomorphic in θ (see Theorem 3.7.14 (1)). From the unitary equivalence of Theorem 3.7.14 (2) and the fact that $\{H^\theta\}_\theta$ is a self-adjoint family, (1) follows immediately.

In a similar way, (2) follows from the fact that an eigenvalue of H^θ depends holomorphically on θ since $(H^\theta + 1)^{-1}$ depends holomorphically on θ (cf. [K66, Thm. VII.1.8]). In order to prove (3) and (4) as in [RS80], we need the notion of *analytic vectors* with respect to the unitary group U^θ. Note that $A := \mathrm{id}_{\mathscr{H}_{\mathrm{int}}} \oplus A_{\mathrm{ext}}$ with $A_{\mathrm{ext}} := (s\partial_s + \partial_s s)\mathrm{i}/2 \otimes \mathrm{id}$ is the self-adjoint generator of the unitary group $\theta \mapsto U^\theta$. The subspace of analytic vectors is defined by

$$\mathscr{A} := \Big\{ f \in \bigcap_{k\in\mathbb{N}} \operatorname{dom} A^k \;\Big|\; \sum_n \frac{\tau^n}{n!} \|A^n f\| < \infty \Big\}$$

for some $\tau \geq \vartheta/2$ (see e.g. [RS80, Sect. X.6] or [GG91, Sect. 2.5]). It follows that

$$\begin{cases} U^\theta(\mathscr{A}) & \text{is dense in } \mathscr{H} \text{ and} \\ \theta \mapsto U^\theta f & \text{extends holomorphically to a map } \mathrm{S}_\vartheta \to \mathscr{H} \end{cases} \tag{3.88}$$

for all $f \in \mathscr{A}$ using (3.87). The holomorphic extension is then given by

$$f^\theta := U^\theta f = \sum_n \frac{\theta^n}{n!} (\mathrm{i}A)^n f.$$

To prove (5) we just note that a meromorphic continuation of Ψ_f is given by

$$\Psi_f^\theta(z) = \langle f^{\overline{\theta}}, (H^\theta - z)^{-1} f^\theta \rangle$$

since we have $\Psi_f^0(z) = \Psi_f^\theta(z)$ a priori only for real θ but by holomorphy also for $\theta \in \mathrm{S}_\vartheta$. For (6) we argue as follows: If g is an eigenvector of H^θ with eigenvalue λ then let $f := U^{-\theta} g$. Again, by holomorphy, we have $U^\theta f = g$ not only for real, but also for complex θ. In particular, Ψ_f has a pole at λ. On the other side, if $\Psi_f = \Psi_f^\theta$ has a pole at λ, then $\langle f^\theta, \mathbb{1}_{\{\lambda\}} f^\theta \rangle \neq 0$ and in particular, f^θ is an eigenvector of H^θ. □

Motivated by Theorem 3.8.2 (2) and (3), we make the following definition.

Definition 3.8.3. A *resonance* of H is a discrete eigenvalue of the dilated operator H^θ for some $\theta \in \mathrm{S}_\vartheta$ and $0 < \vartheta < \pi/2$.

Strictly speaking, a resonance is only a *non-real* eigenvalue of H^θ. In order to avoid notation like "the embedded eigenvalues and the resonances" we use the convention to call the embedded (real) eigenvalues also resonances.

Let us now prove a theorem analogous to Theorem 3.4.45:

Theorem 3.8.4. *Assume that K and $H_{\mathrm{int}}^{\mathrm{D}}$ have purely discrete spectrum. Assume in addition that the coupled boundary triple $(\Gamma, \Gamma', \mathscr{G})$ is elliptic (see Theorem 3.6.5 for conditions assuring this). Then the following assertions are true:*

1. *Let $z \in \mathbb{C} \setminus \sigma(H^{\mathrm{D},\theta})$, then*

$$\ker(H^\theta - z) = S^\theta(z) \ker \Lambda(z), \tag{3.89a}$$

where $\Lambda(z): \mathscr{G}^{3/2} \longrightarrow \mathscr{G}^{1/2}$ is the (undilated) Dirichlet-to-Neumann operator, and the $S^\theta(z)$ is bijective as map between the kernels.

2. *For $\lambda \in \mathbb{C} \setminus \sigma(H^{\mathrm{D},\theta})$, we have the spectral relation*

$$z \text{ is a resonance of } H \quad \Leftrightarrow \quad 0 \in \sigma(\Lambda(z)). \tag{3.89b}$$

Moreover, the multiplicity of a resonance z is given by $\dim \ker \Lambda(z)$.

3. *Assume that we have the natural embedding* $\operatorname{dom} H_{\mathrm{int}}^{\mathrm{D}} \hookrightarrow \operatorname{dom} H^0$ *given by extending a function f_{int} trivially by 0 (such functions are in general* not *in the domain of* $\operatorname{dom} H^0$*), then all eigenvalues of $H_{\mathrm{int}}^{\mathrm{D}}$ are (real-valued) resonances, i.e., $\sigma(H_{\mathrm{int}}^{\mathrm{D}})$ belongs to the set of resonances of H.*

Proof. (1) For the inclusion "$\subset$", let $h \in \ker(H^\theta - z) \subset \mathscr{W}^{2,\theta}$ and set $\varphi := \Gamma^\theta h$. Note that $\varphi \in \mathscr{G}^{3/2}$: Let $\widetilde{h} = h_{\mathrm{int}} \oplus S_{\mathrm{ext}}^0(z) \Gamma_{\mathrm{int}} h$, then $\Gamma^0 \widetilde{h} \in \mathscr{G}^{3/2}$ since the undilated boundary triple is elliptic. Moreover, $\Gamma^0 \widetilde{h} = \Gamma^\theta h = \varphi$, hence $\varphi \in \mathscr{G}^{3/2}$. Now, by Proposition 3.7.7, we have

$$\Lambda(z)\varphi = \Gamma'^\theta S^\theta \varphi = \Gamma'^\theta h = 0 \tag{3.90}$$

since $h \in \operatorname{dom} H^\theta$ (Proposition 3.7.5).

For the opposite inclusion "$\supset$", let $\varphi \in \ker \Lambda(z)$. By definition, $\varphi \in \mathscr{G}^{3/2}$, so that $h = S^\theta(z)\varphi \in \mathscr{W}^{2,\theta}$ by Proposition 3.7.2. It follows from (3.90) and again Proposition 3.7.5, that $h \in \operatorname{dom} H^\theta$ and therefore $h \in \ker(H^\theta - z)$. The set equality now shows the surjectivity of $S^\theta(z)$ as map between the kernels. The injectivity is clear, since $S^\theta(z)$ has a left inverse, namely Γ^θ.

(2) Let z be a resonance of H, i.e., there is a θ with $|\operatorname{Im}\theta|$ large enough such that $\ker(H^\theta - z) \neq \{0\}$. By (1), this is equivalent to $\ker \Lambda(z) \neq \{0\}$.

(3) This is a trivial observation, since an eigenfunction f_{int} of $H^{\mathrm{D}}_{\mathrm{int}}$ can be extended to an eigenfunction $f = f_{\mathrm{int}} \oplus 0 \in \operatorname{dom} H^0$. Note that $\Gamma'_{\mathrm{int}} f = \Gamma'_{\mathrm{ext}} f = 0$ so that we also have $f \in \operatorname{dom} H^\theta$. □

Remark 3.8.5. Let us comment on the Dirichlet eigenvalues: The situation of Theorem 3.8.4 (3) actually happens for quantum graphs with Kirchhoff conditions, as it can easily be seen (cf. the loop graph example below. This fact is usually related to the violation of the *unique continuation principle*, stating that a solution of the equation $(H - z)u = 0$ which vanishes on an open set, is identically 0. Since the unique continuation principle is true for elliptic partial differential operators, the situation of Theorem 3.8.4 (3) can never appear on a manifold.

Nevertheless, we have the phenomena of a *Helmholtz resonator* as explained in Remark 1.4.6: If the manifold problem converges to a *decoupled problem*, the Dirichlet eigenvalues of the limit problem are seen on the manifold as resonances close to the Dirichlet eigenvalues (see also Corollaries 7.2.8 and 7.2.13).

Theorem 3.8.4 (2) actually gives a simple way to calculate resonances on quantum graphs with standard ("Kirchhoff") vertex conditions:

Example 3.8.6. Let us present here a simple example of a metric graph with resonances. Let G_{int} be a loop graph, i.e., the graph with one vertex v and one edge e_0 such that $\partial_- e_0 = \partial_+ e_0 = v$ and $\ell_{e_0} = 1$. Moreover, we attach $m = \deg_{\mathrm{ext}} v$ exterior edges to v, so that G_{ext} is a star graph consisting of one vertex v and m half-lines attached to v (see Fig. 1.4 for the case $m = 1$). The corresponding combinatorial Laplacian of the interior graph is $\dddot{\Delta}^{\mathrm{comb}}_{\mathrm{int}} = 0$ on $\mathscr{G} = \mathbb{C}$ (see Example 3.6.14 for the definition of $\dddot{\Delta}^{\mathrm{comb}}_{\mathrm{int}}$). We assume a standard vertex condition at v, i.e., $\mathscr{V}_v = \mathbb{C}(1, \dots, 1)$. The corresponding quantum graph Laplacian on the coupled graph $G = G_{\mathrm{int}} \mathbin{\overline{\uplus}} G_{\mathrm{ext}}$ with $G_{\mathrm{int}} \cap G_{\mathrm{ext}} = \{v\}$ is the Kirchhoff Laplacian Δ_G acting as $(\Delta_G f)_e = -f_e''$ for functions $f = \{f_e\}_e$ with $f_e \in \mathsf{H}^2(\mathbb{R}_+)$ for $e \in E_{\mathrm{ext}}$ resp. $f_{e_0} \in \mathsf{H}^2(0,1)$ on the loop. Moreover,

$$f_e(0) = f_{e_0}(0) = f_{e_0}(1) \quad \text{and} \quad f'_{e_0}(1) - f'_{e_0}(0) = \sum_{e \in E_{\mathrm{ext}}} f'_e(0).$$

From (3.73) for the Dirichlet-to-Neumann map of the coupled boundary triple we obtain

$$\Lambda(z) = -\frac{\sqrt{z}}{\sin\sqrt{z}}\Big(1 - \cos\sqrt{z} + \mathrm{i}(\sin\sqrt{z})m\Big).$$

According to Theorem 3.8.4 (2), we have a resonance $z \notin \sigma(H)^{\mathrm{D}} = [0, \infty)$ iff

$$1 - \cos\sqrt{z} + \mathrm{i}(\sin\sqrt{z})m \quad \Leftrightarrow \quad (1 - m/2)w^2 - 2w + (1 + m/2) = 0$$
$$\Leftrightarrow \quad w = 1 \quad \text{or} \quad w = -\frac{m+2}{m-2},$$

where $w = \mathrm{e}^{\mathrm{i}z}$ (the latter fraction is a solution only for $m \neq 2$). Note that $w = 1$ leads to the (formal) solutions $z = (2\pi k)^2$, $k = 1, 2, \ldots$, corresponding to the Dirichlet spectrum of the interior loop graph. By Theorem 3.8.4 (3), they are actually real-valued resonances, since we can extend a Dirichlet eigenfunction on the loop graph by 0 to an eigenfunction of the entire graph.

The other solution corresponds to

$$z = (2\pi k - \mathrm{i}\log 3)^2 \text{ for } m = 1 \quad \text{resp.} \quad z = \Big((2k+1)\pi - \mathrm{i}\log\frac{m+2}{m-2}\Big)^2 \text{ for } m \geq 3,$$

where $k \in \mathbb{Z}$. In particular, the graph with $m = 1$ and $m \geq 3$ exterior ends has non-real resonances of multiplicity 1, whereas the graph with $m = 2$ exterior edges (a line with a loop attached) has no non-real resonance. The appearance or non-appearance of resonances has been discussed in [DEL10] (see also references therein). Note that we considered the case $m = 1$ already in Sect. 1.4.2.

Let us finally comment on the independence of the definition of resonances of various model settings (for the notion $X = X_{\mathrm{int}} \overline{\cup} X_{\mathrm{ext}}$ see Sect. 5.1.2). If $\mathscr{H} = \mathsf{L}_2(X)$ and $X = X_{\mathrm{int}} \overline{\cup} X_{\mathrm{ext}}$ is a manifold (up to a discrete set of points, like the vertices in a metric graph) having a cylindrical end $X_{\mathrm{ext}} = \mathbb{R}_+ \times Y$, we can view the dilation U^θ as induced by the *flow*

$$\Phi^\theta(x) = \begin{cases} x & \text{if } x \in X_{\mathrm{int}}, \\ (\mathrm{e}^\theta s, y) & \text{if } x = (s, y) \in X_{\mathrm{ext}}, \end{cases}$$

i.e. $U^\theta u = \det D\Phi^\theta \cdot (u \circ \Phi^\theta)$ or

$$(U^\theta u)(x) = \begin{cases} u(x) & \text{if } x \in X_{\mathrm{int}}, \\ \mathrm{e}^{\theta/2}(\mathrm{e}^\theta s, y) & \text{if } x = (s, y) \in X_{\mathrm{ext}}. \end{cases}$$

Here, $\{\Phi^\theta\}_{\theta \in \mathbb{R}}$ is a *non-smooth flow*, i.e. Φ^θ is continuous and

$$\Phi^{\theta_1 + \theta_2} = \Phi^{\theta_1} \circ \Phi^{\theta_2} \qquad \text{and} \qquad \Phi^0 = \mathrm{id}_{\mathbb{R}_+}. \tag{3.91}$$

We say that a unitary group $\{U^\theta\}_{\theta \in \mathbb{R}}$ is *generated by a flow* $\{\Phi^\theta\}_{\theta \in \mathbb{R}}$ if

$$U^\theta u := u_{\mathrm{int}} \oplus \big((s, y) \mapsto \big|(\Phi^\theta)'(s)\big|^{1/2} \cdot u(\Phi^\theta(s), y)\big).$$

The non-smoothness comes from the fact that at the boundary between the interior and exterior part, we have a jump in the derivative. Another example of a non-smooth flow Φ_{s_0} is given by

$$\Phi_{s_0}^\theta(s) = \begin{cases} s, & \text{if } 0 \le s < s_0, \\ e^\theta (s - s_0)s, & \text{if } s_0 \le s, \end{cases}$$

i.e. we use a cut into an interior and exterior part at $s_0 \ge 0$.

Remark 3.8.7. We can as well allow *smooth* flows, as done for example in [HiS89]. We call the flow *smooth* if Φ is smooth on $(0, \infty)$ and if $(\Phi^\theta)^{(k)}(0) = 0$ for all $k \in \mathbb{N}_0$. The advantage is a somehow simpler operator domain: The dilated operator and the undilated operator have the same (θ-independent!) domain, but the price is a less natural operator. For example, on a metric graph, the jump condition can be implemented into the vertex condition. Similar as in the following proposition, it can be seen that the definition of a resonance is independent of the concrete flow implementation also for smooth flows, see [HM87].

We can now assure that our definition of resonances does not depend on what (non-smooth) flow $\{\Phi_{s_0}^\theta\}_\theta$ we use:

Proposition 3.8.8. *Assume that the dilation operator $U_{s_0}^\theta$ is generated by the non-smooth flow Φ_{s_0} with cut into an interior and exterior part at $s_0 \ge 0$. Then the definition of resonances (as discrete eigenvalues of the dilated operator $H_{s_0}^\theta = U_{s_0}^\theta H U_{s_0}^{-\theta}$) is independent of s_0.*

Proof. Denote by U^θ the exterior dilation operators associated with the flows Φ^θ and $\Phi_{s_0}^\theta$ defined above, respectively. The main point is to show that there exists a subspace $\mathscr{A}$ which satisfies (3.88) for *both* U^θ *and* $U_{s_0}^\theta$. Note that $A_{s_0,\mathrm{ext}} = ((s - s_0)\partial_s + \partial_s(s - s_0))\mathrm{i}/2 = A_{\mathrm{ext}} - \mathrm{i}s_0\partial_s$ for the generator; in particular, we let $\mathscr{A}$ be the space of analytic functions on $\mathbb{R}_+$ with exponential decay; this space fulfils (3.88) for the generators $A_{s_0} := 0 \oplus A_{\mathrm{ext}}$ and $A_0 = 0 \oplus -\mathrm{i}\partial_s$. Moreover, an eigenvalue of the dilated operator with respect to U^θ or $U_{s_0}^\theta$ is a pole of the meromorphic continuation of $\Psi_f(z) = \langle f, (H - z)^{-1} f \rangle$ for some $f \in \mathscr{A}$, and the latter definition is clearly independent of s_0. □

3.9 Boundary Maps and Triples Coupled via Graphs

Assume that $G = (V, E, \partial)$ is a graph without exterior edges (for the notation related to a graph, we refer to Sect. 2.1.1, especially to Definitions 2.1.1 and 2.1.2; as usual, the graph is locally finite, i.e. $\deg v = |E_v| < \infty$ for all $v \in V$). We start with the definition of a coupled system via boundary maps. Assume that for each $v \in V$ there is a boundary map $\Gamma_v \colon \mathscr{H}_v^1 \longrightarrow \mathscr{G}_v$ associated with the non-negative quadratic

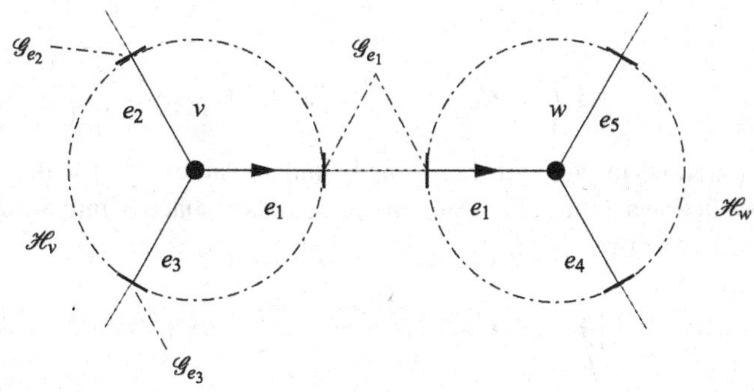

Fig. 3.3 The coupled boundary map $\Gamma : \mathscr{H}^1 \longrightarrow \mathscr{G}$ obtained from the boundary maps $\Gamma_v : \mathscr{H}_v^1 \longrightarrow \mathscr{G}_v$ and $\Gamma_w : \mathscr{H}_w^1 \longrightarrow \mathscr{G}_w$. Here, the boundary space at the vertex v consists of the three boundary spaces at the end-points, i.e. $\mathscr{G}_v = \mathscr{G}_{e_1} \oplus \mathscr{G}_{e_2} \oplus \mathscr{G}_{e_3}$

form $\mathfrak{h}_v$ in the Hilbert space $\mathscr{H}_v$ (see Definition 3.4.1). We assure the following compatibility with the graph structure:

Definition 3.9.1. We say that the family of boundary maps $\{\Gamma_v : \mathscr{H}_v^1 \longrightarrow \mathscr{G}_v\}_{v \in V}$ is *compatible* with the graph $G = (V, E, \partial)$ if for each $e \in E$, there is a Hilbert space $\mathscr{G}_e$, such that

$$\mathscr{G}_v = \bigoplus_{e \in E_v} \mathscr{G}_e$$

and that the boundary map decomposes as

$$\Gamma_v f = \bigoplus_{e \in E_v} \Gamma_{v,e} f$$

for $f \in \mathscr{H}_v^1$, where $\Gamma_{v,e} : \mathscr{H}_v^1 \longrightarrow \mathscr{G}_e$.

For an illustration of the above situation see Fig. 3.3.

Let us now define a *boundary map coupled via the graph G* as follows: Set

$$\mathscr{H} := \bigoplus_{v \in V} \mathscr{H}_v, \qquad \mathscr{H}^{1,\mathrm{dec}} := \bigoplus_{v \in V} \mathscr{H}_v^1 \quad \text{and} \quad \mathscr{G} := \bigoplus_{e \in E} \mathscr{G}_e. \tag{3.92a}$$

The *graph-coupled space of order* 1 is defined as

$$\mathscr{H}^1 := \{ f \in \mathscr{H}^{1,\mathrm{dec}} \mid \Gamma_{\partial_- e, e} f = \Gamma_{\partial_+ e, e} f \ \forall e \in E \}. \tag{3.92b}$$

The *graph-coupled quadratic form* $\mathfrak{h}$ is defined as the restriction of $\bigoplus_v \mathfrak{h}_v$ onto $\mathscr{H}^1$. Moreover, we define

$$\Gamma : \mathscr{H}^1 \longrightarrow \mathscr{G}, \qquad (\Gamma f)_e = \Gamma_{\partial_\pm e, e} f \tag{3.92c}$$

for $f \in \mathscr{H}^1$.

We denote by

$$\mathscr{H}^1_{v,e} := \{ f \in \mathscr{H}^1_v \mid \Gamma_{v,e'} f = 0 \ \forall e' \in E_v \setminus \{e\} \} \tag{3.92d}$$

the space of functions in $\mathscr{H}^1_v$ with vanishing boundary values $\Gamma_{v,e'}$ for all e' except e.

Finally, we define a natural inclusion of the boundary space $\mathscr{G}$ into the *decoupled* boundary space $\mathscr{G}^{\text{dec}}$ by

$$\iota \colon \mathscr{G} = \bigoplus_{e \in E} \mathscr{G}_e \longrightarrow \mathscr{G}^{\text{dec}} := \bigoplus_{v \in V} \mathscr{G}_v, \qquad (\iota\varphi)_v := \varphi_v, \tag{3.92e}$$

where $\varphi_v = \{\varphi_e\}_{e \in E_v}$.

Proposition 3.9.2. *Assume that* $\{\Gamma_v \colon \mathscr{H}^1_v \longrightarrow \mathscr{G}_v\}_{v \in V}$ *is compatible with the graph* $G = (V, E, \partial)$, *that*

$$\sup_{v \in V} \|\Gamma_v\|^2_{1 \to 0} < \infty \tag{3.93a}$$

and that

$$\mathscr{G}^{1/2}_e := \bigcap_{v \in \partial e} \Gamma_{v,e}(\mathscr{H}^1_{v,e}) = \Gamma_{\partial_- e, e}(\mathscr{H}^1_{\partial_- e, e}) \cap \Gamma_{\partial_+ e, e}(\mathscr{H}^1_{\partial_+ e, e}) \tag{3.93b}$$

is dense in $\mathscr{G}_e$ *for all* $e \in E$. *Then the following assertions hold:*

1. *The space* $\mathscr{H}^1$ *is closed in* $\mathscr{H}^{1,\text{dec}}$, *hence* $\mathfrak{h}$ *is a closed non-negative quadratic form in* $\mathscr{H}$.
2. *The operator* $\Gamma \colon \mathscr{H}^1 \longrightarrow \mathscr{G}$ *constructed as above, is a boundary map in the sense of Definition 3.4.1, called the* boundary map coupled via the graph G from the boundary maps Γ_v.
3. *The associated Dirichlet operator is* decoupled *and given by* $H^{\mathrm{D}} = \bigoplus_{v \in V} H^{\mathrm{D}}_v$.
4. *For* $z \notin \sigma(H^{\mathrm{D}}) = \overline{\bigcup_v \sigma(H^{\mathrm{D}}_v)}$, *the associated Dirichlet solution operator* $S(z)$ *is given by* $S(z) = S^{\text{dec}}(z)\iota$, *where* $S^{\text{dec}}(z) := \bigoplus_v S_v(z)$ *and where* $S_v(z)$ *is the Dirichlet solution operator associated with* Γ_v, *i.e.,*

$$S(z)\varphi = \bigoplus_{v \in V} S_v(z)\varphi_v, \qquad \varphi_v = \{\varphi_e\}_{e \in E_v}.$$

5. *The associated Dirichlet-to-Neumann operator* Λ *is given by* $\iota^* \Lambda^{\text{dec}} \iota$, *where* $\Lambda^{\text{dec}} := \bigoplus_v \Lambda_v$, *and where* Λ_v *is the Dirichlet-to-Neumann associated with* Γ_v, *i.e.,*

$$(\Lambda\varphi)_e = \sum_{v \in \partial e} (\Lambda_v \varphi_v)_e.$$

where $(\psi)_e$ *denotes the* e*-th component of* $\psi \in \mathscr{G}_v$.

Proof. (1) Consider the decoupled operator $\Gamma^{\mathrm{dec}} := \bigoplus_v \Gamma_v : \mathscr{H}^{1,\mathrm{dec}} \longrightarrow \mathscr{G}^{\mathrm{dec}}$. By the uniform upper bound (3.93a), Γ^{dec} is bounded. Moreover, $\mathscr{H}^1$ is the preimage of the closed subspace $\iota\mathscr{G} \subset \mathscr{G}^{\mathrm{dec}}$ under Γ^{dec}, hence itself closed in $\mathscr{H}^{1,\mathrm{dec}}$.

(2)–(3) We first check that Γ is a boundary map: The operator Γ is bounded since

$$\begin{aligned}\|\Gamma f\|^2 &= \sum_{e\in E} \|\Gamma_{\partial_\pm e,e} f\|^2 = \frac{1}{2}\sum_{e\in E}\sum_{v\in\partial e} \|\Gamma_{v,e} f\|^2 \\ &= \frac{1}{2}\sum_{v\in V}\sum_{e\in E_v} \|\Gamma_{v,e} f\|^2 = \frac{1}{2}\sum_{v\in V} \|\Gamma_v f\|^2,\end{aligned}$$

using (2.1), and the latter is bounded by $(1/2)\sup_v \|\Gamma_v\|^2_{1\to 0}\|f\|^2_{\mathscr{H}^1}$. Moreover,

$$\mathscr{H}^{1,\mathrm{D}} := \ker \Gamma = \{\, f \in \mathscr{H}^{1,\mathrm{dec}} \mid \Gamma_{v,e} f = 0 \ \forall e \in E_v \ \forall v \in V \,\} = \bigoplus_{v\in V} \mathscr{H}_v^{1,\mathrm{D}},$$

and the latter space is dense in $\mathscr{H}$ since each $\mathscr{H}_v^{1,\mathrm{D}}$ is. Moreover, the space is a direct sum, showing assertion (3). Finally, for the density of $\operatorname{ran}\Gamma$, let $\varphi = \{\varphi_e\}_{e\in E} \in \mathscr{G}$. By the density assumption, there exists $\varphi_{e,n} \in \mathscr{G}_e^{1/2}$ such that $\varphi_{e,n} \to \varphi_e$ in $\mathscr{G}_e$. By definition of $\mathscr{H}_e^{1/2}$, there exist $f_{v,e,n} \in \mathscr{G}^1_{v,e}$ such that $\Gamma_{v,e} f_{v,e,n} = \varphi_{e,n}$. Set $f_{v,n} := \sum_{e'\in E_v} f_{v,e',n}$, then by definition of $\mathscr{H}^1_{v,e}$, we have $\Gamma_{v,e} f_{v,n} = \varphi_{e,n}$ independently of $v = \partial_\pm e$. In particular, $f_n := \{f_{v,n}\}_v \in \mathscr{H}^1$ and $\Gamma f_n = \{\varphi_{e,n}\}_e =: \varphi_n \in \Gamma(\mathscr{H}^1)$. Since φ_n converges componentwise, we have $\varphi_n \to \varphi$, and the density of $\Gamma(\mathscr{H}^1)$ in $\mathscr{G}$ is shown.

(4) Let $h := S^{\mathrm{dec}}(z)\iota\varphi$. Clearly, $\Gamma h = \varphi$, and h is a weak solution, since it is a weak solution on each component.

(5) It is easily seen that the adjoint of ι is given by $(\iota^*\psi)_e = \sum_{v\in\partial e} \psi_{v,e}$ for $\psi \in \mathscr{G}^{\mathrm{dec}}$. The assertion on Λ follows then by Proposition 3.4.11, namely $\Lambda = S^{*1}S$ and a similar assertion for Λ^{dec}. □

For completeness, we also explain the concept of boundary triples coupled via graphs: This actually allows us to give an interpretation of the associated Neumann operator in terms of the boundary map Γ'.

Definition 3.9.3. Assume that $\{\Gamma_v : \mathscr{H}_v^1 \longrightarrow \mathscr{G}_v\}_{v\in V}$ is compatible with the graph $G = (V, E, \partial)$ and that $(\Gamma_v, \Gamma_v', \mathscr{G}_v)$ is a boundary triple associated with $\mathfrak{h}_v$ on $\mathscr{H}_v$. We say that the family of boundary triples $(\Gamma_v, \Gamma_v', \mathscr{G}_v)_{v\in V}$ is *compatible* with the graph $G = (V, E, \partial)$ if the boundary maps $\Gamma_v' : \mathscr{W}_v^2 \longrightarrow \mathscr{G}_v$ decompose as

$$\Gamma_v' f = \bigoplus_{e\in E_v} \Gamma_{v,e}' f$$

for $f \in \mathscr{W}_v^2$, where $\Gamma_{v,e}' : \mathscr{W}_v^2 \longrightarrow \mathscr{G}_e$.

We now define a *boundary triple coupled via a graph* as follows: Set

$$\mathscr{W}^2 := \mathscr{W}^{2,\mathrm{dec}} \cap \mathscr{H}^1, \quad \text{where} \quad \mathscr{W}^{2,\mathrm{dec}} := \bigoplus_{v\in V} \mathscr{W}_v^2, \tag{3.94a}$$

and

$$\Gamma' : \mathscr{W}^2 \longrightarrow \mathscr{G}, \qquad (\Gamma' f)_e = \sum_{v\in\partial e} \Gamma'_{v,e} f = \Gamma'_{\partial_- e,e} f + \Gamma'_{\partial_+ e,e} f \tag{3.94b}$$

for $f \in \mathscr{W}^2$.

The next proposition shows that $(\Gamma, \Gamma', \mathscr{G})$ is indeed a boundary triple, called the *boundary triple coupled via the graph G from the boundary triples* $(\Gamma_v, \Gamma'_v, \mathscr{G}_v)$.

Proposition 3.9.4. *Assume that* $(\Gamma_v, \Gamma'_v, \mathscr{G}_v)_{v\in V}$ *is compatible with the graph* $G = (V, E, \partial)$ *and that, in addition,* (3.93a), (3.93b),

$$\sup_{v\in V} \|\Gamma'_v\|^2_{\mathscr{W}_v^2\to\mathscr{G}_v} < \infty \qquad \text{and} \qquad \sup_{v\in V} \|\check{H}_v\|^2_{\mathscr{W}_v^2\to\mathscr{G}_v} < \infty \tag{3.95a}$$

are fulfilled. Then the following assertions hold:

1. $(\Gamma, \Gamma', \mathscr{G})$*, as constructed above, is a boundary triple with* $\check{H}$ *being the restriction of* $\check{H}^{\mathrm{dec}} := \bigoplus_v \check{H}_v$ *to* $\mathscr{W}^2$*.*
2. *The associated Neumann operator, denoted by* $H = H^{\mathrm{N}}$*, has*
$$\mathscr{W}^{2,\mathrm{N}} := \{\, f \in \mathscr{W}^2 \mid \Gamma_{\partial_- e,e} f = \Gamma_{\partial_+ e,e} f, \quad \Gamma'_{\partial_- e,e} f + \Gamma'_{\partial_+ e,e} f = 0 \ \forall e \in E \,\} \tag{3.95b}$$
in its domain, i.e., $\mathscr{W}^{2,\mathrm{N}} \subset \operatorname{dom} H$*. Moreover,* $Hf = \bigoplus_v \check{H}_v f$ *for* $f \in \mathscr{W}^{2,\mathrm{N}}$*. In particular, the Neumann operator is* coupled, *i.e,* H^{N} *cannot be written as direct sum of the individual Neumann operators.*
3. *Assume that* $(\Gamma, \Gamma', \mathscr{G})$ *is an elliptic boundary triple, and let* $z \in \mathbb{C} \setminus \sigma(H^{\mathrm{D}})$*. Then the associated Dirichlet-to-Neumann operator* $\Lambda(z) : \mathscr{G}^{3/2} \longrightarrow \mathscr{G}^{1/2}$ *can be written as*
$$(\Lambda(z)\varphi)_e = \sum_{v\in\partial e} (\Lambda_v(z)\varphi_v)_e .$$

Proof. (1) A similar argument as for Proposition 3.9.2 (1) shows that $\mathscr{W}^2$ is closed in $\mathscr{W}^{2,\mathrm{dec}}$ (using the fact that $\|f\|_{\mathscr{W}^2} \le \|f\|_{\mathscr{H}^1}$), hence itself a Hilbert space. Moreover, $\Gamma' : \mathscr{W}^2 \longrightarrow \mathscr{G}$ is bounded, since

$$\|\Gamma' f\|^2 \le 2 \sum_{e\in E} \sum_{v\in\partial e} \|\Gamma'_{v,e} f\|^2 = 2 \sum_{v\in V} \sum_{e\in E_v} \|\Gamma'_{v,e} f\|^2 = 2 \sum_{v\in V} \|\Gamma'_v f\|^2,$$

using (2.1), and the latter is bounded by $2 \sup_v \|\Gamma'_v\|^2_{\mathscr{W}_v^2\to\mathscr{G}_v} \|f\|^2_{\mathscr{W}^2}$. The second assumption of (3.95a) assures that $\check{H}$ is bounded as operator $\mathscr{W}^2 \to \mathscr{H}$. Green's formula follows from

$$\begin{aligned}\langle f, \check{H} g\rangle &= \sum_{v\in V}\langle f, \check{H} g\rangle_v = \sum_{v\in V}\big(\mathfrak{h}_v(f,g) - \langle \Gamma_v f, \Gamma_v' g\rangle_{\mathscr{G}_v}\big)\\ &= \mathfrak{h}(f,g) - \sum_{v\in V}\sum_{e\in E_v}\langle \Gamma_{v,e} f, \Gamma'_{v,e} g\rangle_{\mathscr{G}_e}\\ &= \mathfrak{h}(f,g) - \sum_{e\in E}\Big\langle \Gamma_{\partial_\pm e,e} f, \sum_{v\in\partial e}\Gamma'_{v,e} g\Big\rangle_{\mathscr{G}_e}\\ &= \mathfrak{h}(f,g) - \langle \Gamma f, \Gamma' g\rangle_{\mathscr{G}}\end{aligned}$$

for all $f \in \mathscr{H}^1$ and $g \in \mathscr{W}^2$, where we used Green's formula for each v in the second equality, the reordering (2.1) in the third and the fact that $\Gamma_{\partial_- e,e} f = \Gamma_{\partial_+ e,e} f$ in the fourth equation. Since $\mathring{\mathscr{W}}_v^2$ is dense in $\mathscr{H}_v$ and since $\bigoplus_{v\in V}\mathring{\mathscr{W}}_v^2 \subset \mathring{\mathscr{W}}^2 = \ker\Gamma \cap \ker\Gamma'$, the density of $\mathring{\mathscr{W}}^2$ in $\mathscr{H}$ follows.

(2) follows from Proposition 3.4.5.

(3) From the ellipticity of $(\Gamma, \Gamma', \mathscr{G})$ it follows that $h = S(z)\varphi \in \mathscr{W}^2$ and also $h_v \in \mathscr{W}_v^2$ by Proposition 3.9.2 (4). Now we have

$$(\Lambda(z)\varphi)_e = (\Gamma' S(z)\varphi)_e = \sum_{v\in\partial e}\Gamma'_{v,e} S_v(z)\varphi_e = \sum_{v\in\partial e}(\Lambda_v(z)\varphi_v)_e$$

by Proposition 3.4.13. The ellipticity of $(\Gamma, \Gamma', \mathscr{G})$ ensures that $\Lambda(z)$ is bounded as operator $\mathscr{G}^{3/2} \to \mathscr{G}^{1/2}$. □

Remark 3.9.5.

1. We do not provide general criteria assuring the ellipticity of the coupled boundary triple from assumptions on the boundary triples $(\Gamma_v, \Gamma_v', \mathscr{G}_v)$. We will treat such questions in a forthcoming publication. Note that we used the ellipticity of $(\Gamma, \Gamma', \mathscr{G})$ in Proposition 3.9.4 (3) only for the fact that the Dirichlet solution $h = S(z)\varphi$ is in $\mathscr{W}^2$, and that $\Lambda(z): \mathscr{G}^{3/2} \longrightarrow \mathscr{G}^{1/2}$. For the formula for $\Lambda(z)$, the fact $h \in \mathscr{W}^2$ is sufficient.
2. Actually, we only need the coupling of boundary maps and not of boundary triples, as for example in Sect. 4.8 and in Chap. 7. In particular, the coupling is only used in order to pass from the estimates of quadratic forms on star-shaped graph and its corresponding manifold to a general graph and its corresponding graph-like manifold.

Example 3.9.6. Let us illustrate the previous result on a simple example. Denote by G the graph consisting of one edge e and its two boundary vertices $v_\pm = \partial_\pm e$ only. Then $\mathscr{G} = \mathscr{G}_- \oplus \mathscr{G}_+$, where $(\Gamma_\pm, \Gamma'_\pm, \mathscr{G}_\pm)$ denote the boundary triples associated with the vertex $v_\pm$, and by assumption, $\mathscr{G}_\pm = \mathscr{G}_e$. This is precisely the situation of an "interior" and "exterior" boundary triple coupled as in Sect. 3.6.1. In particular, $\iota: \mathscr{G}_e \longrightarrow \mathscr{G}_e \oplus \mathscr{G}_e$ is the diagonal map $\iota\varphi = \varphi \oplus \varphi$ with adjoint $\iota^*(\psi_1 \oplus \psi_2) = \psi_1 + \psi_2$. Therefore, the Dirichlet-to-Neumann map is given by $\Lambda = \Lambda_- + \Lambda_+$, as we have already seen in Proposition 3.6.2.

Chapter 4
The Functional Analytic Part: Two Operators in Different Hilbert Spaces

In this chapter, we develop a general framework in order to deal with operators acting in different Hilbert spaces, but being "close" to each other (in a sense to be made precise). In particular, we generalise the concept of *norm resolvent convergence*. The considerations are quite general, since we have further applications in mind (not all of them included in this work). To our knowledge, such a theory of *norm* convergence of (functions of) operators acting in different Hilbert spaces is new.

In Sect. 4.1, we start our analysis of operators in different spaces with some results on *identification operators*, generalising *quantitatively* the notion of unitarity. In particular, we introduce the notion "δ-quasi-unitary" (resp. the notion "δ-partial isometric" implying the former) for operators $J: \mathscr{H} \longrightarrow \widetilde{\mathscr{H}}$ and $J': \widetilde{\mathscr{H}} \longrightarrow \mathscr{H}$, "identifying" the Hilbert spaces $\mathscr{H}$ and $\widetilde{\mathscr{H}}$ in the sense that, for $\delta = 0$, J and J' are unitary. The notion of "δ-quasi-unitarity" and "δ-partial isometry" is usually related to scales of Hilbert spaces $\mathscr{H}^k(\Delta)$ and $\mathscr{H}^k(\widetilde{\Delta})$ associated with operators Δ and $\widetilde{\Delta}$ on $\mathscr{H}$ and on $\widetilde{\mathscr{H}}$, respectively. Note that for $\delta > 0$, J and J' need not to be bijective, as in many of our examples.

In Sect. 4.2, we analyse non-negative operators $\Delta \geq 0$ and $\widetilde{\Delta} \geq 0$ on $\mathscr{H}$ and $\widetilde{\mathscr{H}}$. For the "untilded" objects, the reader should have a quantum graph $(G, \mathscr{V})$ with $\mathscr{H} = \mathsf{L}_2(G)$ and $\Delta = \Delta_{(G,\mathscr{V})}$ in mind (see Sect. 2.2), the "tilded" objects are associated with a corresponding graph-like manifold X_ε with $\widetilde{\mathscr{H}} = \mathsf{L}_2(X_\varepsilon)$ and Laplacian $\widetilde{\Delta} = \Delta_{X_\varepsilon}$ (see Sect. 5.1). The above mentioned identification operators allow us to define the notion "δ-quasi-unitary equivalence" of Δ and $\widetilde{\Delta}$, generalising at the same time the notion of "unitary equivalent" (if $\delta = 0$) and the notion of "norm resolvent convergence" $\|(\Delta + 1)^{-1} - (\widetilde{\Delta} + 1)^{-1}\| \to 0$ if $\delta \to 0$ in the case that $\mathscr{H} = \widetilde{\mathscr{H}}$, $J = \mathrm{id}$, $J' = \mathrm{id}$ and that $\widetilde{\Delta}$ depends on δ. From this notion, we can deduce norm estimates on the difference of $\varphi(\Delta)$ and $\varphi(\widetilde{\Delta})$, e.g. of the heat operators ($\varphi_t(\lambda) = \mathrm{e}^{-t\lambda}$, Theorem 4.2.9), of spectral projectors ($\varphi = \mathbb{1}_I$, Theorem 4.2.10) or of the unitary evolutions ($\varphi_t(\lambda) = \mathrm{e}^{\mathrm{i}t\lambda}$ Theorem 4.2.15).

In Sect. 4.3 we conclude the convergence of the spectrum, namely the discrete, essential and entire spectrum and of related eigenfunctions, for operators Δ and Δ_n being δ_n-quasi-unitarily equivalent with $\delta_n \to 0$. Sect. 4.4 extends the convergence

O. Post, *Spectral Analysis on Graph-Like Spaces*, Lecture Notes in Mathematics 2039, DOI 10.1007/978-3-642-23840-6_4, © Springer-Verlag Berlin Heidelberg 2012

of non-negative operators to their associated quadratic forms $\mathfrak{d}$ and $\widetilde{\mathfrak{d}}$. Do to so, we need also identification operators "of order 1", respecting the quadratic form domains $\mathscr{H}^1$ and $\widetilde{\mathscr{H}}^1$.

Sects. 4.5–4.7 contain the corresponding theory of generalised norm resolvent convergence, spectral convergence and convergence of sesquilinear forms for closed and, in general, non-self-adjoint operators. Since for such operators H, there is usually no easily expressed upper estimate on the norm of the resolvent $\|(H - z)^{-1}\|$ like $\|(\Delta - z)^{-1}\| = \operatorname{dist}(z, \sigma(\Delta))^{-1}$ for a self-adjoint operator Δ, we assume such an estimate here as in input (called *resolvent norm profile*, see Sect. 3.3).

The general closed operators we have in mind here, arise from the complex dilation method (see Sects. 3.6–3.8). The analysis is more complicated than for non-negative operators, and the results are less strong (e.g. we only show convergence of the discrete spectrum). Indeed, this is enough for our main application, since we are interested in the convergence of resonances, defined as discrete eigenvalues of a complexly dilated operator, shown in Sect. 4.9. For the special case of *sectorial* operators, we developed a convergence theory of sectorial operators acting in different Hilbert spaces, in [MNP10].

Finally, in Sect. 4.8 we provide results how to pass from local estimates to global estimates using the concept of boundary maps coupled via a graph (see Sect. 3.9). This is used in our analysis of graph-like spaces from the passage of star-shaped graphs and their graph-like manifolds to general graphs and their associated graph-like manifolds.

Note that our abstract convergence of operators in different Hilbert spaces also applies to other situations, where the underlying space is perturbed. For example, we can remove small balls or other submanifolds from a given manifolds and compare the original Laplacian on a manifold with the Dirichlet or Neumann Laplacian on the perturbed manifold. Moreover, we can add small handles to the manifold and again compare the two Laplacians. Such problems have been treated e.g. in [RT75, CF78, CF81, A87]. Moreover, we can treat conformal perturbations of the manifold (see e.g. [CdV86, P03b]). For more details, we refer to Sect. 1.6.1.

The reader only interested in the applications to graph-like spaces may just consult Proposition 4.4.13 which contains a summary of the assumptions for "δ-partial isometry" for non-negative quadratic forms. This notion implies the convergence of the spectrum and functions of the operators. In particular, in most of the results of Chaps. 6 and 7, we prove the δ-partial isometry of the quadratic forms associated with a metric graph and a graph-like space.

4.1 Quasi-Unitary Identification Operators

The aim of the present section is to give a quantitative generalisation of unitary operators. The deviation of being unitary is measured by a constant $\delta \geq 0$; the case $\delta = 0$ leads to the unitary case (see Lemma 4.1.4). We often refer to the constant δ itself as *error* or *deviation*.

In order to compare two spaces and operators, we need the following identification operators. Recall the definition of "C-densely embedded" in Definition 3.1.4.

Definition 4.1.1. Let $\delta \geq 0$, $C > 0$ and let

$$J \colon \mathscr{H} \longrightarrow \widetilde{\mathscr{H}}, \qquad\qquad J' \colon \widetilde{\mathscr{H}} \longrightarrow \mathscr{H}$$

be two bounded operators, sometimes called *identification operators*, and assume that $\mathscr{D} \subset \mathscr{H}$, $\widetilde{\mathscr{D}} \subset \widetilde{\mathscr{H}}$ are C-densely embedded subspaces.

1. We say that J and J' are *δ-quasi-unitary* with respect to $\mathscr{D}$ and $\widetilde{\mathscr{D}}$, if the following estimates are fulfilled:[1]

$$\|J\| \leq 2, \tag{4.1a}$$

$$\|J - J'^*\| \leq \delta, \tag{4.1b}$$

$$\|\mathrm{id} - J'J\|_{\mathscr{D} \to \mathscr{H}} \leq \delta \qquad \text{and} \qquad \|\widetilde{\mathrm{id}} - JJ'\|_{\widetilde{\mathscr{D}} \to \widetilde{\mathscr{H}}} \leq \delta. \tag{4.1c}$$

2. We say that J is *δ-partial isometric* with respect to $\mathscr{D}$ and $\widetilde{\mathscr{D}}$, if

$$J^*J = \mathrm{id} \qquad \text{and} \qquad \|\widetilde{\mathrm{id}} - JJ^*\|_{\widetilde{\mathscr{D}} \to \widetilde{\mathscr{H}}} \leq \delta. \tag{4.1d}$$

3. If $\mathscr{D} = \mathscr{H}^k(H)$ and $\widetilde{\mathscr{D}} = \mathscr{H}^k(\widetilde{H})$ for some $0 \leq k \leq 2$ (see Sect. 3.3), we also say that J is *δ-quasi-unitary (δ-partial isometric) of order k with respect to H and $\widetilde{H}$* (or w.r.t. the sesquilinear forms $\mathfrak{h}$ and $\widetilde{\mathfrak{h}}$ if $k = 1$).
4. If $\mathscr{D} = \mathscr{H}^k$ and $\widetilde{\mathscr{D}} = \widetilde{\mathscr{H}}^k$ for some $k \geq 0$ are the scales of order k associated with the non-negative operators Δ and $\widetilde{\Delta}$ (or w.r.t. the quadratic forms $\mathfrak{d}$ and $\widetilde{\mathfrak{d}}$ if $k = 1$, see Sect. 3.2), then we say that J is *δ-quasi-unitary (δ-partial isometric) of order k*, without referring explicitly to Δ and $\widetilde{\Delta}$.

Remark 4.1.2.

1. $\|A\|_{\mathscr{D} \to \mathscr{H}}$ refers to the operator norm of the operator $A \colon \mathscr{D} \longrightarrow \mathscr{H}$. For example, if $\mathscr{D} = \mathscr{H}^2(H)$, then we can express the norm as

$$\|A\|_{2,H \to 0} = \|AR\|, \tag{4.2a}$$

where $R = R(z_0)$ is the resolvent of H in z_0, and if $\mathscr{D} = \mathscr{H}^1(H)$, then

$$\|A\|_{1,H \to 0} = \|AT^{-1}(\Delta + 1)^{-1/2}\| \tag{4.2b}$$

[1] Here and in the sequel, id refers to the identity on $\mathscr{H}$ as well as the embedding $\iota \colon \mathscr{D} \hookrightarrow \mathscr{H}$, and similarly for $\widetilde{\mathrm{id}}$.

and similarly on $\widetilde{\mathscr{H}}$. If $\mathscr{D} = \mathscr{H}^k$ is the scale associated with $\Delta \geq 0$, then

$$\|A\|_{k\to 0} = \|A(\Delta+1)^{-k/2}\|. \tag{4.2c}$$

2. Any other constant larger than 1 as estimate in (4.1a) would do the same, but for simplicity we choose 2. In most applications, we can show that $\|J\| \leq 1 + \mathrm{O}(\sqrt{\delta})$, or even $\|J\| = 1$, e.g. if J is δ-partial isometric, as we will see below.
3. We mostly omit the dependence on C in the notation, especially if $C = 1$ as in the case of non-negative operators.

We deduce the following simple estimates:

Lemma 4.1.3.

1. Assume that J and J' are δ-quasi-unitary, then

$$\|J'\| \leq \|J\| + \delta \leq 2 + \delta, \tag{4.3a}$$

$$\big|\|Jf\| - \|f\|\big| \leq (2C+1)\delta\|f\|_{\mathscr{D}}, \tag{4.3b}$$

$$\big|\|J'u\| - \|u\|\big| \leq (2C+1)\delta\|u\|_{\widetilde{\mathscr{D}}}, \tag{4.3c}$$

$$\|\widetilde{\mathrm{id}} - J'^* J^*\|_{\widetilde{\mathscr{D}}\to\widetilde{\mathscr{H}}} \leq \big(1 + 2(2+\delta)C\big)\delta \tag{4.3d}$$

2. If J is δ-partial isometric, then

$$\|Jf\| = \|f\| \qquad \textit{and} \qquad \big|\|J^*u\| - \|u\|\big| \leq \delta\|u\|_{\widetilde{\mathscr{D}}}. \tag{4.3e}$$

Proof. For the estimate on J' note that

$$\|J'\| \leq \|J' - J^*\| + \|J^*\| \leq \delta + \|J\|$$

and we are done. Moreover, we have

$$\begin{aligned}
\big|\|Jf\|^2 - \|f\|^2\big| &= \big|\langle (J^*J - \mathrm{id})f, f\rangle\big| \\
&\leq \big|\langle (J^* - J')Jf, f\rangle\big| + \big|\langle (J'J - \mathrm{id})f, f\rangle\big| \\
&\leq \|J^* - J'\|\|Jf\|\|f\| + \|J'J - \mathrm{id}\|_{\mathscr{D}\to\mathscr{H}}\|f\|_{\mathscr{D}}\|f\| \\
&\leq \big(2C\delta + \delta\big)\|f\|_{\mathscr{D}}\|f\|.
\end{aligned}$$

If $f \neq 0$, then

$$\big|\|Jf\| - \|f\|\big| = \frac{\big|\|Jf\|^2 - \|f\|^2\big|}{\|Jf\| + \|f\|} \leq \frac{\big|\|Jf\|^2 - \|f\|^2\big|}{\|f\|} \leq (2C+1)\delta\|f\|_{\mathscr{D}}.$$

Estimate (4.3c) follows similarly. For (4.3d) we use

$$\begin{aligned}\|u - J'^* J^* u\| &\le \|u - JJ'u\| + \|(J - J'^*)J'u\| + \|J'^*(J' - J^*)u\| \\ &\le \delta\big(\|u\|_{\widetilde{\mathscr{D}}} + 2\|J'\|\,\|u\|\big) \le \delta\big(1 + 2(2+\delta)C\big)\|u\|_{\widetilde{\mathscr{D}}}.\end{aligned}$$

If J is a δ-partial isometry, then

$$\|Jf\|^2 = \langle f, J^*Jf\rangle = \|f\|^2,$$

and the second estimate can be seen as before.

□

We have the following simple consequences, justifying the names:

Lemma 4.1.4.

1. *If J and J' are 0-quasi-unitary, then $J' = J^*$, and J resp. J' are unitary operators.*
2. *If J is δ-partial isometric, then $\widetilde{P} = JJ^*$ is an orthogonal projection in $\widetilde{\mathscr{H}}$, $\mathscr{H}$ is isometrically embedded into $\widetilde{\mathscr{H}}$ via J, and J, J^* are δ-quasi-unitary.*
3. *If J is δ-partial isometric, then an orthonormal basis $\{f_n\}_n$ of $\mathscr{H}$ is an orthonormal basis of $J(\mathscr{H}) = \operatorname{ran} J = \operatorname{ran} \widetilde{P} \subset \widetilde{\mathscr{H}}$.*

Proof. For the first assertion, note that $J = J'^*$, $\mathrm{id} = J'J$ and $\widetilde{\mathrm{id}} = JJ'$. The second assertion follows from

$$\widetilde{P}^2 = JJ^*JJ^* = JJ^* = \widetilde{P} \qquad \text{and} \qquad \widetilde{P}^* = \widetilde{P}.$$

Moreover, $\|Jf\| = \|f\|$ implies $\|J\| = 1$ and therefore also $\|J^*\| = 1$. The last assertion is easily seen. □

Let us comment on the convergence of orthogonal projections. In our applications, the projections will be spectral projections on bounded intervals or domains.

Theorem 4.1.5. *Let $\alpha \ge 0$, $\delta \ge 0$ and $C > 0$. Assume that $\mathscr{D}$ and $\widetilde{\mathscr{D}}$ are C-densely embedded in $\mathscr{H}$ and $\widetilde{\mathscr{H}}$, and that P, $\widetilde{P}$ are projections in $\mathscr{H}$ and $\widetilde{\mathscr{H}}$, respectively, such that*

$$\|JP - \widetilde{P}J\| \le \alpha\delta.$$

1. *Assume in addition that $\|\mathrm{id} - J'J\|_{\mathscr{D} \to \mathscr{H}} \le \delta$ and that $P\colon \mathscr{H} \longrightarrow \mathscr{D}$ is bounded. Then*[2]

   $$\dim \operatorname{ran} P \le \dim \operatorname{ran} \widetilde{P}$$

   provided $0 < \delta < \delta_0$, where

[2]The case $\dim \operatorname{ran} P = \infty$ is included.

$$\delta_0 := \bigl((2C+1)\|P\|_{\mathscr{H}\to\mathscr{D}} + \alpha\bigr)^{-1}.$$

Moreover, we have

$$\|\widetilde{P}Jf\| \ge \Bigl(1 - \frac{\delta}{\delta_0}\Bigr)\|f\| \tag{4.4}$$

for all $f \in \operatorname{ran} P$.

2. *Assume that* J *and* J' *are* δ*-quasi-unitary w.r.t.* $\mathscr{D}$ *and* $\widetilde{\mathscr{D}}$ *and that*

$$P\colon \mathscr{H} \longrightarrow \mathscr{D} \qquad \text{and} \qquad \widetilde{P}\colon \widetilde{\mathscr{H}} \longrightarrow \widetilde{\mathscr{D}}$$

are bounded. Then

$$\dim \operatorname{ran} P = \dim \operatorname{ran} \widetilde{P}$$

provided $0 < \delta < \delta_0$, *where*

$$\delta_0 := \bigl((2C+1)\max\{\|P\|_{\mathscr{H}\to\mathscr{D}}, \|\widetilde{P}\|_{\widetilde{\mathscr{H}}\to\widetilde{\mathscr{D}}}\} + \alpha\bigr)^{-1}.$$

3. *If, in addition to* (2), $\operatorname{ran} P = \mathbb{C}\psi$ *with* $\|\psi\| = 1$, *then there exists a normalised vector* $\widetilde{\psi} \in \widetilde{\mathscr{H}}$ *generating* $\operatorname{ran} \widetilde{P}$ *such that*

$$\|J\psi - \widetilde{\psi}\| \le C_1\delta \qquad \text{and} \qquad \|J'\widetilde{\psi} - \psi\| \le \widetilde{C}_1\delta,$$

where

$$C_1 := 2\alpha + (2C+1)\|P\|_{\mathscr{H}\to\mathscr{D}} \qquad \text{and} \qquad \widetilde{C}_1 = 3C_1 + \|P\|_{\mathscr{H}\to\mathscr{D}}$$

provided $\delta \le 1$ *for the second estimate.*

Proof. (1) Let us first show the inequality $\dim \operatorname{ran} P \le \dim \operatorname{ran} \widetilde{P}$: Suppose $f \in \operatorname{ran} P = P(\mathscr{H})$. Then

$$\|f\| - \|Jf\| \le (2C+1)\delta\|f\|_{\mathscr{D}} = (2C+1)\delta\|Pf\|_{\mathscr{D}} \le (2C+1)\delta\|P\|_{\mathscr{H}\to\mathscr{D}}\|f\|$$

by Lemma 4.1.3, $f = Pf$ and the assumption. Therefore, we have

$$\begin{aligned}
\|\widetilde{P}Jf\| &\ge \|JPf\| - \|(\widetilde{P}J - JP)f\| \\
&\ge \|Jf\| - \alpha\delta\|f\| \ge \bigl(1 - ((2C+1)\|P\|_{\mathscr{H}\to\mathscr{D}} + \alpha)\delta\bigr)\|f\| \\
&= (1 - \delta/\delta_0)\|f\|
\end{aligned}$$

using the assumption. In particular, if $0 < \delta < \delta_0$, then $\widetilde{P}J \restriction_{P(\mathscr{H})}$ is injective. Moreover, if $f_1, \dots, f_d$ are linear independent in $P(\mathscr{H})$, the same is true for $\widetilde{P}Jf_1, \dots, \widetilde{P}Jf_d$ in $\widetilde{P}(\widetilde{\mathscr{H}})$. If $P(\mathscr{H})$ is infinite-dimensional so is $\widetilde{P}(\widetilde{\mathscr{H}})$. Thus we have shown $\dim \operatorname{ran} P \le \dim \operatorname{ran} \widetilde{P}$.

(2) The remaining point to show is the opposite inequality on the dimension of the projection ranges. The same argument as in (1) with J replaced by J' and P, $\widetilde{P}$ interchanged leads to the result.

(3) For the convergence of the ranges in the one-dimensional case note that

$$\widetilde{\psi} = \frac{1}{\|\widetilde{P} J \psi\|} \widetilde{P} J \psi$$

since $\widetilde{P}$ is a one-dimensional projection and since $\|\widetilde{P} J \psi\| > 0$ for $0 < \delta < \delta_0$ by (4.4). Now we have

$$\begin{aligned}
\|J\psi - \widetilde{\psi}\| &= \Big\| JP\psi - \frac{1}{\|\widetilde{P} J \psi\|} \widetilde{P} J \psi \Big\| \\
&\le \|(JP - \widetilde{P} J)\psi\| + \Big| 1 - \frac{1}{\|\widetilde{P} J \psi\|} \Big| \|\widetilde{P} J \psi\| \\
&= \|(JP - \widetilde{P} J)\psi\| + \big| \|\widetilde{P} J \psi\| - 1 \big| \\
&\le 2\|(JP - \widetilde{P} J)\psi\| + \big| \|JP\psi\| - 1 \big| \\
&= 2\|(JP - \widetilde{P} J)\psi\| + \big| \|J\psi\| - \|\psi\| \big| \\
&\le \big(2\alpha + (2C+1)\|P\|_{\mathscr{H} \to \mathscr{D}}\big)\delta =: C_1 \delta
\end{aligned}$$

since $\psi = P\psi$ and $\|\psi\| = 1$ and using (4.3b). The second estimate follows immediately from

$$\|J'\widetilde{\psi} - \psi\| \le \|J'(\widetilde{\psi} - J\psi)\| + \|(\mathrm{id} - J'J)\psi\| \le ((2+\delta)C_1 + \|P\|_{\mathscr{H} \to \mathscr{D}})\delta$$

using (4.1c) and (4.3a). □

For simplicity, we only considered the case of a one-dimensional projection. In principle, one can also formulate the convergence of the projection ranges for higher dimensions, cf. [K66, G08a] for the convergence of linear subspaces.

Let us finally derive some simple relations between different orders. We start with the scales of Δ and $\widetilde{\Delta}$:

Lemma 4.1.6. *Assume that J is δ-quasi-unitary (δ-partial isometric) of order $k \ge 0$, then J is also δ-quasi-unitary (δ-partial isometric) of order $m \ge k$.*

Proof. The assertion follows from the estimate $\|f\|_k \le \|f\|_m$ and similarly on $\widetilde{\mathscr{H}}$. □

Recall the definition of C-compatibility of order 1 in Definition 3.3.10 (i.e. $H, \widetilde{H} \in \mathscr{R}_1(C, z_0)$):

Lemma 4.1.7. *Assume that $H, \widetilde{H} \in \mathscr{R}_1(C, z_0)$ and that J is δ-quasi-unitary (δ-partial isometric) of order 1 w.r.t. H and $\widetilde{H}$, then J is $C^2\delta$-quasi-unitary ($C^2\delta$-partial isometric) of order 2 w.r.t. H and $\widetilde{H}$.*

Proof. The assertion follows from $\|f\|_{1,H} \le C^2\|f\|_{2,H}$ (see Lemma 3.3.13) and similarly on $\widetilde{\mathscr{H}}$. □

Usually, in the applications we have in mind, it is easier to establish the existence of quasi-unitary maps J and J' for the scale of Hilbert spaces associated with the non-negative operators Δ and $\widetilde{\Delta}$. For the notion of compatibility operators, see Definition 3.3.8. The following result shows how to obtain the quasi-unitarity of J and J' for the operators H and $\widetilde{H}$ (and vice versa):

Lemma 4.1.8. *Assume that $H, \widetilde{H} \in \mathscr{R}_1(C, z_0)$ and that J and J' intertwine the corresponding compatibility operators T and $\widetilde{T}$, i.e.*

$$JT = \widetilde{T}J \qquad \text{and} \qquad J'\widetilde{T} = TJ'. \tag{4.5}$$

1. *If J and J' are δ-quasi-unitary (δ-partial isometric) of order 1 (with respect to Δ and $\widetilde{\Delta}$), then J and J' are $C\delta$-quasi-unitary ($C\delta$-partial isometric) of order 1 with respect to H and $\widetilde{H}$.*
2. *Similarly, if J and J' are δ-quasi-unitary (δ-partial isometric) of order 1 with respect to H and $\widetilde{H}$, then J and J' are $C\delta$-quasi-unitary ($C\delta$-partial isometric) of order 1 (with respect to Δ and $\widetilde{\Delta}$).*

Proof. The only conditions to be verified are

$$\begin{aligned} \|\mathrm{id} - J'J\|_{1,H\to 0} &= \|(\mathrm{id} - J'J)T^{-1}\|_{1\to 0} = \|T^{-1}(\mathrm{id} - J'J)\|_{1\to 0} \\ &\le \|T^{-1}\|\,\|\mathrm{id} - J'J\|_{1\to 0} \le C\delta \end{aligned}$$

and the corresponding statement on $\widetilde{\mathscr{H}}$. Here, we used the intertwining property of T and $\widetilde{T}$ for the second equality. The second part follows similarly. □

Combining the last two lemmata, we obtain:

Corollary 4.1.9. *Assume that $H, \widetilde{H} \in \mathscr{R}_1(C, z_0)$ and that J and J' intertwine the corresponding compatibility operators T and $\widetilde{T}$, If J and J' are δ-quasi-unitary (δ-partial isometric) of order 1 (with respect to Δ and $\widetilde{\Delta}$), then J and J' are $C^3\delta$-quasi-unitary ($C^3\delta$-partial isometric) of order 2 w.r.t. H and $\widetilde{H}$.*

4.2 Convergence of Self-Adjoint Operators in Different Hilbert Spaces

Assume that we have two Hilbert spaces $\mathscr{H}$ and $\widetilde{\mathscr{H}}$, each endowed with a non-negative operator Δ and $\widetilde{\Delta}$. We denote the associated scales of Hilbert spaces by $\mathscr{H}^k$ and $\widetilde{\mathscr{H}}^k$, respectively (see Sect. 3.2). For example, the norm of an operator

$A: \mathscr{H}^k \longrightarrow \widetilde{\mathscr{H}}^{-m}$ is

$$\|A\|_{k\to -m} := \sup_{u\in\mathscr{H}^k} \frac{\|Au\|_{-m}}{\|u\|_k} = \|\widetilde{R}^{m/2} A R^{k/2}\|_{0\to 0},$$

where

$$R := (\Delta + 1)^{-1} \quad \text{and} \quad \widetilde{R} := (\widetilde{\Delta} + 1)^{-1}.$$

By A^* we denote the adjoint of A with respect to the (extended) inner product on $\mathscr{H}$, i.e. $A^*: \widetilde{\mathscr{H}}^m \longrightarrow \mathscr{H}^{-k}$ with norm

$$\|A^*\|_{m\to -k} = \|A\|_{k\to -m}.$$

In order to compare the two spaces and operators Δ and $\widetilde{\Delta}$, we need an identification operator $J: \mathscr{H} \longrightarrow \widetilde{\mathscr{H}}$ of order 0:

Definition 4.2.1. Let $\delta \geq 0$.

1. We say that Δ and $\widetilde{\Delta}$ are *δ-close* (of order 0) with identification operator J if

$$\|\widetilde{\Delta} J - J\Delta\|_{2\to -2} = \|\widetilde{R} J - JR\|_{0\to 0} \leq \delta. \tag{4.6a}$$

2. We say that Δ and $\widetilde{\Delta}$ are *δ-close of order $m \geq 0$* with identification operator J if

$$\|\widetilde{\Delta} J - J\Delta\|_{2+m\to -2} = \|\widetilde{R} J - JR\|_{m\to 0} = \|(\widetilde{R} J - JR)R^{m/2}\| \leq \delta. \tag{4.6b}$$

Remark 4.2.2.

1. Note that estimate (4.6b) is equivalent with

$$\begin{aligned} &\big|\langle Jf, \widetilde{\Delta} u\rangle - \langle J\Delta f, u\rangle\big| \\ &\qquad \leq \delta \|f\|_{2+m} \|u\|_2 = \delta \|(\Delta+1)^{1+m/2} f\| \, \|(\widetilde{\Delta}+1)u\| \end{aligned} \tag{4.6b'}$$

for all $f \in \mathscr{H}^{2+m}$ and $u \in \widetilde{\mathscr{H}}^2$.
2. The notion of δ-closeness alone does not say a lot about the operators Δ and $\widetilde{\Delta}$, e.g., if $J = 0$, then trivially, any pair of operators Δ and $\widetilde{\Delta}$ is 0-close. The important additional information is the δ-quasi-unitarity of J. Nevertheless it is sometimes useful to have this weaker notion.
3. Note that the operators are Δ and $\widetilde{\Delta}$ are 0-close (of order m) w.r.t. J iff J intertwines R and $\widetilde{R}$, i.e. $JR = \widetilde{R} J$.

Next we define when two operators are "unitarily equivalent" up to an error:

Definition 4.2.3. Let $\delta \geq 0$.

1. We say that Δ and $\widetilde{\Delta}$ are *δ-quasi-unitarily equivalent (of order $m \geq 0$)* if there are δ-quasi-unitary operators J and J' of order $k \leq 2$ for which Δ and $\widetilde{\Delta}$ are δ-close (of order m).
2. We say that Δ and $\widetilde{\Delta}$ are *δ-partial isometrically equivalent (of order $m \geq 0$)* if there is a δ-partial isometric operator J of order $k \leq 2$ for which Δ and $\widetilde{\Delta}$ are δ-close (of order m).

The next lemma shows that we indeed generalised the usual notion of "unitary equivalence". Note that the notion "partial isometric equivalent" may be a bit misleading, since Δ and $\widetilde{\Delta}$ are only δ-close and *not* 0-close. In the latter case, we can say more (see also Proposition 4.9.10 for a special case):

Proposition 4.2.4.

1. *The operators Δ, $\widetilde{\Delta}$ are 0-quasi-unitarily equivalent iff Δ and $\widetilde{\Delta}$ are unitarily equivalent.*
2. *Assume that Δ and $\widetilde{\Delta}$ are δ-partial isometrically equivalent and 0-close with respect to the same identification operator J (i.e. $J^*J = \mathrm{id}$, $\|(\widetilde{\mathrm{id}} - JJ^*)\widetilde{R}\| \leq \delta$ and $\widetilde{R}J = JR$), then Δ is unitarily equivalent with $\widetilde{P}\widetilde{\Delta} = \widetilde{\Delta}\widetilde{P}$, where $\widetilde{P} = JJ^*$. In particular, the operators Δ and $\widetilde{\Delta}\widetilde{P}$ have the same spectral properties.*

Proof. The first assertion is obvious (see Lemma 4.1.4 (1)). For the second assertion, we have already seen in Lemma 4.1.4 (2) that $\widetilde{P} = JJ^*$ is an orthogonal projection in $\widetilde{\mathscr{H}}$ and $J : \mathscr{H} \longrightarrow \operatorname{ran} \widetilde{P}$ is unitary. Moreover, since $\widetilde{R}J = JR$, the projection $\widetilde{P}$ commutes with $\widetilde{R}$ and the result follows. □

Let us now show the *transitivity* of δ-quasi-unitary resp. δ-partial isometric equivalence of non-negative operators as follows. Assume that $\mathscr{H}$, $\widetilde{\mathscr{H}}$ and $\widehat{\mathscr{H}}$ are three Hilbert spaces with non-negative operators Δ, $\widetilde{\Delta}$ and $\widehat{\Delta}$, respectively. Moreover, assume that

$$J : \mathscr{H} \longrightarrow \widetilde{\mathscr{H}}, \quad \widetilde{J} : \widetilde{\mathscr{H}} \longrightarrow \widehat{\mathscr{H}}, \quad \widetilde{J}' : \widehat{\mathscr{H}} \longrightarrow \widetilde{\mathscr{H}} \quad \text{and} \quad J' : \widetilde{\mathscr{H}} \longrightarrow \mathscr{H} \tag{4.7}$$

are bounded operators. We define

$$\hat{J} := \widetilde{J} J : \mathscr{H} \longrightarrow \widehat{\mathscr{H}} \qquad \text{and} \qquad \hat{J}' := J' \widetilde{J}' : \widehat{\mathscr{H}} \longrightarrow \mathscr{H}. \tag{4.8}$$

For the next proposition, let us replace condition (4.1a) in the definition of δ-quasi-unitarity by

$$\|J\| \leq 1 + \delta \qquad \text{resp.} \qquad \|\widetilde{J}\| \leq 1 + \widetilde{\delta}. \tag{4.9}$$

Note that then $\|J'\| \leq 1 + 2\delta$ and $\|\widetilde{J}'\| \leq 1 + 2\widetilde{\delta}$ by (4.3a).

Proposition 4.2.5. *Assume that $0 \leq \delta, \widetilde{\delta} \leq 1$.*

1. *Assume that Δ and $\widetilde{\Delta}$ are δ-quasi-unitarily equivalent with identification operators J and J', and that $\widetilde{\Delta}$ and $\widehat{\Delta}$ are $\widetilde{\delta}$-quasi-unitarily equivalent with*

identification operators $\widetilde{J}$ and $\widetilde{J}'$ (both of order 0). Then Δ and $\widehat{\Delta}$ are $\hat{\delta}$-quasi-unitarily equivalent with identification operators $\hat{J}$ and $\hat{J}'$ of order 0, where $\hat{\delta} = 22\delta + 43\widetilde{\delta}$.

2. *Assume that Δ and $\widetilde{\Delta}$ are δ-partial isometrically equivalent with identification operator J, and that $\widetilde{\Delta}$ and $\widehat{\Delta}$ are $\widetilde{\delta}$-partial isometrically equivalent with identification operator $\widetilde{J}$. Then Δ and $\widehat{\Delta}$ are $(\delta + 3\widetilde{\delta})$-partial isometrically equivalent with identification operator $\hat{J}$.*

The asymmetry in δ and $\widetilde{\delta}$ is due to the fact that the assumptions of quasi-unitarity are not symmetric.

Proof. Let us first show that $\hat{J}$ and $\hat{J}'$ are $\underline{\delta}$-quasi-unitary of order 2 (see Definition 4.1.1 (1) and (4)). Obviously, $\|\hat{J}\| \le \|\widetilde{J}\|\|J\| \le (1+\widetilde{\delta})(1+\delta) \le 1+2(\delta+\widetilde{\delta})$. Moreover,

$$\|\hat{J} - \hat{J}'^*\| \le \|(\widetilde{J} - \widetilde{J}'^*)J\| + \|\widetilde{J}'^*(J - J'^*)\| \le 2\widetilde{\delta} + 3\delta.$$

since $\delta, \widetilde{\delta} \le 1$. In addition,

$$\begin{aligned}
\|\mathrm{id} - \hat{J}'\hat{J}\|_{2\to 0} &= \|(\mathrm{id} - J'\widetilde{J}'\widetilde{J}J)R\| \\
&= \big\|(\mathrm{id} - J'J)R + J'(\widetilde{\mathrm{id}} - \widetilde{J}'\widetilde{J})\big(\widetilde{R}J + (JR - \widetilde{R}J)\big)\big\| \\
&\le \|(\mathrm{id} - J'J)R\| \\
&\quad + \|J'\|\big(\|(\widetilde{\mathrm{id}} - \widetilde{J}'\widetilde{J})\widetilde{R}\|\|J\| + (1 + \|\widetilde{J}'\|\|\widetilde{J}\|)\|JR - \widetilde{R}J\|\big) \\
&\le \delta + 3(2\widetilde{\delta} + 7\delta) = 22\delta + 6\widetilde{\delta},
\end{aligned}$$

using (4.6a), where $R = (\Delta + 1)^{-1}$, $\widetilde{R} = (\widetilde{\Delta} + 1)^{-1}$ and $\widehat{R} = (\widehat{\Delta} + 1)^{-1}$. Similarly, we have

$$\|\widehat{\mathrm{id}} - \hat{J}\hat{J}'\|_{2\to 0} = \|(\widehat{\mathrm{id}} - \widetilde{J}JJ'\widetilde{J}')\widehat{R}\| \le \widetilde{\delta} + 2(3\delta + 7\cdot 3\widetilde{\delta}) = 6\delta + 43\widetilde{\delta}$$

using

$$\|\widetilde{J}'\widehat{R} - \widetilde{R}\widetilde{J}'\| \le \|\widetilde{J}'\|(\|\widehat{R}\| + \|\widetilde{R}\|) + \|\widetilde{J}^*\widehat{R} - \widetilde{R}\widetilde{J}^*\| \le 3\widetilde{\delta}.$$

In particular, we have shown that $\hat{J}$ and $\hat{J}'$ are $\hat{\delta}$-quasi-unitary of order 2. Finally, we have

$$\|\widehat{R}\hat{J} - \hat{J}R\| \le \|(\widehat{R}\widetilde{J} - \widetilde{J}\widetilde{R})J\| + \|\widetilde{J}(\widetilde{R}J - JR)\| \le 2\widetilde{\delta} + 2\delta$$

and the $\hat{\delta}$-quasi-unitary equivalence of Δ and $\widehat{\Delta}$ follows.

The $(\delta + 3\widetilde{\delta})$-partial isometric equivalence follows similarly: Obviously, $\hat{J}^*\hat{J} = \widetilde{J}^*J^*J\widetilde{J} = \mathrm{id}$, since $J^*J = \mathrm{id}$ and $\widetilde{J}^*\widetilde{J} = \widetilde{\mathrm{id}}$. Moreover, we can show as before that

$$\|\widehat{\mathrm{id}} - \hat{J}\hat{J}^*\|_{2\to 0} = \|(\widehat{\mathrm{id}} - \widetilde{J}JJ^*\widetilde{J}^*)\widehat{R}\| \le \widetilde{\delta} + (\delta + 2\widetilde{\delta}) = \delta + 3\widetilde{\delta}$$

using now $\|J\| = \|J\|^* = 1$ etc. Finally, $\|\widehat{R}\hat{J} - \hat{J}R\| \le \widetilde{\delta} + \delta$, and the assertion follows. □

Let us now define the notion of convergence of non-negative operators acting in different Hilbert spaces:

Definition 4.2.6. Let Δ_n and Δ be non-negative operators in the Hilbert spaces $\mathscr{H}_n$ and $\mathscr{H}$, respectively. We say that Δ_n *converges to* Δ *(of order* $m \ge 0$*) (in the generalised norm resolvent sense)* ($\Delta_n \to \Delta$) if there is a sequence $\delta_n \to 0$ such that Δ_n and Δ are δ_n-quasi-unitarily equivalent (of order $m \ge 0$). We call δ_n the *convergence speed* of $\Delta_n \to \Delta$.

Again, we have some simple consequences, showing that we generalised existing concepts of convergence of self-adjoint operators:

Lemma 4.2.7.

1. *If* $\mathscr{H} = \widetilde{\mathscr{H}}$ *and* $J = J' = \mathrm{id}$, *then the* δ*-closeness of* Δ *and* $\widetilde{\Delta}$ *is equivalent with the fact that* $\|R - \widetilde{R}\| \le \delta$. *In particular, the norm resolvent convergence* $\|R_n - R\| \to 0$ *is equivalent to the convergence* $\Delta_n \to \Delta$ *in the generalised norm resolvent sense (of order* 0*).*
2. *More generally, assume that* $\mathscr{H}$ *is a closed subspace of* $\widetilde{\mathscr{H}}$. *If* Δ *is a non-negative operator in* $\mathscr{H}$ *and* Δ_n *are non-negative operators in* $\widetilde{\mathscr{H}}$, *we can define a convergence by*[3]

$$\left\|(\Delta_n + 1)^{-1} - (\Delta + 1)^{-1} \oplus 0\right\| \to 0, \tag{4.10}$$

also called generalised norm resolvent convergence. *Setting* $Jf := f \oplus 0$ *and* $J' := J^*$, *then* Δ_n *converges to* Δ *in the generalised norm resolvent sense iff* $\Delta_n \to \Delta$ *in the sense of Definition 4.2.6.*

Proof. The first assertion is obvious. For the second assertion, it is clear that $\mathrm{id} = J'J$ and $JJ' = \widetilde{P}$, where $\widetilde{P}$ is the projection onto $\mathscr{H}$ in $\widetilde{\mathscr{H}}$. Note that Δ_n converges to Δ in the sense of (4.10) iff all three norms

$$\|\widetilde{P}(\Delta_n + 1)^{-1}\widetilde{P} - (\Delta + 1)^{-1}\|, \quad \|\widetilde{P}^{\perp}(\Delta_n + 1)^{-1}\widetilde{P}\|, \quad \|\widetilde{P}^{\perp}(\Delta_n + 1)^{-1}\widetilde{P}^{\perp}\|$$

converge to 0, where $\widetilde{P}^{\perp} := 1 - \widetilde{P}$. □

Let us now derive some consequences of the fact that the operators Δ and $\widetilde{\Delta}$ are δ-close of order m:

Lemma 4.2.8. *Assume that* Δ *and* $\widetilde{\Delta}$ *are* δ*-close of order* $m \ge 0$ *with identification operator* J, *then*

---

[3] The operator $(\Delta + 1)^{-1} \oplus 0$ is the *quasi-inverse* of $\Delta + 1$ viewed as *non-densely* defined operator in $\widetilde{\mathscr{H}}$ (see e.g. [W84]). We can think of $\Delta + 1$ being ∞ on $\mathscr{H}^{\perp} \subset \widetilde{\mathscr{H}}$, so that its inverse is 0 there.

$$\|\widetilde{R}^j J - JR^j\|_{m\to 0} = \|(\widetilde{R}^j J - JR^j)R^{m/2}\| \le j\delta \tag{4.11}$$

for all $j \in \mathbb{N}$.

Proof. We use the resolvent identity

$$\widetilde{R}^j J - JR^j = \sum_{i=0}^{j-1} \widetilde{R}^{j-1-i}(\widetilde{R}J - JR)R^i,$$

and conclude

$$\|(\widetilde{R}^j J - JR^j)R^{m/2}\| \le \sum_{i=0}^{j-1} \|\widetilde{R}^{j-1-i}\| \|(\widetilde{R}J - JR)R^{m/2}\| \|R^i\| \le j\delta$$

using the estimate for $j = 1$ and $\|R\|, \|\widetilde{R}\| \le 1$. □

We want to extend our results to more general functions $\varphi(\Delta)$ of the operator Δ and similarly for $\widetilde{\Delta}$. Here, we essentially use the fact that we have the functional calculus for *self-adjoint* operators. We start with continuous functions on $\mathbb{R}_+ := [0, \infty)$ such that $\lim_{\lambda\to\infty} \varphi(\lambda)$ exist, i.e. with functions continuous on $\overline{\mathbb{R}}_+ := [0, \infty]$. We denote this space by $\mathsf{C}(\overline{\mathbb{R}}_+)$.

Theorem 4.2.9. *Let* $\delta \ge 0$ *and assume that* Δ *and* $\widetilde{\Delta}$ *are* δ*-close of order* $m \ge 0$ *with identification operator* J *and that* $\|J\| \le 2$, *then*

$$\|\varphi(\widetilde{\Delta})J - J\varphi(\Delta)\|_{m\to 0} \le C_\varphi \delta \tag{4.12}$$

for all $\varphi \in \mathsf{C}(\overline{\mathbb{R}}_+)$, *where* $C_\varphi > 0$ *depends only on* φ *and* $\|J\|$.

Proof. Let $\varphi \in \mathsf{C}(\overline{\mathbb{R}}_+)$. By the Stone-Weierstrass theorem there exists a "polynomial"

$$p(\lambda) := \sum_{k=0}^{n} a_k(\lambda + 1)^{-k}$$

in $(\lambda + 1)^{-1}$ such that $\|p - \varphi\|_\infty \le \delta$. Then we have

$$\begin{aligned}
\|\varphi(\widetilde{\Delta})J - J\varphi(\Delta)\|_{m\to 0} &\le \|(\varphi - p)(\widetilde{\Delta})\| \|JR^{m/2}\| + \|J\| \|(\varphi - p)(\Delta)R^{m/2}\| \\
&\quad + \sum_{k=0}^{n} |a_k| \|(\widetilde{R}^k J - JR^k)R^{m/2}\| \\
&\le 4\|\varphi - p\|_\infty + \sum_{k=0}^{n} |a_k| k\delta \\
&\le \Big(4 + \sum_{k=0}^{n} |a_k| k\Big)\delta =: C_\varphi \delta
\end{aligned}$$

using the assumptions, Lemma 4.2.8 and the functional calculus, where $\|\varphi\|_\infty$ denotes the supremum norm of φ. □

In a second step we extend the previous result to certain bounded measurable functions $\psi: \overline{\mathbb{R}}_+ \longrightarrow \mathbb{C}$. We start with the easier case $m = 0$:

Theorem 4.2.10. *Suppose that $U \subset \mathbb{R}_+$ is open and that $\psi: \overline{\mathbb{R}}_+ \longrightarrow \mathbb{C}$ is a measurable, bounded function, continuous on U such that $\lim_{\lambda\to\infty} \psi(\lambda)$ exists. Then there exists a constant $C_\psi > 0$ depending only on ψ such that*

$$\|\psi(\widetilde{\Delta})J - J\psi(\Delta)\| \le C_\psi \delta \tag{4.13}$$

for all operators Δ and $\widetilde{\Delta}$ being δ-close with identification operator J such that $\|J\| \le 2$, provided

$$\sigma(\Delta) \subset U \qquad \text{or} \qquad \sigma(\widetilde{\Delta}) \subset U.$$

Proof. Let χ_1 be a continuous function on $\overline{\mathbb{R}}_+$ satisfying $0 \le \chi_1 \le 1$, $\chi_1 = 1$ on $\sigma(\Delta) \cup \{\infty\}$ (resp. $\chi_1 = 1$ on $\sigma(\widetilde{\Delta}) \cup \{\infty\}$ if U is a neighbourhood of $\sigma(\widetilde{\Delta})$) and $\operatorname{supp} \chi_1 \subset U$. Then $\chi_1\psi$ and $\chi_2 = 1 - \chi_1$ are continuous functions on $\overline{\mathbb{R}}_+$ and

$$\begin{aligned}\|\psi(\widetilde{\Delta})J - J\psi(\Delta)\| &\le \|(\chi_1\psi)(\widetilde{\Delta})J - J(\chi_1\psi)(\Delta)\| \\ &\quad + \|(\chi_2\psi)(\widetilde{\Delta})J - J(\chi_2\psi)(\Delta)\|.\end{aligned}$$

If U is a neighbourhood of $\sigma(\Delta)$ we can estimate the norm with χ_2 by

$$\|\psi\|_\infty \|\chi_2(\widetilde{\Delta})J - J\chi_2(\Delta)\|$$

using the fact that $(\chi_2\psi)(\Delta) = \chi_2(\Delta) = 0$ since $\chi_2 = 0$ on $\sigma(\Delta)$. If U is a neighbourhood of $\sigma(\widetilde{\Delta})$ then we use

$$\|\psi(\widetilde{\Delta})J - J\psi(\Delta)\| = \|\psi(\Delta)J^* - J^*\psi(\widetilde{\Delta})\|,$$

and we argue as in the case where $\sigma(\Delta) \subset U$ with the roles of Δ and $\widetilde{\Delta}$ interchanged.

Applying the preceding theorem twice (in each of the above cases), we obtain the error estimate $C_\psi := C_{\chi_1\psi} + \|\psi\|_\infty C_{\chi_2}$, where C_φ denotes the constant for the *continuous* function φ. □

If $m \ge 1$, then we can still prove the following result, under the stronger assumption of δ-quasi-unitarity:

Theorem 4.2.11. *Suppose that $m \ge 1$, that $U \subset \mathbb{R}_+$ is open and that $\psi: \overline{\mathbb{R}}_+ \longrightarrow \mathbb{C}$ is a measurable, bounded function, continuous on U such that $\lim_{\lambda\to\infty} \psi(\lambda)$ exists. Then there exists a constant C_ψ depending only on ψ such that*

$$\|\psi(\widetilde{\Delta})J - J\psi(\Delta)\|_{m\to 0} \le C_\psi \delta \tag{4.14}$$

for all operators Δ *and* $\widetilde{\Delta}$ *being* δ*-close with the* δ*-quasi-unitary identification operators* J *and* J' *of order* $k \le m$ *w.r.t.* Δ *and* $\widetilde{\Delta}$*, provided*

$$\sigma(\Delta) \subset U \qquad or \qquad \sigma(\widetilde{\Delta}) \subset U$$

and $0 < \delta \le 1$.

Proof. As in the previous proof, we have

$$\begin{aligned}
&\|\psi(\widetilde{\Delta})J - J\psi(\Delta)\|_{m\to 0} \\
&\quad \le \|(\chi_1\psi)(\widetilde{\Delta})J - J(\chi_1\psi)(\Delta)\|_{m\to 0} + \|(\chi_2\psi)(\widetilde{\Delta})J - J(\chi_2\psi)(\Delta)\|_{m\to 0}.
\end{aligned}$$

If U is a neighbourhood of $\sigma(\Delta)$ we estimate the norm with χ_2 by

$$\|\psi\|_\infty \|\chi_2(\widetilde{\Delta})J - J\chi_2(\Delta)\|_{m\to 0}$$

again using the fact that $(\chi_2\psi)(\Delta) = \chi_2(\Delta) = 0$ since $\chi_2 = 0$ on $\sigma(\Delta)$. If U is a neighbourhood of $\sigma(\widetilde{\Delta})$ then we estimate the norm with χ_2 by

$$\|J(\chi_2\psi)(\Delta)\|_{m\to 0} \le \|J\|\,\|(\chi_2\psi)(\Delta)\|_{m\to 0} \le 2\|\psi\|_\infty \|\chi_2(\Delta)\|_{m\to 0}$$

again using the fact that $(\chi_2\psi)(\widetilde{\Delta}) = \chi_2(\widetilde{\Delta}) = 0$ since $\chi_2 = 0$ on $\sigma(\widetilde{\Delta})$. Now

$$\|\chi_2(\Delta)\|_{m\to 0} \le \|\mathrm{id} - J'J\|_{k\to 0}\|\chi_2(\Delta)\|_{m\to k} + \|J'\|\,\|J\chi_2(\Delta) - \chi_2(\widetilde{\Delta})J\|_{m\to 0}.$$

Note that $\|\chi_2(\Delta)\|_{m\to k} \le 1$ since $m \ge k$. Applying Theorem 4.2.9 twice (in each of the above cases), we have the error estimate

$$C_{\chi_1\psi} + 2\|\psi\|_\infty\big(1 + (1+2\delta)C_{\chi_2}\big) \le C_{\chi_1\psi} + 2\|\psi\|_\infty(1 + 3C_{\chi_2}) =: C_\psi$$

using the fact that $\delta \le 1$. □

Denote by $\partial_+ I = \overline{I} \cap \overline{\mathbb{R}_+ \setminus I}$ the boundary of I in the *relative* topology $\mathbb{R}_+ = [0,\infty)$. For example, $\partial[0,\lambda] = \{\lambda\}$. As a consequence of the preceding theorem, we can ensure the convergence of the spectral projections:

Corollary 4.2.12. *Assume that* I *is an interval in* $\mathbb{R}_+$*. Then the spectral projections satisfy*

$$\|\mathbb{1}_I(\widetilde{\Delta})J - J\mathbb{1}_I(\Delta)\|_{m\to 0} \le C_{\mathbb{1}_I}\delta \tag{4.15}$$

provided $\partial_+ I \cap \sigma(\Delta) = \emptyset$ *or* $\partial_+ I \cap \sigma(\widetilde{\Delta}) = \emptyset$.

Proof. We apply either Theorem 4.2.10 if $m = 0$ or Theorem 4.2.11 if $m > 0$ with $\psi = \mathbb{1}_I$, the indicator function of I with the corresponding hypothesises. Note that $\mathbb{1}_I$ is continuous on $U = \mathbb{R}_+ \setminus \partial_+ I$ and that $\lim_{\lambda\to\infty} \mathbb{1}_I(\lambda) \in \{0,1\}$ exists. □

The error estimate is not quite explicit. Namely, it is difficult to control the constant $C_{\mathbb{1}_I}$ in terms of η (or λ) for $I = (\lambda - \eta, \lambda + \eta)$. For this purpose, the holomorphic calculus is more adopted (see Corollary 4.5.15).

Combining the above estimates with a δ-unitary map, we obtain:

Lemma 4.2.13. *Assume that J is δ-quasi-unitary of order k w.r.t. Δ and $\widetilde{\Delta}$ and that*

$$\|\varphi(\widetilde{\Delta})J - J\varphi(\Delta)\|_{m\to 0} = \eta$$

for some function φ, $m \geq 0$ and some constant $\eta > 0$. Then we have

$$\|\varphi(\Delta)J' - J'\varphi(\widetilde{\Delta})\|_{0\to -m} \leq 2\|\varphi\|_\infty \delta + \eta, \tag{4.16a}$$

$$\|\varphi(\Delta) - J'\varphi(\widetilde{\Delta})J\|_{m\to 0} \leq \|\widehat{\varphi}^{k-m}\|_\infty \delta + (2+\delta)\eta, \tag{4.16b}$$

$$\|\varphi(\widetilde{\Delta}) - J\varphi(\Delta)J'\|_{0\to 0} \leq 5\|\widehat{\varphi}^k\|_\infty \delta + 2\eta, \tag{4.16c}$$

where $\widehat{\varphi}^k(\lambda) = (1+\lambda)^{k/2}\varphi(\lambda)$. For the last estimate, we have to assume $m = 0$.

If J is δ-partial isometric and $m = 0$, then

$$\|\varphi(\Delta)J^* - J^*\varphi(\widetilde{\Delta})\| = \|\varphi(\Delta) - J^*\varphi(\widetilde{\Delta})J\| = \eta. \tag{4.16d}$$

Proof. The first estimate follows from

$$\|\varphi(\Delta)J' - J'\varphi(\widetilde{\Delta})\|_{0\to -m} \leq 2\|\varphi\|_\infty \|J' - J^*\| + \|\varphi(\Delta)J^* - J^*\varphi(\widetilde{\Delta})\|_{0\to -m};$$

the second from

$$\begin{aligned}\|\varphi(\Delta) - J'\varphi(\widetilde{\Delta})J\|_{m\to 0} &\leq \|\mathrm{id} - J'J\|_{k\to 0}\|\varphi(\Delta)\|_{m\to k}\\ &\quad + \|J'\|\|J\varphi(\Delta) - \varphi(\widetilde{\Delta})J\|_{m\to 0}\end{aligned}$$

and the third from

$$\|\varphi(\widetilde{\Delta}) - J\varphi(\Delta)J'\| \leq \|\widetilde{\mathrm{id}} - JJ'\|_{k\to 0}\|\varphi(\widetilde{\Delta})\|_{0\to k} + \|J\|\|J'\varphi(\widetilde{\Delta}) - \varphi(\Delta)J'\|$$

together with (4.16a). □

We immediately obtain:

Theorem 4.2.14. *Assume that Δ and $\widetilde{\Delta}$ are δ-quasi-unitarily equivalent of order $m \geq 0$, and that $\varphi \in \mathsf{C}(\overline{\mathbb{R}}_+)$ with $\varphi(\lambda) = \mathrm{O}((\lambda+1)^{-1})$ as $\lambda \to \infty$ then*

$$\|\varphi(\Delta)J' - J'\varphi(\widetilde{\Delta})\|_{0\to -m} \leq C_1\delta, \tag{4.17a}$$

$$\|\varphi(\Delta) - J'\varphi(\widetilde{\Delta})J\|_{m\to 0} \leq C_2\delta, \tag{4.17b}$$

$$\|\varphi(\widetilde{\Delta}) - J\varphi(\Delta)J'\|_{0\to 0} \leq C_3\delta, \tag{4.17c}$$

where C_i depend only on φ and m. Again we have to assume $m = 0$ for the last estimate.

Note that we cannot apply the previous functional calculus to $\varphi_t(\lambda) = \mathrm{e}^{\mathrm{i}t\lambda}$ since $\lim_{\lambda\to\infty} \varphi_t(\lambda)$ does not exist for $t > 0$. Nevertheless we have the following result comparing the time evolution $\mathrm{e}^{\mathrm{i}t\Delta}$ with $\mathrm{e}^{\mathrm{i}t\widetilde{\Delta}}$:

Theorem 4.2.15. *Suppose that Δ and $\widetilde{\Delta}$ are δ-close of order m with identification operator J, then*

$$\|\mathrm{e}^{\mathrm{i}t\widetilde{\Delta}} J - J\mathrm{e}^{\mathrm{i}t\Delta}\|_{2+m\to -2} \le t\delta. \tag{4.18}$$

In particular, if $t = \mathrm{O}(\delta^{-\gamma})$, $0 < \gamma < 1$, then the norm of $\mathrm{e}^{\mathrm{i}t\widetilde{\Delta}} J - J\mathrm{e}^{\mathrm{i}t\Delta}$ as bounded operator from $\mathscr{H}_{2+m}$ to $\widetilde{\mathscr{H}}_{-2}$ is of order $\delta^{1-\gamma}$.

Proof. We have

$$\mathrm{e}^{\mathrm{i}t\widetilde{\Delta}} J\mathrm{e}^{-\mathrm{i}t\Delta} - J = \mathrm{i}\int_0^t \mathrm{e}^{\mathrm{i}\tau\widetilde{\Delta}}(\widetilde{\Delta}J - J\Delta)\mathrm{e}^{-\mathrm{i}\tau\Delta}\mathrm{d}\tau$$

where $\widetilde{\Delta}J - J\Delta$ is a bounded operator from $\mathscr{H}_{2+m}$ to $\widetilde{\mathscr{H}}_{-2}$. Multiplying this equality from the left by $\widetilde{R}$ and on the right by $\mathrm{e}^{\mathrm{i}t\Delta}R = R\,\mathrm{e}^{\mathrm{i}t\Delta}$ yields

$$\widetilde{R}(\mathrm{e}^{\mathrm{i}t\widetilde{\Delta}} J - J\mathrm{e}^{\mathrm{i}t\Delta})R = \mathrm{i}\int_0^t \mathrm{e}^{\mathrm{i}\tau\widetilde{\Delta}}\widetilde{R}(\widetilde{\Delta}J - J\Delta)R\,\mathrm{e}^{\mathrm{i}(t-\tau)\Delta}\mathrm{d}\tau.$$

Therefore,

$$\begin{aligned}\|(\mathrm{e}^{\mathrm{i}t\widetilde{\Delta}} J - J\mathrm{e}^{\mathrm{i}t\Delta})\|_{2+m\to -2} &= \|\widetilde{R}(\mathrm{e}^{\mathrm{i}t\widetilde{\Delta}} J - J\mathrm{e}^{\mathrm{i}t\Delta})R\|_{m\to 0}\\ &\le \int_0^t \|\widetilde{R}(\widetilde{\Delta}J - J\Delta)R\|_{m\to 0}\mathrm{d}\tau = t\|\widetilde{\Delta}J - J\Delta\|_{2+m\to -2} \le t\delta\end{aligned}$$

using $\|\mathrm{e}^{\mathrm{i}t\Delta}\|_{m\to m} = 1$ and the closeness of order m, see (4.6b). □

We also obtain the convergence in the usual operator norm $0 \to 0$ if we use an energy cut at $\lambda_0 > 0$, i.e. if we restrict the functions to the spectral interval $I = [0, \lambda_0]$:

Theorem 4.2.16. *Let $\lambda_0 > 0$ not belong to the spectrum of Δ. Assume in addition that Δ and $\widetilde{\Delta}$ are δ-close with identification operator J, then*

$$\|(\mathrm{e}^{\mathrm{i}t\widetilde{\Delta}} J - J\mathrm{e}^{\mathrm{i}t\Delta})P\| \le ((\lambda_0 + 1)^2 t + 2C_{\mathbb{1}_I})\delta, \tag{4.19a}$$

where $P = \mathbb{1}_I(\Delta)$. If Δ and $\widetilde{\Delta}$ are δ-quasi-unitarily equivalent, i.e. if in addition, J and J' are δ-quasi-unitary (of order 2), then

$$\|(\mathrm{e}^{\mathrm{i}t\widetilde{\Delta}} - J\mathrm{e}^{\mathrm{i}t\Delta}J')\widetilde{P}\| \le (2(\lambda_0 + 1)^2 t + 12C_{\mathbb{1}_I} + \lambda_0 + 9)\delta, \tag{4.19b}$$

where $\widetilde{P} = \mathbb{1}_I(\widetilde{\Delta})$. *In particular, if* $t = \mathrm{O}(\delta^{-\gamma})$, $0 < \gamma < 1$, *then the operator norms in* (4.19a)–(4.19b) *are of order* $\delta^{1-\gamma}$.

Proof. Denote by $\widetilde{P} := \mathbb{1}_I(\widetilde{\Delta})$ the corresponding spectral projection of $\widetilde{\Delta}$ and set $\widetilde{P}^{\perp} := \widetilde{\mathrm{id}} - \widetilde{P}$. Then we have

$$\begin{aligned}\|(\mathrm{e}^{\mathrm{i}t\widetilde{\Delta}}J - J\mathrm{e}^{\mathrm{i}t\Delta})P\| &\le \|\widetilde{P}(\mathrm{e}^{\mathrm{i}t\widetilde{\Delta}}J - J\mathrm{e}^{\mathrm{i}t\Delta})P\| + \|\widetilde{P}^{\perp}(\mathrm{e}^{\mathrm{i}t\widetilde{\Delta}}J - J\mathrm{e}^{\mathrm{i}t\Delta})P\| \\ &\le \|\widetilde{P}(\widetilde{\Delta}+1)\|\,\|\mathrm{e}^{\mathrm{i}t\widetilde{\Delta}}J - J\mathrm{e}^{\mathrm{i}t\Delta})P\|_{2\to -2}\|P(\Delta+1)\| \\ &\quad + 2\|\widetilde{P}^{\perp}JP\|\end{aligned}$$

using the fact that $\mathrm{e}^{\mathrm{i}t\Delta}$ and P commute and similarly for $\widetilde{\Delta}$. The first summand can be estimated by $(\lambda_0 + 1)^2\delta t$ by the continuous functional calculus and Theorem 4.2.15. For the second summand, note that

$$\|\widetilde{P}^{\perp}JP\| \le \|\widetilde{P}^{\perp}\widetilde{P}J\| + \|\widetilde{P}^{\perp}(JP - \widetilde{P}J)\| \le 0 + C_{\mathbb{1}_I}\delta$$

using Theorem 4.2.10 and the assumption $\lambda_0 \notin \sigma(\Delta)$.

For the second assertion, we have

$$\begin{aligned}\|(\mathrm{e}^{\mathrm{i}t\widetilde{\Delta}} - J\mathrm{e}^{\mathrm{i}t\Delta}J')\widetilde{P}\| &\le \|(\mathrm{e}^{\mathrm{i}t\widetilde{\Delta}}(\widetilde{\mathrm{id}} - JJ')\widetilde{P}\| + \|\mathrm{e}^{\mathrm{i}t\widetilde{\Delta}}J(J'\widetilde{P} - PJ')\| \\ &\quad + \|(\mathrm{e}^{\mathrm{i}t\widetilde{\Delta}}J - J\mathrm{e}^{\mathrm{i}t\Delta})PJ'\| + \|J\mathrm{e}^{\mathrm{i}t\Delta}(PJ' - J'\widetilde{P})\|.\end{aligned}$$

The second and forth term can be estimated by $2(2 + C_{\mathbb{1}_I})\delta$ using Lemma 4.2.13 and Theorem 4.2.10. The first term is bounded by $(\lambda_0 + 1)\delta$ by the δ-quasi-unitarity. Finally, the third term can be estimated by twice the bound obtained in (4.19a). □

4.3 Spectral Convergence for Non-negative Operators

In this section, we show the convergence of the discrete and essential spectra provided Δ_n converges to Δ in the generalised norm resolvent sense (see Definition 4.2.6). Recall that $\partial_+ I = \overline{I} \cap \mathbb{R}_+ \setminus I$ is the boundary of I in the relative topology on $\mathbb{R}_+ = [0, \infty)$.

Proposition 4.3.1. *Let* $I \subset \mathbb{R}_+$ *be a bounded measurable set and assume that* Δ, $\widetilde{\Delta}$ *are* δ*-quasi-unitarily equivalent (of order* 0*) such that*

$$\sigma(\Delta) \cap \partial_+ I = \emptyset \qquad \text{or} \qquad \sigma(\widetilde{\Delta}) \cap \partial_+ I = \emptyset.$$

Then

$$\dim \operatorname{ran} \mathbb{1}_I(\Delta) = \dim \operatorname{ran} \mathbb{1}_I(\widetilde{\Delta})$$

provided $0 < \delta < \delta_0$, *where*

$$\delta_0 := \bigl(3C_{2,I}^+ + C_{\mathbb{1}_I}\bigr)^{-1} \qquad \textit{and} \qquad C_{k,I}^+ := \sup_{\lambda \in I}(1+\lambda)^{k/2}.$$

If, in addition, $\lambda \in I$ *is a simple eigenvalue with normalised eigenvector* ψ*, then there exists a normalised eigenvector* $\widetilde{\psi} \in \widetilde{\mathscr{H}}$ *generating* ran $\widetilde{P}$ *such that*

$$\|J\psi - \widetilde{\psi}\| \le C_1\delta \qquad \textit{and} \qquad \|J'\widetilde{\psi} - \psi\| \le \widetilde{C}_1\delta,$$

where $C_1 := 2C_{\mathbb{1}_I} + 3(\lambda+1)$ *and* $\widetilde{C}_1 = 3C_1 + (\lambda+1)$ *provided* $\delta \le 1$ *for the second estimate.*

Proof. We apply Theorem 4.1.5 with $C = 1$ and $\mathscr{D} = \mathscr{H}^2$, so that $\|P\|_{\mathscr{H}\to\mathscr{D}} \le C_{2,I}^+$ by the spectral calculus Theorem 3.2.2, and similarly on $\widetilde{\mathscr{H}}$. The convergence of the projections $P = \mathbb{1}_I(\Delta)$ and $\widetilde{P} = \mathbb{1}_I(\widetilde{\Delta})$ is ensured by Theorem 4.2.10 and the assumption that either the spectrum of Δ or $\widetilde{\Delta}$ is disjoint from $\partial_+ I$, the discontinuity points of $\mathbb{1}_I$ and we have $\alpha = C_{\mathbb{1}_I}$. □

If Δ and $\widetilde{\Delta}$ are only δ-quasi-unitarily equivalent of order $m \ge 1$, we have to work a little bit more due to the asymmetry in the norm $\|\widetilde{P}J - JP\|_{m\to 0}$:

Proposition 4.3.2. *Let* $I \subset \mathbb{R}_+$ *be a bounded measurable set and assume that* Δ, $\widetilde{\Delta}$ *are* δ*-quasi-unitarily equivalent of order* $m \ge 1$ *such that*

$$\sigma(\Delta) \cap \partial_+ I = \emptyset \qquad \textit{or} \qquad \sigma(\widetilde{\Delta}) \cap \partial_+ I = \emptyset.$$

Then

$$\dim \operatorname{ran} \mathbb{1}_I(\Delta) = \dim \operatorname{ran} \mathbb{1}_I(\widetilde{\Delta})$$

provided $0 < \delta < \delta_0$ *for some* $\delta_0 = \delta_0(m, I) > 0$*. If, in addition,* $\lambda \in I$ *is a simple eigenvalue with normalised eigenvector* ψ*, then there exists a normalised eigenvector* $\widetilde{\psi} \in \widetilde{\mathscr{H}}$ *generating* ran $\widetilde{P}$ *such that*

$$\|J\psi - \widetilde{\psi}\| \le C_1\delta \qquad \textit{and} \qquad \|J'\widetilde{\psi} - \psi\| \le \widetilde{C}_1\delta,$$

where $C_1 := 2C_{\mathbb{1}_I}(\lambda+1)^{m/2} + 3(\lambda+1)$ *and* $\widetilde{C}_1 = 3C_1 + (\lambda+1)$ *provided* $\delta \le 1$ *for the second estimate.*

Proof. As before, denote by $P = \mathbb{1}_I(\Delta)$ the spectral projection on I and similarly for $\widetilde{P}$. For the inequality $\dim \operatorname{ran} P \le \dim \operatorname{ran} \widetilde{P}$, we can argue as in the previous proposition, but with $\mathscr{D} = \mathscr{H}^{m+2}$.

The opposite inequality is more difficult due to the asymmetry in the norm $\|\cdot\|_{m\to 0}$. Suppose that $u \in \operatorname{ran} \widetilde{P}$. In addition, let $\chi_1, \chi_2 \in \mathsf{C}(\overline{\mathbb{R}}_+)$ with $\chi_1 + \chi_2 = \mathbb{1}$, supp χ_1 compact and supp $\chi_2 \cap I = \emptyset$. Then

$$\|\chi_1(\Delta)f\|_{-m} = \|R^{m/2}\chi_1(\Delta)f\| \ge \|R^{m/2}\mathbb{1}_I(\Delta)f\| \ge C_{m,I}^-\|f\|$$

by the functional calculus, where

$$C^-_{m,I} := \inf_{\lambda \in I} (1+\lambda)^{-m/2} > 0$$

since I is bounded. Hence, we have the estimate (with $f = J'u$)

$$\begin{aligned}
\|PJ^*u\|_{-m} &\geq \|J^*\widetilde{P}u\|_{-m} - \|(\widetilde{P}J^* - J^*\widetilde{P})u\|_{-m} \\
&\geq \|\chi_1(\Delta)J^*u\|_{-m} - \|\chi_2(\Delta)J^*u\|_{-m} - \|\widetilde{P}J - JP\|_{m\to 0}\|u\| \\
&\geq C^-_{m,I}\|J^*u\| - \|(\chi_2(\Delta)J^* - J^*\chi_2(\widetilde{\Delta}))u\|_{-m} - \delta C_{\mathbb{1}_I}\|u\| \\
&\geq C^-_{m,I}\|J^*u\| - \delta(C_{\chi_2} + C_{\mathbb{1}_I})\|u\|
\end{aligned}$$

by Theorems 4.2.9 and 4.2.11, using the fact that

$$\chi_2(\widetilde{\Delta})u = \chi_2(\widetilde{\Delta})\widetilde{P}u = (\chi_2\mathbb{1}_I)(\widetilde{\Delta})u = 0$$

in the third line since the supports of χ_2 and I are disjoint. Moreover,

$$\|J^*u\| \geq \|J'u\| - \|(J^* - J')u\| \geq (1-\delta)\|u\| - 3\delta\|u\|_2 \geq \big(1 - \delta(1 + 3C^+_{2,I})\big)\|u\|$$

by Lemma 4.1.3 and (4.1b). Finally, we have shown that

$$\|PJ^*u\|_{-m} \geq \Big(C^-_{m,I}\big(1 - \delta(1+3C^+_{2,I})\big) - \delta(C_{\chi_2} + \chi_{\mathbb{1}_I})\Big)\|u\|.$$

In particular, if

$$0 < \delta < \Big(1 + 3C^+_{2,I} + \frac{C_{\chi_2} + C_{\mathbb{1}_I}}{C^-_{m,I}}\Big)^{-1} =: \delta_0,$$

then PJ^* restricted to ran $\widetilde{P}$ is injective, and dim ran $P \geq$ dim ran $\widetilde{P}$ follows as in Theorem 4.1.5.

The assertion on the convergence of the eigenvectors follows by a similar reasoning as in Theorem 4.1.5, noting that now ψ has to be estimated in the norm $\|\psi\|_m = (\lambda+1)^{m/2}$. □

We will now show that the spectrum of the non-negative operators Δ and $\widetilde{\Delta}$ are close in the Hausdorff distance. Since we cannot control all energies $\lambda \in [0,\infty)$ *uniformly*, we introduce the *weighted* distance

$$\overline{d}(a,b) := |\varphi(a) - \varphi(b)|, \qquad \varphi(\lambda) := \frac{1}{1+\lambda}. \tag{4.20}$$

It is easy to see that $\overline{d}$ is a metric on $[0,\infty)$. Moreover, extending the metric onto the space $[0,\infty]$ by setting $\varphi(\infty) := 0$ we obtain a compact metric space. Note that φ is a homeomorphism from $[0,\infty]$ onto $[0,1]$. Since our operators are unbounded, ∞

is always an accumulation point of the spectrum, and we consider it as belonging to the spectrum. In this sense, the spectrum of an unbounded operator is also compact in the metric space $([0,\infty], \overline{d})$. Denote the open ball around λ with radius $\eta > 0$ w.r.t. the metric $\overline{d}$ by $B_\eta(\lambda)$.

For subsets $A, B \subset [0,\infty]$, denote by $\overline{d}_\pm(A, B)$ and $\overline{d}(A, B)$ the maximal outside/inside and Hausdorff distance with respect to the metric $\overline{d}$ on $[0,\infty]$, see Definition A.1.1 in Sect. A.1. We sometimes call this (Hausdorff) distance also *weighted*. We need the weighted distance in order to *compactify* the space $[0,\infty)$.

Let us formulate the continuous dependence of the spectrum for non-negative operators. For the notion of convergence of non-negative operators, see Definition 4.2.6 and for the convergence of sets, we refer to Definition A.1.4.

Theorem 4.3.3. *Assume that $\Delta_n \to \Delta$ of order $m \geq 0$, then $\sigma(\Delta_n) \to \sigma(\Delta)$ in the weighted Hausdorff distance, i.e. $\overline{d}(\sigma(\Delta_n), \sigma(\Delta)) \to 0$.*

Proof. We use the element-wise characterisation of set convergence in Proposition A.1.6. We start with the convergence from outside $\sigma(\Delta_n) \searrow \sigma(\Delta)$: Let $\lambda \notin \sigma(\Delta)$. Since $\sigma(\Delta)$ is closed, there exists $\eta > 0$ such that $B_\eta(\lambda) \cap \sigma(\Delta) = \emptyset$. By Proposition 4.3.1,

$$0 = \dim \operatorname{ran} \mathbb{1}_{B_\eta(\lambda)}(\Delta) = \dim \operatorname{ran} \mathbb{1}_{B_\eta(\lambda)}(\Delta_n)$$

provided n is large enough (such that $\delta_n \leq \delta_0 = \delta_0(\eta, \lambda)$). In particular, $B_\eta(\lambda) \cap \sigma(\Delta_n) = \emptyset$ *eventually*.

In order to show the convergence from inside $\sigma(\Delta_n) \nearrow \sigma(\Delta)$, let $\lambda \in \sigma(\Delta)$ and $\eta > 0$. We have two cases:

Case A: If there exists an open interval $I \subset B_\eta(\lambda)$ with $\lambda \in I$ such that the boundary points of I are not in $\sigma(\Delta)$, then again by Proposition 4.3.1,

$$0 < \dim \operatorname{ran} \mathbb{1}_I(\Delta) = \dim \operatorname{ran} \mathbb{1}_I(\Delta_n)$$

eventually, and in particular, there exists $\lambda_n \in \sigma(\Delta_n)$ with $\overline{d}(\lambda_n, \lambda) \leq \eta$.

Case B: If no such interval exists, then $B_\eta(\lambda) \subset \sigma(\Delta)$. If in this case, $B_\eta(\lambda) \cap \sigma(\Delta_n) = \emptyset$ eventually, we apply once more Proposition 4.3.1, but with the roles of Δ and Δ_n interchanged, using the fact that $\partial_+ B_\eta(\lambda) \cap \sigma(\Delta_n) = \emptyset$. As a result, $B_\eta(\lambda) \cap \sigma(\Delta) = \emptyset$, a contradiction. In particular, we can find $\lambda_n \in \sigma(\Delta_n)$ with $\overline{d}(\lambda_n, \lambda) \leq \eta$. □

For the essential spectrum, we also have continuity:

Theorem 4.3.4. *Assume that $\Delta_n \to \Delta$ of order $m \geq 0$, then $\sigma_{\mathrm{ess}}(\Delta_n) \to \sigma_{\mathrm{ess}}(\Delta)$.*

Proof. We use again the characterisation Proposition A.1.6. In order to show the convergence from outside $\sigma_{\mathrm{ess}}(\Delta_n) \searrow \sigma_{\mathrm{ess}}(\Delta)$, let $\lambda \notin \sigma_{\mathrm{ess}}(\Delta)$. The essential spectrum is closed, therefore there exists $\eta > 0$ such that $B_\eta(\lambda)$ has no point in common with the essential spectrum. Since $B_\eta(\lambda)$ contains at most countably many points from the discrete spectrum of Δ, there exists an open interval $I \subset B_\eta(\lambda)$ with

$\partial_+ I \cap \sigma(\Delta) = \emptyset$. Applying now Proposition 4.3.1 if $m = 0$ resp. Proposition 4.3.2 if $m > 0$ we conclude that

$$\infty > \dim \operatorname{ran} \mathbb{1}_{B_\eta(\lambda)}(\Delta) = \dim \operatorname{ran} \mathbb{1}_{B_\eta(\lambda)}(\Delta_n)$$

so that $I \cap \sigma_{\mathrm{ess}}(\Delta_n) = \emptyset$ eventually.

For the convergence from inside, note that Case A of the preceding proof ($\partial_+ I \cap \sigma(\Delta) = \emptyset$) is literally the same, using

$$\infty = \dim \operatorname{ran} \mathbb{1}_I(\Delta) = \dim \operatorname{ran} \mathbb{1}_I(\Delta_n)$$

eventually. Case B is treated similarly: If no such interval exists, then $B_\eta(\lambda) \subset \sigma_{\mathrm{ess}}(\Delta)$. If in this case, $B_\eta(\lambda) \cap \sigma_{\mathrm{ess}}(\Delta_n) = \emptyset$ eventually, we can find an open bounded interval $I \subset B_\eta(\lambda)$ containing λ such that $\partial_+ I \cap \sigma(\Delta_n) = \emptyset$ for all n, since the discrete spectrum of $\sigma(\Delta_n)$ inside $B_\eta(\lambda)$ is at most countable. Applying Proposition 4.3.1 once more with the roles of Δ and Δ_n interchanged, we obtain

$$\infty > \dim \operatorname{ran} \mathbb{1}_I(\Delta_n) = \dim \operatorname{ran} \mathbb{1}_I(\Delta)$$

eventually, a contradiction to the fact that $\lambda \in \sigma_{\mathrm{ess}}(\Delta)$ and $\lambda \in I$. □

We can only show the lower semi-continuity of the discrete spectrum :

Theorem 4.3.5. *Assume that $\Delta_n \to \Delta$, then $\sigma_{\mathrm{disc}}(\Delta_n) \nearrow \sigma_{\mathrm{disc}}(\Delta)$. Moreover, the multiplicity is preserved, i.e. if λ is of multiplicity μ, then there exist μ eigenvalues $\lambda_{j,n}$, $j = 1, \dots, \mu$, (not necessarily mutually distinct) such that $\lambda_{j,n} \to \lambda$ as $n \to \infty$.*

If, in addition, $\mu = 1$, i.e. if there exists a simple eigenvalue $\lambda \in \sigma_{\mathrm{disc}}(\Delta)$ with normalised eigenvector ψ, then there exists a sequence of normalised eigenvectors $\psi_n \in \mathscr{H}_n$ of Δ_n such that

$$\|J_n \psi - \psi_n\| \to 0 \qquad \text{and} \qquad \|J'_n \psi_n - \psi\| \to 0,$$

where J_n and J'_n are the associated identification operators.

Proof. The proof is similar to the proof of the second part of Theorem 4.3.3: Let $\lambda \in \sigma_{\mathrm{disc}}(\Delta)$. Then by definition of the discrete spectrum, there exists $\eta > 0$ such that $B_\eta(\lambda) \cap \sigma_{\mathrm{disc}}(\Delta) = \{\lambda\}$. In particular, by Proposition 4.3.1 resp. Proposition 4.3.2,

$$0 < \mu := \dim \operatorname{ran} \mathbb{1}_{B_\eta(\lambda)}(\Delta) = \dim \operatorname{ran} \mathbb{1}_{B_\eta(\lambda)}(\Delta_n) < \infty$$

eventually, showing the existence of μ sequences $\{\lambda_{j,n}\}_n$ converging to λ. Moreover, Case B cannot occur here. □

Remark 4.3.6. Note that we cannot expect the upper semi-continuity of $\sigma_{\mathrm{disc}}(\cdot)$ as the following simple example shows: Let T be a bounded operator in $\mathscr{H}$ with purely discrete spectrum spectrum and set $T_n := n^{-1} T$. Then $T_n \to 0$. Moreover, 0 is in the essential spectrum of the limit operator, but not in the discrete one. But it is impossible to find a neighbourhood around 0 not containing discrete spectrum, i.e. the characterisation of convergence from outside in Proposition A.1.6 is not fulfilled. Although the limit operator is not the resolvent of an unbounded operator (in the usual sense), it is easy to construct also examples where $\Delta_n \to \Delta$ does not imply the upper semi-continuity of $\sigma_{\mathrm{disc}}(\cdot)$.

Remark 4.3.7. Using the min-max characterisation of eigenvalues Theorem 3.2.3, we obtain a more precise estimate for the discrete spectrum *below* the essential spectrum (see Corollary 4.4.19). The problem in determine the convergence velocity of the eigenvalues is the complicated dependence of the constant $\delta_0 = \delta_0(m, I)$ on η for $I = (\lambda - \eta, \lambda + \eta)$, see Propositions 4.3.1 and 4.3.2. Nevertheless, the previous result on the entire discrete spectrum also allows to show the convergence of a discrete eigenvalue *inside* a spectral gap. Note that we obtain a more explicit error estimate using the holomorphic functional calculus, see Sect. 4.6.

4.4 Convergence of Quadratic Forms in Different Hilbert Spaces

In this section, we define quasi-unitary equivalence for two *quadratic forms*. Even if the assumptions are stronger than the assumptions for δ-quasi-unitary equivalence of *operators*, it is in general easier to deal with the first order domains (as it is the case in our applications).

Assume that we have two Hilbert spaces $\mathscr{H}$ and $\widetilde{\mathscr{H}}$, each endowed with a non-negative (unbounded) quadratic form $\mathfrak{d}$ and $\widetilde{\mathfrak{d}}$. We denote the corresponding operators by Δ and $\widetilde{\Delta}$, and the associated scales of Hilbert spaces by $\mathscr{H}^k$ and $\widetilde{\mathscr{H}}^k$, respectively, see Theorem 3.1.1 and Sect. 3.2. Note that $\mathscr{H}^1 = \operatorname{dom} \mathfrak{d}$ and $\widetilde{\mathscr{H}}^1 = \operatorname{dom} \widetilde{\mathfrak{d}}$.

Let us now introduce the following identification operators acting on the scale of order 1.

Definition 4.4.1. Let $\delta \geq 0$ and

$$J^1 : \mathscr{H}^1 \longrightarrow \widetilde{\mathscr{H}}^1 \qquad \text{and} \qquad J'^1 : \widetilde{\mathscr{H}}^1 \longrightarrow \mathscr{H}^1$$

be bounded linear operators, called *identification operators of order* 1 associated with the quadratic forms $\mathfrak{d}$ and $\widetilde{\mathfrak{d}}$.

1. We say that J^1 and J'^1 are *δ-adjoint* with respect to $\mathfrak{d}$ and $\widetilde{\mathfrak{d}}$, if

$$\|J^1 - J^{-1}\|_{1 \to -1} = \|\widetilde{R}^{1/2}(J^1 - J^{-1})R^{1/2}\| \leq \delta \tag{4.21a}$$

is fulfilled, where

$$J^{-1} := (J'^1)^* : \mathscr{H}^{-1} \longrightarrow \widetilde{\mathscr{H}}^{-1}.$$

2. We say that J^1 is $(1+\delta)$-*bounded* if
$$\|J^1\|_{1\to 1} \le 1+\delta. \tag{4.21b}$$
3. We say that J'^1 is a *left-inverse* for J^1, if
$$J'^1 J^1 = \mathrm{id}_1 . \tag{4.21c}$$
4. We say that the quadratic forms $\mathfrak{d}$ and $\widetilde{\mathfrak{d}}$ are δ-*close* with first order identification operators J^1 and J'^1 if J^1 and J'^1 are δ-adjoint and if
$$\|\widetilde{\Delta} J^1 - J^{-1}\Delta\|_{1\to -1} \le \delta. \tag{4.21d}$$
5. We say that J^1 and J'^1 are δ-*partial isometric* w.r.t. $\mathfrak{d}$ and $\widetilde{\mathfrak{d}}$, if $\mathfrak{d}$ and $\widetilde{\mathfrak{d}}$ are δ-close and J'^1 is a left-inverse for J^1, i.e. if (4.21a), (4.21c) and (4.21d) are fulfilled.

Remark 4.4.2.

1. Note that J^{-1} does not denote the inverse of J.
2. Estimate (4.21a) is equivalent with
$$\big|\langle J^1 f, u\rangle - \langle f, J'^1 u\rangle\big| \le \delta \|f\|_1 \|u\|_1 \tag{4.21a'}$$
for all $f \in \mathscr{H}^1$ and $u \in \widetilde{\mathscr{H}}^1$. Similarly, condition (4.21d) for the δ-closeness can be expressed as
$$\big|\widetilde{\mathfrak{d}}(J^1 f, u) - \mathfrak{d}(f, J'^1 u)\big| \le \delta \|f\|_1 \|u\|_1. \tag{4.21d'}$$
3. We do not need in general the boundedness of J^1 or J'^1 in quantitative estimates. Note, that the boundedness is needed in order to assure that $J^{-1} f$ is a *bounded* anti-linear form on $\widetilde{\mathscr{H}}^1$. For partial isometric identification operators, the $(1+\delta)$-boundedness follows a posteriori, cf. Lemma 4.4.3. Moreover, the boundedness of J'^1 is needed if one wants to understand the expression individually, e.g., if $J^{-1}\Delta$ is understood as the concatenation of $\Delta\colon \mathscr{H}^1 \longrightarrow \mathscr{H}^{-1}$ and $J^{-1}\colon \mathscr{H}^{-1} \longrightarrow \widetilde{\mathscr{H}}^{-1}$.
   Nevertheless, we can define expressions like $(J^1 - J^{-1})f$ and $(\widetilde{\Delta} J^1 - J^{-1}\Delta) f$ via the sesquilinear forms $\mathscr{H}^1 \times \widetilde{\mathscr{H}}^1 \to \mathbb{C}$,
$$(f,u) \mapsto \langle J^1 f, u\rangle - \langle f, J'^1 u\rangle \quad \text{resp.} \quad (f,u) \mapsto \widetilde{\mathfrak{d}}(J^1 f, u) - \mathfrak{d}(f, J'^1 u)$$
   without referring to the boundedness of J^1 or J'^1.
4. We will give below an easy way how to show the δ-adjointness using also the identification operators J and J' of order 0 (see Definition 4.4.9).
5. We sometimes say that Δ, $\widetilde{\Delta}$ are δ-close of order -1, if $\mathfrak{d}$ and $\widetilde{\mathfrak{d}}$ are δ-close although this is slightly inconsistent with Definition 4.2.1 (2).

6. We do not assume the analogue statement of $\|u - JJ'u\|_{1\to 0} \le \delta$ for the first order identification operators.

Let us state some consequences of the definitions, explaining in particular the name "δ-partial isometry":

Lemma 4.4.3. *Let $\delta \ge 0$.*

1. Assume that J^1 is $(1+\delta)$-bounded and that J^1 and J'^1 are δ'-close, then J'^1 is $(1+\delta+2\delta')$-bounded.
2. Assume that J^1 and J'^1 are δ-partial isometric w.r.t. $\mathfrak{d}$ and $\widetilde{\mathfrak{d}}$, then

$$\big|\|J^1 f\|_1 - \|f\|_1\big| \le 2\delta\|f\|_1. \tag{4.22a}$$

In particular, J^1 is $(1+2\delta)$-bounded and J'^1 is $(1+3\delta)$-bounded.

Proof. The first assertion follows from

$$\begin{aligned}\|J'^1 u\|_1^2 &= \langle u, J'^{1*}(\Delta+1)J'^1 u\rangle_{1,-1}\\ &= \langle u, (J^{-1}\Delta - \widetilde{\Delta}J^1)J'^1 u\rangle_{1,-1} + \langle u, (\widetilde{\Delta}+1)J^1 J'^1 u\rangle_{1,-1}\\ &\quad + \langle u, (J^{-1} - J^1)J'^1 u\rangle_{1,-1}\\ &\le \|u\|_1\big(2\delta' + \|J^1\|_{1\to 1}\big)\|J'^1 u\|_1.\end{aligned}$$

Here, $\langle\cdot,\cdot\rangle_{1,-1}$ stands for the sesquilinear pairing of $\mathscr{H}^1 \times \mathscr{H}^{-1}$. For the second assertion, note that

$$\begin{aligned}(J^1)^*(\widetilde{\Delta}+1)J^1 &= (\Delta+1)J'^1 J^1 + [(J^1)^*(\widetilde{\Delta}+1) - (\Delta+1)J'^1]J^1\\ &= (\Delta+1) + [(\widetilde{\Delta}J^1 - J^{-1}\Delta)^* + (J^1 - J^{-1})^*]J^1\end{aligned}$$

using (4.21c) in the second equality. It follows that

$$\begin{aligned}\big|\|J^1 f\|_1^2 - \|f\|_1^2\big| &= \big|\langle f, [(J^1)^*(\widetilde{\Delta}+1)J^1 - (\Delta+1)]f\rangle_{1,-1}\big|\\ &\overset{\text{CS}}{\le} \|f\|_1\big(\|\widetilde{\Delta}J^1 - J^{-1}\Delta\|_{1\to -1} + \|J^1 - J^{-1}\|_{1\to -1}\big)\|J^1 f\|_1\\ &\le 2\delta\|f\|_1\|J^1 f\|_1\end{aligned}$$

using the δ-closeness and δ-adjointness. Now, if $f \ne 0$, then $J^1 f \ne 0$, and

$$\big|\|J^1 f\|_1 - \|f\|_1\big| = \frac{\big|\|J^1 f\|_1^2 - \|f\|_1^2\big|}{\|J^1 f\|_1 + \|f\|_1} \le \frac{2\delta\|f\|_1\|J^1 f\|_1}{\|J^1 f\|_1} = 2\delta\|f\|_1.$$

The $(1+2\delta)$-boundedness of J^1 is now obvious, and the $(1+3\delta)$-boundedness of J'^1 follows from the first assertion. □

As in Sect. 4.2, we derive the convergence of powers of the resolvent: Note that

$$\|A\|_{-1\to 1} = \|(\widetilde{\Delta}+1)^{1/2} A (\Delta+1)^{1/2}\|.$$

Lemma 4.4.4. *Assume that* $\mathfrak{d}$ *and* $\widetilde{\mathfrak{d}}$ *are* δ*-close with identification operators* J^1 *and* J'^1, *then*

$$\|\widetilde{R}^k J^{-1} - J^1 R^k\|_{-1\to 1} \le (3k-1)\delta \tag{4.23}$$

for $k \ge 1$.

Proof. We start with $k = 1$ and note that

$$(\widetilde{\Delta}+1)^{1/2}(\widetilde{R}J^{-1} - J^1 R)(\Delta+1)^{1/2} = \widetilde{R}^{1/2}\big[(J^{-1}\Delta - \widetilde{\Delta}J^1) + (J^{-1} - J^1)\big]R^{1/2}$$

from which the estimate $\|\widetilde{R}J^{-1} - J^1 R\|_{-1\to 1} \le 2\delta$ follows. For $k \ge 2$, we write

$$\widetilde{R}^k J^{-1} - J^1 R^k = \sum_{j=0}^{k-1} \widetilde{R}^{k-1-j}(\widetilde{R}J^{-1} - J^1 R)R^j + \sum_{j=1}^{k-1} \widetilde{R}^{k-j}(J^1 - J^{-1})R^j.$$

We estimate the norm of the first sum by $2k\delta$ and the second by $(k-1)\delta$. □

For continuous functions $\varphi \in \mathsf{C}(\overline{\mathbb{R}}_+)$ having fast enough decay we have:

Theorem 4.4.5. *Let* $\varphi\colon [0,\infty) \longrightarrow \mathbb{C}$ *be a continuous function such that* $\lim_{\lambda\to\infty}(\lambda+1)\varphi(\lambda)$ *exists. Suppose in addition that* $\mathfrak{d}$ *and* $\widetilde{\mathfrak{d}}$ *are* δ*-close with* J^1 *and* J'^1, *then*

$$\|\varphi(\widetilde{\Delta})J^{-1} - J^1\varphi(\Delta)\|_{-1\to 1} \le C_\varphi \delta, \tag{4.24}$$

where C_φ *depends only on* φ.

Proof. Let $\widehat{\varphi}(\lambda) = (\lambda+1)\varphi(\lambda)$. By assumption, $\widehat{\varphi} \in \mathsf{C}(\overline{\mathbb{R}}_+)$ and therefore by the Stone-Weierstrass theorem, there exists a polynomial

$$\widehat{p}(\lambda) := \sum_{k=0}^{n} a_k (\lambda+1)^{-k}$$

in $(\lambda+1)^{-1}$ such that $\|\widehat{p} - \widehat{\varphi}\|_\infty \le \delta$. Then $p(\lambda) := (\lambda+1)^{-1}\widehat{p}(\lambda)$ is a polynomial without constant term and we have

$$\begin{aligned}\|\varphi(\widetilde{\Delta})J^{-1} - J^1\varphi(\Delta)\|_{-1\to 1} &\le \|(\varphi - p)(\widetilde{\Delta})\|_{-1\to 1}\|J^{-1}\|_{-1\to -1}\\ &\quad + \|J^1\|_{1\to 1}\|(\varphi - p)(\Delta)\|_{-1\to 1}\\ &\quad + \sum_{k=0}^{n} |a_k|\,\|\widetilde{R}^{k+1}J^{-1} - J^1 R^{k+1}\|_{-1\to 1}\end{aligned}$$

$$\leq 4\|\widehat{\varphi} - \widehat{p}\|_\infty + \sum_{k=0}^{n} |a_k|(3k+2)\delta$$

$$\leq \Big(4 + \sum_{k=0}^{n} |a_k|(3k+2)\Big)\delta =: C_\varphi \delta$$

using $\|(\varphi - p)(\Delta)\|_{-1\to 1} = \|(\widehat{\varphi} - \widehat{p})(\Delta)\| \leq \|\widehat{\varphi} - \widehat{p}\|_\infty$ and a similar estimate for $\widetilde{\Delta}$, our assumptions and Lemma 4.4.4. □

Similarly as for zeroth order identification operators (see Theorem 4.2.10), we can extend the previous result to certain bounded measurable functions $\psi: \overline{\mathbb{R}}_+ \longrightarrow \mathbb{C}$.

Theorem 4.4.6. *Suppose that $U \subset \mathbb{R}_+$ and that $\psi: \overline{\mathbb{R}}_+ \longrightarrow \mathbb{C}$ is a measurable, bounded function, continuous on U such that $\lim_{\lambda\to\infty}(\lambda + 1)\psi(\lambda)$ exists. Then there exists a constant $C_\psi > 0$ depending only on ψ such that*

$$\|\psi(\widetilde{\Delta})J^{-1} - J^1\psi(\Delta)\|_{-1\to 1} \leq C_\psi \delta \tag{4.25}$$

for all non-negative quadratic forms $\mathfrak{d}$ and $\widetilde{\mathfrak{d}}$ being δ-close with J^1 and J'^1, provided

$$\sigma(\Delta) \subset U \qquad \text{or} \qquad \sigma(\widetilde{\Delta}) \subset U.$$

Proof. Let χ_1 be a continuous function on $\overline{\mathbb{R}}_+$ satisfying $0 \leq \chi_1 \leq 1$, $\chi_1 = 1$ on $\sigma(\Delta) \cup \{\infty\}$ and $\operatorname{supp} \chi_1 \subset U$. Then $\chi_1\psi$ and $\chi_2 = 1 - \chi_1$ are continuous functions on $\overline{\mathbb{R}}_+$ and

$$\|\psi(\widetilde{\Delta})J^{-1} - J^1\psi(\Delta)\|_{-1\to 1} \leq \|(\chi_1\psi)(\widetilde{\Delta})J^{-1} - J^1(\chi_1\psi)(\Delta)\|_{-1\to 1} + \|(\chi_2\psi)(\widetilde{\Delta})J^{-1} - J^1(\chi_2\psi)(\Delta)\|_{-1\to 1}.$$

If U is a neighbourhood of $\sigma(\Delta)$ we can estimate the norm with χ_2 by

$$\|\psi\|_\infty \|\chi_2(\widetilde{\Delta})J^{-1} - J^1\chi_2(\Delta)\|_{-1\to 1}$$

using the fact that $(\chi_2\psi)(\Delta) = \chi_2(\Delta) = 0$ since $\chi_2 = 0$ on $\sigma(\Delta)$. Applying the preceding theorem twice, we have the error estimate

$$C_\psi := C_{\chi_1\psi} + \|\psi\|_\infty C_{\chi_2}.$$

Note that in both cases, the functions $\chi_1\psi$ and χ_2 are continuous and still lie in $\mathsf{C}(\overline{\mathbb{R}}_+)$ when multiplied with $(\lambda + 1)$. If U is a neighbourhood of $\sigma(\widetilde{\Delta})$ we obtain the result by duality (as in the proof of Theorem 4.2.10). □

As a consequence of the preceding theorem, we can ensure the convergence of the spectral projections:

Corollary 4.4.7. *Assume that I is a* bounded *interval in $\mathbb{R}_+$. Then the spectral projections satisfy*

$$\|\mathbb{1}_I(\widetilde{\Delta})J^{-1} - J^1\mathbb{1}_I(\Delta)\|_{-1\to 1} \le C_{\mathbb{1}_I}\delta. \tag{4.26}$$

provided $\partial_+ I \cap \sigma(\Delta) = \emptyset$ or $\partial_+ I \cap \sigma(\widetilde{\Delta}) = \emptyset$.

Proof. We apply the preceding theorem with $\psi = \mathbb{1}_I$, the indicator function of I. Note that $\mathbb{1}_I$ is continuous on $U = \mathbb{R}_+ \setminus \partial_+ I$ and that $\lim_{\lambda\to\infty}(\lambda+1)\mathbb{1}_I(\lambda) = 0$ exists since I is bounded. □

Let us derive another result on the unitary time evolution (cf. Theorems 4.2.15 and 4.2.16), expressed in terms of the associated forms:

Theorem 4.4.8. *Suppose that*

$$\left|\widetilde{\mathfrak{d}}(J^1\psi(t), u) - \mathfrak{d}(\psi(t), J'^1 u)\right| \le \mu(t)\|u\|_1$$

for all $u \in \widetilde{\mathscr{H}}_1$, where $\psi(t) := \mathrm{e}^{\mathrm{i}t\Delta}\psi(0)$ is the time evolution of a state $\psi(0) \in \mathscr{H}$ w.r.t. Δ and that J^1, J'^1 are δ-quasi-unitary. Then

$$\|(\mathrm{e}^{\mathrm{i}t\widetilde{\Delta}}J^1 - J^{-1}\mathrm{e}^{\mathrm{i}t\Delta})\psi(0)\|_{-1} \le 6\delta\|\psi(0)\|_1 + \int_0^t \mu(\tau)\mathrm{d}\tau.$$

In particular, if $\mathfrak{d}$ and $\widetilde{\mathfrak{d}}$ are δ-close with J^1 and J'^1, then $\mu(t) = \delta\|\psi(0)\|_1$ and

$$\|\mathrm{e}^{\mathrm{i}t\widetilde{\Delta}}J^1 - J^{-1}\mathrm{e}^{\mathrm{i}t\Delta}\|_{1\to -1} \le (6+t)\delta.$$

Proof. Applying the operator calculus to the formula

$$\mathrm{e}^{\mathrm{i}t\lambda} = \mathrm{i}\int_0^t \lambda \mathrm{e}^{\mathrm{i}\tau\lambda}\mathrm{d}\tau + 1$$

with λ replaced by Δ and $\widetilde{\Delta}$, we arrive at the operator equality

$$\mathrm{e}^{\mathrm{i}t\widetilde{\Delta}}J^1 - J^{-1}\mathrm{e}^{\mathrm{i}t\Delta} = \mathrm{i}\int_0^t \mathrm{e}^{\mathrm{i}\tau\widetilde{\Delta}}(\widetilde{\Delta}J^1 - J^{-1}\widetilde{\Delta})\mathrm{e}^{\mathrm{i}\tau\Delta}\mathrm{d}\tau + (J^1 - J^{-1}),$$

where the integral exist in the operator norm topology on $\mathscr{H}_1 \to \widetilde{\mathscr{H}}_{-1}$. Therefore,

$$\begin{aligned}
&\|(\mathrm{e}^{\mathrm{i}t\widetilde{\Delta}}J^1 - J^{-1}\mathrm{e}^{\mathrm{i}t\Delta})\psi(0)\|_{-1}\\
&\le \int_0^t \|(\widetilde{\Delta}J^1 - J^{-1}\widetilde{\Delta})\psi(\tau)\|_{-1}\mathrm{d}\tau + \|J^1 - J^{-1}\|_{1\to -1}\|\psi(0)\|_1\\
&\le \int_0^t \mu(\tau)\mathrm{d}\tau + \delta\|\psi(0)\|_1
\end{aligned}$$

The last assertion follows then immediately. □

Let us now relate the first order identification operators with the ones defined on the entire Hilbert spaces, here called zeroth order identification operators (see Definition 4.1.1):

Definition 4.4.9. Assume that J, J' are zeroth order identification operators. We say that the linear operators

$$J^1 \colon \mathscr{H}^1 \longrightarrow \widetilde{\mathscr{H}}^1, \qquad J'^1 \colon \widetilde{\mathscr{H}}^1 \longrightarrow \mathscr{H}^1$$

are δ-*compatible* with the zeroth order identification operators J and J' if

$$\|J - J^1\|_{1\to 0} \le \delta \qquad \text{and} \qquad \|J' - J'^1\|_{1\to 0} \le \delta. \tag{4.27}$$

A simple consequence is the following:

Lemma 4.4.10. *Assume that J^1 and J'^1 are δ-compatible with the zeroth order identification operators J and J' such that $\|J' - J^*\| \le \delta'$, then J^1 and J'^1 are $(2\delta + \delta')$-adjoint (see Definition 4.4.1 (1)).*

Proof. We have

$$J^{-1} - J^1 = (J'^1 - J')^* + (J'^* - J) + (J - J^1)$$

and the estimate follows from (4.27) and the assumptions. □

Finally, we define the following property of quadratic forms to be "unitarily equivalent" up to an error:

Definition 4.4.11. Let $\delta \ge 0$.

1. We say that the quadratic forms $\mathfrak{d}$ and $\widetilde{\mathfrak{d}}$ are δ-*quasi-unitarily equivalent* if there exist first and zeroth order identification operators J^1, J'^1 and J, J', such that J^1 and J'^1 are bounded,
   a. $\mathfrak{d}$ and $\widetilde{\mathfrak{d}}$ are δ-close,
   b. J^1 and J'^1 are δ-compatible with J and J', and
   c. J and J' are δ-quasi-unitary of order 1.
2. We say that the quadratic forms $\mathfrak{d}$ and $\widetilde{\mathfrak{d}}$ are δ-*partial isometrically equivalent* if there exist first and zeroth order identification operators J^1, J'^1 and J, such that J^1 and J'^1 are bounded,
   a. $\mathfrak{d}$ and $\widetilde{\mathfrak{d}}$ are δ-close,
   b. J^1 and J'^1 are δ-compatible with J and J^*,
   c. J is δ-partial isometric of order 1, and
   d. J'^1 is a left-inverse for J^1.

Let us give a self-contained summary of all estimates necessary in order to assure the δ-quasi-unitary resp. δ-partial isometric equivalence of quadratic forms. Recall that $\mathscr{H}^1 = \operatorname{dom} \mathfrak{d}$ and $\widetilde{\mathscr{H}}^1 = \operatorname{dom} \widetilde{\mathfrak{d}}$ with norms defined by

$$\|f\|_1^2 = \mathfrak{d}(f) + \|f\|^2 \quad \text{and} \quad \|u\|_1^2 = \widetilde{\mathfrak{d}}(u) + \|u\|^2.$$

Proposition 4.4.12. *Let $\delta \geq 0$ and assume that*

$$J: \mathscr{H} \longrightarrow \widetilde{\mathscr{H}}, \qquad J': \widetilde{\mathscr{H}} \longrightarrow \mathscr{H},$$
$$J^1: \mathscr{H}^1 \longrightarrow \widetilde{\mathscr{H}}^1, \qquad J'^1: \widetilde{\mathscr{H}}^1 \longrightarrow \mathscr{H}^1$$

are linear operators. If J^1 is bounded and if the estimates

$$\|Jf\|^2 \leq 4\|f\|^2, \tag{4.28a}$$
$$|\langle J'u, f\rangle - \langle u, Jf\rangle|^2 \leq \delta^2\|f\|^2\|u\|^2, \tag{4.28b}$$
$$\|f - J'Jf\|^2 \leq \delta^2\|f\|_1^2, \qquad \|u - JJ'u\|^2 \leq \delta^2\|u\|_1^2 \tag{4.28c}$$
$$\|J^1 f - Jf\|^2 \leq \delta^2\|f\|_1^2, \qquad \|J'^1 u - J'u\|^2 \leq \delta^2\|u\|_1^2, \tag{4.28d}$$
$$\left|\widetilde{\mathfrak{d}}(J^1 f, u) - \mathfrak{d}(f, J'^1 u)\right|^2 \leq \delta^2\|f\|_1^2\|u\|_1^2 \tag{4.28e}$$

are fulfilled for all $f \in \mathscr{H}^1$ and $u \in \widetilde{\mathscr{H}}^1$, then the quadratic forms $\mathfrak{d}$ and $\widetilde{\mathfrak{d}}$ are (3δ)-quasi-unitarily equivalent.

On the other hand, if $\mathfrak{d}$ and $\widetilde{\mathfrak{d}}$ are δ-quasi-unitarily equivalent, then the estimates (4.28) *are fulfilled.*

Proof. Almost everything is obvious from Definitions 4.1.1, 4.4.1 and 4.4.9. Note that we obtain 3δ due to Lemma 4.4.10, and since δ-adjointness is part of the definition of δ-closeness of $\mathfrak{d}$ and $\widetilde{\mathfrak{d}}$. The boundedness of J'^1 follows from Lemma 4.4.3. □

Proposition 4.4.13. *Let $\delta \geq 0$ and assume that*

$$J: \mathscr{H} \longrightarrow \widetilde{\mathscr{H}},$$
$$J^1: \mathscr{H}^1 \longrightarrow \widetilde{\mathscr{H}}^1, \qquad J'^1: \widetilde{\mathscr{H}}^1 \longrightarrow \mathscr{H}^1$$

are linear operators. If the estimates

$$J^*Jf = f, \qquad \|u - JJ^*u\|^2 \leq \delta^2\|u\|_1^2 \tag{4.29a}$$
$$\|J^1 f - Jf\|^2 \leq \delta^2\|f\|_1^2, \qquad \|J'^1 u - J^*u\|^2 \leq \delta^2\|u\|_1^2, \tag{4.29b}$$
$$J'^1 J^1 f = f, \tag{4.29c}$$
$$\left|\widetilde{\mathfrak{d}}(J^1 f, u) - \mathfrak{d}(f, J'^1 u)\right|^2 \leq \delta^2\|f\|_1^2\|u\|_1^2 \tag{4.29d}$$

are fulfilled for all $f \in \mathscr{H}^1$ and $u \in \widetilde{\mathscr{H}}^1$, then the quadratic forms $\mathfrak{d}$ and $\widetilde{\mathfrak{d}}$ are (2δ)-partial isometrically equivalent.

On the other hand, if $\mathfrak{d}$ and $\widetilde{\mathfrak{d}}$ are δ-partial isometrically equivalent, then the estimates (4.29) *are fulfilled.*

Remark 4.4.14. Note that we do not need a quantitative bound on $\|J^1\|_{1\to 1}$. The boundedness of the identification operators in the partial isometric case follow by Lemmata 4.1.3 and 4.4.3.

Let us now relate the quasi-unitary resp. partial isometric equivalence (see Definition 4.2.3) of operators with the one of quadratic forms:

Proposition 4.4.15. *If $\mathfrak{d}$ and $\widetilde{\mathfrak{d}}$ are δ-quasi-unitarily equivalent (δ-partial isometrically equivalent), then the associated operators Δ and $\widetilde{\Delta}$ are δ'-quasi-unitarily equivalent (δ'-partial isometrically equivalent) of order $m = 0$, where $\delta' = 4\delta$. In particular, the results of Sects. 4.2 and 4.3 apply.*

Proof. The δ-quasi-unitarity resp. δ-partial isometry of the zeroth order identification operators is already part of Definition 4.4.11. We only have to show the δ-closeness (of order 0) of the operators, namely

$$\begin{aligned}\|\widetilde{\Delta}J - J\Delta\|_{2\to -2} &\le \|\widetilde{\Delta}(J - J^1)\|_{2\to -2} + \|\widetilde{\Delta}J^1 - J^{-1}\Delta\|_{2\to -2}\\ &\quad + \|(J'^1 - J')^*\Delta\|_{2\to -2} + \|(J'^* - J)\Delta\|_{2\to -2} \le 4\delta\end{aligned}$$

using $\|\widetilde{\Delta}\|_{0\to -2} \le 1$, $\|J - J^1\|_{2\to 0} \le \|J - J^1\|_{1\to 0} \le \delta$ etc. □

As in Proposition 4.2.5, we can show the *transitivity* of δ-quasi-unitary resp. δ-partial isometric equivalence of *quadratic forms*. Assume that $\mathscr{H}$, $\widetilde{\mathscr{H}}$ and $\widehat{\mathscr{H}}$ are three Hilbert spaces with non-negative operators $\mathfrak{d}$, $\widetilde{\mathfrak{d}}$ and $\widehat{\mathfrak{d}}$, respectively. Moreover, assume that

$$J\colon \mathscr{H} \longrightarrow \widetilde{\mathscr{H}}, \quad \widetilde{J}\colon \widetilde{\mathscr{H}} \longrightarrow \widehat{\mathscr{H}}, \quad \widetilde{J}'\colon \widehat{\mathscr{H}} \longrightarrow \widetilde{\mathscr{H}} \quad \text{and} \quad J'\colon \widetilde{\mathscr{H}} \longrightarrow \mathscr{H},$$

$$J^1\colon \mathscr{H}^1 \longrightarrow \widetilde{\mathscr{H}}^1,\ \widetilde{J}^1\colon \widetilde{\mathscr{H}}^1 \longrightarrow \widehat{\mathscr{H}}^1,\ \widetilde{J}'^1\colon \widehat{\mathscr{H}}^1 \longrightarrow \widetilde{\mathscr{H}}^1 \text{ and } J'^1\colon \widetilde{\mathscr{H}}^1 \longrightarrow \mathscr{H}^1$$

are bounded operators. We define

$$\begin{aligned}\hat{J} &:= \widetilde{J}J\colon \mathscr{H} \longrightarrow \widehat{\mathscr{H}}, & \hat{J}' &:= J'\widetilde{J}'\colon \widehat{\mathscr{H}} \longrightarrow \mathscr{H},\\ \hat{J}^1 &:= \widetilde{J}^1 J^1\colon \mathscr{H}^1 \longrightarrow \widehat{\mathscr{H}}^1, & \hat{J}'^1 &:= J'^1\widetilde{J}'^1\colon \widehat{\mathscr{H}}^1 \longrightarrow \mathscr{H}^1.\end{aligned}$$

Let us use the equivalent characterisation of "δ-quasi-unitary equivalence" in (4.28) as definition, except that we replace condition (4.28a) in the definition of δ-quasi-unitarity by

$$\|J\| \le 1 + \delta \qquad \text{resp.} \qquad \|\widetilde{J}\| \le 1 + \widetilde{\delta}.$$

Moreover, let us assume that the first order identification operators J^1 and $\widetilde{J}^1$ in Definition 4.4.11 are $(1+\delta)$- resp. $(1+\widetilde{\delta})$-bounded, i.e.

$$\|J^1\|_{1\to 1} \le 1 + \delta \qquad \text{resp.} \qquad \|\widetilde{J}^1\|_{1\to 1} \le 1 + \widetilde{\delta}.$$

Note that J^1 and J^{-1} are 3δ-adjoint (Lemma 4.4.10), since we used (4.28) as definition. In particular, $\|J'\| \le 1 + 2\delta$, $\|J'^1\|_{1\to 1} \le 1 + 5\delta$ by (4.3a) and the proof of Lemma 4.4.3. A similar remark holds for $\widetilde{J}'$ and $\widetilde{J}'^1$.

Similarly, we use the equivalent characterisation of of "δ-partial isometric equivalence" in (4.29) as definition. Note that here, J^1 and J^{-1} are 2δ-adjoint. We do not need extra conditions on the bounds of the identification operators, since $\|J\| = \|J^*\| = 1$ resp. $\|\widetilde{J}\| = \|\widetilde{J}^*\| = 1$ and

$$\|J^1\|_{1\to 1} \le 1 + 3\delta \qquad \text{resp.} \qquad \|\widetilde{J}^1\|_{1\to 1} \le 1 + 3\widetilde{\delta},$$
$$\|J'^1\|_{1\to 1} \le 1 + 6\delta \qquad \text{resp.} \qquad \|\widetilde{J}'^1\|_{1\to 1} \le 1 + 6\widetilde{\delta}$$

(see the proof of Lemma 4.4.3).

Proposition 4.4.16. *Assume that* $0 \le \delta, \widetilde{\delta} \le 1$.

1. *Assume that* $\mathfrak{d}$ *and* $\widetilde{\mathfrak{d}}$ *are* δ*-quasi-unitarily equivalent with identification operators* J, J^1, J' *and* J'^1, *and that* $\widetilde{\mathfrak{d}}$ *and* $\widehat{\mathfrak{d}}$ *are* $\widetilde{\delta}$*-quasi-unitarily equivalent with identification operators* $\widetilde{J}$, $\widetilde{J}^1$, $\widetilde{J}'$ *and* $\widetilde{J}'^1$. *Then* $\mathfrak{d}$ *and* $\widehat{\mathfrak{d}}$ *are* $\hat{C}(\delta + \widetilde{\delta})$*-quasi-unitarily equivalent with identification operators* $\hat{J}$, $\hat{J}^1$, $\hat{J}'$ *and* $\hat{J}'^1$, *where* $\hat{C}$ *is a universal constant.*
2. *Assume that* $\mathfrak{d}$ *and* $\widetilde{\mathfrak{d}}$ *are* δ*-partial isometrically equivalent with identification operators* J, J^1 *and* J'^1, *and that* $\widetilde{\mathfrak{d}}$ *and* $\widehat{\mathfrak{d}}$ *are* $\widetilde{\delta}$*-partial isometrically equivalent with identification operators* $\widetilde{J}$, $\widetilde{J}^1$ *and* $\widetilde{J}'^1$. *Then* $\mathfrak{d}$ *and* $\widehat{\mathfrak{d}}$ *are* $\hat{C}'(\delta + \widetilde{\delta})$*-partial isometrically equivalent with identification operator* $\hat{J}$, $\hat{J}^1$ *and* $\hat{J}'^1$, *where again* $\hat{C}$ *is a universal constant.*

Proof. We set $R = (\Delta + 1)^{-1}$, $\widetilde{R} = (\widetilde{\Delta} + 1)^{-1}$ and $\widehat{R} = (\widehat{\Delta} + 1)^{-1}$ and define $\hat{J}^{-1} := (\hat{J}'^1)^* = \widetilde{J}^{-1} J^{-1}$.

Let us start with the quasi-unitary equivalence. We have already shown in the proof of Proposition 4.2.5 that

$$\|\hat{J}\| \le 1 + 2(\delta + \widetilde{\delta}) \qquad \text{and} \qquad \|\hat{J} - \hat{J}'^*\| \le 2\widetilde{\delta} + 3\delta$$

using $\delta, \widetilde{\delta} \le 1$. Moreover,

$$\begin{aligned}
\|\mathrm{id} - \hat{J}'\hat{J}\|_{1\to 0} &= \|(\mathrm{id} - J'\widetilde{J}'\widetilde{J}J)R^{1/2}\| \\
&= \big\|(\mathrm{id} - J'J)R^{1/2} + J'(\widetilde{\mathrm{id}} - \widetilde{J}'\widetilde{J})\big(\widetilde{R}^{1/2} + (JR^{1/2} - \widetilde{R}^{1/2}J)\big)\big\| \\
&\le \|(\mathrm{id} - J'J)R^{1/2}\| + \|J'\|\big(\|(\widetilde{\mathrm{id}} - \widetilde{J}'\widetilde{J})\widetilde{R}^{1/2}\|\,\|J\| \\
&\qquad + (1 + \|\widetilde{J}'\|\,\|\widetilde{J}\|)\|JR^{1/2} - \widetilde{R}^{1/2}J\|\big) \\
&\le \delta + 3(2\widetilde{\delta} + 7C_\varphi\delta) = 21(C_\varphi + 1)\delta + 6\widetilde{\delta},
\end{aligned}$$

using Theorem 4.2.9 with $\varphi(\lambda) = (\lambda + 1)^{-1/2}$. Similarly, we have

$$\|\widehat{\mathrm{id}} - \hat{J}\hat{J}'\|_{1\to 0} \le \widetilde{\delta} + 2(3\delta + 7(2 + C_\varphi)\widetilde{\delta}) = 6\delta + (29 + 14C_\varphi)\widetilde{\delta}$$

using additionally (4.17a). In particular, we have shown that $\hat{J}$ and $\hat{J}'$ are $\hat{\delta}$-quasi-unitary of order 1. For the compatibility of the zeroth and first order operators, we estimate

$$\begin{aligned}\|\hat{J}^1 - \hat{J}\|_{1\to 0} &= \big\|(\widetilde{J}^1 - \widetilde{J})J^1 + \widetilde{J}(J^1 - J)\big\|_{1\to 0}\\ &\le \|\widetilde{J}^1 - \widetilde{J}\|_{1\to 0}\|J^1\|_{1\to 1} + \|\widetilde{J}\|\,\|J^1 - J\|_{1\to 0} \le 2(\widetilde{\delta} + \delta).\end{aligned}$$

Finally, we have

$$\begin{aligned}\|\widehat{\Delta}\hat{J}^1 - \hat{J}^{-1}\Delta\|_{1\to -1} &= \big\|(\widehat{\Delta}\widetilde{J}^1 - \widetilde{J}^{-1}\widetilde{\Delta})J^1 + \widetilde{J}^{-1}(\widetilde{\Delta}J^1 - J^{-1}\Delta)\big\|_{1\to -1}\\ &\le \|\widehat{\Delta}\widetilde{J}^1 - \widetilde{J}^{-1}\widetilde{\Delta}\|_{1\to -1}\|J^1\|_{1\to 1}\\ &\quad + \|\widetilde{J}^{-1}\|_{-1\to -1}\|\widetilde{\Delta}J^1 - J^{-1}\Delta\|_{1\to -1}\\ &\le 2\widetilde{\delta} + 5\delta,\end{aligned}$$

and the $\hat{\delta}$-quasi-unitary equivalence of $\mathfrak{d}$ and $\widehat{\mathfrak{d}}$ follows with $\hat{C} := 21C_\varphi + 29$.

The partial isometric equivalence follows similarly: Obviously, $\hat{J}^*\hat{J} = \mathrm{id}$ and $\hat{J}'^1\hat{J}^1 = \mathrm{id}$. Moreover, we can show as before that

$$\begin{aligned}&\|\widehat{\mathrm{id}} - \hat{J}\hat{J}^*\|_{1\to 0} \le \widetilde{\delta} + (\delta + 2C_\varphi\widetilde{\delta}) = \delta + (1 + 2C_\varphi)\widetilde{\delta} \quad \text{and}\\ &\|\hat{J}^1 - \hat{J}\|_{1\to 0} \le \|\widetilde{J}^1 - \widetilde{J}\|_{1\to 0}\|J^1\|_{1\to 1} + \|\widetilde{J}\|\,\|J^1 - J\|_{1\to 0} \le 4\widetilde{\delta} + \delta.\end{aligned}$$

Finally, $\|\widehat{\Delta}\hat{J}^1 - \hat{J}^{-1}\Delta\|_{1\to -1} \le 4\widetilde{\delta} + 6\delta$, and the assertion follows with $\hat{C}' = 6 + 2C_\varphi$. □

Let us now show the following estimates from the ones already considered:

Proposition 4.4.17. *Suppose that J^1 and J'^1 are first order identification operators and that φ is a measurable bounded function, then*

$$\|\varphi(\Delta)J'^{-1} - J'^1\varphi(\widetilde{\Delta})\|_{-1\to 1} = \|\varphi(\widetilde{\Delta})J^{-1} - J^1\varphi(\Delta)\|_{-1\to 1},$$

where $J'^{-1} = (J^1)^*$. Moreover, if J'^1 is a left-inverse for J^1 and if J'^1 is $(1+\delta)$-bounded (see Definition 4.4.1) then

$$\|\varphi(\Delta) - J'^1\varphi(\widetilde{\Delta})J^{-1}\|_{-1\to 1} \le (1+\delta)\|\varphi(\widetilde{\Delta})J^{-1} - J^1\varphi(\Delta)\|_{-1\to 1}.$$

If in particular, ψ is a measurable function, continuous on $U \subset \mathbb{R}_+$ such that $\sigma(\Delta) \subset U$ and $\psi(\lambda) = \mathrm{O}(\lambda^{-1})$ as $\lambda \to \infty$, then

$$\|\psi(\Delta) - J'^1\psi(\widetilde{\Delta})J^{-1}\|_{-1\to 1} \le C_\psi(1+\delta)\delta\,,$$

provided J^1 and J'^1 are δ-isometric.

Proof. The first estimate is obvious from taking the adjoint. The second follows from $\varphi(\Delta) = J'^1 J^1 \varphi(\Delta)$ and $\|J'^1\|_{1\to 1} \le 1+\delta$. The last statement is an immediate consequence of Theorem 4.4.6. □

We conclude this section with the convergence of eigenvalues. If the operators Δ and $\widetilde{\Delta}$ associated with the quadratic forms $\mathfrak{d}$ and $\widetilde{\mathfrak{d}}$ have *purely discrete* spectrum, or more generally, have discrete spectrum below the essential spectrum, then we can directly show the convergence of the eigenvalues using the variational min-max characterisation of the eigenvalues, denoted by λ_k and $\widetilde{\lambda}_k$, $k \in \mathbb{N}$, written in the ascending order and repeated according to multiplicity. This estimate is more explicit than the one given in Theorem 4.3.5.

We start with a simple application of the min-max principle. Let $\mathfrak{d}$ be a non-negative quadratic form on $\mathscr{H}$ with domain $\mathscr{H}^1$ and similarly $\widetilde{\mathfrak{d}}$ a non-negative quadratic form on $\widetilde{\mathscr{H}}$ with domain $\widetilde{\mathscr{H}}^1$. Denote by λ_k resp. $\widetilde{\lambda}_k$ the corresponding (discrete) spectrum.

Proposition 4.4.18. *Assume that we have a first order identification operator*

$$J^1 \colon \mathscr{H}^1 \longrightarrow \widetilde{\mathscr{H}}^1$$

such that there exists $\delta_0, \delta_1 \ge 0$ and $\delta'_0, \delta'_1 \ge 0$ with

$$\|J^1 f\|^2 \ge \|f\|^2 - \delta_0 \|f\|^2 - \delta_1 \mathfrak{d}(f) \tag{4.30a}$$

$$\widetilde{\mathfrak{d}}(J^1 f) \le \mathfrak{d}(f) + \delta'_0 \|f\|^2 + \delta'_1 \mathfrak{d}(f) \tag{4.30b}$$

for all $f \in \mathscr{H}^1$. If $\delta_0 + \delta_1 \lambda_k < 1$ then

$$\widetilde{\lambda}_k \le \frac{1+\delta'_1}{1-\delta_0-\delta_1\lambda_k} \cdot \lambda_k + \frac{\delta'_0}{1-\delta_0-\delta_1\lambda_k}$$

for all $k \in \mathbb{N}$. In particular, if $\delta'_0 = \delta'_1 = 0$, then

$$\frac{1-\delta_0}{1+\delta_1\widetilde{\lambda}_k} \cdot \widetilde{\lambda}_k \le \lambda_k.$$

Proof. Let $f \in E_k$, where E_k denotes the space generated by the first k eigenfunctions associated with $\mathfrak{d}$. Let $f \in E_k$, then

$$\|J^1 f\|^2 \ge \bigl(1-\delta_0-\delta_1\lambda_k\bigr)\|f\|^2 \qquad \text{and} \qquad \widetilde{\mathfrak{d}}(J^1 f) \le \bigl(\lambda_k + \delta'_0 + \delta'_1\lambda_k\bigr)\|f\|^2$$

provided $f \neq 0$, and therefore

$$\frac{\widetilde{\mathfrak{d}}(J^1 f)}{\|J^1 f\|^2} \le \frac{\lambda_k + \delta'_0 + \delta'_1\lambda_k}{1-\delta_0-\delta_1\lambda_k}. \tag{4.31}$$

If $\delta_0 + \delta_1 \lambda_k < 1$, then $Jf = 0$ implies $f = 0$, and in particular, J^1 restricted to E_k is injective and $J^1(E_k)$ is a k-dimensional subspace of $\widetilde{\mathscr{H}}^1$. From the min-max principle Theorem 3.2.3 we deduce from inequality (4.31) that

$$\widetilde{\lambda}_k = \inf_{\widetilde{D}_k \subset \widetilde{\mathscr{H}}^1} \sup_{u \in \widetilde{D}_k \setminus \{0\}} \frac{\widetilde{\mathfrak{d}}(u)}{\|u\|^2} \leq \sup_{f \in E_k \setminus \{0\}} \frac{\widetilde{\mathfrak{d}}(J^1 f)}{\|J^1 f\|^2} \leq \frac{\lambda_k + \delta_0' + \delta_1' \lambda_k}{1 - \delta_0 - \delta_1 \lambda_k}$$

choosing the k-dimensional space $\widetilde{D}_k = J^1(E_k)$. □

A straightforward consequence is the following:

Corollary 4.4.19. *Assume that we have first order identification operators*

$$J^1 : \mathscr{H}^1 \longrightarrow \widetilde{\mathscr{H}}^1 \quad \text{and} \quad J'^1 : \widetilde{\mathscr{H}}^1 \longrightarrow \mathscr{H}^1$$

such that there exists $\delta, \delta' \geq 0$ *with*

$$\|J^1 f\|^2 \geq \|f\|^2 - \delta\|f\|^2 - \delta \mathfrak{d}(f) \tag{4.32a}$$

$$\widetilde{\mathfrak{d}}(J^1 f) \leq \mathfrak{d}(f) + \delta\|f\|^2 + \delta \mathfrak{d}(f) \tag{4.32b}$$

$$\|J'^1 u\|^2 \geq \|u\|^2 - \delta'\|u\|^2 - \delta' \widetilde{\mathfrak{d}}(u) \tag{4.32c}$$

$$\mathfrak{d}(J'^1 u) \leq \widetilde{\mathfrak{d}}(u) + \delta'\|u\|^2 + \delta' \widetilde{\mathfrak{d}}(u) \tag{4.32d}$$

for all $f \in \mathscr{H}^1$ *and* $u \in \widetilde{\mathscr{H}}^1$. *If* $\delta + \delta' < (1 + \lambda_k)^{-1}$ *then*

$$\frac{1 - \delta'}{1 + (1 + \lambda_k)\delta'} \cdot \lambda_k - \frac{\delta'}{1 + (1 + \lambda_k)\delta'} \leq \widetilde{\lambda}_k \leq \frac{1 + \delta}{1 - (1 + \lambda_k)\delta} \cdot \lambda_k + \frac{\delta}{1 - (1 + \lambda_k)\delta}$$

for all $k \in \mathbb{N}$. *In particular,* $|\lambda_k - \widetilde{\lambda}_k| = \mathrm{O}(\delta) + \mathrm{O}(\delta')$, *where the error only depends on* λ_k, δ *and* δ'.

4.5 Convergence of Non-self-adjoint Operators

In this section we are going to prove closeness results (in terms of norm estimates) for (in general) non-self-adjoint, closed operators H and $\widetilde{H}$ in the Hilbert spaces $\mathscr{H}$ and $\widetilde{\mathscr{H}}$, respectively. The main difference with the self-adjoint case is that we do not have the functional calculus for *all* continuous functions, but only for *holomorphic* functions. We therefore have to modify some arguments. In particular, we have to provide upper estimates on the norm of the resolvents as quantitative input. Note that for a self-adjoint operator $A = A^*$, we have

$$\|(A - z)^{-1}\| = \frac{1}{\operatorname{dist}(\sigma(A), z)}$$

for $z \in \rho(A)$ by the spectral calculus Theorem 3.2.2. Here, the upper estimate "$\le$" is only fulfilled for *self-adjoint* or, more generally, for *normal* operators ($A^* A = AA^*$).

Denote by $\mathscr{H}^k(H)$ and $\mathscr{H}^k(\widetilde{H})$ the corresponding scales of Hilbert spaces with norms $\|\cdot\|_{k,H}$, and $\|\cdot\|_{k,\widetilde{H}}$ for $k = -2, 0, 2$, respectively, defined in Sect. 3.3.1. Moreover, denote by $\rho(H)$ the *resolvent set* of H. Let

$$z_0 \in \rho(H) \cap \rho(\widetilde{H})$$

be a fixed reference point in the resolvent set and let

$$R := (H - z_0)^{-1} \qquad \text{and} \qquad \widetilde{R} := (\widetilde{H} - z_0)^{-1}. \tag{4.33}$$

In most cases, we can set $z_0 = -1$, and we often omit the dependence on z_0 in the notation.

Recall the definition of the resolvent norm profile in Definition 3.3.3, i.e. $H \in \mathscr{R}(\gamma, U)$, if U is part of the resolvent set $\rho(H)$ and if $\|R(z)\| \le \gamma(z)$ for all $z \in U$. Typically, the resolvent estimate will be fulfilled for classes like ϑ-sectorial operators. In this case, U can be chosen as the complement of a sector, i.e, $U = \{ z \in \mathbb{C} \mid |\arg z| > \vartheta \}$, and $\gamma(z) = C_{\le\vartheta}/|z|$.

Definition 4.5.1. Let $\delta \ge 0$ and let $\gamma: U \longrightarrow (0, \infty)$ be a continuous function on $U \subset \mathbb{C}$.

1. We say that H and $\widetilde{H}$ are *δ-close* (with zeroth order identification operator J) if

$$\|\widetilde{H} J - JH\|_{2,H\to -2,\widetilde{H}} = \|\widetilde{R} J - JR\| \le \delta. \tag{4.34}$$

   is fulfilled.
2. We say that H and $\widetilde{H}$ are *(δ, γ, U)-close* with the zeroth order identification operator J if $H, \widetilde{H} \in \mathscr{R}(\gamma, U)$ (i.e. H, $\widetilde{H}$ have resolvent norm profile γ) and if (4.34) if fulfilled.

Remark 4.5.2. Note that (δ, γ, U)-closeness in particular implies that U belongs to the resolvent set of *both* operators H *and* $\widetilde{H}$. In particular, we need some a priori information on *both* spectra.

Let us now define the quasi-unitary equivalence:

Definition 4.5.3. Let $\delta \ge 0$. We say that H and $\widetilde{H}$ are *δ-quasi-unitarily (-partial isometrically) equivalent* if there are δ-quasi-unitary (-partial isometric) operators J and J' of order 2 for which H and $\widetilde{H}$ are δ-close.

Sometimes it is convenient to include the resolvent norm profile in the definition, and to assume the conditions on J and J' also for the adjoints H^* and $\widetilde{H}^*$:

Definition 4.5.4. Let $\delta \ge 0$ and $\gamma: U \longrightarrow (0, \infty)$ be continuous.

1. We say that H and $\widetilde{H}$ are *(δ, γ, U)-quasi-unitarily (-partial isometrically) equivalent* if there are δ-quasi-unitary (-partial isometric) operators J and J' of order 2 for which H and $\widetilde{H}$ are (δ, γ, U)-close.

2. We say that H and $\widetilde{H}$ are $(\delta, \gamma, U)^*$*-quasi-unitarily (-partial isometrically) equivalent* if there are δ-quasi-unitary (-partial isometric) operators J and J' of order 2 for which H and $\widetilde{H}$ as well as H^* and $\widetilde{H}^*$ are (δ, γ, U)-close.
3. If $U = \{z_0\}$ and $C = \gamma(z_0) > 0$ has only a single value, then we also say that $H, \widetilde{H}$ are (δ, C, z_0)*-quasi-unitarily (resp. -partial isometrically) equivalent*, and similarly for $(\delta, C, z_0)^*$-quasi-unitarily (resp. -partial isometric) equivalence.

Remark 4.5.5. Let us summarise the necessary conditions for convenience here:

1. The operators H and $\widetilde{H}$ are δ*-quasi-unitarily equivalent* iff there are identification operators J and J' such that the following conditions are fulfilled:

$$\|J\| \le 2, \qquad \|J' - J^*\| \le \delta, \tag{4.35a}$$

$$\|(\mathrm{id} - J'J)R\| \le \delta, \qquad \|(\widetilde{\mathrm{id}} - JJ')\widetilde{R}\| \le \delta, \tag{4.35b}$$

$$\|\widetilde{R}J - JR\| \le \delta, \tag{4.35c}$$

where $R = (H - z_0)^{-1}$.

If, in addition, U belongs to the resolvent set of H and $\widetilde{H}$, and if

$$\|R(z)\| \le \gamma(z), \qquad \|\widetilde{R}(z)\| \le \gamma(z), \tag{4.35d}$$

where $R(z) = (H - z)^{-1}$ and $\widetilde{R}(z) = (\widetilde{H} - z)^{-1}$, then $H, \widetilde{H}$ are (δ, γ, U)*-quasi-unitarily equivalent*. The stared version of $(\delta, \gamma, U)^*$-quasi-unitary equivalence means that estimates (4.35b)–(4.35d) are also true for R^* and $\widetilde{R}^*$.
2. The operators H and $\widetilde{H}$ are δ*-partial isometrically equivalent* iff there is an identification operator J such that the following conditions are fulfilled:

$$\mathrm{id} = J^*J, \qquad \|(\widetilde{\mathrm{id}} - JJ^*)\widetilde{R}\| \le \delta, \tag{4.35e}$$

$$\|\widetilde{R}J - JR\| \le \delta. \tag{4.35f}$$

If, in addition, (4.35d) is fulfilled, then the operators H and $\widetilde{H}$ are (δ, γ, U)*-partial isometrically equivalent*. The stared version of $(\delta, \gamma, U)^*$-partial isometric equivalence means that estimates (4.35e)–(4.35f) are also true for R^* and $\widetilde{R}^*$.
3. If $U = \{z_0\}$ and $C = \gamma(z_0)$, then the resolvent estimate reduces to $\|R\| \le C$ and $\|\widetilde{R}\| \le C$.

Let us now define the notion of convergence of closed operators acting in different Hilbert spaces:

Definition 4.5.6. Let H_n and H be closed operators in the Hilbert spaces $\mathscr{H}_n$ and $\mathscr{H}$, respectively. We say that H_n *converges to* H *(in the generalised norm resolvent sense)* ($H_n \to H$) if there is a constant C independent of n, $z_0 \in \rho(H) \cap \rho(H_n)$ for all n and a sequence $\delta_n \to 0$ such that H_n and H are $(\delta_n, C, z_0)^*$-quasi-unitarily equivalent. We call δ_n the *convergence speed* of $H_n \to H$.

As in the self-adjoint case, we have the following simple consequences, showing that we generalised existing concepts of convergence of closed operators:

Lemma 4.5.7. *If* $\mathscr{H} = \widetilde{\mathscr{H}}$ *and* $J = J' = \mathrm{id}$, *then* δ*-closeness of* H *and* $\widetilde{H}$ *is equivalent with the fact that* $\|R - \widetilde{R}\| \leq \delta$. *In particular, the norm resolvent convergence* $\|R_n - R\| \to 0$ *is equivalent with the convergence* $H_n \to H$ *in the generalised norm resolvent sense.*

Let us now derive some consequences of the fact that the operators H and $\widetilde{H}$ are (δ, γ, U)-close.

Lemma 4.5.8. *Let* $C \geq 1$ *and assume that* $H, \widetilde{H}$ *are* (δ, C, z_0)*-close, then*

$$\|\widetilde{R}^k J - JR^k\| \leq kC^{k-1}\delta \tag{4.36}$$

for $k \geq 1$.

Proof. We use the resolvent identity

$$\widetilde{R}^k J - JR^k = \sum_{i=0}^{k-1} \widetilde{R}^{k-1-i}(\widetilde{R}J - JR)R^i,$$

and conclude

$$\|\widetilde{R}^k J - JR^k\| \leq \sum_{i=0}^{k-1} \|\widetilde{R}^{k-1-i}\| \|\widetilde{R}J - JR\| \|R^i\| \leq kC^{k-1}\delta$$

using the estimate for $k = 1$ and the assumption $\|R\|, \|\widetilde{R}\| \leq C$. □

A simple argument allows us to deal with all z in $\rho(H)$ *and* $\rho(\widetilde{H})$, the price is a norm estimate on the resolvents $R(z)$ *and* $\widetilde{R}(z)$:

Lemma 4.5.9. *Given* $z \in \rho(H) \cap \rho(\widetilde{H})$, *then*

$$V(z) = \big[\mathrm{id} + (z - z_0)\widetilde{R}(z)\big] V(z_0) \big[\mathrm{id} + (z - z_0) R(z)\big], \tag{4.37a}$$

where $V(z) := \widetilde{R}(z)J - JR(z)$. *In particular, if* $\gamma : U \longrightarrow (0, \infty)$ *is continuous and* $H, \widetilde{H} \in \mathscr{R}(\gamma, U)$, *then*

$$\|\widetilde{R}(z)J - JR(z)\| \leq \gamma_0(z)^2 \|\widetilde{R}(z_0)J - JR(z_0)\|, \tag{4.37b}$$

is fulfilled for all $z \in U$, *where*

$$\gamma_0(z) = 1 + |z - z_0| \gamma(z) \geq 1$$

is continuous on U.

Proof. We have

$$\begin{aligned} V(z) &= V(z_0) + (z - z_0)\big(\widetilde{R}(z)\widetilde{R}(z_0)J - JR(z)R(z_0)\big) \\ &= V(z_0) + (z - z_0)\big(\widetilde{R}(z)V(z_0) + V(z)R(z_0)\big) \end{aligned}$$

where we have used the second resolvent identity. Reordering the terms yields

$$V(z)\big[\mathrm{id} - (z - z_0)R(z_0)\big] = \big[\mathrm{id} + (z - z_0)\widetilde{R}(z)\big]V(z_0).$$

Since $\mathrm{id} + (z - z_0)R(z)$ is the inverse of $\mathrm{id} - (z - z_0)R(z_0)$, we obtain (4.37a). The norm estimate is then obvious. □

We can now easily extend the convergence results to a suitable class of holomorphic functions of the operators: Remember that H and $\widetilde{H}$ being (δ, γ, U)-close implies that $H, \widetilde{H} \in \mathscr{R}(\gamma, U)$, and in particular that

$$\sigma(H) \cap U = \emptyset \qquad \text{and} \qquad \sigma(\widetilde{H}) \cap U = \emptyset. \tag{4.38}$$

If $D \subset \mathbb{C}$ is a domain with $\partial D \subset U$, then

$$\mathsf{L}_1(\partial D, \gamma_0(z)^2 \mathrm{d}|z|) \subset \mathsf{L}_1(\partial D, \mathrm{d}|z|)$$

since $\gamma_0(z) \geq 1$ (see Lemma 4.5.9). The next lemma allows us to extend the convergence result to holomorphic functions having fast enough decay along ∂D if D is unbounded:

Theorem 4.5.10. *Let $\delta \geq 0$, $\gamma: U \longrightarrow (0, \infty)$ be a continuous function. Assume that D is an open set in $\mathbb{C}$ with piecewise smooth boundary $\partial D \subset U$, and that φ is holomorphic in a neighbourhood of $\overline{D}$. Assume in addition that $\varphi \in \mathsf{L}_1(\partial D, \gamma_0(z)^2 \mathrm{d}|z|)$, then*

$$\|\varphi_D(\widetilde{H})J - J\varphi_D(H)\| \leq C_D(\varphi, \gamma_0^2)\delta \tag{4.39}$$

for all H, $\widetilde{H}$ being (δ, γ, U)-close, where $C_D(\varphi, \gamma_0^2)$ depends only on D, φ and γ. Moreover, the integrability condition on φ is satisfied if the curve ∂D is compact.

Proof. Since ∂D is contained in the resolvent set of both operators and due to our integrability assumption on φ, the holomorphic functional calculus Theorem 3.3.6 applies,

$$\varphi_D(H) = \frac{1}{2\pi \mathrm{i}} \oint_{\partial D} \frac{\varphi(z)}{z - H} \mathrm{d}z,$$

and a similar claim is valid for $\widetilde{H}$. Hence

$$J\varphi_D(H) - \varphi_D(\widetilde{H})J = -\frac{1}{2\pi \mathrm{i}} \oint_{\partial D} \big(JR(z) - \widetilde{R}(z)J\big)\varphi(z)\mathrm{d}z,$$

and therefore,

$$\|J\varphi_D(H) - \varphi_D(\widetilde{H})J\| \le \frac{\delta}{2\pi}\oint_{\partial D} \gamma_0(z)^2|\varphi(z)|\mathrm{d}|z| =: C_D(\varphi, \gamma_0^2)\,\delta.$$

Since $\gamma_0(z)^2$ depends continuously on z, the right-hand side is in particular finite if ∂D is compact. □

Let $z \in \rho(H)$. Once we know a priori that also $z \in \rho(\widetilde{H})$, we obtain a *quantitative* estimate on the norm of $\widetilde{R}(z)$:

Lemma 4.5.11. *Assume that $z, z_0 \in \rho(H) \cap \rho(\widetilde{H})$ and $C > 0$. Then there is a constant*

$$\delta_1(z) := \min\Big\{1, \frac{1}{2}\Big(|z - z_0|\big(1 + 3(2C + 1 + |z - z_0|\,\|R(z)\|)\big)\Big)^{-1}\Big\} \tag{4.40a}$$

such that

$$\|\widetilde{R}(z)\| \le 12\big(1 + C + (|z - z_0| + 1)\|R(z)\|\big) \tag{4.40b}$$

for all $\widetilde{H}$ being $(\delta, C, z_0)^*$-unitarily equivalent with H provided $0 \le \delta \le \delta_1(z)$. Moreover,

$$\|V(z)\| = \|\widetilde{R}(z)J - JR(z)\| \le 12\big(1 + |z - z_0|\,\|R(z)\|\big)^2\delta. \tag{4.40c}$$

Proof. Note first that

$$\begin{aligned}
\|\widetilde{R}(z)(\widetilde{\mathrm{id}} - JJ')\| &= \|(\widetilde{\mathrm{id}} - J'^*J^*)\widetilde{R}(z)^*\| \\
&= \|(\widetilde{\mathrm{id}} - J'^*J^*)\widetilde{R}(z_0)^*(\widetilde{\mathrm{id}} + (\bar{z} - \bar{z}_0)\widetilde{R}(z)^*\| \\
&\le \|(\widetilde{\mathrm{id}} - J'^*J^*)\widetilde{R}(z_0)^*\|(1 + |z - z_0|\,\|\widetilde{R}(z)\|) \\
&\le (1 + 2(2 + \delta)C)(1 + |z - z_0|\,\|\widetilde{R}(z)\|)\delta
\end{aligned}$$

using (4.3d). Now we have

$$\begin{aligned}
\|\widetilde{R}(z)\| &\le \|\widetilde{R}(z)(\widetilde{\mathrm{id}} - JJ')\| + \|(\widetilde{R}(z)J - JR(z))J'\| + \|JR(z)J'\| \\
&\le \delta\Big[\big(1 + |z - z_0|\,\|\widetilde{R}(z)\|\big)\big(1 + 2(2 + \delta)C\big) \\
&\quad + \big(1 + |z - z_0|\,\|R(z)\|\big)\big(1 + |z - z_0|\,\|\widetilde{R}(z))\|(2 + \delta)\Big] \\
&\quad + 2(2 + \delta)\|R(z)\|
\end{aligned}$$

using also (4.37a). Bringing the $\|\widetilde{R}(z)\|$-terms on the LHS and dividing by the resulting coefficient yields

$$\|\widetilde{R}(z)\| \le \frac{2(2 + \delta)\|R(z)\| + \delta\big[1 + 2(2 + \delta)C + (2 + \delta)(1 + |z - z_0|\,\|R(z)\|)\big]}{1 - \delta|z - z_0|\big(1 + (2 + \delta)(2C + 1 + |z - z_0|\,\|R(z)\|)\big)},$$

leading to the desired assertion on $\|\widetilde{R}(z)\|$ using $\delta \le 1$. Note that we estimated the denominator from below by $1/2$ since $0 \le \delta \le \delta_1(z)$. Similarly,

$$1 + |z - z_0| \|\widetilde{R}(z)\| \le \frac{1 + 2(2+\delta)|z - z_0| \|R(z)\|}{1 - \delta|z - z_0|\bigl(1 + (2+\delta)(2C + 1 + |z - z_0| \|R(z)\|)\bigr)}$$
$$\le 12\bigl(1 + |z - z_0| \|R(z)\|\bigr).$$

Finally, we have

$$\|V(z)\| \le \bigl(1 + |z - z_0| \|R(z)\|\bigr)\bigl(1 + |z - z_0| \|\widetilde{R}(z)\|\bigr)\delta \le 12\bigl(1 + |z - z_0| \|R(z)\|\bigr)^2\delta$$

again using (4.37a). □

Remark 4.5.12. It is a priori not clear whether our assumptions of δ-quasi unitary equivalence or even δ-partial isometry are strong enough to assure that $z \in \rho(H)$ implies $z \in \rho(\widetilde{H})$. Note that we used the existence (and boundedness) of the resolvent at z in the previous proof e.g. in the first step (when using the fact $\|A\| = \|A^*\|$ for *bounded* operators).

Due to this lack of information, we always have to include *a priori* information of the spectrum of H *and* $\widetilde{H}$, and we cannot omit the restriction "provided $\partial D \subset \rho(\widetilde{H})$" e.g. in Theorem 4.5.14, Corollary 4.5.15, Corollary 4.5.18 etc.

From the preceding lemma, we can easily deduce a quantitative resolvent estimate for $\widetilde{H}$ from a resolvent estimate for H. We only need an estimate (by C) on the norm of the resolvents in a common reference point z_0. This fact is part of the definition of $(\delta, C, z_0)^*$-quasi-unitary equivalence, see Definition 4.5.4 (3):

Proposition 4.5.13. *Let $z_0 \in \mathbb{C}$ and assume that D is an open set in $\mathbb{C}$ with compact, piecewise smooth boundary ∂D. Suppose that $\gamma \colon \partial D \cup \{z_0\} \longrightarrow (0,\infty)$ is continuous and set $C := \gamma(z_0)$. Then there exist a constant $\delta_1 = \delta_1(\gamma, C, z_0) > 0$ and a continuous function $\widetilde{\gamma} \colon \partial D \longrightarrow (0,\infty)$ such that*

$$\|R(z)\| \le \gamma(z) \qquad \Rightarrow \qquad \|\widetilde{R}(z)\| \le \widetilde{\gamma}(z), \qquad z \in \partial D$$

for all closed operators $H, \widetilde{H}$ being $(\delta, C, z_0)^*$-quasi-unitarily equivalent and all $0 \le \delta \le \delta_1$ provided $\partial D \subset \rho(H)$ and $\partial D \subset \rho(\widetilde{H})$. *The constant δ_1 and the function $\widetilde{\gamma}$ are given by*

$$\delta_1 := \min_{z\in\partial D}\Bigl\{1, \frac{1}{2}\Bigl(|z - z_0|\bigl(1 + 3(2C + 1 + |z - z_0|\gamma(z))\bigr)\Bigr)^{-1}\Bigr\} > 0 \quad \textit{and}$$

$$\widetilde{\gamma}(z) := 12\bigl(1 + C + (|z - z_0| + 1)\gamma(z)\bigr) = 12(C + \gamma_0(z) + \gamma(z)).$$

In particular, we can choose $\gamma(z) := \|R(z)\|$ (note that $\|R(\cdot)\|$ is continuous, see Lemma 3.3.2).

Similarly as in Theorem 4.5.10, we can prove the following result for holomorphic functions of the operators, but now only with a resolvent estimate on H:

Theorem 4.5.14. *Let $z_0 \in \mathbb{C}$. Assume that D is an open set in $\mathbb{C}$ with compact, piecewise smooth boundary ∂D, and that φ is holomorphic in a neighbourhood of $\overline{D}$. Let $H \in \mathscr{R}(\gamma, \partial D \cup \{z_0\})$ for some continuous function $\gamma\colon \partial D \cup \{z_0\} \longrightarrow (0,\infty)$ (e.g., $\partial D \subset \rho(H)$ and $\gamma(z) := \|R(z)\|$) and set $C := \gamma(z_0)$. Then there exist constants $\delta_1 = \delta_1(\gamma, C, z_0) > 0$ and $C' = C'(\varphi, \gamma, z_0) > 0$ such that*

$$\|\varphi_D(\widetilde{H})J - J\varphi_D(H)\| \le C'\delta \tag{4.41a}$$

for all closed operators $\widetilde{H}$, where H, $\widetilde{H}$ are $(\delta, C, z_0)^*$-quasi-unitarily equivalent and $0 \le \delta \le \delta_1$ provided $\partial D \subset \rho(\widetilde{H})$. The constant C' is given by

$$C'(\varphi, \gamma, z_0) := C_D(\varphi, 12\gamma_0^2) = \frac{12}{2\pi}\oint_{\partial D} \big(1 + |z - z_0|\gamma(z)\big)^2 |\varphi(z)|\,\mathrm{d}|z|. \tag{4.41b}$$

Proof. The assertion follows immediately from

$$\begin{aligned}\|J\varphi_D(H) - \varphi_D(\widetilde{H})J\| &\le \frac{\delta}{2\pi}\oint_{\partial D} \|V(z)\|\,|\varphi(z)|\,\mathrm{d}|z| \\ &\le \frac{12\delta}{2\pi}\oint_{\partial D} \big(1 + |z - z_0|\gamma(z)\big)^2 |\varphi(z)|\,\mathrm{d}|z|\end{aligned}$$

using (4.40c). □

Applying the previous theorem with $\varphi = \mathbb{1}$ we obtain the following assertion on spectral projectors:

Corollary 4.5.15. *Let $z_0 \in \mathbb{C}$. Assume that D is an open set in $\mathbb{C}$ with compact, piecewise smooth boundary ∂D. Let $H \in \mathscr{R}(\gamma, \partial D \cup \{z_0\})$ for some continuous function $\gamma\colon \partial D \cup \{z_0\} \longrightarrow (0,\infty)$ and set $C := \gamma(z_0)$. Then there exist constants $\delta_1 = \delta_1(\gamma, C, z_0) > 0$ and $C' = C'(\gamma, z_0) > 0$ such that the spectral projections satisfy*

$$\|\mathbb{1}_D(\widetilde{H})J - J\mathbb{1}_D(H)\| \le C'\delta \tag{4.42}$$

for all closed operators $\widetilde{H}$, where H, $\widetilde{H}$ are $(\delta, C, z_0)^*$-quasi-unitarily equivalent and $0 \le \delta \le \delta_1$ provided $\partial D \subset \rho(\widetilde{H})$.

Remark 4.5.16. We might drop the condition of compactness of ∂D provided $\|R(z)\|$ decays fast enough along ∂D. In particular, we need the integrability condition $C'(\varphi, \gamma) < \infty$ (cf. (4.41b). Moreover, we have to assure that $\inf_{z\in\partial D} \delta_1(z)$ (see the definition in (4.40a)) remains *strictly* positive.

The next proposition shows how to extend the functional calculus to other combinations of identification operators:

Proposition 4.5.17. *Assume that J is δ-quasi-unitary of order 2 w.r.t. H and $\widetilde{H}$ and that*

$$\|\varphi(\widetilde{H})J - J\varphi(H)\| = \eta$$

for some function φ and some constant $\eta > 0$. Then we have

$$\|\varphi(H)J' - J'\varphi(\widetilde{H})\| \le 2C(\varphi)\delta + \eta, \tag{4.43a}$$

$$\|\varphi(H) - J'\varphi(\widetilde{H})J\| \le C(\widehat{\varphi})\delta + (2+\delta)\eta, \tag{4.43b}$$

$$\|\varphi(\widetilde{H}) - J\varphi(H)J'\| \le \big(C(\widehat{\varphi}) + 4C(\varphi)\big)\delta + 2\eta, \tag{4.43c}$$

where $\widehat{\varphi}(z) = (z - z_0)\varphi(z)$ and where $C(\varphi)$ is an upper bound of $\|\varphi(H)\|$ and $\|\varphi(\widetilde{H})\|$.

If J is δ-partial isometric, then

$$\|\varphi(H)J^* - J^*\varphi(\widetilde{H})\| = \|\varphi(H) - J^*\varphi(\widetilde{H})J\| = \eta. \tag{4.43d}$$

Proof. The first estimate follows from

$$\begin{aligned}\|\varphi(H)J' - J'\varphi(\widetilde{H})\| &\le (\|\varphi(H)\| + \|\varphi(\widetilde{H})\|)\|J' - J^*\| \\ &\quad + \|\varphi(H)J^* - J^*\varphi(\widetilde{H})\| \\ &\le 2C(\varphi)\delta + \eta;\end{aligned}$$

the second from

$$\begin{aligned}\|\varphi(H) - J'\varphi(\widetilde{H})J\| &\le \|\mathrm{id} - J'J\|_{2,H\to 0}\|\varphi(H)\|_{0\to 2,H} \\ &\quad + \|J'\|\,\|J\varphi(H) - \varphi(\widetilde{H})J\| \\ &\le \delta C(\widehat{\varphi}) + (2+\delta)\eta\end{aligned}$$

and the third from

$$\begin{aligned}\|\varphi(\widetilde{H}) - J\varphi(H)J'\| &\le \|\widetilde{\mathrm{id}} - JJ'\|_{2,\widetilde{H}\to 0}\|\varphi(\widetilde{H})\|_{0\to 2,\widetilde{H}} \\ &\quad + \|J\|\,\|J'\varphi(\widetilde{H}) - \varphi(H)J'\|\end{aligned}$$

together with (4.43a). □

As a consequence of the preceding proposition and Theorem 4.5.14 we obtain:

Corollary 4.5.18. *Let $z_0 \in \mathbb{C}$. Assume that D is an open set in $\mathbb{C}$ with compact, piecewise smooth boundary ∂D, and that φ is holomorphic in a neighbourhood* *of $\overline{D}$. Let $H \in \mathscr{R}(\gamma, \partial D \cup \{z_0\})$ for some continuous function $\gamma \colon \partial D \cup \{z_0\} \longrightarrow (0,\infty)$ (e.g., $\partial D \subset \rho(H)$ and $\gamma(z) := \|R(z)\|$) and set $C := \gamma(z_0)$. Then there exist constants $\delta_1 = \delta_1(\gamma, C, z_0) > 0$ and $C_i' = C_i'(\varphi,\gamma,z_0) > 0$ such that*

$$\|\varphi_D(H)J' - J'\varphi_D(\widetilde{H})\| \le C_1'\delta, \tag{4.44a}$$

$$\|\varphi_D(H) - J'\varphi_D(\widetilde{H})J\| \le C_2'\delta, \tag{4.44b}$$

$$\|\varphi_D(\widetilde{H}) - J\varphi_D(H)J'\| \le C_3'\delta \tag{4.44c}$$

for all closed operators $\widetilde{H}$, where H, $\widetilde{H}$ are $(\delta, C, z_0)^*$-quasi-unitarily equivalent and $0 \le \delta \le \delta_1$ provided $\partial D \subset \rho(\widetilde{H})$.

4.6 Spectral Convergence for Non-self-adjoint Operators

If H and $\widetilde{H}$ are δ-quasi-unitarily equivalent, and if $z \in \rho(H)$, then we do not know in general that $z \in \rho(\widetilde{H})$, even for small δ (see Remark 4.5.12). Therefore, we assume that we know already the essential spectra of *both* operators, and make a convergence statement on the discrete spectrum only. This restriction does not matter in our applications since the essential spectrum of *both* operators is known.

We start with an assertion on the dimensions of the spectral projections provided the operators are quasi-unitarily equivalent (see Definition 4.5.4):

Lemma 4.6.1. *Assume that $z_0 \in \mathbb{C}$ and that D is an open set in $\mathbb{C}$ with compact, piecewise smooth boundary ∂D. Let $H \in \mathscr{R}(\gamma, \partial D \cup \{z_0\})$ for some continuous function $\gamma \colon \partial D \cup \{z_0\} \longrightarrow (0,\infty)$ (e.g., $\partial D \subset \rho(H)$ and $\gamma(z) := \|R(z)\|$) and set $C := \gamma(z_0)$. Then there exists a constant $\delta_1' = \delta_1'(\gamma, C, z_0) > 0$ such that*

$$\dim \operatorname{ran} \mathbb{1}_D(H) = \dim \operatorname{ran} \mathbb{1}_D(\widetilde{H})$$

for all operators $\widetilde{H}$, where $H, \widetilde{H}$ are $(\delta, C, z_0)^*$-quasi-unitarily equivalent and $0 \le \delta \le \delta_1'$ provided $\partial D \subset \rho(\widetilde{H})$.

If, in addition, $\dim \operatorname{ran} \mathbb{1}_D(H) = 1$, *i.e. if there exists a simple eigenvalue $\lambda \in D$ with normalised eigenvector ψ, then there exists a normalised eigenvector $\widetilde{\psi} \in \widetilde{\mathscr{H}}$ generating* $\operatorname{ran} \mathbb{1}_D(\widetilde{H})$ *such that*

$$\|J\psi - \widetilde{\psi}\| \le C_1\delta \qquad \text{and} \qquad \|J'\widetilde{\psi} - \psi\| \le \widetilde{C}_1\delta,$$

provided $0 \le \delta \le \delta_1'$. Here, C_1 and $\widetilde{C}_1$ depend only on γ, C and z_0.

Proof. We apply Theorem 4.1.5 with C as above and $\mathscr{D} = \mathscr{H}^2(H)$, then $\mathscr{H}^2(H)$ is C-densely embedded in $\mathscr{H}$, and similarly for $\widetilde{\mathscr{D}} = \mathscr{H}^2(\widetilde{H})$. Moreover, we set $P = \mathbb{1}_D(H)$ and $\widetilde{P} = \mathbb{1}_D(\widetilde{H})$, hence

$$\|P\|_{\mathscr{H}\to\mathscr{D}} \le C_D(\cdot - z_0, \gamma) \qquad \text{and} \qquad \|\widetilde{P}\|_{\widetilde{\mathscr{H}}\to\widetilde{\mathscr{D}}} \le C_D(\cdot - z_0, \widetilde{\gamma})$$

by the holomorphic functional calculus Theorem 3.3.6 and Proposition 4.5.13. The norm estimate on the projection difference is ensured by Corollary 4.5.15, so that $\alpha := C_D(\mathbb{1}, 12\gamma_0^2)$ in the notation of Theorem 4.1.5 and (4.41b) and the assertions follow with $\delta_1' := \min\{\delta_0, \delta_1\}$ where δ_0 is defined in Theorem 4.1.5 (2) and δ_1 is defined in Proposition 4.5.13. □

Let us give a quantitative version in the case that $D = B_\eta(\lambda)$ is a ball of radius η around the discrete eigenvalue λ. This quantitative version is useful when determine the convergence speed of discrete eigenvalues (see Theorem 4.6.4):

Corollary 4.6.2. *Assume that H is a closed operator with $\lambda \in \sigma_{\text{disc}}(H)$ and $z_0 \in \rho(H)$. Choose $\eta_0 > 0$ such that $B_{\eta_0}(\lambda) \cap \sigma(H) = \{\lambda\}$. Moreover, assume that there exists $C, C_2 > 0$ and $\beta \geq 1$ such that $\|R(z_0)\| \leq C$ and*

$$\gamma_\lambda(\eta) := \max_{z \in \partial B_\eta(\lambda)} \|R(z)\| \leq C_2 \eta^{-\beta} \tag{4.45}$$

for all $\eta \in (0, \eta_0]$. Then for all $\eta \in (0, \eta_0]$, there exists a constant $\delta_1(\eta) = \delta_1(\eta, C, z_0, C_2) > 0$ such that

$$\dim \operatorname{ran} \mathbb{1}_{B_\eta(\lambda)}(H) = \dim \operatorname{ran} \mathbb{1}_{B_\eta(\lambda)}(\widetilde{H})$$

for all operators $\widetilde{H}$, where $H, \widetilde{H}$ are $(\delta, C, z_0)^*$-quasi-unitarily equivalent and $0 \leq \delta \leq \delta_1(\eta)$ provided $\partial B_\eta(\lambda) \subset \rho(\widetilde{H})$. Moreover, $\delta_1(\eta) = \mathrm{O}(\eta^{2\beta-1})$.

In particular, if H is self-adjoint (or normal, i.e. H and H^* commute), then we can choose $C_2 = 1$ and $\beta = 1$, i.e. $\delta_1(\eta) = \mathrm{O}(\eta)$.

Proof. Let us give a careful analysis of the constants involved in the previous proof. Set $d := d(\lambda, z_0) := \max_{z \in \partial B_\eta(\lambda)} |z - z_0|$. Then we have the estimate

$$\begin{aligned}
(\delta_0)^{-1} &:= (2C+1)\max\{\|P\|_{\mathscr{H}\to\mathscr{D}}, \|\widetilde{P}\|_{\widetilde{\mathscr{H}}\to\widetilde{\mathscr{D}}}\} + \alpha \\
&\leq (2C+1)C_D(\cdot - z_0, \widetilde{\gamma}) + C_D(\mathbb{1}, 12\gamma_0) \\
&\leq 12(2C+1)\eta\big[C + \big(1 + (d+1)\gamma_\lambda(\eta)\big)^2\big] = \mathrm{O}(\eta^{1-2\beta})
\end{aligned}$$

on the constant δ_0 using the definition of $C_D(\varphi, \gamma)$ in Theorem 3.3.6. Moreover, for δ_1 we have

$$(\delta_1)^{-1} \leq \max\{1, 2d[1 + 3(2C + 1 + d\gamma_\lambda(\eta))]\} = \mathrm{O}(\eta^{-\beta})$$

where the errors depend only on C, z_0, d and C_2. Since $\beta \geq 1$, the dominant error is given by $\mathrm{O}(\eta^{1-2\beta})$.

Note that for self-adjoint or normal operators H, we have the equality $\|R(z)\| = d(z, \sigma(H))^{-1} = 1/\eta$ if $z \in \partial_\lambda B(\eta)$ for η small enough. □

For the notion of convergence of closed operators, we refer to Definition 4.5.6. Recall that $H_n \to H$ already implies (by definition) that $z_0 \in \rho(H_n) \cap \rho(H)$

for all n with norm of the resolvents bounded by some constant C, independent of n.

Theorem 4.6.3. *Assume that H is a closed operator in $\mathscr{H}$ and that $\lambda \in \sigma_{\mathrm{disc}}(H)$ with multiplicity μ. Let H_n be a sequence of closed operators in $\mathscr{H}_n$ such that $H_n \to H$ and $B_{\eta_0}(\lambda) \cap \sigma_{\mathrm{ess}}(H_n) = \emptyset$ eventually for some $\eta_0 > 0$, then for each n there exist μ eigenvalues $\lambda_{j,n}$, $j = 1, \ldots, \mu$, (not necessarily mutually distinct) such that $\lambda_{j,n} \to \lambda$ as $n \to \infty$.*

If, in addition, $\mu = 1$, i.e. if the eigenvalue $\lambda \in \sigma_{\mathrm{disc}}(H)$ is simple with normalised eigenvector ψ, then there exists a sequence of normalised eigenvectors $\psi_n \in \mathscr{H}_n$ of H_n such that

$$\|J_n\psi - \psi_n\| \to 0 \qquad \text{and} \qquad \|J_n'\psi_n - \psi\| \to 0,$$

where J_n and J_n' are the associated identification operators.

Proof. Let $\lambda \in \sigma_{\mathrm{disc}}(H) \cap D$. Then by definition of the discrete spectrum, there exists $0 < \eta_1 \le \eta_0$ such that $B_{\eta_1}(\lambda) \cap \sigma_{\mathrm{disc}}(H) = \{\lambda\}$. Let $0 < \eta \le \eta_1$. Since

$$\bigcup_n \sigma(H_n) \cap B_\eta(\lambda) = \bigcup_n \sigma_{\mathrm{disc}}(H_n) \cap B_\eta(\lambda)$$

by assumption, and since the latter set is at most countable, we can always assume (by slightly modifying η) that

$$\partial B_\eta(\lambda) \subset \rho(H_n)$$

for all n. In particular, by the first part of Lemma 4.6.1,

$$0 < \mu := \dim \operatorname{ran} \mathbb{1}_{B_\eta(\lambda)}(H) = \dim \operatorname{ran} \mathbb{1}_{B_\eta(\lambda)}(H_n) < \infty$$

eventually, showing the existence of μ sequences $\{\lambda_{j,n}\}_n$ converging to λ. The eigenvector convergence follows immediately from the second part of Lemma 4.6.1. □

Note that we have shown the convergence of the discrete spectrum in the maximal outer distance on bounded subsets (see Definition A.1.1 and Proposition A.1.6), i.e.

$$\sigma_{\mathrm{disc}}(H_n) \cap D \nearrow \sigma_{\mathrm{disc}}(H) \cap D$$

for all balls $D \subset \mathbb{C}$.

If we have a resolvent estimate of the operator H near an eigenvalue $\lambda \in \sigma_{\mathrm{disc}}(H)$, we get a quantitative estimate on the eigenvalues of $\widetilde{H}$ close to λ:

Theorem 4.6.4. *Let $C > 0$. Assume that H is a closed operator such that $\lambda \in \sigma_{\mathrm{disc}}(H)$ is a discrete eigenvalue of multiplicity μ and that $z_0 \in \rho(H)$. Moreover, assume that there exists $C_2 > 0$ and $\beta \ge 1$ such that $\|R(z_0)\| \le C$ and*

$$\gamma_\lambda(\eta) := \max_{z \in \partial D_\eta(\lambda)} \|R(z)\| \le C_2 \eta^{-\beta}$$

for all $0 < \eta \le \eta_0$. *Let* $\widetilde{H}$ *be a closed operator, such that* $H, \widetilde{H}$ *are* $(\delta, C, z_0)^*$-*quasi-unitarily equivalent for* $\delta \ge 0$ *small enough and that* $\lambda \notin \sigma_{\mathrm{ess}}(\widetilde{H})$. *Then there exist* μ *eigenvalues* $\widetilde{\lambda}_j$, $j = 1, \dots, \mu$, *(not necessarily mutually distinct) such that*

$$|\widetilde{\lambda}_j - \lambda| = \mathrm{O}(\delta^{\frac{1}{2\beta - 1}}).$$

In particular, if H *is self-adjoint, then the error is* $\mathrm{O}(\delta)$, *provided* H *and* $\widetilde{H}$ *are* δ-*quasi-unitarily equivalent (see Definition 4.2.3).*

If, in addition, $\mu = 1$, *i.e. if there exists a simple eigenvalue* $\lambda \in \sigma_{\mathrm{disc}}(H)$ *with normalised eigenvector* ψ, *then there exists a normalised eigenvector* $\widetilde{\psi} \in \widetilde{\mathscr{H}}$ *of* $\widetilde{H}$ *such that*

$$\|J\psi - \widetilde{\psi}\| = \mathrm{O}(\delta) \qquad \text{and} \qquad \|J'\widetilde{\psi} - \psi\| = \mathrm{O}(\delta),$$

where J *and* J' *are the associated identification operators. All errors depend only on* C, z_0, C_2 *and* β.

Proof. The proof is similar to the previous one, using now the quantitative information of Corollary 4.6.2. Note that since $\sigma_{\mathrm{ess}}(\widetilde{H})$ is closed, there exists $\eta_1 > 0$ such that $B_{\eta_1}(\lambda) \cap \sigma_{\mathrm{ess}}(\widetilde{H}) = \emptyset$, and the existence of a ball of radius $0 < \eta \le \eta_1$ such that $\partial B_\eta \subset \rho(\widetilde{H})$ is ensured. □

Remark 4.6.5. Assume that $H_n \to H$ with convergence speed $\delta_n \to 0$. Once we have a resolvent estimate for the operator H of the type (4.45) (or a similar estimate), we can conclude the convergence speed of the discrete eigenvalues.

4.7 Convergence of Non-symmetric Forms

As in the previous section we are going to prove convergence results for (in general) non-self-adjoint, closed operators H and $\widetilde{H}$ in the Hilbert spaces $\mathscr{H}$ and $\widetilde{\mathscr{H}}$, respectively. Here in contrast, we use sesquilinear forms $\mathfrak{h}$ and $\widetilde{\mathfrak{h}}$ associated with H and $\widetilde{H}$ (see Definition 3.3.16), since in many applications, it is easier to deal with the first-order domains.

Denote by $\mathscr{H}^k(H)$ and $\mathscr{H}^k(\widetilde{H})$ the corresponding scales of Hilbert spaces with norms denoted by $\|\cdot\|_{k,H}$ and $\|\cdot\|_{k,\widetilde{H}}$. For *even* k ($k = 0, \pm 2$), they are defined naturally via the operators H, $\widetilde{H}$ and the resolvents $R = (H - z_0)^{-1}$ and $\widetilde{R} := (\widetilde{H} - z_0)^{-1}$ for $z_0 \in \rho(H) \cap \rho(\widetilde{H})$ in both resolvents sets, see Sect. 3.3.1. For *odd* k ($k = \pm 1$), we defined a natural scale together with norms $\|\cdot\|_{1,H}$, $\|\cdot\|_{1,\widetilde{H}}$ in Definition 3.3.8 associated with the non-negative operators Δ, $\widetilde{\Delta}$, respectively. Recall that

$$\|f\|_{1,H}^2 = \|(\Delta + 1)^{1/2} T f\|^2 \qquad \text{and} \qquad \|f\|_{1,H^*}^2 = \|(\Delta + 1)^{1/2} T^* f\|^2$$

and that $\|(\Delta + 1)^{1/2} T f\|^2 = \|Tf\|^2 + \mathfrak{d}(Tf)$. Moreover, $\mathscr{H}^{-1}(H) = (\mathscr{H}^1(H^*))^*$ is defined with the adjoint operator H^*. Similar remarks hold for the scale on $\widetilde{\mathscr{H}}$.

We introduce the following terminology for identification operators of order 1, similarly as in Definition 4.4.1.

Definition 4.7.1. Let $\delta \geq 0$ and

$$J^1 \colon \mathscr{H}^1(H) \longrightarrow \mathscr{H}^1(\widetilde{H}) \qquad \text{and} \qquad \bar{J}'^1 \colon \mathscr{H}^1(\widetilde{H}^*) \longrightarrow \mathscr{H}^1(H^*)$$

be bounded linear operators, called *identification operators of order* 1 associated with the sesquilinear forms $\mathfrak{h}$ and $\widetilde{\mathfrak{h}}$.

1. We say that J^1 and $\bar{J}'^1$ are δ*-adjoint* with respect to $\mathfrak{h}$ and $\widetilde{\mathfrak{h}}$, if

$$\|J^1 - J^{-1}\|_{1,H\to -1,\widetilde{H}} \leq \delta \tag{4.46a}$$

is fulfilled, where

$$J^{-1} := (\bar{J}'^1)^* \colon \mathscr{H}^{-1}(H) \longrightarrow \mathscr{H}^{-1}(\widetilde{H}).$$

2. Let $C_1 \geq 1$. We say that J^1 is C_1*-bounded* if

$$\|J^1\|_{1,H\to 1,\widetilde{H}} \leq C_1. \tag{4.46b}$$

3. We say that the sesquilinear forms $\mathfrak{h}$ and $\widetilde{\mathfrak{h}}$ are δ*-close* with first order with first order identification operators J^1 and $\bar{J}'^1$, if J^1 and $\bar{J}'^1$ are δ-adjoint and if

$$\|\widetilde{H} J^1 - J^{-1} H\|_{1,H\to -1,\widetilde{H}} \leq \delta \tag{4.46c}$$

Remark 4.7.2.

1. The boundedness of $\bar{J}'^1$ is needed a priori only to define the adjoint J^{-1}, and in most cases we do not need a quantitative bound.
2. If we consider the adjoints H^* and $\widetilde{H}^*$, we also have to introduce the identification operators

$$\bar{J}^1 \colon \mathscr{H}^1(H^*) \longrightarrow \mathscr{H}^1(\widetilde{H}^*) \qquad \text{and} \qquad J'^1 \colon \mathscr{H}^1(\widetilde{H}) \longrightarrow \mathscr{H}^1(H)$$

associated with the *adjoint* sesquilinear forms $\mathfrak{h}^*$ and $\widetilde{\mathfrak{h}}^*$, where $\mathfrak{h}^*(f,g) := \overline{\mathfrak{h}(g,f)}$. Moreover, we need J'^1 in order to define partial isometry in Definition 4.7.12.

In our applications, we will have $H = H^\theta$ for some complex parameter, where $\{H^\theta\}_\theta$ is a self-adjoint family of operators, i.e. $(H^\theta)^* = H^{\overline{\theta}}$. In this situation, $J^1 = J^{1,\theta}$ and $\bar{J}^1 = J^{1,\overline{\theta}}$, and assumptions on J^1 are automatically fulfilled for $\bar{J}^1$ etc.

3. The norm $\|\cdot\|_{1,H\to -1,\widetilde{H}}$ can be expressed as

$$\|V\|_{1,H\to -1,\widetilde{H}} = \|(\widetilde{\Delta}+1)^{-1/2}\widetilde{T}^{-1}VT^{-1}(\Delta+1)^{-1/2}\|. \tag{4.47}$$

4. Estimate (4.46a) is equivalent wit

$$\left|\langle u, J^1 f\rangle - \langle \bar{J}'^1 u, f\rangle\right| \le \delta \|u\|_{1,\widetilde{H}^*}\|f\|_{1,H} \tag{4.46a'}$$

for all $u \in \mathscr{H}^1(\widetilde{H}^*)$ and $f \in \mathscr{H}^1(H)$. Similarly, (4.46c) is equivalent with

$$\left|\widetilde{\mathfrak{h}}(u, J^1 f) - \mathfrak{h}(\bar{J}'^1 u, f)\right| \le \delta \|u\|_{1,\widetilde{H}^*}\|f\|_{1,H} \tag{4.46c'}$$

for $u \in \mathscr{H}^1(\widetilde{H}^*)$ and $f \in \mathscr{H}^1(H)$.

It is sometimes also useful to include the resolvent norm profile (of order 1) in the closeness definition. Recall the definition of $\mathscr{R}_1(\gamma, U)$ in Definition 3.3.10:

Definition 4.7.3. Let $\delta \ge 0$ and let $\gamma \colon U \longrightarrow (0,\infty)$ be a continuous function. We say that the sesquilinear forms $\mathfrak{h}$ and $\widetilde{\mathfrak{h}}$ are *(δ, γ, U)-close* with first order identification operators J^1 and $\bar{J}'^1$, if $\mathfrak{h}$ and $\widetilde{\mathfrak{h}}$ are δ-close with J^1 and $\bar{J}'^1$ and H, $\widetilde{H}^*$ have resolvent norm profile γ of order 1 on $U \subset \rho(H) \cap \rho(\widetilde{H}^*)$, i.e. (4.46a), (4.46c) and $H, \widetilde{H}^* \in \mathscr{R}_1(\gamma, U)$ are fulfilled. If $U = \{z_0\}$ and $\gamma(z_0) = C$, we also say that $\mathfrak{h}$ and $\widetilde{\mathfrak{h}}$ are *(δ, C, z_0)-close.*

We can now develop a functional calculus similarly as in Sect. 4.5:

Lemma 4.7.4. *Let $\delta \ge 0$ and $C \ge 1$. Assume that $\mathfrak{h}$ and $\widetilde{\mathfrak{h}}$ are (δ, C, z_0)-close, then*

$$\|\widetilde{R}^k J^{-1} - J^1 R^k\|_{-1,H\to 1,\widetilde{H}} \le \bigl(k\bigl(1 + C^3(1+|z_0|)\bigr) - 1\bigr)C^{3k-4}\delta \tag{4.48}$$

for $k \ge 1$.

Proof. We start with $k = 1$ and note that

$$\widetilde{R}J^{-1} - J^1 R = \widetilde{R}\bigl((J^{-1}H - \widetilde{H}J^1) + z_0(J^{-1} - J^1)\bigr)R.$$

Then we can estimate the norm of the last expression by

$$\begin{aligned}\|\widetilde{R}J^{-1} - J^1 R\|_{-1,H\to 1,\widetilde{H}} &\le \|\widetilde{R}\|_{-1,\widetilde{H}\to 1,\widetilde{H}}\bigl(\|J^{-1}H - \widetilde{H}J^1\|_{1,H\to -1,\widetilde{H}} \\ &+ |z_0|\,\|J^{-1} - J^1\|_{1,H\to -1,\widetilde{H}}\bigr)\|R\|_{-1,H\to 1,H} \le (1+|z_0|)C^2\delta\end{aligned}$$

using the norm resolvent estimates and the assumptions. For $k \ge 2$, we have

$$\widetilde{R}^k J^{-1} - J^1 R^k = \sum_{j=0}^{k-1} \widetilde{R}^{k-1-j}(\widetilde{R}J^{-1} - J^1 R)R^j + \sum_{j=1}^{k-1} \widetilde{R}^{k-j}(J^1 - J^{-1})R^j.$$

We estimate the norm of the first sum by $kC^{3(k-1)}(1+|z_0|)C^2\delta$ using Lemma 3.3.15 for the estimate $\|R\|_{-1,H\to-1,H} \le C^3$ and $\|\widetilde{R}\|_{1,H\to 1,H} \le C^3$, and using the estimate for $k = 1$. Moreover, the second sum can be estimated by $(k-1)C^2C^{3(k-2)}\delta$ using the assumption $\|R\|_{-1,H\to 1,H} \le C$ and similarly for $\widetilde{R}$, and using again Lemma 3.3.15 as well as the assumption of δ-adjointness. □

A simple argument allows us to deal with all z in $\rho(H)$ *and* $\rho(\widetilde{H})$, the proof is analogous to the one of Lemma 4.5.9:

Lemma 4.7.5. *Let $\delta \ge 0$. Assume that $z \in \rho(H)\cap\rho(\widetilde{H})$ and that there is a constant $\gamma(w) > 0$ such that $\|R(w)\|_{-1,H\to 1,H}, \|\widetilde{R}(w)\|_{-1,\widetilde{H}\to 1,\widetilde{H}} \le \gamma(w)$ for $w = z$ and $w = z_0$. If in addition, $\mathfrak{h}$ and $\widetilde{\mathfrak{h}}$ are δ-close then*

$$\|\widetilde{R}(z)J^{-1} - J^1R(z)\|_{-1,H\to 1,\widetilde{H}} \le \gamma_1(z)\delta, \tag{4.49}$$

where

$$\gamma_1(z) := (2+|z_0|)\gamma(z_0)^2\widetilde{\gamma}_0(z)^2 \ge 1$$

and $\widetilde{\gamma}_0(z) = 1 + |z-z_0|\gamma(z_0)^2\gamma(z)$.

In particular, if $\gamma\colon U \longrightarrow (0,\infty)$ is continuous, and if $\mathfrak{h}$ and $\widetilde{\mathfrak{h}}$ are (δ,γ,U)-close, then estimate (4.49) *is fulfilled for all $z \in U$ and γ_1 is continuous on U.*

Proof. We set $V^1(z) := \widetilde{R}(z)J^{-1} - J^1R(z)$ and have

$$\begin{aligned} V^1(z) = V^1(z_0) &+ (z-z_0)\bigl(\widetilde{R}(z)V^1(z_0) \\ &+ \bigl[\mathrm{id} + (z-z_0)\widetilde{R}(z)\bigr]\widetilde{R}(z_0)(J^{-1}-J^1)R(z_0) + V^1(z)R(z_0)\bigr) \end{aligned}$$

where we have used the second resolvent identity. Reordering the terms yields

$$V^1(z)\bigl[\mathrm{id} - (z-z_0)R(z_0)\bigr] = \bigl[\mathrm{id} + (z-z_0)\widetilde{R}(z)\bigr]\bigl(V^1(z_0) + \widetilde{R}(z_0)(J^{-1}-J^1)R(z_0)\bigr).$$

Since $\mathrm{id} + (z-z_0)R(z)$ is the inverse of $\mathrm{id} - (z-z_0)R(z_0)$, we obtain

$$\begin{aligned} V^1(z) = \Bigl(&\bigl[\mathrm{id}_{\mathscr{H}^1(\widetilde{H})} + (z-z_0)\widetilde{R}(z)\bigr] \\ &\times \bigl(V(z_0) + \widetilde{R}(z_0)(J^{-1}-J^1)R(z_0)\bigr)\Bigr)\bigl[\mathrm{id}_{\mathscr{H}^{-1}(H)} + (z-z_0)R(z)\bigr]. \end{aligned}$$

Now, we use Lemma 4.7.4 for the norm bound on $V(z_0)$. Moreover, we estimate the expressions $[\ldots]$ by $\widetilde{\gamma}_0(z)$ and the resolvents in z_0 by $\gamma(z_0)$. □

We can now easily extend the convergence results to a suitable class of holomorphic functions of the operators. The proof is the same as for Theorem 4.5.10 using Theorem 3.3.20 instead of Theorem 3.3.6. Recall that $\mathfrak{h}$ and $\widetilde{\mathfrak{h}}$ being (δ,γ,U)-close implies that $H, \widetilde{H} \in \mathscr{R}_1(\gamma, U)$, and in particular, that

$$\sigma(H)\cap U = \emptyset \quad \text{and} \quad \sigma(\widetilde{H})\cap U = \emptyset.$$

Moreover, if $D \subset \mathbb{C}$ is open with $\partial D \subset U$ then

$$\mathsf{L}_1(\partial D, \gamma_1(z)\mathrm{d}|z|) \subset \mathsf{L}_1(\partial D, \mathrm{d}|z|)$$

since $\gamma_1(z) \geq 1$.

Theorem 4.7.6. *Let $\delta \geq 0$, $\gamma: U \longrightarrow (0, \infty)$ be a continuous function. Assume that D is open in $\mathbb{C}$ with piecewise smooth boundary $\partial D \subset U$, and that φ is holomorphic in a neighbourhood of $\overline{D}$. Assume in addition that $\varphi \in \mathsf{L}_1(\partial D, \gamma_1(z)\mathrm{d}|z|)$, then*

$$\|\varphi_D(\widetilde{H})J^{-1} - J^1\varphi_D(H)\|_{-1,H\to 1,\widetilde{H}} \leq C_D(\varphi, \gamma_1)\delta \tag{4.50}$$

for all $\mathfrak{h}$, $\widetilde{\mathfrak{h}}$ being (δ, γ, U)-close, where $C_D(\varphi, \gamma_1)$ depends only on D, φ and γ. Moreover, the integrability condition on φ is satisfied if the curve ∂D is compact.

Let us now relate the identification operators acting on the scale of order 1 with the ones acting on the scale of order 0:

Definition 4.7.7. Let $\delta \geq 0$ and $C \geq 1$.

1. We say that the identification operator

$$J^1: \mathscr{H}^1(H) \longrightarrow \mathscr{H}^1(\widetilde{H})$$

is (δ, C, z_0)-*compatible* with the zeroth order identification operator $J: \mathscr{H} \longrightarrow \widetilde{\mathscr{H}}$, if $H, \widetilde{H} \in \mathscr{R}_1(C, z_0)$ and

$$\|J - J^1\|_{1,H\to 0} \leq \delta, \tag{4.51a}$$

$$\widetilde{T}J = JT, \tag{4.51b}$$

where T and $\widetilde{T}$ are the compatibility operators of the scales $\mathscr{H}^1(H)$ and $\mathscr{H}^1(\widetilde{H})$, respectively (see Definition 3.3.8).

2. Similarly,

$$\bar{J}'^1: \mathscr{H}^1(\widetilde{H}^*) \longrightarrow \mathscr{H}^1(H^*)$$

is called (δ, C, z_0)-*compatible* with the identification operator $J': \widetilde{\mathscr{H}} \longrightarrow \mathscr{H}$ of order 0, if $H, \widetilde{H} \in \mathscr{R}_1(C, z_0)$ and

$$\|J' - \bar{J}'^1\|_{1,\widetilde{H}^*\to 0} \leq \delta, \tag{4.51c}$$

$$T^*J' = J'\widetilde{T}^*. \tag{4.51d}$$

As in Lemma 4.4.10, we have:

Lemma 4.7.8. *Let $\delta \geq 0$ and $C \geq 1$. Assume that J^1 is (δ, C, z_0)-compatible with the zeroth order identification operator J, and that $\bar{J}'^1$ is (δ, C, z_0)-compatible with the zeroth order identification operator J'. If, in addition, $\|J' - J^*\| \leq \delta'$, then J^1 and $\bar{J}'^1$ are δ''-adjoint, where $\delta'' = 2C\delta + C^2\delta'$.*

Proof. We have

$$J^{-1} - J^1 = (\bar{J}'^1 - J')^* + (J'^* - J) + (J - J^1)$$

and the estimate follows from (4.51a), (4.51c) and Lemma 3.3.13. □

Let us now define the following property of sesquilinear forms to be "unitarily equivalent" up to an error:

Definition 4.7.9. Let $\delta \geq 0$ and let $\gamma : U \longrightarrow (0, \infty)$ be a continuous function, and set $C := \gamma(z_0)$. We say that the sesquilinear forms $\mathfrak{h}$ and $\widetilde{\mathfrak{h}}$ are (δ, γ, U)-*quasi-unitarily equivalent* if there exist first and zeroth order identification operators J^1, $\bar{J}'^1$ and J, J', such that

1. $\mathfrak{h}$ and $\widetilde{\mathfrak{h}}$ are (δ, γ, U)-close,
2. J^1 and $\bar{J}'^1$ are bounded; moreover, J^1 is (δ, C, z_0)-compatible with J, and $\bar{J}'^1$ is (δ, C, z_0)-compatible with J',
3. J and J' are δ-quasi-unitary of order 1 w.r.t. H and $\widetilde{H}$.

Remark 4.7.10. It does not really matter what scale of order 1 of Hilbert spaces we use for the quasi-unitary operators J and J': A priori, we defined it with respect to $\mathscr{D} = \mathscr{H}^1(H)$ and $\widetilde{\mathscr{D}} = \mathscr{H}^1(\widetilde{H})$, but we have seen in Lemma 4.1.8, that we could equivalently use the "free" scale $\mathscr{H}^1$ and $\widetilde{\mathscr{H}}^1$ associated with Δ and $\widetilde{\Delta}$, passing to a different δ.

Let us give a self-contained summary of all estimates necessary in order to assure the δ-quasi-unitary equivalence. Recall that $\mathscr{H}^1(H) = T^{-1}(\mathscr{H}^1)$ and $\mathscr{H}^1(\widetilde{H}) = \widetilde{T}^{-1}(\widetilde{\mathscr{H}}^1)$, and $\mathscr{H}^1 = \operatorname{dom} \mathfrak{d}$ and $\widetilde{\mathscr{H}}^1 = \operatorname{dom} \widetilde{\mathfrak{d}}$, where $\mathfrak{d}$ and $\widetilde{\mathfrak{d}}$ are the free forms (see Definition 3.3.8). As norms, we have set

$$\|f\|_{1,H}^2 = \mathfrak{d}(Tf) + \|Tf\|^2 \qquad \text{and} \qquad \|u\|_{1,H^*}^2 = \widetilde{\mathfrak{d}}(T^*u) + \|T^*u\|^2,$$

$$\|f\|_1^2 = \mathfrak{d}(f) + \|f\|^2 \qquad \text{and} \qquad \|u\|_1^2 = \widetilde{\mathfrak{d}}(u) + \|u\|^2$$

for $f \in \mathscr{H}^1(H)$ and $u \in \mathscr{H}^1(\widetilde{H})$ resp. $f \in \mathscr{H}^1$ and $u \in \widetilde{\mathscr{H}}^1$.

Proposition 4.7.11. *Let $\delta \geq 0$, $\gamma : U \longrightarrow (0, \infty)$ be continuous, $U \subset \mathbb{C}$, $z_0 \in U$ and set $C := \gamma(z_0)$. Assume that*

$$J : \mathscr{H} \longrightarrow \widetilde{\mathscr{H}}, \qquad J' : \widetilde{\mathscr{H}} \longrightarrow \mathscr{H},$$

$$J^1 : \mathscr{H}^1(H) \longrightarrow \mathscr{H}^1(\widetilde{H}), \qquad \bar{J}'^1 : \mathscr{H}^1(\widetilde{H}^*) \longrightarrow \mathscr{H}^1(H^*)$$

are linear operators. If J^1 and $\bar{J}'^1$ are bounded, and if the estimates

$$\|Jf\|^2 \leq 4\|f\|^2, \tag{4.52a}$$

$$|\langle J'u, f\rangle - \langle u, Jf\rangle|^2 \leq \delta^2 \|f\|^2 \|u\|^2, \tag{4.52b}$$

$$\|f - J'Jf\|^2 \le \delta^2 \|f\|_1^2, \qquad \|u - JJ'u\|^2 \le \delta^2 \|u\|_1^2 \tag{4.52c}$$

$$\|J^1 f - Jf\|^2 \le \delta^2 \|f\|_{1,H}^2, \qquad \|\bar{J}'^1 u - J'u\|^2 \le \delta^2 \|u\|_{1,\widetilde{H}^*}^2, \tag{4.52d}$$

$$\left|\widetilde{\mathfrak{h}}(u, J^1 f) - \mathfrak{h}(\bar{J}'^1 u, f)\right|^2 \le \delta^2 \|u\|_{1,\widetilde{H}^*}^2 \|f\|_{1,H}^2, \tag{4.52e}$$

$$\widetilde{T} J = JT, \qquad T^* J' = J' \widetilde{T}^*, \tag{4.52f}$$

$$\|T\| \le C, \ \|T^{-1}\| \le C, \ \|\widetilde{T}\| \le C, \ \|\widetilde{T}^{-1}\| \le C, \tag{4.52g}$$

$$U \subset \rho(H) \cap \rho(\widetilde{H}), \tag{4.52h}$$

$$\|(H - z)^{-1}\|_{-1,H \to 1,H} \le \gamma(z), \qquad \|(\widetilde{H} - z)^{-1}\|_{-1,\widetilde{H} \to 1,\widetilde{H}} \le \gamma(z) \tag{4.52i}$$

are fulfilled for all $f \in \mathscr{H}$ *and* $u \in \widetilde{\mathscr{H}}$ *in the appropriate subspaces, then the sesquilinear forms* $\mathfrak{h}$ *and* $\widetilde{\mathfrak{h}}$ *are* $(C(2 + C)\delta, \gamma, U)$*-quasi-unitarily equivalent.*

On the other hand, if $\mathfrak{h}$ *and* $\widetilde{\mathfrak{h}}$ *are* δ*-quasi-unitarily equivalent, then the estimates* (4.52) *are fulfilled with* δ *replaced by* $C\delta$.

Proof. The assertions follow immediately from Definitions 4.1.1, 4.7.1 and 4.7.7. Note that we obtain $C(2 + C)\delta$ due to Lemma 4.7.8, and since δ-adjointness is part of the δ-closeness of $\mathfrak{h}$ and $\widetilde{\mathfrak{h}}$. Moreover, passing from the scale $\|\cdot\|_1$ to $\|\cdot\|_{1,H}$ and vice versa (cf. Lemma 4.1.8) gives an error $C\delta \le C(2 + C)\delta$. □

The notion of "δ-partial isometry" needs some more notation. Basically, we need an identification operator $J'^1 : \mathscr{H}^1(\widetilde{H}) \longrightarrow \mathscr{H}^1(H)$ associated with the operators H and $\widetilde{H}$, not their adjoints as for $\bar{J}'^1$.

Definition 4.7.12. Let $\delta \ge 0$. We say that the identification operators

$$J^1 : \mathscr{H}^1(H) \longrightarrow \mathscr{H}^1(\widetilde{H}) \qquad \text{and} \qquad J'^1 : \mathscr{H}^1(\widetilde{H}) \longrightarrow \mathscr{H}^1(H)$$

are *δ-partial isometries* if the following conditions are fulfilled:

1. J'^1 is a left-inverse for J^1
2. The (non-negative) quadratic forms $\mathfrak{d} \circ T$ and $\widetilde{\mathfrak{d}} \circ \widetilde{T}$ are δ-close in $(\mathscr{H}, \|T\cdot\|)$ and $(\widetilde{\mathscr{H}}, \|\widetilde{T}\cdot\|)$, i.e.

$$\left|\langle \widetilde{T} u, \widetilde{T} J^1 f\rangle - \langle T J'^1 u, Tf\rangle\right| \le \delta \|u\|_{1,\widetilde{H}} \|f\|_{1,H} \tag{4.53a}$$

$$\left|\widetilde{\mathfrak{d}}(\widetilde{T} u, \widetilde{T} J^1 f) - \mathfrak{d}(T J'^1 u, Tf)\right| \le \delta \|u\|_{1,\widetilde{H}} \|f\|_{1,H} \tag{4.53b}$$

for all $u \in \mathscr{H}^1(\widetilde{H})$ and $f \in \mathscr{H}^1(H)$.

Remark 4.7.13.

1. The conditions (4.53a)–(4.53b) are equivalent with

$$\|\widetilde{T}^* \widetilde{T} J^1 - (TJ'^1)^* T\|_{1,H \to -1,\widetilde{H}^*} \le \delta,$$

$$\|\widetilde{T}^* \widetilde{\Delta}(\widetilde{T} J^1) - (TJ'^1)^* \Delta T\|_{1,H \to -1,\widetilde{H}^*} \le \delta,$$

the first one being the δ-adjointness of J^1 and J'^1 in the Hilbert spaces with norms $f \to \|Tf\|$ and $u \mapsto \|\widetilde{T}u\|$, see Remark 3.3.9 (3). These conditions are only needed for Lemma 4.7.14.

2. One has to distinguish between J'^1 and $\bar{J}'^1$, the first one being the identification operator associated with H and $\widetilde{H}$ used e.g. in the definition of δ-partial isometry (see Definition 4.7.12), the second is associated with the *adjoints* H^* and $\widetilde{H}^*$ used e.g. for the scale of order -1 in Definition 4.7.1, see also Remark 4.7.2 (2).

We deduce the following estimates, explaining the name "partial isometry":

Lemma 4.7.14. *Let $\delta \geq 0$ and $C_1 > 0$.*

1. *Assume that J^1 is C_1-bounded, that J^1 and J'^1 fulfil* (4.53a)–(4.53b) *with $\delta \geq 0$, then J'^1 is $(C_1 + 2\delta)$-bounded.*
2. *Assume that J^1 and J'^1 are δ-partial isometric, then*

$$\big| \|J^1 f\|_{1,\widetilde{H}} - \|f\|_{1,H} \big| \leq 2\delta \|f\|_{1,H}. \tag{4.54}$$

In particular, J^1 is $(1 + 2\delta)$-bounded and J'^1 is $(1 + 4\delta)$-bounded.

Proof. The first assertion follows similarly as in the proof of Lemma 4.4.3 from

$$\begin{aligned}
\|J'^1 u\|^2_{1,H} &= \langle u, (J'^1)^* T^* (\Delta + 1) T J'^1 u \rangle_{1,-1} \\
&= \langle u, ((J'^1)^* T^* \Delta T - \widetilde{T}^* \widetilde{\Delta} \widetilde{T} J^1) J'^1 u \rangle_{1,-1} \\
&\quad + \langle u, \widetilde{T}^* (\widetilde{\Delta} + 1) \widetilde{T} J^1 J'^1 u \rangle_{1,-1} \\
&\quad + \langle u, ((J'^1)^* T^* T - \widetilde{T}^* \widetilde{T} J^1) J'^1 u \rangle_{1,-1} \\
&\leq \|u\|_1 \big(2\delta + \|J^1\|_{1,H \to 1,\widetilde{H}} \big) \|J'^1 u\|_{1,H}.
\end{aligned}$$

Here, $\langle \cdot, \cdot \rangle_{1,-1}$ stands for the sesquilinear pairing of $\mathscr{H}^1(H) \times \mathscr{H}^{-1}(H^*)$. For the second assertion, we have

$$\begin{aligned}
\big| \|J^1 f\|^2_{1,\widetilde{H}} - \|f\|^2_{1,H} \big| &= \big| \langle f, [(\widetilde{T} J^1)^* (\widetilde{\Delta} + 1) \widetilde{T} J^1 - T^* (\Delta + 1) T] f \rangle \big| \\
&\leq \big| \widetilde{\mathfrak{d}}(\widetilde{T} u, \widetilde{T} J^1 f) - \mathfrak{d}(T J'^1 u, T f) \big| + \big| \langle \widetilde{T} u, \widetilde{T} J^1 f \rangle \big| \\
&\quad - \langle T J'^1 u, T f \rangle \leq 2\delta \|u\|_{1,\widetilde{H}} \|f\|_{1,H}
\end{aligned}$$

using $J'^1 u = J'^1 J^1 f = f$ and (4.53a)–(4.53b), where $u = J^1 f$. Now, if $f \neq 0$, then $J^1 f \neq 0$, and

$$\begin{aligned}
\big| \|J^1 f\|_{1,\widetilde{H}} - \|f\|_{1,H} \big| &= \frac{\big| \|J^1 f\|^2_{1,H} - \|f\|^2_{1,H} \big|}{\|J^1 f\|_{1,\widetilde{H}} + \|f\|_{1,H}} \leq \frac{2\delta \|f\|_{1,H} \|J^1 f\|_{1,\widetilde{H}}}{\|J^1 f\|_{1,\widetilde{H}}} \\
&= 2\delta \|f\|_{1,H}.
\end{aligned}$$

The assertion on the boundedness of J^1 is obvious, and the boundedness of J'^1 follows by the first part. □

Let us now define the following property of sesquilinear forms to be "partial isometrically equivalent" up to an error:

Definition 4.7.15. Let $\delta \ge 0$ and let $\gamma : U \longrightarrow (0,\infty)$ be a continuous function, and set $C := \gamma(z_0)$. We say that the sesquilinear forms $\mathfrak{h}$ and $\widetilde{\mathfrak{h}}$ are (δ,γ,U)*-partial isometrically equivalent* if there exist first and zeroth order identification operators J^1, J'^1, $\bar{J}'^1$ and J, such that:

1. $\mathfrak{h}$ and $\widetilde{\mathfrak{h}}$ are (δ,γ,U)-close,
2. J^1 is (δ, C, z_0)-compatible with J, and $\bar{J}'^1$ is bounded and (δ, C, z_0)-compatible with J^*,
3. J is δ-partial isometric of of order 1 w.r.t. H and $\widetilde{H}$,
4. J^1 and J'^1 are δ-partial isometries.

Remark 4.7.16. Note that for the partial isometric equivalence, we need both identification operators J'^1 and $\bar{J}'^1$ for the operators *and* their adjoints.

Let us give a self-contained summary of all estimates necessary in order to assure the δ-partial isometric equivalence as in Proposition 4.7.11.

Proposition 4.7.17. *Let $\delta \ge 0$ and assume that*

$$J : \mathscr{H} \longrightarrow \widetilde{\mathscr{H}}, \qquad J^1 : \mathscr{H}^1(H) \longrightarrow \mathscr{H}^1(\widetilde{H}),$$
$$J'^1 : \mathscr{H}^1(\widetilde{H}) \longrightarrow \mathscr{H}^1(H), \qquad \bar{J}'^1 : \mathscr{H}^1(\widetilde{H}^*) \longrightarrow \mathscr{H}^1(H^*)$$

are linear operators, and that $\bar{J}'^1$ is bounded. If the estimates

$$J^*Jf = f, \qquad J'^1J^1 f = f, \tag{4.55a}$$

$$\|u - JJ^*u\|^2 \le \delta^2 \|u\|_1^2, \tag{4.55b}$$

$$\|J^1 f - Jf\|^2 \le \delta^2 \|f\|_{1,H}^2, \qquad \|\bar{J}'^1 \bar{u} - J^*\bar{u}\|^2 \le \delta^2 \|\bar{u}\|^2_{1,\widetilde{H}^*}, \tag{4.55c}$$

$$\left|\widetilde{\mathfrak{h}}(u, J^1 f) - \mathfrak{h}(\bar{J}'^1 u, f)\right|^2 \le \delta^2 \|u\|^2_{1,\widetilde{H}^*} \|f\|^2_{1,H}, \tag{4.55d}$$

$$\widetilde{T}J = JT, \qquad \widetilde{T}^*J = JT^*, \tag{4.55e}$$

$$\|T\| \le C,\ \|T^{-1}\| \le C,\ \|\widetilde{T}\| \le C,\ \|\widetilde{T}^{-1}\| \le C, \tag{4.55f}$$

$$U \subset \rho(H) \cap \rho(\widetilde{H}), \tag{4.55g}$$

$$\|(H-z)^{-1}\|_{-1,H\to 1,H} \le \gamma(z), \qquad \|(\widetilde{H}-z)^{-1}\|_{-1,\widetilde{H}\to 1,\widetilde{H}} \le \gamma(z) \tag{4.55h}$$

$$\left|\langle \widetilde{T}u, \widetilde{T}J^1 f\rangle - \langle TJ'^1 u, Tf\rangle\right|^2 \le \delta^2 \|u\|^2_{1,\widetilde{H}} \|f\|^2_{1,H}, \tag{4.55i}$$

$$\left|\widetilde{\mathfrak{d}}(\widetilde{T}u, \widetilde{T}J^1 f) - \mathfrak{d}(TJ'^1 u, Tf)\right|^2 \le \delta^2 \|u\|^2_{1,\widetilde{H}} \|f\|^2_{1,H}, \tag{4.55j}$$

are fulfilled for all $f \in \mathscr{H}$ *and* $u \in \widetilde{\mathscr{H}}$ *in the appropriate subspaces, then the sesquilinear forms* $\mathfrak{h}$ *and* $\widetilde{\mathfrak{h}}$ *are* $(2C\delta, \gamma, U)$*-partial isometrically equivalent.*

On the other hand, if $\mathfrak{h}$ *and* $\widetilde{\mathfrak{h}}$ *are* (δ, γ, U)*-partial isometrically equivalent, then the estimates* (4.55) *are fulfilled with* δ *replaced by* $C\delta$*.*

Let us finally show that the equivalence of the sesquilinear forms implies the equivalence of the operators (cf. Definition 4.5.4): We use the above list of properties, although the constants may change from the original definition.

Lemma 4.7.18. *Assume that* $\mathfrak{h}$ *and* $\widetilde{\mathfrak{h}}$ *are* (δ, γ, U)*-quasi-unitary (resp. -partial isometrically) equivalent in the sense that* (4.52) *(resp.* (4.55)*) are fulfilled, then the corresponding operators* H *and* $\widetilde{H}$ *are* (δ', γ', U)*-quasi-unitary (resp. -partial isometrically) equivalent, where*

$$\delta' := \bigl(3C^2(1 + |z_0|C^3) + C^4\bigr)\delta, \qquad \gamma'(z) = C^2\gamma(z) \quad \text{and} \quad C = \gamma(z_0).$$

In particular, the results of Sects. 4.5 and 4.6 apply.

Proof. For the closeness of operators, we need to verify (4.34), namely,

$$\begin{aligned}\|\widetilde{H}J - JH\|_{2,H\to -2,\widetilde{H}} \le \|\widetilde{H}(J - J^1)\|_{2,H\to -2,\widetilde{H}} + \|\widetilde{H}J^1 - J^{-1}H\|_{2,H\to -2,\widetilde{H}} \\ + \|(J'^1 - J')^* H\|_{2,H\to -2,\widetilde{H}} + \|(J'^* - J)H\|_{2,H\to -2,\widetilde{H}}.\end{aligned} \tag{4.56}$$

Now,

$$\begin{aligned}\|\widetilde{H}(J - J^1)\|_{2,H\to -2,\widetilde{H}} &\le \|\widetilde{H}\|_{0\to -2,\widetilde{H}} \|J - J^1\|_{2,H\to 0} \\ &\le (1 + |z_0|C^3)C^2 \|J - J^1\|_{1,H\to 0} \le (1 + |z_0|C^3)C^2\delta\end{aligned}$$

using Lemmata 3.3.5 and 3.3.12 for the estimate $\|\widetilde{H}\|_{0\to -2,\widetilde{H}} \le 1 + |z_0|C^3$ and Lemma 3.3.13 for the estimate on the inclusion $\mathscr{H}^2(H) \hookrightarrow \mathscr{H}^1(H)$. The third and forth term of (4.56) can be estimated similarly. The second term can be estimated by $C^4\delta$, using again Lemma 3.3.13 for the estimate of the inclusion of order 2 into 1. Similarly, we have shown in Corollary 4.1.9 that δ-quasi-unitarity (resp. -partial isometrically) of order 1 implies $C^3\delta$-quasi-unitarity (resp. -partial isometrically) of order 2 w.r.t. H and $\widetilde{H}$. The resolvent norm profile estimate of order 0 follows from Lemma 3.3.12. □

4.8 Closeness of Coupled Boundary Maps

In this section we show the closeness of a system coupled via a graph from the closeness of its building blocks associated with each vertex.

Let us first define the closeness of boundary maps: Remember that an operator $\Gamma: \mathscr{H}^1 = \operatorname{dom} \mathfrak{h} \longrightarrow \mathscr{G}$ is a *boundary map* associated with a closed quadratic form

$\mathfrak{h} \geq 0$ in $\mathscr{H}$ if Γ is bounded, $\ker \Gamma$ is dense in $\mathscr{H}$ and $\operatorname{ran} \Gamma$ is dense in $\mathscr{G}$ (see Definition 3.4.1).

Definition 4.8.1. Assume that

$$\Gamma : \mathscr{H}^1 \longrightarrow \mathscr{G} \qquad \text{and} \qquad \widetilde{\Gamma} : \widetilde{\mathscr{H}}^1 \longrightarrow \widetilde{\mathscr{G}}.$$

are boundary maps associated with $\mathfrak{h}$ and $\widetilde{\mathfrak{h}}$ in $\mathscr{H}$ and $\widetilde{\mathscr{H}}$, respectively. We say that Γ and $\widetilde{\Gamma}$ are *δ-close* with the first order identification operators

$$J^1 : \mathscr{H}^1 \longrightarrow \widetilde{\mathscr{H}}^1 \qquad \text{and} \qquad J'^1 : \widetilde{\mathscr{H}}^1 \longrightarrow \mathscr{H}^1$$

and the boundary identification operator $I : \mathscr{G} \longrightarrow \widetilde{\mathscr{G}}$ if $\|I\| \leq 1$,

$$\widetilde{\Gamma} J^1 = I\Gamma \qquad \text{and} \qquad \|\Gamma J'^1 - I^* \widetilde{\Gamma}\|_{1\to 0} \leq \delta.$$

Note that we work on the scale of Hilbert spaces associated with the *Neumann* operator H^{N} and $\widetilde{H}^{\mathrm{N}}$, since these are the operators associated with the forms $\mathfrak{h}$ and $\widetilde{\mathfrak{h}}$.

Remark 4.8.2.

1. Here, $\|A\|_{1\to 0}$ denotes the norm of an operator $A : \widetilde{\mathscr{H}}^1 \longrightarrow \mathscr{G}$. More precisely, we have
$$\Gamma J'^1 : \widetilde{\mathscr{H}}^1 \longrightarrow \mathscr{G}^{1/2} \qquad \text{and} \qquad I^* \widetilde{\Gamma} : \widetilde{\mathscr{H}}^1 \longrightarrow \mathscr{G}.$$
If, for example, I intertwines the Dirichlet-to-Neumann operator, i.e. $I\Lambda = \widetilde{\Lambda} I$, then we also have $I^*(\widetilde{\mathscr{G}}^{1/2}) \subset \mathscr{G}^{1/2}$, and therefore, the difference of the above operators maps into $\mathscr{G}^{1/2}$. In general, we will not expect this to be true, so we loose some regularity here.
2. For simplicity, we do not consider the more general case $\|\widetilde{\Gamma} J^1 - I\Gamma\|_{1\to 0} \leq \delta$ instead of $\widetilde{\Gamma} J^1 = I\Gamma$.
3. If $\delta = 0$, then the δ-closeness becomes
$$\widetilde{\Gamma} J^1 = I\Gamma \qquad \text{and} \qquad \Gamma J'^1 = I^* \widetilde{\Gamma}.$$
4. Note that we only need the *boundary maps* for the coupling in the following, not a corresponding *boundary triple* (see Remark 3.9.5 (2)).

Let us now show the closeness of the boundary maps for the coupled system. We recall the notation of Sect. 3.9. Assume that $G = (V, E, \partial)$ is a (locally finite) graph and that for each vertex $v \in V$, there are two boundary maps

$$\Gamma_v : \mathscr{H}_v^1 \longrightarrow \mathscr{G}_v \qquad \text{and} \qquad \widetilde{\Gamma}_v : \widetilde{\mathscr{H}}_v^1 \longrightarrow \widetilde{\mathscr{G}}_v$$

associated with $\mathfrak{h}_v$ and $\widetilde{\mathfrak{h}}_v$, such that the boundary maps are both compatible with the graph G (see Definition 3.9.1). Recall, that the compatibility means that the boundary spaces can be decomposed into

$$\mathscr{G}_v = \bigoplus_{e \in E_v} \mathscr{G}_e \qquad \text{and} \qquad \widetilde{\mathscr{G}}_v = \bigoplus_{e \in E_v} \widetilde{\mathscr{G}}_e$$

and

$$\Gamma_v f = \bigoplus_{e \in E_v} \Gamma_{v,e} f \qquad \text{and} \qquad \widetilde{\Gamma}_v u = \bigoplus_{e \in E_v} \widetilde{\Gamma}_{v,e} u$$

with $\Gamma_{v,e}\colon \mathscr{G}_v \longrightarrow \mathscr{G}_e$ and $\widetilde{\Gamma}_{v,e}\colon \widetilde{\mathscr{G}}_v \longrightarrow \widetilde{\mathscr{G}}_e$. Moreover, we assume that

$$\sup_{v \in V} \|\Gamma_v\|_{1\to 0} < \infty \qquad \text{and} \qquad \sup_{v \in V} \|\widetilde{\Gamma}_v\|_{1\to 0} < \infty \tag{4.57a}$$

and that

$$\mathscr{G}_e^{1/2} := \bigcap_{v \in \partial e} \Gamma_{v,e}(\mathscr{H}^1_{v,e}) \qquad \text{and} \qquad \widetilde{\mathscr{G}}_e^{1/2} := \bigcap_{v \in \partial e} \widetilde{\Gamma}_{v,e}(\widetilde{\mathscr{H}}^1_{v,e}) \tag{4.57b}$$

are dense in $\mathscr{G}_e$ resp. $\widetilde{\mathscr{G}}_e$ for all $e \in E$ (see (3.92d) for the definition of $\mathscr{H}^1_{v,e}$). Then we defined a *coupled* boundary map $\Gamma\colon \mathscr{H}^1 \longrightarrow \mathscr{G}$ with $\mathscr{G} := \bigoplus_{e \in E} \mathscr{G}_e$ and $\mathscr{H}^1$ as well as a coupled quadratic form $\mathfrak{h}$ (see Proposition 3.9.2), and similarly for the objects with a tilde $\widetilde{\cdot}$.

Proposition 4.8.3. *Assume that the boundary maps Γ_v and $\widetilde{\Gamma}_v$ are 0-close with first order identification operators J^1_v and J'^1_v and boundary identification operator $I_v\colon \mathscr{G}_v \longrightarrow \widetilde{\mathscr{G}}_v$. Moreover, assume that*

$$I_v = \bigoplus_{e \in E_v} I_e, \qquad \textit{where} \qquad I_e\colon \mathscr{G}_e \longrightarrow \widetilde{\mathscr{G}}_e, \tag{4.58}$$

and define the first order identification operators as the restrictions

$$J^1 := \bigoplus_{v \in V} J^1_v \restriction_{\mathscr{H}^1} \qquad \textit{and} \qquad J'^1 := \bigoplus_{v \in V} J'^1_v \restriction_{\widetilde{\mathscr{H}}^1}.$$

Then the following assertions are true:

1. *The operators J^1 and J'^1 map into the* coupled *spaces $\widetilde{\mathscr{H}}^1$ and $\mathscr{H}^1$, respectively.*
2. *The coupled boundary maps Γ and $\widetilde{\Gamma}$ are 0-close with J^1 and J'^1 and the boundary identification operator $I := \bigoplus_{e \in E} I_e$.*
3. *Assume that for each $v \in V$, the operators J^1_v and J'^1_v are δ_v-adjoint, then J^1 and J'^1 are δ-adjoint with $\delta := \sup_v \delta_v$.*
4. *Assume that for each $v \in V$, the operators J^1_v and J'^1_v are δ_v-compatible with the zeroth order identification operators J_v and J'_v, then J^1 and J'^1 are δ-compatible with the zeroth order operators*

$$J := \bigoplus_{v \in V} J_v\colon \mathscr{H} \longrightarrow \widetilde{\mathscr{H}} \qquad \textit{and} \qquad J' := \bigoplus_{v \in V} J'_v\colon \widetilde{\mathscr{H}} \longrightarrow \mathscr{H}$$

with $\delta := \sup_v \delta_v$.

5. *Assume that for each $v \in V$, the forms $\mathfrak{h}_v$ and $\widetilde{\mathfrak{h}}_v$ are δ_v-close, then the coupled forms $\mathfrak{h}$ and $\widetilde{\mathfrak{h}}$ are δ-close with $\delta := \sup_v \delta_v$.*
6. *The statement in* (5) *remains true if "closeness" is replaced by "quasi-unitary equivalence".*
7. *The statement in* (5) *remains true if "closeness" is replaced by "partial isometric equivalence".*

Proof. Note first that the 0-closeness and (4.58) imply that

$$\widetilde{\Gamma}_{v,e} J_v^1 = I_e \Gamma_{v,e}. \tag{4.59}$$

In order to show (1) we have to assure that $J^1 f \in \widetilde{\mathscr{H}}^1$ if $f \in \mathscr{H}^1$. But

$$\widetilde{\Gamma}_{\partial_\pm e,e} J^1 f = I_e \Gamma_{\partial_\pm e,e} f,$$

the latter expression depends only on $v = \partial_\pm e$, and not on the sign by definition of $f \in \mathscr{H}^1$. In particular, $J^1 f \in \widetilde{\mathscr{H}}^1$. The assertion for J'^1 follows similarly.

(2) From (4.59) we conclude

$$(\widetilde{\Gamma} J^1 f)_e = \widetilde{\Gamma}_{v,e} J_v^1 f_v = I_e \Gamma_{v,e} f_v = (I \Gamma f)_e$$

for $f \in \mathscr{H}^1$, and similarly $\Gamma J'^1 = I^* \widetilde{\Gamma}$.

(3) We have

$$\begin{aligned} \big|\langle J'^1 u, f\rangle - \langle f, J^1 f\rangle\big| &= \sum_{v \in V} \big|\langle J_v'^1 u, f\rangle_v - \langle f, J_v^1 f\rangle_v\big| \\ &\le \sum_{v \in V} \delta_v \|u\|_{1,v} \|f\|_{1,v} \overset{\text{CS}}{\le} \sup_v \delta_v \|u\|_1 \|f\|_1 \end{aligned}$$

and (5) follows similarly. For (4), we observe that

$$\|(J^1 - J)f\|^2 = \sum_{v \in V} \|(J_v^1 - J_v) f\|_v^2 \le \sum_{v \in V} \delta_v^2 \|f\|_{v,1}^2 \le \delta^2 \|f\|_1^2,$$

and similarly for $J'^1 - J'$. (6) follows from (4) and (5) and the fact that J, J' are δ-quasi-unitary provided all J_v, J_v' are δ_v-quasi-unitary; e.g., we have

$$\|u - JJ'u\|^2 = \sum_{v \in V} \|u - J_v J_v' u\|_v^2 \le \sum_{v \in V} \delta_v^2 \|u\|_{v,1}^2 \le \delta^2 \|u\|_1^2$$

and similarly for the other conditions of δ-quasi-unitarity. (7) follows in the same way, e.g. we have $J^* J f = \bigoplus_{v \in V} J_v^* J_v f = \bigoplus_{v \in V} f_v = f$. □

Remark 4.8.4. Note that the 0-closeness allows us to define J'^1 simply as the restriction of the direct sum to $\widetilde{\mathscr{H}}^1$. If we would have started with δ_v-closed boundary maps, one has to modify the first order identification operator (see e.g. Proposition 4.9.2), but we do not need this more general case in our examples.

4.9 Convergence of Resonances

In this section we provide a criterion for the convergence of resonances. To to so, we show the closeness of coupled dilated systems starting from the closeness of the interior systems and some natural conditions on the half-line models.

Assume that

$$\mathscr{H} = \mathscr{H}_{\mathrm{int}} \oplus \mathscr{H}_{\mathrm{ext}} \qquad \text{and} \qquad \widetilde{\mathscr{H}} = \widetilde{\mathscr{H}}_{\mathrm{int}} \oplus \widetilde{\mathscr{H}}_{\mathrm{ext}} \tag{4.60}$$

and that there are boundary triples $(\Gamma_{\mathrm{int}}, \Gamma'_{\mathrm{int}}, \mathscr{G})$ and $(\widetilde{\Gamma}_{\mathrm{int}}, \widetilde{\Gamma}'_{\mathrm{int}}, \widetilde{\mathscr{G}})$ associated with $\mathfrak{h}_{\mathrm{int}}$ and $\widetilde{\mathfrak{h}}_{\mathrm{int}}$, resp.

Moreover, we assume that we have half-line boundary triples $(\Gamma_{\mathrm{ext}}, \Gamma'_{\mathrm{ext}}, \mathscr{G})$ and $(\widetilde{\Gamma}_{\mathrm{ext}}, \widetilde{\Gamma}'_{\mathrm{ext}}, \widetilde{\mathscr{G}})$ on $\mathscr{H}_{\mathrm{ext}} = \mathsf{L}_2(\mathbb{R}_+) \otimes \mathscr{G}$ and $\widetilde{\mathscr{H}}_{\mathrm{ext}} = \mathsf{L}_2(\mathbb{R}_+) \otimes \widetilde{\mathscr{G}}$ associated with the transversal operators K and $\widetilde{K}$ on $\mathscr{G}$ and $\widetilde{\mathscr{G}}$, respectively, constructed in Sect. 3.5.

Let $I: \mathscr{G} \longrightarrow \widetilde{\mathscr{G}}$ be a boundary identification operator with $\|I\| \le 1$. As identification operator, we set

$$J_{\mathrm{ext}}: \mathscr{H}_{\mathrm{ext}} \longrightarrow \widetilde{\mathscr{H}}_{\mathrm{ext}}, \qquad J_{\mathrm{ext}} := \mathrm{id}_{\mathbb{R}_+} \otimes I.$$

We assume that K and $\widetilde{K}$ have purely discrete spectrum. Remember that then $\mathscr{G}^k_{\mathrm{ext}} = \operatorname{dom} K^{k/2}$ and $\mathscr{G}^1_{\mathrm{ext}} = \operatorname{dom} \mathfrak{k}$ (cf. Corollary 3.5.5), where $\mathscr{G}^k_{\mathrm{ext}}$ is the scale of Hilbert spaces associated with the exterior Dirichlet-to-Neumann map $\Lambda_{\mathrm{ext}} = \sqrt{K+1}$. Similar assertions hold on $\widetilde{\mathscr{G}}$.

For convenience of the reader, we recall the definition of the dilated quadratic form $\mathfrak{h}^\theta_{\mathrm{ext}}$ here. For more details, we refer to Sect. 3.5.2. Namely, we have

$$\mathfrak{h}^\theta_{\mathrm{ext}}(f) = \mathrm{e}^{-2\theta} \|f'\|^2 + \mathfrak{h}^\perp(f) = \int_{\mathbb{R}_+} \bigl(\mathrm{e}^{-2\theta} \|f'(s)\|^2_{\mathscr{G}} + \mathfrak{k}(f(s))\bigr) \mathrm{d}s$$

for $f \in \mathscr{H}^1_{\mathrm{ext}}$ and similarly for $\widetilde{\mathfrak{h}}^\theta_{\mathrm{ext}}$. Here and in the sequel, $\theta \in \mathrm{S}_\vartheta$, where $0 \le \vartheta < \vartheta_0 < \pi/2$ and

$$\mathrm{S}_\vartheta = \bigl\{ \theta \in \mathbb{C} \bigm| |\operatorname{Im} \theta| \le \vartheta/2 \bigr\} \tag{4.61}$$

denotes the strip of thickness ϑ. We have the following result on the identification operators for the exterior spaces:

Proposition 4.9.1. *Assume that $I(\mathscr{G}^1_{\mathrm{ext}}) \subset \widetilde{\mathscr{G}}^1_{\mathrm{ext}}$ and $I^*(\widetilde{\mathscr{G}}^1_{\mathrm{ext}}) \subset \mathscr{G}^1_{\mathrm{ext}}$. Then the following assertions hold:*

1. *J_{ext} is δ-partial isometric w.r.t. $\mathfrak{h}_{\mathrm{ext}}$ iff I is δ-partial isometric w.r.t. $\mathfrak{k}$.*
2. *Assume that $I: \mathscr{G}^1_{\mathrm{ext}} \longrightarrow \widetilde{\mathscr{G}}^1_{\mathrm{ext}}$ is bounded, then*

$$J_{\mathrm{ext}}(\mathscr{H}^1_{\mathrm{ext}}) \subset \widetilde{\mathscr{H}}^1_{\mathrm{ext}} \qquad \text{and} \qquad \|J_{\mathrm{ext}}\|_{1\to 1} \le \max\{\|I\|, \|I\|_{1\to 1}\}.$$

Similarly, if $I^*: \widetilde{\mathscr{G}}^1_{\text{ext}} \longrightarrow \mathscr{G}^1_{\text{ext}}$ *is bounded, then*

$$J^*_{\text{ext}}(\widetilde{\mathscr{H}}^1_{\text{ext}}) \subset \mathscr{H}^1_{\text{ext}} \quad \textit{and} \quad \|J^*_{\text{ext}}\|_{1\to 1} \le \max\{\|I\|, \|I^*\|_{1\to 1}\}.$$

3. *Assume that* $I: \mathscr{G}^1_{\text{ext}} \longrightarrow \widetilde{\mathscr{G}}^1_{\text{ext}}$ *is bounded, then the forms* $\mathfrak{h}^\theta_{\text{ext}}$ *and* $\widetilde{\mathfrak{h}}^\theta_{\text{ext}}$ *are 0-close with identification operator* J_{ext} *iff* $\mathfrak{k}$ *and* $\widetilde{\mathfrak{k}}$ *are 0-close with identification operator* I.
4. *If* $I(\mathscr{G}^2_{\text{ext}}) \subset \widetilde{\mathscr{G}}^2_{\text{ext}}$ *and* $IK = \widetilde{I}K$, *then* $\mathfrak{k}$ *and* $\widetilde{\mathfrak{k}}$ *as well as* $\mathfrak{h}^\theta_{\text{ext}}$ *and* $\widetilde{\mathfrak{h}}^\theta_{\text{ext}}$ *are 0-close. In particular,* $\|I\|_{k\to k} = \|I^*\|_{k\to k} = \|I\|$.

Proof. For the first assertion, note that $J^*_{\text{ext}} J_{\text{ext}} = \text{id}$ iff $I^* I = \text{id}_{\mathscr{G}}$. Moreover, if $\|\widetilde{\text{id}} - JJ^*\|^2_{1\to 0} \le \delta$ then for $u = v \otimes \psi \in \widetilde{\mathscr{H}}^1_{\text{ext}}$, we have

$$\|v \otimes (\psi - II^*\psi)\|^2 = \|v\|^2 \|\psi - II^*\psi\|^2 \le \delta\big(\|v'\|^2\|\psi\|^2 + \|v\|^2(\widetilde{\mathfrak{k}}(\psi) + \|\psi\|^2)\big)$$

implying

$$\|\psi - II^*\psi\|^2 \le \delta^2\Big(\frac{\|v'\|^2}{\|v\|^2}\|\psi\|^2 + (\widetilde{\mathfrak{k}}(\psi) + \|\psi\|^2)\Big)$$

for all $v \in \mathsf{H}^1(\mathbb{R}_+)$, $v \ne 0$. Since the infimum over all such v of the quotient is $0 = \inf \sigma(-\partial^{\mathrm{N}}_{ss})$ (the spectrum of the Neumann Laplacian on $\mathbb{R}_+$), the desired estimate $\|\text{id} - II^*\|_{1\to 0} \le \delta$ follows. On the other hand, if I is δ-partial isometric, then

$$\begin{aligned}
\|u - J_{\text{ext}} J^*_{\text{ext}} u\|^2 &= \int_{\mathbb{R}_+} \|u(s) - II^*u(s)\|^2_{\widetilde{\mathscr{G}}} \mathrm{d}s \\
&\le \delta^2 \int_{\mathbb{R}_+} \big(\widetilde{\mathfrak{k}}(u(s)) + \|u(s)\|^2\big)\mathrm{d}s \le \delta \|u\|^2_{\widetilde{1}}.
\end{aligned}$$

For the second assertion, we estimate

$$\begin{aligned}
\|J_{\text{ext}}(g \otimes \varphi)\|^2_{\widetilde{1}} &= \|g'\|^2\|I\varphi\|^2 + \|g\|^2\big(\widetilde{\mathfrak{k}}(I\varphi) + \|I\varphi\|^2\big) \\
&= \|g'\|^2\|I\varphi\|^2 + \|g\|^2\|I\varphi\|^2_{\widetilde{1}} \\
&\le \|g'\|^2\|I\|^2\|\varphi\|^2 + \|g\|^2\|I\|^2_{1\to 1}\|\varphi\|^2_1 \\
&\le \max\{\|I\|^2, \|I\|^2_{1\to 1}\}\|g \otimes \varphi\|^2_1
\end{aligned}$$

using the fact that $I\varphi \in \operatorname{dom} \widetilde{\mathfrak{k}}$ if $\varphi \in \operatorname{dom} \mathfrak{k}$. In particular, we have $J(g \otimes \varphi) \in \widetilde{\mathscr{H}}^1_{\text{ext}}$. Since the space $\mathscr{H}^1_{\text{ext}}$ is the closure of linear combinations of elementary tensors $g \otimes \varphi$ under the norm defined as in (3.50c), we have shown $J(\mathscr{H}^1_{\text{ext}}) \subset \widetilde{\mathscr{H}}^1_{\text{ext}}$. The second inclusion follows similarly.

The third assertion is a consequence of

$$\begin{aligned}\mathfrak{h}(f, J_{\mathrm{ext}}^* u) - \widetilde{\mathfrak{h}}(J_{\mathrm{ext}} f, u) &= \int_{\mathbb{R}_+} \big(\langle f'(s), I^* u'(s)\rangle + \mathfrak{k}(f(s), I^* u(s)) \\ &\qquad - \langle I f'(s), u'(s)\rangle - \widetilde{\mathfrak{k}}(I f(s), u(s))\big) \mathrm{d}s \\ &= \int_{\mathbb{R}_+} \big(\mathfrak{k}(f(s), I^* u(s)) - \widetilde{\mathfrak{k}}(I f(s), u(s))\big) \mathrm{d}s\end{aligned}$$

using $(If(s))' = If'(s)$ since I is bounded and similarly for I^*. If now $\mathfrak{h}_{\mathrm{ext}}$ and $\widetilde{\mathfrak{h}}_{\mathrm{ext}}$ are 0-close then the previous equation with $f = g \otimes \varphi$ and $u = v \otimes \psi$ yields

$$0 = \int_{\mathbb{R}_+} \overline{g(s)} v(s) \big(\mathfrak{k}(\varphi, I^* \psi) - \widetilde{\mathfrak{k}}(I\varphi, \psi)\big) \mathrm{d}s,$$

and therefore the 0-closeness of $\mathfrak{k}$ and $\widetilde{\mathfrak{k}}$. The other direction can be seen similarly.

The first statement of the fourth assertion is obvious. Moreover,

$$\|I\|_{k\to k} = \|(\widetilde{K}+1)^{k/2} I (K+1)^{-k/2}\| = \|(\widetilde{K}+1)^{k/2} (\widetilde{K}+1)^{-k/2} I\| = \|I\|$$

using Corollary 3.5.5, $IK = \widetilde{K} I$ and the functional calculus. The statement on I^* follows from $\|I^*\|_{k\to k} = \|I\|_{-k\to -k}$. □

Recall the definition of the (complexly) dilated coupled sesquilinear forms and operators (see Sect. 3.7 and especially Definition 3.7.13): The complexly dilated sesquilinear form is given by

$$\begin{aligned}\mathfrak{h}^\theta(f, g) &= \mathfrak{h}_{\mathrm{int}}(f, g) + \mathfrak{h}_{\mathrm{ext}}^\theta(f, g) \\ &= \mathfrak{h}_{\mathrm{int}}(f, g) + \int_{\mathbb{R}_+} \big(\mathrm{e}^{-2\theta} \langle f'(s), g'(s)\rangle_{\mathscr{G}} + \mathfrak{k}(f(s), g(s))\big) \mathrm{d}s\end{aligned}$$

for $f \in \mathscr{H}^{1,\overline{\theta}}$ and $g \in \mathscr{H}^{1,\theta}$, where

$$\mathscr{H}^{1,\theta} = \{\, f \in \mathscr{H}^1_{\mathrm{int}} \oplus \mathscr{H}^1_{\mathrm{ext}} \mid \Gamma_{\mathrm{ext}} f = \mathrm{e}^{\theta/2} \Gamma_{\mathrm{int}} f \,\}$$

and $\theta \in \mathrm{S}_\vartheta$ (see (4.61)). We equip $\mathscr{H}^{1,\theta}$ with the norm given by

$$\|f\|_{1,\theta}^2 := \|T^\theta f\|_1^2 = \mathfrak{h}(T^\theta f) + \|T^\theta f\|^2,$$

where $\mathfrak{h} := \mathfrak{h}^0$ is the undilated coupled (non-negative) form, and where

$$T^\theta : \mathscr{H} \longrightarrow \mathscr{H}, \qquad T^\theta u := u_{\mathrm{int}} \oplus \mathrm{e}^{-\theta/2} u_{\mathrm{ext}}$$

serves as compatibility operator, turning the restriction $T^\theta : \mathscr{H}^{1,\theta} \longrightarrow \mathscr{H}^1$ into a unitary operator.

Moreover, the corresponding dilated operator H^θ is given by

$$H^\theta f = \check{H}_{\mathrm{int}} f \oplus \left(-\mathrm{e}^{-2\theta} f''_{\mathrm{ext}} + (\mathrm{id}_{\mathbb{R}_+} \otimes K) f_{\mathrm{ext}}\right)$$

for $f \in \mathscr{H}^{2,\theta}$, where

$$\mathscr{H}^{2,\theta} = \left\{ u \in \mathscr{W}^2_{\mathrm{int}} \oplus \mathscr{W}^2_{\mathrm{ext}} \,\middle|\, \Gamma_{\mathrm{ext}} u = \mathrm{e}^{\theta/2} \Gamma_{\mathrm{int}} u, \quad \Gamma'_{\mathrm{ext}} u = \mathrm{e}^{3\theta/2} \Gamma'_{\mathrm{int}} u \right\}.$$

In addition, H^θ has spectrum in the sector $\Sigma_{\vartheta_0} = \{ z \in \mathbb{C} \mid |\arg z| \le \vartheta_0 \}$, where $0 \le \vartheta < \vartheta_0 < \pi/2$, and resolvent norm profile γ^1 of order 1, i.e.

$$\begin{aligned} &\| R(z) \|_{-1,\theta \to 1,\theta} \\ &\quad \le \gamma^1(z) := \left(\mathrm{e}^{|\mathrm{Re}\,\theta|} + \left(1 + C_{\le\vartheta}(z,-1) C^{\le\vartheta}(-1,z)\right)^2\right) C_{\le\vartheta}(z,-1) \end{aligned} \tag{4.62}$$

for $z \in U_{\vartheta_0} := \mathbb{C} \setminus \Sigma_{\vartheta_0}$, see Theorem 3.7.14 (4). For a definition of the constants $C_{\le\vartheta}(z,w)$ and $C^{\le\vartheta}(z,w)$ we refer to Lemma 3.5.13. In particular $H^\theta \in \mathscr{R}_1(\gamma^1, U_{\vartheta_0})$. Finally, we have shown that H^θ depends holomorphically on θ, i.e. that $(H^\theta - z)^{-1} \colon \mathscr{H} \longrightarrow \mathscr{H}$ is holomorphic in θ, provided the coupled boundary triple is elliptic, what we will assume for the rest of this section. If K has discrete spectrum, we have shown that the discrete eigenvalues of H^θ are (locally) constant in θ (see Theorem 3.8.2). In particular, we defined a *resonance* of H as a discrete eigenvalue of H^θ (see Definition 3.8.3) for θ having large enough imaginary part.

Similarly, we have the above objects on $\widetilde{\mathscr{H}}$, endowed notationally with a tilde $\widetilde{\cdot}$. The aim of the following is to show the quasi-unitarity of the coupled dilated operators H^θ and $\widetilde{H}^\theta$, starting from some quasi-unitarity information on the interior part.

We begin with the definition of the first order identification operators and show their adjointness. In order to do so, we need the following piecewise affine-linear, continuous cut-off function

$$\rho \colon \mathbb{R}_+ \longrightarrow [0,1], \qquad \rho(s) = 1 - s, \quad 0 \le s < 1, \qquad \rho(s) = 0, \quad s \ge 1. \tag{4.63}$$

As a consequence, $\rho(0) = 1$, $\|\rho\|^2 = 1/2 \le 1$ and $\|\rho'\|^2 = 1$. In particular, we set

$$J^{1,\theta} f := J^1_{\mathrm{int}} f \oplus (\mathrm{id}_{\mathbb{R}_+} \otimes I) f_{\mathrm{ext}} \tag{4.64a}$$

$$J'^{1,\overline{\theta}} u := J'^1_{\mathrm{int}} u \oplus \left[(\mathrm{id}_{\mathbb{R}_+} \otimes I^*) u_{\mathrm{ext}} + \rho \otimes \mathrm{e}^{\overline{\theta}/2} \left(\Gamma_{\mathrm{int}} J'^1_{\mathrm{int}} u - I^* \widetilde{\Gamma}_{\mathrm{int}} u \right) \right]. \tag{4.64b}$$

Recall the notion of adjointness for first order identification operators associated with non-negative quadratic forms (Definition 4.4.1 (1)) and general sesquilinear forms (Definition 4.7.1 (1)):

Proposition 4.9.2. *Let $\delta \geq 0$. Suppose the following assumptions:*

1. *The first order identification operators J^1_{int} and J'^1_{int} are δ-adjoint with respect to the scales $\mathscr{H}^1_{\mathrm{int}}$ and $\widetilde{\mathscr{H}}^1_{\mathrm{int}}$.*
2. *The interior boundary maps Γ_{int} and $\widetilde{\Gamma}_{\mathrm{int}}$ are δ-close with identification operators J^1_{int}, J'^1_{int} and $I : \mathscr{G} \longrightarrow \widetilde{\mathscr{G}}$.*
3. *The boundary identification operator $I : \mathscr{G} \longrightarrow \widetilde{\mathscr{G}}$ intertwines the (exterior) transversal operators K and $\widetilde{K}$, i.e. $\operatorname{ran} I \subset \operatorname{dom} \widetilde{K}$ and $\widetilde{K} I = I K$.*
4. *The exterior boundary triple $(\Gamma_{\mathrm{ext}}, \Gamma'_{\mathrm{ext}}, \mathscr{G})$ is bounded, i.e, the transversal operator K is bounded (see Corollary 3.5.5 and Proposition 3.4.11).*

Then the maps $J^{1,\theta}$ and $J'^{1,\overline{\theta}}$ are bounded as operators

$$J^{1,\theta} : \mathscr{H}^{1,\theta} \longrightarrow \widetilde{\mathscr{H}}^{1,\theta} \qquad \text{and} \qquad J'^{1,\overline{\theta}} : \widetilde{\mathscr{H}}^{1,\overline{\theta}} \longrightarrow \mathscr{H}^{1,\overline{\theta}}, \tag{4.65}$$

where ρ is given in (4.63). Moreover, $J^{1,\theta}$ and $J'^{1,\overline{\theta}}$ are δ'-adjoint, where $\delta' = (1 + \mathrm{e}^{\operatorname{Re}\theta})\delta$.

Proof. First, we have to show that the operators map into the right spaces: Assume that $f \in \mathscr{H}^{1,\theta}$. Then

$$\widetilde{\Gamma}_{\mathrm{ext}} u = u_{\mathrm{ext}}(0) = I f_{\mathrm{ext}}(0) = \mathrm{e}^{\theta/2} I \Gamma_{\mathrm{int}} f = \mathrm{e}^{\theta/2} \widetilde{\Gamma}_{\mathrm{int}} J^1_{\mathrm{int}} f = \mathrm{e}^{\theta/2} \widetilde{\Gamma}_{\mathrm{int}} u,$$

i.e. $u = J^{1,\theta} f \in \widetilde{\mathscr{H}}^{1,\theta}$ using (2) for the fourth equality. Next, assume that $u \in \widetilde{\mathscr{H}}^{1,\overline{\theta}}$ and set $f = J'^{1,\overline{\theta}} u$. Then $\Gamma_{\mathrm{int}} J'^1_{\mathrm{int}} u - I^* \widetilde{\Gamma}_{\mathrm{int}} u$, a priori only in $\mathscr{G}$ already belongs to $\mathscr{G}^1_{\mathrm{ext}} = \operatorname{dom} \mathfrak{k} = \mathscr{G}$ by (4). In particular, $f_{\mathrm{ext}} \in \mathscr{H}^1_{\mathrm{ext}}$. Moreover,

$$\begin{aligned} \Gamma_{\mathrm{ext}} f &= f_{\mathrm{ext}}(0) + \mathrm{e}^{\overline{\theta}/2} \big(\Gamma_{\mathrm{int}} J'^1_{\mathrm{int}} u - I^* \widetilde{\Gamma}_{\mathrm{int}} u \big) \\ &= I^* u_{\mathrm{ext}}(0) + \mathrm{e}^{\overline{\theta}/2} \big(\Gamma_{\mathrm{int}} J'^1_{\mathrm{int}} u - I^* \widetilde{\Gamma}_{\mathrm{int}} u \big) \\ &= \mathrm{e}^{\overline{\theta}/2} \Gamma_{\mathrm{int}} J'^1_{\mathrm{int}} u = \mathrm{e}^{\overline{\theta}/2} \Gamma_{\mathrm{int}} f, \end{aligned}$$

i.e. $f = J'^{1,\overline{\theta}} u \in \mathscr{H}^{1,\overline{\theta}}$ using again (2) for the third equality. Moreover,

$$\begin{aligned} \| J^{1,\theta} f \|^2_{1,\theta} &\leq \| J^1_{\mathrm{int}} \|^2_{1 \to 1} \| f \|^2_{1,\mathrm{int}} + \| I \|^2_{1 \to 1} \| \mathrm{e}^{-\theta/2} f \|^2_{1,\mathrm{ext}} \\ &\leq \max\{ \| J^1_{\mathrm{int}} \|^2_{1 \to 1}, 1 \} \| f \|^2_{1,\theta} \end{aligned}$$

using the boundedness of J^1_{int}, and $\| I \|_{1 \to 1} \leq 1$ following from (3) and Proposition 4.9.1 (4). Moreover, we have

$$\begin{aligned} &\big| \langle J'^{1,\overline{\theta}} u, f \rangle - \langle u, J^{1,\theta} f \rangle \big| \\ &\qquad \leq \big| \langle J'^1_{\mathrm{int}} u, f \rangle_{\mathrm{int}} - \langle u, J^1_{\mathrm{int}} f \rangle_{\mathrm{int}} \big| \\ &\qquad\qquad + \big| \langle \rho \otimes \mathrm{e}^{\overline{\theta}/2} (\Gamma_{\mathrm{int}} J'^1_{\mathrm{int}} u - I^* \widetilde{\Gamma}_{\mathrm{int}} u), f \rangle_{\mathrm{ext}} \big| \end{aligned}$$

$$\leq \delta \|u\|_{1,\mathrm{int}} \bigl(\|f\|_{1,\mathrm{int}} + \|\rho\| \mathrm{e}^{\operatorname{Re}\theta} \|\mathrm{e}^{-\theta/2} f\|_{\mathrm{ext}} \bigr)$$

$$\overset{\mathrm{CS}}{\leq} \delta (1 + \mathrm{e}^{\operatorname{Re}\theta}) \|f\|_{1,\theta} \|u\|_{1,\overline{\theta}}$$

using (1) and (2) and $\|\rho\| \leq 1$. □

Recall the notion of closeness for quadratic forms (Definition 4.4.1 (4)) and for general sesquilinear forms (Definition 4.7.1 (3)):

Proposition 4.9.3. *Let $\delta \geq 0$ and suppose the following assumptions:*

1. *The quadratic forms $\mathfrak{h}_{\mathrm{int}}$ and $\widetilde{\mathfrak{h}}_{\mathrm{int}}$ are δ-close with the identification operators J^1_{int} and J'^1_{int}.*
2. *The boundary maps Γ_{int} and $\widetilde{\Gamma}_{\mathrm{int}}$ are δ-close.*
3. *We have* $\operatorname{ran} I \subset \operatorname{dom} \widetilde{K}$ *and $\widetilde{K} I = I K$.*
4. *The (exterior) transversal operator K is bounded.*

Then the coupled dilated sesquilinear forms $\mathfrak{h}^\theta$ and $\widetilde{\mathfrak{h}}^\theta$ are δ'-close with $\delta' = (1 + \mathrm{e}^{-\operatorname{Re}\theta} + \mathrm{e}^{\operatorname{Re}\theta} \|K\|)\delta$.

Proof. The arguments are similar to the ones of the previous proof, namely,

$$\begin{aligned}
&\bigl|\mathfrak{h}^\theta(J'^{1,\overline{\theta}} u, f) - \widetilde{\mathfrak{h}}^\theta(u, J^{1,\theta} f)\bigr| \\
&\quad \leq \bigl|\mathfrak{h}_{\mathrm{int}}(J'^1 u, f) - \widetilde{\mathfrak{h}}_{\mathrm{int}}(u, J^1 f)\bigr| \\
&\qquad + \mathrm{e}^{-2\operatorname{Re}\theta} \bigl| \langle \rho' \otimes \mathrm{e}^{\overline{\theta}/2} (\Gamma_{\mathrm{int}} J'^1_{\mathrm{int}} u - I^* \widetilde{\Gamma}_{\mathrm{int}} u), f' \rangle_{\mathrm{ext}} \bigr| \\
&\qquad + \bigl| \mathfrak{h}^\perp_{\mathrm{ext}} (\rho \otimes \mathrm{e}^{\overline{\theta}/2} (\Gamma_{\mathrm{int}} J'^1_{\mathrm{int}} u - I^* \widetilde{\Gamma}_{\mathrm{int}} u), f) \bigr| \\
&\quad \leq \delta \|u\|_{1,\mathrm{int}} \bigl(\|f\|_{1,\mathrm{int}} + \mathrm{e}^{-\operatorname{Re}\theta} \|\rho'\| \|\mathrm{e}^{-\theta/2} f'\|_{\mathrm{ext}} + \mathrm{e}^{\operatorname{Re}\theta} \|\rho\| \|K\| \|\mathrm{e}^{-\theta/2} f\|_{\mathrm{ext}} \bigr) \\
&\quad \overset{\mathrm{CS}}{\leq} \delta \|u\|_{1,\overline{\theta}} \bigl(1 + \mathrm{e}^{-\operatorname{Re}\theta} + \mathrm{e}^{\operatorname{Re}\theta} \|K\|\bigr) \|f\|_{1,\theta}
\end{aligned}$$

since

$$\mathfrak{h}^\theta_{\mathrm{ext}}((\mathrm{id} \otimes I^*) u, f) = \widetilde{\mathfrak{h}}^\theta_{\mathrm{ext}}(u, (\mathrm{id} \otimes I) f)$$

due to (3) and Proposition 4.9.1 (4). Moreover, we used the closeness of the interior forms and (2). The δ'-adjointness of the identification operators has already been shown in Proposition 4.9.2 with an error smaller than the one given here. □

Let us now define the zeroth order identification operators starting from zeroth order identification operators J_{int} and J'_{int} on the interior part:

$$J f := J_{\mathrm{int}} f \oplus (\mathrm{id}_{\mathbb{R}_+} \otimes I) f_{\mathrm{ext}} \qquad \text{and} \qquad J' u := J'_{\mathrm{int}} u \oplus (\mathrm{id}_{\mathbb{R}_+} \otimes I^*) u_{\mathrm{ext}}. \tag{4.66}$$

Recall the notion of δ-compatibility for identification operators of order 1 in Definition 4.4.9 for non-negative quadratic forms and in Definition 4.7.7 for general sesquilinear forms:

Proposition 4.9.4. *Let $\delta \geq 0$ and assume that both coupled boundary triples are elliptic. Moreover, we suppose the following assumptions:*

1. *The first-order identification operators J^1_{int} and J'^1_{int} are δ-compatible with J_{int} and J'_{int}, (of order 1 w.r.t. $\mathfrak{h}_{\mathrm{int}}$ and $\widetilde{\mathfrak{h}}_{\mathrm{int}}$).*
2. *The boundary maps Γ_{int} and $\widetilde{\Gamma}_{\mathrm{int}}$ are δ-close.*
3. *We have* $\operatorname{ran} I \subset \operatorname{dom} \widetilde{K}$ *and* $\widetilde{K} I = IK$.
4. *The (exterior) transversal operator K is bounded.*

Then the coupled first-order operator $J^{1,\theta}$ is $(\delta, C, -1)$-compatible with J and $J'^{1,\overline{\theta}}$ is $(\delta', C, -1)$-compatible with J', where $\delta' = (1 + \mathrm{e}^{\operatorname{Re}\theta/2})\delta$ and $C = \gamma^1(-1)$.

Proof. We have

$$\|(J^{1,\theta} - J)f\|^2 = \|(J^1_{\mathrm{int}} - J_{\mathrm{int}})f\|^2_{\mathrm{int}} \leq \delta^2 \|f\|^2_{1,\mathrm{int}} \leq \delta^2 \|f\|^2_{1,\theta}$$

and

$$\begin{aligned}\|(J'^{1,\overline{\theta}} - J')u\|^2 &= \|(J'^1_{\mathrm{int}} - J'_{\mathrm{int}})u\|^2_{\mathrm{int}} + \|\rho \otimes \mathrm{e}^{\theta/2}\big(\Gamma_{\mathrm{int}} J'^1_{\mathrm{int}} u - I^* \widetilde{\Gamma}_{\mathrm{int}} u_{\mathrm{int}}\big)\|^2_{\mathrm{ext}} \\ &\leq \delta^2\big(1 + \|\rho\|^2 \mathrm{e}^{\operatorname{Re}\theta}\big)\|u\|^2_{1,\mathrm{int}} \leq \delta^2(1 + \mathrm{e}^{\operatorname{Re}\theta/2})^2 \|u\|^2_{1,\mathrm{int}}.\end{aligned}$$

The fact that H^θ and $\widetilde{H}^\theta \in \mathscr{R}_1(C, -1)$ was already shown in Theorem 3.7.14 (4) using the ellipticity. Moreover, it is easily seen that

$$\widetilde{T}^\theta J = J T^\theta \qquad \text{and} \qquad T^{\overline{\theta}} J' = \widetilde{T}^{\overline{\theta}} J'$$

and we are done. □

Recall the notion of δ-quasi-unitarity and δ-partial isometry in Definition 4.1.1:

Proposition 4.9.5. *Let $\delta \geq 0$. Suppose the following assumptions:*

1. *The zeroth order identification operators J_{int} and J'_{int} are δ-quasi-unitary (resp. J is δ-partial isometric) of order 1 w.r.t. $\mathfrak{h}_{\mathrm{int}}$ and $\widetilde{\mathfrak{h}}_{\mathrm{int}}$.*
2. *The boundary identification operator I is δ-partial isometric of order 1 w.r.t. $\mathfrak{k}$ and $\widetilde{\mathfrak{k}}$, i.e. that $I^* I = \mathrm{id}_{\mathscr{G}}$ and*

$$\|\psi - II^*\psi\|^2 \leq \delta^2 \|\psi\|^2_{\tilde{1}} = \delta^2\, \widetilde{\mathfrak{k}}(\psi) + \|\psi\|^2.$$

Then the coupled zeroth order operators J and J' defined in (4.66) are δ'-quasi-unitary (resp. J^θ is δ'-partial isometric) (of order 1 w.r.t. $\mathfrak{h}^\theta$ and $\widetilde{\mathfrak{h}}^\theta$), where $\delta' = \max\{1, \mathrm{e}^{\operatorname{Re}\theta/2}\}\delta$.

Proof. We have

$$\|Ju\|^2 = \|J_{\mathrm{int}} u\|^2_{\mathrm{int}} + \|(\mathrm{id} \otimes I)u\|^2_{\mathrm{ext}} \leq 4\|u\|^2_{\mathrm{int}} + \|u\|^2_{\mathrm{ext}} \leq 4\|u\|^2$$

using $\|I\| \leq 1$. Similarly,

$$\|(J' - J^*)u\|^2 \leq \delta^2 \|u\|^2_{\mathrm{int}} + 0 \leq \delta^2 \|u\|^2.$$

Finally,

$$\begin{aligned}\|u - JJ'u\|^2 &= \|u - J_{\mathrm{int}} J'_{\mathrm{int}} u\|^2_{\mathrm{int}} + \|u - (\mathrm{id}_{\mathbb{R}_+} \otimes I I^*) u\|^2_{\mathrm{ext}} \\ &\le \delta^2 \|u\|^2_{1,\mathrm{int}} + \delta^2 \|u\|^2_{1,\mathrm{ext}} \le \max\{1, \mathrm{e}^{\mathrm{Re}\,\theta}\} \delta^2 \|u\|^2_{1,\theta}\end{aligned}$$

using Proposition 4.9.1 and similarly for the remaining estimate. The partial isometry follows similarly. □

Recall Definition 4.4.1 (5) and Definition 4.7.12 for the notion of partial isometry of the first order operators associated with non-negative quadratic forms and general sesquilinear forms.

Proposition 4.9.6. *Let $\delta \ge 0$. Suppose the following assumptions:*

1. *The first order identification operators J^1_{int} and J'^1_{int} are δ-partial isometric w.r.t. $\mathfrak{h}_{\mathrm{int}}$ and $\widetilde{\mathfrak{h}}_{\mathrm{int}}$.*
2. *The boundary maps Γ_{int} and $\widetilde{\Gamma}_{\mathrm{int}}$ are δ-close.*
3. *We have* $\operatorname{ran} I \subset \operatorname{dom} \widetilde{K}$ *and $\widetilde{K} I = I K$.*
4. *The (exterior) transversal operator K is bounded.*
5. *The boundary identification operator I is δ-partial isometric of order* 1 *w.r.t. $\mathfrak{k}$ and $\widetilde{\mathfrak{k}}$.*

Then the coupled first-order operators $J^{1,\theta}$ and $J'^{1,\theta}$ defined in (4.64) *are δ'-partial isometries, where $\delta' = (2 + \|K\|)\delta$.*

Note that we used θ instead of $\overline{\theta}$ for the operator $J'^{1,\theta}$, see Remark 4.7.13 (2).

Proof. The condition $J'^{1,\theta} J^{1,\theta} = \mathrm{id}$ is obvious. Moreover, we have

$$\begin{aligned}&\big|\langle T^\theta J'^{1,\theta} u, T^\theta f\rangle - \langle \widetilde{T}^\theta u, \widetilde{T}^\theta J^{1,\theta} f\rangle\big| \\ &\quad \le \big|\langle J'^1_{\mathrm{int}} u, f\rangle_{\mathrm{int}} - \langle u, J^1_{\mathrm{int}} f\rangle_{\mathrm{int}}\big| + \big|\langle \rho \otimes (\Gamma_{\mathrm{int}} J'^1_{\mathrm{int}} u - I^* \widetilde{\Gamma}_{\mathrm{int}} u), \mathrm{e}^{-\theta/2} f\rangle_{\mathrm{ext}}\big| \\ &\quad \le \delta \|u\|_{1,\mathrm{int}} \big(\|f\|_{1,\mathrm{int}} + \|\rho\|\, \|\mathrm{e}^{-\theta/2} f\|_{\mathrm{ext}}\big) \\ &\quad \overset{\mathrm{CS}}{\le} 2\delta \|u\|_{1,\theta} \|f\|_{1,\theta}\end{aligned}$$

using (1) and (2) and $\|\rho\| \underset{\sim}{\le} 1$. The last condition to be verified is on the undilated coupled forms $\mathfrak{h} = \mathfrak{h}^0$ and $\widetilde{\mathfrak{h}} = \widetilde{\mathfrak{h}}^0$, namely,

$$\begin{aligned}&\big|\mathfrak{h}(T^\theta J'^{1,\theta} u, T^\theta f) - \widetilde{\mathfrak{h}}(\widetilde{T}^\theta u, \widetilde{T}^\theta J^{1,\theta} f)\big| \\ &\quad \le \big|\mathfrak{h}_{\mathrm{int}}(J'^1 u, f) - \widetilde{\mathfrak{h}}_{\mathrm{int}}(u, J^1 f)\big| + \big|\langle \rho' \otimes (\Gamma_{\mathrm{int}} J'^1_{\mathrm{int}} u - I^* \widetilde{\Gamma}_{\mathrm{int}} u), \mathrm{e}^{-\theta/2} f'\rangle_{\mathrm{ext}}\big| \\ &\quad + \big|\mathfrak{h}^{\perp}_{\mathrm{ext}}(\rho \otimes (\Gamma_{\mathrm{int}} J'^1_{\mathrm{int}} u - I^* \widetilde{\Gamma}_{\mathrm{int}} u), \mathrm{e}^{-\theta/2} f)\big| \\ &\quad \le \|u\|_{1,\mathrm{int}} \delta \big(\|f\|_{1,\mathrm{int}} + \|\rho'\|\, \|\mathrm{e}^{-\theta/2} f'\|_{\mathrm{ext}} + \|\rho\|\, \|K\|\, \|\mathrm{e}^{-\theta/2} f\|_{\mathrm{ext}}\big) \\ &\quad \overset{\mathrm{CS}}{\le} \delta (2 + \|K\|) \|u\|_{1,\theta} \|f\|_{1,\theta}\end{aligned}$$

since

$$\mathfrak{h}^\theta_{\mathrm{ext}}((\mathrm{id} \otimes I^*)u, f) = \widetilde{\mathfrak{h}}^\theta_{\mathrm{ext}}(u, (\mathrm{id} \otimes I) f)$$

due to (3) and Proposition 4.9.1 (4), and again using (1) and (2). Moreover, we used the closeness of the interior forms and (2). The δ'-adjointness of the identification operators has already been shown in Proposition 4.9.2 with an error smaller than the one given here. □

We can now state the main result of this section. Recall the notion of δ-quasi-unitary (resp. -partial isometric) equivalence of non-negative quadratic forms (Definition 4.4.11) and of general sesquilinear forms (Definition 4.7.9):

Theorem 4.9.7. *Let $\delta \geq 0$ and $0 \leq \vartheta < \vartheta_0 < \pi/2$. Assume that both coupled boundary triples are elliptic. Moreover, we suppose the following assumptions:*

1. *The quadratic forms $\mathfrak{h}_{\mathrm{int}}$ and $\widetilde{\mathfrak{h}}_{\mathrm{int}}$ are δ-quasi-unitarily (resp. -partial isometrically) equivalent.*
2. *The boundary maps Γ_{int} and $\widetilde{\Gamma}_{\mathrm{int}}$ are δ-close.*
3. *We have* $\operatorname{ran} I \subset \operatorname{dom} \widetilde{K}$ *and* $\widetilde{K} I = I K$.
4. *The (exterior) transversal operator K is bounded.*
5. *Assume in addition that I is δ-partial isometric of order* 1 *w.r.t. $\mathfrak{k}$ and $\widetilde{\mathfrak{k}}$.*

Then the coupled sesquilinear forms $\mathfrak{h}^\theta$ and $\widetilde{\mathfrak{h}}^\theta$ are $(\delta', \gamma^1, U_{\vartheta_0})$-quasi-unitarily (resp. -partial isometrically) equivalent for $\theta \in S_\vartheta$, where $\delta' = \bigl(1 + \mathrm{e}^{|\operatorname{Re}\theta|}(1 + \|K\|)\bigr)\delta$, where γ^1 is given in (4.62) *and where $U_{\vartheta_0} := \mathbb{C} \setminus \Sigma_{\vartheta_0}$.*

Proof. The result follows from Propositions 4.9.3, 4.9.4 and 4.9.5 and Proposition 4.9.6. Moreover, the resolvent norm profile is shown in Theorem 3.7.14 (4). □

As a consequence, we obtain:

Theorem 4.9.8. *Let $\delta \geq 0$ and $0 \leq \vartheta < \vartheta_0 < \pi/2$. Assume that both coupled boundary triples are elliptic. Moreover, we assume:*

1. *The quadratic forms $\mathfrak{h}_{\mathrm{int}}$ and $\widetilde{\mathfrak{h}}_{\mathrm{int}}$ are δ-quasi-unitarily (resp. -partial isometrically) equivalent.*
2. *The boundary maps Γ_{int} and $\widetilde{\Gamma}_{\mathrm{int}}$ associated with $\mathfrak{h}_{\mathrm{int}}$ and $\widetilde{\mathfrak{h}}_{\mathrm{int}}$ are δ-close.*
3. *We have* $\operatorname{ran} I \subset \operatorname{dom} \widetilde{K}$ *and* $\widetilde{K} I = I K$.
4. *The (exterior) transversal operator K is bounded.*
5. *The boundary identification operator I is δ-partial isometric of order* 1 *w.r.t. $\mathfrak{k}$ and $\widetilde{\mathfrak{k}}$.*

Then the coupled dilated operators H^θ and $\widetilde{H}^\theta$ are $(\delta'', C^2\gamma^1, U_{\vartheta_0})^*$-quasi-unitarily (resp. -partial isometrically) equivalent for all $\theta \in S_\vartheta$, where

$$\delta'' := 7C^5\bigl(1 + \mathrm{e}^{|\operatorname{Re}\theta|}(1 + \|K\|)\bigr)\delta, \qquad C := \gamma^1(-1) \geq 1,$$

$U_{\vartheta_0} := \mathbb{C} \setminus \Sigma_{\vartheta_0}$ and where γ^1 is given in (4.62).

Proof. The quasi-unitary (resp. -partial isometric) equivalence for the sesquilinear forms was proven in Theorem 4.9.7, and the quasi-unitary (resp. -partial isometric) equivalence for operators follows then from Lemma 4.7.18; note that $z_0 = -1$ here. Since $(H^\theta)^* = H^{\overline{\theta}}$, the same statements are also true for the adjoints, and the result follows. □

Recall the holomorphic functional calculus of Theorem 3.3.6. We can conclude all the results of Sects. 4.5 and 4.7 having quasi-unitarity as assumption. In particular, from Theorem 4.5.14 we obtain:

Corollary 4.9.9. *Assume that $D \subset \mathbb{C}$ is open with piecewise smooth compact boundary ∂D. Suppose that φ is holomorphic in a neighbourhood of $\overline{D}$. Under the assumptions of the previous theorem, we have*

$$\|\varphi_D(\widetilde{H}^\theta) - J\varphi_D(H^\theta)J'\| \leq C'\delta$$

provided $\partial D \subset \rho(H^\theta) \cap \rho(\widetilde{H}^\theta)$ and $\delta > 0$ is small enough, where C' depends only on ∂D, φ, ϑ, ϑ_0, $\operatorname{Re}\theta$ and $\|K\|$.

In particular, the preceding corollary applies to spectral projections by choosing $\varphi = \mathbb{1}$. Let us now state some more assertions on the spectrum:

Proposition 4.9.10. *Let $\delta \geq 0$ and assume the following:*

1. *The boundary space $\mathscr{G}$ is finite-dimensional,*
2. *We have $\operatorname{ran} I \subset \operatorname{dom} \widetilde{K}$ and $\widetilde{K}I = IK$.*
3. *The operator I is δ-partial isometric of order 1 w.r.t. $\mathfrak{k}$ and $\widetilde{\mathfrak{k}}$.*
4. *The operator $\widetilde{K}$ has purely discrete spectrum.*

Then

$$\lambda_k(K) = \lambda_k(\widetilde{K}), \quad k = 1, \dots, k_0 \qquad \text{and} \qquad \lambda_k(\widetilde{K}) \geq \frac{1}{\delta^2} - 1, \quad k > k_0,$$

where $k_0 = \dim \mathscr{G}$.

Proof. Note that I is a partial isometry ($I^*I = \mathrm{id}$) from $\mathscr{G}$ onto $\operatorname{ran} \widetilde{P} \subset \widetilde{\mathscr{G}}$, where $\widetilde{P} := II^*$ is an orthogonal projection. Moreover, from $\widetilde{K}I = IK$ it follows that K is unitarily equivalent with $\widetilde{K}\widetilde{P} = \widetilde{P}\widetilde{K}$, showing the first statement. Finally, the condition $\|(\mathrm{id} - II^*)(\widetilde{K} + 1)^{1/2}\| \leq \delta$ of the δ-partial isometry is equivalent with

$$\frac{1}{\delta^2}\widetilde{P}^\perp \leq \widetilde{P}^\perp(\widetilde{K} + 1)\widetilde{P}^\perp$$

where $\widetilde{P}^\perp = 1 - \widetilde{P}$, from which the latter eigenvalue estimate follows. □

Let us now state the convergence of the essential spectrum of the coupled dilated operators H^θ and $\widetilde{H}^\theta$. Recall the explicit form of the essential spectrum in Proposition 3.8.1:

Theorem 4.9.11. *Let $\delta \geq 0$ and $0 \leq \vartheta < \vartheta_0 < \pi/2$. Assume that the boundary space $\mathscr{G}$ is finite-dimensional, and that both coupled boundary triples are elliptic. Moreover, we assume:*

1. *The quadratic forms $\mathfrak{h}_{\mathrm{int}}$ and $\widetilde{\mathfrak{h}}_{\mathrm{int}}$ are δ-quasi-unitarily equivalent.*
3. *We have* $\operatorname{ran} I \subset \operatorname{dom} \widetilde{K}$ *and* $\widetilde{K} I = I K$.
4. *The operator I is δ-partial isometric of order* 1 *w.r.t.* $\mathfrak{k}$ *and* $\widetilde{\mathfrak{k}}$.
5. *The operator $\widetilde{K}$ has purely discrete spectrum.*

Then for $\theta \in \mathrm{S}_\vartheta$ with $\operatorname{Im}\theta \neq 0$, *the first* $k_0 = \dim \mathscr{G}$ *branches*

$$B_k = \lambda_k(K) + \mathrm{e}^{-2\theta}[0, \infty), \quad k = 1, \ldots, k_0$$

of the essential spectra of the coupled dilated operators H^θ and $\widetilde{H}^\theta$ agree. Moreover, the remaining branches B_k, $k > k_0$, have a base point $\lambda_k(\widetilde{K}) \geq \delta^{-2} - 1$, i.e. they tend to ∞ as $\delta \to 0$.

Finally, if the interior Dirichlet-to-Neumann map $\widetilde{\Lambda}_{\mathrm{int}}$ has purely discrete spectrum and if $\operatorname{Im}\theta \neq 0$, *then also the real essential spectra*

$$\sigma_{\mathrm{ess}}(H^\theta) \cap \mathbb{R}_+ = \sigma_{\mathrm{ess}}(H^{\mathrm{D}}_{\mathrm{int}}) \qquad \text{and} \qquad \sigma_{\mathrm{ess}}(\widetilde{H}^\theta) \cap \mathbb{R}_+ = \sigma_{\mathrm{ess}}(\widetilde{H}^{\mathrm{D}}_{\mathrm{int}})$$

converge, i.e. there is a function $\eta(\delta) \to 0$ as $\delta \to 0$ depending only on $\mathfrak{h}_{\mathrm{int}}$ such that

$$\overline{d}\,(\sigma_{\mathrm{ess}}(H^\theta) \cap \mathbb{R}_+, \sigma_{\mathrm{ess}}(\widetilde{H}^\theta) \cap \mathbb{R}_+) \leq \eta(\delta),$$

where $\overline{d}$ denotes the weighted Hausdorff distance (cf. Definition A.1.1 and (4.20)*).*

Proof. Since we assumed that $\mathscr{G}$ is finite-dimensional, the transversal operator K is obviously bounded. From Proposition 3.8.1 and Proposition 4.9.10, the statements on the non-real essential spectra follows. For the real essential spectra note that $\widetilde{R}^{\mathrm{N}}_{\mathrm{int}} - \widetilde{R}^{\mathrm{D}}_{\mathrm{int}} = \widetilde{B}^*_{\mathrm{int}} \widetilde{\Lambda}^{-1}_{\mathrm{int}} \widetilde{B}_{\mathrm{int}}$ by Theorem 3.4.44, and the latter operator is compact by assumption, so that the interior Neumann and Dirichlet operators have the same essential spectrum. But the convergence of the interior Neumann essential spectrum follows by the first assumption, Proposition 4.4.15 and Theorem 4.3.4. The function $\eta(\delta)$ is just the modulus of continuity of the continuity stated in Theorem 4.3.4. □

Recall the definition of a resonance of H (Definition 3.8.3) (i.e. an eigenvalue of H^θ for some θ with large enough imaginary part). In particular, we show that the resonances of $\widetilde{H}$ are close to the resonances of H:

Theorem 4.9.12. *Let $\delta \geq 0$ and $\vartheta \in [0, \pi/2)$. Assume that the boundary space $\mathscr{G}$ is finite-dimensional, and that both coupled boundary triples are elliptic. Moreover, we assume:*

1. *$\mathfrak{h}_{\mathrm{int}}$ and $\widetilde{\mathfrak{h}}_{\mathrm{int}}$ are δ-quasi-unitarily equivalent;*
2. *The boundary maps Γ_{int} and $\widetilde{\Gamma}_{\mathrm{int}}$ associated with $\mathfrak{h}_{\mathrm{int}}$ and $\widetilde{\mathfrak{h}}_{\mathrm{int}}$ are δ-close;*
3. $\operatorname{ran} I \subset \operatorname{dom} \widetilde{K}$ *and* $\widetilde{K} I = I K$;

4. *I is δ-partial isometric of order* 1 *w.r.t.* $\mathfrak{k}$ *and* $\widetilde{\mathfrak{k}}$*;*
5. $\widetilde{K}$ *has purely discrete spectrum.*

Let $\theta \in S_\vartheta$ and $\lambda \in \sigma_{\mathrm{disc}}(H^\theta)$ be a discrete eigenvalue with multiplicity μ. Then there exist a function $\eta(\delta) \to 0$ as $\delta \to 0$ and μ eigenvalues $\widetilde{\lambda}_j \in \sigma_{\mathrm{disc}}(\widetilde{H}^\theta)$, $j = 1, \dots, \mu$, (not necessarily mutually distinct) such that $|\widetilde{\lambda}_j - \lambda| \le \eta(\delta)$.

If, in addition, $\mu = 1$, i.e. if λ is a simple eigenvalue with normalised eigenvector ψ, then there exists a normalised eigenvector $\widetilde{\psi}$ of $\widetilde{H}^\theta$ such that

$$\|J\psi - \widetilde{\psi}\| \le C_{1,\vartheta}\delta \qquad \text{and} \qquad \|J'\widetilde{\psi} - \psi\| \le \widetilde{C}_{1,\vartheta}\delta.$$

Finally, η, $C_{1,\vartheta}$ and $\widetilde{C}_{1,\vartheta}$ depend only on ϑ, $\operatorname{Re}\theta$, $\|K\|$ *and* H^θ.

Proof. The theorem is a result of Theorem 4.9.8, Theorem 4.6.3 and Lemma 4.6.1. Again, the function $\eta(\delta)$ is the modulus of continuity of the continuity stated in Theorem 4.6.3. □

Remark 4.9.13. It would be interesting to see whether we can obtain a resolvent estimate around $\lambda \in \sigma_{\mathrm{disc}}(H^\theta)$ as in (4.45) from the abstract setting here. Such an estimate would allow us to say something about the rate of convergence of the resonances (see Theorem 4.6.4). For example, in [HiM91, Hi92] it was shown for a so-called *Helmholtz resonator* that each discrete eigenvalue of the compact part of the decoupled model is exponentially close to a resonance, i.e. the error is of the form $O(e^{-c/\varepsilon})$ for some constant $c > 0$. For a similar phenomena on graph-like spaces, see Remark 1.4.6.

Chapter 5
Manifolds, Tubular Neighbourhoods and Their Perturbations

In this chapter, we collect some general facts about Riemannian manifolds and tubular neighbourhoods of a one-dimensional space. Moreover, we provide convergence results for small perturbations of a metric on a manifold. In the subsequent Chap. 6, we mostly consider so-called "graph-like manifolds" with exact product structure for an edge neighbourhood and with exact scaling behaviour of the vertex neighbourhoods, simplifying the reduction to the metric graph significantly. The results in this chapter are then used to briefly discuss the more general case of a graph G embedded in $\mathbb{R}^2$ and a neighbourhood X_ε shrinking to G as $\varepsilon \to 0$ and other perturbation of the product structure.

Section 5.1 contains some basic notation for manifolds with or without boundary, manifolds constructed from building blocks and the definition of the associated Laplacian via its quadratic form. When dealing with graph-like manifolds shrinking to a metric graph (see Chap. 6), basically all convergence results for the identification operators and the Laplacians follow from a few fundamental inequalities, namely the min-max eigenvalue estimate (Propositions 5.1.1 and 5.1.2) and the Sobolev trace estimate (5.6). We finally collect some facts about the scaling of a metric $g_\varepsilon = \varepsilon^2 g$ on a manifold M using the convenient short-hand notation εM for the Riemannian manifold (M, g_ε).

In Sect. 5.2, we consider perturbations $\widetilde{g}$ of a metric g on a general (not necessarily compact) manifold, and show that the associated quadratic forms and Laplacians are δ-quasi-unitarily equivalent in Theorem 5.2.6 (see Chap. 4 for the concept and the implications such as spectral convergence). The error δ is expressed purely intrinsically in terms of $\widetilde{g}$ relative to g only. This result is of course well-known, but shows the strength of our two-Hilbert space approach of Chap. 4 and how natural it can be applied in this situation. Moreover, using quadratic forms only, we can avoid the use of derivatives of the perturbation $\widetilde{g}$ w.r.t. g.

Sections 5.3–5.6 contains material on tubular neighbourhoods of an interval. Such tubular neighbourhoods appear in a graph-like manifold as a neighbourhood of an edge. The considerations in this chapter allow us to consider later on only a pure product structure $X_{\varepsilon,e} = I_e \times \varepsilon Y_e$ as neighbourhood of an edge e with associated interval $I = I_e = [0, \ell_e]$ in the metric graph and transversal manifold

O. Post, *Spectral Analysis on Graph-Like Spaces*, Lecture Notes in Mathematics 2039, DOI 10.1007/978-3-642-23840-6_5,

$Y = Y_e$, and to treat other situations such as the ε-neighbourhood of an embedded graph in $\mathbb{R}^2$ as a perturbation of the product structure. In Sect. 5.3, we consider different aspects like general perturbations of the natural product structure on a tubular neighbourhood and the effect of a small shortening of the underlying base space, the interval. Section 5.4 treats the particular case of the interval I and the tubular neighbourhood being embedded in $\mathbb{R}^2$ as a curve and its neighbourhood. Finally, Sects. 5.5–5.6 treat the thin radius limit of quadratic forms associated with tubular neighbourhoods with transversal boundary (i.e., $\partial Y \neq \emptyset$) and Neumann resp. Dirichlet conditions on $I \times \partial Y$.

5.1 Manifolds

Let us first fix some notation. Denote by X a d-dimensional smooth manifold, with or without boundary. Let g be a Riemannian metric on X. We often refer to the Riemannian manifold (X, g) simply as X, if the metric is clear from the context. We assume that g is smooth as section into the bundle of positive sesquilinear forms on the tangent bundle TX. When constructing manifolds from building blocks, it is convenient to allow a weaker regularity, namely, that the coordinate change maps of X are only Lipschitz continuous, and that the metric is only continuous, see Sect. 5.1.2. For more details on Lipschitz manifolds see e.g. [KSh03] or [Tel83] and references therein.

The canonical Riemannian measure will be denoted by $\mathrm{d}X$ or $\mathrm{d}(X, g)$ or mostly by $\mathrm{d}x$ for the integration variable $x \in X$. When speaking about the volume of a measurable subset $U \subset X$, we also write $\operatorname{vol} U = \operatorname{vol}_d U$.

5.1.1 *Manifolds with Boundary*

If X has (total) boundary ∂X, we use the convention $X = \mathring{X} \,\dot\cup\, \partial X$, i.e. X is the disjoint union of the *interior* $\mathring{X}$ and the *boundary* ∂X. We assume that the boundary is (piecewise) smooth with metric induced by X. Moreover, we denote the corresponding Riemannian measure by $\mathrm{d}\partial X$ or simply by $\mathrm{d}y$ for $y \in \partial X$. For the volume we write $\operatorname{vol} \partial X = \operatorname{vol}_m \partial X$, where $m := d - 1$ is the dimension.

For a subset $U \subset X$, denote by

$$\mathring{\partial} U := \overline{U} \cap \overline{X \setminus U} \cap \mathring{X} \qquad \text{and} \qquad \partial U := \mathring{\partial} U \cup \overline{U} \cap \partial X$$

the *internal* resp. *total* boundary of U, where $\overline{(\cdot)}$ denotes the closure in the topological space X. Note that we can interpret $\mathring{\partial}$ as (topological) boundary operator in the topological space $\mathring{X}$ (e.g., $\mathring{\partial}\mathring{X} = \emptyset$), but ∂ is *not* the boundary operator of the topological space X, since in general $\partial X \neq \emptyset$.

When introducing boundary conditions or boundary triples, it is convenient to decompose the (total) boundary ∂X into two closed subsets, namely we assume that there are two subsets $\partial_0 X$ and $\partial_1 X$ of the boundary of ∂X such that

$$\partial X = \partial_0 X \,\overline{\cup}\, \partial_1 X$$

(i.e. the union is disjoint up to $(d-1)$-dimensional measure 0). We sometimes call $\partial_0 X$ and $\partial_1 X$ the *transversal* resp. *longitudinal* boundary of X, referring to the fact that we fix boundary conditions on the transversal boundary, and use the longitudinal boundary for boundary triples in order to couple different spaces via the longitudinal boundary in Chap. 7.

Having chosen such a decomposition, we denote by

$$\partial_0 U := \overline{U} \cap \partial_0 X \qquad \text{and} \qquad \partial_1 U := \overline{U} \cap \partial_1 X$$

the *(transversal and longitudinal) boundary* of $U \subset X$, i.e. the boundary part of U belonging to the boundary component $\partial_0 X$ and $\partial_1 X$, respectively. In particular, we have

$$\partial U = \mathring{\partial} U \,\overline{\cup}\, \partial_0 U \,\overline{\cup}\, \partial_1 U$$

(disjoint union up to $m = (d-1)$-dimensional measure 0), see Fig. 5.1.

5.1.2 *Manifolds Constructed from Building Blocks*

Usually, we consider manifolds of class C^∞, but sometimes, we allow the manifold or its boundary to be only Lipschitz, i.e. the coordinate change maps are Lipschitz continuous. As for the manifold, we sometimes allow a weaker regularity for the Riemannian metric. The minimal regularity of the metric (viewed as section in the

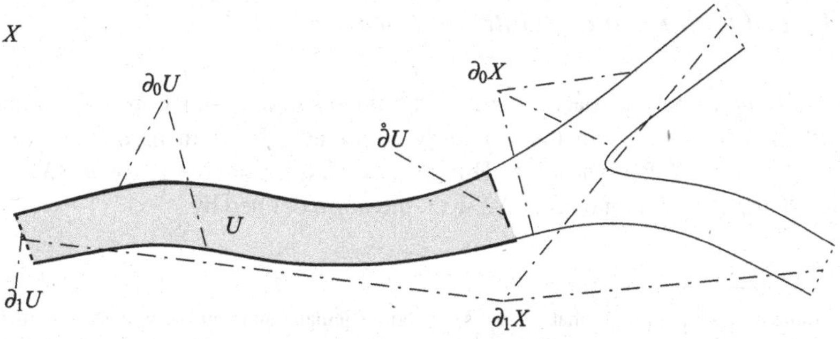

Fig. 5.1 A typical example we have in mind for the different boundary components of a manifold X, here embedded in $\mathbb{R}^2$. Note that $X \subset \mathbb{R}^2$ is closed, $\partial X = \partial_0 X \cup \partial_1 X$ is the total boundary (being here the boundary of X as subset in $\mathbb{R}^2$). The total boundary ∂U consists of the internal boundary $\mathring{\partial} U$ and the transversal and longitudinal boundary parts $\partial_0 U$ and $\partial_1 U$ of ∂X, respectively

bundle of sesquilinear forms on TX) is continuity. A particular example of a non-smooth manifold may arise from the following construction of a manifold X by a number of components X_i: For each index i, let X_i be a d-dimensional manifold with Lipschitz boundary ∂X_i. For each pair (i, j) with $i \neq j$ we identify a *smooth*, relatively open subset $\partial_j X_i$ of ∂X_i with a *smooth*, relatively open subset $\partial_i X_j$ of ∂X_j (the subsets may be empty for certain pairs (i, j)). Let us denote the identified parts by $\partial X_i \cap \partial X_j \cong \partial_j X_i \cong \partial_i X_j$. If the manifolds X_i and X_j carry a metric, we assume that the identification is an *isometry*. We denote the resulting space by

$$X = \overline{\bigcup_i} X_i$$

as short hand for the above construction, without explicitly referring to the underlying identifications. We remark that the notion is consistent with the notion "disjoint up to measure 0", denoted by the same symbol. Moreover, the property of a subset $S \subset X$ to have measure 0 w.r.t. the canonical volume measure in the Riemannian manifold (X, g) is independent of the metric g. The total manifold X has boundary

$$\partial X = \bigcup_i \partial X_i \setminus \bigcup_{i \neq j} (\partial X_i \cap \partial X_j),$$

which we *assume* to be Lipschitz. We may define Lipschitz continuous coordinate charts near $\partial X_i \cap \partial X_j$ via charts from X_i and X_j near the *smooth* boundary parts $\partial_j X_i$ and $\partial_i X_j$. But since we deal with L_2-spaces only, it is enough to cover the manifold with charts up to sets of measure 0. In particular, charts for X_i are enough, since $\partial X_i \cap \partial X_j$ has d-dimensional measure 0 in X. The identification of the boundary parts $\partial_j X_i$ and $\partial_i X_j$ usually arises naturally from the construction (if, e.g., the total manifold X is a priori given as a subset of $\mathbb{R}^\nu$). Otherwise, the identification is tacitly included in the notation $X = \dot{\bigcup}_i X_i$ (see Fig. 5.2).

5.1.3 *Laplacians and Quadratic Forms*

Let $\mathsf{C}_c^\infty(X, \partial_0 X)$ be the space of complex-valued smooth[1] function with compact support in X *disjoint* from the boundary component $\partial_0 X$. In particular, we set $\mathsf{C}_c^\infty(X) := \mathsf{C}_c^\infty(X, \emptyset)$. The basic Hilbert space $\mathsf{L}_2(X) = \mathsf{L}_2(X, g)$ on (X, g) is defined as the completion of $\mathsf{C}_c^\infty(X)$ w.r.t. the norm defined by

[1] We assume here for simplicity that (X, g) is a smooth Riemannian manifold with smooth metric. For Lipschitz manifolds or metrics which are only continuous, one has to replace the attribute "smooth" by "Lipschitz" for the quadratic form domain. In this case, functions in the domain of the corresponding operator may no longer be twice weakly differentiable. Nevertheless, by Theorem 3.1.1, there always exists an associated operator, and it is straightforward to calculate the corresponding domain in the non-smooth case.

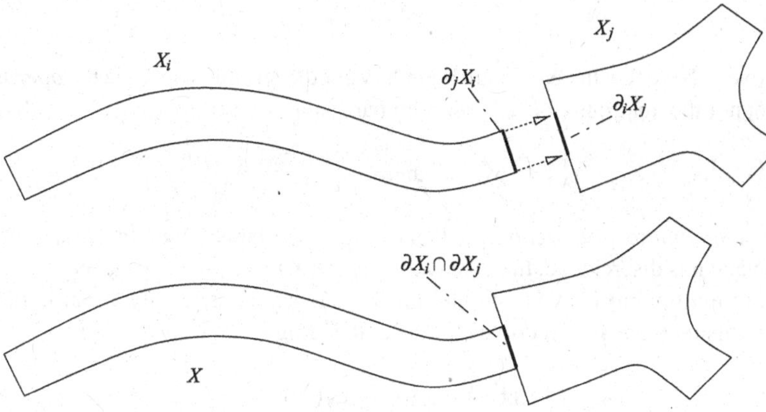

Fig. 5.2 The construction of a manifold from two building blocks. The boundary of the resulting manifold X may have additional "corners"

$$\|u\|_X^2 = \|u\|^2_{\mathsf{L}_2(X)} := \int_X |u|^2 \mathrm{d}X. \tag{5.1}$$

We often use other intuitive notation like $\|u\|_X = \|u\|_g$ etc. The Sobolev space $\mathsf{H}^1(X, \partial_0 X) = \mathsf{H}^1(X, \partial_0 X, g)$ of order 1 is defined as the completion of $\mathsf{C}_c^\infty(X, \partial_0 X)$ w.r.t. the norm given by

$$\|u\|^2_{\mathsf{H}^1(X)} := \|u\|_X^2 + \|\mathrm{d}u\|_X^2 = \int_X \left(|u|^2 + |\mathrm{d}u|_g^2\right)\mathrm{d}X, \tag{5.2}$$

where $\mathrm{d}u$ is the exterior derivative on X and $|\cdot|_g$ is the norm defined via the canonical metric on T^*X. Namely, in local coordinates, we have

$$|\mathrm{d}u|_g^2 = \sum_{i,j=1}^{d} g^{ij}\, \partial_i \overline{u}\, \partial_j u,$$

where $\{g^{ij}\}_{ij}$ is the inverse of the matrix $\{g_{ij}\}_{ij} = \{g(\partial_i, \partial_j)\}_{ij}$. In particular, we set

$$\mathsf{H}^1(X) := \mathsf{H}^1(X, \emptyset) \qquad \text{and} \qquad \mathring{\mathsf{H}}^1(X) := \mathsf{H}^1(X, \partial X).$$

The *Laplacian* $\Delta_X^{\partial_0 X} = \Delta_{(X,g)}^{\partial_0 X}$ on (X, g) is defined via its quadratic form $\mathfrak{d}^{\partial_0 X}$, namely

$$\mathfrak{d}^{\partial_0 X}(u) = \|\mathrm{d}u\|_X^2 \qquad \text{and} \qquad \operatorname{dom} \mathfrak{d}^{\partial_0 X} = \mathsf{H}^1(X, \partial_0 X)$$

(see Theorem 3.1.1). Using Green's formula, one can see that functions u in the domain of $\Delta_X^{\partial_0 X}$ fulfil Dirichlet boundary conditions on $\partial_0 X$ and Neumann boundary conditions on $\partial_1 X$, i.e.

$$u \restriction_{\partial_0 X} = 0 \qquad \text{and} \qquad \partial_n u \restriction_{\partial_1 X} = 0,$$

respectively. Note that the Neumann boundary condition enters only via the *operator* domain, not the quadratic form domain. In particular, we set

$$\Delta_X := \Delta_X^{\emptyset} \qquad \text{and} \qquad \Delta_X^{\mathrm{D}} := \Delta_X^{\partial X}$$

for the *Neumann* resp. *Dirichlet* Laplacian. If X is compact, then the spectrum of the Laplacian is discrete, i.e. the eigenvalues $\lambda_k^{\partial_0 X}(X) = \lambda_k^{\partial_0 X}(X, g)$ (repeated with respect to multiplicity) have finite multiplicity and accumulate only at ∞. In order to show this fact, one has to prove that the embedding

$$\mathsf{H}^1(X, \partial_0 X) \hookrightarrow \mathsf{L}_2(X)$$

is compact (see e.g. [Ros97, Thm. 1.22]). This follows from the Rellich-Kondarachov compactness theorem for the form domain $\mathring{\mathsf{H}}^1(X)$ of the Dirichlet Laplacian. For the Neumann Laplacian (or mixed boundary conditions), we need additionally a (local) Sobolev extension operator. Such an operator exists in our setting, since we assumed that the boundary ∂X is Lipschitz. We will occasionally omit the label $(\cdot)^{\partial_0 X}$ indicating where we put the Dirichlet boundary condition, if it is clear from the context.

5.1.4 Basic Estimates

Let M be a compact manifold. We define the *averaging operator* $\fint_M$ by

$$\fint_M u := \frac{1}{\operatorname{vol} M} \int_M u \, \mathrm{d}M \tag{5.3}$$

for $u \in \mathsf{L}_2(M)$, where $\operatorname{vol} M = \operatorname{vol}_d M$ denotes the volume of the manifold M. In particular, we have the following result, a consequence of the Min-max principle Theorem 3.2.3:

Proposition 5.1.1. *We have*

$$\|u - \fint_M u\|_M^2 \leq \frac{1}{\lambda_2(M)} \|\mathrm{d}u\|_M^2$$

for $u \in \mathsf{H}^1(M)$, *where* $\lambda_2(M)$ *denotes the second (first non-zero)* Neumann *eigenvalue of the Laplacian on* M, *i.e. the Laplacian associated with the form* $\mathfrak{d}_M(u) := \|\mathrm{d}u\|_M^2$ *with domain* $\operatorname{dom} \mathfrak{d}_M := \mathsf{H}^1(M)$.

Proof. Apply the min-max characterisation and note that $\fint_M u$ is the projection of u onto the first (constant) eigenfunction. □

Assume that M has boundary $\partial M = \partial_0 M \,\dot\cup\, \partial_1 M$. We impose Dirichlet boundary conditions on $\partial_0 M$ and denote by φ_1 the first eigenfunction associated with the Laplacian $\Delta_M^{\partial_0 M}$ with Dirichlet conditions on $\partial_0 M$ (and Neumann conditions on $\partial_1 M$). Denote by $\lambda_k = \lambda_k^{\partial_0 M}(M)$ the corresponding sequence of eigenvalues.

Proposition 5.1.2. *Assume that M is connected and compact, then we have*

$$\left\| u - \langle \varphi_1, u\rangle \varphi_1 \right\|_M^2 \le \frac{1}{\lambda_2 - \lambda_1}\left(\|du\|_M^2 - \lambda_1 \|u\|_M^2\right) \tag{5.4a}$$

for $u \in \mathsf{H}^1(M, \partial_0 M)$. *In particular, if* $\partial_0 M \neq \emptyset$, *we have* $\lambda_1 > 0$ *and*

$$\|u\|_M^2 \le \frac{1}{\lambda_1}\|du\|_M^2 \tag{5.4b}$$

for $u \in \mathsf{H}^1(M, \partial_0 M)$.

Proof. For the first estimate, we start with the observation

$$\left\| u - \langle \varphi_1, u\rangle \varphi_1 \right\|^2 = \|u\|^2 - \left|\langle \varphi_1, u\rangle\right|^2, \tag{5.5a}$$

$$\left\| du - \langle \varphi_1, u\rangle d\varphi_1 \right\|^2 = \|du\|^2 - \lambda_1\left|\langle \varphi_1, u\rangle\right|^2 \tag{5.5b}$$

using $\langle du, d\varphi_1\rangle = \langle u, d^* d\varphi_1\rangle = \lambda_1 \langle u, \varphi_1\rangle$ in the second equation. Now,

$$\begin{aligned}\left\| u - \langle \varphi_1, u\rangle \varphi_1 \right\|^2 &\le \frac{1}{\lambda_2}\left(\|du\|^2 - \lambda_1\left|\langle \varphi_1, u\rangle\right|^2\right)\\ &= \frac{1}{\lambda_2}\left(\|du\|^2 - \lambda_1\|u\|^2\right) + \frac{\lambda_1}{\lambda_2}\left(\|u\|^2 - \left|\langle \varphi_1, u\rangle\right|^2\right)\\ &= \frac{1}{\lambda_2}\left(\|du\|^2 - \lambda_1\|u\|^2\right) + \frac{\lambda_1}{\lambda_2}\left\| u - \langle \varphi_1, u\rangle \varphi_1 \right\|^2\end{aligned}$$

using the min-max characterisation and the equalities (5.5). Since M is connected, $\lambda_1 < \lambda_2$ and we can bring the last term on the LHS. Dividing by $(1 - \lambda_1/\lambda_2) > 0$ yields the result. The second estimate is a direct consequence of the min-max characterisation and the fact that $\lambda_1 > 0$: Otherwise, φ_1 would be constant, and $\varphi_1 \restriction_{\partial_0 M} = 0$ would imply $\varphi_1 = 0$. □

Note that Proposition 5.1.1 is a special case of (5.4a), where $\partial_0 M = \emptyset$ and therefore $\lambda_1 = 0$ and φ_1 is constant.

The next proposition compares different average operators. As assumption, we need a Sobolev trace estimate, usually fulfilled in our applications.

Proposition 5.1.3. *Assume that M is a d-dimensional manifold and that* $\{M_i\}_i$ *consists of subsets of M with* $\operatorname{vol}_d M_i > 0$. *Moreover, we assume that* $\dot\bigcup_i M_i \subset M$ *(i.e.* M_i *are mutually disjoint up to measure 0). In addition, let* $\Gamma = \dot\bigcup_i \Gamma_i \subset \partial M$ *where* $\operatorname{vol}_m(\partial \Gamma_i) > 0$ *and* $\Gamma_i \subset \partial M_i$ *(see Fig. 5.3) such that there exists* $a_i > 0$ *with*

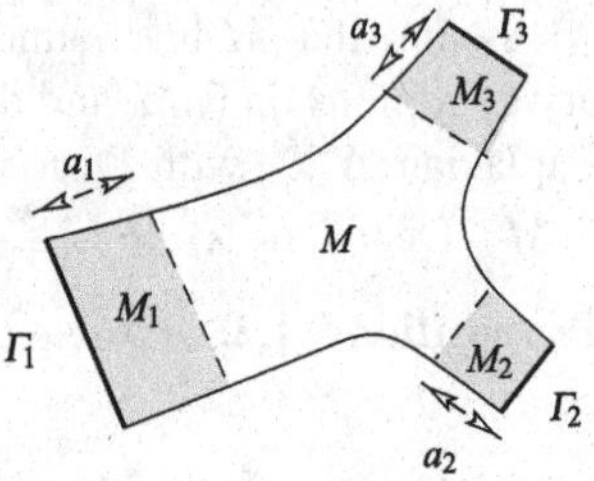

Fig. 5.3 A typical situation in Proposition 5.1.3

$$\|u\|^2_{\Gamma_i} \le a\|du\|^2_{M_i} + \frac{2}{a}\|u\|^2_{M_i} \tag{5.6}$$

for all $0 < a \le a_i$ *and all* i. *Then we have*

$$\begin{aligned}|f_M u - f_\Gamma u|^2 &\le \frac{1}{\mathrm{vol}_m \Gamma}\sum_i \mathrm{vol}_m \Gamma_i |f_M u - f_{\Gamma_i} u|^2\\ &\le \frac{1}{\mathrm{vol}_m \Gamma}\Big(a + \frac{2}{a\lambda_2(M)}\Big)\|du\|^2_M\end{aligned}$$

for all $0 < a \le a_- := \inf_i a_i$.

Proof. The proof follows from

$$\begin{aligned}|f_M u - f_\Gamma u|^2 &= \Big|\frac{1}{\mathrm{vol}\,\Gamma}\sum_i \mathrm{vol}\,\Gamma_i (f_M u - f_{\Gamma_i} u)\Big|^2\\ &\stackrel{\mathrm{CS}}{\le} \frac{1}{\mathrm{vol}\,\Gamma}\sum_i \mathrm{vol}\,\Gamma_i |f_M u - f_{\Gamma_i} u|^2\\ &\stackrel{\mathrm{CS}}{\le} \frac{1}{\mathrm{vol}\,\Gamma}\sum_i \|f_M u - u\|^2_{\Gamma_i}\\ &\le \frac{1}{\mathrm{vol}\,\Gamma}\sum_i \Big(a\|du\|^2_{M_i} + \frac{2}{a}\|f_M u - u\|^2_{M_i}\Big)\\ &\le \frac{1}{\mathrm{vol}\,\Gamma}\Big(a\|du\|^2_M + \frac{2}{a}\|f_M u - u\|^2_M\Big)\end{aligned}$$

noting that $df_M u = 0$. The estimate follows now from Proposition 5.1.1. □

5.1.5 Some Scaling Behaviour

Let M be a d-dimensional Riemannian manifold with metric g and let $\varepsilon > 0$. Denote the volume form by dM. Moreover, denote by εM the Riemannian manifold with the same underlying space, but with metric $g_\varepsilon := \varepsilon^2 g$ (the *ε-homothetic version of M*). We have the following scaling behaviour:

$$\mathrm{d}(\varepsilon M) = \varepsilon^d \mathrm{d}M, \qquad \|u\|^2_{\varepsilon M} = \varepsilon^d \|u\|^2_M, \tag{5.7a}$$

$$|\mathrm{d}u|^2_{g_\varepsilon} = \frac{1}{\varepsilon^2}|\mathrm{d}u|^2_g, \qquad \|\mathrm{d}u\|^2_{\varepsilon M} = \varepsilon^{d-2}\|du\|^2_M, \tag{5.7b}$$

$$\Delta_{\varepsilon M} = \frac{1}{\varepsilon^2}\Delta_M, \qquad f_{\varepsilon M} u = f_M u \tag{5.7c}$$

$$\sigma(\Delta_{\varepsilon M}) = \frac{1}{\varepsilon^2}\sigma(\Delta_M), \qquad \lambda_k(\varepsilon M) = \frac{1}{\varepsilon^2}\lambda_k(M) \tag{5.7d}$$

Remark 5.1.4. Note that by definition, the scaled spaces $Y_{\varepsilon,e} = \varepsilon Y_e$, $X_{\varepsilon,v} = \varepsilon^\alpha X_v$ etc. are independent of ε as manifolds, the parameter $\varepsilon > 0$ only enters through their metric. Moreover, the various norms of the L_2- and Sobolev spaces depend on ε, but as vector spaces, the scaled and unscaled versions agree, e.g., $\mathsf{L}_2(Y_{\varepsilon,e}) = \mathsf{L}_2(Y_e)$.

If $\Gamma \subset M$ is of dimension $m = d - 1$, then the Sobolev trace estimate (5.6) scales as

$$\|u\|^2_{\varepsilon\Gamma} \le a_\varepsilon \|\mathrm{d}u\|^2_{\varepsilon M} + \frac{2}{a_\varepsilon}\|u\|^2_{\varepsilon M} \tag{5.8}$$

where $a_\varepsilon = \varepsilon a$, i.e. a_ε scales like a *length*.

Let us introduce another intuitive notation: Let M be a differentiable (or Lipschitz) manifold with two metrics g and $\widetilde{g}$. Abusing slightly the notation, we write M and $\widetilde{M}$ for the Riemannian manifolds (M, g) and $(M, \widetilde{g})$, respectively. If $0 < r_1 \le r_2$ we use the notation

$$r_1 M \le \widetilde{M} \le r_2 M \tag{5.9}$$

as shortcut for

$$r_1^2 g_x(v, v) \le \widetilde{g}_x(v, v) \le r_2^2 g_x(v, v)$$

for all $v \in T_x M$, $x \in M$. In particular, if M_ε denotes the manifold M with metric g_ε for $0 < \varepsilon \le 1$ and $M := M_1$, then the inequalities

$$\varepsilon^{\alpha_2} M \le M_\varepsilon \le \varepsilon^{\alpha_1} M, \qquad 0 < \alpha_1 \le \alpha_2 \tag{5.10}$$

imply

$$\varepsilon^{\alpha_2 d}\|u\|^2_M \le \|u\|^2_{M_\varepsilon} \le \varepsilon^{\alpha_1 d}\|u\|^2_M, \tag{5.11a}$$

$$\varepsilon^{\alpha_2 d - 2\alpha_1}\|\mathrm{d}u\|^2_M \le \|\mathrm{d}u\|^2_{M_\varepsilon} \le \varepsilon^{\alpha_1 d - 2\alpha_2}\|\mathrm{d}u\|^2_M, \tag{5.11b}$$

$$\varepsilon^{\alpha_2 d}\,\mathrm{vol}\, M \le \mathrm{vol}\, M_\varepsilon \le \varepsilon^{\alpha_1 d}\,\mathrm{vol}\, M, \tag{5.11c}$$

$$\varepsilon^{d(\alpha_2-\alpha_1)-2\alpha_1}\lambda_2(M) \le \lambda_2(M_\varepsilon) \le \varepsilon^{-d(\alpha_2-\alpha_1)-2\alpha_2}\lambda_2(M). \tag{5.11d}$$

Let us reformulate the above Propositions 5.1.1, 5.1.2 and 5.1.3 in the case of a scaled manifold:

Corollary 5.1.5. *For the scaled manifold εM we have*

$$\|u - f_{\varepsilon M} u\|^2_{\varepsilon M} \le \frac{\varepsilon^2}{\lambda_2(M)} \|du\|^2_{\varepsilon M}.$$

Corollary 5.1.6. *For the scaled d-dimensional manifold εM and Dirichlet boundary conditions on $\partial_0 M$, we have*

$$\|u - \langle \varphi_1, u\rangle_M \varphi_1\|^2_{\varepsilon M} \le \frac{\varepsilon^2}{\lambda_2^{\partial_0 M}(M) - \lambda_1^{\partial_0 M}(M)} \Big(\|du\|^2_{\varepsilon M} - \frac{\lambda_1}{\varepsilon^2} \|u\|^2_{\varepsilon M} \Big),$$

$$\|u\|^2_{\varepsilon M} \le \frac{\varepsilon^2}{\lambda_1^{\partial_0 M}(M)} \|du\|^2_{\varepsilon M}$$

for $u \in \mathsf{H}^1(\varepsilon M, \partial_0 M)$, provided that $\partial_0 M \neq \emptyset$ in the latter case.

Corollary 5.1.7. *Assume that $\overline{\dot{\bigcup}}_i M_i \subset M$ and that $\Gamma = \overline{\dot{\bigcup}}_i \Gamma_i \subset \partial M$ where $\operatorname{vol}_d(M_i) > 0$, $\operatorname{vol}_m(\partial \Gamma_i) > 0$ and $\Gamma_i \subset \partial M_i$ such that (5.6) is fulfilled for $a_i > 0$ and all i or that the scaled version (5.8) is fulfilled for Γ_i and M_i with $a_\varepsilon = \varepsilon a_i$. Then we have*

$$\begin{aligned} \big|f_M u - f_\Gamma u\big|^2 &\le \frac{1}{\operatorname{vol}_m \Gamma} \sum_i \operatorname{vol}_m \Gamma_i \big|f_M u - f_{\Gamma_i} u\big|^2 \\ &\le \frac{\varepsilon^2}{\operatorname{vol}_m \Gamma} \Big(a + \frac{2}{a \lambda_2(M)} \Big) \|du\|^2_{\varepsilon M} \end{aligned}$$

for all $0 < a \le a_- := \inf_i a_i$.

5.2 Perturbations of the Metric

Let $g, \widetilde{g}$ be two metrics on the d-dimensional manifold X. The manifold is not necessarily assumed to be compact. Abusing the notation slightly, we also write X for the Riemannian manifold (X, g) and $\widetilde{X}$ for the Riemannian manifold $(X, \widetilde{g})$. We think of $\widetilde{g}$ as being a small perturbation of the metric g. In order to define the perturbation intrinsically, we introduce a section A into the bundle $\mathscr{L}(TX)$ with fibres consisting of endomorphisms of $T_x X$.

Definition 5.2.1. We call the positive linear operator A_x on $T_x X$, defined via

$$\widetilde{g}_x(v, v) = g_x(v, A_x v) \qquad \forall\, v \in T_x X, \quad x \in X, \tag{5.12}$$

the *relative distortion* of $\widetilde{g}$ w.r.t. g. Moreover, we denote by

$$\rho(x) := \det A_x^{1/2}$$

the *relative density* of $\widetilde{g}$ w.r.t g.

In a coordinate chart, A_x can be expressed as a $d \times d$-matrix. Moreover, $x \mapsto A_x$ has the same regularity as $x \mapsto g_x$ and $x \mapsto \widetilde{g}_x$ (e.g. smooth or Lipschitz).

The Riemannian measures are related via

$$\mathrm{d}\widetilde{X} = \rho \mathrm{d}X.$$

justifying the name "relative density" for ρ. Similarly, for the cotangent bundle, we define a section A^* in $\mathscr{L}(T^*X)$ relating the canonical metrics g^* and $\widetilde{g}^*$ on T^*X corresponding to g and $\widetilde{g}$, respectively, via

$$\widetilde{g}_x^*(\xi,\xi) = g_x^*(A_x^*)\xi,\xi) \qquad \forall\, \xi \in T_x^*X, \quad x \in X. \tag{5.13}$$

Note that

$$|A_x^*|_{\mathscr{L}(T_x^*X,g_x^*)} = |A_x|^{-1}_{\mathscr{L}(T_xX,g_x)}, \tag{5.14}$$

where

$$|A_x|_{\mathscr{L}(T_xX,g_x)} := \sup_{v\in T_xX\setminus\{0\}} \frac{g_x(v, A_x v)}{g_x(v,v)}$$

is the canonical operator norm in (T_xX, g_x) and similarly we define the norm of A_x^* as operator in (T_x^*X, g_x^*). By a slight abuse of notation, we also write A_x^{-1} for the section A_x^*, having the relation (5.14) in mind. Moreover, we often use the same symbol g for the metric on TX *and* on T^*X (or any other derived tensor bundle).

We first provide conditions assuring that $\mathsf{L}_2(X, g)$ and $\mathsf{H}^1(X, g)$ are independent of the metric as vector space and have equivalent norms. We set

$$\|\rho\|_\infty := \sup_x |\rho(x)| \qquad \text{and} \qquad \|A\|_\infty := \sup_x |A_x|_{\mathscr{L}(T_xX,g_x)}$$

using the pointwise operator norm in the second case. Let us make some comments on the norm estimates of A, ρ and related expressions: Denote by

$$0 < \alpha_1(x) \le \alpha_2(x) \le \cdots \le \alpha_d(x)$$

the d eigenvalues of A_x, ordered and repeated with respect to multiplicity. Then

$$\|A\|_\infty = \|\alpha_d\|_\infty, \qquad \|\rho\|_\infty^2 = \|\alpha_1 \cdot \ldots \cdot \alpha_d\|_\infty \le \|\alpha_d\|_\infty^d$$

$$\|A^{-1}\|_\infty = \|\alpha_1^{-1}\|_\infty, \qquad \|\rho^{-1}\|_\infty^2 = \|\alpha_1^{-1} \cdot \ldots \cdot \alpha_d^{-1}\|_\infty \le \|\alpha_1^{-1}\|_\infty^d.$$

Moreover,

$$\|\rho A^{-1}\|_\infty^2 = \|\alpha_1^{-1}\alpha_2 \cdot \ldots \cdot \alpha_d\|_\infty \le \|\alpha_1^{-1}\|_\infty \|\alpha_d\|_\infty^{d-1},$$
$$\|\rho^{-1} A\|_\infty^2 = \|\alpha_1^{-1} \cdot \ldots \cdot \alpha_{d-1}^{-1}\alpha_d\|_\infty \le \|\alpha_1^{-1}\|_\infty^{d-1} \|\alpha_d\|_\infty.$$

In particular, $\|A\|_\infty < \infty$ implies $\|\rho\|_\infty < \infty$ and $\|A^{-1}\|_\infty < \infty$ implies $\|\rho^{-1}\|_\infty < \infty$.

Proposition 5.2.2. *Assume that $\|A^{\pm 1}\|_\infty < \infty$, then we have*

$$\frac{1}{\|\rho^{-1}\|_\infty} \|u\|^2_{\mathsf{L}_2(X,g)} \le \|u\|^2_{\mathsf{L}_2(X,\widetilde{g})} \le \|\rho\|_\infty \|u\|^2_{\mathsf{L}_2(X,g)}, \tag{5.15a}$$

$$\frac{1}{\|\rho^{-1} A\|_\infty} \|\mathrm{d}u\|^2_{\mathsf{L}_2(X,g)} \le \|\mathrm{d}u\|^2_{\mathsf{L}_2(X,\widetilde{g})} \le \|\rho A^{-1}\|_\infty \|u\|^2_{\mathsf{L}_2(X,g)}. \tag{5.15b}$$

In particular, the spaces $\mathsf{L}_2(X,g)$ and $\mathsf{L}_2(X,\widetilde{g})$ resp. $\mathsf{H}^1(X,g)$ and $\mathsf{H}^1(X,\widetilde{g})$ agree as vector spaces and have equivalent norms.

Proof. The proof of the second estimate follows from

$$\|\mathrm{d}u\|^2_{\mathsf{L}_2(X,\widetilde{g})} = \int_X \langle \rho A^{-1}\mathrm{d}u, \mathrm{d}u\rangle_g \mathrm{d}(X,g)$$

and similarly for the first estimate. □

For the notation $r_- X \le \widetilde{X} \le r_+ X$ we refer to (5.9). As a corollary we obtain:

Corollary 5.2.3. *Assume that $r_- X \le \widetilde{X} \le r_+ X$. Then*

$$r_-^d \|u\|^2_{\mathsf{L}_2(X,g)} \le \|u\|^2_{\mathsf{L}_2(X,\widetilde{g})} \le r_+^d \|u\|^2_{\mathsf{L}_2(X,g)}, \tag{5.16a}$$

$$r_-^{d-1} r_+^{-1} \|\mathrm{d}u\|^2_{\mathsf{L}_2(X,g)} \le \|\mathrm{d}u\|^2_{\mathsf{L}_2(X,\widetilde{g})} \le r_+^{d-1} r_-^{-1} \|u\|^2_{\mathsf{L}_2(X,g)}. \tag{5.16b}$$

Proof. Note that the estimate on the manifold, i.e. $r_-^2 g_x \le \widetilde{g}_x \le r_+^2 g_x$ in the sense of quadratic forms, implies that the relative distortion A_x of $\widetilde{g}$ w.r.t. g fulfils

$$r_-^2 \mathbb{1}_x \le A_x \le r_+^2 \mathbb{1}_x$$

where $\mathbb{1}_x$ denotes the identity operator on $T_x X$. □

On a compact manifold, *continuity* of A (and therefore of ρ) is sufficient in order to assure that the above spaces agree and have equivalent norms. Moreover we have the following result:

Proposition 5.2.4. *Assume that X is a compact manifold with continuous metrics g, $\widetilde{g}$. Denote by $\lambda_k(X,g)$ and $\lambda_k(X,\widetilde{g})$ the k-th eigenvalue of the Laplacian associated with (X,g) and $(X,\widetilde{g})$, respectively. Then*

$$\begin{aligned} \left|\lambda_k(X,\widetilde{g}) - \lambda_k(X,g)\right| &\le \|\rho^{-1}\|_\infty\big(\|\rho A^{-1} - \mathbb{1}\|_\infty + \|\mathbb{1} - \rho\|_\infty\big)\lambda_k(X,g) \\ &\le 8\delta\lambda_k(X,g), \end{aligned} \tag{5.17}$$

provided

$$\delta := \max\big\{\|\mathbb{1} - \rho\|_\infty, \|\mathbb{1} - A\|_\infty\big\} \le \frac{1}{2}$$

in the latter estimate.

Proof. Note first that due to the compactness of X, the spectra of the Laplace operators with respect to g and $\widetilde{g}$ are purely discrete. Moreover, the estimates of Proposition 5.2.2 hold with strictly positive and finite constants. In particular, $\mathsf{H}^1(X,g) = \mathsf{H}^1(X,\widetilde{g})$. Comparing the *Rayleigh quotients* $R(u,g) := \|\mathrm{d}u\|^2_{(X,g)}/\|u\|^2_{(X,g)}$ for g and $\widetilde{g}$ yield

$$\begin{aligned} \left|R(u,\widetilde{g}) - R(u,g)\right| &= \left|\frac{\langle \rho A^{-1}\mathrm{d}u, \mathrm{d}u\rangle_X}{\langle \rho u, u\rangle_X} - \frac{\langle \mathrm{d}u, \mathrm{d}u\rangle_X}{\langle u, u\rangle_X}\right| \\ &= \left|\frac{\langle(\rho A^{-1} - \mathbb{1})\mathrm{d}u, \mathrm{d}u\rangle_X}{\langle \rho u, u\rangle_X} + R(u,g)\frac{\langle(\mathbb{1} - \rho)u, u\rangle_X}{\langle \rho u, u\rangle_X}\right| \\ &\le \underbrace{\|\rho^{-1}\|_\infty\Big(\|\rho A^{-1} - \mathbb{1}\|_\infty + \|\mathbb{1} - \rho\|_\infty\Big)}_{=:\eta} R(u,g). \end{aligned}$$

Applying the min-max characterisation Theorem 3.2.3 to the inequality

$$(1 - \eta)R(u,g) \le R(u,\widetilde{g}) \le (1 + \eta)R(u,g)$$

gives the corresponding inequality for the eigenvalues. Finally, it is easy to see that $\eta \le 8\delta$ if $\delta \le 1/2$. □

The eigenvalue estimate can be seen similarly using the more abstract setting of Corollary 4.4.19. The above lemma shows that it is enough to consider only *continuous* metrics, since no derivative of g is needed in the estimates.

We now want to compare the two resolvents associated with the Laplacians on (X,g) and $(X,\widetilde{g})$ using the abstract theory developed in Sect. 4.4. Let $\mathscr{H} := \mathsf{L}_2(X,g)$ and let $\mathscr{H}^k$ be the scale of Hilbert spaces associated with the Laplacian $\Delta_{(X,g)}$, see Sect. 3.2. Note that $\mathscr{H}^1 = \mathsf{H}^1(X,g)$. We denote by $\widetilde{\mathscr{H}}^k$ the corresponding spaces for the metric $\widetilde{g}$. Denote by $\mathfrak{d}$ the quadratic form $\mathfrak{d}(f) := \|\mathrm{d}f\|^2_X$ with domain $\operatorname{dom}\mathfrak{d} := \mathscr{H}^1$ and similarly, $\widetilde{\mathfrak{d}}(u) := \|\mathrm{d}u\|^2_{\widetilde{X}}$ with $\operatorname{dom}\widetilde{\mathfrak{d}} := \widetilde{\mathscr{H}}^1$.

Let $J\colon \mathscr{H} \longrightarrow \widetilde{\mathscr{H}}$, $Jf = f$ and $J'\colon \widetilde{\mathscr{H}} \longrightarrow \mathscr{H}$, $J'u = u$. The next lemma follows immediately from the fact that $\mathrm{d}\widetilde{X} = \rho\mathrm{d}X$:

Lemma 5.2.5. *The adjoint of J is given by $J^*\colon \widetilde{\mathscr{H}} \longrightarrow \mathscr{H}$ with $J^*u = \rho u$.*

We are now able to compare the quadratic forms, from which the resolvent estimates follow (see Corollary 5.2.7). For the notion of quasi-unitary equivalence of quadratic forms, see Definition 4.4.11:

Theorem 5.2.6. *Assume that* $\|A^{\pm 1}\|_\infty < \infty$, *then* $J(\mathscr{H}^1) = \widetilde{\mathscr{H}}^1$ *and* $J'(\widetilde{\mathscr{H}}^1) = \mathscr{H}^1$. *Moreover, the quadratic forms* $\mathfrak{d}$ *and* $\widetilde{\mathfrak{d}}$ *are* δ*-quasi-unitarily equivalent with*

$$\delta := \max\bigl\{\|\rho^{1/2} - \rho^{-1/2}\|_\infty, \|\rho^{-1/2}A^{1/2} - \rho^{1/2}A^{-1/2}\|_\infty\bigr\}.$$

If in addition, $\hat{\delta} := \max\{\|\mathbb{1} - \rho\|_\infty, \|\mathbb{1} - A\|_\infty\} \leq 1/2$ *then the forms* $\mathfrak{d}$ *and* $\widetilde{\mathfrak{d}}$ *are* $4\hat{\delta}$*-quasi-unitarily equivalent with identification operators*

$$J : \mathsf{L}_2(X, g) \longrightarrow \mathsf{L}_2(X, \widetilde{g}), \quad f \mapsto f, \quad \text{and} \quad J' : \mathsf{L}_2(X, g) \longrightarrow \mathsf{L}_2(X, \widetilde{g}), \quad u \mapsto u. \tag{5.18}$$

In particular, the results of Sect. 4.4 apply.

Proof. The first assertion follows from Proposition 5.2.2. Moreover, the identification operators J and J' already respect the spaces of order 1, $J'J = \mathrm{id}$ and $JJ' = \widetilde{\mathrm{id}}$. In addition, $\|Jf\|^2_{\widetilde{X}} \leq \|\rho\|_\infty \|f\|^2_X$ and similarly $\|J'u\|^2_X \leq \|\rho^{-1}\|_\infty \|u\|^2_{\widetilde{X}}$ and we have

$$\bigl|\|\rho^{\pm 1}\|^{1/2}_\infty - 1\bigr| \leq \|\rho^{\pm 1/2} - 1\|_\infty \leq \|\rho^{1/2} - \rho^{-1/2}\|_\infty,$$

i.e. $\|J\|, \|J'\| \leq 2$ provided $\delta \leq 1$. The remaining estimates to show are

$$\|(J' - J^*)u\|^2_X = \|(\mathbb{1} - \rho)\rho^{-1/2}\rho^{1/2}u\|^2_X \leq \|\rho^{1/2} - \rho^{-1/2}\|^2_\infty \|u\|^2_{\widetilde{X}}$$

and

$$\begin{aligned}
\bigl|\mathfrak{d}(J'u, f) - \widetilde{\mathfrak{d}}(u, Jf)\bigr| &= \bigl|\langle \mathrm{d}u, \mathrm{d}f\rangle_X - \langle \mathrm{d}u, \mathrm{d}f\rangle_{\widetilde{X}}\bigr| \\
&= \Bigl|\int_X \bigl\langle (\mathbb{1} - \rho A^{-1})\rho^{-1/2}A^{1/2}\rho^{1/2}A^{-1/2}\mathrm{d}u, \mathrm{d}f\bigr\rangle_g \mathrm{d}X\Bigr| \\
&\leq \|(\mathbb{1} - \rho A^{-1})\rho^{-1/2}A^{1/2}\|_\infty \int_X \rho^{1/2}|\mathrm{d}u|_{\widetilde{g}}|\mathrm{d}f|_g \mathrm{d}X \\
&\overset{\mathrm{CS}}{\leq} \|(\rho^{-1/2}A^{1/2} - \rho^{1/2}A^{-1/2})\|_\infty \|\mathrm{d}u\|_{\widetilde{X}} \|\mathrm{d}f\|_X
\end{aligned}$$

for $f \in \mathscr{H}^1$ and $u \in \widetilde{\mathscr{H}}^1$. The $4\hat{\delta}$-quasi unitary equivalence follows easily from basic estimates. □

For the notion of quasi-unitary equivalence of operators, we refer to Definition 4.2.3). Note that the quasi-unitary equivalence of operators follows form the one of quadratic forms by Proposition 4.4.15.

Corollary 5.2.7. *With the notation of the preceding theorem, the operators* $\Delta_{(X,g)}$ *and* $\Delta_{(X,\widetilde{g})}$ *are* 4δ*-quasi-unitarily equivalent with identification operators given in* (5.18). *In particular,*

$$\|J(\Delta_{(X,g)} + 1)^{-1} - (\Delta_{(X,\tilde{g})} + 1)^{-1} J\| \leq 4\delta. \tag{5.19}$$

Moreover, the results of Sects. 4.2 and 4.3 apply.

Remark 5.2.8. The advantage of choosing J' instead of the adjoint J^* is to avoid assumptions involving the *derivative* of ρ (and therefore derivatives of the metrics). If we would have set $J' := J^*$, then the last estimate in the previous proof would contain an extra term with $\mathrm{d}\rho$.

In particular, we can assume without loss of generality that the metric is only *continuous*, i.e. that $x \mapsto g_x$ is continuous, up to an error. The corresponding Laplacian is defined via its quadratic form; the operator domain may be different from a standard Sobolev space.

Corollary 5.2.9. *Let g be a continuous metric on X, then for any $\delta > 0$ there exists* a smooth *metric such that $\Delta_{(X,g)}$ and $\Delta_{(X,\tilde{g})}$ are 4δ-quasi-unitarily equivalent with identification operators given in* (5.18).

5.3 Tubular Neighbourhoods

5.3.1 Perturbations of the Product Structure

The following considerations are motivated by tubular neighbourhoods of an edge as discussed in a concrete situation in the next section. We start with a slightly more general setting. Let us consider the metric g on $X := I \times Y$ defined by

$$g = \ell^2 \mathrm{d}s^2 + r^2 h, \tag{5.20}$$

where ℓ and r are positive functions on X to be specified later on. Here, $I \subset \mathbb{R}$ is an interval and (Y, h) is a Riemannian manifold of dimension $m \geq 1$. We identify the tangent space $T_{(s,y)}X$ with $\mathbb{R} \times T_y Y$ and write $v = (v_1, v_2) \in T_{(s,y)}X$.

Let $\tilde{g}$ be another metric, close to g given by

$$\tilde{g} = (1 + \mathrm{o}_{11})\ell^2 \mathrm{d}s^2 + 2\ell r\, \mathrm{d}s\, \mathrm{o}_{12} + (1 + \mathrm{o}_{22}) r^2 h, \tag{5.21}$$

where $\mathrm{o}_{11}, \mathrm{o}_{22}$ are real-valued functions on X and $\mathrm{o}_{12,(s,y)} \colon T_y X \longrightarrow \mathbb{R}$. The corresponding relative distortion A of $\tilde{g}$ w.r.t. g (see Definition 5.2.1) in this setting is

$$A = \begin{pmatrix} 1 + \mathrm{o}_{11} & \ell^{-1} r\, \mathrm{o}_{12} \\ \ell r^{-1} \mathrm{o}_{12}^* & (1 + \mathrm{o}_{22})\, \mathrm{id}_{TY} \end{pmatrix}. \tag{5.22}$$

Its d eigenvalues are given by

$$\alpha_k = \begin{cases} 1 + \mathrm{o}_{22} & \text{(multiplicity } m-1\text{)}, \\ 1 + \frac{1}{2}\big(\mathrm{o}_{11} + \mathrm{o}_{22} \pm \sqrt{(\mathrm{o}_{11} - \mathrm{o}_{22})^2 + 4|\mathrm{o}_{12}|^2}\big), \end{cases}$$

where $|\mathrm{o}_{12}|^2 = \mathrm{o}_{12}\mathrm{o}_{12}^*$. Moreover, we have

$$\begin{aligned} \rho^2 = \det A &= (1 + \mathrm{o}_{22})^{m-1}\big((1 + \mathrm{o}_{11})(1 + \mathrm{o}_{22}) - |\mathrm{o}_{12}|^2\big), \\ \|\rho^{1/2} - \rho^{-1/2}\|_\infty &= 2\sinh\Big(\frac{1}{2}\log\|\rho\|_\infty\Big) \\ &= 2\sinh\Big(\frac{m-1}{4}\log(1 + \mathrm{o}_{22}) \\ &\qquad + \frac{1}{4}\log\big((1 + \|\mathrm{o}_{11}\|_\infty)(1 + \|\mathrm{o}_{22}\|_\infty) - \|\mathrm{o}_{12}\|_\infty^2\big)\Big) \\ &= \mathrm{O}(\mathrm{o}_{11}) + \mathrm{O}(\mathrm{o}_{22}) + \mathrm{O}(|\mathrm{o}_{12}|^2), \\ \|\rho^{-1/2}A^{1/2} - \rho^{1/2}A^{-1/2}\|_\infty &= 2\sinh\Big(\frac{1}{2}\log\|\rho^{-1}A\|_\infty\Big) \\ &= \mathrm{O}(\mathrm{o}_{11}) + \mathrm{O}(\mathrm{o}_{22}) + \mathrm{O}(|\mathrm{o}_{12}|^2), \qquad (\mathrm{o}_{ij} \to 0), \end{aligned}$$

where the errors $\mathrm{O}(\mathrm{o}_{ij})$ are short hand for an expression of order ε with $\varepsilon = \|\mathrm{o}_{ij}\|_\infty$, depending only on m and ε, and similarly for $|\mathrm{o}_{12}|^2$.

Denote by $\mathfrak{d}(f) := \|\mathrm{d}f\|^2_{(X,g)}$ and $\widetilde{\mathfrak{d}}(u) := \|\mathrm{d}u\|^2_{(X,\widetilde{g})}$ the quadratic forms associated with the corresponding Riemannian manifolds. Then Theorem 5.2.6 yields:

Proposition 5.3.1. *The quadratic forms $\mathfrak{d}$ and $\widetilde{\mathfrak{d}}$ associated with the spaces (X,g) and $(X,\widetilde{g})$ are δ-quasi-unitarily equivalent, where*

$$\delta := \mathrm{O}(\mathrm{o}_{11}) + \mathrm{O}(\mathrm{o}_{22}) + \mathrm{O}(|\mathrm{o}_{12}|^2) \qquad \textit{as } \mathrm{o}_{ij} \to 0.$$

Moreover, the error terms depend only on $m = \dim Y$ and on the suprema of o_{ij}.

It is sometimes useful to compare the metric $g := \ell^2 \mathrm{d}s^2 + r^2 h$ with a warped product metric (see Fig. 5.4) provided the function r depends on $s \in I$ only:

Definition 5.3.2. Let I and Y be two Riemannian manifolds with metrics $\mathrm{d}s^2$ and h. Moreover, let $r \colon I \longrightarrow (0,\infty)$ be a continuous map. We call the metric

$$g_0 := \mathrm{d}s^2 + r^2 h$$

a *warped product metric*. We denote the Riemannian manifold $X = I \times Y$ with metric g_0 by

$$X = I \times_r Y,$$

the *warped product* of I and Y with *radius function* r.

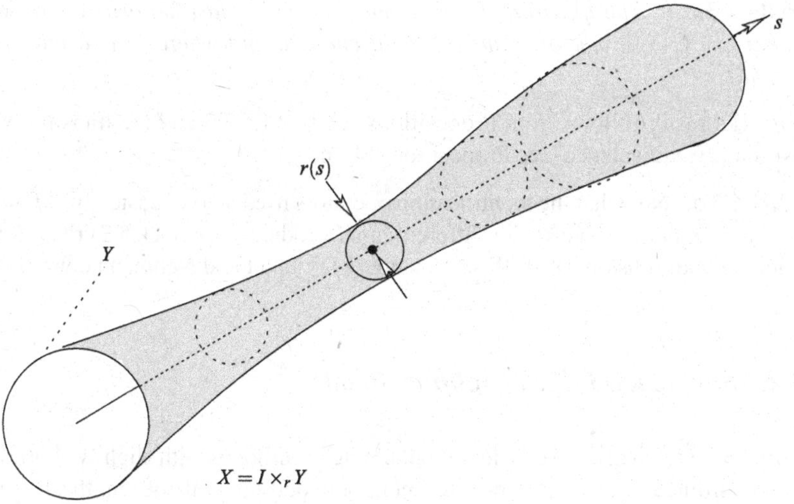

Fig. 5.4 A warped product space without transversal boundary. Here, the transversal manifold (Y, h) is a 1-sphere and the manifold $X = I \times_r Y$, i.e. the product $X = I \times Y$ with metric $g_0 = ds^2 + r(s)^2 h$ consists of the rotation surface

The relative distortion A_0 of g w.r.t. g_0 and the relative density $\rho_0 := (\det A_0)^{1/2}$ are given by

$$A_0 = \begin{pmatrix} \ell^2 & 0 \\ 0 & 1 \end{pmatrix} \qquad \text{and} \qquad \rho_0 = \ell.$$

In particular, we have

$$\begin{aligned} \|\rho_0^{1/2} - \rho_0^{-1/2}\|_\infty &= 2\sinh\Big(\frac{1}{2}\log\|\ell\|_\infty\Big) \\ \leq \|\rho_0^{-1/2} A_0^{1/2} - \rho_0^{1/2} A_0^{-1/2}\|_\infty &= 2\sinh\Big(\frac{1}{2}\log\|\rho_0^{-1} A_0\|_\infty\Big) \\ &= 2\sinh\Big(\frac{1}{2}\big|\log\|\ell\|_\infty\big|\Big), \end{aligned}$$

since $\rho_0^{-1} A_0$ has the eigenvalues $\ell^{\pm 1}$. Denote by $\mathfrak{d}_0(f) := \|df\|^2_{(X,g_0)}$ the quadratic form on (X, g_0). Comparing (X, g_0) with (X, g), we obtain again from Theorem 5.2.6:

Proposition 5.3.3. *The quadratic forms* $\mathfrak{d}$ *and* $\mathfrak{d}_0$ *associated with the spaces* (X, g) *and* (X, g_0) *are* δ_0*-quasi-unitarily equivalent, where* $\delta_0 = 2\sinh(|\log\|\ell\|_\infty|/2) = O(\|\ell\|_\infty - 1)$.

Combining Propositions 5.3.1 and 5.3.3 together with the transitivity of quasi-unitary equivalence (Proposition 4.4.16) yields:

Corollary 5.3.4. *The quadratic forms $\widetilde{\mathfrak{d}}$ and $\mathfrak{d}_0$ are δ'-quasi-unitarily equivalent, where $\delta' = \hat{C}(\delta + \delta_0)$ for δ and δ_0 small enough, and where $\hat{C}$ is a universal constant.*

Proof. The result follows from Propositions 5.3.1 and 5.3.3 and the transitivity of quasi-unitary equivalence (see Proposition 4.4.16). □

Remark 5.3.5. Note that the identification operators used above (basically, $Jf = f$ and $J'u = u$, but in spaces with different inner products, see Sect. 5.2) respect the boundary conditions on ∂Y if $\partial Y \neq \emptyset$, namely Dirichlet and Neumann conditions.

5.3.2 Shortened Edge Neighbourhoods

We sometimes also need to consider tubular neighbourhoods with slightly shortened length coordinate. Let $\hat{I} \subset I$ be two open, bounded intervals of length $(1 - \tau)a$ and a, respectively, together with an affine-linear transformation $\varphi\colon I \longrightarrow \hat{I}$. If we assume that φ is increasing, then φ is uniquely determined by the intervals. Moreover, $\varphi'(s) = 1 - \tau$. By our assumptions, $0 \le \tau < 1$. Furthermore, we need that ℓ and r are uniformly continuous functions.

Let us introduce the following notation in order to quantify the error terms in the next proposition:

Definition 5.3.6. Let $\eta\colon I \longrightarrow \mathbb{R}$ be a function and $B_\delta(s_0) := \{\, s \in I \mid |s - s_0| < \delta \,\}$ be the (open) ball of radius δ around s_0 in I. The *local module of continuity* of η in s_0 is defined as

$$\operatorname{moc}_{s_0}(\eta, \delta) := \sup_{s \in B_\delta(s_0)} \bigl|\eta(s) - \eta(s_0)\bigr|.$$

The *global module of continuity* of η is defined as

$$\operatorname{moc}(\eta, \delta) := \sup_{s_0 \in I} \operatorname{moc}_{s_0}(\eta, \delta).$$

If η depends also on y, i.e. $\eta\colon I \times Y \longrightarrow \mathbb{R}$ then we define the *uniform global module of continuity* of η with respect to the first variable as

$$\operatorname{moc}(\eta, \delta) := \sup_{y \in Y} \operatorname{moc}(\eta(\cdot, y), \delta).$$

If η is Lipschitz (i.e. weakly differentiable with derivative in L_∞ w.r.t. the first variable s) with derivative denoted by η', then

$$\operatorname{moc}_{s_0}(\eta, \delta) \le \delta \sup_{s \in B_\delta(s_0)} |\eta'(s)| \qquad \text{and} \qquad \operatorname{moc}(\eta, \delta) \le \delta \|\eta'\|_\infty$$

where $\|\eta'\|_\infty$ is the supremum of $\eta'(s)$ resp. $\eta'(s, y)$, $s \in I$, $y \in Y$.

Set $X := I \times Y$ with metric $g = \ell^2 ds^2 + r^2 h$ as in (5.20) and $\hat{X} := \hat{I} \times Y$ with the same metric restricted to $\hat{X}$. Recall that $m = \dim Y$. We want to compare the quadratic forms $\mathfrak{d}(f) := \|df\|^2_X$ and $\hat{\mathfrak{d}}(u) := \|du\|^2_{\hat{X}}$:

Proposition 5.3.7. *The quadratic forms* $\mathfrak{d}$ *and* $\hat{\mathfrak{d}}$ *are* $\hat{\delta}$*-quasi-unitarily equivalent where*

$$\hat{\delta}^2 := \frac{4}{1-\tau}\Big(\tau^2 e^{\mu} + 4\sinh^2(\mu/2)\Big)$$

and

$$\mu := \operatorname{moc}(\log \ell, \tau a) + m \operatorname{moc}(\log r, \tau a).$$

Moreover, if the functions ℓ *and* r *are Lipschitz continuous in the first variable* s*, then*

$$\mu \le \tau a \cdot \big(\|(\log \ell)'\|_\infty + m\|(\log r)'\|_\infty\big)$$

and in particular, $\hat{\delta} = O(\tau)$. *Finally, if* ℓ *and* r *are constant functions, then* $\mu = 0$.

Proof. Let $Jf(\hat{s}, y) := f(\varphi^{-1}(\hat{s}), y)$ and $J'u(s, y) := u(\varphi(s), y)$. Obviously, these identification operators fulfil $J'Jf = f$ and $JJ'u = u$. Denote the density of the metric g by $\rho = \ell r^m$. Then we have

$$\begin{aligned}
\|Jf\|^2_{\hat{X}} &= \int_{\hat{I}\times Y} \big|f(\varphi^{-1}(\hat{s}), y)\big|^2 \rho(\hat{s}, y) d\hat{s} dy \\
&= \int_{I\times Y} |f(s, y)|^2 \frac{\rho(\varphi(s), y)}{\rho(s, y)} \rho(s, y)(1-\tau) ds dy \\
&\le (1-\tau) e^{\operatorname{moc}(\log \rho,\, \tau a)} \|f\|^2_X
\end{aligned}$$

since $0 \le s - \varphi(s) \le \tau a$. In particular, $\|J\| \le e^{\operatorname{moc}(\log \rho, \tau a)/2}$. Moreover,

$$J^* u(s, y) = \frac{\rho(\varphi(s), y)}{\rho(s, y)}(1-\tau) u(\varphi(s), y)$$

and therefore

$$\begin{aligned}
\|(J^* - J')u\|^2_X &= \int_{I\times Y} \Big|\frac{\rho(\varphi(s), y)}{\rho(s, y)}(1-\tau) - 1\Big|^2 \big|u(\varphi(s), y)\big|^2 \rho(s, y) ds dy \\
&= \frac{1}{1-\tau} \int_{\hat{I}\times Y} \Big|\Big(\frac{\rho(\hat{s}, y)}{\rho(\varphi^{-1}(\hat{s}), y)}\Big)^{1/2}(1-\tau) - \Big(\frac{\rho(\varphi^{-1}(\hat{s}), y)}{\rho(\hat{s}, y)}\Big)^{1/2}\Big|^2 \\
&\quad \times \big|u(\varphi^{-1}(\hat{s}), y)\big|^2 \rho(\hat{s}, y) d\hat{s} dy \\
&\le \frac{2}{1-\tau}\Big(\tau^2 e^{\operatorname{moc}(\log \rho,\, \tau a)} + 4\sinh^2(\operatorname{moc}(\log \rho, \tau a)/2)\Big)\|u\|^2_{\hat{X}}.
\end{aligned}$$

Now,

$$\operatorname{moc}(\log \rho, \tau a) \leq \operatorname{moc}(\log \ell, \tau a) + m \operatorname{moc}(\log r, \tau a) =: \mu$$

since $\rho = \ell r^m$. The estimate for the quadratic forms is similar up to a factor 2. □

5.4 Embedded Tubular Neighbourhoods

Let us illustrate the error estimates of the preceding section in the case of an embedded curve, keeping the notation introduced there. In Fig. 5.5 we sketched the different metric spaces associated with the tubular neighbourhood.

Suppose that we have a curve $\psi : I \longrightarrow \mathbb{R}^2$ in $\mathbb{R}^2$ in arc length parametrisation. Then

$$\Psi_\varepsilon(s, y) := \psi(s) + \varepsilon r(s) y \mathrm{n}(s), \qquad (s, y) \in I \times Y =: X$$

defines a tubular neighbourhood with variable width r of the curve $\psi(I)$ as subset of $\mathbb{R}^2$. Here, $Y = [-1/2, 1/2]$ and n is a continuous unit vector field normal to the tangent vector field ψ' and $r(s)$ defines the radius of the neighbourhood. Denote by

$$\kappa := \psi_1' \psi_2'' - \psi_1'' \psi_2' \tag{5.23}$$

the signed curvature of ψ. In order to avoid self-intersections of the neighbourhood, we assume that

$$\ell_\varepsilon(s, y) := 1 + \varepsilon \kappa(s) r(s) y > 0 \tag{5.24}$$

for all $(s, y) \in I \times Y$. Now $U_\varepsilon := \Psi_\varepsilon(X)$ together with the Euclidean metric on $\mathbb{R}^2$ defines a metric $\widetilde{g}_\varepsilon$ on X having the matrix representation

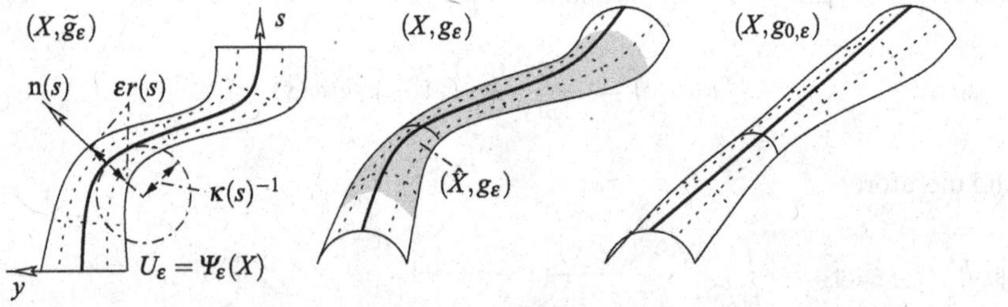

Fig. 5.5 The different tubular neighbourhoods. On the *left*, the original (*flat*) tubular neighbourhood with variable radius $r(s)$ and curvature $\kappa(s)$ of the generating curve ψ (the thick line). In the *middle*, we sketched a manifold with metric $g_\varepsilon = \ell_\varepsilon^2 ds^2 + \varepsilon^2 r(s)^2 h$. On the *right*, a manifold with warped product metric $g_{0,\varepsilon} = ds^2 + \varepsilon^2 r(s)^2 h$ is shown. Note that the coordinate lines in (X, g_ε) and $(X, g_{0,\varepsilon})$ are orthogonal, whereas in $(X, \widetilde{g}_\varepsilon) \cong U_\varepsilon \subset \mathbb{R}^2$, they are only orthogonal if $r(s)$ is constant. The *shaded part* in the middle corresponds to the shortened neighbourhood $(\hat{X}, g_\varepsilon)$ (see Proposition 5.4.4)

$$\widetilde{g}_\varepsilon \cong \begin{pmatrix} \ell_\varepsilon^2 + \varepsilon^2 y^2 (r')^2 & \varepsilon^2 r r' y \\ \varepsilon^2 r r' y & \varepsilon^2 r^2 \end{pmatrix}. \tag{5.25}$$

with respect to the basis ∂_s and ∂_y of $T_{(s,y)}X$. Let

$$g_\varepsilon = \ell_\varepsilon^2 \mathrm{d}s^2 + \varepsilon^2 r^2 \mathrm{d}y^2$$

be the reference metric as in (5.20). Then the errors in the relative distortion A of $\widetilde{g}$ w.r.t. g defined in (5.22) are given by

$$\mathrm{o}_{11} = \frac{\varepsilon^2 y^2 (r')^2}{\ell_\varepsilon^2}, \qquad \mathrm{o}_{12} = \frac{\varepsilon y r'}{\ell_\varepsilon} \qquad \text{and} \qquad \mathrm{o}_{22} = 0.$$

Moreover, the relative density is given by

$$\rho^2 = \det A = (1 + \mathrm{o}_{11}) - \mathrm{o}_{12}^2 = 1. \tag{5.26}$$

Let us summarise the above observations in the following proposition. The assertion is a special case of Proposition 5.3.1:

Proposition 5.4.1. *The quadratic forms $\mathfrak{d}_\varepsilon$ and $\widetilde{\mathfrak{d}}_\varepsilon$ associated with the manifolds (X, g_ε) and $(X, \widetilde{g}_\varepsilon)$, respectively, are δ_ε-quasi-unitarily equivalent, where*

$$\delta_\varepsilon = C_m\Big(\varepsilon \Big\| \frac{r'}{\ell_\varepsilon} \Big\|_\infty\Big) \le C_m\Big(\frac{\varepsilon \|r'\|_\infty}{1 - \varepsilon \|\kappa\|_\infty \|r\|_\infty / 2}\Big).$$

Here, $C_m(\cdot)$ is a monotonously increasing function depending only on m and $C_m(\tau) = \mathrm{O}(\tau^2)$. In particular, $\delta_\varepsilon = \mathrm{O}(\varepsilon^2)$.

Let us now compare the metric $\widetilde{g}_\varepsilon$ with the exact warped product metric

$$g_{0,\varepsilon} = \mathrm{d}s^2 + \varepsilon^2 r^2 \mathrm{d}y^2.$$

We first compare g_ε with $g_{0,\varepsilon}$. Denote by A_0 the relative distortion of g_ε w.r.t. $g_{0,\varepsilon}$. Then

$$A_0 = \begin{pmatrix} \ell_\varepsilon^2 & 0 \\ 0 & 1 \end{pmatrix},$$

and the relative density is

$$\rho_0 = (\det A)^{1/2} = \ell_\varepsilon$$

using (5.24). Now, from Corollary 5.3.4 we conclude:

Proposition 5.4.2. *The quadratic forms $\widetilde{\mathfrak{d}}_\varepsilon$ and $\mathfrak{d}_{0,\varepsilon}$ associated with the perturbed warped product $(X, \widetilde{g}_\varepsilon)$ and the exact warped product $(X, g_{0,\varepsilon})$ are δ'_ε-quasi-unitarily equivalent, where*

$$\delta'_\varepsilon = \hat{C}_m\Big(\varepsilon\Big\|\frac{r'}{\ell_\varepsilon}\Big\|_\infty\Big) + \hat{C}_0(\varepsilon\|\kappa r\|_\infty)$$
$$\leq \hat{C}_m\Big(\frac{\varepsilon^2\|r'\|_\infty}{1-\varepsilon\|\kappa\|_\infty\|r\|_\infty/2}\Big) + \hat{C}_0(\varepsilon\|\kappa\|_\infty\|r\|_\infty).$$

Moreover, $\hat{C}_m(\tau) = \mathrm{O}(\tau^2)$ *and* $\hat{C}_0(\tau) = \mathrm{O}(\tau)$ *are universal monotone functions. In particular,* $\delta'_\varepsilon = \mathrm{O}(\varepsilon)$. *The error depends only on the suprema of the curvature* $\|\kappa\|_\infty$, *of the radius* $\|r\|_\infty$ *and its derivative* $\|r'\|_\infty$.

Remark 5.4.3. Sometimes the metric g_ε is better adopted to the problem than the (exact) warped product metric $g_{0,\varepsilon}$: The error in the quasi-unitary equivalence comparing $\widetilde{g}_\varepsilon$ with g_ε is of order $\mathrm{O}(\varepsilon^2)$ instead of $\mathrm{O}(\varepsilon)$ when $\widetilde{g}_\varepsilon$ is compared with $g_{0,\varepsilon}$. This is of particular importance in the case of Dirichlet boundary conditions, see Sects. 5.6 and 6.11.4, where one needs cancellation of divergent terms.

Let us finally compare the metric $g_\varepsilon = \ell_\varepsilon^2 \mathrm{d}s^2 + \varepsilon^2 r^2 h$ on $X = I \times Y$ with the same metric restricted to the smaller space $\hat{X} = \hat{I} \times Y$, where $I = (0, a)$ and $\hat{I}$ is an open subinterval of length $(1-\tau)a$ as in Sect. 5.3.2. Proposition 5.3.7 reads in this case:

Proposition 5.4.4. *The quadratic forms* $\mathfrak{d}_\varepsilon$ *and* $\hat{\mathfrak{d}}_\varepsilon$ *associated with* (X, g_ε) *and* $(\hat{X}, g_\varepsilon)$ *are* $\hat{\delta}_\varepsilon$*-quasi-unitarily equivalent where*

$$\hat{\delta}_\varepsilon^2 := \frac{4}{1-\tau}\Big(\tau^2 \mathrm{e}^{\mu_\varepsilon} + 4\sinh^2(\mu_\varepsilon/2)\Big)$$

and

$$\mu_\varepsilon := \tau a\Big(\frac{\varepsilon\|\kappa' r + \kappa r'\|_\infty}{1-\varepsilon\|\kappa r\|_\infty} + m\Big\|\frac{r'}{r}\Big\|_\infty\Big).$$

In particular, if the length error is $\tau = \varepsilon$, *then* $\hat{\delta}_\varepsilon = \mathrm{O}(\varepsilon)$ *and the error depends only on the suprema of* κ, κ', r, r^{-1} *and* r'.

5.5 Reduction to a One-Dimensional Problem: Tubular Neighbourhoods with Neumann Boundary Conditions

In the next two sections, we show that the Laplacian on a tubular neighbourhood $X = I \times Y$ with certain ε-depending metrics g_ε can be reduced to a one-dimensional problem on I in the limit $\varepsilon \to 0$. We have to distinguish the cases where $\partial Y = \emptyset$ or we impose Neumann boundary conditions on ∂Y and the case where we impose Dirichlet boundary conditions on ∂Y. We start with the Neumann (or boundaryless) case and treat the Dirichlet case in the next section.

Let us begin with a general result which is useful in order to get rid of first order derivatives $\mathrm{d}u$:

Lemma 5.5.1. *Let M be a smooth manifold with boundary ∂M Then*

$$\int_M \operatorname{Re}\langle u\sigma \mathrm{d}\tau, \mathrm{d}u\rangle \mathrm{d}M = \frac{1}{2}\int_M \mathrm{d}^*(\sigma \mathrm{d}\tau)|u|^2 \mathrm{d}M + \frac{1}{2}\int_{\partial M} \sigma(\mathrm{d}\tau)_{\mathrm{n}}|u|^2 \mathrm{d}\partial M$$

for bounded, smooth functions σ, τ with bounded derivatives and $u \in \mathsf{H}^1(M)$. Here, d^* denotes the formal adjoint of d.

Proof. The proof follows from

$$\begin{aligned}\int_M \operatorname{Re}\langle u\sigma \mathrm{d}\tau, \mathrm{d}u\rangle \mathrm{d}M &= \int_M \operatorname{Re}(\mathrm{d}^*(\overline{u}\sigma \mathrm{d}\tau)u)\mathrm{d}M + \int_{\partial M} \sigma(\mathrm{d}\tau)_{\mathrm{n}}|u|^2 \mathrm{d}\partial M \\ &= \int_M \mathrm{d}^*(\sigma \mathrm{d}\tau)|u|^2 \mathrm{d}M - \int_M \operatorname{Re}\langle \mathrm{d}u, u\sigma \mathrm{d}\tau\rangle \mathrm{d}M \\ &\quad + \int_{\partial M} \sigma(\mathrm{d}\tau)_{\mathrm{n}}|u|^2 \mathrm{d}\partial M.\end{aligned}$$

The second integral equals the one with which we have started, and the result follows. □

Let us now fix the tubular neighbourhood. Assume that I is a one-dimensional space, i.e. I is either an interval (bounded or unbounded) or I is a 1-sphere $\mathbb{S}^1$. Moreover, assume that (Y, h) is a *connected*, compact Riemannian manifold of dimension $m \geq 1$. We will consider the Laplacian on Y; if $\partial Y \neq \emptyset$, we impose Neumann boundary conditions. In any case, the first eigenvalue is 0, i.e. $\lambda_1(Y) = 0$, whereas the second is non-zero, $\lambda_2(Y) > 0$ since we assumed Y to be connected. Let $X := I \times Y$ with the *warped product metric*

$$g_\varepsilon := \mathrm{d}s^2 + \varepsilon^2 r(s)^2 h$$

for some bounded and continuous function

$$r: I \longrightarrow (0, \infty).$$

We use the notation $X_\varepsilon := I \times_r \varepsilon Y$ as short hand for the Riemannian manifold (X, g_ε).

The aim of the following is to show that the quadratic form

$$\widetilde{\mathfrak{d}}(u) := \|\mathrm{d}u\|^2_{X_\varepsilon} \qquad \text{with} \qquad \operatorname{dom}\widetilde{\mathfrak{d}} := \widetilde{\mathscr{H}}^1 := \mathsf{H}^1(X_\varepsilon)$$

in the Hilbert space $\widetilde{\mathscr{H}} := \mathsf{L}_2(X_\varepsilon)$ converges to a 1-dimensional problem on I. In particular, we set

$$\mathscr{H} := \mathsf{L}_2(I, w\mathrm{d}s),$$

where $\mathsf{L}_2(I, w\mathrm{d}s)$ is the L_2-space with norm given by

$$\|f\|^2_{(I,w\mathrm{d}s)} := \int_I |f(s)|^2 w(s)\mathrm{d}s, \qquad w(s) := r(s)^m.$$

As quadratic form on $\mathscr{H}$, we set

$$\mathfrak{d}(f) := \|f'\|^2_{\mathscr{H}} = \int_I |f'(s)|^2 w(s)\mathrm{d}s \qquad \text{with} \qquad \operatorname{dom}\mathfrak{d} := \mathscr{H}^1 := \mathsf{H}^1(I, w\mathrm{d}s),$$

where $f \in \mathsf{H}^1(I, w\mathrm{d}s)$ iff $f, f' \in \mathsf{L}_2(I, w\mathrm{d}s)$. The operator D corresponding to the non-negative quadratic form $\mathfrak{d}$ (see Theorem 3.1.1) is given by

$$Df = -\frac{1}{w}(wf')' = -f'' - m(\log r)' f' \tag{5.27}$$

for suitable f's.

In order to compare the forms $\widetilde{\mathfrak{d}}$ and $\mathfrak{d}$ resp. the operators Δ_{X_ε} and D, we introduce the identification operator

$$J \colon \mathsf{L}_2(I, w\mathrm{d}s) \longrightarrow \mathsf{L}_2(X_\varepsilon), \qquad f \mapsto f \otimes \mathbb{1}_\varepsilon,$$

where $\mathbb{1}_\varepsilon = \varepsilon^{-m/2}\mathbb{1}$ and $\mathbb{1}$ denote the first (constant) normalised eigenfunctions on εY and Y, respectively. Denote by

$$\widetilde{P} := JJ^*, \qquad \text{i.e.} \qquad (\widetilde{P}u)(s,\cdot) = \langle \mathbb{1}, u(s,\cdot)\rangle_Y \mathbb{1},$$

the projection onto the transversally constant functions. We have the following approximation result:

Proposition 5.5.2.

1. *The identification operator J is δ_ε-partial isometric of order* 1 *with error given by $\delta_\varepsilon = \varepsilon(\lambda_2(Y))^{-1/2}$.*
2. *The forms $\mathfrak{d}$ and $\widetilde{\mathfrak{d}}$ are* 0*-close with the identification operators J and J^*.*
3. *The forms $\mathfrak{d}$ and $\widetilde{\mathfrak{d}}$ are δ_ε-partial isometrically equivalent.*
4. *The operator D is unitary equivalent with $\Delta_{X_\varepsilon}\widetilde{P} = \widetilde{P}\Delta_{X_\varepsilon}$.*

Proof. We have $(J^*u)(s) = \varepsilon^{m/2}\langle \mathbb{1}, u(s,\cdot\rangle_Y$. Then obviously, the spaces of order 1 are respected by J and J^*, i.e. J and J^* are also first order identification operators. Moreover, $J^*J = \mathrm{id}$ on $\mathscr{H}$ and

$$\begin{aligned}\|u - JJ^*u\|^2_{X_\varepsilon} &= \int_I \|u(s,\cdot) - \langle \mathbb{1}, u(s,\cdot)\rangle_Y \mathbb{1}\|^2_Y \varepsilon^m w(s)\mathrm{d}s \\ &\le \frac{\varepsilon^2}{\lambda_2(Y)} \int_I \|\mathrm{d}_Y u(s,\cdot)\|^2_Y \varepsilon^m w(s)\mathrm{d}s = \frac{\varepsilon^2}{\lambda_2(Y)} \|\mathrm{d}_Y u\|^2_{X_\varepsilon} \le \frac{\varepsilon^2}{\lambda_2(Y)} \|\mathrm{d}u\|^2_{X_\varepsilon}\end{aligned}$$

by Proposition 5.1.1. In particular, J is δ_ε-partial isometric of order 1 (see Definition 4.1.1 (4)). The forms $\mathfrak{d}$ and $\widetilde{\mathfrak{d}}$ are 0-close (see Definition 4.4.1 (4)) since

$$\widetilde{\mathfrak{d}}(u, Jf) = \varepsilon^{m/2} \int_{I \times Y} \overline{u}' f' \mathrm{d}Y \, w \mathrm{d}s = \mathfrak{d}(J^* u, f)$$

and $\mathrm{d}_Y(f \otimes \mathbb{1}) = 0$. The third statement follows easily from (1) and (2), and the last one is a consequence of Proposition 4.2.4 (2). □

It is sometimes convenient to work in an L_2-space without weight: Set

$$U : \mathsf{L}_2(I, w\mathrm{d}s) \longrightarrow \mathsf{L}_2(I), \qquad Uf := w^{1/2} f$$

where $\mathsf{L}_2(I) = \mathsf{L}_2(I, \mathrm{d}s)$ is the (unweighted) standard L_2-space. Obviously, U is unitary. Moreover, the quadratic form $\mathfrak{d}$ turns into a quadratic form $\mathfrak{q}$ on $\mathsf{L}_2(I)$ and is given in the next proposition. For simplicity, we assume that the "longitudinal" boundary terms on ∂I vanish, and that r is sufficiently regular. In the following proposition, we use the notation

$$[F]_{\partial I} := F(b) - F(a) \tag{5.28}$$

for a continuous function $s \mapsto F(s)$ depending on $s \in \mathbb{R}$ such that $F(\pm\infty) := \lim_{s \to \pm\infty} F(s) = 0$, where I is an interval with endpoints $a < b$. If $I = \mathbb{S}^1$, we use the convention $[F]_{\partial I} = 0$. Recall that $w = r^m$.

Proposition 5.5.3. *Assume that r', r'' exist and that r, r^{-1}, r', r'' are bounded on I. Then the transformed quadratic form*

$$\mathfrak{q}(\widetilde{f}) := \mathfrak{d}(U^{-1}\widetilde{f}), \qquad \operatorname{dom} \mathfrak{q} := U(\operatorname{dom} \mathfrak{d}) = \mathsf{H}^1(I),$$

is given by

$$\mathfrak{q}(\widetilde{f}) = \int_I (|\widetilde{f}'(s)|^2 + Q|\widetilde{f}(s)|^2) \mathrm{d}s - \frac{1}{2}\Big[\frac{w'}{w}|\widetilde{f}|^2\Big]_{\partial I}, \tag{5.29}$$

where

$$Q := q^2 + q' = \frac{1}{4}\Big(\frac{w'}{w}\Big)^2 + \frac{1}{2}\Big(\frac{w'}{w}\Big)', \qquad q := \frac{(\log w)'}{2} = \frac{m(\log r)'}{2} = \frac{mr'}{2r}.$$

In particular, if $r' \restriction_{\partial I} = 0$ or $\partial I = \emptyset$, then the boundary term vanishes.

Proof. We have

$$\mathfrak{q}(\widetilde{f}) = \int_I \big|w^{-1/2}\widetilde{f}' + (w^{-1/2})'\widetilde{f}\big|^2 w\mathrm{d}s = \int_I \Big(|\widetilde{f}'|^2 + \frac{1}{4}\Big(\frac{w'}{w}\Big)^2 - \frac{w'}{w}\operatorname{Re}\big(\overline{\widetilde{f}'}\widetilde{f}\big)\Big)\mathrm{d}s$$

and the result follows from Lemma 5.5.1 with $M := I$, $\sigma = w^{-1}$ and $\tau = w$. Note that the boundary term vanishes if $r' = 0$ on ∂I or $\partial I = \emptyset$. □

In particular, we can replace the form $\mathfrak{d}$ by $\mathfrak{q}$ in Proposition 5.5.2. For example, we have:

Corollary 5.5.4. *The forms* $\mathfrak{q}$ *and* $\widetilde{\mathfrak{d}}$ *defined as above are* δ_ε*-partial isometrically equivalent, where* $\delta_\varepsilon = \varepsilon\lambda_2(Y)^{-1/2}$.

Remark 5.5.5. We can also use Dirichlet boundary conditions on ∂I in Proposition 5.5.3, i.e. replace $\mathfrak{d}$ by the form $\mathring{\mathfrak{d}}$ defined on $\operatorname{dom}\mathring{\mathfrak{d}} = \mathring{\mathsf{H}}^1(I)$. Note that the unitary transformation U respects the Dirichlet domain, i.e. $f = 0$ on ∂I iff $Uf = 0$ on ∂I. Moreover, the boundary term of the quadratic form $\mathfrak{q}$ vanishes.

5.6 Reduction to a One-Dimensional Problem: Tubular Neighbourhoods with Dirichlet Boundary Conditions

In this section, we compare the Dirichlet Laplacian on $X = I \times Y$ with a metric g_ε, "shrinking" the transversal compact m-dimensional manifold (Y, h) to a point, with a one-dimensional problem on I (an interval or a sphere). We impose *Dirichlet* boundary conditions on $\partial Y \neq \emptyset$.

Let us start with a metric of the form

$$g = \ell^2 \mathrm{d}s^2 + r^2 h \tag{5.30}$$

having the additional function ℓ as longitudinal coefficient. We introduce the parameter ε later on. We assume that r is a function of $s \in I$ only. In the Dirichlet case, we do not assume that the "longitudinal" coefficient $\ell: I \times Y \longrightarrow (0, \infty)$ equals 1 as in the Neumann case in the previous section. We will see that some information of $\ell = \ell_\varepsilon$ remains in the limit, due to the fact that in the Dirichlet case, we have to substract the *divergent* first transversal eigenmode $\lambda_1^{\mathrm{D}}(Y)/\varepsilon^2$ in order to expect a convergent limit.

The Dirichlet Laplacian acts in the Hilbert space $\mathsf{L}_2(X, g) = \mathsf{L}_2(I \times Y, \rho \mathrm{d}s \mathrm{d}Y)$ with density $\rho = \ell r^m$. Let us calculate the unitarily equivalent operator on the L_2-space without density, namely on $\mathsf{L}_2(I \times Y, \mathrm{d}s\mathrm{d}Y)$, using the following unitary operator

$$U: \mathsf{L}_2(X, g) \longrightarrow \mathsf{L}_2(I \times Y, \mathrm{d}s\mathrm{d}Y), \qquad Uu = \rho^{1/2} u.$$

Let $\widetilde{\mathfrak{d}}$ be the quadratic form defined by

$$\widetilde{\mathfrak{d}}(u) := \|\mathrm{d}u\|_X^2 \qquad \text{and} \qquad \operatorname{dom}\widetilde{\mathfrak{d}} = \mathsf{H}^1(X, I \times \partial Y, g).$$

The corresponding Laplacian is the Laplacian with Dirichlet boundary conditions on $\partial_0 X := I \times \partial Y$ and Neumann boundary conditions on $\partial_1 X := \partial I \times Y$. For the notation $[F]_{\partial I}$ in the following lemma see (5.28).

Lemma 5.6.1. *Assume that r and ℓ are functions differentiable up to order 2. In addition, assume (for simplicity) that r, r^{-1}, r', r'' and ℓ, ℓ^{-1}, ℓ', ℓ'' are bounded. Then the transformed quadratic form*

$$\mathfrak{q}(\widetilde{u}) := \widetilde{\mathfrak{d}}(U^{-1}\widetilde{u}), \qquad \operatorname{dom}\mathfrak{q} := U(\operatorname{dom}\widetilde{\mathfrak{d}}),$$

is given by $\operatorname{dom}\mathfrak{q} = \mathsf{H}^1(X, I \times \partial Y, \mathrm{d}s\mathrm{d}Y)$ *and*

$$\mathfrak{q}(\widetilde{u}) = \int_{I\times Y} (\ell^{-2}|\widetilde{u}'|^2 + r^{-2}|\mathrm{d}_Y\widetilde{u}|_h^2 + K|\widetilde{u}|^2)\mathrm{d}s\mathrm{d}Y - \Big[\int_Y \frac{Q}{\ell^2}|\widetilde{u}|^2\Big]_{\partial I}$$

where

$$K := -\frac{|\mathrm{d}_Y\ell|_h^2}{4r^2\ell^2} - \frac{\Delta_Y\ell}{2r^2\ell} + \frac{Q^2+Q'}{\ell^2} - \frac{2\ell'Q}{\ell^3}, \qquad Q := \frac{1}{2}\Big(\frac{\ell'}{\ell} + m\frac{r'}{r}\Big)$$

Proof. The claim follows from a straightforward but long calculation, using the formula in Lemma 5.5.1. □

Let us now fix the function $\ell = \ell_\varepsilon$, motivated by the embedded edge neighbourhood of Sect. 5.4. In particular, we

$$\ell_\varepsilon(s, y) = 1 + r_\varepsilon(s)\kappa(s)b(y) \tag{5.31a}$$

with the following assumptions on b, r_ε and κ: the functions $b\colon Y \longrightarrow \mathbb{R}$, $r_1\colon I \longrightarrow (0,\infty)$ and $\kappa\colon I \longrightarrow \mathbb{R}$ are twice differentiable and bounded with bounded derivatives. Moreover, r_1^{-1} is bounded and b is harmonic, i.e.

$$\Delta_Y b = 0, \qquad \text{and} \qquad r_\varepsilon(s) = \varepsilon r_1(s). \tag{5.31b}$$

Then the metric is given by

$$g_\varepsilon = \ell_\varepsilon^2\mathrm{d}s^2 + r_\varepsilon(s)^2h = \big(1 + r_\varepsilon(s)\kappa(s)b(y)\big)^2\mathrm{d}s^2 + \varepsilon^2r_1(s)^2h.$$

Remark 5.6.2. In Sect. 5.4, we analysed the special case $Y = [-1/2, 1/2]$ and $b(y) = y$. The metric g_ε does not arise from an embedded tubular neighbourhood with variable radius function r_1 as the metric $\widetilde{g}_\varepsilon$ does. Note that $\widetilde{g}_\varepsilon$ has "off-diagonal terms", see Sect. 5.4. If the function r_1 is constant, then $\widetilde{g}_\varepsilon = g_\varepsilon$, i.e. the off-diagonal terms vanish. In principle, one can also consider tubular neighbourhoods and *variable* radius, together with the corresponding Dirichlet Laplacian.[2]

Note that κ can be interpreted as a sort of "(extrinsic) curvature" of the curve generating the tubular neighbourhood as in Sect. 5.4. We denote by X_ε the corresponding Riemannian manifold (X, g_ε) with quadratic form associated with the Laplacian with Dirichlet boundary conditions on $\partial_0 X = I \times \partial Y$ given by

[2]For a curve embedded in $\mathbb{R}^3$ see e.g. [KP88, Sect. 4].

$$\mathfrak{d}_\varepsilon(u) := \|du\|^2_{X_\varepsilon} \qquad \text{and} \qquad \operatorname{dom} \mathfrak{d}_\varepsilon = \mathsf{H}^1(X_\varepsilon, I \times \partial Y, g_\varepsilon).$$

Again, we want to use the unitarily equivalent quadratic form in the Hilbert space $\mathsf{L}_2(X, \mathrm{d}s\mathrm{d}Y)$ without density, defined by

$$\mathfrak{q}_\varepsilon(\widetilde{u}) := \mathfrak{d}_\varepsilon(U_\varepsilon^{-1}\widetilde{u}), \qquad \operatorname{dom} \mathfrak{q}_\varepsilon := U_\varepsilon(\operatorname{dom} \mathfrak{d}_\varepsilon),$$

where

$$U_\varepsilon \colon \mathsf{L}_2(X_\varepsilon) \longrightarrow \mathsf{L}_2(X, \mathrm{d}s\mathrm{d}Y), \qquad U_\varepsilon u := \varepsilon^{m/2} r_1^{m/2} \ell_\varepsilon^{1/2} u$$

is unitary.

Using an asymptotic expansion for $K = K_\varepsilon$ given in Lemma 5.6.1, we obtain the following result:

Lemma 5.6.3. *The transformed quadratic form $\mathfrak{q}_\varepsilon$ is given by*

$$\mathfrak{q}_\varepsilon(\widetilde{u}) = \int_{I \times Y} \Big(\frac{1}{\ell_\varepsilon^2} |\widetilde{u}'|^2 + \frac{1}{\varepsilon^2 r^2} |\mathrm{d}_Y \widetilde{u}|^2_h + K_\varepsilon |\widetilde{u}|^2 \Big) \mathrm{d}s\mathrm{d}Y - \frac{1}{2} \Big[\int_Y \frac{1}{\ell_\varepsilon^2} \Big(\frac{\ell_\varepsilon'}{\ell_\varepsilon} + m \frac{r_1'}{r_1} \Big) |\widetilde{u}|^2 \Big]_{\partial I},$$

where

$$K_\varepsilon = -\frac{1}{4} |\mathrm{d}_Y b|^2_h \kappa^2 + q^2 + q' + R_\varepsilon, \qquad q := \frac{m r_1'}{2 r_1},$$

and $R_\varepsilon = \mathrm{O}(\varepsilon)$ depends only on $\|b\|_\infty$, $\|\kappa^{(j)}\|_\infty$ and $\|r^{(j)}\|_\infty$, $j = 0, 1, 2$.

Remark 5.6.4. Let us make a comment on why for example the curvature term κ^2 is only seen in the *Dirichlet case*: Formally, one could also consider the boundaryless case, but on a manifold Y without boundary, all harmonic functions are constant (on each connected component), so that $\ell = \ell(s)$ is independent of y; in particular, the potential K_ε does not contain the curvature term κ^2. Otherwise, if we drop the condition of harmonicity of b, we obtain an additional term of the form

$$-\frac{\Delta_Y \ell_\varepsilon}{2 r_\varepsilon^2 \ell_\varepsilon} = -\frac{\kappa \Delta_Y b}{2 \varepsilon r_1 \ell_\varepsilon} \to \infty$$

which is *divergent* as $\varepsilon \to 0$.

Moreover, the transformation $Uf = \rho^{1/2} f$ respects the Dirichlet condition on $I \times \partial Y$, but *not* the Neumann (or Robin) condition, since the normal derivative will be transformed in a more complicated expression involving the density function ρ; in particular, the transformation is useless in the Neumann case, since then, the effect of the potential κ^2 and the complicated boundary condition cancel each other.

Denote by $\lambda_1 = \lambda_1^{\mathrm{D}}(Y) > 0$ the first Dirichlet eigenvalue with corresponding normalised eigenfunction φ_1. Without loss of generality, we may assume that φ_1 is real-valued. Moreover, denote by $\lambda_2 = \lambda_2^{\mathrm{D}}(Y)$ the second Dirichlet eigenvalue. If Y is connected, then $\lambda_2 > \lambda_1$.

Let us now compare the form $\mathfrak{d}_\varepsilon$ defined on X_ε given as below with a 1-dimensional problem on $\mathsf{L}_2(I)$. For simplicity, we assume that the radius function r_1 is constant on I, and that we impose also Dirichlet boundary conditions on the "longitudinal" boundary $\partial I \times Y$. In particular, we have no boundary term in the transformed quadratic form $\mathfrak{q}_\varepsilon$. As quadratic form in $\mathsf{L}_2(I)$ we define

$$\mathfrak{h}_0(f) := \|f'\|_I^2 + \langle f, K_0 f\rangle_I, \qquad \operatorname{dom} \mathfrak{h}_0 := \mathring{\mathsf{H}}^1(I),$$

where we imposed Dirichlet conditions on ∂I. The potential is given by

$$K_0 := -\frac{\|\varphi_1 \mathrm{d}_Y b\|_Y^2}{4}\kappa^2 \le 0.$$

Note that the potential contains the extrinsic curvature κ of the curve embedded in some ambient space as in the concrete situation of Sect. 5.4.

In order to get a convergence result, we need to rescale the form $\mathfrak{d}_\varepsilon$: Denote by

$$\mathfrak{h}_\varepsilon(u) := \mathfrak{d}_\varepsilon(u) - \frac{\lambda_1}{\varepsilon^2}\|u\|_{X_\varepsilon}^2, \qquad \operatorname{dom} \mathfrak{h}_\varepsilon := \mathring{\mathsf{H}}^1(X_\varepsilon)$$

the rescaled quadratic form with Dirichlet condition on the entire boundary of X_ε.

Formally, we defined the partial isometric equivalence in Definition 4.4.11 only for *non-negative* quadratic forms. Since the additional potentials K_0 resp. K_ε in $\mathfrak{h}_0$ resp. $\mathfrak{q}_\varepsilon - \varepsilon^{-2}\lambda_1 \mathbb{1} \cong \mathfrak{h}_\varepsilon$ are bounded, we can pass to the forms $\widetilde{\mathfrak{h}}_\varepsilon := \mathfrak{h}_\varepsilon + c\mathbb{1}$ and $\widetilde{\mathfrak{h}}_0 := \mathfrak{h}_0 + c\mathbb{1}$, which are non-negative for some $c \ge 0$ large enough. Note that $c \ge \|K_0\|_\infty, \|K_\varepsilon\|_\infty$ is sufficient.

Proposition 5.6.5. *Assume that Y is connected and that the radius function r_1 is constant, w.l.o.g. $r_1(s) = 1$, and that b is harmonic. Then the quadratic forms $\mathfrak{h}_\varepsilon = \mathfrak{d}_\varepsilon - (\varepsilon^{-2}\lambda_1)\mathbb{1}$ and $\mathfrak{h}_0$ are $\mathrm{O}(\varepsilon)$-partial isometrically equivalent, where the error depends only on $\|b\|_\infty$, $\|\varphi_1 \mathrm{d}_Y b\|_Y$, $\|\kappa^{(j)}\|_\infty$, $j = 0, 1, 2$ and $\lambda_2 - \lambda_1$.*

Proof. We set $\mathscr{H} := \mathsf{L}_2(I)$, $\mathscr{H}^1 := \mathring{\mathsf{H}}^1(I)$, $\widetilde{\mathscr{H}} := \mathsf{L}_2(X_\varepsilon)$ and $\widetilde{\mathscr{H}}^1 := \mathring{\mathsf{H}}^1(X_\varepsilon)$. The identification operator $J : \mathscr{H} \longrightarrow \widetilde{\mathscr{H}}$ is defined as $Jf := U_\varepsilon^{-1}(f \otimes \varphi_1)$. Its adjoint is given by $(J^*u)(s) = \langle \varphi_1, \widetilde{u}(s, \cdot)\rangle$ where $\widetilde{u} = U_\varepsilon u$. Moreover,

$$\begin{aligned}
\|u - JJ^*u\|^2 &= \int_I \left\|\widetilde{u}(s) - \langle \varphi_1, \widetilde{u}(s)\rangle \varphi_1\right\|^2 \mathrm{d}s \\
&\le \frac{\varepsilon^2}{\lambda_2 - \lambda_1} \int_{I \times Y} \frac{1}{\varepsilon^2}\left(|\mathrm{d}_Y \widetilde{u}|_h^2 - \lambda_1 |\widetilde{u}|^2\right) \mathrm{d}Y \,\mathrm{d}s \\
&\le \frac{\varepsilon^2}{\lambda_2 - \lambda_1}\left(\mathfrak{q}_\varepsilon(\widetilde{u}) - \langle \widetilde{u}, K_\varepsilon \widetilde{u}\rangle\right) \\
&\le \frac{\varepsilon^2}{\lambda_2 - \lambda_1}\left(\mathfrak{h}_\varepsilon(u) + c\|u\|^2\right)
\end{aligned}$$

by Proposition 5.1.2, provided $c \geq \|K_\varepsilon\|_\infty$. Obviously, the identification operators J and J^* respect the quadratic form domains. The closeness of $\mathfrak{h}_\varepsilon$ and $\mathfrak{h}_0$ follows from

$$\begin{aligned}\mathfrak{h}_\varepsilon(Jf,u) - \mathfrak{h}_0(f,J^*u) &= \mathfrak{q}_\varepsilon(f\otimes\varphi_1,\widetilde{u}) - \frac{\lambda_1}{\varepsilon^2}\langle f\otimes\varphi_1,\widetilde{u}\rangle_{I\times Y}\\ &\quad - \mathfrak{h}_0(f, s\mapsto\langle\varphi_1,\widetilde{u}(s,\cdot)\rangle)\\ &= \int_{I\times Y}\Big[\Big(\frac{1}{\ell_\varepsilon}-\ell_\varepsilon\Big)\overline{f}'\otimes\varphi_1\cdot\ell_\varepsilon^{-1}\widetilde{u}' + (K_\varepsilon(s,y)\\ &\quad - K_0(s))\overline{f}\otimes\varphi_1\cdot\widetilde{u}\Big]\mathrm{d}Y\,\mathrm{d}s,\end{aligned}$$

where we used partial integration on Y and the eigenvalue equation $\Delta_Y\varphi = \lambda_1\varphi$ in order to get rid of the y-derivatives. Moreover, using the expansion of K_ε in Lemma 5.6.3 gives

$$\begin{aligned}\int_{I\times Y}(K_\varepsilon(s,y)-K_0(s))\overline{f}\otimes\varphi_1\cdot\widetilde{u}\mathrm{d}Y\,\mathrm{d}s &= \int_{I\times Y} R_\varepsilon(s,y)\overline{f}(s)\varphi_1(y)\widetilde{u}(s,y)\mathrm{d}s\mathrm{d}y\\ &\quad - \int_I\frac{\kappa(s)^2}{4}\overline{f}(s)\int_Y\big(|\mathrm{d}_Y b(y)|_h^2\\ &\quad - \|\varphi_1\mathrm{d}_Y b\|_Y^2\big)\widetilde{u}(s,y)\mathrm{d}y\mathrm{d}s.\end{aligned}$$

The integral over Y can be estimated by

$$\int_Y\Big(|\mathrm{d}_Y b|_h^2 - \|\varphi_1\mathrm{d}_Y b\|_Y^2\Big)\varphi_1\cdot v\mathrm{d}Y = \langle w - \langle\varphi_1,w\rangle\varphi_1, v\rangle_Y = \langle w, v - \langle\varphi_1,v\rangle\varphi_1\rangle_Y,$$

where $w := |\mathrm{d}_Y b|_h^2\varphi_1$ and $v := \widetilde{u}(s,\cdot)$ using the fact that the projection $v\mapsto v - \langle\varphi_1,v\rangle\varphi_1$ onto $\varphi_1^\perp$ is self-adjoint. In particular, the latter integral can be estimated by

$$\Big|\int_Y\Big(|\mathrm{d}_Y b|_h^2 - \|\varphi_1\mathrm{d}_Y b\|_Y^2\Big)\varphi_1\cdot v\mathrm{d}Y\Big|^2 \leq \frac{\|w\|_Y^2}{\lambda_2-\lambda_1}\big(\|\mathrm{d}_Y v\|_Y^2 - \lambda_1\|v\|_Y^2\big)$$

using again Proposition 5.1.2. Altogether, we obtain

$$\begin{aligned}&\big|\mathfrak{h}_\varepsilon(Jf,u) - \mathfrak{h}_0(f,J^*u)\big|^2\\ &\overset{\text{CS}}{\leq} \delta_\varepsilon^2(\|f'\|^2+\|f\|^2)\Big(\|\ell_\varepsilon^{-1}\widetilde{u}'\|^2 + \frac{1}{\varepsilon^2}(\|\mathrm{d}_Y\widetilde{u}\|^2 - \lambda_1\|\widetilde{u}\|^2) + \|\widetilde{u}\|^2\Big)\\ &\leq \delta_\varepsilon^2\big(\mathfrak{h}_0(f) + (\|K_0\|_\infty+1)\|f\|^2\big)\big(\mathfrak{h}_\varepsilon(u) + (\|K_\varepsilon\|_\infty+1)\|\widetilde{u}\|^2\big)\\ &\leq \delta_\varepsilon^2\big(\mathfrak{h}_0(f) + (c+1)\|f\|^2\big)\big(\mathfrak{h}_\varepsilon(u) + (c+1)\|\widetilde{u}\|^2\big),\end{aligned}$$

where $c \geq \|K_0\|_\infty, \|K_\varepsilon\|_\infty$. The total error of the partial isometry is given by

$$\delta_\varepsilon^2 := \max\Big\{3\|\ell_\varepsilon^{-1} - \ell_\varepsilon\|_\infty^2, 3\|R_\varepsilon\|_\infty^2, \frac{3\varepsilon^2}{\lambda_2 - \lambda_1} \max\Big\{\frac{\|\kappa\|_\infty^2 \|\varphi_1 \mathrm{d}_Y b\|_Y^2}{4}, 1\Big\}\Big\}.$$

Note that $\delta_\varepsilon = \mathrm{O}(\varepsilon)$, since

$$\|\ell_\varepsilon^{-1} - \ell_\varepsilon\|_\infty = 2\|\sinh \log \ell_\varepsilon\|_\infty \leq 2 \sinh\big(\varepsilon \|\kappa\|_\infty \|b\|_\infty\big) = \mathrm{O}(\varepsilon)$$

and $\|R_\varepsilon\|_\infty = \mathrm{O}(\varepsilon)$ by Lemma 5.6.3. □

Similarly, we can prove the above result if the boundary term in $\mathfrak{q}_\varepsilon$ does not vanish: Let us state the following result for Neumann boundary conditions on $\partial I \times Y$, i.e. we have no condition in the quadratic form domain on $\partial I \times Y$.

Proposition 5.6.6. *Assume that Y is connected and that the radius function r_1 is constant, w.l.o.g. $r_1(s) = 1$, and that b is harmonic. Then the quadratic forms $\mathfrak{h}_\varepsilon = \mathfrak{d}_\varepsilon - \varepsilon^{-2}\lambda_1 \mathbb{1}$ on $\operatorname{dom} \mathfrak{h}_\varepsilon := \mathsf{H}^1(X_\varepsilon, I \times \partial Y)$ and $\mathfrak{h}_0$ on $\operatorname{dom} \mathfrak{h}_0 = \mathsf{H}^1(I)$ are $\mathrm{O}(\varepsilon)$-partial isometrically equivalent, where the error depends only on $\|b\|_\infty$, $\|\varphi_1 \mathrm{d}_Y b\|_Y$, $\|\kappa^{(j)}\|_\infty$, $j = 0, 1, 2$, $\min\{\operatorname{len}(I), 1\}$ and $\lambda_2 - \lambda_1$.*

Proof. In addition to the preceding proposition, we have to estimate the additional boundary term in $\mathfrak{q}_\varepsilon$, which can be done as follows: Note first that $\widetilde{u} = \varepsilon^{m/2} \ell_\varepsilon^{1/2} u$ and

$$\begin{aligned} &\Big| -\frac{1}{2} \Big[\int_Y \frac{1}{\ell_\varepsilon^2} \cdot \frac{\ell_\varepsilon'}{\ell_\varepsilon} \overline{f(\cdot)\varphi_1 \widetilde{u}} \mathrm{d}Y \Big]_{\partial I} \Big|^2 \\ &\leq \big\|\ell_\varepsilon^{-3/2} (\log \ell_\varepsilon)'\big\|_\infty^2 \sum_{s \in \partial I} |f(s)|^2 \varepsilon^m \int_Y |u(s, \cdot)|^2 \mathrm{d}Y \\ &\leq \rho_\infty \big\|\ell_\varepsilon^{-3/2} (\log \ell_\varepsilon)'\big\|_\infty^2 \Big(a \|f'\|_I^2 + \frac{2}{a}\|f\|_I^2\Big) \Big(a \|\mathrm{d}u\|_{X_\varepsilon}^2 + \frac{2}{a} \|u\|_{X_\varepsilon}^2\Big) \end{aligned}$$

for $0 < a \leq \operatorname{len}(I)$ using Corollary A.2.8 twice, where $\rho_\infty = (\|\ell_\varepsilon\|_\infty \|\ell_\varepsilon^{-1}\|_\infty)^d$ for the estimate involving u. The value of the relative distortion ρ_∞ (see Definition A.2.2) follows from

$$\frac{1}{\|\ell_\varepsilon^{-1}\|_\infty^2} \Big(\mathrm{d}s^2 + \frac{1}{\varepsilon^2} h\Big) \leq g_\varepsilon = \ell_\varepsilon^2 \mathrm{d}s^2 + \varepsilon^{-2} h \leq \|\ell_\varepsilon\|_\infty^2 \Big(\mathrm{d}s^2 + \frac{1}{\varepsilon^2} h\Big).$$

In particular, we get an extra error term of order $\mathrm{O}(\varepsilon)$. □

Chapter 6
Plumber's Shop: Estimates for Star Graphs and Related Spaces

In this chapter, we consider the reduction of a graph-like manifold to the underlying quantum graph model. We treat here only star-shaped graphs., i.e., a graph with one central vertex v and $\deg v$-many edges $e \in E_v = E$ attached. The case of general graphs and their graph-like manifolds will be treated in Chap. 7. Note that any graph G can be decomposed into star-shaped graphs G_v by cutting each interior edge (i.e., each edge of finite length) in the middle.

Let us now consider such a single building block G_v. For ease of notation, we often omit the dependence on the central vertex[1] v of G_v, and we use ℓ_e as length of the emanating edges instead of $\ell_e/2$. A *graph-like manifold* associated with the star-shaped graph $G = G_v$ consists of a space with cylindrical ends $X_{\varepsilon,e} = I_e \times \varepsilon Y_e$ with radius ε of the transversal (or cross-sectional) manifold Y_e for each edge $e \in E = E_v$ and a central manifold $X_{\varepsilon,v}$. Strictly speaking, a *graph-like manifold* is a *family* of Riemannian manifolds $\{X_\varepsilon\}_{0<\varepsilon\le\varepsilon_0}$ for a certain $\varepsilon_0 > 0$, but in general, no confusion should occur. If we want to stress the ε-dependence of X_ε, we also speak of a *scaled graph-like manifold*..

In the limit, X_ε "converges" to the underlying metric graph $G = G_v$. We show that the Laplacian on X_ε is δ_ε-partial isometrically equivalent with a Laplace-like operator on the metric graph and we provide precise error estimates for $\delta_\varepsilon \to 0$. The limit operator depends on the choice of boundary conditions on the "transversal" boundary $\partial_0 X_\varepsilon$. Note that the transversal boundary $\partial_0 X_\varepsilon$ is non-trivial if the cross-section Y_e has non-trivial boundary.

We first treat the Neumann (or boundaryless) case, since in this case the lowest transversal eigenmode is 0 and the corresponding eigenfunction is constant. The limit operator is basically determined by the vertex neighbourhood scaling. Of particular interest are the non-decoupling limit operators of Sects. 6.4, 6.6 and 6.10. Similar results have been proven in [RuS01a, KuZ01, KuZ03, EP05]

[1]Having the decomposition of G into star-shaped graphs G_v in mind, we do not consider the free ends of the edges of G_v with finite length as vertices, but only as boundary points (see Remark 2.2.17).

O. Post, *Spectral Analysis on Graph-Like Spaces*, Lecture Notes in Mathematics 2039, DOI 10.1007/978-3-642-23840-6_6, © Springer-Verlag Berlin Heidelberg 2012

(see also Sects. 1.2.1–1.2.2). The slowly decaying and borderline case (see below) were introduced in [KuZ03] (see also [EP05]) showing the convergence of the spectrum for compact graphs and manifolds. The notion "plumber's shop" has been adopted from the article of Rubinstein and Schatzman [RuS01a], where the authors used it for the necessary local estimates of the identification operators from the graph to the graph-like manifold and vice versa. We extend the analysis here to non-compact spaces and show in particular the δ_ε-quasi unitary equivalence (resp. δ_ε-partial isometry) implying e.g. the convergence of resolvents and the convergence of the entire spectrum. At the end of this chapter, we describe a special situation, where we deal with Dirichlet boundary conditions (see Sect. 6.11), cf. also [P05, MV07, G08b, G08a].

Let us briefly describe the content of the subsequent sections: Sect. 6.1 contains the different graph models for the limit space, depending on the scaling behaviour of the vertex neighbourhood $X_{\varepsilon,v}$. In Sect. 6.2 we construct graph-like manifolds, starting with a simple example, namely, that of a vertex neighbourhood scaling as $X_{\varepsilon,v} = \varepsilon X_v$. If the scaling is different like e.g. $X_{\varepsilon,v} = \varepsilon^\alpha X_v$ for some $\alpha \in (0,1)$, we have to assure that the vertex neighbourhood still can be glued to the edge neighbourhood $X_{\varepsilon,e} = I_e \times \varepsilon Y_e$. This needs some homogeneity of the transversal manifold Y_e allowing an isometric embedding $\varepsilon Y_e \hookrightarrow \varepsilon^\alpha Y_e$, treated in detail in Sect. 6.2.2. As a simple example, the reader should have a Euclidean ball as transversal manifold Y_e in mind. We also introduce the associated quadratic forms for the Laplacians and boundary triples. The latter are used in Chap. 7 in order to pass from star-shaped graphs to general graphs. Section 6.3 contains some necessary estimates on functions on $X_{\varepsilon,v}$.

Sections 6.4–6.6 treat the δ_ε-partial isometry of the models with vertex neighbourhood $X_{\varepsilon,v} = \varepsilon^\alpha X_v$ depending on $\alpha \in (0,1]$ for the Neumann Laplacian. Section 6.7 treats a star-shaped graph and its neighbourhood embedded in $\mathbb{R}^2$. Section 6.8 contains a different way how to join the scaled vertex neighbourhood $X_{\varepsilon,v} = \varepsilon^\alpha X_v$ for $\alpha \in (0,1)$ with the edge neighbourhoods $X_{\varepsilon,e}$, namely by joining both with a suitable truncated cone. This situation allows us to treat e.g. boundaryless transversal manifolds Y_e like a sphere $\mathbb{S}^1$. Sections 6.9–6.10 contain then the reduction to the corresponding graph models. Finally, in Sect. 6.11, we consider a special situation of a Laplacian with Dirichlet boundary condition and the associated (decoupled) graph model.

6.1 The Graph Models for Neumann Boundary Conditions

Let us start with a simple example of a star-shaped metric graph $G = G_v$ having only one vertex v and $\deg v$-many adjacent edges e of length $\ell_e \in (0,\infty]$. We identify the (metric) edge e with the interval $I_e := [0,\ell_e]$ (resp. $I_e = [0,\infty)$ for exterior edges, i.e. $\ell_e = \infty$) oriented in such a way that 0 corresponds to the vertex v. Moreover, the metric graph $G = G_v$ is given by the abstract space $G_v := \bigcup_e I_e/\sim$ where $\sim$ identifies the points $0 \in I_e$ with the vertex v. The star-shaped metric graph

G has boundary ∂G consisting of the (free) endpoints $\ell_e \in I_e$ of the interior edges, i.e. the edges of finite length. Recall that the free endpoints are not considered as vertices of the graph G (see Remark 2.2.17).

We will define different types of Laplacians on G via their quadratic forms. The different types of Laplacians depend on a parameter α, which is the length scale rate of the vertex neighbourhood, i.e. $\operatorname{vol} X_{\varepsilon,v} \approx \varepsilon^{\alpha d} \operatorname{vol} X_v$ and the total transversal volume $\operatorname{vol} Y_\varepsilon$. Here,

$$Y_\varepsilon := \dot{\bigcup_e} Y_{\varepsilon,e} \tag{6.1}$$

denotes the total scaled transversal manifold consisting of the disjoint union of the transversal manifolds $Y_{\varepsilon,e}$, and $Y := Y_1$ the corresponding unscaled manifold. In particular

$$\operatorname{vol} Y_\varepsilon = \varepsilon^m \operatorname{vol} Y, \qquad \text{where} \qquad m := d - 1.$$

The spaces are described in detail in Sects. 6.2 and 6.8. The operators in the limit $\varepsilon \to 0$ are determined by the ratio

$$\operatorname{vol}(\varepsilon, v) := \frac{\operatorname{vol}_d X_{\varepsilon,v}}{\operatorname{vol}_m Y_\varepsilon} \cong \varepsilon^{\alpha d - m} \frac{\operatorname{vol}_d X_v}{\operatorname{vol}_m Y}. \tag{6.2}$$

We will distinguish three cases, depending on the limit $\operatorname{vol}(0, v) := \lim_{\varepsilon \to 0} \operatorname{vol}(\varepsilon, v)$:

- **Fast decaying vertex volume:** $\frac{m}{d} < \alpha \le 1$ $(\operatorname{vol}(0, v) = 0)$,
- **The borderline case:** $\alpha = \frac{m}{d}$ $(\operatorname{vol}(0, v) \in (0, \infty))$,
- **Slowly decaying vertex volume:** $0 < \alpha < \frac{m}{d}$ $(\operatorname{vol}(0, v) = \infty)$.

6.1.1 Fast Decaying Vertex Volume

In the fast decaying case, the limit space is the simple quantum graph $(G, \mathscr{V})$ together with its associated Laplacian operator. For the notation of metric and quantum graphs and associated operators, we refer to Sect. 2.2. Here, the vertex space $\mathscr{V} = \mathscr{V}_v$ is a weighted standard vertex space at the central vertex, i.e.

$$\mathscr{V}_v := \mathbb{C} p(v),$$

where $p(v) = \{p_e(v)\}_{e \in E_v}$ consists of positive entries $p_e(v) > 0$ only. Since our graph has only one central vertex, we also write $p = \{p_e\}_{e \in E}$, where $p = p(v)$, $p_e = p_e(v)$ and $E = E_v$. We interpret $p_e > 0$ as a weight, determined by the corresponding transversal manifold Y_e, i.e. $p_e = (\operatorname{vol} Y_e)^{1/2}$ (see Sects. 6.4–6.6).

Let us recall the definitions in our situation here. The basic Hilbert space associated with $G = G_v$ is

$$\mathscr{H} = \mathscr{H}_v := \mathsf{L}_2(G) = \bigoplus_{e \in E_v} \mathsf{L}_2(I_e).$$

We define the Laplacian by its quadratic form

$$\mathfrak{d}(f) = \mathfrak{d}_v(f) := \|f'\|_G^2 = \sum_{e \in E} \|f_e'\|_{I_e}^2, \qquad \operatorname{dom} \mathfrak{d} := \mathsf{H}_p^1(G),$$

where $\mathsf{H}_p^1(G)$ denotes the Sobolev space associated with the weight p and is given by

$$\mathscr{H}^1 = \mathscr{H}_v^1 := \mathsf{H}_p^1(G) = \{\, f \in \mathsf{H}_{\max}^1(G) = \bigoplus_e \mathsf{H}^1(I_e) \mid \underline{f} \in \mathbb{C}p \,\}. \tag{6.3}$$

Moreover, $\underline{f} := \{f_e(0)\}_e \in \mathbb{C}^{E_v} \cong \mathbb{C}^{\deg v}$ is the evaluation vector of f at the vertex v. We use the notation

$$\underline{f} = f(v)p, \qquad \text{i.e.} \qquad f_e = f(v)p_e \tag{6.4}$$

for all $e \in E$. In particular, we have

$$|\underline{f}|^2 = \sum_{e \in E_v} |f_e(0)|^2 = |f(v)|^2 |p|^2.$$

A simple consequence of the Sobolev trace estimate in Corollary A.2.8 resp. Corollary A.2.18 is the following:

Lemma 6.1.1. *We have*

$$|f(v)|^2 \le |p|^{-2}\Big(a\|f'\|_G^2 + \frac{2}{a}\|f\|_G^2\Big) \le |p|^{-2} \max\Big\{a, \frac{2}{a}\Big\} \|f\|_{\mathsf{H}^1(G)}^2,$$

for all $f \in \mathsf{H}_p^1(G)$ and for all $0 < a \le \min_{e \in E} \ell_e$, where $\|f\|_{\mathsf{H}^1(G)}^2 = \|f'\|_G^2 + \|f\|_G^2$.

As in the previous lemma, we often need a lower bound on the length ℓ_e:

Definition 6.1.2. We call $\ell_-(v) := \min_{e \in E_v}\{\ell_e, 1\}$ the *(modified) minimal length* at the vertex v.

It is convenient to have an upper bound on the minimal length, say $\ell_-(v) \le 1$, since then, we have the estimate

$$\max\Big\{a, \frac{2}{a}\Big\} \le \frac{2}{\ell_-(v)}$$

for $0 < a \le \min_e \ell_e$. Nevertheless, it is sometimes useful to keep the original formulation including the parameter a, e.g., when showing that a quadratic form is relatively bounded w.r.t. another quadratic form (see e.g. [EP09]).

The following result is a special case of Corollary 2.2.11. For convenience, we formulate it in our case here:

Proposition 6.1.3. *The quadratic form $\mathfrak{d}$ with* $\operatorname{dom}\mathfrak{d} = \mathsf{H}^1_p(G)$ *is closed. The associated non-negative operator $\Delta = \Delta_v$ is given by $(\Delta f)_e = -f_e''$ with domain $\mathscr{H}^2 = \operatorname{dom}\Delta$, where $f \in \mathscr{H}^2$ iff $f \in \mathsf{H}^2_{\max}(G)$ and*

$$\frac{f_{e_1}(v)}{p_{e_1}} = \frac{f_{e_2}(v)}{p_{e_2}} =: f(v) \quad \forall\, e_1, e_2 \in E_v, \qquad \sum_{e \in E_v} p_e f_e'(0) = 0 \tag{6.5a}$$

$$\text{and} \quad f_e'(\ell_e) = 0 \quad \text{for all } e \in E \text{ such that } \ell_e < \infty. \tag{6.5b}$$

Proof. By Lemma 6.1.1, $\mathsf{H}^1_p(G)$ is a closed subspace of $\mathsf{H}^1_{\max}(G)$. Since the norm associated with the quadratic form $\mathfrak{d}$ is precisely the Sobolev norm $\|\cdot\|_{\mathsf{H}^1(G)}$, the closeness of $\mathfrak{d}$ follows. The assertion on the associated operator is straightforward. □

In order to glue different spaces $G = G_v$ together according to a given graph, we use the notion of boundary triples introduced in Sect. 3.4. Recall that the boundary of G_v consists of the "free" endpoints of the edges of finite length, i.e.

$$\partial G = \partial G_v = \dot{\bigcup_{e \in E_{v,\mathrm{int}}}} \{\ell_e\},$$

where $E_{\mathrm{int}} = E_{v,\mathrm{int}} = \{\, e \in E \mid \ell_e < \infty \,\}$ denotes the set of interior edges. We associate a boundary triple $(\Gamma_v, \Gamma_v', \mathscr{G}_v)$ to G_v as follows: Set

$$\mathscr{G}_v := \mathbb{C}^{E_{v,\mathrm{int}}} = \bigoplus_{e \in E_{v,\mathrm{int}}} \mathscr{G}_e$$

with $\mathscr{G}_e = \mathbb{C}$, and the boundary maps are given by

$$\Gamma_v f = \bigoplus_{e \in E_{v,\mathrm{int}}} \Gamma_{v,e} f \qquad \text{and} \qquad \Gamma_v' f = \bigoplus_{e \in E_{v,\mathrm{int}}} \Gamma_{v,e}' f,$$

and where

$$\Gamma_{v,e}\colon \mathscr{H}_v^1 \longrightarrow \mathbb{C} \qquad\qquad \Gamma_{v,e}'\colon \mathscr{W}_v^2 \longrightarrow \mathbb{C},$$
$$f \mapsto f_e(\ell_e), \qquad\qquad f \mapsto f_e'(\ell_e).$$

Moreover,

$$\mathscr{W}_v^2 = \Big\{ f \in \mathsf{H}^2_{\max}(G_v) \,\Big|\, \frac{f_{e_1}(v)}{p_{e_2}} = \frac{f_{e_1}(v)}{p_{e_2}} \quad \forall\, e_1, e_2 \in E_v, \quad \sum_{e \in E_v} p_e f_e'(0) = 0 \Big\}$$

is the Sobolev space of order 2, and the (Sobolev-)maximal operator is defined by $(\check{\Delta} f)_e = -f_e''$ for $f \in \mathscr{W}_v^2$.

For the notation $\sin_{z,e,\pm}$, $\tan_z$ and $\cot_z$ we refer to (2.17).

Proposition 6.1.4.

1. *The triple* $(\Gamma_v, \Gamma'_v, \mathscr{G}_v)$ *is a bounded, elliptic boundary triple associated with* $\mathfrak{d} = \mathfrak{d}_v$ *and*
$$\|\Gamma_v\|_{1\to 0} \le (2/\ell_-(v))^{1/2} = \Big(\frac{2}{\min_e\{\ell_e, 1\}}\Big)^{1/2}$$
2. *The corresponding Neumann operator is* Δ_v *with domain given in* (6.5b). *The corresponding Dirichlet operator has the vertex condition as in* (6.5a) *at* v *and Dirichlet conditions* $f_e(\ell_e) = 0$, $e \in E_{v,\mathrm{int}}$.
3. *Let* $F = \{F_e\}_{e\in E_{v,\mathrm{int}}} \in \mathscr{G}_v$ *and* $z \in \mathbb{C} \setminus \mathbb{R}_+$. *Then the Dirichlet solution operator is given by* $h = S_v(z)F$, *where*
$$h_e(s) = \begin{cases} h(v)p_e \sin_{z,e,-}(s) + F_e \sin_{z,e,+}(s), & \text{if } e \in E_{v,\mathrm{int}} \\ h(v)p_e \mathrm{e}^{\mathrm{i}\sqrt{z}s}, & \text{if } e \in E_{v,\mathrm{ext}} \end{cases}$$
with
$$h(v) = \frac{1}{C(z)} \sum_{e\in E_{v,\mathrm{int}}} \frac{p_e}{\sin_z \ell_e} \cdot F_e \quad \text{and} \quad C(z) = \sum_{e\in E_{v,\mathrm{int}}} p_e^2 \cot_z \ell_e - \mathrm{i} \sum_{e\in E_{v,\mathrm{ext}}} p_e^2.$$
4. *The Dirichlet-to-Neumann map is represented by the matrix*
$$\Lambda_v(z)_{e_1,e_2} = \begin{cases} -\dfrac{\sqrt{z}}{C(z)} \cdot \dfrac{p_{e_1} p_{e_2}}{\sin_z \ell_{e_1} \cdot \sin_z \ell_{e_2}}, & \text{if } e_1 \ne e_2 \\ -\dfrac{\sqrt{z}}{C(z)} \cdot \dfrac{p_{e_1}^2}{(\sin_z \ell_{e_1})^2} + \sqrt{z}\cot_z \ell_{e_1}, & \text{if } e_1 = e_2 \end{cases}$$
for $e_1, e_2 \in E_{v,\mathrm{int}}$.

Proof. (1) follows from Proposition 2.2.18, where we interpret the boundary ∂G of G as vertices with associated vertex space $\mathbb{C}$ (see Remark 2.2.17). (2) is obvious. Assertions (3)–(4) follow from a straightforward calculation. Note that $s \mapsto \mathrm{e}^{\mathrm{i}\sqrt{z}s}$ is in $\mathsf{L}_2(\mathbb{R}_+)$ since $\operatorname{Im}\sqrt{z} > 0$. □

Let us calculate the Dirichlet solution and Dirichlet-to-Neumann map in the special case of an equilateral star graph $\ell_e = 1$ for all $e \in E_v$:

Corollary 6.1.5. *Assume that* $\ell_e = 1$ *for all* $e \in E_v$ *(in particular, there are no exterior edges).*

1. *The spectrum of the (Neumann) and Dirichlet operator* Δ_v *and* Δ_v^{D} *is given by*
$$\sigma(\Delta_v^{\mathrm{N}}) = \Big\{\frac{\pi^2 k^2}{4} \,\Big|\, k = 0, 1, 2, \dots\Big\} \quad \text{and} \quad \sigma(\Delta_v^{\mathrm{D}}) = \Big\{\frac{\pi^2 k^2}{4} \,\Big|\, k = 1, 2, \dots\Big\},$$
respectively.

2. *Let* $F = \{F_e\}_{e \in E_v} \in \mathscr{G}_v$ *and* $z \in \mathbb{C} \setminus \mathbb{R}_+$. *Then the Dirichlet solution operator is given by* $h = S_v(z)F$, *where*

$$h_e(s) = h(v) p_e \sin_{z,e,-}(s) + F_e \sin_{z,e,+}(s) \quad \textit{with}$$

$$h(v) = \frac{1}{\cos\sqrt{z}} \cdot \frac{1}{|p|^2} \sum_{e \in E_v} p_e F_e.$$

3. *The Dirichlet-to-Neumann map is given by*

$$\Lambda_v(z) = \frac{\sqrt{z}}{\sin\sqrt{z} \cdot \cos\sqrt{z}} \left(-P + (\cos\sqrt{z})^2 \mathbb{1}\right),$$

where P *is the orthogonal projection onto* $\mathbb{C}p$ *in* $\mathbb{C}^{E_v}$. *The eigenvalues of* $\Lambda(z)$ *are* $-\sqrt{z}\tan\sqrt{z}$ *with eigenvector* p *and* $\sqrt{z}\cot\sqrt{z}$ *with eigenspace* $p^\perp$ *and multiplicity* $\deg v - 1$. *In particular, for* $z = -\kappa^2$ *and* $\kappa > 0$, *the eigenvalues of* $\Lambda(-\kappa^2)$ *are positive and given by* $\kappa \tanh\kappa$ *and* $\kappa \coth\kappa$.

We can use the above formula for $\Lambda(z)$ in order to obtain a meromorphic continuation. Note that the poles of $\Lambda(z)$ (i.e. the poles of the eigenvalues) are precisely given by the Dirichlet spectrum, and the zeros of $\Lambda(z)$ (i.e. the values $z \in \mathbb{R}_+$ for which $\Lambda(z)$ is not invertible) are given by the Neumann spectrum. In Chap. 7, we use such "building blocks" consisting of a star graph at the vertex v in order to construct a coupled boundary triple as in Sect. 3.9. The (global) Laplacian on the entire graph is then the Neumann operator associated with this coupled boundary triple.

6.1.2 Slowly Decaying Vertex Volume

In the slowly decaying case, the limit space is an *extended* quantum graph $(G, \mathbb{C}p, L)$ with weighted standard vertex space $\mathbb{C}p$ and trivial operator $L = 0$ (see Definition 2.3.1) together with the associated Laplacian. Let us recall the definitions. The basic Hilbert space is the extended Hilbert space

$$\mathscr{H} = \mathscr{H}_v := \mathsf{L}_2(G) := \bigoplus_{e \in E} \mathsf{L}_2(I_e) \oplus \mathbb{C}, \tag{6.6}$$

i.e, we have an additional state $F = F(v) \in \mathbb{C}$ at the vertex v. The quadratic form is the extended quadratic form defined by

$$\mathfrak{d}(\hat{f}) = \mathfrak{d}_v(\hat{f}) := \|f'\|^2_G = \sum_e \|f'_e\|^2_{I_e},$$

where $\hat{f} = (f, F) \in \operatorname{dom}\mathfrak{d} = \mathscr{H}^1 \subset \mathsf{L}_2(G) \oplus \mathbb{C}$ and

$$\mathscr{H}^1 = \mathscr{H}^1_v := \bigoplus_e \mathsf{H}^1_0(I_e) \oplus \mathbb{C}.$$

Note that in Sect. 2.3, we used the notation $\hat{\mathfrak{d}}_0$ for $\mathfrak{d}$. Moreover, $\mathsf{H}^1_0(I_e) = \mathsf{H}^1(I_e, 0) = \{ f \in \mathsf{H}^1(I_e) \mid f_e(0) = 0\}$ is the Sobolev space with Dirichlet condition in 0. Obviously, the quadratic form $\mathfrak{d}$ is *decoupled*, i.e. $\mathfrak{d} = \bigoplus_e \mathfrak{d}^{\mathrm{D}}_e \oplus \mathfrak{o}$, where $\mathfrak{d}^{\mathrm{D}}_e$ is the quadratic form on $\mathsf{H}^1_0(I_e)$ with Dirichlet condition at 0 and $\mathfrak{o}$ is the trivial quadratic form on $\mathbb{C}$ (see also Corollary 2.3.4). The corresponding operator

$$\Delta = \Delta_v = \oplus \Delta^0_{I_e} \oplus 0,$$

is also decoupled, where $\Delta^0_{I_e}$ is the Laplacian with Dirichlet condition at 0 and Neumann condition at ℓ_e (if $\ell_e < \infty$).

Moreover, the boundary triple $(\Gamma_v, \Gamma'_v, \mathscr{G}_v)$ associated with $\mathfrak{d}_v$ is given by $\mathscr{G}_v := \mathbb{C}^{E_{v,\mathrm{int}}}$ and

$$\Gamma_v : \mathscr{H}^1_v \longrightarrow \mathbb{C}^{E_{v,\mathrm{int}}}, \qquad \Gamma'_v : \mathscr{W}^2_v \longrightarrow \mathbb{C}^{E_{v,\mathrm{int}}},$$

$$\Gamma_v \hat{f} := \{f_e(\ell_e)\}_{e \in E_{v,\mathrm{int}}}, \qquad \Gamma'_v \hat{f} := \{f'_e(\ell_e)\}_{e \in E_{v,\mathrm{int}}},$$

where

$$\mathscr{W}^2_v := \bigoplus_e \mathsf{H}^2_0(I_e) \oplus \mathbb{C} \qquad \text{and} \qquad \mathsf{H}^2_0(I_e) = \mathsf{H}^2(I_e) \cap \mathsf{H}^1_0(I_e)$$

with the corresponding maximal operator $\check{\Delta}$. Note that the boundary triple is *decoupled*, i.e. $(\Gamma_v, \Gamma'_v, \mathscr{G}_v)$ is the direct sum of the boundary triples $(\Gamma_e, \Gamma'_e, \mathscr{G}_e)$ associated with the form $\mathfrak{d}^{\mathrm{D}}_e$, where

$$\Gamma_e f_e := f_e(\ell_e), \qquad \Gamma'_e f_e := f'_e(\ell_e), \qquad \mathscr{G}_e := \mathbb{C}.$$

The following result is now obvious:

Proposition 6.1.6.

1. *The triple $(\Gamma_v, \Gamma'_v, \mathscr{G}_v)$ is a bounded, elliptic boundary triple associated with $\mathfrak{d} = \mathfrak{d}_v$ with $\|\Gamma_v\|_{1\to 0} \le (2/\ell_-(v))^{1/2}$.*
2. *The corresponding Neumann and Dirichlet operators are*

$$\Delta_v = \Delta^{\mathrm{N}}_v = \bigoplus_{e \in E_v} \Delta^0_{I_e} \oplus 0 \quad \textit{and} \quad \Delta^{\mathrm{D}}_v = \bigoplus_{e \in E_v} \Delta^{\partial I_e}_{I_e} \oplus 0,$$

where $\Delta^{\partial I_e}_{I_e}$ is the Laplacian with Dirichlet conditions on ∂I_e. Moreover,

$$\sigma(\Delta^{\mathrm{N}}_v) = \Big\{ \frac{(2k-1)^2}{(2\ell_e)^2} \,\Big|\, e \in E,\ k \in \mathbb{N} \Big\} \quad \textit{and} \quad \sigma(\Delta^{\mathrm{D}}_v) = \Big\{ \frac{k^2}{\ell_e^2} \,\Big|\, e \in E,\ k \in \mathbb{N} \Big\}.$$

3. *Let $F = \{F_e\}_{e \in E_{v,\mathrm{int}}} \in \mathscr{G}_v$ and $z \in \mathbb{C} \setminus \mathbb{R}_+$. Then the Dirichlet solution operator is given by $h = S_v(z) F$, where $h_e = F_e \sin_{z,e,+}$ if $e \in E_{v,\mathrm{int}}$ and $h_e = 0$ if $e \in E_{v,\mathrm{ext}}$.*

4. *The Dirichlet-to-Neumann map is given by*

$$\Lambda_v(z) = \bigoplus_{e \in E_{v,\mathrm{int}}} \Lambda_e(z), \quad \textit{where} \quad \Lambda_e(z) = \sqrt{z}\cot(\sqrt{z}\,\ell_e).$$

6.1.3 The Borderline Case

In the borderline case, the limit space is again an *extended* quantum graph $(G, \mathbb{C}p, L)$ with weighted standard vertex space $\mathbb{C}p$, but now with non-trivial operator L. Since the vertex space $\mathbb{C}p$ is one-dimensional, $L = L(v)$ is the multiplication with a real number. Let us recall the definitions from Sect. 2.3. The underlying Hilbert space is the same as in (6.6). The quadratic form is given by

$$\mathfrak{d}(\hat{f}) = \mathfrak{d}_v(\hat{f}) := \|f'\|_G^2 = \sum_e \|f_e'\|_{I_e}^2$$

with domain $\operatorname{dom}\mathfrak{d} = \mathscr{H}^1$ and

$$\mathscr{H}^1 = \mathscr{H}_v^1 := \{\, \hat{f} = (f, F) \in \mathsf{H}_p^1(G) \oplus \mathbb{C} \,|\, f(v) = LF \,\}$$

coupling the extra state with the values of f at the vertex v. In the notation of Sect. 2.3, we have $\mathfrak{d} = \hat{\mathfrak{d}}_L$ and $\mathscr{H}^1 = \hat{\mathscr{H}}_L^1$. We will fix the quantities $p_e(v) > 0$ and $L = L(v) > 0$ in Sect. 6.6. The corresponding operator acts as

$$\Delta_v \hat{f} = \Big(-f'', -\frac{L}{|p|^2} \sum_{e \in E_v} p_e f_e'(0)\Big) \tag{6.7a}$$

for $\hat{f} = (f, F) \in \operatorname{dom}\Delta_v = \mathscr{H}^2$, where

$$\mathscr{H}^2 := \{\, (f, F) \in \mathsf{H}_p^2(G) \oplus \mathbb{C} \,|\, f(v) = LF, \;\; f_e'(\ell_e) = 0 \;\forall e \in E_{\mathrm{int}} \,\} \tag{6.7b}$$

(see Corollary 2.3.5).

The boundary triple $(\Gamma_v, \Gamma_v', \mathscr{G}_v)$ associated with $\mathfrak{d}_v$ in this situation is given by $\mathscr{G}_v := \mathbb{C}^{E_{v,\mathrm{int}}}$ and

$$\Gamma_v : \mathscr{H}_v^1 \longrightarrow \mathbb{C}^{E_{v,\mathrm{int}}}, \qquad \Gamma_v' : \mathscr{W}_v^2 \longrightarrow \mathbb{C}^{E_{v,\mathrm{int}}},$$
$$\Gamma_v \hat{f} := \{f_e(\ell_e)\}_{e \in E_{v,\mathrm{int}}}, \qquad \Gamma_v' \hat{f} := \{f_e'(\ell_e)\}_{e \in E_{v,\mathrm{int}}},$$

where

$$\mathscr{W}_v^2 := \{\, \hat{f} = (f, F) \in \mathsf{H}_p^2(G) \oplus \mathbb{C} \,|\, f(v) = LF \,\}.$$

is the Sobolev space of order 2, and the (Sobolev-)maximal operator is defined formally as in (6.7a), but with domain $\mathscr{W}_v^2$.

The proof of the following result is very similar to the one of Proposition 6.1.4:

Proposition 6.1.7.

1. *The triple* $(\Gamma_v, \Gamma_v', \mathscr{G}_v)$ *is a bounded, elliptic boundary triple associated with* $\mathfrak{d} = \mathfrak{d}_v$ *and* $\|\Gamma_v\|_{1\to 0} \le (2/\ell_-(v))^{1/2}$.
2. *The corresponding Neumann operator is* Δ_v *with domain given in (6.7b). The corresponding Dirichlet operator has the same domain, but with the conditions* $f_e(\ell_e) = 0$ *at the finite ends,* $e \in E_{v,\mathrm{int}}$.
3. *Let* $F = \{F_e\}_{e\in E_{v,\mathrm{int}}} \in \mathscr{G}_v$ *and* $z \in \mathbb{C} \setminus \mathbb{R}_+$. *Then the Dirichlet solution operator is given by* $\hat{h} = S_v(z)F$, *where* $\hat{h} = (h, H)$ *and*

$$h_e(s) = \begin{cases} h(v)p_e \sin_{z,e,-}(s) + F_e \sin_{z,e,+}(s), & \text{if } e \in E_{v,\mathrm{int}} \\ h(v)p_e e^{\mathrm{i}\sqrt{z}s}, & \text{if } e \in E_{v,\mathrm{ext}} \end{cases}$$

with

$$h(v) = \frac{1}{\widetilde{C}(z)} \sum_{e\in E_{v,\mathrm{int}}} \frac{p_e}{\sin_z \ell_e} \cdot F_e \quad \text{and}$$

$$\widetilde{C}(z) = \sum_{e\in E_{v,\mathrm{int}}} p_e^2 \cot_z \ell_e - \mathrm{i} \sum_{e\in E_{v,\mathrm{ext}}} p_e^2 - \frac{\sqrt{z}}{L^2}.$$

4. *The Dirichlet-to-Neumann map is represented by the matrix*

$$\Lambda_v(z)_{e_1,e_2} = \begin{cases} -\dfrac{\sqrt{z}}{\widetilde{C}(z)} \cdot \dfrac{p_{e_1} p_{e_2}}{\sin_z \ell_{e_1} \cdot \sin_z \ell_{e_2}}, & \text{if } e_1 \ne e_2 \\ -\dfrac{\sqrt{z}}{\widetilde{C}(z)} \cdot \dfrac{p_{e_1}^2}{(\sin_z \ell_{e_1})^2} + \sqrt{z}\cot_z \ell_{e_1}, & \text{if } e_1 = e_2 \end{cases}$$

for $e_1, e_2 \in E_{v,\mathrm{int}}$.

Let us calculate the Dirichlet solution and Dirichlet-to-Neumann map in the special case of an equilateral star graph $\ell_e = 1$ for all $e \in E_v$:

Corollary 6.1.8. *Assume that* $\ell_e = 1$ *for all* $e \in E_v$ *(in particular, there are no exterior edges).*

1. *The spectrum of the (Neumann) and Dirichlet operator* Δ_v *and* Δ_v^{D} *is given by*

$$\sigma(\Delta_v^{\mathrm{N}}) = \Big\{ \frac{(2k-1)^2\pi^2}{4} \,\Big|\, k \in \mathbb{N} \Big\} \cup \{ \lambda > 0 \,|\, D(\sqrt{\lambda}) = 0 \} \quad \text{and}$$

$$\sigma(\Delta_v^{\mathrm{D}}) = \{ \pi^2 k^2 \,|\, k \in \mathbb{N} \} \cup \{ \lambda > 0 \,|\, D(\sqrt{\lambda}) = 0 \}$$

respectively, where

$$D(w) := \cos w - \frac{1}{|p|^2 L^2} \cdot w \sin w.$$

2. *Let* $F = \{F_e\}_{e \in E_v} \in \mathscr{G}_v$ *and* $z \in \mathbb{C} \setminus \mathbb{R}_+$. *Then the Dirichlet solution operator is given by* $h = S_v(z)F$, *where*

$$h_e(s) = h(v) p_e \sin_{z,e,-}(s) + F_e \sin_{z,e,+}(s)$$

with

$$h(v) = \frac{1}{D(\sqrt{z})|p|^2} \sum_{e \in E_v} p_e F_e$$

3. *The Dirichlet-to-Neumann map is given by*

$$\Lambda_v(z) = \sqrt{z}\Big(\frac{-1}{D(\sqrt{z}) \sin \sqrt{z}} \cdot P + \cot \sqrt{z} \cdot \mathbb{1}\Big),$$

where P *is the orthogonal projection onto* $\mathbb{C}p$ *in* $\mathbb{C}^{E_v}$. *The eigenvalues of* $\Lambda(z)$ *are*

$$-\sqrt{z}((D(\sqrt{z}) \sin \sqrt{z})^{-1} - \cot \sqrt{z}) \quad \text{and} \quad \sqrt{z} \cot \sqrt{z}$$

with eigenvector p *and eigenspace* $p^\perp$ *of dimension* $\deg v - 1$, *respectively. In particular, for* $z = -\kappa^2$ *and* $\kappa > 0$, *the eigenvalues of* $\Lambda(-\kappa^2)$ *are positive and given by*

$$\kappa\Big(\coth \kappa - \frac{1}{(\cosh \kappa + (|p|^2 L)^{-2} \kappa \sinh \kappa) \sinh \kappa}\Big) \qquad \text{and} \qquad \kappa \coth \kappa.$$

Note that formally, $L = 0$ corresponds to the slowly decaying case and $L = \infty$ to the fast decaying case (in the sense that $L \to \infty$ enforces $F = 0$ and $\sum_e p_e f_e'(0) = 0$). Moreover, $D(w) = D_L(w) \to \cos w$ as $L \to \infty$, and we obtain the formula for the Dirichlet-to-Neumann map as in the fast decaying case.

6.2 The Manifold Models

6.2.1 *A Simple Graph-Like Manifold*

Edge and Vertex Neighbourhoods

In this section, we construct a building block consisting roughly of the neighbourhood of the star-graph $G = G_v$. Let us first treat the case when all length ℓ_e are *finite*; for the general case see Sect. 6.2.4. Assume that $X = X_v^+$ is a d-dimensional

Riemannian manifold having a finite number of cylindrical ends X_e, labelled by $e \in E = E_v$. More precisely, let $X_e, X_v \subset X$ be d-dimensional, compact, connected submanifolds, such that

$$X = X_v^+ = X_v \,\overline{\dot\cup}\, \overline{\dot{\bigcup_{e \in E}}} X_e, \tag{6.8}$$

i.e. the union is disjoint up to sets of d-dimensional measure 0. For the cylindrical ends, we assume that

$$X_e = I_e \times Y_e$$

with product metric, where $I_e := [0, \ell_e]$ and where Y_e is a compact connected Riemannian manifold of dimension $m := d - 1$.

Definition 6.2.1. We call the manifold X a *graph-like manifold* associated with the metric star graph G. Moreover, we call X_v and X_e the *vertex neighbourhood* and *edge neighbourhood*, respectively. Finally, we call the manifold Y_e the *transversal manifold* associated with the edge e.

Let us now carefully describe the different boundary components. For the general notion we refer to Sect. 5.1.2.

The Boundaryless Case

We first treat the case when $Y_e = \emptyset$ for all edges e. Then the boundary of X consists of

$$\partial X = \partial_1 X = \dot{\bigcup_e} \{\ell_e\} \times Y_e \qquad \text{and} \qquad \partial_0 X = \emptyset,$$

i.e. the total boundary equals the *longitudinal boundary*, the *transversal boundary* $\partial_0 X$ is empty.

The boundary of the subset $X_v \subset X$ consists of internal boundary only, i.e.

$$\partial X_v = \mathring{\partial} X_v = \dot{\bigcup_e} \partial_e X_v,$$

where $\partial_e X_v \cong Y_e$ is the component meeting the edge neighbourhood X_e. The boundary of the cylindrical end X_e is given by

$$\partial X_e = \mathring{\partial} X_e \,\dot\cup\, \partial_1 X_e,$$

where

$$\partial_1 X_e = \{\ell_e\} \times Y_e \qquad \text{and} \qquad \mathring{\partial} X_e = \partial_v X_e = \{0\} \times Y_e \tag{6.9}$$

denotes the longitudinal and internal boundary, respectively. In this situation, all boundary components are *smooth* submanifolds (provided all manifolds Y_e are smooth).

The Case of Transversal Boundaries

Let us now treat the case when *all* transversal manifolds Y_e have non-trivial boundary. The mixed case can be treated similarly. In this situation, the boundary of X consists of

$$\partial X = \partial_1 X \,\overline{\dot\cup}\, \partial_0 X,$$

where

$$\partial_1 X = \dot{\bigcup_e} \{\ell_e\} \times Y_e \qquad \text{and} \qquad \partial_0 X = \overline{\partial X \setminus \partial_1 X}$$

denotes the *longitudinal boundary* and *transversal boundary*, respectively. Moreover,

$$\partial X_v = \mathring{\partial} X_v \,\overline{\dot\cup}\, \partial_0 X_v$$

is the decomposition into the internal and transversal boundary of the vertex neighbourhood X_v, where

$$\mathring{\partial} X_v = \dot{\bigcup_e} \partial_e X_v,$$

and $\partial_e X_v \cong Y_e$ is the component meeting the edge neighbourhood X_e, and where $\partial_0 X_v = \partial_0 X \cap X_v$. In addition,

$$\partial X_e = \partial_v X_e \,\overline{\dot\cup}\, \partial_0 X_e \,\overline{\dot\cup}\, \partial_1 X_e$$

is the decomposition into the internal, transversal and longitudinal boundary of the edge neighbourhood X_e, respectively. Here,

$$\partial_0 X_e = I_e \times \partial Y_e,$$

and $\partial_1 X_e$ and $\mathring{\partial} X_e$ are given as in (6.9). Note that in both cases, we have

$$\partial_v X_e = \partial_e X_v \cong Y_e \qquad \text{and} \qquad \partial_1 X_e \cong Y_e$$

isometrically.

A Further Decomposition of the Vertex Neighbourhood

In order to show Sobolev trace estimates as in (6.20), we need to decompose the vertex neighbourhood X_v into further components, namely,

$$X_v = X_v^- \,\overline{\dot\cup}\, \overline{\dot{\bigcup_{e \in E}}} X_{v,e} \tag{6.10}$$

where $X_v^-, X_{v,e} \subset X_v$ are again closed subsets of X and d-dimensional compact connected submanifolds, and where $X_{v,e}$ is a *collar neighbourhood* of the boundary component $\partial_e X_v$. More precisely, we assume that

$$X_{v,e} = I_{v,e} \times Y_e, \qquad I_{v,e} = [0, \ell_{v,e}],$$

together with the product metric. We assume that

$$\tau \min_e \ell_e \le \ell_{v,e} \le \ell_{\max} \qquad \forall\, e \in E_v,$$

where $0 < \tau \le 1$ and $\ell_{\max} > 0$. For convenience, we choose $\tau = 1$ and $\ell_{\max} = 1$, so that

$$\ell_-(v) \le \min_e \ell_e \le \min_e \ell_{v,e} \le 1, \tag{6.11}$$

but any other value for τ and $\ell_{\max}$ would work as well, as far as it is independent of v. Sometimes it is useful to keep the independent lengths scale; for example when dealing with approximations of other vertex conditions as in [EP09].

Remark 6.2.2. It is easy to see that the decomposition (6.10) always exists if X has smooth transversal boundary $\partial_0 X$ (or if $\partial_0 X = \emptyset$). If X_v does not have long enough cylindrical ends $X_{v,e}$, we start with another decomposition of X into $\hat{X}_e$ and $\hat{X}_v$. More precisely, $\hat{X}_v$ consists of X_v and a cylinder of length $\tau\ell_e$ taken from the (old) edge neighbourhood X_e for each edge. The new edge neighbourhood $\hat{X}_e$ has now length $(1-\tau)\ell_e$. In order to obtain an edge neighbourhood of full length ℓ_e, we need to stretch the length coordinate $\hat{s}_e \in \hat{I}_e = [\tau\ell_e, \ell_e]$ into $s_e \in I_e = [0, \ell_e]$ by an affine-linear transformation.

We have seen in Proposition 5.3.7 that such a rescaling of the longitudinal direction leads to $\hat{\delta}$-quasi-unitary equivalent quadratic forms on each edge, where $\hat{\delta} = 2\tau(1-\tau)^{-1/2}$. Note that we have $r(s) = 1$ and $\ell(s, y) = 1$ in the notation of Proposition 5.3.7. Moreover, it is easy to see that the quadratic forms on the full space $\hat{X}_v \mathbin{\dot\cup} \dot{\bigcup}_e \hat{X}_e$ resp. $\hat{X}_v \mathbin{\dot\cup} \dot{\bigcup}_e X_e$ (cf. Sect. 6.2.3 below) are again $\hat{\delta}$-quasi-unitarily equivalent. In particular, as identification operators we define $J := \mathrm{id}_{\hat{X}_v} \oplus \bigoplus_e J_e$ (recall that $(J_e f)(\hat{s}_e, y_e) = f(s_e, y_e)$ as in the proof of Proposition 5.3.7) and similarly for J'. These operators respect the spaces of order 0 and 1 (but not of higher order).

In the scaled version X_ε of the manifold X defined below, we can choose $\tau = \mathrm{O}(\varepsilon)$.

The Scaled Graph-Like Manifold

In what follows, we need a shrinking parameter $0 < \varepsilon \le 1$. Set

$$X_{\varepsilon,e} := I_e \times \varepsilon Y_e \qquad \text{and} \qquad X_{\varepsilon,v} := \varepsilon X_v; \tag{6.12}$$

i.e. $X_{\varepsilon,e}$ and $X_{\varepsilon,v}$ are the manifolds $X_e = I_e \times Y_e$ and X_v with metrics

$$g_{\varepsilon,e} = \mathrm{d}s_e^2 + \varepsilon^2 h_e \qquad \text{and} \qquad g_{\varepsilon,v} = \varepsilon^2 g_v,$$

where h_e and g_v are metrics on Y_e and X_v, respectively. Moreover, we decompose the vertex neighbourhood $X_{\varepsilon,v}$ as above into

$$X_{\varepsilon,v} = X^-_{\varepsilon,v} \,\overline{\dot\cup}\, \overline{\dot\bigcup_{e\in E}} X_{\varepsilon,v,e},$$

where

$$X_{\varepsilon,v,e} = \varepsilon(I_{v,e} \times Y_e) \qquad \text{and} \qquad X^-_{\varepsilon,v} = \varepsilon X^-_v.$$

The regularity of the scaled manifold X_ε is as good as the regularity of the unscaled manifold X, i.e. if the edge and vertex neighbourhoods X_v and X_e fit smoothly (resp. Lipschitz continuous) together, then $X_{\varepsilon,v}$ and $X_{\varepsilon,e}$ fit smoothly (resp. Lipschitz continuous) to a manifold

$$X_\varepsilon = X_{\varepsilon,v} \,\overline{\dot\cup}\, \overline{\dot\bigcup_{e\in E}} X_{\varepsilon,e}$$

(see Figs. 6.1 and 6.2).

Definition 6.2.3. We call the Riemannian manifold X_ε constructed above the *scaled graph-like manifold* associated with the metric star graph G with vertex neighbourhood scaling ε.

We will frequently refer to X_ε simply as *graph-like manifold*, although more precisely, we should speak of the *family* $\{X_\varepsilon\}_\varepsilon$ of graph-like manifolds. However, no confusion should occur.

Moreover, we use again the short hand notation X_ε for the Riemannian manifold (X, g_ε), where g_ε is given by $g_{\varepsilon,e}$ and $g_{\varepsilon,v}$ on the components $X_{\varepsilon,e}$ and $g_{\varepsilon,v}$, respectively.

For the different boundary components we employ a similar notation as in the unscaled case. In particular, we have the scaling behaviour

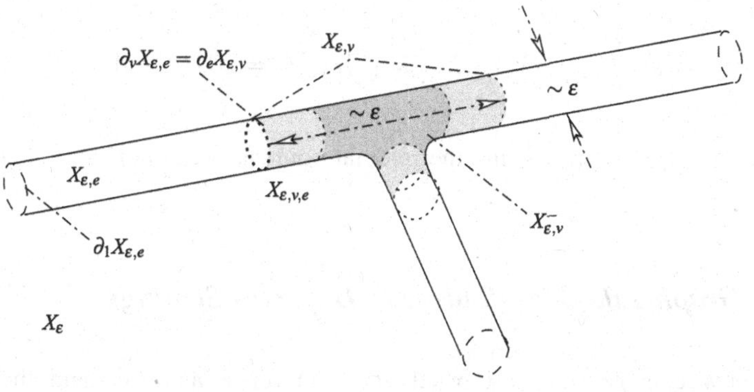

Fig. 6.1 A smooth 2-dimensional graph-like manifold with vertex neighbourhood scaling ε: The transversal manifolds Y_e are 1-dimensional spheres, and have no boundary. Moreover, the total boundary ∂X_ε of X_ε equals the longitudinal boundary $\partial_1 X_\varepsilon$ and consists of three 1-spheres here (*dashed line*). The internal boundaries $\mathring{\partial} X_{\varepsilon,e}$, $\mathring{\partial} X_{\varepsilon,v}$ and $\mathring{\partial} X^-_{\varepsilon,v}$ are drawn by *dotted lines*

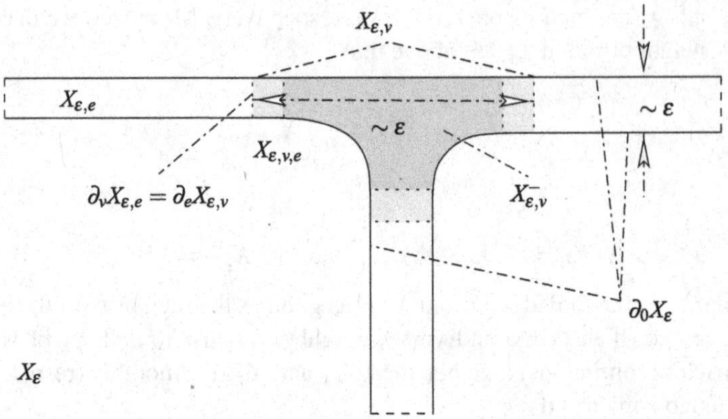

Fig. 6.2 A 2-dimensional graph-like manifold with smooth transversal boundary $\partial_0 X_\varepsilon$ (*solid line*) with vertex neighbourhood scaling ε. The transversal manifolds Y_e are 1-dimensional intervals. The longitudinal boundary $\partial_1 X_\varepsilon$ is drawn *dashed*, and the internal boundaries $\mathring{\partial} X_{\varepsilon,e}$, $\mathring{\partial} X_{\varepsilon,v}$ and $\mathring{\partial} X^-_{\varepsilon,v}$ are drawn *dotted*

$$\partial_1 X_\varepsilon = \varepsilon \partial_1 X$$

for the longitudinal boundary of the entire manifold. Moreover, for the edge neighbourhood, we have

$$\mathring{\partial} X_{\varepsilon,e} = \partial_v X_{\varepsilon,e} = \varepsilon \partial_v X_e = \varepsilon \mathring{\partial} X_e,$$

and

$$\partial_0 X_{\varepsilon,e} = I_e \times \varepsilon \partial Y_e \qquad \text{and} \qquad \partial_1 X_{\varepsilon,e} = \varepsilon \partial_1 X_e;$$

recall that $\partial_0 X_e = I_e \times \partial Y_e$. For the vertex neighbourhood, we have behaviour

$$\partial_1 X_{\varepsilon,v} = \varepsilon \partial_1 X_v, \quad \mathring{\partial} X_{\varepsilon,v} = \dot{\bigcup_e} \partial_e X_{\varepsilon,v} = \dot{\bigcup_e} \varepsilon \partial_e X_v = \varepsilon \mathring{\partial} X_v \quad \text{and} \quad \partial_e X_{\varepsilon,v} = \varepsilon \partial_e X_v$$

for the transversal boundary and the internal boundary (components) of $X_{v,e}$ (see Fig. 6.2).

6.2.2 *Graph-Like Manifolds with Different Scalings*

Let us now consider the case when the transversal manifold and the vertex neighbourhood have a different scaling. Namely, we assume that the edge and vertex neighbourhoods are defined as

$$X_{\varepsilon,e} := I_e \times \varepsilon Y_e \qquad \text{and} \qquad X_{\varepsilon,v} := \varepsilon^\alpha X_v; \tag{6.13}$$

i.e. $X_{\varepsilon,e}$ and $X_{\varepsilon,v}$ are given by the manifolds $I_e \times Y_e$ and X_v with metrics

$$g_{\varepsilon,e} = \mathrm{d}s_e^2 + \varepsilon^2 h_e \qquad \text{and} \qquad g_{\varepsilon,v} = \varepsilon^{2\alpha} g_v,$$

respectively. Here, the vertex neighbourhood scales with a *slower* rate

$$0 < \alpha < 1$$

than the transversal manifolds Y_e.

In order that the boundary components $\partial_v X_{\varepsilon,e} \cong \varepsilon Y_e$ of the edge neighbourhood and the corresponding component $\varepsilon^\alpha \partial_e X_v \cong \varepsilon^\alpha Y_e$ of the vertex neighbourhood fit together, we have to assume a sort of homogeneity on the transversal manifold Y_e. In particular, we assume that ∂Y_e is non-trivial for all e, and that for each edge e, there is an isometric embedding

$$\varphi_{\varepsilon,e} : \varepsilon Y_e \hookrightarrow Y_e \tag{6.14}$$

for all $0 < \varepsilon \leq 1$, i.e. there is a differentiable map

$$\varphi_{\varepsilon,e} : Y_e \hookrightarrow Y_e$$

such that

$$h_e(y)\big(\mathrm{d}\varphi_{\varepsilon,e}(y).v, \mathrm{d}\varphi_{\varepsilon,e}(y).v\big) = \varepsilon^2 h_e(y)(v, v)$$

for all $v \in T_y Y_e$ and $y \in Y_e$. We assume that the longitudinal derivative $\varphi'_{\varepsilon,e}$ defined as

$$\varphi'_{\varepsilon,e}(y) := \partial_\varepsilon \varphi_{\varepsilon,e}(y) \in T_{\widetilde{y}} Y_e, \qquad \widetilde{y} = \varphi_{\varepsilon,e}(y),$$

is bounded by some constant $C_e > 0$ depending only on Y_e, i.e. that

$$|\varphi'_{\varepsilon,e}(y)|^2_{h_e(\widetilde{y})} := h_e\big(\varphi'_{\varepsilon,e}(y), \varphi'_{\varepsilon,e}(y)\big) \leq C_e^2 \tag{6.15}$$

for all $y \in Y_e$ and $0 < \varepsilon \leq 1$.

The basic example we have in mind is $Y_e = \overline{B}(0, r_e) \subset \mathbb{R}^m$, i.e. a closed ball in Euclidean space, together with

$$\varphi_{\varepsilon,e}(y) := \varepsilon y \tag{6.16}$$

on the unscaled space Y_e (see Fig. 6.3). Here, we have

$$\varphi'_{\varepsilon,e}(y) = y \qquad \text{and} \qquad |\varphi'_{\varepsilon,e}(y)| = |y| \leq r_e =: C_e,$$

i.e. the constant C_e depends only on the radius of Y_e. Note that the embedding $\varphi_{\varepsilon,e}$ is not unique. We could alternatively define $\varphi_{\varepsilon,e}(y) := \varepsilon y - y_0$ for some point $y_0 \in B(0, r_e)$ provided ε is small enough.

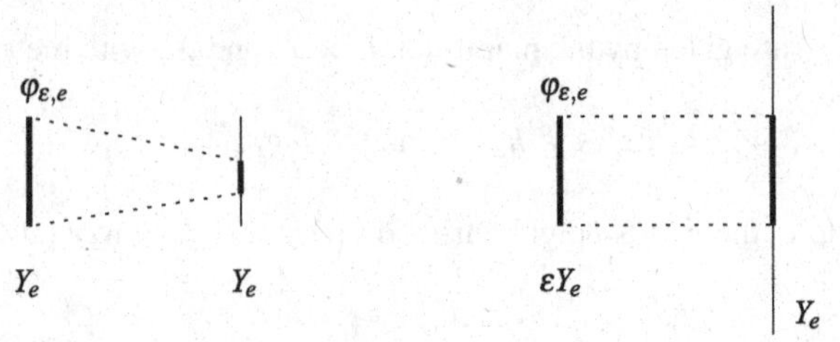

Fig. 6.3 The embedding $\varphi_{\varepsilon,e}$ for Y_e being an interval, on the left the unscaled manifold (on which we formally define the map $\varphi_{\varepsilon,e}$), on the right the isometric picture of the scaled version

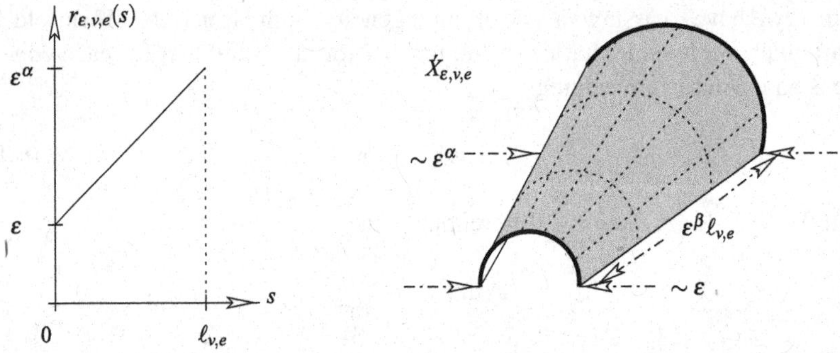

Fig. 6.4 The radius function $r_{\varepsilon,v,e}$ and the truncated cone $\acute{X}_{\varepsilon,v,e}$ with underlying space $X_{v,e} = I_{v,e} \times Y_e$ and warped product metric $\acute{g}_{\varepsilon,e} = \varepsilon^{2\beta} ds^2 + r_{\varepsilon,v,e}(s)^2 h_e$. The truncated cone has length $\varepsilon^\beta \ell_{v,e}$ and two ends εY_e and $\varepsilon^\alpha Y_e$. Here, the transversal manifold Y_e is an interval. Note that the coordinate lines (*dotted*) of the underlying space are orthogonal

Denote by

$$X_{\varepsilon,v,e} = \varepsilon^\alpha (I_{v,e} \times Y_e) \qquad \text{and} \qquad X^-_{\varepsilon,v} = \varepsilon^\alpha X^-_v$$

the scaled subsets of the decomposition (6.10); recall that $I_{v,e} = [0, \ell_{v,e}]$, and that $X_{\varepsilon,v,e} = \varepsilon^\alpha X_{v,e}$ is a short hand notation for the manifold

$$X_{v,e} := I_{v,e} \times Y_e \quad \text{with product metric} \quad g_{\varepsilon,v,e} := \varepsilon^{2\alpha}(ds^2 + h_e).$$

In the following, we need a truncated cone defined by

$$\acute{X}_{\varepsilon,v,e} := \varepsilon^\beta I_{v,e} \times_{r_{\varepsilon,v,e}} Y_e. \tag{6.17}$$

Here, $r_{\varepsilon,v,e}$ is the affine-linear function given by $r_{\varepsilon,v,e}(0) = \varepsilon$ and $r_{\varepsilon,v,e}(\ell_{v,e}) = \varepsilon^\alpha$ (see Fig. 6.4), and $\acute{X}_{\varepsilon,v,e}$ is the corresponding warped product (see Definition 5.3.2), i.e. $\acute{X}_{\varepsilon,v,e}$ is the Riemannian manifold with underlying (ε-independent!) space $X_{v,e}$ with warped product metric

$$\acute{g}_{\varepsilon,v,e} = \varepsilon^{2\beta} \mathrm{d}s_e^2 + r_{\varepsilon,v,e}(s_e)^2 h_e,$$

where

$$\beta > \alpha$$

is an additional scaling parameter to be fixed later on. The homogeneity of Y_e allows us to embed the manifold $\acute{X}_{\varepsilon,v,e}$ into $X_{\varepsilon,v,e} = \varepsilon^\alpha X_{v,e}$ as follows. For simplicity, we omit the labels e and v occasionally, e.g., we write $\psi_\varepsilon = \psi_{\varepsilon,v,e}$ for the embedding and $r_\varepsilon = r_{\varepsilon,v,e}$ and for the radius function. We define the embedding by

$$\psi_\varepsilon \colon \acute{X}_{\varepsilon,v,e} \hookrightarrow X_{\varepsilon,v,e}, \qquad \psi_\varepsilon(s,y) = (\varepsilon^{\beta-\alpha} s, \varphi_{\widetilde{r}_\varepsilon(s),e}(y)) \in I_{v,e} \times Y_e$$

(see Fig. 6.5),
where

$$\widetilde{r}_\varepsilon(s) = \varepsilon^{-\alpha} r_\varepsilon(s) = \varepsilon^{1-\alpha} + \frac{(1-\varepsilon^{1-\alpha})s}{\ell_{v,e}}.$$

Note that the additional factor $\varepsilon^{-\alpha}$ comes from the fact that the underlying space of the Riemannian manifold $X_{\varepsilon,v,e} = \varepsilon^\alpha X_{v,e}$ is the ε-independent space $X_{v,e} = I_{v,e} \times Y_e$.

In general, the embedding is *not* isometric: For example the coordinate lines of the embeddings are not all orthogonal as it is for the coordinate lines of the truncated cone $\acute{X}_{\varepsilon,v,e}$ (see Figs. 6.4 and 6.5). Let us therefore compare the warped product metric $\acute{g}_{\varepsilon,v,e}$ on the truncated cone with the pull-back

$$\widetilde{g}_{\varepsilon,v,e} := \psi_\varepsilon^* g_{\varepsilon,v,e}$$

of the product metric on the product $X_{\varepsilon,v,e}$. We use the short hand notation

$$\widetilde{X}_{\varepsilon,v,e} \quad \text{for the Riemannian manifold} \quad (X_{v,e}, \widetilde{g}_{\varepsilon,v,e}).$$

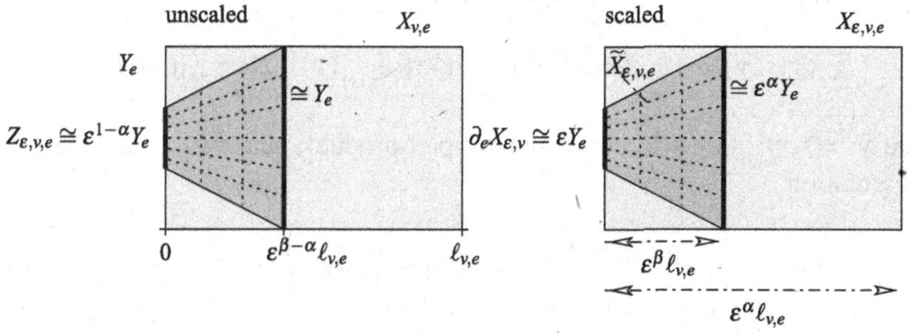

Fig. 6.5 The truncated cone (*dark grey*) embedded in the product $X_{\varepsilon,v,e}$. On the *left*, we sketched the unscaled manifold $X_{v,e}$ in which ψ_ε formally maps, the range of ψ_ε is the dark grey area. On the *right*, we have the isometric picture. Here, the dark grey area is isometric to $\widetilde{X}_{\varepsilon,v,e}$, i.e. the truncated cone with pull-back metric $\widetilde{g}_{\varepsilon,v,e} = \psi_\varepsilon^* g_{\varepsilon,v,e} = \varepsilon^{2\alpha}\psi_\varepsilon^*(\mathrm{d}s^2 + h_e)$. The coordinate lines (*dotted*) of the truncated cone are in general not orthogonal, and the pull-back metric $\widetilde{g}_{\varepsilon,v,e}$ on the cone therefore differs from the warped product metric $\acute{g}_{\varepsilon,v,e}$

Note that $\widetilde{X}_{\varepsilon,v,e}$ is (by definition) isometrically embedded in the scaled product $X_{\varepsilon,v,e} = \varepsilon^\alpha X_{v,e}$ (see Fig. 6.5). We occasionally identify $\widetilde{X}_{\varepsilon,v,e}$ with its isometric image in $X_{\varepsilon,v,e}$.

For the notion of *relative distortion* and *relative density* we refer to Definition 5.2.1.

Lemma 6.2.4. *Assume that* $0 < \alpha < 1$ *and* $0 < \varepsilon \le 1$. *Denote by* $A_\varepsilon = A_{\varepsilon,v,e}$ *(resp.* ρ_ε*) the relative distortion (resp. density) of the pull-back metric* $\widetilde{g}_{\varepsilon,v,e}$ *with respect to the warped product metric* $\acute{g}_{\varepsilon,v,e}$. *Then the relative density is trivial, i.e.* $\rho_\varepsilon = \det A_\varepsilon^{1/2} = \mathbb{1}$ *is constant on* $X_{v,e}$. *In other words, the embedding* ψ_ε *is volume-preserving. Moreover, the eigenvalues of* A_ε *are* 1 *(with multiplicity* $m-1$*),* $0 < \alpha_- \le 1$ *and* $1 \le \alpha_+ = 1/\alpha_-$*, where*

$$\alpha_+ = 1 + \frac{1}{2}\Big(\mathrm{o}_{11} + \sqrt{\mathrm{o}_{11}^2 + 4\mathrm{o}_{11}}\Big) \quad \textit{and}$$

$$\mathrm{o}_{11}(s,y) = \varepsilon^{-2\beta}(r_\varepsilon'(s))^2 |\varphi'_{\widetilde{r}_\varepsilon(s)}(y)|^2_{h_e(y)} \le \varepsilon^{-2(\beta-\alpha)} \cdot \frac{C_e^2}{\ell_{v,e}^2}.$$

Finally,

$$1 + \mathrm{o}_{11} \le \alpha_+ \le 2 + \mathrm{o}_{11} = \mathrm{O}(\varepsilon^{-2(\beta-\alpha)}).$$

Proof. We choose coordinates y_i in a neighbourhood of a point $y \in Y$. Without loss of generality, we can assume that $\{\partial_{y_1}, \dots, \partial_{y_m}\}$ is an orthonormal basis of $(T_y Y_e, h_e(y))$ *at the point* y *only*. In this way, $\{\partial_s, \partial_{y_1}, \dots, \partial_{y_m}\}$ becomes an orthonormal basis of $T_{(s,y)} X_{v,e} \cong \mathbb{R} \oplus T_y Y_e$ with metric $\mathrm{d}s^2 + h_e$.

The pull-back metric $\widetilde{g}_{\varepsilon,v,e}$ is represented by the matrix

$$\widetilde{G}_\varepsilon = \widetilde{G}_{\varepsilon,v,e} = \begin{pmatrix} \varepsilon^{2\beta} + r_\varepsilon'(s)^2 |\varphi'_{\widetilde{r}_\varepsilon(s)}(y)|^2_h & \widetilde{g}_{\varepsilon,v,e}(s,y)(\partial_s, \partial_{y_i}) \\ \widetilde{g}_{\varepsilon,v,e}(s,y)(\partial_{y_j}, \partial_s) & r_\varepsilon(s)^2 \delta_{ij} \end{pmatrix}$$

with respect to the orthonormal basis $\{\partial_s, \partial_{y_1}, \dots, \partial_{y_m}\}$, where

$$\widetilde{g}_{\varepsilon,v,e}(s,y)(\partial_s, \partial_{y_i}) = \varepsilon^{2\alpha} \widetilde{r}_\varepsilon'(s) h_e(\widetilde{y})\big(\varphi'_{\widetilde{r}_\varepsilon(s)}(y)), D\varphi_{\widetilde{r}_\varepsilon(s)}(y).\partial_{y_i}\big),$$

where $\widetilde{y} = \varphi_{\widetilde{r}_\varepsilon(s)}(y)$. Moreover, the warped product metric $\acute{g}_{\varepsilon,v,e}$ has the matrix representation

$$\acute{G}_\varepsilon = \acute{G}_{\varepsilon,v,e} = \begin{pmatrix} \varepsilon^{2\beta} & 0 \\ 0 & r_\varepsilon(s)^2 \delta_{ij} \end{pmatrix}.$$

The relative distortion $A_\varepsilon = A_{\varepsilon,v,e} = \acute{G}_\varepsilon^{-1} \widetilde{G}_\varepsilon$ of $\widetilde{g}_{\varepsilon,v,e}$ with respect to $\acute{g}_{\varepsilon,v,e}$ is represented by

$$A_\varepsilon = \begin{pmatrix} 1 + \mathrm{o}_{11} & \varepsilon^{-\beta} r_\varepsilon(s) \mathrm{o}_{12} \\ \varepsilon^\beta r_\varepsilon(s)^{-1} \mathrm{o}_{12}^* & \mathrm{id}_{TY} \end{pmatrix}$$

(for the notation see (5.22)), where

$$\mathrm{o}_{11}(s,y) = \varepsilon^{-2\beta}(r_\varepsilon'(s))^2 \big|\varphi'_{\widetilde{r}_\varepsilon(s)}(y)\big|^2_{h_e(\widetilde{y})} \in \mathbb{R} \quad \text{and}$$
$$\mathrm{o}_{12}(s,y) = \varepsilon^{-\beta} r_\varepsilon(s)^{-1} \widetilde{g}_\varepsilon(\partial_s,\cdot) \in T_y^* Y_e.$$

Moreover, we have

$$|\mathrm{o}_{12}(s,y)|^2_{(T_y^*Y,h_e(y))} = \sum_{i=1}^m \varepsilon^{-2(\beta-\alpha)}\big(\widetilde{r}_\varepsilon'(s)\big)^2 \Big|h_e(\widetilde{y})\big(\varphi'_{\widetilde{r}_\varepsilon(s)}(y), \widetilde{e}_i\big)\Big|^2 = \mathrm{o}_{11},$$

where

$$\widetilde{e}_i := \frac{1}{\widetilde{r}_\varepsilon(s)} \cdot D\varphi_{\widetilde{r}_\varepsilon(s)}(y).\partial_{y_i}.$$

Here, we used the fact that $\{\widetilde{e}_i\}_{i=1,\dots,m}$ is an orthonormal basis of $(T_{\widetilde{y}}Y_e, h_e(\widetilde{y}))$ in $\widetilde{y} = \varphi_{\widetilde{r}_\varepsilon(s)}(y)$ since $\varphi_{\widetilde{r}_\varepsilon(s)} : \widetilde{r}_\varepsilon(s) Y_e \longrightarrow Y_e$ is an isometry.

We have calculated the relative density and eigenvalues of A_ε in Sect. 5.3.1. We conclude that $\rho_\varepsilon = \det A_\varepsilon^{1/2} = \mathbb{1}$ is constant on $X_{v,e}$, since $|\mathrm{o}_{12}|^2 = \mathrm{o}_{11}$ and $\mathrm{o}_{22} = 0$ (in the notation of Sect. 5.3.1). In other words, the embedding ψ_ε is volume-preserving. In particular, $\alpha_- = 1/\alpha_+$. Finally, the estimates on α_+ are obvious, and the estimate on o_{11} follows from combining (6.15) with the fact that $r_\varepsilon'(s) = \varepsilon^\alpha(1-\varepsilon^{1-\alpha})/\ell_{v,e} \le \varepsilon^\alpha/\ell_{v,e}$. □

We can now state the main conclusion needed in further estimates. Basically, the next estimate says that we can replace the norm contribution of the derivative on the truncated cone with warped product metric by the pull-back metric up to a constant. We will see in Lemma 6.3.2 that the divergent constant can be compensated. Recall that $u'(s,y) = \partial_s u(s,y)$ is a short hand notation for the partial derivative in the first (i.e. longitudinal) variable.

Corollary 6.2.5. *We have*

$$\|u'\|^2_{\acute{X}_{\varepsilon,v,e}} \le \|du\|^2_{\acute{X}_{\varepsilon,v,e}} \le \Big(2 + \varepsilon^{-2(\beta-\alpha)} \cdot \frac{C_e^2}{\ell_{v,e}^2}\Big) \cdot \|du\|^2_{\widetilde{X}_{\varepsilon,v,e}}$$

for $u \in \mathsf{H}^1(X_{v,e})$.

Proof. The first inequality is obvious, since we have the pointwise estimate

$$|u'\mathrm{d}s|^2_{\acute{g}^*_{\varepsilon,v,e}} = \varepsilon^{-2\beta}|u'|^2 \le \varepsilon^{-2\beta}|u'|^2 + \frac{1}{r_\varepsilon^2} h_e(\mathrm{d}_{Y_e}u, \mathrm{d}_{Y_e}u) = |\mathrm{d}u|^2_{\acute{g}^*_{\varepsilon,v,e}}.$$

The second inequality follows from $\rho_\varepsilon = \mathbb{1}$ and

$$\acute{g}^*_{\varepsilon,v,e}(\xi,\xi) = \widetilde{g}_{\varepsilon,v,e}(\xi,(A^*)^{-1}\xi) \le \alpha_+ \widetilde{g}^*_{\varepsilon,v,e}(\xi,\xi)$$

for $\xi \in T^*X_{v,e}$ using (5.14). □

Let us finally fix some notation for boundary components of $X_{\varepsilon,v,e}$ and $X_{\varepsilon,v}$: The boundary of the truncated cone $\acute{X}_{\varepsilon,v,e}$ (or of $\tilde{X}_{\varepsilon,v,e}$, which is the same) has two components isometric to εY_e and $\varepsilon^\alpha Y_e$, a "smaller" and "larger" end. Let $Z_{\varepsilon,v,e} := \psi_{\varepsilon,v,e}(\{0\} \times Y_e) \cong \varphi_{\varepsilon^{1-\alpha}}(Y_e)$ be the range of the "smaller" boundary component of the truncated cone $\acute{X}_{\varepsilon,v,e} = (X_{v,e}, \acute{g}_{\varepsilon,v,e})$ under the embedding $\psi_{\varepsilon,v,e}$ into the product $X_{\varepsilon,v,e} = (X_{v,e}, g_{\varepsilon,v,e}) = (I_{v,e} \times Y_e, \varepsilon^{2\alpha} g_{v,e})$, where $g_{v,e} = \mathrm{d}s^2 + h_e$. Note that here the space $Z_{\varepsilon,v,e}$ is ε-dependent, and

$$(Z_{\varepsilon,v,e}, g_{v,e} \restriction_{Z_{\varepsilon,v,e}}) \cong (Y_e, \varepsilon^{2(1-\alpha)}) = \varepsilon^{1-\alpha} Y_e.$$

The space $Z_{\varepsilon,v,e}$ equipped with the (restricted) scaled product metric $g_{\varepsilon,v,e}$ is isometric with εY_e, i.e.

$$(Z_{\varepsilon,v,e}, g_{\varepsilon,v,e} \restriction_{Z_{\varepsilon,v,e}}) \cong (Y_e, \varepsilon^2 h_e) = \varepsilon Y_e.$$

Here, we have to break our convention, that the underlying spaces are always ε-independent, since we deal with boundary components with scaling behaviour εY_e embedded into the space $X_{\varepsilon,v,e} = \varepsilon^\alpha X_{v,e}$. We use the short hand notation

$$\partial_e X_{\varepsilon,v} := (Z_{\varepsilon,v,e}, g_{\varepsilon,v,e} \restriction_{Z_{\varepsilon,v,e}}) \tag{6.18}$$

for the Riemannian manifold corresponding to the shorter end of the truncated cone embedded into the scaled manifold $X_{\varepsilon,v,e}$ (see Fig. 6.5). The product $X_{\varepsilon,v,e}$ has a boundary component $(\{0\} \times Y_e, \varepsilon^{2\alpha} h_e) \subset X_{\varepsilon,v,e}$, denoted by $\varepsilon^\alpha \partial_e X_v$. Note that the isometric embedding

$$\partial_e X_{\varepsilon,v} \hookrightarrow \varepsilon^\alpha \partial X_v$$

is not surjective provided $\alpha < 1$ and $\varepsilon < 1$.

Identifying the smaller end $\partial_e X_{\varepsilon,v} \cong \varepsilon Y_e$ of the truncated cone with the end $\partial_v X_{\varepsilon,e} = \varepsilon \partial_x X_e = \{0\} \times \varepsilon Y_e$ of the edge neighbourhood, we obtain the total space X_ε by

$$X_\varepsilon := X_{\varepsilon,v} \,\overline{\cup}\, \overline{\dot{\bigcup_e}} X_{\varepsilon,e} = \varepsilon^\alpha X_v \,\overline{\cup}\, \overline{\dot{\bigcup_e}} I_e \times \varepsilon Y_e,$$

where we used the notation of Sect. 5.1.2 (see Figs. 6.6 and 6.7). Note that X_ε can be equipped with Lipschitz-continuous charts. In addition, we assume that X_ε has Lipschitz boundary ∂X_ε.

Now, the *internal boundary* of $X_{\varepsilon,v}$ as subset of X_ε is given by

$$\mathring{\partial} X_{\varepsilon,v} = \dot{\bigcup_e} \partial_e X_{\varepsilon,v},$$

isometric to the disjoint union of εY_e, $e \in E_v$. The remaining (closed) part of $\partial X_{\varepsilon,v}$, i.e. the *transversal boundary*, is then given by

$$\partial_0 X_{\varepsilon,v} := \overline{\partial X_{\varepsilon,v} \setminus \mathring{\partial} X_{\varepsilon,v}}$$

(see Fig. 6.6).

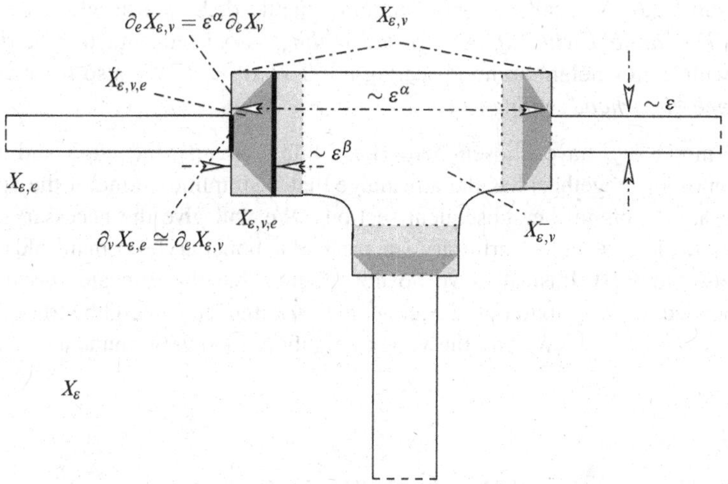

Fig. 6.6 A 2-dimensional graph-like manifold with non-smooth boundary. The shrinking rate at the vertex neighbourhood is $0 < \alpha < 1$. The transversal manifolds Y_e are 1-dimensional intervals. Note that the boundary part $\partial_v X_{\varepsilon,e} \cong \varepsilon Y_e$ (*short thick line*) is identified with the "small" end of the truncated cone $\acute{X}_{\varepsilon,v,e}$ isometric to εY_e. The transversal boundary $\partial_0 X_{\varepsilon,v}$ of $X_{\varepsilon,v}$ is plotted by a *thin solid line*

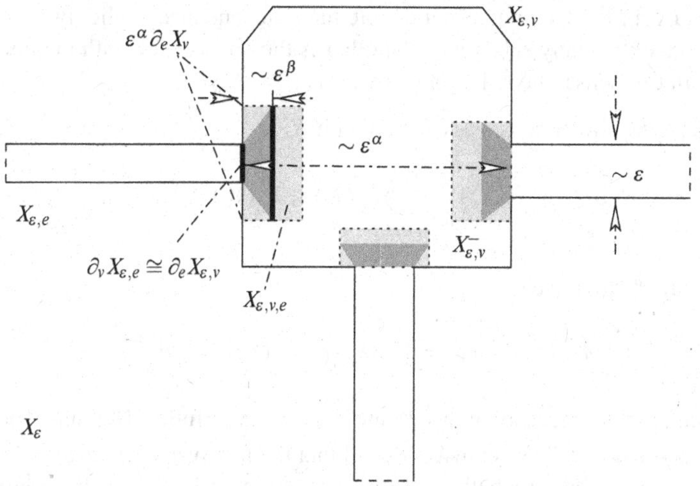

Fig. 6.7 Another 2-dimensional graph-like manifold with non-smooth boundary. Here the product sets $X_{\varepsilon,v,e}$ (*light and dark grey*) are really "inside" the vertex neighbourhood $X_{\varepsilon,v}$. The transversal boundary $\partial_0 X_{\varepsilon,v}$ of $X_{\varepsilon,v}$ is plotted by a *thin solid line*

Definition 6.2.6. We call the (non-smooth) manifold X_ε constructed above the *graph-like manifold with different vertex scaling* associated with the metric star graph with vertex neighbourhood scaling ε^α, $0 < \alpha < 1$. We also refer to α as *vertex neighbourhood scaling rate*.

The model we have chosen here (i.e. gluing directly the edge and vertex neighbourhoods together) has the advantage that it simplifies some of the approximation arguments in the subsequent sections. We will give the necessary modifications in order to allow arbitrary (in particular boundaryless) manifolds Y_e in Sects. 6.9 and 6.10. Basically, we do not assume that the truncated cone $\acute{X}_{\varepsilon,v,e}$ is embedded in $X_{\varepsilon,v}$, but put the cone *between* the edge neighbourhood with boundary $\partial_v X_{\varepsilon,e} = \varepsilon \partial_v X_e$ and the vertex neighbourhood with boundary $\partial_e X_{\varepsilon,v} = \varepsilon^\alpha X_v$.

6.2.3 The Associated Quadratic Form, Operator and Boundary Triple

Let us now define the Laplacian associated with the space X_ε. We start with the Neumann Laplacian (in the case that $\partial X_\varepsilon \neq \emptyset$). The Dirichlet case under certain special conditions will be treated in Sect. 6.11. It is convenient to define the Laplacian via its quadratic form.

Remark 6.2.7. Here and in the sequel, we denote the ε-depending spaces and operators by a tilde $\widetilde{(\cdot)}$, and do not denote the ε-dependence explicitly. One reason is not to have too many subscripts, the other is the consistency with the notation in Chap. 4. In this sense, the notation $\widetilde{(\cdot)}$ reads as $(\cdot)_\varepsilon$.

The underlying Hilbert space associated with X_ε is

$$\widetilde{\mathscr{H}} := \mathsf{L}_2(X_\varepsilon) = \bigoplus_{e \in E} \mathsf{L}_2(I_e) \otimes \mathsf{L}_2(Y_e) \oplus \mathsf{L}_2(X_v). \tag{6.19}$$

As quadratic form $\widetilde{\mathfrak{d}}$, we set

$$\widetilde{\mathscr{H}}^1 := \operatorname{dom} \widetilde{\mathfrak{d}} := \mathsf{H}^1(X_\varepsilon), \qquad \widetilde{\mathfrak{d}}(u) := \|du\|^2_{X_\varepsilon}.$$

In particular, the corresponding operator $\widetilde{\Delta} = \Delta_{X_\varepsilon}$ fulfils Neumann boundary conditions on $\partial X_\varepsilon = \partial_0 X_\varepsilon \,\bar{\cup}\, \partial_1 X_\varepsilon$. Recall that the *transversal boundary* $\partial_0 X_\varepsilon$ may be empty, depending on whether all transversal manifolds Y_e are boundaryless or not.

In order to pass from a star graph G_v and its associated graph-like manifold $X^+_{\varepsilon,v} = X_\varepsilon$ to more complicated graphs, we need the notion of boundary triples introduced in Sect. 3.4. In Chap. 7, we use these "building blocks" associated with a vertex v in order to construct a global boundary triple as in Sect. 3.9.

Define $\widetilde{\mathfrak{d}}_v := \widetilde{\mathfrak{d}}$,

$$\mathscr{H}_v := \mathsf{L}_2(X^+_{\varepsilon,v}), \qquad \mathscr{H}^1_v := \mathsf{H}^1(X^+_{\varepsilon,v}) \qquad \text{and} \qquad \mathscr{W}^2_v := \mathsf{H}^2(X^+_{\varepsilon,v}),$$

where $\mathsf{H}^2(X^+_{\varepsilon,v})$ is the usual Sobolev space of order 2.

Remark 6.2.8. We developed a theory in Sect. 3.4 *avoiding* quantitative estimates involving the second order norm, so the precise definition (possibly depending on ε) of the norm $\|\cdot\|_{\mathsf{H}^2(X^+_{\varepsilon,v})}$ is of no importance. We only need quantitative estimates on "first order" spaces, like the norm estimate on $\widetilde{\Gamma}_v$ in the proposition below.

As boundary space, we choose the L_2-space on the *longitudinal boundary* $\partial_1 X^+_{\varepsilon,v} \cong Y_{\varepsilon,v}$, consisting of the "free" cylindrical ends $\{\ell_e\} \times Y_{\varepsilon,e}$, i.e.

$$\widetilde{\mathscr{G}}_v := \mathsf{L}_2(Y_{\varepsilon,v}) = \bigoplus_{e \in E_v} \mathsf{L}_2(Y_{\varepsilon,e}) = \bigoplus_{e \in E_v} \widetilde{\mathscr{G}}_e, \qquad \widetilde{\mathscr{G}}_e = \mathsf{L}_2(Y_{\varepsilon,e}),$$

where

$$Y_{\varepsilon,v} := \dot{\bigcup_{e \in E_v}} Y_{\varepsilon,e}$$

denotes the disjoint union of all transversal cross-sections $Y_{\varepsilon,e}$ with edge e adjacent to v (see (6.1)). The boundary maps

$$\widetilde{\Gamma}_v : \mathscr{H}^1_v \longrightarrow \widetilde{\mathscr{G}}_v \qquad \text{and} \qquad \widetilde{\Gamma}'_v : \mathscr{W}^2_v \longrightarrow \widetilde{\mathscr{G}}_v$$

are defined by

$$\widetilde{\Gamma}_v u = \bigoplus_{e \in E_v} \widetilde{\Gamma}_{v,e} u \qquad \text{and} \qquad \widetilde{\Gamma}'_v u = \bigoplus_{e \in E_v} \widetilde{\Gamma}'_{v,e} u,$$

where

$$\widetilde{\Gamma}_{v,e} : \mathscr{H}^1_v \longrightarrow \widetilde{\mathscr{G}}_e, \qquad \widetilde{\Gamma}'_{v,e} : \mathscr{W}^2_v \longrightarrow \widetilde{\mathscr{G}}_e,$$
$$u \mapsto u_e(\ell_e, \cdot), \qquad u \mapsto u'_e(\ell_e, \cdot).$$

Here, $u'_e(\ell_e, \cdot)$ denotes the outer normal derivative on $\partial_1 X_{\varepsilon,e}$. Let us summarise the facts already proven in Sect. 3.4:

Proposition 6.2.9. *The triple* $(\widetilde{\Gamma}_v, \widetilde{\Gamma}'_v, \widetilde{\mathscr{G}}_v)$ *is an elliptic boundary triple associated with the quadratic form* $\widetilde{\mathfrak{d}}_v$. *Moreover,*

$$\|\widetilde{\Gamma}_v\|_{1 \to 0} \le \frac{2}{\ell_-(v)} = \frac{2}{\min_e\{\ell_e, 1\}},$$

where $\|\widetilde{\Gamma}_v\|_{1\to 0}$ denotes the norm of $\widetilde{\Gamma}_v$ as operator $\widetilde{\mathscr{H}}_v \to \widetilde{\mathscr{G}}_v$. In particular, the norm bound is independent of ε. Let $z \notin \sigma(\Delta^{\mathrm{D}}_{X^+_{\varepsilon,v}})$. The associated Dirichlet-to-Neumann map $\widetilde{\Lambda}_v(z)$ in $\widetilde{\mathscr{G}}_v$ is the usual Dirichlet-to-Neumann map, i.e. $\psi = \widetilde{\Lambda}_v(z)\varphi$ is defined by $\psi = \partial_{\mathrm{n}} h \restriction_{\partial_1 X^+_{\varepsilon,v}}$, where $h = S(z)\varphi$ is the unique solution of the Dirichlet problem

$$(\Delta_{X^+_{\varepsilon,v}} - z)h = 0, \qquad h \restriction_{\partial_1 X_\varepsilon} = \varphi.$$

Moreover, the scale of boundary Hilbert spaces agrees as vector space with the corresponding Sobolev spaces, i.e.

$$\widetilde{\mathscr{G}}^k_v = \mathsf{H}^k(Y_{\varepsilon,v}) = \bigoplus_{e \in E_v} \mathsf{H}^k(Y_{\varepsilon,e}).$$

Proof. The proof follows from elliptic theory for partial differential operators, see Theorem 3.4.39 if the transversal manifolds do not have boundary and Theorem 3.4.40 for the case with boundary. Note that the ends have a cylindrical structure as required in Theorem 3.4.40. The norm estimate on $\widetilde{\Gamma}_v$ is a consequence of Corollary A.2.12, similarly as for the Sobolev trace estimate (6.21). □

6.2.4 Manifolds with Infinite Ends

For ends of infinite length, i.e. for edges $e \in E_{\mathrm{ext}} = E \setminus E_{\mathrm{int}}$, we do not have a longitudinal boundary. In particular, the longitudinal boundary of the graph-like manifold X is given by

$$\partial_1 X = \bigcup_{e \in E_{\mathrm{int}}} \partial_1 X_e = \bigcup_{e \in E_{\mathrm{int}}} \{\ell_e\} \times Y_e$$

and similarly for the scaled graph-like manifold X_ε. Consequently, the boundary space associated with X_ε is given by

$$\mathscr{G} := \bigoplus_{e \in E_{\mathrm{int}}} \mathsf{L}_2(Y_{\varepsilon,e}),$$

i.e. we only have to replace the edge set E by E_{int} in Sect. 6.2.3.

6.3 Some Vertex Neighbourhoods Estimates

Let us now provide estimates for functions on the vertex neighbourhood in terms of the natural scale of Hilbert space of order one, associated with the Neumann Laplacian on X or subsets of X. We follow closely the arguments of [EP05, P06]. Similar arguments have been used in [RuS01a, KuZ01, KuZ03].

We tried to obtain estimates which are as geometric as possible, i.e. which depend only on intrinsic quantities like the ratio $\operatorname{vol} X_v / \operatorname{vol} \mathring{\partial} X_v$ or on the second (first non-zero Neumann) eigenvalue $\lambda_2(X_v)$. In some applications, it is essential to have a control over these parameters: For example, one can approximate other vertex conditions in the limit than the weighted standard ones by Schrödinger-type operators on the manifold. Here, the minimal length parameter $\ell_-(v) = \min_e\{\ell_e, 1\}$ at a vertex is important, see [EP09].

Let us start with two fundamental estimates, which are shown in a general setting in Sect. A.2. Let us briefly recall the notation as needed here: The interval I is one of the intervals $I_e = [0, \ell_e]$ or $I_{v,e} = [0, \ell_{v,e}]$. For the fibre space we set $\mathscr{H}(s) = \mathscr{H}_0 = \mathsf{L}_2(Y_e)$ and $\|\cdot\|_{\mathscr{H}(s)} = \|\cdot\|_{\mathscr{H}_0}$, i.e. the distortion functions $\rho_\pm$ are constant, namely $\rho_\pm = 1$, and the fibred space $\mathscr{H}(I) \cong \mathsf{L}_2(I) \otimes \mathsf{L}_2(Y_e) = \mathsf{L}_2(I \times Y_e)$ is a product. Then from Corollary A.2.12 we obtain the following *Sobolev trace estimates*

$$\|u\|^2_{\partial_e X_v} \le \widetilde{a}\|\mathrm{d}u\|^2_{X_{v,e}} + \frac{2}{\widetilde{a}}\|u\|^2_{X_{v,e}} \tag{6.20}$$

$$\|u\|^2_{\partial_v X_e} \le a\|u'\|^2_{X_e} + \frac{2}{a}\|u\|^2_{X_e} \tag{6.21}$$

for $0 < a \le \ell_e$ and $0 < \widetilde{a} \le \ell_{v,e}$.

Remark 6.3.1. Consider the restriction as operator $\Gamma_{v,e} \colon \mathsf{H}^1(X_{v,e}) \longrightarrow \mathsf{L}_2(\partial_e X_v)$, then the norm can be estimated as

$$\|\Gamma_{v,e}\|_{1\to 0} \ge \frac{\|u\|^2_{\partial_e X_v}}{\|\mathrm{d}u\|^2_{X_{v,e}} + \|u\|^2_{X_{v,e}}} = \frac{\operatorname{vol} \partial_e X_v}{\ell_{v,e} \operatorname{vol} Y_e} = \frac{1}{\ell_{v,e}},$$

where we inserted the constant function $u = \mathbb{1}$. In particular, we cannot expect a better coefficient of $\|u\|^2_{X_{v,e}}$ in estimate (6.20) in terms of $\widetilde{a}$ up to the universal constant 2. Note that this constant comes from the Cauchy-Young inequality (see Sect. A.2), and can be pushed down to any constant $1 < \tau \le 2$, the price being a larger coefficient of $\|\mathrm{d}u\|^2_{X_{v,e}}$. A similar remark holds for (6.21).

In order to define identification operators from the graph-like manifold onto the graph, we need the following averaging operators

$$f_v u := \fint_{X_v} u \,\mathrm{d}x_v \qquad \text{and} \qquad f_e u(s) := \fint_{Y_e} u(s,\cdot)\,\mathrm{d}y_e \tag{6.22}$$

for $u \in \mathsf{L}_2(X)$, where $\fint_{X_v}$ resp. $\fint_{Y_e}$ denote the normalised integral, see (5.3). By Cauchy-Schwarz, we also have

$$\operatorname{vol} Y_e \left|f_e u(0)\right|^2 \overset{\mathrm{CS}}{\le} \|u\|^2_{\partial_e X_v} = \|u\|^2_{\partial_v X_e}.$$

and similarly for $⨍_e u(s)$. In particular, by Fubini, $⨍_e u(s)$ is defined and finite for almost all $s \in I_e$. Note that

$$⨍_e u(0) = ⨍_{\partial_e X_{\varepsilon,v}} u \tag{6.23}$$

since $\partial_e X_{\varepsilon,v}$ and $\partial_v X_e = \{0\} \times Y_e$ are identified by assumption. Recall that if the scaling rate α of the vertex neighbourhood fulfils $0 < \alpha < 1$ and if $0 < \varepsilon < 1$, then $\partial_e X_{\varepsilon,v} \cong \varepsilon Y_e$ is strictly embedded in the entire boundary component $\varepsilon^\alpha \partial_e X_v \cong \varepsilon^\alpha Y_e$ (see Fig. 6.6). In this case, the two averaging operators

$$⨍_{\partial_e X_{\varepsilon,v}} u \qquad \text{and} \qquad ⨍_{\varepsilon^\alpha \partial_e X_v} u = ⨍_{\partial_e X_v} u$$

differ, since the first operator averages over a smaller set than the second. Nevertheless, we can compare the two averaging operators. Recall the definition of C_e and p_{m-2} in (6.15) and Definition A.2.13, respectively.

Lemma 6.3.2. *Assume that* $0 < \alpha < 1$ *and* $\beta > \alpha$*, then*

$$\varepsilon^m \operatorname{vol} Y_e \Big|⨍_{\partial_e X_{\varepsilon,v}} u\Big|^2 \le \delta^2_{\varepsilon,v,e} \|du\|^2_{X_{\varepsilon,v,e}} + 4\varepsilon^m \operatorname{vol} Y_e \Big|⨍_{\partial_e X_v} u\Big|^2 \quad \textit{and}$$

$$\varepsilon^m \operatorname{vol} Y_e \Big|⨍_{\partial_e X_v} u\Big|^2 \le \delta^2_{\varepsilon,v,e} \|du\|^2_{X_{\varepsilon,v,e}} + 4\varepsilon^m \operatorname{vol} Y_e \Big|⨍_{\partial_e X_{\varepsilon,v}} u\Big|^2,$$

where

$$\delta^2_{\varepsilon,v,e} := 4\varepsilon^{1-(\beta-\alpha)} p_{m-2}(\varepsilon^{1-\alpha})\Big(2\varepsilon^{2(\beta-\alpha)} + \frac{C_e^2}{\ell^2_{v,e}}\Big) + 4\varepsilon^{m(1-\alpha)+\beta}\ell_{v,e}$$
$$= \mathrm{O}\big(\varepsilon^{1-(\beta-\alpha)}[\log \varepsilon]\big).$$

Here, the term $[\log \varepsilon]$ *occurs only if* $m = 1$.

Proof. For the first estimate, we have

$$\begin{aligned}\varepsilon^m \operatorname{vol} Y_e \Big|⨍_{\partial_e X_{\varepsilon,v}} u\Big|^2 &\le 2a p_{m-2}(\varepsilon^{1-\alpha})\varepsilon^{1-\alpha} \|u'\|^2_{\acute{X}_{\varepsilon,v,e}} + 2\varepsilon^m \operatorname{vol} Y_e \Big|⨍_{\{a\}\times Y_e} u\Big|^2\\ &\le 2a p_{m-2}(\varepsilon^{1-\alpha})\varepsilon^{1-\alpha}\Big(2 + \varepsilon^{-2(\beta-\alpha)} \cdot \frac{C_e^2}{\ell^2_{v,e}}\Big)\|du\|^2_{\acute{X}_{\varepsilon,v,e}} + 2\varepsilon^m \operatorname{vol} Y_e \Big|⨍_{\{a\}\times Y_e} u\Big|^2\\ &= 2\varepsilon^{1-(\beta-\alpha)} p_m(\varepsilon^{1-\alpha})\Big(2\varepsilon^{2(\beta-\alpha)} + \frac{C_e^2}{\ell^2_{v,e}}\Big)\|du\|^2_{\acute{X}_{\varepsilon,v,e}} + 2\varepsilon^m \operatorname{vol} Y_e \Big|⨍_{\{a\}\times Y_e} u\Big|^2\end{aligned}$$

using Proposition A.2.17 with $a = \varepsilon^\beta \ell_{v,e}$, $r_0 = \varepsilon$ and $r_1 = \varepsilon^\alpha$ on the warped product space $\acute{X}_{\varepsilon,v,e}$. Here, $\{a\} \times Y_e$ denotes the transversal manifold in $X_{v,e}$ at $s = a$. Moreover, we used Corollary 6.2.5 in the second estimate. Similarly, we can express the averaging operator over $\{a\} \times Y_e$ by the averaging operator over $\partial_e X_v = \{0\} \times Y_e$, now via the product space $X_{\varepsilon,v,e}$, namely,

$$\varepsilon^{\alpha m} \operatorname{vol} Y_e \left| f_{\{a\} \times Y_e} u \right|^2 \le 2a \|u'\|^2_{\varepsilon^\alpha([0,a] \times Y_e)} + 2\varepsilon^{m\alpha} \operatorname{vol} Y_e \left| f_{\partial_e X_v} u \right|^2$$

using again Proposition A.2.17 with $r_0 = r_1 = \varepsilon^\alpha$. Altogether, we obtain the desired estimate using the fact that $\widetilde{X}_{\varepsilon,v,e}$ and $\varepsilon^\alpha([0,a] \times Y_e)$ are isometric subsets of $X_{\varepsilon,v,e}$. Finally, it is easy to see that the error $\varepsilon^{1-(\beta-\alpha)}$ is dominant, since we assumed $\beta > \alpha$. Moreover $p_{m-2}(\varepsilon^{1-\alpha}) \le 1$ if $m \ge 2$ and $p_{-1}(\varepsilon^{1-\alpha}) = \mathrm{O}(|\log \varepsilon|)$ by Lemma A.2.14. The second estimate follows similarly. □

Let us now come back to the unscaled case. The following lemma is an application of Proposition 5.1.3 together with the Sobolev trace estimate (6.21) as input. Recall that $\mathring{\partial} X_v = \bigcup_e \partial_e X_v$ denotes the internal boundary of X_v (see Sect. 6.2.1) and that $\ell_-(v) := \min_e\{\ell_e, 1\}$:

Lemma 6.3.3. *We have*

$$\operatorname{vol} \mathring{\partial} X_v \left| f_{\mathring{\partial} X_v} u - f_v u \right|^2 \overset{\text{CS}}{\le} \sum_{e \in E} \operatorname{vol} Y_e \left| f_{\partial_e X_v} u - f_v u \right|^2 \le \frac{1}{\ell_-(v)} \Bigl(1 + \frac{2}{\lambda_2(X_v)}\Bigr) \|du\|^2_{X_v}$$

for $u \in \mathsf{H}^1(X_v)$.

We immediately obtain in the scaled case:

Corollary 6.3.4. *We have*

$$\begin{aligned}\varepsilon^{\alpha m} \operatorname{vol} \mathring{\partial} X_v \left| f_{\mathring{\partial} X_v} u - f_v u \right|^2 &\overset{\text{CS}}{\le} \varepsilon^{\alpha m} \sum_{e \in E} \operatorname{vol} Y_e \left| f_{\partial_e X_v} u - f_v u \right|^2 \\ &\le \frac{\varepsilon^\alpha}{\ell_-(v)} \Bigl(1 + \frac{2}{\lambda_2(X_v)}\Bigr) \|du\|^2_{X_{\varepsilon,v}}\end{aligned}$$

for $u \in \mathsf{H}^1(X_{\varepsilon,v})$.

If $\alpha = 1$, then we can immediately apply the previous result to the graph-like manifolds. If $0 < \alpha < 1$, then the latter averaging operators are not compatible with the ones on the edge neighbourhoods. We can compensate this fact by Lemma 6.3.2 (recall that $\mathring{\partial} X_{\varepsilon,v} = \bigcup_e \partial_e X_{\varepsilon,v} \cong \bigcup_e \varepsilon Y_e$ is the boundary part where the edge neighbourhoods are attached):

Proposition 6.3.5. *Assume that* $0 < \alpha \le 1$, $\beta > \alpha$ *and* $0 < \varepsilon < 1$, *then*

$$\operatorname{vol} \mathring{\partial} X_{\varepsilon,v} \left| f_{\mathring{\partial} X_{\varepsilon,v}} u - f_v u \right|^2 \overset{\text{CS}}{\le} \varepsilon^m \sum_{e \in E} \operatorname{vol} Y_e \left| f_e u(0) - f_v u \right|^2 \le \hat{\delta}^2_{\varepsilon,v} \|du\|^2_{X_{\varepsilon,v}}$$

for $u \in \mathsf{H}^1(X_{\varepsilon,v})$, *where*

$$\hat{\delta}^2_{\varepsilon,v} := \frac{\varepsilon}{\ell_-(v)} \Bigl(1 + \frac{2}{\lambda_2(X_v)}\Bigr) = \mathrm{O}(\varepsilon)$$

if $\alpha = 1$ *and*

$$\hat{\delta}^2_{\varepsilon,v} := \max_e \delta^2_{\varepsilon,v,e} + 4\varepsilon^{2\alpha+m-\alpha d} \frac{1}{\ell_-(v)}\Big(1 + \frac{2}{\lambda_2(X_v)}\Big) = \mathrm{O}(\varepsilon^{1-(\beta-\alpha)}[\log \varepsilon])$$

if $0 < \alpha < 1$.

The proposition is only needed in the cases $m/d < \alpha \le 1$ and $\alpha = m/d$.

Proof. The case $\alpha = 1$ follows immediately from Corollary 6.3.4 and (6.23). For $0 < \alpha < 1$ we have the chain of estimates

$$\begin{aligned}\operatorname{vol} \mathring{\partial} X_{\varepsilon,v} \big|f_{\mathring{\partial} X_{\varepsilon,v}} u - f_v u\big|^2 &\overset{\text{CS}}{\le} \varepsilon^m \sum_{e\in E} \operatorname{vol} Y_e \big|f_{\partial_e X_{\varepsilon,v}} u - f_v u\big|^2 \\ &\le \max_e \delta^2_{\varepsilon,v,e} \|du\|^2_{X_{\varepsilon,v}} + 4\varepsilon^m \sum_e \operatorname{vol} Y_e \big|f_{\partial_e X_v} u \\ &\quad - f_v u\big|^2 \le \hat{\delta}^2_{\varepsilon,v} \|du\|^2_{X_{\varepsilon,v}}\end{aligned}$$

using Lemma 6.3.2 with u replaced by $u - f_v u$ and (6.23) for each edge, and Corollary 6.3.4 in the last estimate. Again, the error $\mathrm{O}(\varepsilon^{1-(\beta-\alpha)})$ is dominant, since

$$(2\alpha + m - \alpha d) - (1 - (\beta - \alpha)) = (m-1)(1-\alpha) + (\beta - \alpha) > 0$$

and we are done. □

We also need an estimate over the vertex neighbourhood. On the scaled manifold X_ε, the estimate assures that no family of eigenfunctions $\{u_\varepsilon\}_\varepsilon$ with eigenvalues lying in a bounded interval can concentrate on $X_{\varepsilon,v}$ as $\varepsilon \to 0$. We start with an estimate on the unscaled manifold (i.e. we set $\varepsilon = 1$):

Lemma 6.3.6. *We have*

$$\begin{aligned}\|u\|^2_{X_v} &\le 4\Big[\frac{1}{\lambda_2(X_v)} + \frac{\operatorname{vol} X_v}{\operatorname{vol} \mathring{\partial} X_v}\Big(\widetilde{a} + \frac{2}{\widetilde{a}\lambda_2(X_v)}\Big)\Big] \|du\|^2_{X_v} + 4 \operatorname{vol} X_v \big|f_{\mathring{\partial} X_v} u\big|^2 \\ &\le 4\Big[\frac{1}{\lambda_2(X_v)} + \frac{\operatorname{vol} X_v}{\operatorname{vol} \mathring{\partial} X_v}\Big(\widetilde{a} + \frac{2}{\widetilde{a}\lambda_2(X_v)}\Big)\Big] \|du\|^2_{X_v} \\ &\quad + 4\frac{\operatorname{vol} X_v}{\operatorname{vol} \mathring{\partial} X_v}\Big[a\|u'\|^2_{X_E} + \frac{2}{a}\|u\|^2_{X_E}\Big]\end{aligned}$$

for $u \in \mathsf{H}^1(X)$, $0 < a \le \min_e \ell_e$ *and* $0 < \widetilde{a} \le \min_e \ell_{v,e}$, *where* $X_E := \dot{\bigcup}_e X_e$ *denotes the union of all edge neighbourhoods.*

Proof. We start with the estimate

$$\|u\|^2_{X_v} \le 2\Big(\|u - f_v u\|^2_{X_v} + \|f_v u\|^2_{X_v}\Big) \le 2\Big(\frac{2}{\lambda_2(X_v)}\|du\|^2_{X_v} + \operatorname{vol} X_v |f_v u|^2\Big)$$

using Lemma 6.3.3 and the fact that $f_v u$ is constant. Moreover, the last term has an upper bound given by

$$\begin{aligned}\operatorname{vol} \mathring{\partial} X_v |f_v u|^2 &\le 2 \operatorname{vol} \mathring{\partial} X_v \Big(|f_v u - f_{\mathring{\partial} X_v} u|^2 + |f_{\mathring{\partial} X_v} u|^2 \Big) \\ &\le 2 \Big(\widetilde{a} + \frac{2}{\widetilde{a} \lambda_2(X_v)} \|du\|^2_{X_v} + \sum_e \operatorname{vol} \partial_e X_v |f_{\partial_e X_v} u|^2 \Big)\end{aligned}$$

using (6.20). Since $\varepsilon = 1$, we have the equality of the averagings of the vertex and edge neighbourhood boundaries (6.23), and we can estimate the latter sum by

$$\sum_e \operatorname{vol} \partial_v X_e |f_{\partial_e X_v} u|^2 \le \sum_e \|\partial_v X_e\|^2 u \le a \|u'\|^2_{X_E} + \frac{2}{a} \|u\|^2_{X_E}$$

for all $0 < a \le \min_e \ell_e$ due to (6.20) on each edge e. □

Let us now treat the case of a scaling rate $0 < \alpha \le 1$ at the vertex neighbourhood:

Proposition 6.3.7. *Assume that* $0 < \alpha \le 1$ *and* $\alpha < \beta < 1 + \alpha$, *then we have*

$$\|u\|^2_{X_{\varepsilon,v}} \le \delta^2_{\varepsilon,v} \big(\|du\|^2_{X_\varepsilon} + \|u\|^2_{X_\varepsilon} \big),$$

where

$$\delta^2_{\varepsilon,v} := \max\Big\{ \frac{8\varepsilon}{\ell_-(v)} \cdot \frac{\operatorname{vol} X_v}{\operatorname{vol} \mathring{\partial} X_v}, \varepsilon^2 C_v^2 \Big\} = \mathrm{O}(\varepsilon) \qquad \text{if} \qquad \alpha = 1,$$

$$\delta^2_{\varepsilon,v} := \max\Big\{ \frac{32 \varepsilon^{\alpha d - m}}{\ell_-(v)} \cdot \frac{\operatorname{vol} X_v}{\operatorname{vol} \mathring{\partial} X_v}, \varepsilon^{2\alpha} C_v^2 + 4 \varepsilon^{\alpha d - m} \max_e \delta^2_{\varepsilon,v,e} \cdot \frac{\operatorname{vol} X_v}{\operatorname{vol} \mathring{\partial} X_v} \Big\}$$

with $\delta^2_{\varepsilon,v} = \mathrm{O}(\varepsilon^{\alpha d - m})$ *if* $0 < \alpha < 1$, *and*

$$C_v^2 := \frac{4}{\lambda_2(X_v)} + \frac{\operatorname{vol} X_v}{\operatorname{vol} \mathring{\partial} X_v} \cdot \frac{4}{\ell_-(v)} \Big(1 + \frac{2}{\lambda_2(X_v)} \Big). \tag{6.24}$$

Proof. The case $\alpha = 1$ follows immediately from Lemma 6.3.6. For the case $0 < \alpha < 1$, we obtain

$$\begin{aligned}\|u\|^2_{X_{\varepsilon,v}} &\le \varepsilon^{2\alpha} C_v^2 \|du\|^2_{X_{\varepsilon,v}} + 4 \varepsilon^{\alpha d} \operatorname{vol} X_v |f_{\mathring{\partial} X_v} u|^2 \\ &\overset{\text{CS}}{\le} \varepsilon^{2\alpha} C_v^2 \|du\|^2_{X_{\varepsilon,v}} + 4 \varepsilon^{\alpha d} \frac{\operatorname{vol} X_v}{\operatorname{vol} \mathring{\partial} X_v} \sum_e \operatorname{vol} Y_e |f_{\partial_e X_v} u|^2\end{aligned}$$

$$\leq \Big(\varepsilon^{2\alpha} C_v^2 + 4\varepsilon^{\alpha d - m} \max_e \delta^2_{\varepsilon,v,e} \cdot \frac{\operatorname{vol} X_v}{\operatorname{vol} \mathring{\partial} X_v}\Big) \|du\|^2_{X_{\varepsilon,v}}$$

$$+ 16\varepsilon^{\alpha d} \frac{\operatorname{vol} X_v}{\operatorname{vol} \mathring{\partial} X_v} \sum_e \operatorname{vol} Y_e \big|f_e u(0)\big|^2$$

using Lemma 6.3.6 with $\widetilde{a} = \ell_-(v) \leq 1$ and the appropriate scaling behaviour in the first inequality. Moreover, we employed Lemma 6.3.2 and (6.23) in the third inequality. Now, the latter sum over the averaging operators can be estimated by the Sobolev trace estimate (6.21) on each edge neighbourhood with $a = \ell_-(v) \leq 1$, leading to the final expression for $\delta^2_{\varepsilon,v}$. For the error estimate, we note that $\varepsilon^{\alpha d - m}\delta^2_{\varepsilon,v,e}$ is dominant in the second term of $\delta^2_{\varepsilon,v}$, since

$$2\alpha - \big((\alpha d - m) + 1 - (\beta - \alpha)\big) > (d-2)(1-\alpha) \geq 0$$

using $\beta > \alpha$. Since also $\beta < 1 + \alpha$, the total error is of order $\varepsilon^{\alpha d - m}$. □

Let us make a comment on the optimality of the previous vertex neighbourhood estimate in the case $\alpha = 1$:

Remark 6.3.8. Assume that X_ε is compact, i.e. all length ℓ_e are finite and bounded by $0 < \ell_- \leq \ell_e \leq \ell_+$. Then the constant function $u = \mathbb{1}$ belongs to $\mathsf{H}^1(X)$, and we have

$$\frac{\|u\|^2_{X_{\varepsilon,v}}}{\|u\|^2_{X_\varepsilon} + \|du\|^2_{X_\varepsilon}} = \frac{\|u\|^2_{X_{\varepsilon,v}}}{\|u\|^2_{X_\varepsilon}} = \varepsilon \frac{\operatorname{vol} X_v}{\operatorname{vol} X_E + \varepsilon \operatorname{vol} X_v} \geq \frac{\varepsilon}{\ell_+} \cdot \frac{\operatorname{vol} X_v}{\operatorname{vol} Y + \varepsilon \ell_+^{-1} \operatorname{vol} X_v},$$

where

$$\operatorname{vol} X_E = \sum_e \operatorname{vol} X_e = \sum_e \ell_e \operatorname{vol} Y_e. \tag{6.25}$$

In particular, the coefficient of the norm contribution $\|u\|^2_{X_{\varepsilon,E}}$ in the previous vertex neighbourhood estimate is at least of order $\ell_+^{-1}\varepsilon$ (see also Lemma 6.3.6). Moreover, if the lengths do not vary too much, the coefficient of $\|u\|^2_{X_{\varepsilon,E}}$ has to be $\varepsilon \operatorname{vol} X_v (\operatorname{vol} \mathring{\partial} X_v)^{-1} \ell_-^{-1}$ up to some universal factor (8 in our case). In this sense, the scaled version of the estimate in Lemma 6.3.6 is optimal.

In the fast decaying vertex case, we can also consider a globally smooth manifold X_ε with *non-homogeneous* vertex neighbourhood scaling, namely[2] we assume that X_ε decomposes as

$$X_\varepsilon = X_{\varepsilon,v} \,\overline{\dot\cup}\, \overline{\dot\bigcup_e} X_{\varepsilon,e}, \tag{6.26}$$

[2]For the notation $\varepsilon X_v \leq X_{\varepsilon,v} \leq \varepsilon^\alpha X_v$ see (5.9).

where

$$X_{\varepsilon,e} = I_e \times \varepsilon Y_e, \qquad \varepsilon X_v \le X_{\varepsilon,v} \le \varepsilon^\alpha X_v \qquad \text{and} \qquad \mathring{\partial} X_{\varepsilon,v} = \varepsilon \mathring{\partial} X_v \qquad (6.27)$$

for some $\alpha \in (m/d, 1)$. In particular, we may assume that the manifold X_ε near $\partial_v X_{\varepsilon,e} \cong \partial_e X_{\varepsilon,v}$ is a cylinder of radius ε, and the scaling rate on $X_{\varepsilon,v}$ changes only away from $\partial_e X_{\varepsilon,e}$. Therefore, (6.23) holds in this case. The next corollary follows then immediately from Lemma 6.3.6 and the scaling estimates (5.11) with $\alpha_1 = \alpha$ and $\alpha_2 = 1$:

Corollary 6.3.9. *Let $u \in \mathsf{H}^1(X_{\varepsilon,v})$, then*

$$\|u\|^2_{X_{\varepsilon,v}} \le \delta^2_{\varepsilon,v}\big(\|du\|^2_{X_\varepsilon} + \|u\|^2_{X_\varepsilon}\big),$$

where $\delta^2_{\varepsilon,v}$ is given in this case by

$$\delta^2_{\varepsilon,v} := \max\Big\{\frac{8\varepsilon^{\alpha d - m}}{\ell_-(v)} \cdot \frac{\operatorname{vol} X_v}{\operatorname{vol} \mathring{\partial} X_v}, \varepsilon^{\alpha(d+2)-d} C_v^2\Big\}.$$

In particular, if $\alpha < m/d = (d-1)/m$, then $\delta_{\varepsilon,v} = \mathrm{O}(\varepsilon^{(\alpha d - m)/2})$.

Proof. For the error estimate, note that in the fast decaying case, we have

$$\alpha(d+2) - d = \alpha d - m + (2\alpha - 1) > 2\alpha - 1 > 1 - \frac{2}{d} \ge 0,$$

i.e. the term $\varepsilon^{\alpha d - m}$ is dominant. □

In the case that the vertex volume decays slowly, i.e. $0 < \alpha < m/d$, we need a different estimate:

Proposition 6.3.10. *Assume that $0 < \alpha < m/d = (d-1)/d$ and $\alpha < \beta < 1 + \alpha$, then we have*

$$\varepsilon^m \sum_{e \in E} \operatorname{vol} Y_e \big|f_e u(0)\big|^2 \le \widetilde{\delta}^2_{\varepsilon,v}\Big(\|du\|^2_{X_{\varepsilon,v}} + \|u\|^2_{X_{\varepsilon,v}}\Big)$$

for $u \in \mathsf{H}^1(X_{\varepsilon,v})$, where

$$\widetilde{\delta}^2_{\varepsilon,v} := \max\Big\{\max_e \delta^2_{\varepsilon,v,e} + 4\varepsilon^{m - \alpha d + 2\alpha}, \varepsilon^{m-\alpha d}\frac{8}{\ell_-(v)}\Big\} = \mathrm{O}(\varepsilon^{m - \alpha d}).$$

Proof. We have

$$\begin{aligned}\varepsilon^m \sum_e \operatorname{vol} Y_e \big|f_e u(0)\big|^2 &= \varepsilon^m \sum_e \operatorname{vol} Y_e \big|f_{\partial_e X_{\varepsilon,v}} u\big|^2 \\ &\le \max_e \delta^2_{\varepsilon,v,e}\|du\|^2_{X_{\varepsilon,v}} + 4\varepsilon^m \sum_e \operatorname{vol} Y_e \big|f_{\partial_e X_v} u\big|^2\end{aligned}$$

$$\overset{\text{CS}}{\le} \max_e \delta^2_{\varepsilon,v,e} \|du\|^2_{X_{\varepsilon,v}} + 4\varepsilon^m \sum_e \Big(\|u'\|^2_{X_{v,e}} + \frac{2}{\ell_-(v)} \|u\|^2_{X_{v,e}} \Big)$$

$$\le \big(\max_e \delta^2_{\varepsilon,v,e} + 4\varepsilon^{m-\alpha d+2\alpha}\big) \|du\|^2_{X_{\varepsilon,v}} + \varepsilon^{m-\alpha d} \frac{8}{\ell_-(v)} \|u\|^2_{X_{\varepsilon,v}}$$

using (6.23) and Lemma 6.3.2. Moreover, we employed (6.20) with $\widetilde{a} = \ell_-(v)$ and the scaling behaviour of the norms. For the error estimate, it is easy to see that $\varepsilon^{m-\alpha d}$ dominates the error of $\delta^2_{\varepsilon,v,e} = \mathrm{O}(\varepsilon^{1-(\beta-\alpha)}[\log \varepsilon])$, since $\beta < 1+\alpha$ and $\alpha d < m$. □

6.4 Fast Decaying Vertex Neighbourhoods: Reduction to the Graph Model

Let us now prove one of the main results of this chapter, namely the partial isometric equivalence of the Laplacian on X_ε and the (weighted standard) Laplacian on the star-shaped graph G. More precisely, we prove the partial isometric equivalence of the corresponding quadratic forms (see Definition 4.4.11 and Sect. 4.4). We first consider the case when the transversal volume $\operatorname{vol} Y_{\varepsilon,e} = \varepsilon^m \operatorname{vol} Y_e$ dominates the vertex neighbourhood volume $\operatorname{vol} X_{\varepsilon,v} = \varepsilon^{\alpha d} \operatorname{vol} X_v$, i.e. we are in the fast decaying case

$$\frac{m}{d} = \frac{d-1}{d} < \alpha \le 1. \tag{6.28}$$

In this situation, the vertex neighbourhoods are negligible, and in the limit, we obtain the (weighted) standard ("Kirchhoff") vertex condition at the vertex v. For the notation and the definition of the spaces, we refer to Sects. 6.1.1 and 6.2.1.

We now define the zeroth and first order identification operators. Recall that we have set

$$\mathscr{H} := \mathsf{L}_2(G), \qquad \mathscr{H}^1 := \mathsf{H}^1_p(G), \qquad \mathfrak{d}(f) := \|f'\|^2, \tag{6.29a}$$

$$\widetilde{\mathscr{H}} := \mathsf{L}_2(X_\varepsilon), \qquad \widetilde{\mathscr{H}}^1 := \mathsf{H}^1(X_\varepsilon), \qquad \widetilde{\mathfrak{d}}(u) := \|du\|^2_{X_\varepsilon}. \tag{6.29b}$$

Moreover, we need the weights of the vertex condition, namely we set

$$p_e := (\operatorname{vol} Y_e)^{1/2}. \tag{6.30}$$

Let $J_\varepsilon \colon \mathscr{H} \longrightarrow \widetilde{\mathscr{H}}$ be given by

$$J_\varepsilon f := \bigoplus_{e \in E} (f_e \otimes \mathbb{1}_{\varepsilon,e}) \oplus 0 = \varepsilon^{-m/2} \bigoplus_{e \in E} (f_e \otimes \mathbb{1}_e) \oplus 0 \tag{6.31}$$

with respect to the decomposition (6.19). Here, $\mathbb{1}_e$ is the normalised eigenfunction of Y_e associated with the lowest eigenvalue 0, i.e. $\mathbb{1}_e(y) = (\operatorname{vol}_{d-1} Y_e)^{-1/2}$ and

$\mathbb{1}_{\varepsilon,e} = \varepsilon^{-m/2}\mathbb{1}_e$ the corresponding normalised eigenfunction of the scaled manifold $Y_{\varepsilon,e} = \varepsilon Y_e$. In order to respect the Sobolev spaces of order 1 we need a first order identification operator $J_\varepsilon^1 : \mathscr{H}^1 \longrightarrow \widetilde{\mathscr{H}}^1$, which we define by

$$J_\varepsilon^1 f := \bigoplus_{e\in E}(f_e \otimes \mathbb{1}_{\varepsilon,e}) \oplus \varepsilon^{-m/2} f(v)\mathbb{1}_v = \varepsilon^{-m/2}\Bigl(\bigoplus_{e\in E}(f_e \otimes \mathbb{1}_e) \oplus f(v)\mathbb{1}_v\Bigr). \tag{6.32}$$

Here, $\mathbb{1}_v$ is the constant function on X_v with value 1. Note that J_ε^1 is well defined:

$$(J_\varepsilon^1 f)_e(0,y) = \varepsilon^{-m/2} p_e^{-1} f_e(0) = \varepsilon^{-m/2} f(v) = (J_\varepsilon^1 f)_v$$

due to (6.30) and (6.4), i.e. the function $J_\varepsilon^1 f$ matches along the components $\partial_v X_{\varepsilon,e}$ and $\partial_e X_{\varepsilon,v}$, independently of the embedding (6.14) (if $\alpha < 1$), and therefore $J_\varepsilon^1 f \in \mathsf{H}^1(X)$. Moreover, $f(v)$ is defined for $f \in \mathsf{H}^1(G)$ (see Lemma 6.1.1). The definition of J_ε^1 is quite natural, since we do not expect the eigenfunctions u_ε of the Laplacian on X_ε with eigenvalues in a bounded interval (independent of ε) to vary very much on $X_{\varepsilon,v}$.

The adjoint $J_\varepsilon^* : \widetilde{\mathscr{H}} \longrightarrow \mathscr{H}$ of J_ε is given by

$$(J_\varepsilon^* u)_e(s) = \langle \mathbb{1}_{\varepsilon,e}, u_e(s,\cdot)\rangle_{Y_{\varepsilon,e}} = \varepsilon^{m/2}(\operatorname{vol} Y_e)^{1/2} \underline{f}_e u(s). \tag{6.33}$$

Furthermore, we define the corresponding first order operator $J_\varepsilon'^1 : \widetilde{\mathscr{H}}^1 \longrightarrow \mathscr{H}^1$ by

$$(J_\varepsilon'^1 u)_e(s) := \varepsilon^{m/2}(\operatorname{vol} Y_e)^{1/2}\Bigl[\underline{f}_e u(s) + \chi_e(s)\bigl(\underline{f}_v u - \underline{f}_e u(0)\bigr)\Bigr]. \tag{6.34}$$

Here χ_e is a continuous, piecewise smooth cut-off function such that $\chi_e(0) = 1$ and $\chi_e(\ell_e) = 0$. If we choose the function χ_e to be piecewise affine-linear with $\chi_e(0) = 1$, $\chi_e(a) = 0$ and $\chi_e(\ell_e) = 0$, where $a := \ell_-(v) \le 1$, then $\|\chi_e\|^2_{I_e} = a/3 \le 1$ and $\|\chi_e'\|^2_{I_e} = a^{-1} = \ell_-(v)^{-1}$. Moreover, $(J_\varepsilon'^1 u)_e(0) = \varepsilon^{m/2} p_e \underline{f}_v u$ so that $f := J_\varepsilon'^1 u$ fulfils $\underline{f}(0) \in \mathbb{C}p$ and therefore $f \in \mathsf{H}^1_p(G)$.

Now we are in position to demonstrate that the two quadratic forms $\mathfrak{d}$ and $\widetilde{\mathfrak{d}}$ associated with the (Neumann-)Laplacian on X_ε and the (weighted standard) Laplacian on G are δ-partial isometrically equivalent (see Definition 4.4.11 and Proposition 4.4.13):

Theorem 6.4.1. *Assume that $\alpha \in (m/d, 1]$. Then the quadratic forms $\widetilde{\mathfrak{d}}$ and $\mathfrak{d}$ are δ_ε-partial isometrically equivalent with identification operators J_ε, J_ε^1 and $J_\varepsilon'^1$, where*

$$\delta_\varepsilon^2 := \max\Bigl\{\frac{8\varepsilon}{\ell_-(v)} \cdot \frac{\operatorname{vol} X_v}{\operatorname{vol} \mathring{\partial} X_v},\ \frac{\varepsilon}{\ell_-(v)^2}\Bigl(1 + \frac{2}{\lambda_2(X_v)}\Bigr),\ \frac{\varepsilon^2}{\min_e \lambda_2(Y_e)},\ \varepsilon^2 C_v^2\Bigr\} = \mathrm{O}(\varepsilon)$$

$$\delta_\varepsilon^2 := \max\Bigl\{\delta_{\varepsilon,v}^2,\ \frac{1}{\ell_-(v)} \cdot \hat{\delta}_{\varepsilon,v}^2,\ \frac{\varepsilon^2}{\min_e \lambda_2(Y_e)}\Bigr\} = \mathrm{O}(\varepsilon^{\alpha d - m}).$$

and where the first line is valid for $\alpha = 1$ *and the second line for* $m/d < \alpha < 1$. *In particular, the error depends only on an upper estimate of* $\operatorname{vol} X_v / \operatorname{vol} \mathring{\partial} X_v$, *and lower estimates on* $\lambda_2(X_v)$, $\lambda_2(Y_e)$ *and* ℓ_e $(e \in E)$, *if* $\alpha = 1$, *and additionally on upper estimates on* C_e, *if* $\alpha \in (m/d, 1)$.

Recall the definition of C_v and $\delta_{\varepsilon,v}$ in Proposition 6.3.7 and the definition of $\hat{\delta}_{\varepsilon,v}$ in Proposition 6.3.5. The constant C_e was given in (6.15). If $\alpha \in (m/d, 1)$ then we have set $\beta = 1$ in $\delta_{\varepsilon,v,e}$ (see Lemma 6.3.2). Note that $\alpha < 1 < 1 + \alpha$.

Proof. The first condition in (4.29a), i.e. $J_\varepsilon^* J_\varepsilon f = f$, is easily seen to be fulfilled. The second estimate in (4.29a) is more involved. Here, we have

$$\|u - J_\varepsilon J_\varepsilon^* u\|_{X_\varepsilon}^2 = \sum_e \|u - \mathit{f}_e u\|_{X_{\varepsilon,e}}^2 + \|u\|_{X_{\varepsilon,v}}^2.$$

The first term can be estimated by

$$\begin{aligned}\|u - \mathit{f}_e u\|_{X_{\varepsilon,e}}^2 &= \int_{I_e} \|u(s) - \mathit{f}_e u(s)\|_{Y_{\varepsilon,e}}^2 \, ds \\ &\le \frac{\varepsilon^2}{\lambda_2(Y_e)} \int_{I_e} \|d_{Y_e} u(s)\|_{Y_{\varepsilon,e}}^2 \, ds = \frac{\varepsilon^2}{\lambda_2(Y_e)} \|d_{Y_e} u\|_{X_{\varepsilon,e}}^2\end{aligned}$$

using Proposition 5.1.1, where $u(s) := u(s, \cdot)$ and $\lambda_2(Y_e)$ denotes the second (first non-vanishing) eigenvalue of the Laplacian on Y_e. Note that $\lambda_2(Y_{\varepsilon,e}) = \varepsilon^{-2} \lambda_2(Y_e)$. The second term can be estimated by Proposition 6.3.7, so that

$$\delta_\varepsilon^2 \ge \max\Big\{ \frac{\varepsilon^2}{\min_e \lambda_2(Y_e)}, \delta_{\varepsilon,v}^2 \Big\}$$

is enough for the estimate (4.29a). Moreover, the first estimate in condition (4.29b) reads as

$$\|J_\varepsilon f - J_\varepsilon^1 f\|_{X_\varepsilon}^2 = \varepsilon^{\alpha d - m} \operatorname{vol} X_v |f(v)|^2 \le \frac{2\varepsilon^{\alpha d - m}}{\ell_-(v)} \cdot \frac{\operatorname{vol} X_v}{\operatorname{vol} \mathring{\partial} X_v} \big(\|f'\|_G^2 + \|f\|_G^2 \big)$$

using Lemma 6.1.1 with $a = \ell_-(v)$, the scaling $\operatorname{vol} X_{\varepsilon,v} = \varepsilon^{\alpha d} \operatorname{vol} X_v$ and the fact that $|p|^2 = \operatorname{vol} \mathring{\partial} X_v$ due to (6.30). Note that the error term obtained here is already contained in $\delta_{\varepsilon,v}$ (see Proposition 6.3.7). The second estimate in (4.29b) is here

$$\|J_\varepsilon^* u - J_\varepsilon'^1 u\|_G^2 = \varepsilon^m \sum_{e \in E} \|\chi_e\|_{I_e}^2 \operatorname{vol} Y_e \big| \mathit{f}_v u - \mathit{f}_e u(0) \big|^2 \le \hat{\delta}_{\varepsilon,v}^2 \|du\|_{X_{\varepsilon,v}}^2,$$

using Proposition 6.3.5. Condition (4.29c), i.e. $J_\varepsilon'^1 J_\varepsilon^1 f = f$, is easily seen. Moreover, for estimate (4.29d) we have

$$\left|\mathfrak{d}(J_\varepsilon'^1 u, f) - \widetilde{\mathfrak{d}}(u, J_\varepsilon^1 f)\right|^2 \le \varepsilon^m \left|\sum_e (\operatorname{vol} Y_e)^{1/2} \left(f_v \overline{u} - f_e \overline{u}(0)\right) \langle \chi_e', f' \rangle_{I_e}\right|^2$$

$$\overset{\text{CS}}{\le} \frac{\hat{\delta}^2_{\varepsilon,v}}{\ell_-(v)} \|du\|^2_{X_{\varepsilon,v}} \|f'\|^2_G$$

using again Proposition 6.3.5. Note that the derivative terms cancel on the edges due to the product structure of the metric and the fact that $d_{Y_e} 1\!\!1_e = 0$. Moreover, the vertex contribution vanishes since $d_{X_v} 1\!\!1 = 0$. If $\alpha < 1$ then we fix $\beta = 1$, and we have

$$0 < \alpha d - m < 1 - \beta + \alpha = \alpha < 1,$$

and in particular, $\delta_\varepsilon^2 = O(\varepsilon^{\alpha d - m})$. Recall that if $\alpha < 1$, then

$$\delta^2_{\varepsilon,v} = O(\varepsilon^{\alpha d - m}) \qquad \text{and} \qquad \hat{\delta}^2_{\varepsilon,v} = O(\varepsilon^{1-(\beta-\alpha)}[\log \varepsilon]) = O(\varepsilon^\alpha [\log \varepsilon]),$$

where the term $[\log \varepsilon]$ appears only if $m = 1$.

□

Let us again make a remark on the optimality of the errors:

Remark 6.4.2. Assume that we have given the identification operators as above. If we chose $f_e(s) = p_e$ for $s \in I_e$ on each edge, then $f \in H^1_p(G)$ and

$$\|J_\varepsilon^1 - J_\varepsilon\|^2_{1\to 0} \ge \frac{\|J_\varepsilon^1 f - J_\varepsilon f\|^2}{\|f\|_1^2} = \frac{\varepsilon^{\alpha d - m} \operatorname{vol} X_v}{\operatorname{vol} X_E} = O(\varepsilon^{\alpha d - m}),$$

where $\operatorname{vol} X_E = \sum_e \ell_e \operatorname{vol} Y_e$. Moreover, if $u = 1\!\!1$ is constant, then

$$\|\widetilde{\mathrm{id}} - J_\varepsilon J_\varepsilon^*\|^2_{1\to 0} \ge \frac{\|u - J_\varepsilon J_\varepsilon^* u\|^2}{\|u\|_1^2} = \frac{\|u\|^2_{X_{\varepsilon,v}}}{\|u\|_1^2} = \frac{\varepsilon^{\alpha d - m} \operatorname{vol} X_v}{\operatorname{vol} X_E + \varepsilon^{\alpha d - m} \operatorname{vol} X_v}$$
$$= O(\varepsilon^{\alpha d - m}).$$

In particular, we cannot expect that the deviation δ_ε from being unitarily equivalent is better than $\delta_\varepsilon = O(\varepsilon^{(\alpha d - m)/2})$. Moreover, if $\alpha = 1$, we only get $\delta_\varepsilon = O(\varepsilon^{1/2})$ with our choice of identification operators. We obtain the same error estimate if we directly compare the eigenvalues using Corollary 4.4.19. Note that Grieser [G08a] found an eigenvalue expansion in terms of ε, and therefore gets the better estimate $\lambda_k(\varepsilon) = \lambda_k(0) + O(\varepsilon)$, where $\lambda_k(\varepsilon)$ and $\lambda_k(0)$ denote the k-th eigenvalue of Δ_{X_ε} ($\varepsilon > 0$) and Δ, respectively.

Alternatively, we can consider a globally smooth manifold X_ε with *non-homogeneous* vertex neighbourhood scaling, namely[3] we assume that X_ε decomposes as in (6.26) such that

$$X_{\varepsilon,e} = I_e \times \varepsilon Y_e, \qquad \varepsilon X_v \preceq X_{\varepsilon,v} \preceq \varepsilon^\alpha X_v \qquad \text{and} \qquad \mathring{\partial} X_{\varepsilon,v} = \varepsilon \mathring{\partial} X_v$$

for some $\alpha \in (m/d, 1)$. In this case, we obtain the following result:

Theorem 6.4.3. *Assume that the graph-like manifold scales as in* (6.27). *Then the metric graph quadratic form* $\mathfrak{d}$ *on* $\mathsf{L}_2(G_v)$ *is* δ_ε*-partial isometrically equivalent with the manifold quadratic form* $\widetilde{\mathfrak{d}}$ *on* $\mathsf{L}_2(X_\varepsilon)$*, where*

$$\delta_\varepsilon^2 := \max\Big\{\frac{8\varepsilon^{\alpha d - m}}{\ell_-(v)} \cdot \frac{\operatorname{vol} X_v}{\operatorname{vol} \mathring{\partial} X_v}, \frac{\varepsilon^{2\alpha-1}}{\ell_-(v)^2}\Big(1 + \frac{2}{\lambda_2(X_v)}\Big), \frac{\varepsilon^2}{\min_e \lambda_2(Y_e)}, \varepsilon^{\alpha(d+2)-d} C_v^2\Big\}$$

and $\ell_-(v) := \min_{e \in E_v}\{1, \ell_e\}$. *In particular,* $\delta_\varepsilon = \mathrm{O}(\varepsilon^{(\alpha d - m)/2})$, *and the error depends only on an upper estimate of* $\operatorname{vol} X_v / \operatorname{vol} \mathring{\partial} X_v$, *and lower estimates on* $\lambda_2(X_v)$, $\lambda_2(Y_e)$ *and* ℓ_e *for* $e \in E$.

Recall the definition of C_v in (6.24).

Proof. The proof is similar to the proof of Theorem 6.4.1, using now Lemma 6.3.3 together with the scaling inequalities (5.11) and Corollary 6.3.9. Note that the error is indeed of order $\mathrm{O}(\varepsilon^{\alpha d - m})$ since $\alpha d - m \le \alpha(d+2) - d \le 2\alpha - 1 \le 2$. □

We also obtain a closeness result for the associated boundary triples $(\Gamma_v, \Gamma'_v, \mathscr{G}_v)$ and $(\widetilde{\Gamma}_v, \widetilde{\Gamma}'_v, \widetilde{\mathscr{G}}_v)$ defined in Sects. 6.1.1 and 6.2.3. For the notion of δ-closeness of two boundary maps Γ_v and $\widetilde{\Gamma}_v$ we refer to Definition 4.8.1. We define the *boundary identification operator* as

$$I_\varepsilon : \mathscr{G}_v = \mathbb{C}^{E_{v,\mathrm{int}}} \longrightarrow \widetilde{\mathscr{G}}_v = \mathsf{L}_2(Y_{\varepsilon,\mathrm{int}}) = \bigoplus_{e \in E_{v,\mathrm{int}}} \mathsf{L}_2(Y_{\varepsilon,e}),$$

$$I_\varepsilon F := \{\varepsilon^{-m/2} F_e \mathbb{1}_e\}_e, \tag{6.35}$$

i.e. we extend the single value F_e to a constant normalised function on Y_e. Its adjoint is given by

$$(I_\varepsilon^* \psi)_e = \varepsilon^{m/2} \langle \mathbb{1}_e, \psi_e \rangle_{Y_e} = \varepsilon^{m/2} (\operatorname{vol} Y_e)^{1/2} \textstyle\fint_e \psi.$$

Moreover, we set

$$\mathfrak{k}(F) := 0 \qquad \text{and} \qquad \widetilde{\mathfrak{k}}(\psi) = \|\mathrm{d}_Y \psi\|^2_{Y_\varepsilon}$$

[3] For the notation $\varepsilon X_v \preceq X_{\varepsilon,v} \preceq \varepsilon^\alpha X_v$ see (5.9).

with $\operatorname{dom}\widetilde{\mathfrak{e}} = \mathsf{H}^1(Y_\varepsilon) = \mathscr{G}_v^1$. The corresponding transversal operators are $K = 0$ and $\widetilde{K} = \Delta_{Y_{\varepsilon,\mathrm{int}}}$; the latter has Neumann boundary condition, if $\partial Y = \bigcup_e \partial Y_e \neq \emptyset$.

Theorem 6.4.4. *The boundary maps*

$$\Gamma_v : \mathsf{H}^1_p(G) \longrightarrow \mathbb{C}^{E_{v,\mathrm{int}}}, \qquad \widetilde{\Gamma}_v : \mathsf{H}^1(X_\varepsilon) \longrightarrow \mathsf{L}_2(Y_{\varepsilon,\mathrm{int}}),$$

$$\Gamma_v f := \{f_e(\ell_e)\}_e, \qquad \widetilde{\Gamma}_v u := \{u_e(\ell_e, \cdot)\}_e$$

associated with the forms $\mathfrak{d}$ and $\widetilde{\mathfrak{d}}$ are 0-close with the identification operators J^1_ε, J'^1_ε and the boundary identification operator I_ε. Moreover, I_ε is a $\hat{\delta}_\varepsilon$-partial isometry w.r.t. $\mathfrak{k}$ and $\widetilde{\mathfrak{k}}$, where $\hat{\delta}_\varepsilon = \varepsilon(\min_e(\lambda_2(Y_e))^{-1/2}$. Finally, $\widetilde{K} I_\varepsilon = I_\varepsilon K (= 0)$.

Proof. It follows straightforward from the definitions that

$$(\widetilde{\Gamma} J^1_\varepsilon f)_e = \varepsilon^{-m/2} f_e(\ell_e) \mathbb{1}_e = (I \Gamma f)_e \quad \text{and}$$

$$(\Gamma J'^1_\varepsilon u)_e = \varepsilon^{m/2} (\operatorname{vol} Y_e)^{1/2} \fint_e u(\ell_e) = (I^* \widetilde{\Gamma} u)_e.$$

In particular, the 0-closeness follows. For the partial isometry of I_ε, the condition $I^*_\varepsilon I_\varepsilon F = F$ is obvious. Moreover,

$$\|\psi - I_\varepsilon I^*_\varepsilon \psi\|^2_{Y_{\varepsilon,\mathrm{int}}} = \varepsilon^m \sum_{e \in E} \|\psi_e - \langle \mathbb{1}_e, \psi_e \rangle \mathbb{1}_e\|^2_{Y_e} \le \frac{\varepsilon^2}{\min_e(\lambda_2(Y_e)} \widetilde{\mathfrak{k}}(\psi)$$

for $\psi \in \operatorname{dom}\widetilde{\mathfrak{k}} = \mathsf{H}^1(Y_{\varepsilon,\mathrm{int}})$ using Proposition 5.1.1 on each manifold Y_e. The intertwining property $\widetilde{K} I_\varepsilon = I_\varepsilon K$ of I_ε is obvious, since $\widetilde{K}\{\mathbb{1}_e\}_e = 0$. □

6.5 Slowly Decaying Vertex Neighbourhoods: Reduction to the Graph Model

In this section, we prove the partial isometric equivalence of the Laplacian on X_ε and the decoupled Dirichlet Laplacian on the graph G (more precisely, of the associated quadratic forms). Here, we assume that the vertex neighbourhood volume $\operatorname{vol} X_{\varepsilon,v} = \varepsilon^{\alpha d} \operatorname{vol} X_v$ dominates the transversal volume $\operatorname{vol} Y_{\varepsilon,e} = \varepsilon^m \operatorname{vol} Y_e$, i.e.

$$0 < \alpha < \frac{m}{d} = \frac{d-1}{d}. \tag{6.36}$$

In this case, the vertex neighbourhood serves as big obstacle. In the limit, the obstacle enforces a Dirichlet condition on each edge at the vertex v, and leads to an extra state associated with the vertex. We first treat the simpler situation where the differently scaled manifolds $X_{\varepsilon,e} = I_e \times \varepsilon Y_e$ and $X_{\varepsilon,v} = \varepsilon^\alpha X_v$ are directly glued together; the general case will be treated in Sect. 6.9. For the notation and the definition of the manifold, we refer to Sect. 6.2.1.

Let us now define the zeroth and first order identification operators. Remember that

$$\mathscr{H} := \mathsf{L}_2(G) \oplus \mathbb{C}, \quad \mathscr{H}^1 := \bigoplus_e \mathsf{H}^1_0(I_e) \oplus \mathbb{C}, \quad \mathfrak{d}(\hat{f}) := \|f'\|^2, \tag{6.37a}$$

$$\widetilde{\mathscr{H}} := \mathsf{L}_2(X_\varepsilon), \quad \widetilde{\mathscr{H}}^1 := \mathsf{H}^1(X_\varepsilon), \quad \widetilde{\mathfrak{d}}(u) := \|\mathrm{d}u\|^2_{X_\varepsilon} \tag{6.37b}$$

as in Sects. 6.1.2 and 6.2.3.

Let $J_\varepsilon \colon \mathscr{H} \longrightarrow \widetilde{\mathscr{H}}$ be given by

$$\begin{aligned} J_\varepsilon \hat{f} &:= \bigoplus_{e\in E} (f_e \otimes \mathbb{1}_{\varepsilon,e}) \oplus F(v)\mathbb{1}_{\varepsilon,v} \\ &= \varepsilon^{-m/2}\Big(\bigoplus_{e\in E}(f_e \otimes \mathbb{1}_e) \oplus \varepsilon^{(m-\alpha d)/2}F(v)\mathbb{1}_v\Big), \end{aligned} \tag{6.38}$$

where $\mathbb{1}_v$ resp. $\mathbb{1}_{\varepsilon,v}$ is the normalised constant function on X_v resp. $X_{\varepsilon,v} = \varepsilon^\alpha X_v$. For the spaces of order 1, we define $J^1_\varepsilon \colon \mathscr{H}^1 \longrightarrow \widetilde{\mathscr{H}}^1$ by

$$\begin{aligned} J^1_\varepsilon \hat{f} &:= \bigoplus_{e\in E}\Big(f_e + \Big(\frac{\operatorname{vol} Y_{\varepsilon,e}}{\operatorname{vol} X_{\varepsilon,v}}\Big)^{1/2} \chi_e F(v)\Big) \otimes \mathbb{1}_{\varepsilon,e} \oplus F(v)\mathbb{1}_{\varepsilon,v} \\ &= \varepsilon^{-m/2}\Big(\bigoplus_{e\in E}\Big(f_e + \varepsilon^{(m-\alpha d)/2}\Big(\frac{\operatorname{vol} Y_e}{\operatorname{vol} X_v}\Big)^{1/2} \chi_e F(v)\Big) \otimes \mathbb{1}_e \oplus \\ &\qquad \oplus \varepsilon^{(m-\alpha d)/2} F(v)\mathbb{1}_v\Big), \end{aligned} \tag{6.39}$$

where χ_e is the continuous, piecewise affine-linear cut-off function with $\chi_e(0) = 1$, $\chi_e(a) = 0$ and $\chi_e(\ell_e) = 0$ with $a := \ell_-(v) \le 1$. The operator J^1_ε is well defined:

$$(J^1_\varepsilon \hat{f})_e(0, y) = (\operatorname{vol} X_{\varepsilon,v})^{-1/2} F(v) = (J^1_\varepsilon \hat{f})(x), \quad x \in X_v,$$

since $f_e(0) = 0$, i.e. the function $J^1_\varepsilon \hat{f}$ matches along the components $\partial_v X_{\varepsilon,e}$ and $\partial_e X_{\varepsilon,v}$ (again independently of the embedding (6.14)), and $J_\varepsilon \hat{f} \in \mathsf{H}^1(X)$. Note that the identification operator J^1_ε reflects the fact that we expect an additional, almost constant eigenfunction on the big vertex neighbourhood as $\varepsilon \to 0$.

The adjoint $J^*_\varepsilon \colon \widetilde{\mathscr{H}} \longrightarrow \mathscr{H}$ of J_ε is given by

$$(J^*_\varepsilon u)_e(s) = \langle \mathbb{1}_{\varepsilon,e}, u_e(s,\cdot)\rangle_{Y_{\varepsilon,e}} = \varepsilon^{m/2}(\operatorname{vol} Y_e)^{1/2} ⨍_e u(s), \tag{6.40a}$$

$$(J^*_\varepsilon u)(v) = \langle \mathbb{1}_{\varepsilon,v}, u_v\rangle_{X_{\varepsilon,v}} = \varepsilon^{\alpha d/2}(\operatorname{vol} X_v)^{1/2} ⨍_v u, \tag{6.40b}$$

where the averaging operators $⨍_v u$ and $⨍_e u(s)$ where introduced in (6.22). Note that these operators do not see the scaling, i.e. $⨍_{\varepsilon,v} u = ⨍_v u$ and $⨍_{\varepsilon,e} u(s) = ⨍_e u(s)$, but recall that there are two different averaging operators associated with the boundary component $\partial_e X_v$, see Lemma 6.3.2. Furthermore, we define $J_\varepsilon'^1 : \widetilde{\mathscr{H}}^1 \longrightarrow \mathscr{H}^1$ by

$$(J_\varepsilon'^1 u)_e(s) := \varepsilon^{m/2} (\operatorname{vol} Y_e)^{1/2} \Big(\fint_e u(s) - \chi_e(s) \fint_e u(0) \Big), \tag{6.41a}$$

$$(J_\varepsilon'^1 u)(v) := \varepsilon^{\alpha d/2} (\operatorname{vol} X_v)^{1/2} \fint_v u. \tag{6.41b}$$

To see that the operator $J_\varepsilon'^1$ indeed maps into the given space, let $\hat{f} = (f, F) := J_\varepsilon'^1 u$ then $f_e(0) = 0$ and therefore $f_e \in \mathsf{H}_0^1(I_e)$ and $\hat{f} \in \mathscr{H}^1$.

Now we are in position to demonstrate that the two quadratic forms $\mathfrak{d}$ and $\widetilde{\mathfrak{d}}$ are δ-partial isometrically equivalent (see Definition 4.4.11 and Proposition 4.4.13):

Theorem 6.5.1. *The quadratic forms $\widetilde{\mathfrak{d}}$ and $\mathfrak{d}$ are δ_ε-partial isometrically equivalent with identification operators J_ε, J_ε^1 and $J_\varepsilon'^1$, where*

$$\delta_\varepsilon^2 := \max\left\{ \frac{2\varepsilon^{m-\alpha d}}{\ell_-(v)} \cdot \frac{\operatorname{vol} \mathring{\partial} X_v}{\operatorname{vol} X_v}, \frac{2\widetilde{\delta}_{\varepsilon,v}^2}{\ell_-(v)}, \frac{\varepsilon^2}{\min_e \lambda_2(Y_e)}, \frac{\varepsilon^{2\alpha}}{\lambda_2(X_v)} \right\} = \mathrm{O}(\varepsilon^{\min\{m-\alpha d, 2\alpha\}}).$$

In particular, the error depends only on upper estimates of $\operatorname{vol} \partial X_v / \operatorname{vol} X_v$ *and C_e (see* (6.15)*) and on lower estimates of $\lambda_2(X_v)$, $\lambda_2(X_e)$ and ℓ_e.*

Recall the definition of $\widetilde{\delta}_{\varepsilon,v}$ in Proposition 6.3.10. Note that the fastest decay (in ε) of the error term is achieved for $m - \alpha_0 d = 2\alpha_0$, i.e. for $\alpha_0 = m/(d+2) < m/d$. If $0 < \alpha < \alpha_0$, then $\delta_\varepsilon = \mathrm{O}(\varepsilon^\alpha)$, if $\alpha_0 < \alpha < m/d$, then $\delta_\varepsilon = \mathrm{O}(\varepsilon^{(m-\alpha d)/2})$.

Proof. The first condition in (4.29a), i.e. $J_\varepsilon^* J_\varepsilon \hat{f} = \hat{f}$, is easily seen to be fulfilled. For the second condition in (4.29a), we estimate

$$\begin{aligned} \|u - J_\varepsilon J_\varepsilon^* u\|_{X_\varepsilon}^2 &= \sum_e \|u - \fint_e u\|_{X_{\varepsilon,e}}^2 + \|u - \fint_v u\|_{X_{\varepsilon,v}}^2 \\ &\le \max_{e \in E} \left\{ \frac{\varepsilon^2}{\lambda_2(Y_e)}, \frac{\varepsilon^{2\alpha}}{\lambda_2(X_v)} \right\} \|du\|_{X_\varepsilon}^2 \end{aligned}$$

by Proposition 5.1.1 and the scaling behaviour of the spectrum. Moreover, the first condition in (4.29b) is here

$$\|J_\varepsilon^1 \hat{f} - J_\varepsilon \hat{f}\|_{X_\varepsilon}^2 = \varepsilon^{m-\alpha d} \sum_e \frac{\operatorname{vol} Y_e}{\operatorname{vol} X_v} \|\chi_e\|_{I_e}^2 |F(v)|^2 \le \varepsilon^{m-\alpha d} \frac{\operatorname{vol} \mathring{\partial} X_v}{\operatorname{vol} X_v} |F(v)|^2,$$

using the fact that $\|\chi_e\|_{I_e}^2 \le 1$. The second estimate in (4.29b) can be seen by

$$\|J_\varepsilon^* u - J_\varepsilon'^1 u\|_{\mathscr{H}}^2 = \varepsilon^m \sum_{e \in E} \operatorname{vol} Y_e \|\chi_e\|_{I_e}^2 |\fint_e u(0)|^2 \le \widetilde{\delta}_{\varepsilon,v}^2 \Big(\|du\|_{X_{\varepsilon,v}}^2 + \|u\|_{X_{\varepsilon,v}}^2 \Big)$$

using Proposition 6.3.10. Moreover, equation (4.29c) is again easily seen to be fulfilled. For the form estimate, we have

$$\begin{aligned}
\left|\mathfrak{d}(J_\varepsilon'^1 u, \hat f) - \widetilde{\mathfrak{d}}(u, J_\varepsilon^1 \hat f)\right|^2 &= \Big|\sum_e \Big[\varepsilon^{m/2} (\operatorname{vol} Y_e)^{1/2} \fint_e \overline{u(0)} \langle \chi_e', f_e' \rangle_{I_e} \\
&\quad + \varepsilon^{(m-\alpha d)/2} \Big(\frac{\operatorname{vol} Y_e}{\operatorname{vol} X_v}\Big)^{1/2} \langle u_e', \chi_e' \otimes \mathbb{1}_e \rangle_{X_e} F(v)\Big]\Big|^2 \\
&\overset{\text{CS}}{\le} \frac{2}{\ell_-(v)} \Big[\delta_{\varepsilon,v}^2 \Big(\|du\|^2_{X_{\varepsilon,v}} + \|u\|^2_{X_{\varepsilon,v}}\Big) \|f'\|^2_G \\
&\quad + \varepsilon^{m-\alpha d} \frac{\operatorname{vol} \mathring{\partial} X_v}{\operatorname{vol} X_v} \|u'\|^2_{X_{\varepsilon,E}} \big|F(v)\big|^2\Big]
\end{aligned}$$

using Proposition 6.3.10 again and $\|\chi_e'\|_{I_e} = \ell_-(v)^{-1}$. Here, $X_{\varepsilon,E} := \dot{\bigcup}_e X_{\varepsilon,e}$. Note that $\delta_{\varepsilon,v}^2 = \mathrm{O}(\varepsilon^{m-\alpha d})$ if $\alpha \in (0, m/d)$. □

As in the fast decaying case, we obtain a closeness result for the associated boundary triples $(\Gamma_v, \Gamma_v', \mathscr{G}_v)$ and $(\widetilde{\Gamma}_v, \widetilde{\Gamma}_v', \widetilde{\mathscr{G}}_v)$ defined in Sects. 6.1.2 and 6.2.3. The boundary identification operator $I_\varepsilon \colon \mathscr{G}_v \longrightarrow \widetilde{\mathscr{G}}_v$ is the same as in (6.35). Moreover, we use the same transversal operators $K = 0$ and $\widetilde{K} = \Delta_{Y_{\varepsilon,\mathrm{int}}}$ as in Sect. 6.4:

Theorem 6.5.2. *The boundary maps*

$$\Gamma_v \colon \mathscr{H}_v^1 \longrightarrow \mathbb{C}^{E_{v,\mathrm{int}}}, \qquad \widetilde{\Gamma}_v \colon \mathsf{H}^1(X_\varepsilon) \longrightarrow \mathsf{L}_2(Y_{\varepsilon,\mathrm{int}}),$$
$$\Gamma_v \hat f := \{f_e(\ell_e)\}_e, \qquad \widetilde{\Gamma}_v u := \{u_e(\ell_e, \cdot)\}_e$$

associated with the forms $\mathfrak{d}$ and $\widetilde{\mathfrak{d}}$ are 0-close with the identification operators J_ε^1, $J_\varepsilon'^1$ and the boundary identification operator I_ε. Moreover, I_ε is a $\hat\delta_\varepsilon$-partial isometry w.r.t. $\mathfrak{k}$ and $\widetilde{\mathfrak{k}}$, where $\hat\delta_\varepsilon = \varepsilon(\min_e(\lambda_2(Y_e))^{-1/2}$. Finally, $\widetilde{K} I_\varepsilon = I_\varepsilon K (= 0)$.

Proof. It follows straightforward from the definitions that

$$(\widetilde{\Gamma} J_\varepsilon^1 \hat f)_e = \varepsilon^{-m/2} f_e(\ell_e) \mathbb{1}_e = (I \Gamma \hat f)_e \quad \text{and}$$
$$(\Gamma J_\varepsilon'^1 u)_e = \varepsilon^{m/2} (\operatorname{vol} Y_e)^{1/2} \fint_e u(\ell_e) = (I^* \widetilde{\Gamma} u)_e,$$

where $(I^*\psi)_e = \varepsilon^{m/2} (\operatorname{vol} Y_e)^{1/2} \fint_e \psi$. In particular, the 0-closeness follows. The remaining assertions follow as in Theorem 6.4.4. □

6.6 The Borderline Case: Reduction to the Graph Model

In this section, we consider the case when the transversal volume $\operatorname{vol} Y_{\varepsilon,e} = \varepsilon^m \operatorname{vol} Y_e$ and the vertex neighbourhood volume $\operatorname{vol} X_{\varepsilon,v} = \varepsilon^{\alpha d} \operatorname{vol} X_v$ are of the same order, i.e. if

$$\alpha = \frac{m}{d} = \frac{d-1}{d}. \tag{6.42}$$

Here, we obtain a *non-trivial* limit operator, different from the usual (weighted) standard Laplacian. Namely, the limit operator is the corresponding extended Laplacian, extended at the vertex v by the vertex space $\mathbb{C}p(v)$ and the "operator" $L(v) > 0$ (see Sect. 2.3 for details). We first present the simpler situation where the differently scaled manifolds $X_{\varepsilon,e} = I_e \times \varepsilon Y_e$ and $X_{\varepsilon,v} = \varepsilon^\alpha X_v$ are directly glued together. This construction is only possible for certain homogeneous transversal manifolds like $Y_e = \overline{B}(0, r_e)$; the general case will be treated in Sect. 6.10. For the notation and the definition of the manifold, we refer to Sect. 6.2.1.

As usual, we start with defining the zeroth and first order identification operators. Remember that in this situation,

$$\mathscr{H} := \mathsf{L}_2(G) \oplus \mathbb{C}, \quad \mathscr{H}^1 := \{\, \hat{f} \in \mathsf{H}^1_p(G) \oplus \mathbb{C} \,|\, f(v) = L(v)F(v) \,\}, \tag{6.43a}$$

$$\widetilde{\mathscr{H}} := \mathsf{L}_2(X_\varepsilon), \qquad \widetilde{\mathscr{H}}^1 := \mathsf{H}^1(X_\varepsilon), \tag{6.43b}$$

$\mathfrak{d}(\hat{f}) := \|f'\|^2$ and $\widetilde{\mathfrak{d}}(u) := \|du\|^2_{X_\varepsilon}$ (see Sects. 6.1.3 and 6.2.3), where $\hat{f} = (f, F)$. Moreover, by

$$L(v) = (\operatorname{vol} X_v)^{-1/2} \quad \text{and} \quad p_e := (\operatorname{vol} Y_e)^{1/2}, \tag{6.44}$$

we relate the different graph and manifold parameters. Let $J_\varepsilon \colon \mathscr{H} \longrightarrow \widetilde{\mathscr{H}}$ be given by

$$J_\varepsilon \hat{f} := \bigoplus_{e \in E} (f_e \otimes \mathbb{1}_{\varepsilon,e}) \oplus F(v)\mathbb{1}_{\varepsilon,v} = \varepsilon^{-m/2}\Big(\bigoplus_{e \in E} (f_e \otimes \mathbb{1}_e) \oplus F(v)\mathbb{1}_v\Big) \tag{6.45}$$

using (6.42). For the spaces of order 1, we use the same operator restricted to $\mathscr{H}^1$, i.e. $J^1_\varepsilon \hat{f} := J_\varepsilon \hat{f}$ if $\hat{f} \in \mathscr{H}^1$. Note that

$$\begin{aligned}(J^1_\varepsilon \hat{f})_e(0, y) &= \varepsilon^{-m/2} p_e^{-1} f_e(0) = \varepsilon^{-m/2} f(v)\\ &= \varepsilon^{-m/2} (\operatorname{vol} X_v)^{-1/2} F(v) = (J^1_\varepsilon \hat{f})(x), \quad x \in X_v\end{aligned}$$

due to (6.44) and (6.4), i.e. the function $J^1_\varepsilon \hat{f}$ matches along the components $\partial_v X_{\varepsilon,e}$ and $\partial_e X_{\varepsilon,v}$ (again independently of the embedding (6.14)), and therefore $J^1_\varepsilon \hat{f} \in \mathsf{H}^1(X_\varepsilon)$.

The adjoint $J^*_\varepsilon \colon \widetilde{\mathscr{H}} \longrightarrow \mathscr{H}$ of J_ε is given by

$$(J^*_\varepsilon u)_e(s) = \langle \mathbb{1}_{\varepsilon,e}, u_e(s, \cdot) \rangle_{Y_{\varepsilon,e}} = \varepsilon^{m/2} (\operatorname{vol} Y_e)^{1/2} {\textstyle\int_e} u(s), \tag{6.46a}$$

$$(J^*_\varepsilon u)(v) = \langle \mathbb{1}_{\varepsilon,v}, u_v \rangle_{X_{\varepsilon,v}} = \varepsilon^{m/2} (\operatorname{vol} X_v)^{1/2} {\textstyle\int_v} u \tag{6.46b}$$

using again (6.42). Furthermore, we define $J'^1_\varepsilon : \widetilde{\mathscr{H}}^1 \longrightarrow \mathscr{H}^1$ by

$$(J'^1_\varepsilon u)_e(s) := \varepsilon^{m/2}(\operatorname{vol} Y_e)^{1/2}\Bigl(\underline{f}_e u(s) + \chi_e(s)\bigl(\underline{f}_v u - \underline{f}_e u(0)\bigr)\Bigr), \tag{6.47a}$$

$$(J'^1_\varepsilon u)(v) := \varepsilon^{m/2}(\operatorname{vol} X_v)^{1/2}\underline{f}_v u. \tag{6.47b}$$

As usual, χ_e is the continuous, piecewise affine-linear cut-off function with $\chi_e(0) = 1$, $\chi_e(a) = 0$ and $\chi_e(\ell_e) = 0$, where $a = \ell_-(v)$. To see that J'^1_ε maps indeed into the given space, let $\hat{f} = (f, F) := J'^1_\varepsilon u$. Then $f_e(0) = \varepsilon^{m/2} p_e \underline{f}_v u$ so that $f(v) = \varepsilon^{m/2}\underline{f}_v u$, and $\underline{f} \in \mathbb{C}p$, i.e. $f \in \mathsf{H}^1_p(G)$. Moreover, $F(v) = (\operatorname{vol} X_v)^{1/2} f(v)$ and finally, $\hat{f} \in \mathscr{H}^1$.

Now we are again in position to demonstrate that the two quadratic forms $\mathfrak{d}$ and $\widetilde{\mathfrak{d}}$ are δ-partial isometrically equivalent (see Definition 4.4.11 and Proposition 4.4.13):

Theorem 6.6.1. *The quadratic forms $\widetilde{\mathfrak{d}}$ and $\mathfrak{d}$ are δ_ε-partial isometrically equivalent with identification operators J_ε, J^1_ε and J'^1_ε, where*

$$\delta^2_\varepsilon := \max\Bigl\{\frac{\hat{\delta}^2_{\varepsilon,v}}{\ell_-(v)}\Bigl(1 + \frac{2}{\lambda_2(X_v)}\Bigr), \frac{\varepsilon^2}{\min_e \lambda_2(Y_e)}, \frac{\varepsilon^{2\alpha}}{\lambda_2(X_v)}\Bigr\}.$$

In particular, the error is of order[4] *$\delta_\varepsilon = \mathrm{O}(\varepsilon^{1/2-})$ and depends only on an upper estimate on C_e and on lower estimates on $\lambda_2(X_v)$, $\lambda_2(Y_e)$ and ℓ_e.*

Recall the definition of $\hat{\delta}_{\varepsilon,v}$ in Proposition 6.3.5.

Proof. The first condition in (4.29a) is easily seen to be fulfilled. For the second condition in (4.29a) we estimate

$$\begin{aligned}\|u - J_\varepsilon J^*_\varepsilon u\|^2_{X_\varepsilon} &= \sum_e \|u - \underline{f}_e u\|^2_{X_{\varepsilon,e}} + \|u - \underline{f}_v u\|^2_{X_{\varepsilon,v}} \\ &\le \max_{e\in E}\Bigl\{\frac{\varepsilon^2}{\lambda_2(Y_e)}, \frac{\varepsilon^{2\alpha}}{\lambda_2(X_v)}\Bigr\}\|du\|^2_{X_\varepsilon}\end{aligned}$$

by Proposition 5.1.1 and the scaling behaviour. The first estimate in (4.29b) is trivially fulfilled, even with equality. The second estimate in (4.29b) reads as

$$\|J^*_\varepsilon u - J'^1_\varepsilon u\|^2_{\mathscr{H}} = \varepsilon^m \sum_{e\in E} \|\chi_e\|^2_{I_e} \operatorname{vol} Y_e \bigl|\underline{f}_v u - \underline{f}_e u(0)\bigr|^2 \le \hat{\delta}^2_{\varepsilon,v}\|du\|^2_{X_{\varepsilon,v}}$$

[4] Here, $\delta_\varepsilon = \mathrm{O}(\varepsilon^{1/2-})$ means that for any $\eta > 0$ small enough, we have $\delta_\varepsilon = \mathrm{o}(\varepsilon^{1/2-\eta})$, i.e. $\delta_\varepsilon \varepsilon^{-(1/2-\eta)} \to 0$ as $\varepsilon \to 0$.

using Proposition 6.3.5. Moreover, (4.29c) is easily seen to be fulfilled. For the estimate on the forms, we have

$$\begin{aligned}\left|\mathfrak{d}(J'^1_\varepsilon u, \hat{f}) - \widetilde{\mathfrak{d}}(u, J^1_\varepsilon \hat{f})\right|^2 &= \varepsilon^m \Big|\sum_e (\operatorname{vol} Y_e)^{1/2}(f_v \overline{u} - f_e \overline{u}(0))\langle \chi'_e, f'\rangle_{I_e}\Big|^2 \\ &\overset{\text{CS}}{\leq} \frac{\hat{\delta}^2_{\varepsilon,v}}{\ell_-(v)} \|du\|^2_{X_{\varepsilon,v}} \|f'\|^2_G\end{aligned}$$

using again Proposition 6.3.5 and $\|\chi'_e\|_{I_e} = \ell_-(v)^{-1}$. For the total error recall that $\hat{\delta}^2_{\varepsilon,v} = \mathrm{O}(\varepsilon^{1-(\beta-\alpha)}[\log \varepsilon])$, where $[\log \varepsilon]$ occurs only if $m = 1$. Since $2\alpha > (1 - \beta + \alpha)$ in the borderline case, the dominant error in δ^2_ε is of the same order. Choosing $\beta = \alpha + 2\eta$ for $0 < 2\eta < 1$, then $\delta^2_\varepsilon = \mathrm{O}(\varepsilon^{1-2\eta}[\log \varepsilon]) = \mathrm{o}(\varepsilon^{1-\eta})$, as we claimed. □

As in the fast and slowly decaying case, we obtain a closeness result for the associated boundary triples $(\Gamma_v, \Gamma'_v, \mathscr{G}_v)$ and $(\widetilde{\Gamma}_v, \widetilde{\Gamma}'_v, \widetilde{\mathscr{G}}_v)$ defined in Sects. 6.1.3 and 6.2.3. The boundary identification operator $I_\varepsilon : \mathscr{G}_v \longrightarrow \widetilde{\mathscr{G}}_v$ is the same as in (6.35). Moreover, we use the same transversal operators $K = 0$ and $\widetilde{K} = \Delta_{Y_{\varepsilon,\mathrm{int}}}$ as in Sect. 6.4:

Theorem 6.6.2. *The boundary maps*

$$\Gamma_v : \mathscr{H}^1_v \longrightarrow \mathbb{C}^{E_{v,\mathrm{int}}}, \qquad \widetilde{\Gamma}_v : \mathsf{H}^1(X_\varepsilon) \longrightarrow \mathsf{L}_2(Y_{\varepsilon,\mathrm{int}}),$$

$$\Gamma_v \hat{f} := \{f_e(\ell_e)\}_e, \qquad \widetilde{\Gamma}_v u := \{u_e(\ell_e, \cdot)\}_e$$

associated with the forms $\mathfrak{d}$ and $\widetilde{\mathfrak{d}}$ are 0-close with the identification operators J^1_ε, J'^1_ε and the boundary identification operator I_ε. Moreover, I_ε is a $\hat{\delta}_\varepsilon$-partial isometry w.r.t. $\mathfrak{k}$ and $\widetilde{\mathfrak{k}}$, where $\hat{\delta}_\varepsilon = \varepsilon(\min_e(\lambda_2(Y_e))^{-1/2}$. Finally, $\widetilde{K} I_\varepsilon = I_\varepsilon K (= 0)$.

The proof is literally the same as for Theorem 6.4.4.

6.7 The Embedded Case

We show in this section that certain neighbourhoods of an embedded (star) graph fall into the class of graph-like manifolds as described above up to some error estimates. In particular, we obtain similar limit theorems as for graph-like manifolds.

6.7.1 Embedded Graph-Like Spaces

Assume that $G \subset \mathbb{R}^2$ is a metric star graph embedded in $\mathbb{R}^2$, and that $\hat{X}_\varepsilon \subset \mathbb{R}^2$ is a closed neighbourhood of G with decomposition

$$\hat{X}_\varepsilon = \hat{X}_{\varepsilon,v} \mathbin{\overline{\dot\cup}} \overline{\dot{\bigcup_{e \in E}}} \hat{X}_{\varepsilon,e}. \tag{6.48}$$

For simplicity only, we assume that all length ℓ_e are finite, and therefore G and $\hat{X}_\varepsilon$ are compact. Let us give a more precise description of $\hat{X}_\varepsilon$ in terms of the parametrisation

$$\psi_e = (\psi_{e,1}, \psi_{e,2}) \colon I_e \longrightarrow \mathbb{R}^2$$

of the edge e as curve in $\mathbb{R}^2$, see also Sect. 5.4 for the case of a single curve. We assume that ψ_e has sufficient regularity, e.g., $\psi_e \in \mathsf{C}^3$ is enough. The vertex v corresponds to the point $\psi_e(0) \in \mathbb{R}^2$, and the latter is independent of $e \in E_v$. Denote by

$$\kappa_e := \psi'_{e,1}\psi''_{e,2} - \psi''_{e,1}\psi'_{e,2} \tag{6.49}$$

the curvature of the embedded edge e. Assume that for each edge, there is a Lipschitz continuous function $r_e \colon I_e \longrightarrow (0,\infty)$. The εr_e*-tubular neighbourhood* of the embedded graph is defined via the parametrisation

$$\Psi_{\varepsilon,e}(s,y) := \psi_e(s) + \varepsilon r_e(s) y \mathrm{n}_e(s), \qquad (s,y) \in I_e \times Y,$$

where n_e denotes a continuous unit vector field normal to the tangent vector field ψ'_e, and where $Y := [-1/2, 1/2]$. In order to avoid self-intersections, we assume that

$$0 < \varepsilon \le \varepsilon_0 := \min_e \frac{1}{\|\kappa_e\|_\infty \|r_e\|_\infty}.$$

Then

$$1 + \varepsilon\kappa_e(s) r_e(s) y \ge \frac{1}{2} > 0$$

for all $y \in Y$, $s \in I_e$, $0 < \varepsilon \le \varepsilon_0$ and all edges e (see condition (5.24)).

Let $0 < \alpha \le 1$ and $0 < \varepsilon \le \varepsilon_\alpha := \varepsilon_0^{1/\alpha}$. We assume that the components of $\hat{X}_\varepsilon$ are given by

$$\hat{X}_{\varepsilon,e} := \Psi_{\varepsilon,e}([\ell_e\varepsilon^\alpha, \ell_e] \times Y]) \qquad \text{and} \qquad \hat{X}_{\varepsilon,v} := \bigcup_{e \in E_v} \Psi_{\varepsilon^\alpha,e}([0, \ell_e\varepsilon^\alpha] \times Y]),$$

i.e. $\hat{X}_{\varepsilon,e}$ consists of the εr_e-neighbourhood of the curve ψ_e away from the vertex, and $\hat{X}_{\varepsilon,v}$ denotes a tubular neighbourhood of radius of order ε^α of the "ε^α-star graph" $G_\varepsilon = \overline{\dot\bigcup}_e \psi_e([0, \ell_e\varepsilon^\alpha])$.

The associated quadratic form on $\hat{X}_\varepsilon$ is defined by

$$\hat{\mathfrak{d}}(u) := \|\nabla u\|^2_{\hat{X}_\varepsilon}, \qquad \operatorname{dom}\hat{\mathfrak{d}} := \mathsf{H}^1(\hat{X}_\varepsilon).$$

Let us compare the quadratic form $\hat{\mathfrak{d}}$ with the quadratic form $\widetilde{\mathfrak{d}}$ associated with the scaled graph-like manifold $X_\varepsilon = X_{\varepsilon,v} \,\overline{\dot\cup}\, \overline{\bigcup}_e X_{\varepsilon,e}$ (see Definition 6.2.6), where

$$X_{\varepsilon,v} = \varepsilon^\alpha X_v \qquad \text{and} \qquad X_{\varepsilon,e} = I_e \times_{\varepsilon r_e} Y = I_e \times_{r_e} \varepsilon Y$$

(for the notation of the warped product, see Definition 5.3.2). Here, the transversal space, for which we need an embedding of different scalings as in (6.14), is given by an interval of length $r_e(0)$. Moreover, we use the embedding (6.16), and obtain $C_e = r_e(0)$ as bound on the derivative of the embedding.

The concept of δ-quasi-unitarily equivalence of quadratic forms was introduced in Sect. 4.4 (see Definition 4.4.11). We consider three cases of the embeddings:

- **Straight edges, constant radius:** Assume that r_e is constant and that $\kappa_e = 0$. Then $\hat{X}_{\varepsilon,e}$ is isometric to $I_{\varepsilon,e} \times Y_{\varepsilon,e}$ where $I_{\varepsilon,e}$ is an interval of length $(1-\varepsilon^\alpha)\ell_e$ and $Y_{\varepsilon,e}$ is an interval of length $\varepsilon r_e > 0$. Moreover, the vertex neighbourhood $\hat{X}_{\varepsilon,v}$ is ε^α-homothetic to a fixed subset $X_v \subset \mathbb{R}$, i.e. $\hat{X}_{\varepsilon,v}$ is isometric to $X_{\varepsilon,v} = \varepsilon^\alpha X_v$.

  In order to compare $\hat{X}_{\varepsilon,e}$ with $X_{\varepsilon,e}$, we need to enlarge slightly the longitudinal coordinate. From Proposition 5.4.4 (with $r_e(s) = r_e$ and $\ell_e(s,y) = 1$ being constant) it follows that the quadratic forms $\hat{\mathfrak{d}}$ and $\widetilde{\mathfrak{d}}$ are $\hat{\delta}_\varepsilon$-quasi-unitarily equivalent, where

  $$\hat{\delta}_\varepsilon = 2\varepsilon^\alpha(1-\varepsilon^\alpha)^{-1/2} = \mathrm{O}(\varepsilon^\alpha)$$

  depends only on α.
- **Curved edges, constant radius:** Assume that r_e is again constant, but the curvature κ_e may be non-trivial (see Fig. 6.8).

  The error from the edge neighbourhood is $\mathrm{O}(\varepsilon^\alpha)$, since we have again the correction of the longitudinal coordinate (Proposition 5.4.4). The error depends now on $\|\kappa_e\|_\infty$, $\|\kappa_e'\|_\infty$ and r_e. Moreover, the curvature produces a perturbation from the product structure, so by Proposition 5.4.2, we get another error term of order $\mathrm{O}(\varepsilon)$, depending on $\|\kappa_e\|_\infty$ and r_e.

  If the curvature κ_e vanishes near the vertex, we can choose $\hat{X}_{\varepsilon,v}$ such that $\hat{X}_{\varepsilon,v}$ is ε-homothetic with a fixed set $X_v \subset \mathbb{R}^2$. If the curvature κ_e does not vanish near the vertex v, the vertex neighbourhood differs slightly from an *exact* ε-homothetic one. The error made in the quadratic forms is of order $\mathrm{O}(\varepsilon^\alpha)$. This can be seen by cutting the vertex neighbourhood into deg v-many pieces which can be treated as embedded tubular neighbourhoods as in Sect. 5.4 away from the vertex (see the grey-shaded parts of $\hat{X}_{\varepsilon,v}$ in Fig. 6.8). The radius of the tubular neighbourhoods may be chosen to be $r_e(0)$ since $r_e(s) \approx r_e(0)$ for $s = \mathrm{O}(\varepsilon^\alpha)$. "Straightening" the edge leads to an error term of order $\mathrm{O}(\varepsilon^\alpha)$ as in Proposition 5.4.2.

  Altogether, the quadratic forms $\hat{\mathfrak{d}}$ and $\widetilde{\mathfrak{d}}$ are $\mathrm{O}(\varepsilon^\alpha)$-quasi-unitarily equivalent, where the error depends only on $\|\kappa_e\|_\infty$, $\|\kappa_e'\|_\infty$, r_e and α.
- **Curved edges, variable radius:** In the most general case, we allow a variable radius r_e and non-trivial curvature κ_e.

  For the edge neighbourhood, the error is $\mathrm{O}(\varepsilon^\alpha)$ from the shortened edge (Proposition 5.4.4) depending on r_e and κ_e and its first derivatives. In addition,

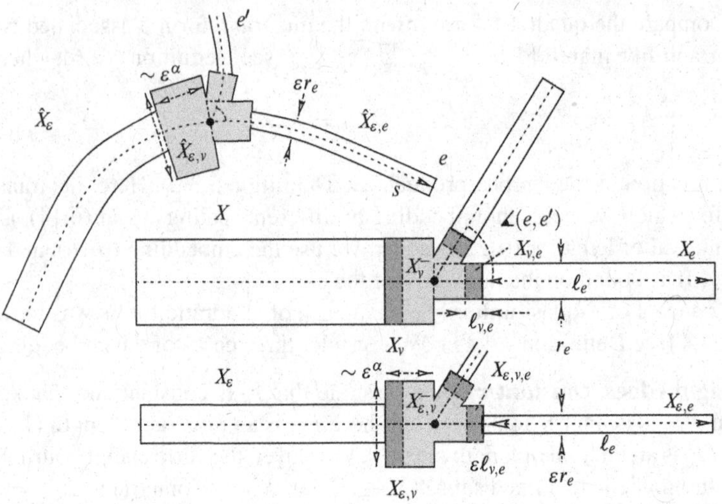

Fig. 6.8 A 2-dimensional closed tubular neighbourhood $\hat{X}_\varepsilon$ of an embedded star graph with curved edges in $\mathbb{R}^2$. Here, $\hat{X}_\varepsilon$ has non-smooth boundary and constant radius on each edge. Below, we sketched the corresponding unscaled and scaled graph-like manifolds X and X_ε (with vertex neighbourhood scaling ε^α), respectively. Note that the edge neighbourhoods X_e and $X_{\varepsilon,e}$ have the same length ℓ_e as the original edge e. We will see later on that the angle between two edges does not directly enter in our model assumptions, only a lower bound on the second Neumann eigenvalue $\lambda_2(X_v)$ is needed

the curvature and the variable radius produce a perturbation from the product structure, so by Proposition 5.4.2, we get another error term of order $O(\varepsilon)$ depending on the same quantities.

Moreover, the vertex neighbourhood $\hat{X}_{\varepsilon,v}$ is ε^α-homothetic up to an error $O(\varepsilon^\alpha)$ of the quadratic forms depending now on r_e and the curvature κ_e by the same arguments as in the previous case.

Altogether, the quadratic forms $\hat{\mathfrak{d}}$ and $\tilde{\mathfrak{d}}$ are $O(\varepsilon^\alpha)$-quasi-unitarily equivalent, where the error depends now on $\|r_e\|_\infty$, $\|r_e^{-1}\|_\infty$, $\|r_e'\|_\infty$, $\|\kappa_e\|_\infty$, $\|\kappa_e'\|_\infty$ and α.

6.7.2 *Reduction to the Graph Model*

Let us now apply the results of Sects. 6.4–6.6 to the situation when the star graph G is embedded in $\mathbb{R}^2$ with curved edges and has a constant tubular radius $r_e > 0$ around each edge, see Fig. 6.8. Here, $d = 2$ and $m = 1$. Basically, the quadratic forms on the embedded space and the abstract graph-like manifold are $\hat{\delta}_\varepsilon$-quasi-unitarily equivalent, where $\hat{\delta}_\varepsilon = O(\varepsilon^\alpha)$ depends on $\|\kappa_e\|_\infty$, $\|\kappa_e'\|_\infty$, r_e and the scaling rate α of the vertex neighbourhood. Combining the results of Sects. 6.4–6.6, we can show that the quadratic form on $\hat{X}_\varepsilon$ is δ_ε-quasi-unitarily equivalent with the corresponding quadratic form on the graph (depending on the scaling rate α of the vertex neighbourhood).

The fast decaying case:

Assume that the scaling rate α of the vertex neighbourhood fulfils $m/d = 1/2 < \alpha \leq 1$, then we have (see Sect. 6.4):

Theorem 6.7.1. *The quadratic form* $\hat{\mathfrak{d}}(u) := \|\nabla u\|^2_{\hat{X}_\varepsilon}$ *is* δ_ε*-quasi-unitarily equivalent with the quadratic form* $\mathfrak{d}$ *associated with the (weighted) standard Laplacian on G as defined in Sect. 6.1.1, where* $\delta_\varepsilon = \mathrm{O}(\varepsilon^{(\alpha d - m)/2}) = \mathrm{O}(\varepsilon^{\alpha - 1/2})$. *The error depends only on lower bounds on* $\lambda_2(X_v)$, ℓ_e *and on upper bounds on* r_e, $\operatorname{vol} X_v / \operatorname{vol} \mathring{\partial} X_v$, $\|\kappa_e\|_\infty$ *and* $\|\kappa'_e\|_\infty$.

The proof follows from Theorem 6.4.1 and the additional error estimates considered in Sect. 6.7.1 below. Note that the error term of Theorem 6.4.1 is dominant, since $\alpha d - m = 2\alpha - 1 \leq \alpha$. Moreover, the upper bound on r_e arises since we need a lower bound on $\lambda_2(Y_e) = \pi^2/r_e^2$. Here, Y_e is an interval of length $r_e > 0$.

The slowly decaying case:

Assume that $0 < \alpha < m/d = 1/2$, then we have (see Sect. 6.5):

Theorem 6.7.2. *The quadratic form* $\hat{\mathfrak{d}}(u) := \|\nabla u\|^2_{\hat{X}_\varepsilon}$ *is* δ_ε*-quasi-unitarily equivalent with the decoupled extended Dirichlet form* $\mathfrak{d}$ *as defined in Sect. 6.1.2, where* $\delta_\varepsilon = \mathrm{O}(\varepsilon^\alpha)$ *if* $0 < \alpha \leq 1/4$ *and* $\delta_\varepsilon = \mathrm{O}(\varepsilon^{1/2-\alpha})$ *if* $1/4 \leq \alpha < 1/2$. *The error depends only on lower bounds on* $\lambda_2(X_v)$, ℓ_e *and on upper bounds on* r_e, $\operatorname{vol} \mathring{\partial} X_v / \operatorname{vol} X_v$, $\|\kappa_e\|_\infty$ *and* $\|\kappa'_e\|_\infty$.

The proof follows from Theorem 6.5.1 and the reasoning in Sect. 6.7.1.

The borderline case:

Assume that $\alpha = m/d = 1/2$, then we have (see Sect. 6.6):

Theorem 6.7.3. *The quadratic form* $\hat{\mathfrak{d}}(u) := \|\nabla u\|^2_{\hat{X}_\varepsilon}$ *is* δ_ε*-quasi-unitarily equivalent with the coupled form* $\mathfrak{d}$ *as defined in Sect. 6.1.3, where* $\delta_\varepsilon = \mathrm{O}(\varepsilon^{1/2-})$. *The error depends only on lower bounds on* $\lambda_2(X_v)$, ℓ_e *and on upper bounds on* r_e, $\|\kappa_e\|_\infty$ *and* $\|\kappa'_e\|_\infty$.

The proof follows similarly from Theorem 6.6.1.

Remark 6.7.4. Although the curvature κ_e of the edge and the angle $\measuredangle(e, e')$ between two adjacent edges (see e.g. Fig. 6.8) are not detectable in the limit, the error

δ_ε depends on κ_e, κ'_e and $\lambda_2(X_v)$. From Theorem 4.6.4 we see that $\lambda_k(\varepsilon) = \lambda_k(0) + \mathrm{O}(\varepsilon^{1/2})$ for the k-th eigenvalue $\lambda_k(\varepsilon)$ and $\lambda_k(0)$ of $\hat{\mathfrak{d}}$ and $\mathfrak{d}$, respectively. Using the eigenvalue comparison Corollary 4.4.19 one can show directly the same result. Note that [G08a, GJ09] obtained an asymptotic expansion $\lambda_k(\varepsilon) = \lambda_k(0) + \varepsilon\lambda_k^1(0) + \varepsilon^2\lambda_k^2(0) + \dots$ for the eigenvalues of *straight* edges. It would be interesting whether one can detect information e.g. on the angles between the edges via the coefficients $\lambda_k^1(0), \lambda_k^2(0), \dots$. For a curved tubular neighbourhood with Dirichlet boundary conditions around a closed curve of length ℓ with positive curvature in $\mathbb{R}^3$, the first eigenvalue expands as

$$\lambda_1^{\mathrm{D}}(\varepsilon) = \frac{\lambda_1}{\varepsilon^2} - \frac{3}{4\ell}\int_0^\ell \kappa_1(s)^2 \mathrm{d}s + \mathrm{O}(\varepsilon)$$

(cf. [KP88, Thm. 4.1]), where λ_1 is the first Dirichlet eigenvalue of the unit disc and κ_1 is the curvature of the curve embedded in $\mathbb{R}^3$.

Remark 6.7.5. If we consider tubular neighbourhoods of an edge e with *variable* radius function $r_e \colon I_e \longrightarrow (0,\infty)$ (see Fig. 6.9), we obtain a different limit form on each edge (see Sect. 5.5). Let us illustrate the necessary arguments in the case of fast decaying vertex volume. The other cases can be treated similarly. As reference transversal manifold we use $Y_e = [-1/2, 1/2]$, so that $\operatorname{vol} Y_e = 1$. We assume that r_e has sufficient regularity, say, $r_e \in \mathsf{C}^2$.

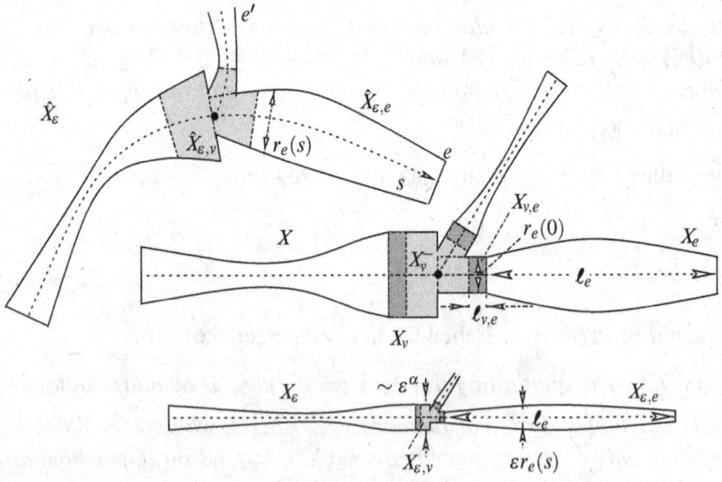

Fig. 6.9 A 2-dimensional closed tubular neighbourhood $\hat{X}_\varepsilon$ with *variable* radius on each edge, here for simplicity with vertex neighbourhood scaling ε. Note that the "straightened" unscaled vertex space X_v has constant radius $r_e(0)$ on the arm associated with the edge e. The scaled graph-like manifold X_ε is drawn for a smaller ε than the original tubular neighbourhood $\hat{X}_\varepsilon$

In the limit, we basically have to replace the measure $\mathrm{d}s = \mathrm{d}s_e$ on I_e by the weighted measure $w_e(s)\mathrm{d}s = r_e(s)^m \mathrm{d}s$ with $m = 1$ here. The corresponding Hilbert space is now given by

$$\mathsf{L}_2(G, r\mathrm{d}s) := \bigoplus_e \mathsf{L}_2(I_e, r_e(s)\mathrm{d}s).$$

Moreover, the limit quadratic form is defined as

$$\mathfrak{d}(f) = \sum_e \|f_e'\|^2_{I_e, r_e \mathrm{d}s} = \int_{I_e} |f_e'(s)|^2 r_e(s) \mathrm{d}s_e$$

with domain

$$\operatorname{dom} \mathfrak{d} = \Big\{ f \in \bigoplus_e \mathsf{H}^1(I_e, r_e(s)\mathrm{d}s_e) \;\Big|\; f \text{ continuous at } v \Big\}.$$

The corresponding operator will act as in (5.27) on each edge. Using the transformation $f_e \mapsto \widetilde{f}_e = r_e^{1/2} f_e$ on each edge, one obtains a unitary operator from the weighted space $\mathsf{L}_2(G, r\mathrm{d}s)$ into the unweighted Hilbert space

$$\mathsf{L}_2(G) = \bigoplus_e \mathsf{L}_2(I_e).$$

Moreover, the quadratic form transforms into a quadratic form $\mathfrak{q}$ given by

$$\mathfrak{q}(\widetilde{f}) = \sum_e \big(\|\widetilde{f}_e'\|^2_{I_e} + \langle \widetilde{f}_e, Q_e \widetilde{f}_e \rangle_{I_e} \big) + L(v)|f(v)|^2$$

with domain

$$\operatorname{dom} \mathfrak{q} = \mathsf{H}^1_{\mathrm{cont}}(G) = \big\{ f \in \mathsf{H}^1_{\max}(G) \,\big|\, \underline{f}(0) = f(v)p \in \mathbb{C}p \big\}$$

where $p = \{p_e\}$ and $p_e = r_e(0)^{1/2}$ (see Proposition 5.5.3). Here, the potentials on the edges and vertices are given by

$$Q_e := q_e^2 + q_e', \qquad q_e := \frac{1}{2}(\log r_e)' \qquad \text{and} \qquad L(v) := \sum_e q_e(0) p_e^2.$$

Note that

$$Q_e = \Big(\frac{r_e'}{2r_e}\Big)^2 + \Big(\frac{r_e'}{2r_e}\Big)' \qquad \text{and} \qquad L(v) = \frac{1}{2} \sum_e r_e'(0).$$

The transformed quadratic form $\mathfrak{q}$ is non-negative since the unitary equivalent form $\mathfrak{d}$ is – although the edge and vertex potentials Q_e and $L(v)$ may achieve negative values.

6.8 Slowly Decaying and Borderline Case for Arbitrary Transversal Manifolds

In the following sections, we provide the necessary modifications in order to treat *arbitrary* transversal manifolds Y_e (and not only "homogeneous" ones as in Sect. 6.2.2) also in the case of slowly decaying vertex volume (Sect. 6.9) or the borderline case (Sect. 6.10). As before, we construct the manifold associated with a star graph G with one vertex v and finitely many edges $e \in E = E_v$ of lengths $\ell_e \in (0, \infty]$. We define the (modified) graph-like manifold and collect some estimates needed for both cases in this section.

6.8.1 *The Enlarged Vertex Neighbourhood with Added Truncated Cones*

If the transversal manifolds Y_e are not homogeneous, i.e. we do not have an embedding as in (6.14), then we have to modify the construction of a graph-like manifold. This is in particular the case for transversal manifolds like $Y_e = \mathbb{S}^1$. For such transversal manifolds, we cannot glue together two parts $X_{\varepsilon,e} = I_e \times \varepsilon Y_e$ and $X_{\varepsilon,v} = \varepsilon^\alpha X_v$ $(\alpha < 1)$ having a different scaling, such that the resulting space has a (Lipschitz) continuous metric. In order to overcome this problem, we add a truncated cone $\acute{X}_{\varepsilon,v,e}$ between the edge neighbourhood $X_{\varepsilon,e}$ and the vertex neighbourhood $X_{\varepsilon,v}$ for each edge (see Figs. 6.10 and 6.11).

The truncated cone is given by the warped product

$$\acute{X}_{\varepsilon,v,e} := \varepsilon^\beta I_{\varepsilon,v} \times_{r_{\varepsilon,v,e}} Y_e, \qquad \beta > \alpha$$

(see Definition 5.3.2), where the radius function is given by the affine-linear function $r_{\varepsilon,v,e} \colon I_{v,e} = [0, \ell_{v,e}] \longrightarrow (0, \infty)$ with

$$r_{\varepsilon,v,e}(0) = \varepsilon \qquad \text{and} \qquad r_{\varepsilon,v,e}(\ell_{v,e}) = \varepsilon^\alpha.$$

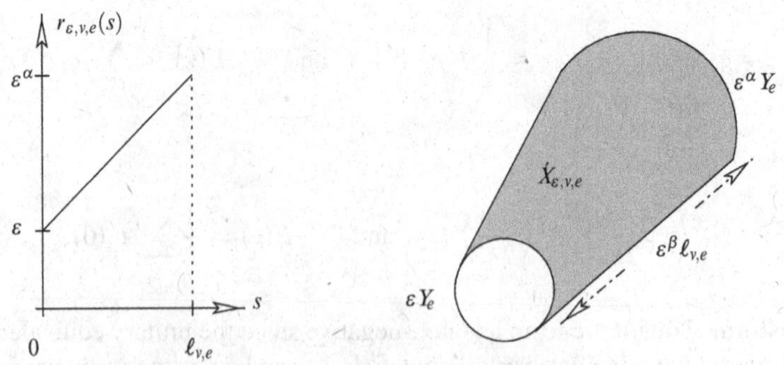

Fig. 6.10 The truncated cone $\acute{X}_{\varepsilon,v,e}$ for the transversal manifold $Y_e = \mathbb{S}^1$

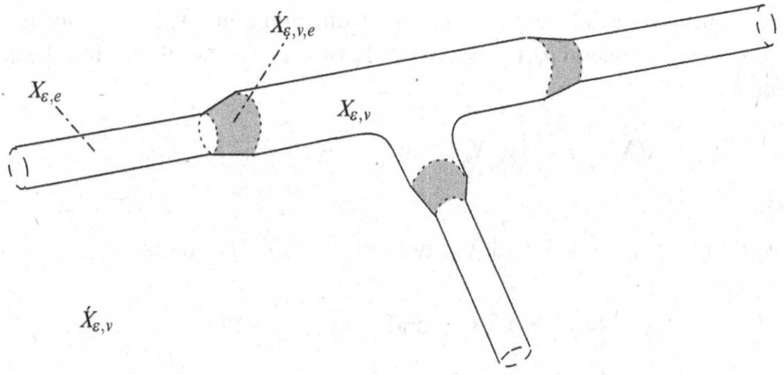

Fig. 6.11 The enlarged vertex neighbourhood $\acute{X}_{\varepsilon,v}$ with an added truncated cone $\acute{X}_{\varepsilon,v,e}$ for each adjacent edge

Recall that $\acute{X}_{\varepsilon,v,e}$ is the manifold $X_{v,e} = I_{v,e} \times Y_e$ with metric

$$\acute{g}_{\varepsilon,v,e} = \varepsilon^{2\beta} ds^2 + r_{\varepsilon,v,e}(s)^2 h_e,$$

where h_e is a given metric on Y_e. Such a truncated cone was already used in Sect. 6.2.2 where we *embedded* the cone into $X_{\varepsilon,v,e} \subset X_{\varepsilon,v}$. Here, in contrast, we put the truncated cones *between* each edge neighbourhood and the central vertex neighbourhood: We therefore define the *enlarged vertex neighbourhood* by

$$\acute{X}_{\varepsilon,v} := X_{\varepsilon,v} \,\overline{\cup}\, \overline{\dot{\bigcup_e}}\, \acute{X}_{\varepsilon,v,e}$$

(see Fig. 6.11),

where again, the central part of $\acute{X}_{\varepsilon,v}$ scales homogeneously, i.e.

$$X_{\varepsilon,v} = \varepsilon^{\alpha} X_v, \qquad \text{for some} \qquad 0 < \alpha \le \frac{m}{d}$$

(recall that $m := \dim Y_e = d - 1$ and $\dim X_v = d$). Note that in the fast decaying case $m/d < \alpha < 1$, we have already treated smooth manifolds in Theorem 6.4.3 allowing a *non-homogeneous* scaling (6.27) of $X_{\varepsilon,v}$. If $\beta \le 1$, then the enlarged vertex neighbourhood $\acute{X}_{\varepsilon,v}$ fulfils (6.27) with underlying unscaled space

$$\acute{X}_v := X_v \,\overline{\cup}\, \overline{\dot{\bigcup_e}}\, X_{v,e}$$

(i.e. we set $\varepsilon = 1$). Note that $X_{v,e} = \acute{X}_{v,e}$, since for $\varepsilon = 1$, the truncated cone becomes a cylinder. The *graph-like manifold (with added cones)* associated with the underlying star graph G is now defined by

$$X_\varepsilon := \acute{X}_{\varepsilon,v} \,\overline{\cup}\, \overline{\dot{\bigcup_e}}\, X_{\varepsilon,e},$$

where we identified each "small" end of the truncated cone $\acute{X}_{\varepsilon,v,e}$ (isometric to εY_e) with the corresponding end $\partial_v X_{\varepsilon,e} = \{0\}\times\varepsilon Y_e$ of the edge neighbourhood. As usual, we denote by

$$\mathring{\partial}\acute{X}_{\varepsilon,v} = \dot{\bigcup_e} \partial_e \acute{X}_{\varepsilon,v} \quad \text{and} \quad \mathring{\partial} X_{\varepsilon,v} = \dot{\bigcup_e} \partial_e X_{\varepsilon,v}$$

the *internal* boundary of $\acute{X}_{\varepsilon,v}$ and $X_{\varepsilon,v}$ as subsets of X_ε. Note that

$$\mathring{\partial}\acute{X}_{\varepsilon,v} = \varepsilon\mathring{\partial}\acute{X}_v \quad \text{and} \quad \mathring{\partial} X_{\varepsilon,v} = \varepsilon^\alpha \mathring{\partial} X_v.$$

We start with a lemma assuring that we do not add "too much" volume by introducing the truncated cones. Recall that $\operatorname{vol} X_{\varepsilon,v} = \varepsilon^{\alpha d} \operatorname{vol} X_v$:

Lemma 6.8.1. *Assume that* $0 < \varepsilon \le 1$, *then we have*

$$\operatorname{vol} X_{\varepsilon,v} \le \operatorname{vol} \acute{X}_{\varepsilon,v} \le \Big(1 + \varepsilon^{\beta-\alpha} \frac{\operatorname{vol} \mathring{\partial} X_v}{\operatorname{vol} X_v}\Big) \operatorname{vol} X_{\varepsilon,v}.$$

In particular, if $\alpha < \beta$, *then the additional truncated cones are negligible, i.e.*

$$\frac{\operatorname{vol} X_{\varepsilon,v}}{\operatorname{vol} \acute{X}_{\varepsilon,v}} \to 1 \quad \textit{as} \quad \varepsilon \to 0.$$

Proof. We apply Proposition A.2.17 with $a = \varepsilon^\beta \ell_{v,e}$, $r_0 = \varepsilon$, $r_1 = \varepsilon^\alpha$ and obtain

$$\frac{\varepsilon^{\beta+\alpha m}}{m+1} \cdot \ell_{v,e} \operatorname{vol} Y_e \le \operatorname{vol} \acute{X}_{\varepsilon,v,e} \le \varepsilon^{\beta+\alpha m} \cdot \ell_{v,e} \operatorname{vol} Y_e \le \varepsilon^{\beta+\alpha m} \cdot \operatorname{vol} Y_e,$$

since $\ell_{v,e} \le 1$. In particular, we have

$$0 \le \operatorname{vol} \acute{X}_{\varepsilon,v} \operatorname{vol} X_{\varepsilon,v} = \sum_e \operatorname{vol} \acute{X}_{\varepsilon,v,e} \le \varepsilon^{\beta+\alpha m} \operatorname{vol} \mathring{\partial} X_v = \varepsilon^{\beta-\alpha} \frac{\operatorname{vol} \mathring{\partial} X_v}{\operatorname{vol} X_v} \cdot \operatorname{vol} X_{\varepsilon,v}$$

noting that $\operatorname{vol} \mathring{\partial} X_v = \operatorname{vol} \mathring{\partial} \acute{X}_v = \sum_e \operatorname{vol} Y_e$. □

6.8.2 Some More Vertex Neighbourhood Estimates

Here, we provide the necessary estimates needed for the reduction to the graph models in Sects. 6.9 and 6.10. Let us start with introducing the following averaging operators

$$\acute{\fint}_{v} u := \fint_{\acute{X}_v} u, \qquad \acute{\fint}_{\varepsilon,v} u := \fint_{\acute{X}_{\varepsilon,v}} u, \tag{6.50a}$$

$$\fint_{v} u := \fint_{X_v} u = \fint_{X_{\varepsilon,v}} u, \qquad \fint_{e} u(s) := \fint_{Y_e} u(s,\cdot) = \fint_{Y_{\varepsilon,e}} u(s,\cdot) \tag{6.50b}$$

for $u \in \mathsf{L}_2(X)$. As before, $\fint_e u(s)$ is finite for almost every $s \in I_e = [0, \ell_e]$, and $\fint_e u \in \mathsf{L}_2(I_e)$. Since $\acute{X}_{\varepsilon,v}$ does not scale homogeneously, we do not have the scale invariance

$$\acute{\fint}_{v} u \neq \acute{\fint}_{\varepsilon,v} u$$

as for $X_{\varepsilon,v} = \varepsilon^\alpha X_v$ and $Y_{\varepsilon,e} = \varepsilon Y_e$. Since we identified the smaller end $\partial_e \acute{X}_{\varepsilon,v} = \varepsilon \partial_e \acute{X}_v \cong \varepsilon Y_e$ of the truncated cone $\acute{X}_{\varepsilon,v,e}$ with one end of the edge neighbourhood $\partial_v X_{\varepsilon,v} = \{0\} \times \varepsilon Y_e$ in the total space we have

$$\fint_e u(0) = \fint_{\partial_e \acute{X}_{\varepsilon,v}} u \tag{6.51}$$

for $u \in \mathsf{H}^1(X_\varepsilon)$.

Let us first compare the averaging operators on $\acute{X}_{\varepsilon,v}$ and $X_{\varepsilon,v}$:

Lemma 6.8.2. *We have*

$$\begin{aligned} &\left|(\operatorname{vol} \acute{X}_{\varepsilon,v})^{1/2} \acute{\fint}_{\varepsilon,v} u - (\operatorname{vol} X_{\varepsilon,v})^{1/2} \fint_{v} u\right|^2 \\ &\le 2\Big(1 - \frac{\operatorname{vol} X_{\varepsilon,v}}{\operatorname{vol} \acute{X}_{\varepsilon,v}}\Big) \|u\|^2_{\acute{X}_{\varepsilon,v}} \le 2\varepsilon^{\beta-\alpha} \frac{\operatorname{vol} \mathring{\partial} X_v}{\operatorname{vol} X_v} \cdot \|u\|^2_{\acute{X}_{\varepsilon,v}} \end{aligned}$$

for $u \in \mathsf{H}^1(\acute{X}_{\varepsilon,v})$.

Proof. We have

$$\begin{aligned} &\left|(\operatorname{vol} \acute{X}_{\varepsilon,v})^{1/2} \acute{\fint}_{\varepsilon,v} u - (\operatorname{vol} X_{\varepsilon,v})^{1/2} \fint_{v} u\right|^2 \\ &= \frac{1}{\operatorname{vol} \acute{X}_{\varepsilon,v}} \left| \int_{\acute{X}_{\varepsilon,v} \setminus X_{\varepsilon,v}} u + \Big(1 - \Big(\frac{\operatorname{vol} \acute{X}_{\varepsilon,v}}{\operatorname{vol} X_{\varepsilon,v}}\Big)^{1/2}\Big) \int_{X_{\varepsilon,v}} u \right|^2 \\ &\overset{\text{CS}}{\le} 2\Big[\frac{\operatorname{vol}(\acute{X}_{\varepsilon,v} \setminus X_{\varepsilon,v})}{\operatorname{vol} \acute{X}_{\varepsilon,v}} \|u\|^2_{\acute{X}_{\varepsilon,v} \setminus X_{\varepsilon,v}} + \Big(1 - \Big(\frac{\operatorname{vol} \acute{X}_{\varepsilon,v}}{\operatorname{vol} X_{\varepsilon,v}}\Big)^{1/2}\Big)^2 \frac{\operatorname{vol} X_{\varepsilon,v}}{\operatorname{vol} \acute{X}_{\varepsilon,v}} \|u\|^2_{X_{\varepsilon,v}}\Big] \\ &\le 2 \max\Big\{1 - \rho, (\sqrt{\rho} - 1)^2\Big\} \|u\|^2_{\acute{X}_{\varepsilon,v}}, \end{aligned}$$

where $\rho := \operatorname{vol} X_{\varepsilon,v} / \operatorname{vol} \acute{X}_{\varepsilon,v} \le 1$. The result follows from the fact that $(\sqrt{\rho} - 1)^2 \le |1 - \rho|$ for all $\rho \ge 0$. For the last estimate, we apply Lemma 6.8.1 and the fact that $1 - \rho \le 1/\rho - 1$ if $\rho \le 1$. □

Next, we need to compare the second (first non-vanishing) Neumann eigenvalues $\lambda_2(\acute{X}_{\varepsilon,v})$ and $\lambda_2(X_{\varepsilon,v}) = \varepsilon^{-2\alpha}\lambda_2(X_v)$ on $\acute{X}_{\varepsilon,v}$ and $X_{\varepsilon,v} = \varepsilon^\alpha X_v$, respectively:

Lemma 6.8.3. *Assume that* $\beta > \alpha$ *and that*

$$0 < \varepsilon \le \varepsilon_0 := \min\Big\{\big(8 + 8\lambda_2^{\mathrm{D}}(X_v)\big)^{-\frac{1}{\beta-\alpha}}, 1\Big\}, \tag{6.52a}$$

where $\lambda_2^{\mathrm{D}}(X_v)$ *denotes the second eigenvalue with Dirichlet conditions on* $\bigcup_e \partial_e X_v$,

$$\frac{1}{\lambda_2(\acute{X}_{\varepsilon,v})} \le \frac{2}{\lambda_2(X_{\varepsilon,v})} = \frac{2\varepsilon^{2\alpha}}{\lambda_2(X_v)}. \tag{6.52b}$$

Proof. The eigenvalue estimate follows from Proposition 4.4.18 with $\mathscr{H} = \mathsf{L}_2(\acute{X}_{\varepsilon,v})$, $\widetilde{\mathscr{H}} = \mathsf{L}_2(X_{\varepsilon,v})$, $\mathfrak{d}(f) := \|\mathrm{d}f\|^2_{\acute{X}_{\varepsilon,v}}$, $\widetilde{\mathfrak{d}}(u) := \|\mathrm{d}u\|^2_{X_{\varepsilon,v}}$ and $J^1 f := f \restriction_{X_{\varepsilon,v}}$. Namely, we have to check the necessary estimates Proposition 4.4.18: The estimate $\widetilde{\mathfrak{d}}(J^1 f) \le \mathfrak{d}(f)$ is trivially fulfilled, i.e. we have $\delta_0' = \delta_1' = 0$. For the estimate on the quadratic forms, we use again the decomposition of the central manifold

$$X_{\varepsilon,v} = X_{\varepsilon,v}^- \,\overline{\dot\cup}\, \overline{\dot{\bigcup_e}}\, X_{\varepsilon,v,e} \quad \text{where} \quad X_{\varepsilon,v}^- = \varepsilon^\alpha X_v^- \quad \text{and} \quad X_{\varepsilon,v,e} = \varepsilon^\alpha X_{v,e}.$$

Applying Proposition A.2.17 with $a = \ell_{v,e}\varepsilon^\beta$, $b = \ell_{v,e}\varepsilon^\alpha$, $r_0 = \varepsilon$ and $r_1 = \varepsilon^\alpha$ (and a change of the longitudinal coordinate $s \mapsto \varepsilon^\beta s$) yields

$$\begin{aligned}\|f\|^2_{\mathscr{H}} - \|J^1 f\|^2_{\widetilde{\mathscr{H}}} &= \sum_e \|f\|^2_{\acute{X}_{\varepsilon,v,e}}\\ &\le 4\varepsilon^{\alpha+\beta}\sum_e \ell_{v,e}^2 \cdot \Big(\|f'\|^2_{X^+_{\varepsilon,v,e}} + \frac{1}{\varepsilon^{2\alpha}\ell_{v,e}^2}\|f\|^2_{X^+_{\varepsilon,v,e}}\Big)\\ &\le \delta_1\mathfrak{d}(f) + \delta_0\|f\|^2,\end{aligned}$$

where $X^+_{\varepsilon,v,e} := \acute{X}_{\varepsilon,v,e} \,\overline{\dot\cup}\, X_{\varepsilon,v,e}$ denotes the union of the truncated cone and the cylinder $X_{\varepsilon,v,e} \subset X_{\varepsilon,v}$, and where $\delta_0 := 4\varepsilon^{\beta-\alpha}$ and $\delta_1 := 4\varepsilon^{\beta+\alpha}$ (recall that $\ell_{v,e} \le 1$). In addition, the condition $\delta_0 + \delta_1\lambda_k < 1$ is fulfilled, since

$$\delta_0 + \delta_1\lambda_k(\acute{X}_{\varepsilon,v}) \le 4\big(1 + \lambda_k^{\mathrm{D}}(X_v)\big)\varepsilon^{\beta-\alpha} \le \frac{1}{2}$$

provided $0 < \varepsilon \le \varepsilon_0$. Here, we used the eigenvalue monotonicity

$$\lambda_k(\acute{X}_{\varepsilon,v}) \le \lambda_k^{\mathrm{D}}(X_{\varepsilon,v}) = \varepsilon^{-2\alpha}\lambda_k^{\mathrm{D}}(X_v)$$

since $X_{\varepsilon,v} \subset \acute{X}_{\varepsilon,v}$. Finally, from Proposition 4.4.18 we obtain

$$\lambda_k(X_{\varepsilon,v}) \le \frac{\lambda_k(\acute{X}_{\varepsilon,v})}{1 - \delta_0 - \delta_1 \lambda_k(\acute{X}_{\varepsilon,v})} \le 2\lambda_k(\acute{X}_{\varepsilon,v}),$$

and the result follows with $k = 2$. □

The following estimate is needed in the slowly decaying case; a similar result has been proven in Proposition 6.3.10. For the definition of p_{m-2}, we refer to Definition A.2.13.

Proposition 6.8.4. *Assume that $\beta > \alpha$ and that $0 < \alpha < m/d = (d-1)/d$, then we have*

$$\varepsilon^m \sum_{e \in E} \operatorname{vol} Y_e \left| f_e u(0) \right|^2 \le \widetilde{\delta}^2_{\varepsilon,v} \Big(\|du\|^2_{\acute{X}_{\varepsilon,v}} + \|u\|^2_{X_{\varepsilon,v}} \Big)$$

for $u \in \mathsf{H}^1(\acute{X}_{\varepsilon,v})$, where

$$\widetilde{\delta}^2_{\varepsilon,v} := \max\Big\{ 2\varepsilon^{1+\beta-\alpha} p_{m-2}(\varepsilon^{1-\alpha}),\ 2\varepsilon^{m-\alpha d+2\alpha},\ \varepsilon^{m-\alpha d} \frac{4}{\ell_-(v)} \Big\}.$$

Here, $\widetilde{\delta}^2_{\varepsilon,v} = \mathrm{O}(\varepsilon^{m-\alpha d})$ provided $\beta > (1-\alpha)m - 1$.

Proof. We have

$$\begin{aligned} \varepsilon^m \operatorname{vol} Y_e \left| f_e u(0) \right|^2 &= \varepsilon^m \operatorname{vol} Y_e \left| f_{\partial_e \acute{X}_{\varepsilon,v}} u \right|^2 \\ &\le 2\varepsilon^\beta \ell_{v,e} p_{m-2}(\varepsilon^{1-\alpha}) \varepsilon^{1-\alpha} \|u'\|^2_{\acute{X}_{\varepsilon,v,e}} + 2\varepsilon^m \operatorname{vol} Y_e \left| f_{\partial_e X_v} u \right|^2 \end{aligned}$$

using (6.51) and Proposition A.2.17 with $a = \varepsilon^\beta \ell_{v,e}$. The latter averaging term can be estimated by

$$\begin{aligned} \varepsilon^m \operatorname{vol} Y_e \left| f_{\partial_e X_v} u \right|^2 &\overset{\text{CS}}{\le} \varepsilon^m \|u\|^2_{\partial_e X_v} \le \varepsilon^m \Big(\ell_{v,e} \|u'\|^2_{X_{v,e}} + \frac{2}{\ell_{v,e}} \|u\|^2_{X_{v,e}} \Big) \\ &= \varepsilon^m \Big(\varepsilon^{-(d-2)\alpha} \ell_{v,e} \|u'\|^2_{X_{\varepsilon,v,e}} + \frac{2\varepsilon^{-d\alpha}}{\ell_{v,e}} \|u\|^2_{X_{\varepsilon,v,e}} \Big) \end{aligned}$$

using (6.20) and the scaling behaviour of the norms. The result follows by summing over e. For the error estimate, it is easy to see that $\varepsilon^{m-\alpha d}$ dominates the error of $\delta^2_{\varepsilon,v}$ for $\beta > (1-\alpha)m - 1$. Recall that $p_{m-1}(\tau) \to 1/(m-1)$ if $m \ge 2$ and $p_{-1}(\tau) = \mathrm{O}(|\log \varepsilon|)$ as $\tau \to 0$ (see Lemma A.2.16). □

The next proposition is similar to Proposition 6.3.5 and used in the borderline case:

Proposition 6.8.5. *Assume that $\beta > \alpha$, then we have*

$$\sum_{e \in E} \operatorname{vol} Y_e \left| f_e u(0) - f_v u \right|^2 \le \hat{\delta}^2_{\varepsilon,v} \|du\|^2_{\acute{X}_{\varepsilon,v}}$$

for $u \in \mathsf{H}^1(\acute{X}_{\varepsilon,v})$, where

$$\hat{\delta}^2_{\varepsilon,v} := \max\Big\{2\varepsilon^{1+\beta-\alpha} p_{m-2}(\varepsilon^{1-\alpha}),\ 2\varepsilon^{2\alpha+m-\alpha d} \cdot \frac{1}{\ell_-(v)}\Big(1 + \frac{2}{\lambda_2(X_v)}\Big)\Big\}.$$

If $\alpha = m/d = (d-1)/d$ then $\hat{\delta}^2_{\varepsilon,v} = \mathrm{O}(\varepsilon^{2\alpha})$.

Proof. As in the previous proof, we apply Proposition A.2.17 with u replaced by $w = u - f_v u$ and (6.51) and obtain

$$\begin{aligned} \varepsilon^m \operatorname{vol} Y_e \left| f_e u(0) - f_v u \right|^2 &= \varepsilon^m \operatorname{vol} Y_e \left| f_{\partial_e \acute{X}_{\varepsilon,v}} w \right|^2 \\ &\le 2\varepsilon^\beta \ell_{v,e} p_{m-2}(\varepsilon^{1-\alpha}) \varepsilon^{1-\alpha} \|u'\|^2_{\acute{X}_{\varepsilon,v,e}} + 2\varepsilon^m \operatorname{vol} Y_e \left| f_{\partial_e X_v} w \right|^2. \end{aligned}$$

Note that $u' = w'$. Using Corollary 6.3.4, we can estimate the sum over the second term by

$$2\varepsilon^m \sum_e \operatorname{vol} Y_e \left| f_{\partial_e X_v} w \right|^2 \le 2\varepsilon^{2\alpha+m-\alpha d} \cdot \frac{1}{\ell_-(v)}\Big(1 + \frac{2}{\lambda_2(X_v)}\Big) \|du\|^2_{X_{\varepsilon,v}},$$

since the central manifold $X_{\varepsilon,v} = \varepsilon^\alpha X_v$ scale homogeneously. If $\alpha = m/d$ then $2\alpha < 1 + \beta - \alpha$ and therefore, $\hat{\delta}^2_{\varepsilon,v} = \mathrm{O}(\varepsilon^{2\alpha})$. □

6.9 Slowly Decaying and Arbitrary Transversal Manifolds: Reduction to the Graph Model

Let us now show that the quadratic form $\widetilde{\mathfrak{d}}$ associated with the (Neumann) Laplacian on the d-dimensional manifold X_ε is δ_ε-partial isometrically equivalent with a quadratic form on the underlying metric graph. Here, the graph-like space associated with the metric star graph G is given by

$$X_\varepsilon = \acute{X}_{\varepsilon,v} \,\overline{\cup}\, \overline{\dot{\bigcup_e}} X_{\varepsilon,e}$$

where $\acute{X}_{\varepsilon,v}$ is the central manifold $X_{\varepsilon,v} = \varepsilon^\alpha X_v$ together with the truncated cones $\acute{X}_{\varepsilon,v,e}$ for $e \in E_v$. In particular, for each edge $e \in E_v$, the truncated cone $\acute{X}_{\varepsilon,v}$ joins the end of radius ε from the edge neighbourhood with the corresponding boundary part of radius ε^α of the central manifold $X_{\varepsilon,v}$. For details we refer to Sect. 6.8.1. Let us assume in this section, that the (d-dimensional) vertex neighbourhood volume dominates the (m-dimensional) transversal volume, i.e. we assume that

$$0 < \alpha < \frac{m}{d} = \frac{d-1}{d}.$$

In this case, we have

$$\frac{\operatorname{vol} \acute{X}_{\varepsilon,v}}{\partial \acute{X}_{\varepsilon,v}} \approx \frac{\operatorname{vol} X_{\varepsilon,v}}{\operatorname{vol} Y_\varepsilon} = \varepsilon^{\alpha d - m} \frac{\operatorname{vol} X_v}{\operatorname{vol} Y} \to 0 \quad \text{as} \quad \varepsilon \to \infty$$

(see Lemma 6.8.1), where

$$\operatorname{vol} Y_\varepsilon = \operatorname{vol} \partial \acute{X}_{\varepsilon,v} = \varepsilon^m \operatorname{vol} Y \qquad \text{and} \qquad \operatorname{vol} Y = \sum_e \operatorname{vol} Y_e$$

denotes the m-dimensional volume of the union of all transversal manifolds $Y_{\varepsilon,e} = \varepsilon Y_e$.

Let us now define the zeroth and first order identification operators. We use the same spaces as in (6.37). Denote by $\mathbb{1}_{\varepsilon,v}$ the normalised constant function on $\acute{X}_{\varepsilon,v}$ and by $\acute{\mathbb{1}}_v$ the constant function on $\acute{X}_v$ with value 1. Let $J_\varepsilon \colon \mathscr{H} \longrightarrow \widetilde{\mathscr{H}} = \mathsf{L}_2(X_\varepsilon)$ be given by

$$\begin{aligned} J_\varepsilon \hat{f} &:= \bigoplus_{e \in E} (f_e \otimes \mathbb{1}_{\varepsilon,e}) \oplus F(v)\mathbb{1}_{\varepsilon,v} \\ &= \varepsilon^{-m/2} \bigoplus_{e \in E} (f_e \otimes \mathbb{1}_e) \oplus (\operatorname{vol} \acute{X}_{\varepsilon,v})^{-1/2} F(v) \acute{\mathbb{1}}_v \end{aligned} \tag{6.53}$$

for $\hat{f} = (f, F) \in \mathscr{H} = \mathsf{L}_2(G) \oplus \mathbb{C}$; similarly as in (6.38), but with respect to the decomposition

$$\mathsf{L}_2(X_\varepsilon) = \bigoplus_{e \in E} \mathsf{L}_2(X_{\varepsilon,e}) \oplus \mathsf{L}_2(\acute{X}_{\varepsilon,v}). \tag{6.54}$$

For the spaces of order 1, we define $J_\varepsilon^1 \colon \mathscr{H}^1 \longrightarrow \widetilde{\mathscr{H}}^1 = \mathsf{H}^1(X_\varepsilon)$ by

$$\begin{aligned} J_\varepsilon^1 \hat{f} &:= \bigoplus_{e \in E} \Big(f_e + \Big(\frac{\operatorname{vol} Y_{\varepsilon,e}}{\operatorname{vol} X_{\varepsilon,v}} \Big)^{1/2} \chi_e F(v) \Big) \otimes \mathbb{1}_{\varepsilon,e} \oplus (\operatorname{vol} X_{\varepsilon,v})^{-1/2} F(v) \acute{\mathbb{1}}_v \qquad (6.55) \\ &= \varepsilon^{-m/2} \Big(\bigoplus_{e \in E} \Big(f_e + \varepsilon^{(m-\alpha d)/2} \Big(\frac{\operatorname{vol} Y_e}{\operatorname{vol} X_v} \Big)^{1/2} \chi_e F(v) \Big) \otimes \mathbb{1}_e \oplus \frac{\varepsilon^{(m-\alpha d)/2}}{(\operatorname{vol} X_v)^{1/2}} \\ &\qquad \times F(v) \acute{\mathbb{1}}_v \Big), \end{aligned}$$

for $\hat{f} = (f, F)$ in the decoupled space $\mathscr{H}^1 = \bigoplus_e \mathsf{H}^1_0(I_e) \oplus \mathbb{C}$, where $\mathsf{H}^1_0(I_e) = \{ f \in \mathsf{H}^1(I_e) \mid f(0) = 0 \}$. Again, we use a similar definition as in (6.39), but with respect to the decomposition (6.54). As in Sect. 6.5, it is easy to see that the function $J^1_\varepsilon \hat{f}$ matches along the different components, i.e. $J^1_\varepsilon \hat{f} \in \mathsf{H}^1(X)$.

The mapping in the opposite direction is given by the adjoint $J^*_\varepsilon \colon \widetilde{\mathscr{H}} \longrightarrow \mathscr{H}$, namely,

$$(J^*_\varepsilon u)_e(s) = \langle \mathbb{1}_{\varepsilon,e}, u_e(s,\cdot)\rangle_{Y_{\varepsilon,e}} = \varepsilon^{m/2}(\operatorname{vol} Y_e)^{1/2} f_e u(s), \tag{6.56a}$$

$$(J^*_\varepsilon u)(v) = \langle \acute{\mathbb{1}}_{\varepsilon,v}, u_v\rangle_{\acute{X}_{\varepsilon,v}} = (\operatorname{vol} \acute{X}_{\varepsilon,v})^{1/2} \acute{f}_{\varepsilon,v} u. \tag{6.56b}$$

Furthermore, we define $J'^1_\varepsilon \colon \widetilde{\mathscr{H}}^1 \longrightarrow \mathscr{H}^1$ by

$$(J'^1_\varepsilon u)_e(s) := \varepsilon^{m/2}(\operatorname{vol} Y_e)^{1/2}\Big(f_e u(s) - \chi_e(s) f_e u(0)\Big), \tag{6.57a}$$

$$(J'^1_\varepsilon u)(v) := \langle \mathbb{1}_{\varepsilon,v}, u_v\rangle_{X_{\varepsilon,v}} = (\operatorname{vol} X_{\varepsilon,v})^{1/2} f_{\varepsilon,v} u = \varepsilon^{\alpha d/2}(\operatorname{vol} X_v)^{1/2} f_v u \tag{6.57b}$$

as in (6.41). Recall the definition of the averaging operators in (6.50). Again, it is easily seen that $(f, F) := J'^1_\varepsilon u \in \mathscr{H}^1$, in particular, that $f_e(0) = 0$.

Now we are in position to demonstrate that the decoupled Dirichlet form on the graph $\mathfrak{d}(f) = \sum_e \|f'_e\|^2_{I_e}$ with $\operatorname{dom} \mathfrak{d} = \mathscr{H}^1$ and the quadratic form $\widetilde{\mathfrak{d}}(u) = \|du\|^2_{X_{\varepsilon,v}}$ with $\operatorname{dom} \widetilde{\mathfrak{d}} = \mathsf{H}^1(X_\varepsilon)$ on the manifold are δ-partial isometrically equivalent (see Proposition 4.4.13 and Definition 4.4.11):

Theorem 6.9.1. *The quadratic forms $\widetilde{\mathfrak{d}}$ and $\mathfrak{d}$ are δ_ε-partial isometrically equivalent with identification operators J_ε, J^1_ε and J'^1_ε, where*

$$\delta^2_\varepsilon := \max\Big\{\big(2\varepsilon^{m-\alpha d} + \varepsilon^{\beta-\alpha}\big)\frac{1}{\ell_-(v)} \cdot \frac{\operatorname{vol} \mathring{\partial} X_v}{\operatorname{vol} X_v}, \widetilde{\delta}^2_{\varepsilon,v} + 2\varepsilon^{\beta-\alpha} \cdot \frac{\operatorname{vol} \mathring{\partial} X_v}{\operatorname{vol} X_v}, \frac{2\widetilde{\delta}^2_{\varepsilon,v}}{\ell_-(v)}, \frac{\varepsilon^2}{\min_e \lambda_2(Y_e)}, \frac{2\varepsilon^{2\alpha}}{\lambda_2(X_v)}\Big\},$$

$\beta > \alpha$, $0 < \varepsilon \le \varepsilon_0$, and where ε_0 is given in (6.52a). In particular, if $\beta > m(1-\alpha)$, then $\delta_\varepsilon = \mathrm{O}(\varepsilon^{\min\{(m-\alpha d)/2, \alpha\}})$, and the error depends only on an upper estimate of $\operatorname{vol} \mathring{\partial} X_v / \operatorname{vol} X_v$, and $\lambda^{\mathrm{D}}_2(X_v)$, and lower estimates on $\lambda_2(X_v)$, $\lambda_2(Y_e)$ and ℓ_e.

Recall the definition of $\widetilde{\delta}_{\varepsilon,v}$ in Proposition 6.8.4.

Proof. As usual, the first condition in (4.29a) is easily seen to be fulfilled. For the second condition, we estimate

$$\|u - J_\varepsilon J^*_\varepsilon u\|^2_{X_\varepsilon} = \sum_e \|u - f_e u\|^2_{X_{\varepsilon,e}} + \|u - \acute{f}_{\varepsilon,v} u\|^2_{\acute{X}_{\varepsilon,v}}$$

$$\le \max_{e \in E}\Big\{\frac{\varepsilon^2}{\lambda_2(Y_e)}, \frac{1}{\lambda_2(\acute{X}_{\varepsilon,v})}\Big\}\|du\|^2_{X_\varepsilon} \le \max_{e \in E}\Big\{\frac{\varepsilon^2}{\lambda_2(Y_e)}, \frac{2\varepsilon^{2\alpha}}{\lambda_2(X_v)}\Big\}\|du\|^2_{X_\varepsilon}$$

by Proposition 5.1.1 and Lemma 6.8.3. The first condition in (4.29b) is here

$$\|J_\varepsilon^1 \hat{f} - J_\varepsilon \hat{f}\|^2_{\mathcal{X}_\varepsilon} = \Big[\varepsilon^{m-\alpha d} \sum_e \frac{\operatorname{vol} Y_e}{\operatorname{vol} X_v} \|\chi_e\|^2_{I_e} + \Big|\Big(\frac{\operatorname{vol} \acute{X}_{\varepsilon,v}}{\operatorname{vol} X_{\varepsilon,v}}\Big)^{1/2} - 1\Big|^2\Big] |F(v)|^2$$
$$\le \big(\varepsilon^{m-\alpha d} + \varepsilon^{\beta-\alpha}\big) \frac{\operatorname{vol} \mathring{\partial} X_v}{\operatorname{vol} X_v} |F(v)|^2$$

using Lemma 6.8.1 together with $|\tau^{1/2} - 1| \le |\tau - 1|$ for $\tau \ge 0$ and the fact that $\|\chi_e\|^2_{I_e} \le 1$. The second estimate in (4.29b) can be seen by

$$\|J_\varepsilon^* u - J_\varepsilon'^1 u\|^2_{\mathscr{H}} = \sum_{e \in E} \operatorname{vol} Y_e \|\chi_e\|^2_{I_e} |f_e u(0)|^2$$
$$+ \Big|(\operatorname{vol} \acute{X}_{\varepsilon,v})^{1/2} \acute{f}_{\varepsilon,v} u - (\operatorname{vol} X_{\varepsilon,v})^{1/2} f_{\varepsilon,v} u\Big|^2$$
$$\le \Big(\widetilde{\delta}^2_{\varepsilon,v} + 2\varepsilon^{\beta-\alpha} \frac{\operatorname{vol} \mathring{\partial} X_v}{\operatorname{vol} X_v}\Big)\big(\|du\|^2_{\acute{X}_{\varepsilon,v}} + \|u\|^2_{\acute{X}_{\varepsilon,v}}\big)$$

applying Proposition 6.8.4 and Lemma 6.8.2. Moreover, (4.29c) is again easily seen to be fulfilled. Finally, (4.29d) follows from

$$\big|\widetilde{\mathfrak{d}}(u, J_\varepsilon^1 \hat{f}) - \mathfrak{d}(J_\varepsilon'^1 u, \hat{f})\big|^2 = \Big|\sum_e \Big[\varepsilon^{(m-\alpha d)/2} \Big(\frac{\operatorname{vol} Y_e}{\operatorname{vol} X_v}\Big)^{1/2} \langle u_e', \chi_e' \otimes \mathbb{1}_{\varepsilon,e}\rangle_{X_{\varepsilon,e}} F(v)$$
$$+ \varepsilon^{m/2} (\operatorname{vol} Y_e)^{1/2} \int_e \overline{u(0)} \langle \chi_e', f_e' \rangle_{I_e}\Big]\Big|^2$$
$$\overset{\text{CS}}{\le} \frac{2}{\ell_-(v)} \Big[\varepsilon^{m-\alpha d} \frac{\operatorname{vol} \mathring{\partial} X_v}{\operatorname{vol} X_v} \|u'\|^2_{X_{\varepsilon,E}} |F(v)|^2$$
$$+ 2\widetilde{\delta}^2_{\varepsilon,v} \big(\|du\|^2_{\acute{X}_{\varepsilon,v}} + \|u\|^2_{\acute{X}_{\varepsilon,v}}\big) \|f'\|^2_{\mathcal{G}}\Big]$$

using Proposition 6.8.4 again and $\|\chi_e'\|_{I_e} \le 1/\ell_-(v)$. For the error estimate on $\delta^2_{\varepsilon,v}$, note that $\beta > m(1-\alpha)$ implies that $\beta - \alpha > (m - \alpha d)$ and $\widetilde{\delta}^2_{\varepsilon,v} = \mathrm{O}(\varepsilon^{m-\alpha d})$. □

We can also allow a non-homogeneous scaling on X_v, namely,

$$\varepsilon^{\alpha_2} X_v \le X_{\varepsilon,v} \le \varepsilon^{\alpha_1} X_v, \qquad 0 < \alpha_1 \le \alpha_2 < \frac{m}{d}. \tag{6.58}$$

Then similarly as in Lemma 6.8.1, we conclude

$$\operatorname{vol} X_{\varepsilon,v} \le \operatorname{vol} \acute{X}_{\varepsilon,v} \le \Big(\varepsilon^{\beta-(\alpha_2-\alpha_1)m-\alpha_2} \frac{\operatorname{vol} \mathring{\partial} X_v}{\operatorname{vol} X_v} + 1\Big) \operatorname{vol} X_{\varepsilon,v}, \quad \text{and}$$
$$\varepsilon^{d(\alpha_2-\alpha_1)-2\alpha_1} \lambda_2(X_v) \le \lambda_2(X_{\varepsilon,v})$$

using (5.9). In particular, if α_1, α_2 and β fulfil

$$\beta > (\alpha_2 - \alpha_1)m + \alpha_2, \qquad (d+2)\alpha_1 > d\alpha_2,$$

then the two quadratic forms are again δ_ε-partial isometrically equivalent as in Theorem 6.9.1 with $\delta_\varepsilon = \mathrm{O}(\varepsilon^\gamma)$ for some $\gamma > 0$ depending on α_1, α_2 and β.

6.10 The Borderline Case with Arbitrary Transversal Manifolds: Reduction to the Graph Model

Here, we treat the case when the vertex neighbourhood volume is of the same order as the transversal volume, i.e. we assume that

$$\alpha = \frac{m}{d} = \frac{d-1}{d}.$$

For the notation of the spaces, see Sect. 6.8.1. The setting is similar to the one in Sect. 6.6; we only use the decomposition (6.54) instead of (6.19). Moreover, we use averagings over different spaces for the zeroth and first order identification operators, in contrast to Sect. 6.5. For the notation of the total spaces on the graph and the manifold, we refer to (6.43). Here, the parameters of the graph and the manifold are related by

$$L(v) = (\operatorname{vol} X_v)^{-1/2} \qquad \text{and} \qquad p_e = (\operatorname{vol} Y_e)^{1/2}, \tag{6.59}$$

i.e. the coupling factor $L(v)$ is given by the unscaled volume of the *smaller* vertex neighbourhood $X_v \subset \acute{X}_v$.

Let us define the zeroth and first order identification operators. The operator $J_\varepsilon \colon \mathscr{H} \longrightarrow \widetilde{\mathscr{H}}$ is given by

$$\begin{aligned} J_\varepsilon f &:= \bigoplus_{e \in E} (f_e \otimes \pm_{\varepsilon,e}) \oplus F(v)\acute{\pm}_{\varepsilon,v} \\ &= \varepsilon^{-m/2} \bigoplus_{e \in E} (f_e \otimes \pm_e) \oplus (\operatorname{vol} \acute{X}_{\varepsilon,v})^{-1/2} F(v)\acute{\mathbb{1}}_v \end{aligned} \tag{6.60}$$

as in (6.45), but with respect to the decomposition (6.54). As before, $\acute{\pm}_{\varepsilon,v}$ is the normalised constant function on $\acute{X}_{\varepsilon,v} = \varepsilon^\alpha \acute{X}_v$ and $\acute{\mathbb{1}}_v$ the constant function on $\acute{X}_v$ with value 1. For the spaces of order 1, we define $J_\varepsilon^1 \colon \mathscr{H}^1 \longrightarrow \widetilde{\mathscr{H}}^1$ by

$$\begin{aligned} J_\varepsilon^1 \hat{f} &:= \bigoplus_{e \in E} (f_e \otimes \pm_{\varepsilon,e}) \oplus \Big(\frac{\operatorname{vol} \acute{X}_{\varepsilon,v}}{\operatorname{vol} X_{\varepsilon,v}}\Big)^{1/2} F(v)\acute{\pm}_{\varepsilon,v} \\ &= \varepsilon^{-m/2} \Big(\bigoplus_{e \in E} (f_e \otimes \pm_e) \oplus (\operatorname{vol} X_v)^{-1/2} F(v)\acute{\mathbb{1}}_v \Big). \end{aligned}$$

Note that the latter operator is well defined:

$$(J_\varepsilon^1 \hat{f})_e(0,y) = \varepsilon^{-m/2} p_e^{-1} f_e(0) = \varepsilon^{-m/2} f(v) = \varepsilon^{-m/2} (\operatorname{vol} X_v)^{-1/2} F(v)$$
$$= (J_\varepsilon^1 \hat{f})(x)$$

for $x \in \acute{X}_v$ due to (6.44) and (6.4), i.e. the function $J_\varepsilon^1 \hat{f}$ matches along the different components and therefore $J_\varepsilon^1 \hat{f} \in \mathsf{H}^1(X)$.

The mapping in the opposite direction is given by the adjoint $J_\varepsilon^* \colon \widetilde{\mathscr{H}} \longrightarrow \mathscr{H}$, i.e.

$$(J_\varepsilon^* u)_e(s) = \langle \mathbb{1}_{\varepsilon,e}, u_e(s,\cdot)\rangle_{Y_{\varepsilon,e}} = \varepsilon^{m/2} (\operatorname{vol} Y_e)^{1/2} \underline{f}_e u(s), \tag{6.61a}$$

$$(J_\varepsilon^* u)(v) = \langle \mathbb{1}_{\varepsilon,v}, u_v \rangle_{\acute{X}_{\varepsilon,v}} = (\operatorname{vol} \acute{X}_{\varepsilon,v})^{1/2} \acute{\underline{f}}_{\varepsilon,v} u. \tag{6.61b}$$

Furthermore, we define $J_\varepsilon'^1 \colon \widetilde{\mathscr{H}}^1 \longrightarrow \mathscr{H}^1$ by

$$(J_\varepsilon'^1 u)_e(s) := \varepsilon^{m/2} (\operatorname{vol} Y_e)^{1/2} \Big[\underline{f}_e u(s) + \chi_e(s) \Big(\underline{f}_v u - \underline{f}_e u(0) \Big) \Big], \tag{6.62a}$$

$$(J_\varepsilon'^1 u)(v) := \langle \mathbb{1}_{\varepsilon,v}, u_v \rangle_{X_{\varepsilon,v}} = (\operatorname{vol} X_{\varepsilon,v})^{1/2} \underline{f}_{\varepsilon,v} u = \varepsilon^{m/2} (\operatorname{vol} X_v)^{1/2} \underline{f}_v u, \tag{6.62b}$$

again as in (6.47), but with respect to the decomposition (6.54). Recall the definition of the averaging operators in (6.50). To see that the operator indeed maps into the given space, let $\hat{f} = (f, F) := J_\varepsilon'^1 u$ then $f_e(0) = \varepsilon^{m/2} p_e \underline{f}_v u$ so that $f(v) = \varepsilon^{m/2} \underline{f}_v u$, and $\underline{f} \in \mathbb{C}p$, i.e. $f \in \mathsf{H}_p^1(G)$. Moreover, $F(v) = (\operatorname{vol} X_v)^{1/2} f(v)$ and finally, $\hat{f} \in \mathscr{H}^1$.

Let us now demonstrate that the two quadratic forms $\mathfrak{d}(\hat{f}) := \sum_e \|f_e'\|_{I_e}^2$, $\hat{f} \in \mathscr{H}^1$ and $\widetilde{\mathfrak{d}}$ are δ-partial isometrically equivalent (see Definition 4.4.11 and Proposition 4.4.13):

Theorem 6.10.1. *In the borderline case $\alpha = m/d$, the quadratic forms $\widetilde{\mathfrak{d}}$ and $\mathfrak{d}$ are δ_ε-partial isometrically equivalent with identification operators J_ε, J_ε^1 and $J_\varepsilon'^1$, where*

$$\delta_\varepsilon^2 := \max\Big\{ 2\varepsilon^{\beta-\alpha} \cdot \frac{\operatorname{vol} \mathring{\partial} X_v}{\operatorname{vol} X_v}, \frac{\hat{\delta}_{\varepsilon,v}^2}{\ell_-(v)}, \frac{\varepsilon^2}{\min_e \lambda_2(Y_e)}, \frac{2\varepsilon^{2\alpha}}{\lambda_2(X_v)} \Big\} = \mathrm{O}(\varepsilon^{2\alpha})$$

provided $0 < \varepsilon \le \varepsilon_0$ (see (6.52a)) and $\beta \ge 3\alpha$. Moreover, the error depends only on an upper estimate of $\operatorname{vol} \mathring{\partial} X_v / \operatorname{vol} X_v$ and $\lambda_2^{\mathrm{D}}(X_v)$, and on lower estimates on $\lambda_2(X_v)$, $\lambda_2(Y_e)$ and ℓ_e.

Recall the definition of $\hat{\delta}_{\varepsilon,v}$ in Proposition 6.8.5.

Proof. The first condition in (4.29a) is easily seen to be fulfilled, and the second estimate follows as in the proof of Theorem 6.9.1. The first condition of the δ_ε-partial isometric equivalence in (4.29b) is here

$$\|J_\varepsilon^1 \hat{f} - J_\varepsilon \hat{f}\|^2_{X_\varepsilon} = \Big[\Big(\frac{\operatorname{vol} \acute{X}_{\varepsilon,v}}{\operatorname{vol} X_{\varepsilon,v}}\Big)^{1/2} - 1\Big]^2 \big|F(v)\big|^2 \le \varepsilon^{\beta-\alpha} \frac{\operatorname{vol} \partial \mathring{X}_v}{\operatorname{vol} X_v} \big|F(v)\big|^2$$

using Lemma 6.8.1. The second estimate in (4.29b) follows from

$$\begin{aligned}\|J_\varepsilon'^1 u - J_\varepsilon^* u\|^2_{\mathscr{H}} &= \varepsilon^m \sum_{e \in E} \|\chi_e\|^2_{I_e} \operatorname{vol} Y_e \big|f_v u - f_e u(0)\big|^2 \\ &\quad + \big|(\operatorname{vol} X_{\varepsilon,v})^{1/2} f_v u - (\operatorname{vol} \acute{X}_{\varepsilon,v})^{1/2} \acute{f}_{\varepsilon,v} u\big|^2 \\ &\le \max\Big\{\hat{\delta}^2_{\varepsilon,v},\ 2\varepsilon^{\beta-\alpha} \frac{\operatorname{vol} \partial \mathring{X}_v}{\operatorname{vol} X_v}\Big\} \big(\|du\|^2_{\acute{X}_{\varepsilon,v}} + \|u\|^2_{\acute{X}_{\varepsilon,v}}\big)\end{aligned}$$

using Lemma 6.8.2 and Proposition 6.8.5. Moreover, (4.29c)is easily seen to be fulfilled. Finally, we have

$$\begin{aligned}\big|\mathfrak{d}(J_\varepsilon'^1 u, \hat{f}) - \widetilde{\mathfrak{d}}(u, J_\varepsilon^1 \hat{f})\big|^2 &= \varepsilon^m \Big|\sum_e p_e \big(f_v \overline{u} - f_e \overline{u}(0)\big) \langle \chi_e', f' \rangle_{I_e}\Big|^2 \\ &\overset{\text{CS}}{\le} \frac{\hat{\delta}^2_{\varepsilon,v}}{\ell_-(v)} \|du\|^2_{\acute{X}_{\varepsilon,v}} \|f'\|^2_G\end{aligned}$$

using Proposition 6.8.5 and $\|\chi_e'\|_{I_e} \le 1/\ell_-(v)$. □

Note that the error we obtain here converges faster than the one in Theorem 6.6.1, since $\alpha = m/d > 1/2$.

6.11 Dirichlet Boundary Conditions: The Decoupling Case

In this section we present an abstract graph-like manifold with non-trivial transversal boundary $\partial_0 X$. In particular, we show that if the vertex neighbourhood is small in a spectral sense, then the rescaled Laplacian with Dirichlet boundary conditions on $\partial_0 X$ converges to a decoupled Laplacian on the underlying graph. The rescaling is necessary since the lowest transversal eigenvalue λ_1 is non-zero in the Dirichlet case. Recently, Molchanov and Vainberg resp. Grieser (see [MV07, G08a] and Sect. 1.2.2) treated the general case (i.e. general vertex neighbourhoods, allowing in principle also other types of boundary conditions like Robin-type ones). The limit problem in the general situation depends on a *scattering* problem on the unscaled manifold (with infinite ends). As in our situation here, one obtains generically only a decoupled limit operator.

6.11.1 The Graph and Manifold Models

Let $G = G_v$ be a metric star graph having only one vertex v and $\deg v$ adjacent edges E of length $\ell_e \in (0, \infty]$. Before defining the limit operator on the graph, we fix the abstract manifold associated with the graph. Assume that $X = X_v^+$ is a d-dimensional Riemannian manifold with boundary ∂X having a finite number of cylindrical ends, labelled by $e \in E = E_v$. More precisely, we assume that

$$X = X_v^+ = X_v \,\overline{\dot\cup}\, \overline{\dot\bigcup_{e \in E}} X_e$$

is a manifold with $|E|$-many ends. Here, $X_e := I_e \times Y_e$, and Y_e is a compact, connected Riemannian manifold of dimension $m := d - 1$, having non-trivial boundary ∂Y_e. Recall that $I_e = [0, \ell_e]$ if $\ell_e < \infty$ and $I_e = [0, \infty)$ if $\ell_e = \infty$. Let us recall some notation for the different boundary components already introduced in Sect. 6.2 (see also Fig. 6.12).

Namely, we denote by $\partial_1 X$ and $\partial_0 X$ the longitudinal and transversal boundary of X i.e.

$$\partial_1 X := \bigcup_{e \in E_{\text{int}}} \{\ell_e\} \times Y_e \qquad \text{and} \qquad \partial_0 X_v = \overline{\partial X \setminus \partial_1 X}.$$

Here, $E_{\text{int}} = \{e \in E \mid \ell_e < \infty\}$ denotes the set of interior edges (i.e. edges of finite length. Moreover, the boundary of X_v decomposes as $\partial X_v = \mathring{\partial} X_v \,\overline{\dot\cup}\, \partial_0 X_v$, i.e. into the internal and transversal boundary, respectively. The internal boundary has $(\deg v)$-many components

$$\mathring{\partial} X_v = \bigcup_e \partial_e X_v, \qquad \text{where} \quad \partial_e X_v = \partial_v X_e = \{0\} \times Y_e$$

is the boundary component of X_v where the edge neighbourhood X_e is attached.

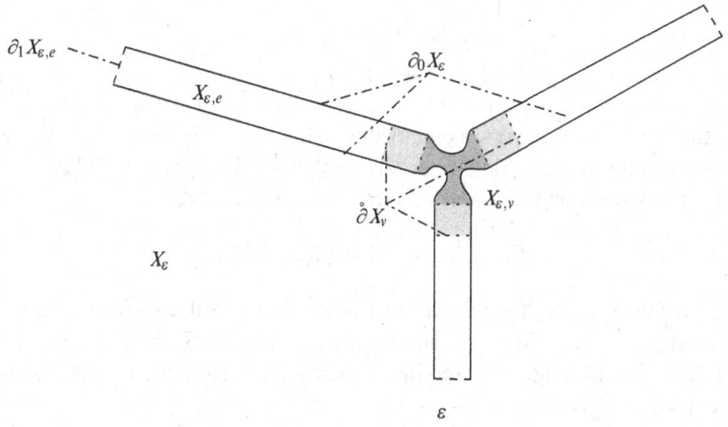

Fig. 6.12 A graph-like manifold X_ε with scaling parameter ε and Dirichlet boundary conditions on the transversal boundary $\partial_0 X_\varepsilon$. In *dark* and *light grey* we pictured the vertex neighbourhood $X_{\varepsilon,v}$

As in Sect. 6.2, we assume that the vertex neighbourhood X_v decomposes as

$$X_v = X_v^- \,\dot\cup\, \dot{\bigcup_{e \in E}} X_{v,e}$$

where $X_{v,e} = I_{v,e} \times Y_e$ and $I_{v,e} = [0, \ell_{v,e}]$ with $\ell_{v,e} > 0$ such that $0 < \ell_-(v) \leq \ell_{v,e} \leq \min\{\ell_e, 1\}$, where $\ell_-(v) = \min_e \ell_e$ is the minimal edge length.

We assume that the metrics on X_e and $X_{v,e}$ have product structure $d^2 s + h_e$, where h_e is a metric on Y_e. From Corollary A.2.7 we obtain the Sobolev trace estimate

$$\|u\|^2_{\partial_e X_v} \leq \widetilde{a} \|u'\|^2_{X_{v,e}} + \frac{2}{\widetilde{a}} \|u\|^2_{X_{v,e}} \tag{6.63}$$

for $0 < \widetilde{a} \leq \ell_{v,e}$ and

$$u \in \mathsf{H}^1(X_v, \partial_0 X_v) = \{u \in \mathsf{H}^1(X_v) \,|\, u\restriction_{\partial_0 X_v} = 0\}.$$

As scaled manifolds we consider

$$Y_{\varepsilon,e} := \varepsilon Y_e, \qquad X_{\varepsilon,e} := I_e \times Y_{\varepsilon,e} \qquad \text{and} \qquad X_{\varepsilon,v} := \varepsilon X_v, \tag{6.64}$$

for $0 < \varepsilon \leq 1$, i.e. the edge neighbourhoods shrink in the transversal direction of order ε, whereas the vertex neighbourhoods shrink in all directions of order ε. Denote by X_ε the corresponding total scaled space.

Denote the k-th eigenvalue of the Dirichlet Laplacian on Y_e by $\lambda_k(Y_e) = \lambda_k^{\mathrm{D}}(Y_e)$. The quadratic form on the manifold X_ε is defined by

$$\widetilde{\mathfrak{h}}(u) := \|du\|^2_{X_\varepsilon} - \frac{\lambda_1}{\varepsilon^2} \|u\|^2_{X_\varepsilon},$$

where

$$\lambda_1 := \min_e \lambda_1^{\mathrm{D}}(Y_e) > 0. \tag{6.65}$$

denotes the smallest transversal eigenvalue. This rescaling is necessary since the first transversal mode is no longer zero. Note that $\lambda_1^{\mathrm{D}}(Y_{\varepsilon,e}) = \varepsilon^{-2} \lambda_1^{\mathrm{D}}(Y_e)$.

The limit operator on the graph survives only on the edges

$$E_0 := \{\, e \in E \,|\, \lambda_1^{\mathrm{D}}(Y_e) = \lambda_1 \,\}, \tag{6.66}$$

i.e. only on those edges, for which the lowest transversal eigenvalue agrees with the minimal eigenvalue. We call these edges *thick*. Let G_0 be the corresponding metric subgraph with edges E_0 and one vertex v. The associated Hilbert space and quadratic form are given by

$$\mathsf{L}_2(G_0) := \bigoplus_{e \in E_0} \mathsf{L}_2(I_e), \quad \mathfrak{h}(f) := \|f'\|^2_{G_0} = \sum_{e \in E_0} \|f_e'\|^2_{I_e}, \quad \operatorname{dom} \mathfrak{h} := \bigoplus_e \mathsf{H}^1_0(I_e),$$

where $\mathsf{H}^1_0(I_e) = \{ f \in \mathsf{H}^1(I_e) \mid f(0) = 0 \}$. In particular, the limit operator (i.e. the operator associated with $\mathfrak{h}$) is the *decoupled* Laplacian

$$\Delta = \bigoplus_{e \in E_0} \Delta^0_{I_e}$$

with Dirichlet condition at each point $0 \in I_e$ of the thick edges.

6.11.2 Some Vertex and Edge Neighbourhoods Estimates

We start with an estimate showing that no eigenfunction concentrates on the vertex neighbourhood:

Lemma 6.11.1. *Assume that $\lambda_1(X_v) > \lambda_1$, then we have*

$$\|u\|^2_{X_v} \le \frac{1}{\lambda_1(X_v) - \lambda_1} \big(\|\mathrm{d}u\|^2_{X_v} - \lambda_1 \|u\|^2_{X_v} \big)$$

for $u \in \mathsf{H}^1(X_v, \partial_0 X_v)$, where $\lambda_k(X_v) = \lambda_k^{\partial_0 X_v}(X_v)$ denote the eigenvalues of the Laplacian with Dirichlet conditions on $\partial_0 X_v$ and Neumann conditions on $\mathring{\partial} X_v$.

Proof. Applying Proposition 5.1.2 we obtain

$$\|u\|^2_{X_v} \le \frac{1}{\lambda_1(X_v)} \|\mathrm{d}u\|^2_{X_v} = \frac{1}{\lambda_1(X_v)} \big(\|\mathrm{d}u\|^2_{X_v} - \lambda_1 \|u\|^2_{X_v} \big) + \frac{\lambda_1}{\lambda_1(X_v)} \|u\|^2_{X_v}.$$

By assumption, $\lambda_1/\lambda_1(X_v) < 1$, so we can bring the last term on the LHS and divide by $(1 - \lambda_1/\lambda_1(X_v)) > 0$. □

Let us make a comment on the spectral condition used in the previous lemma::

Remark 6.11.2. The condition $\lambda_1(X_v) > \lambda_1$ on the vertex neighbourhood eigenvalue $\lambda_1(X_v) = \lambda_1^{\partial_0 X_v}(X_v)$ and the lowest transversal eigenvalue λ_1 is a sort of "smallness condition" on the vertex neighbourhood. In particular, for given transversal manifolds Y_e, the eigenvalue $\lambda_1(X_v)$ has to be large enough. Roughly speaking, this means, that X_v is "small". We give a construction for the existence of such manifolds in the embedded case in Remark 6.11.10 and show as well, that the ε-neighbourhood of an embedded graph does not fulfil the condition, see the discussion taken from [P05] in Sect. 6.11.5.

Under the above spectral condition, we can also assure that the quadratic form $\widetilde{\mathfrak{h}}$ is non-negative:

Proposition 6.11.3. *Assume that $\lambda_1(X_v) \ge \lambda_1$, then $\widetilde{\mathfrak{h}} \ge 0$.*

Proof. We have

$$\begin{aligned}\widetilde{\mathfrak{h}}(u) &= \sum_e \Big(\|du\|^2_{X_{\varepsilon,e}} - \frac{\lambda_1}{\varepsilon^2}\|u\|^2_{X_{\varepsilon,e}}\Big) + \|du\|^2_{X_{\varepsilon,v}} - \lambda_1\|u\|^2_{X_{\varepsilon,v}} \\ &\geq \sum_e \Big(\|d_{Y_e}u\|^2_{X_{\varepsilon,e}} - \frac{\lambda_1(Y_e)}{\varepsilon^2}\|u\|^2_{X_{\varepsilon,e}}\Big) + (\lambda_1(X_v) - \lambda_1)\|u\|^2_{X_{\varepsilon,v}} \geq 0\end{aligned}$$

using Proposition 5.1.2. □

Next, we need to estimate the contribution of a function u on the internal boundary $\mathring{\partial} X_v$ of X_v. Denote by φ_e the first Dirichlet eigenfunction on Y_e.

Lemma 6.11.4. *We have*

$$\sum_e |\langle \varphi_e, u(0,\cdot)\rangle_{Y_e}|^2 \overset{\text{CS}}{\leq} \|u\|^2_{\mathring{\partial} X_v} \leq \Big(\widetilde{a} + \frac{\lambda_1 \widetilde{a}^2 + 2}{\widetilde{a}(\lambda_1(X_v) - \lambda_1)}\Big)(\|du\|^2_{X_v} - \lambda_1\|u\|^2_{X_v})$$

for $u \in \mathsf{H}^1(X_v, \partial_0 X_v)$ *and* $0 < \widetilde{a} \leq \min_e \ell_{v,e}$.

Proof. We have

$$\begin{aligned}\sum_e |\langle \varphi_e, u(0,\cdot)\rangle_{Y_e}|^2 &\overset{\text{CS}}{\leq} \sum_e \|u\|^2_{\partial_e X_v} \leq \sum_e \Big(\widetilde{a}\|du\|^2_{X_{v,e}} + \frac{2}{\widetilde{a}}\|u\|^2_{X_{v,e}}\Big) \\ &\leq \widetilde{a}(\|du\|^2_{X_v} - \lambda_1\|u\|^2_{X_v}) + \Big(\widetilde{a}\lambda_1 + \frac{2}{\widetilde{a}}\Big)\|u\|^2_{X_v} \\ &\leq \Big(\widetilde{a} + \frac{\lambda_1 \widetilde{a}^2 + 2}{\widetilde{a}(\lambda_1(X_v) - \lambda_1)}\Big)(\|du\|^2_{X_v} - \lambda_1\|u\|^2_{X_v})\end{aligned}$$

where we used (6.63) in the second inequality and Lemma 6.11.1 in the last inequality. □

Similarly as in Lemma 6.11.1, we can show that the "thin" edges do not contribute to the limit:

Lemma 6.11.5. *Assume that* $e \in E \setminus E_0$, *i.e.* $\lambda_1(Y_e) > \lambda_1$, *then we have*

$$\|u\|^2_{X_e} \leq \frac{1}{\lambda_1(Y_e) - \lambda_1}(\|du\|^2_{X_e} - \lambda_1\|u\|^2_{X_e})$$

for $u \in \mathsf{H}^1(X_e, \partial_0 X_e)$.

Proof. Applying Proposition 5.1.2 we obtain

$$\begin{aligned}\|u\|^2_{X_e} &= \int_{I_e} \|u(s)\|^2_{Y_e}\, ds \leq \frac{1}{\lambda_1(Y_e)} \int_{I_e} \|d_{Y_e}u(s)\|^2_{Y_e}\, ds \\ &\leq \frac{1}{\lambda_1(Y_e)}\|du\|^2_{X_e} = \frac{1}{\lambda_1(Y_e)}(\|du\|^2_{X_e} - \lambda_1\|u\|^2_{X_e}) + \frac{\lambda_1}{\lambda_1(Y_e)}\|u\|^2_{X_e}.\end{aligned}$$

By assumption, $\lambda_1/\lambda_1(Y_e) < 1$, so we can bring the last term on the LHS and divide by $(1 - \lambda_1/\lambda_1(Y_e)) > 0$. □

6.11.3 Reduction to the Graph Model

We now define the zeroth and first order identification operators. Here,

$$\mathscr{H} := \mathsf{L}_2(G_0) = \bigoplus_{e\in E_0} \mathsf{L}_2(I_e), \qquad \mathscr{H}^1 := \bigoplus_{e\in E_0} \mathsf{H}^1_0(I_e), \tag{6.67a}$$

$$\widetilde{\mathscr{H}} := \mathsf{L}_2(X), \qquad \widetilde{\mathscr{H}}^1 := \mathsf{H}^1(X, \partial_0 X). \tag{6.67b}$$

Let $J_\varepsilon \colon \mathscr{H} \longrightarrow \widetilde{\mathscr{H}}$ be given by

$$J_\varepsilon f := \bigoplus_{e\in E_0} (f_e \otimes \varphi_{\varepsilon,e}) \oplus 0 = \varepsilon^{-m/2} \bigoplus_{e\in E_0} (f_e \otimes \varphi_e) \oplus 0, \tag{6.68}$$

where we set the value 0 on the vertex neighbourhood and on all "thin" edges $E \setminus E_0$. Moreover, $\varphi_{\varepsilon,e}$ is the normalised eigenfunction of $Y_{\varepsilon,e}$ associated with the lowest eigenvalue $\lambda_1(Y_{\varepsilon,e}) = \varepsilon^{-2}\lambda_1(Y_e) > 0$. Note that $\varphi_{\varepsilon,e} = \varepsilon^{-m/2}\varphi_e$ where φ_e is the normalised eigenfunction associated with $\lambda_1(Y_e)) = \lambda_1$. Since functions in $\mathscr{H}^1$ vanish at the vertex, the map $J^1_\varepsilon := J_\varepsilon \restriction_{\mathscr{H}^1}$ already respects the Sobolev spaces of order 1, i.e. $J^1_\varepsilon \colon \mathscr{H}^1 \longrightarrow \widetilde{\mathscr{H}}^1$.

The mapping in the opposite direction is given by the adjoint $J^*_\varepsilon \colon \widetilde{\mathscr{H}} \longrightarrow \mathscr{H}$, i.e. we have

$$(J^*_\varepsilon u)_e(s) = \begin{cases} \langle \varphi_{\varepsilon,e}, u(s,\cdot)\rangle_{Y_{\varepsilon,e}} = \varepsilon^{m/2}\langle \varphi_e, u(s,\cdot)\rangle_{Y_e}, & \text{if } e \in E_0, \\ 0, & \text{otherwise.} \end{cases} \tag{6.69}$$

Furthermore, we define $J'^1_\varepsilon \colon \widetilde{\mathscr{H}}^1 \longrightarrow \mathscr{H}^1$ by

$$\begin{aligned} (J'^1_\varepsilon u)_e(s) &:= \langle \varphi_{\varepsilon,e}, u(s,\cdot)\rangle_{Y_{\varepsilon,e}} - \chi_e(s)\langle \varphi_{\varepsilon,e}, u(0,\cdot)\rangle_{Y_{\varepsilon,e}} \\ &= \varepsilon^{m/2}\Big(\langle \varphi_e, u(s,\cdot)\rangle_{Y_e} - \chi_e(s)\langle \varphi_e, u(0,\cdot)\rangle_{Y_e}\Big) \end{aligned} \tag{6.70}$$

on all thick edges $e \in E_0$, and we set $(J'^1_\varepsilon u)_e = 0$ if $e \in E \setminus E_0$. As usual, χ_e is the piecewise affine-linear with $\chi_e(0) = 1$, $\chi_e(a) = 0$ and $\chi_e(\ell_e) = 0$ for $a = \ell_-(v)$. In particular, $\|\chi_e\|^2_{I_e} = a/3 \le 1$ and $\|\chi'_e\|^2_{I_e} = a^{-1}$. Moreover, $(J'^1_\varepsilon u)_e(0) = 0$ so that $J'^1_\varepsilon u \in \mathscr{H}^1$.

Now we are in position to demonstrate that the two quadratic forms $\mathfrak{h}$ and $\widetilde{\mathfrak{h}}$ are δ-partial isometrically equivalent (see Definition 4.4.11 and Proposition 4.4.13):

Theorem 6.11.6. *Assume that*

$$\lambda_1(X_v) - \lambda_1 > 0, \tag{6.71a}$$

then the quadratic forms $\widetilde{\mathfrak{h}}$ *and* $\mathfrak{h}$ *are* δ_ε*-partial isometrically equivalent with identification operators* J_ε, J_ε^1 *and* $J_\varepsilon'^1$, *where*

$$\delta_\varepsilon^2 := \max\Bigl\{\frac{\varepsilon}{\ell_-(v)^2}\Bigl(1+\frac{\lambda_1+2}{\lambda_1(X_v)-\lambda_1}\Bigr), \frac{\varepsilon^2}{\min_e(\lambda_{2,e}-\lambda_1)}, \frac{\varepsilon^2}{\lambda_1(X_v)-\lambda_1}\Bigr\}$$

and

$$\lambda_{2,e} := \begin{cases} \lambda_2(Y_e), & \text{if } e \in E_0 \text{ (i.e. } \lambda_1(Y_e) = \lambda_1\text{)}, \\ \lambda_1(Y_e), & \text{otherwise.} \end{cases} \tag{6.71b}$$

Moreover, $\lambda_k(Y_e) = \lambda_k^{\mathrm{D}}(Y_e)$ *denotes the* k*-th Dirichlet eigenvalue on* Y_e *and* $\lambda_k(X_v)$ *the* k*-th eigenvalues with Dirichlet condition on* $\partial_0 X_v$ *and Neumann conditions on* $\mathring{\partial} X_v$ *and* $\ell_-(v) := \min_e\{\ell_e, 1\}$.

In particular, $\delta_\varepsilon = \mathrm{O}(\varepsilon^{1/2})$ *where the error depends only on an upper estimate of* $\lambda_1 := \min_e \lambda_1(Y_e)$ *and on lower estimates of* ℓ_e, $\lambda_{2,e} - \lambda_1$ *and* $\lambda_1(X_v) - \lambda_1$.

Note that $\min_e\{\lambda_{2,e} - \lambda_1\}$ is the distance of λ_1 to the remaining spectrum of $\Delta_Y = \bigoplus_{e\in E} \Delta_{Y_e}$.

Proof. The first condition in (4.29a) is easily seen to be fulfilled. The second estimate can be shown as follows:

$$\begin{aligned}
\|u - J_\varepsilon J_\varepsilon^* u\|_{X_\varepsilon}^2 &= \varepsilon^m \sum_{e\in E_0} \int_{I_e} \|u(s,\cdot) - \langle\varphi_e, u(s,\cdot)\rangle \varphi_e\|_{Y_e}^2 \,\mathrm{d}s + \sum_{e\in E\setminus E_0} \|u\|_{X_{\varepsilon,e}}^2 \\
&\quad + \varepsilon^d \|u\|_{X_v}^2 \\
&\le \sum_e \frac{\varepsilon^2}{\lambda_{2,e}-\lambda_1}\Bigl(\|\mathrm{d}u\|_{X_{\varepsilon,e}}^2 - \frac{\lambda_1}{\varepsilon^2}\|u\|_{X_{\varepsilon,e}}^2\Bigr) \\
&\quad + \frac{\varepsilon^2}{\lambda_1(X_v)-\lambda_1}\Bigl(\|\mathrm{d}u\|_{X_{\varepsilon,v}}^2 - \frac{\lambda_1}{\varepsilon^2}\|u\|_{X_{\varepsilon,v}}^2\Bigr) \\
&\le \varepsilon^2 \max_e\Bigl\{\frac{1}{\lambda_{2,e}-\lambda_1}, \frac{1}{\lambda_1(X_v)-\lambda_1}\Bigr\}\Bigl(\|\mathrm{d}u\|_{X_\varepsilon}^2 - \frac{\lambda_1}{\varepsilon^2}\|u\|_{X_\varepsilon}^2\Bigr)
\end{aligned}$$

using Proposition 5.1.2, Lemmata 6.11.1 and 6.11.5 and again the scaling behaviour. The first estimate (4.29b) is trivially fulfilled, here even with equality. The second condition follows from

$$\begin{aligned}
\|J_\varepsilon^* u - J_\varepsilon'^1 u\|_{G_0}^2 &= \varepsilon^m \sum_{e\in E_0} \|\chi_e\|_{I_e}^2 \bigl|\langle\varphi_e, u(0,\cdot)\rangle\bigr|^2 \\
&\le \varepsilon a\Bigl(\widetilde{a} + \frac{\lambda_1\widetilde{a}^2+2}{\widetilde{a}(\lambda_1(X_v)-\lambda_1)}\Bigr)\Bigl(\|\mathrm{d}u\|_{X_{\varepsilon,v}}^2 - \frac{\lambda_1}{\varepsilon^2}\|u\|_{X_{\varepsilon,v}}^2\Bigr)
\end{aligned}$$

using Lemma 6.11.4 for $0 < a \le \min_e \ell_e$ and $0 < \widetilde{a} \le \min_e \ell_{v,e}$ and the scaling behaviour. Moreover, (4.29c) is easily seen to be fulfilled. Finally, we have

$$\begin{aligned}
\bigl|\mathfrak{h}(J_\varepsilon'^1 u, f) - \widetilde{\mathfrak{h}}(u, J_\varepsilon^1 f)\bigr|^2 &= \varepsilon^m \Bigl|\sum_{e\in E_0} \langle \chi_e', f'\rangle_{I_e} \langle \varphi_e, u(0,\cdot)\rangle\Bigr|^2 \\
&\le \frac{\varepsilon}{a}\Bigl(\widetilde{a} + \frac{\lambda_1 \widetilde{a}^2 + 2}{\widetilde{a}(\lambda_1(X_v) - \lambda_1)}\Bigr)\|f'\|_G^2 \\
&\quad \times \Bigl(\|du\|_{X_{\varepsilon,v}}^2 - \frac{\lambda_1}{\varepsilon^2}\|u\|_{X_{\varepsilon,v}}^2\Bigr)
\end{aligned}$$

using Lemma 6.11.4. Note that we also used the fact that $\lambda_1 = \lambda_1^{\mathrm{D}}(Y_e)$ for $e \in E_0$ in the first equality of the latter estimate. The final expression for δ_ε can be obtained by setting $a = \widetilde{a} = \ell_v(v) \le 1$. □

Let us define the associated boundary triples in the Dirichlet case: The *boundary spaces* are given by

$$\mathscr{G}_v := \mathbb{C}^{E_{0,\mathrm{int}}} \qquad \text{and} \qquad \widetilde{\mathscr{G}}_v = \mathsf{L}_2(Y_{\varepsilon,\mathrm{int}}) = \bigoplus_{e\in E_{\mathrm{int}}} \mathsf{L}_2(Y_{\varepsilon,e})$$

where $E_{0,\mathrm{int}} := E_0 \cap E_{\mathrm{int}}$ and $E_{\mathrm{int}} = \{\, e \in E \mid \ell_e < \infty \,\}$. We denote by $(\Gamma_v, \Gamma_v', \mathscr{G}_v)$ and $(\widetilde{\Gamma}_v, \widetilde{\Gamma}_v', \widetilde{\mathscr{G}}_v)$ the boundary triples defined as in Sects. 6.1 and 6.2. Note that the first one is bounded and elliptic (see Proposition 2.2.18) and the second is unbounded and elliptic (see Theorem 3.4.41).

We define the *boundary identification operator* in the Dirichlet case by

$$I_\varepsilon \colon \mathscr{G}_v \longrightarrow \widetilde{\mathscr{G}}_v, \qquad (I_\varepsilon F)_e := \begin{cases} F_e \varphi_{\varepsilon,e}, & \text{if } e \in E_{0,\mathrm{int}}, \\ 0, & \text{otherwise,} \end{cases}$$

i.e. we choose the F_e-multiple of the first Dirichlet eigenfunction on the thick edges. The adjoint is given by $(I^*\psi)_e = \varepsilon^{m/2}\langle \varphi_e, \psi_e\rangle$ for $e \in E_{0,\mathrm{int}}$. The transversal operators here are

$$K = 0 \qquad \text{and} \qquad \widetilde{K} = \Delta_{Y_{\varepsilon,\mathrm{int}}}^{\mathrm{D}} - \frac{\lambda_1}{\varepsilon^2} = \bigoplus_{e\in E}\Bigl(\Delta_{Y_{\varepsilon,e}}^{\mathrm{D}} - \frac{\lambda_1}{\varepsilon^2}\Bigr)$$

with associated quadratic forms $\mathfrak{k} = 0$ on $\mathscr{G}_v$ and $\widetilde{\mathfrak{k}}(\psi) = \|d\psi\|_{Y_{\varepsilon,\mathrm{int}}}^2 - \varepsilon^{-2}\lambda_1\|\psi\|_{Y_{\varepsilon,\mathrm{int}}}^2$.

For the notion of δ-closeness of two boundary maps Γ_v and $\widetilde{\Gamma}_v$ we refer to Definition 4.8.1.

Theorem 6.11.7. *The boundary maps*

$$\Gamma_v \colon \mathsf{H}_p^1(G_0) \longrightarrow \mathbb{C}^{E_{0,\mathrm{int}}}, \qquad \widetilde{\Gamma}_v \colon \mathsf{H}^1(X_\varepsilon) \longrightarrow \mathsf{L}_2(Y_{\varepsilon,\mathrm{int}}),$$

$$\Gamma_v f := \{f_e(\ell_e)\}_e, \qquad \widetilde{\Gamma}_v u := \{u(\ell_e,\cdot)\}_e$$

associated with the forms $\mathfrak{h}$ *and* $\widetilde{\mathfrak{h}}$ *are* 0*-close with the identification operators* J^1_ε*,* J'^1_ε *and the boundary identification operator* I_ε*. Moreover,* I_ε *is a* $\hat{\delta}_\varepsilon$*-partial isometry w.r.t.* $\mathfrak{k}$ *and* $\widetilde{\mathfrak{k}}$*, where* $\hat{\delta}_\varepsilon = \varepsilon(\min_e(\lambda_{2,e} - \lambda_1))^{-1/2}$*. Finally,* $\widetilde{K} I_\varepsilon = I_\varepsilon K (= 0)$*.*

Proof. It follows straightforward from the definitions that

$$(\widetilde{\Gamma} J^1 f)_e = \varepsilon^{-m/2} f_e(\ell_e)\varphi_e = (I_\varepsilon \Gamma f)_e,$$

$$(\Gamma J'^1 u)_e = \varepsilon^{m/2} \langle \varphi_e, u(\ell_e, \cdot)\rangle = (I^*_\varepsilon \widetilde{\Gamma} u)_e$$

for thick edges $e \in E_{0,\mathrm{int}}$. In particular, the 0-closeness follows.

For the partial isometry, $I^*_\varepsilon I_\varepsilon F = F$ is obvious. Moreover,

$$\begin{aligned}\|\psi - I_\varepsilon I^*_\varepsilon \psi\|^2_{Y_{\varepsilon,\mathrm{int}}} &= \varepsilon^m \Big(\sum_{e \in E_0} \|\psi_e - \langle \varphi_e, \psi_e\rangle \varphi_e\|^2_{Y_e} + \sum_{e \in E \setminus E_0} \|\psi_e\|^2_{Y_e} \Big) \\ &\le \frac{\varepsilon^2}{\min_e(\lambda_{2,e} - \lambda_1)} \widetilde{\mathfrak{k}}(\psi)\end{aligned}$$

for $\psi \in \operatorname{dom} \widetilde{\mathfrak{k}} = \mathring{\mathsf{H}}^1(Y_{\varepsilon,\mathrm{int}})$ using Proposition 5.1.2 and an argument similar as in Lemma 6.11.5 on each component. The intertwining property $\widetilde{K} I_\varepsilon = I_\varepsilon K$ of I_ε is obvious, since $\widetilde{K}$ vanishes on the eigenfunctions φ_e of the thick edges e $\in E_0$. □

6.11.4 The Embedded Case

Let us apply the previous results to the situation when the graph G is embedded in $\mathbb{R}^2$ and has a constant tubular radius $r_e > 0$ around each edge. We denote the closed neighbourhood of G by $\hat{X}_\varepsilon \subset \mathbb{R}^2$ (see Sect. 6.7.1 for details). The transversal manifolds Y_e here are just intervals of length r_e with eigenvalues $\lambda^{\mathrm{D}}_k(Y_e) = \pi^2 k^2 r_e^{-2}$. The *thick* edges here are given by

$$E_0 := \{\, e \in E \mid \lambda^{\mathrm{D}}_1(Y_e) = \lambda_1 \,\} = \{\, e \in E \mid r_e = r_+ \,\},$$

where

$$\lambda_1 = \min_e \lambda^{\mathrm{D}}_1(Y_e) = \frac{\pi^2}{r_+^2} \qquad \text{and} \qquad r_+ = \max_e r_e,$$

explaining the name "thick" edge. The quadratic form on $\hat{X}_\varepsilon$ is defined as

$$\hat{\mathfrak{h}}_\varepsilon(u) := \|\nabla u\|^2_{\hat{X}_\varepsilon} - \frac{\lambda_1}{\varepsilon^2}\|u\|^2_{\hat{X}_\varepsilon} = \|\nabla u\|^2_{\hat{X}_\varepsilon} - \frac{\pi^2}{\varepsilon^2 r_+^2}, \qquad \operatorname{dom} \hat{\mathfrak{h}}_\varepsilon = \mathsf{H}^1(\hat{X}_\varepsilon, \partial_0 \hat{X}_\varepsilon).$$

Denote by $\hat{\mathfrak{h}}_{\varepsilon,e}$ the restriction of $\hat{\mathfrak{h}}_\varepsilon$ to the Sobolev space $\mathsf{H}^1(\hat{X}_{\varepsilon,e}, \partial_0 \hat{X}_{\varepsilon,e})$, where $\hat{X}_{\varepsilon,e}$ is given in (6.48) and where $\partial_0 \hat{X}_{\varepsilon,e} = \partial \hat{X}_\varepsilon \cap \hat{X}_{\varepsilon,e}$ is the transversal boundary.

We have seen in Sect. 5.4 that the metric on $\hat{X}_{\varepsilon,e}$ in the intrinsic coordinates $I_{\varepsilon,e} \times Y_e$ is given by

$$g_{\varepsilon,e} = \ell_{\varepsilon,e}^2 \mathrm{d}s^2 + \varepsilon^2 r_e^2 \mathrm{d}y_e^2,$$

i.e. $\hat{\mathfrak{h}}_{\varepsilon,e}$ is unitarily equivalent with the corresponding form $\mathfrak{h}_{\varepsilon,e}$ on $X_{\varepsilon,e} = (I_{\varepsilon,e} \times Y_e, g_{\varepsilon,e})$, where

$$\mathfrak{h}_{\varepsilon,e}(u) := \|\mathrm{d}u\|^2_{X_{\varepsilon,e}} - \frac{\lambda_1}{\varepsilon^2}\|u\|^2_{X_{\varepsilon,e}}, \qquad \operatorname{dom}\mathfrak{h}_{\varepsilon,e} = \mathsf{H}^1(X_{\varepsilon,e}, \partial_0 X_{\varepsilon,e}).$$

Here, we keep the slightly shortened length of the embedded edge neighbourhood, i.e. $I_{\varepsilon,e} = [0, \ell_{\varepsilon,e}]$ is an interval of length $\ell_{\varepsilon,e} = \ell_e(1 - \varepsilon\tau)$ for some $\tau \in (0,1)$. Moreover,

$$\ell_{\varepsilon,e}(s, y) = 1 + \varepsilon\kappa_e(s) r_e y,$$

where κ_e denotes the curvature of the embedded edge (see (6.49)). Note that the function $b(y) = y$ is harmonic on the interval Y_e.

On the interval $I_{\varepsilon,e}$, we define a quadratic form by

$$\mathfrak{q}_{\varepsilon,e}(f) := \|f_e'\|^2_{I_{\varepsilon,e}} + \langle f_e, K_e f_e\rangle_{I_{\varepsilon,e}}, \qquad K_e := -\frac{\kappa_e^2}{4}$$

and $\operatorname{dom}\mathfrak{q}_e := \mathsf{H}^1_0(I_{\varepsilon,e})$. In Proposition 5.6.6 we have shown, that $\mathfrak{h}_{\varepsilon,e}$ (and therefore $\hat{\mathfrak{h}}_{\varepsilon,e}$) is δ_ε-partial isometrically equivalent with the quadratic form $\mathfrak{q}_{\varepsilon,e}$. Note that we used a fixed interval I in Propositions 5.6.5 and 5.6.6, but the explicit formula for δ_ε shows, that the result remains true for ε-depending intervals. The error depends on upper bounds on r_e and $\|\kappa_e^{(j)}\|_\infty$, $j = 0, 1, 2$.

In a last step, we can get rid of the ε-dependence of the interval $I_{\varepsilon,e}$, by an argument similarly as in Proposition 5.3.7: Denote by $\mathfrak{q}_e$ the quadratic form with the same formal expression as $\mathfrak{q}_{\varepsilon,e}$, but now defined on $\operatorname{dom}\mathfrak{q}_e := \mathsf{H}^1_0(I_e)$. Now, it is easy to see that $\mathfrak{q}_{\varepsilon,e}$ and $\mathfrak{q}_e$ are $\mathrm{O}(\varepsilon)$-quasi unitarily equivalent, where the error depends only on $\|\kappa_e\|_\infty$ and $\|\kappa_e'\|_\infty$.

Remark 6.11.8. It is not clear if we could directly compare the quadratic form $\mathfrak{h}_{\varepsilon,e}$ with a similar quadratic form on the *full* space $I_e \times Y_e$ as in Proposition 5.4.4 for the Neumann case. Note here, that the divergent terms with $\lambda_1\varepsilon^{-2}$ coming from the rescaling usually cause trouble, if they do not cancel each other *exactly*.

Putting the decoupled parts together, we can define a quadratic form

$$\mathfrak{q} := \bigoplus_{e\in E_0} \mathfrak{q}_e, \qquad \operatorname{dom} q := \mathscr{H}^1 = \bigoplus_{e\in E_0} \mathsf{H}^1_0(I_e)$$

on the graph. The corresponding operator is

$$\bigoplus_{e\in E_0}\Big(\Delta^0_{I_e} - \frac{\kappa_e^2}{4}\Big),$$

where $\Delta^0_{I_e}$ is the Laplacian on I_e with Dirichlet condition at 0 and Neumann condition at ℓ_e, if $\ell_e < \infty$. Finally, we obtain the following result; its proof being similar to the one of Theorem 6.11.6:

Theorem 6.11.9. *Assume that*

$$\lambda_1(X_v) - \frac{\pi^2}{r_+^2} > 0, \tag{6.72}$$

then the quadratic form $\hat{\mathfrak{h}}_\varepsilon$ is δ_ε-partial isometrically equivalent with the decoupled form $\mathfrak{q}$ as defined above with the curvature induced potentials, where $\delta_\varepsilon = \mathrm{O}(\varepsilon^{1/2})$. The error depends only on lower bounds on ℓ_e and $\lambda_1(X_v) - \pi^2 r_+^{-2}$, and on upper bounds on $\min_{e \in E \setminus E_0}(r_+ - r_e)$, $\|\kappa_e\|_\infty$, $\|\kappa_e'\|_\infty$ and $\|\kappa_e''\|_\infty$.

6.11.5 The Spectral Vertex Neighbourhood Condition

We finish this section with a comment on the smallness condition (6.72) on the vertex neighbourhood X_v, and give examples where this condition holds or fails. To simplify the presentation, we assume that the star graph G is again embedded in $\mathbb{R}^2$.

First, we show, that the condition can always be fulfilled, provided the vertex neighbourhood is small enough:

Remark 6.11.10. Suppose that we start with the closed 1-neighbourhood denoted by $X_v(0)$, i.e. we set $\varepsilon = 1$ and regard the unscaled vertex neighbourhood X_v. For simplicity, we assume that the curvature vanishes near the vertices. Therefore $X_v(0)$ is bounded by straight lines. Then we deform $X_v(0)$ smoothly in order to obtain a family $X_v(\tau)$, $\tau \geq 0$, shrinking to the graph, but fixing the boundary parts $\partial_e X_v(\tau) = \partial_e X_v(0)$, $e \in E_v$, where the edge neighbourhoods touch (cf. Fig. 6.13).

As in [P03a, Sect. 7] we can show that the first eigenvalue of the Laplacian on $X_v(\tau)$ tends to ∞, i.e.

$$\lambda_1(X_v(\tau)) \to \infty \qquad \text{as } \tau \to \infty,$$

where we impose Dirichlet boundary conditions on the transversal boundary $\partial_0 X_v(\tau)$ and Neumann boundary conditions on the internal boundary $\mathring{\partial} X_v(\tau) = \bigcup_e \partial_e X_v(\tau)$ (see Fig. 6.13) In particular, we can always find some $\tau \in (0, \infty)$ for which $X_v := X_v(\tau)$ satisfies (6.72). Fixing this shrinking parameter τ, we obtain an (ε-unscaled) vertex neighbourhood X_v fulfilling the desired estimate.

An Example Not Satisfying the Smallness Assumption

Let us briefly give an example of a vertex neighbourhood not satisfying (6.72). For suitable vertex neighbourhoods (e.g. arising from the ε-neighbourhood of a

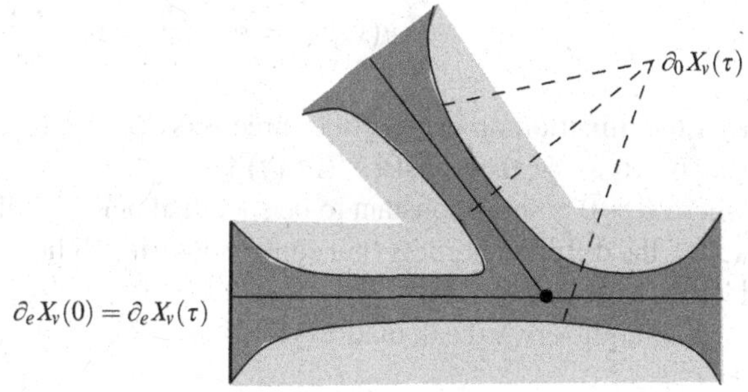

Fig. 6.13 The original vertex neighbourhood $X_v(0)$ (*light grey*) and the shrunken vertex neighbourhood $X_v(\tau)$ (*dark grey*). The internal boundary is drawn by a *thin solid line*, and the boundary component $\partial_e X_v(\tau) = \partial_e X_v(0)$ by a *thick solid line*

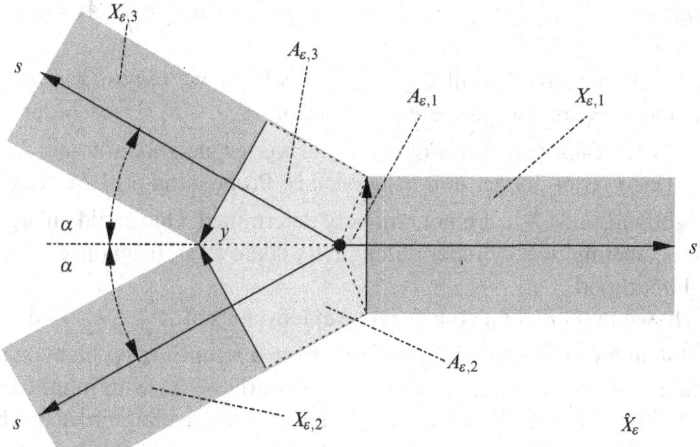

Fig. 6.14 A simple trial function supported in a neighbourhood of the vertex has an eigenvalue below the threshold $\lambda_1/\varepsilon^2 = \pi^2/(r_+\varepsilon)^2$ with $r_+ = 2$

graph) we will show the existence of an eigenvalue below the threshold $\lambda_1/\varepsilon^2 = \pi^2/(r_+\varepsilon^2)$. The case of an ε-neighbourhood of a vertex with four infinite edges emanating (a "cross") is considered in [SRW89] and [ABGM91]. In the former reference one can also find a contour plot of the first eigenfunction. In particular, the rescaled operator $\Delta^{\mathrm{D}}_{\hat{X}_\varepsilon} - \pi^2/(r_+\varepsilon^2)^2$ is no longer non-negative, i.e. (6.72) fails.

The existence of such an eigenvalue below the threshold can easily be established by inserting an appropriate trial function in the Rayleigh quotient. We consider a graph with one vertex and three adjacent edges of length ℓ and denote its ε-neighbourhood by $\hat{X}_\varepsilon$. We decompose $\hat{X}_\varepsilon$ into three rectangles $X_{\varepsilon,e}$ and three sets $A_{\varepsilon,e}$ as in Fig. 6.14.

On the rectangle $X_{\varepsilon,e}$ we use the coordinates $0 < s < \ell$ and $-\varepsilon < y < \varepsilon$ where $s = 0$ corresponds to the common boundary with $A_{\varepsilon,e}$. We extend these coordinates from $X_{\varepsilon,e}$ onto $A_{\varepsilon,e}$ and define

$$u(s, y) := \varepsilon^{-1/2} \chi(s) \cos\Bigl(\frac{\pi y}{2\varepsilon}\Bigr)$$

as a test function on each of the three sets $X_{\varepsilon,e} \cup A_{\varepsilon,e}$. Here, $\chi(s) = 1$ for $s < 0$ (i.e. on $A_{\varepsilon,e}$), $\chi(s) = \cos(\pi s/(2\varepsilon\kappa))$ for $0 \le s \le \kappa\varepsilon$ and $\chi(s) = 0$ for $\varepsilon < s < \ell$ where $\kappa > 0$ is some constant to be specified below. Although u is not differentiable across the different borders (but continuous), it still lies in the quadratic form domain $\mathsf{H}^1(\hat{X}_\varepsilon, \partial_0 \hat{X}_\varepsilon)$.

A straightforward calculation yields

$$\frac{\|du\|^2_{\hat{X}_\varepsilon}}{\|u\|^2_{\hat{X}_\varepsilon}} - \frac{\pi^2}{4\varepsilon^2} = \Bigl(\frac{8\kappa \cos\alpha + 3\pi^2 \sin\alpha - 16\kappa}{((3\pi^2 - 4)\cos\alpha + 3\pi^2\kappa \sin\alpha + 8)\kappa}\Bigr)\frac{\pi^2}{4\varepsilon^2}. \tag{6.73}$$

This quantity is negative for all $0 < \alpha < 0.93\pi$ if we choose e.g. $\kappa = 3$. In particular, there exists a negative eigenvalue of $\Delta^{\mathrm{D}}_{\hat{X}_\varepsilon} - \lambda_1/\varepsilon^2$ of order ε^{-2}, and Condition (6.72) fails here for any choice of vertex neighbourhoods $X_{\varepsilon,v}$, since $\Delta^{\mathrm{D}}_{\hat{X}_\varepsilon} - \pi^2/(2\varepsilon^2)^2$ is no longer non-negative (see Proposition 6.11.3). Note that the vertex neighbourhoods $X_{\varepsilon,v}$ are not uniquely determined. One could enlarge $X_{\varepsilon,v}$ at each edge emanating by a cylinder of length $a\varepsilon$ taken away from the corresponding edge neighbourhood.

It was shown in [G08a] that the negative eigenvalues of $\Delta^{\mathrm{D}}_{\hat{X}_\varepsilon} - \lambda_1/\varepsilon^2$ do not arise from a problem formulated on the graph, but from a *scattering* problem on a related space obtained from the unscaled vertex neighbourhood X_v with infinite cylinders $X_{e,\infty} = [0, \infty) \times Y_e$ attached. The dependence of the leading order on the angle α in (6.73) indicates that the behaviour should depend on the angles of the edges meeting at a vertex.

Chapter 7
Global Convergence Results

In this final chapter, we prove convergence results for general metric graphs and the associated graph-like manifolds. In particular, we show resolvent convergence, convergence of the discrete and essential spectrum and of resonances. The idea is as follows: we decompose a general metric graph G given by (V, E, ∂, ℓ) (cf. Sect. 2.2) and an associated graph-like manifold X_ε into star graphs G_v and their graph-like manifolds $X^+_{\varepsilon,v}$. For such spaces, we already showed the convergence of the associated quadratic forms $\mathfrak{d}_v$ and $\widetilde{\mathfrak{d}}_v$ (more precisely their $\delta_\varepsilon(v)$-partial isometry resp. $\delta_\varepsilon(v)$-unitary equivalence). The corresponding vertex coupling depends on the behaviour at the vertex region, and is encoded in a space $\mathscr{V}_v \subset \mathbb{C}^{E_v}$ at each vertex. For example in the "fast decaying" case (including the case of a ε-homothetic vertex neighbourhood $X_{\varepsilon,v} = \varepsilon X_v$ in the centre), we have $\mathscr{V}_v = \mathbb{C}p(v)$, where $p(v) = \{p_e\}_{e \in E}$ and $p_e = (\operatorname{vol} Y_e)^{1/2}$ is the square root of the transversal volume. The precise model settings are fixed in the previous chapter.

In Sect. 7.1 we define boundary maps coupled via the graph G (cf. Sect. 3.9 for a definition) starting from the boundary maps associated with the star-shaped spaces G_v and $X^+_{\varepsilon,v}$ together with their quadratic forms $\mathfrak{d}_v$ and $\widetilde{\mathfrak{d}}_v$. To do so, we need suitable global assumptions like e.g. a positive global lower bound $\inf_{e \in E} \ell_e > 0$. Note that the Neumann operator associated with the coupled boundary map is the original operator on G resp. on X_ε. We can then use an abstract convergence result for coupled boundary triples $(\Gamma, \Gamma', \mathscr{G})$ and $(\widetilde{\Gamma}, \widetilde{\Gamma}', \widetilde{\mathscr{G}})$ developed in Sect. 4.8 (more precisely, we only need coupled boundary maps $\Gamma \colon \mathscr{H}^1 \longrightarrow \mathscr{G}$ and $\widetilde{\Gamma} \colon \widetilde{\mathscr{H}}^1 \longrightarrow \widetilde{\mathscr{G}}$). This convergence result needs again some global assumptions, assuring that the local error terms $\delta_\varepsilon(v)$ are uniformly convergent, namely that $\sup_v \delta_\varepsilon(v) \to 0$ as $\varepsilon \to 0$. We treat the different cases depending on the scaling behaviour at the vertex neighbourhood resp. the boundary conditions in Sects. 7.1.1–7.1.4. We also comment on the situation, when the metric graph G is embedded in $\mathbb{R}^2$ and X_ε is a small neighbourhood of G in $\mathbb{R}^2$, in Sect. 7.1.5.

In Sect. 7.2, we use the abstract convergence results for resonances developed in Sect. 4.9. We recall some facts, in particular the definition of the complexly dilated operators H^θ and H^θ_ε. Their discrete eigenvalues are (locally) independent of the complex parameter θ and actually poles of a suitable meromorphic continuation of

O. Post, *Spectral Analysis on Graph-Like Spaces*, Lecture Notes in Mathematics 2039, DOI 10.1007/978-3-642-23840-6_7, © Springer-Verlag Berlin Heidelberg 2012

the resolvents $(\Delta_G - z)^{-1}$ and $(\Delta_{X_\varepsilon} - z)^{-1}$, i.e., resonances (see Sects. 3.6–3.8). Note that in the cases, when the limit operator on the graph *decouples* (like in the slowly decaying and the Dirichlet decoupled case), all resonances on the graph are real-valued and embedded eigenvalues in the continuous spectrum. These eigenvalues are seen on the corresponding graph-like manifold as resonances converging to the real eigenvalues. This phenomena is related to a *Helmholtz resonator* (see Remark 1.4.6 and Corollaries 7.2.8 and 7.2.13).

7.1 Spectral Convergence for Graph-Like Spaces

Let us now explain in more detail how the global spaces and the operators can be obtained from the corresponding objects on the star graph. Let G be a metric graph. Metric graphs and its associated operators are introduced in detail in Sects. 2.2 and 2.3. We decompose G into its star-graph components by splitting each edge of finite length (i.e., *interior* edges $e \in E_{\mathrm{int}}$) into two halves. Next, we collect the half-edges (and edges of infinite length, i.e., *exterior* edges $e \in E_{\mathrm{ext}}$) adjacent to a vertex v in order to form a star graph G_v. We similarly decompose the graph-like manifold X_ε into its closed star-graph components $X^+_{\varepsilon,v}$, i.e. we have

$$G = \dot{\bigcup_{v \in V}} G_v \qquad \text{and} \qquad X_\varepsilon = \dot{\bigcup_{v \in V}} X^+_{\varepsilon,v}$$

for the graph and the manifold, respectively.[1] For the precise definition of the star graph neighbourhoods $X^+_{\varepsilon,v}$ associated with the star graph G_v with different scaling behaviour near the vertex we refer to Sects. 6.2 and 6.8. If we consider a metric graph G embedded in some ambient space $\mathbb{R}^d$ and an associated graph-like space $X_\varepsilon \subset \mathbb{R}^d$, we obtain additional error terms, cf. Sects. 6.7.1 and 7.1.5.

We have proven in Chap. 6 in various cases, that the quadratic form $\|du\|^2_{X^+_{\varepsilon,v}}$ is $\delta_\varepsilon(v)$-partial isometrically equivalent with a corresponding quadratic form on the star graph, where $\delta_\varepsilon(v) \to 0$ as $\varepsilon \to 0$. Using the abstract setting of boundary maps coupled via graphs (see Sect. 3.9), we can define a boundary triple associated with the forms on G and X_ε, respectively. The boundary maps (and triples) $\Gamma_v \colon \mathscr{H}^1_v \longrightarrow \mathscr{G}_v$ and $\widetilde{\Gamma}_v \colon \widetilde{\mathscr{H}}^1_v \longrightarrow \widetilde{\mathscr{G}}_v$ associated with G_v and $X^+_{\varepsilon,e}$ were defined in Sects. 6.1.1–6.1.3 and Sects. 6.2.3–6.2.4, respectively. For the Dirichlet case, see Sect. 6.11.3 (in the sequel, for the Dirichlet case, one has to replace $E_{v,\mathrm{int}} = E_v \cap E_{\mathrm{int}}$ by $E_{0,v,\mathrm{int}} = E_0 \cap E_v \cap E_{\mathrm{int}}$, the set of *thick* edges, see (6.66)).

Let us recall the setting here: We have $\mathscr{G}_v = \mathbb{C}^{E_{v,\mathrm{int}}}$ and $(\Gamma_v f)_e = f_e(\ell_e/2)$, i.e., we evaluate f_e on the *midpoint* of the edge $e \in E_{v,\mathrm{int}}$. The space $\mathscr{H}^1_v$ depends on the model setting, but near the midpoints of an edge, f_e is in H^1.

[1]The notation $G = \dot{\bigcup}_v G_v$ means "disjoint union up to measure 0".

On the manifold model $X^+_{\varepsilon,v}$, we have $\widetilde{\mathscr{G}}_v = \mathsf{L}_2(Y_{\varepsilon,\mathrm{int}})$ where $\partial_1 X^+_{\varepsilon,v} := Y_{\varepsilon,\mathrm{int}} = \dot\bigcup_{e\in E_{v,\mathrm{int}}} Y_{\varepsilon,e}$ is the disjoint union of the transversal manifolds of the interior edges adjacent to v (for the boundary notation of manifolds, see Sect. 5.1.1). Here, $(\widetilde{\Gamma}_v u)_e = u_e(\ell_e/2, \cdot)$. Locally, $X^+_{\varepsilon,v}$ is of product structure near the (longitudinal) boundary $\partial_1 X^+_{\varepsilon,v}$.

Let us first check that we can define boundary maps coupled via the graph (V, E, ∂) from the boundary maps. Here, G_{int} denotes the graph obtained from the graph G by deleting all exterior edges (edges of infinite length). For the Dirichlet case, one has to delete in addition the non-thick edges.

Proposition 7.1.1. *Assume that* $\inf_{e\in E} \ell_e > 0$, *then the boundary maps* $\Gamma_v : \mathscr{H}^1_v \longrightarrow \mathscr{G}_v$ *and* $\widetilde{\Gamma}_v : \widetilde{\mathscr{H}}^1_v \longrightarrow \widetilde{\mathscr{G}}_v$ *associated with* G_v *and* $X^+_{\varepsilon,e}$ *are compatible with the graph* G_{int} *(see Definition 3.9.1), and the coupled boundary maps* $\Gamma : \mathscr{H}^1 \longrightarrow \mathscr{G}$ *and* $\widetilde{\Gamma} : \widetilde{\mathscr{H}}^1 \longrightarrow \widetilde{\mathscr{G}}$ *as defined in* (3.92) *exist.*

Proof. The compatibility is obvious from the construction, namely we set $\Gamma_{v,e} f := f_e(\ell_e/2)$ and $\widetilde{\Gamma}_{v,e} u := u_e(\ell_e/2, \cdot)$. Moreover, $\mathscr{G}_e := \mathbb{C}$ and $\widetilde{\mathscr{G}}_e := \mathsf{L}_2(Y_{\varepsilon,e})$. We now have to check the conditions of Proposition 3.9.2. In both cases, we estimated the norm of Γ_v and $\widetilde{\Gamma}_v$ by $2/\inf_e\{\sqrt{\ell_e}, 1\}$ (see Propositions 6.1.4, 6.1.6, 6.1.7 and 6.2.9), which is finite by assumption. Moreover, $f \in \mathscr{H}^1_{v,e}$ iff $f_{e'}(\ell_{e'}/2) \neq 0$ only for $e = e'$, and similarly for $u \in \widetilde{\mathscr{H}}^1_{v,e}$. Now, $\mathscr{G}^{1/2}_e \cong \mathbb{C} = \mathscr{G}_e$ and $\widetilde{\mathscr{G}}^{1/2}_e \cong \mathsf{H}^{1/2}(Y_{\varepsilon,e}) = \mathscr{G}^{1/2}_e$, so that in particular $\mathscr{G}^{1/2}_e$ and $\widetilde{\mathscr{G}}^{1/2}_e$ are dense in $\mathscr{G}_e$ and $\widetilde{\mathscr{G}}_e$. Therefore, the result follows by Proposition 3.9.2. □

It is now easy to see that the Neumann operators of the coupled boundary triples are the original operators on G resp. X_ε: For example, on the manifold, the quadratic form on the coupled system is defined on the domain

$$\widetilde{\mathscr{H}}^1 = \bigl\{ u \in \bigoplus_v \mathsf{H}^1(X^+_{\varepsilon,v}) \bigm| \widetilde{\Gamma}_{\partial_- e,e} u = \widetilde{\Gamma}_{\partial_+ e,e} u, \ \forall\, e \in E \bigr\}.$$

The conditions on the boundaries $\widetilde{\Gamma}_{\partial_- e,e} u = \widetilde{\Gamma}_{\partial_+ e,e}$ just mean continuity, so that $\widetilde{\mathscr{H}}^1$ equals the Sobolev space $\mathsf{H}^1(X_\varepsilon)$ (resp. $\mathring{\mathsf{H}}^1(X_\varepsilon)$ in the Dirichlet case), i.e., the domain of the quadratic form associated with the original Laplacian on X_ε. In particular, the Neumann operator H^{N} of the coupled system is just the Laplacian Δ_{X_ε} (resp. $\Delta^{\mathrm{D}}_{X_\varepsilon}$). A similar remark holds for the operators on the metric graph.

Let us now fix the scaling on the vertex neighbourhoods and the corresponding limit operators.

7.1.1 Fast Decaying Vertex Volume

We start with the fast decaying case. Recall that each star graph neighbourhood $X^+_{\varepsilon,v}$ decomposes as

$$X_{\varepsilon,v}^{+} = X_{\varepsilon,v} \,\dot{\cup}\, \dot{\bigcup_{e \in E_v}} X_{\varepsilon,e}.$$

In the fast decaying case, we assume that each vertex neighbourhood $X_{\varepsilon,v}$ behaves like

$$\varepsilon X_v \le X_{\varepsilon,v} \le \varepsilon^\alpha X_v, \qquad \text{where} \qquad \frac{m}{d} = \frac{d-1}{d} < \alpha \le 1,$$

(see (5.9) for the notation). For simplicity, we assume that the scaling rate α is independent of the vertex v. The convergence for the quadratic forms associated with the star graph G_v was established in Sect. 6.4. Let us now define the global spaces associated with the entire graph G. The space $\mathscr{H}$ and the quadratic form $\mathfrak{d}$ on G are defined by

$$\mathscr{H} := \mathsf{L}_2(G) = \bigoplus_{e \in E} I_e, \qquad \mathfrak{d}(f) := \sum_{e \in E} \|f_e'\|^2_{I_e}$$

$$\mathscr{H}^1 := \operatorname{dom} \mathfrak{d} := \big\{ f \in \mathsf{H}^1_{\max}(G) \,\big|\, \underline{f}(v) \in \mathbb{C}p(v)\ \forall v \in V \big\}.$$

The corresponding operator is the weighted standard Laplacian $\Delta := \Delta_G$ given by $(\Delta_G f)_e = -f_e''$ and domain

$$\operatorname{dom} \Delta_G = \big\{ u \in \mathsf{H}^2_{\max}(G) \,\big|\, \underline{f}(v) \in \mathbb{C}p(v),\ \sum_{e \in E_v} p_e \underline{\widetilde{f}}'_e(v) = 0\ \forall v \in V \big\},$$

see Corollary 2.2.11. The weights of the vertex space $\mathscr{V}_v = \mathbb{C}p(v)$ are given by

$$p_e := (\operatorname{vol} Y_e)^{1/2}, \qquad \text{where} \qquad p(v) = \{p_e\}_{e \in E_v}.$$

For the manifold, we set

$$\widetilde{\mathscr{H}} := \mathsf{L}_2(X_\varepsilon), \qquad \widetilde{\mathfrak{d}}(u) := \|du\|^2_{X_\varepsilon} \qquad \text{and} \qquad \widetilde{\mathscr{H}}^1 = \operatorname{dom} \widetilde{\mathfrak{d}} := \mathsf{H}^1(X_\varepsilon).$$

The associated operator $\widetilde{\Delta} = \Delta_{X_\varepsilon}$ is the Laplace operator with Neumann boundary conditions if $\partial X_\varepsilon \ne \emptyset$.

In order to assure the global convergence, we need some uniformity conditions. Namely, we assume that

$$\ell_- := \inf_{e \in E} \ell_e > 0, \qquad \lambda_2(E) := \inf_{e \in E} \lambda_2(Y_e) > 0, \tag{7.1a}$$

$$\operatorname{vol}_+ := \sup_{v \in V} \frac{\operatorname{vol} X_v}{\operatorname{vol} \mathring{\partial} X_v} < \infty, \qquad \lambda_2(V) := \inf_{v \in V} \lambda_2(X_v) > 0, \tag{7.1b}$$

where $\lambda_2(Y_e)$ and $\lambda_2(X_v)$ denote the second (first non-vanishing) *Neumann* eigenvalue of Y_e resp. X_v.

Theorem 7.1.2. *Assume that the above global assumptions* (7.1) *are fulfilled. Then* $\mathfrak{d}$ *and* $\widetilde{\mathfrak{d}}$ *are* δ_ε*-partial isometrically equivalent, where* $\delta_\varepsilon = \mathrm{O}(\varepsilon^{(\alpha d - m)/2})$ *depends only on the above global constants. In particular, if the vertex neighbourhoods scale of order* ε *(i.e.* $\alpha = 1$*), then* $\delta_\varepsilon = \mathrm{O}(\varepsilon^{1/2})$.

Proof. The $\delta_{\varepsilon,v}$-partial isometry of $\mathfrak{d}_v$ and $\widetilde{\mathfrak{d}}_v$ follows from Theorem 6.4.3. Moreover, the boundary operators Γ_v and $\widetilde{\Gamma}_v$ are 0-close with the corresponding identification operators J_v^1 and $J_v'^1$ by Theorem 6.4.4. The step from the estimates on each star graph G_v to the global graph G follows now from Proposition 4.8.3 (7). Note that we need (7.1) in order to assure that $\delta_\varepsilon := \sup_v \delta_{\varepsilon,v} < \infty$. □

7.1.2 Slowly Decaying Vertex Volume

In the slowly decaying case, we use the decomposition

$$X_{\varepsilon,v}^{+} = \dot{X}_{\varepsilon,v} \,\overline{\dot\cup}\, \overline{\dot\bigcup_{e\in E_v}} X_{\varepsilon,e}$$

where $\dot{X}_{\varepsilon,v}$ is either the homogeneously scaled manifold $X_{\varepsilon,v} = \varepsilon^\alpha X_v$ (see Sect. 6.2) or the homogeneously scaled manifold $X_{\varepsilon,v} = \varepsilon^\alpha X_v$ together with the truncated cones $\acute{X}_{\varepsilon,v,e}$, i.e. $\dot{X}_{\varepsilon,v} = \acute{X}_{\varepsilon,v} = X_{\varepsilon,v} \,\overline{\dot\cup}\, \overline{\dot\bigcup}_{e\in E} \acute{X}_{\varepsilon,v,e}$ (see Sect. 6.8). In both cases, we assume that the scaling rate α fulfils

$$0 < \alpha < \frac{m}{d} = \frac{d-1}{d}.$$

The convergence for the quadratic forms associated with the star graph G_v was established in Sects. 6.5 and 6.9.

Let us now define the global spaces and operators. The space $\mathscr{H}$ and the quadratic form $\mathfrak{d}$ on the (extended) metric graph G are given by

$$\mathscr{H} := \mathsf{L}_2(G) \oplus \ell_2(V), \qquad \mathfrak{d}(\hat{f}) := \sum_{e\in E} \|f_e'\|_{I_e}^2$$

$$\mathscr{H}^1 := \operatorname{dom}\mathfrak{d} := \{\, \hat{f} = (f, F) \in \mathsf{H}^1_{\max}(G) \oplus \ell_2(V) \mid \underline{f}(v) = 0 \ \forall v \in V \,\},$$

Note that the quadratic form is decoupled. The corresponding operator

$$\Delta = \Delta_G^{\mathrm{D}} \oplus 0 = \bigoplus_{e\in E} \Delta_{I_e}^{\mathrm{D}} \oplus 0$$

is as well decoupled. In the terminology of Sect. 2.3, Δ is the extended operator with $L = 0$ (see Corollary 2.3.4). For the manifold we set

$$\widetilde{\mathscr{H}} := \mathsf{L}_2(X_\varepsilon), \qquad \widetilde{\mathfrak{d}}(u) := \|du\|_{X_\varepsilon}^2 \qquad \text{and} \qquad \widetilde{\mathscr{H}}^1 := \operatorname{dom}\widetilde{\mathfrak{d}} := \mathsf{H}^1(X_\varepsilon).$$

Note again that $\widetilde{\Delta} = \Delta_{X_\varepsilon}$ is the Laplacian with Neumann boundary conditions if X_ε has a non-trivial boundary. The global assumptions are

$$\ell_- := \inf_{e \in E} \ell_e > 0, \qquad \lambda_2(E) := \inf_{e \in E} \lambda_2(Y_e) > 0, \tag{7.2a}$$

$$\mathrm{vol}_- := \inf_{v \in V} \frac{\mathrm{vol}\, X_v}{\mathrm{vol}\, \mathring{\partial} X_v} > 0, \qquad \lambda_2(V) := \inf_{v \in V} \lambda_2(X_v) > 0, \tag{7.2b}$$

where $\lambda_2(Y_e)$ and $\lambda_2(X_v)$ denote the second (first non-vanishing) *Neumann* eigenvalue of Y_e resp. X_v. Moreover, we need either

$$\sup C_e < \infty \qquad \text{or} \qquad \lambda_2^{\mathrm{D}}(V) := \sup_{v \in V} \lambda_2^{\mathrm{D}}(X_v) < \infty \tag{7.2c}$$

depending on whether we used the vertex neighbourhood $\dot{X}_{\varepsilon,v} = X_{\varepsilon,v}$ of Sect. 6.2 with homogeneous transversal manifolds or the space $\dot{X}_{\varepsilon,v} = \acute{X}_{\varepsilon,v}$ with attached truncated cones (see Sect. 6.8). Here, C_e is defined in (6.15) as bound on the derivative of the homogeneous embedding $\varphi_e \colon \varepsilon Y_e \hookrightarrow Y_e$, and $\lambda_2^{\mathrm{D}}(X_v)$ is the second eigenvalue of X_v with Dirichlet condition on $\mathring{\partial} X_v$ and Neumann condition on $\partial_0 X_v$ if $\partial_0 X_v = \partial X_v \cap X_v$ is non-empty.

Theorem 7.1.3. *Assume that the above global assumptions* (7.2) *are fulfilled. Then* $\mathfrak{d}$ *and* $\widetilde{\mathfrak{d}}$ *are* δ_ε*-partial isometrically equivalent. Moreover, the error is of order* $\delta_\varepsilon = \mathrm{O}(\varepsilon^{\min\{(m-\alpha d)/2, \alpha\}})$ *and depends only on the above global constants.*

Proof. The proof follows similarly as in Theorem 7.1.2, now using Theorems 6.5.1, 6.5.2 and 6.9.1. □

7.1.3 The Borderline Case

Let us here use the construction of the vertex neighbourhood $\acute{X}_{\varepsilon,v}$ with the truncated cones attached (see Sects. 6.8 and 6.10). Recall that the homogeneously scaled (central) vertex neighbourhood $X_{\varepsilon,v} = \varepsilon^\alpha X_v$ is assumed to scale with rate

$$\alpha = \frac{m}{d} = \frac{d-1}{d}.$$

The convergence for the quadratic forms associated with the star graph G_v was established in Sect. 6.10.

Let us now define the global spaces and operators. The space $\mathscr{H}$ and the quadratic form $\mathfrak{d}$ on the (extended) metric graph G are given by

$$\mathscr{H} := \mathsf{L}_2(G) \oplus \ell_2(V, |p|^2), \qquad \mathfrak{d}(\hat{f}) := \sum_{e \in E} \|f_e'\|^2_{I_e}$$

$$\mathscr{H}^1 := \operatorname{dom} \mathfrak{d} := \{ \hat{f} = (f, F) \in \mathsf{H}^1_p(G) \oplus \ell_2(V, |p|^2) \mid f(v) = L(v)F(v) \forall v \in V \},$$

where

$$p_e := (\operatorname{vol} Y_e)^{1/2} \qquad \text{and} \qquad L(v) = (\operatorname{vol} X_v)^{-1/2}.$$

Recall the notation $\underline{f}(v) = f(v)p(v)$ and

$$\underline{f}(v) = \{f_e(v)\}_{e \in E}, \quad \mathsf{H}^k_p(G) = \{ f \in \mathsf{H}^k_{\max}(G) \mid \underline{f}(v) \in \mathbb{C}p(v) \ \forall v \in V \}.$$

Moreover, $\ell_2(V, |p|^2)$ is the weighted ℓ_2-space with norm given by $\|F\|^2_{\ell_2(V,|p|^2)} = \sum_v |F(v)|^2 |p(v)|^2$. Note that the space and the quadratic form are the extended space and form associated with the standard weighted quantum graph with operator $L = \{L(v)\}$, see Sect. 2.3. In particular, we have $\mathscr{H} = \hat{\mathscr{H}}$, $\mathfrak{d} = \hat{\mathfrak{d}}_L$ and $\mathscr{H}^1 = \hat{\mathscr{H}}^1_L$, where $L = \{L(v)\}$ acts via multiplication on $\ell_2(V, |p|^2)$. The operator associated with the quadratic form $\mathfrak{d}$ is $\Delta = \hat{\Delta}_L$ with domain

$$\operatorname{dom} \Delta = \{ \hat{f} \in \mathsf{H}^2_p(G) \oplus \ell_2(V, |p|^2) \mid f(v) = L(v)F(v) \ \forall v \in V \},$$

and acts as

$$\Delta \hat{f} = \{-f_e''\}_e \oplus \Big\{ \frac{L(v)}{|p(v)|^2} \sum_{e \in E_v} p_e \underline{\widetilde{f}}'_e(v) \Big\}_v$$

(see Corollary 2.3.5).

Remark 7.1.4. The above operator couples the extra state space $\ell_2(V, |p|^2)$ non-trivially. Note that the eigenvalue equation $\Delta \hat{f} = \lambda \hat{f}$ reads as

$$-f_e'' = \lambda f_e, \qquad \frac{1}{\operatorname{vol} \mathring{\partial} X_v} \sum_{e \in E_v} (\operatorname{vol} Y_e)^{1/2} \underline{\widetilde{f}}'_e(v) = \lambda \cdot (\operatorname{vol} X_v) f(v)$$

for each edge e and each vertex v. Considering only the graph part, we can think of this operator as of a delta-interaction with *energy*-depending strength (see Remark 2.4.4 (1)).

For the manifold we set again

$$\widetilde{\mathscr{H}} := \mathsf{L}_2(X_\varepsilon), \qquad \widetilde{\mathfrak{d}}(u) := \|du\|^2_{X_\varepsilon}, \qquad \text{and} \qquad \widetilde{\mathscr{H}}^1 := \operatorname{dom} \widetilde{\mathfrak{d}} := \mathsf{H}^1(X_\varepsilon),$$

and $\widetilde{\Delta} = \Delta_{X_\varepsilon}$ is the (Neumann) Laplacian on X_ε. The global assumptions are

$$\ell_- := \inf_{e \in E} \ell_e > 0, \qquad \lambda_2(E) := \inf_{e \in E} \lambda_2(Y_e) > 0, \tag{7.3a}$$

$$\operatorname{vol}_- := \inf_{v \in V} \frac{\operatorname{vol} X_v}{\operatorname{vol} \mathring{\partial} X_v} > 0, \qquad \lambda_2(V) := \inf_{v \in V} \lambda_2(X_v) > 0, \tag{7.3b}$$

$$\lambda_2^{\mathrm{D}}(V) := \sup_{v \in V} \lambda_2^{\mathrm{D}}(X_v) < \infty, \tag{7.3c}$$

where $\lambda_2(Y_e)$ and $\lambda_2(X_v)$ denote the second (first non-vanishing) *Neumann* eigenvalue of Y_e resp. X_v, and $\lambda_2^{\mathrm{D}}(X_v)$ the second eigenvalue of X_v with Dirichlet condition on $\mathring{\partial} X_v$ and Neumann condition on $\partial_0 X_v$.

Theorem 7.1.5. *Assume that the above global assumptions* (7.3) *are fulfilled. Then* $\mathfrak{d}$ *and* $\widetilde{\mathfrak{d}}$ *are* δ_ε*-partial isometrically equivalent, where* $\delta_\varepsilon = \mathrm{O}(\varepsilon^{(d-1)/d})$ *depends only on the above global constants.*

Proof. The proof follows similarly as in Theorem 7.1.2, now using Theorems 6.6.2 and 6.10.1. □

7.1.4 The Dirichlet Decoupled Case

Let us now treat the case when the Laplacian on X_ε has Dirichlet boundary conditions on $\partial X_\varepsilon \neq \emptyset$. Before defining the limit graph space, we need some more information on the manifold approximation. Let G be a metric graph given by (V, E, ∂, ℓ). To each edge $e \in E$, we associate the transversal manifold Y_e with non-trivial boundary $\partial Y_e \neq \emptyset$. Note that the lowest transversal eigenvalue $\lambda_1^{\mathrm{D}}(Y_e)$ of the Dirichlet operator on Y_e is no longer 0. In particular,

$$\lambda_1^{\mathrm{D}}(Y_{\varepsilon,e}) = \frac{\lambda_1^{\mathrm{D}}(Y_e)}{\varepsilon^2}$$

diverges as $\varepsilon \to 0$. We therefore need a rescaling in order to expect convergence. Denote by

$$\lambda_1 := \inf_{e \in E} \lambda_1^{\mathrm{D}}(Y_e) \tag{7.4}$$

the infimum of all lowest transversal eigenmodes. We have already seen on a star graph in Sect. 6.11 that only those edges survive in the limit operator which have the lowest eigenmode equal to λ_1, namely

$$E_0 := \{\, e \in E \mid \lambda_1^{\mathrm{D}}(Y_e) = \lambda_1 \,\},$$

i.e. E_0 is the set of edges for which the infimum is *achieved*. We assume here that $E_0 \neq \emptyset$.

Let us now define the limit space. The Hilbert space $\mathscr{H}$, the quadratic form $\mathfrak{d}$ and the operator Δ on the metric graph G are given by

$$\mathscr{H} := \mathsf{L}_2(G_0), \qquad \mathfrak{d}(f) := \sum_{e \in E_0} \|f_e'\|^2_{I_e}, \qquad \mathscr{H}^1 := \operatorname{dom} \mathfrak{d} := \bigoplus_{e \in E_0} \mathring{\mathsf{H}}^1(I_e),$$

$$\Delta = \bigoplus_{e \in E_0} \Delta^{\mathrm{D}}_{I_e}, \quad \operatorname{dom} \Delta = \bigoplus_{e \in E_0} \bigl(\mathsf{H}^2(I_e) \cap \mathring{\mathsf{H}}^1(I_e)\bigr).$$

Note that the limit operator is decoupled. However, we do not need an extra discrete space as in the slowly decaying case of Sect. 7.1.2.

For the manifold we set

$$\widetilde{\mathscr{H}} := \mathsf{L}_2(X_\varepsilon), \quad \widetilde{\mathfrak{d}}(u) := \|du\|^2_{X_\varepsilon} - \frac{\lambda_1}{\varepsilon^2}\|u\|^2_{X_\varepsilon} \quad \text{and} \quad \widetilde{\mathscr{H}}^1 := \operatorname{dom}\widetilde{\mathfrak{d}} := \mathring{\mathsf{H}}^1(X_\varepsilon).$$

The corresponding operator is $\widetilde{\Delta} = \Delta^{\mathrm{D}}_{X_\varepsilon} - \frac{\lambda_1}{\varepsilon^2}$, where $\Delta^{\mathrm{D}}_{X_\varepsilon}$ denotes the Laplacian with Dirichlet boundary conditions on ∂X_ε. The global assumptions are

$$\ell_- := \inf_{e \in E} \ell_e > 0, \qquad \lambda_{21}(E) := \inf_{e \in E_0} \bigl(\lambda^{\mathrm{D}}_2(Y_e) - \lambda^{\mathrm{D}}_1(Y_e)\bigr) > 0, \tag{7.5a}$$

$$\lambda_1(E) := \inf_{e \in E \setminus E_0} \bigl(\lambda^{\mathrm{D}}_1(Y_e) - \lambda_1\bigr) > 0, \; \lambda_1(V) := \inf_{v \in V} \bigl(\lambda_1(X_v) - \lambda_1\bigr) > 0, \tag{7.5b}$$

where $\lambda^{\mathrm{D}}_k(Y_e)$ denotes the k-th Dirichlet eigenvalue of Y_e and $\lambda_1(X_v)$ the first *Dirichlet Neumann* eigenvalue of X_v with Dirichlet condition on the boundary part $\partial_0 X_v$ coming from ∂X_ε and Neumann conditions on the parts $\mathring{\partial} X_v = \bigcup_{e \in E_v} \partial_e X_v$, where the edge neighbourhoods are attached. Recall the definition of λ_1 in (7.4).

Note that the condition $\lambda_1(V) > 0$ is a global *smallness* condition on the vertex neighbourhood X_v, see the discussion in Remark 6.11.2 and Sect. 6.11.5.

Theorem 7.1.6. *Assume that the above global assumptions* (7.5) *are fulfilled. Then* $\mathfrak{d}$ *and* $\widetilde{\mathfrak{d}}$ *are* δ_ε*-partial isometrically equivalent, where* $\delta_\varepsilon = \mathrm{O}(\varepsilon^{1/2})$ *depends only on the above global constants.*

Proof. The proof follows similarly as in Theorem 7.1.2, now using Theorems 6.11.6 and 6.11.7. □

7.1.5 The Embedded Case

Let us make a short comment on the case when the graph G is embedded in $\mathbb{R}^2$, and the manifold X_ε is a closed neighbourhood of G in $\mathbb{R}^2$ (see Sects. 6.7.1 and 6.7) with *constant* radius $r_e > 0$. The previous theorems remain true, if we assume additionally that

$$\sup_e r_e < \infty, \qquad \sup_e \|\kappa_e\|_\infty < \infty \qquad \text{and} \qquad \sup_e \|\kappa'_e\|_\infty < \infty.$$

In the Dirichlet case, we need in addition

$$\sup_e \|\kappa''_e\|_\infty < \infty.$$

We allow variable radii $r_e\colon I_e \longrightarrow (0,\infty)$ for the edge neighbourhoods, and set $Y_e = [-1/2, 1/2]$ as fixed transversal manifold. Moreover, we need the additional uniformity assumptions

$$\sup \|r_e\|_\infty < \infty, \qquad \sup \|r_e^{-1}\|_\infty < \infty \qquad \text{and} \qquad \sup \|r_e'\|_\infty < \infty.$$

In the limit, we obtain the operator

$$(Df)_e = -\frac{1}{r_e}(r_e f_e)'$$

(see Remark 6.7.5) in $\mathsf{L}_2(G,r) = \bigoplus_e \mathsf{L}_2(I_e, r_e(s)\mathrm{d}s_e)$ with domain

$$\operatorname{dom} D = \Big\{ f \in \mathsf{H}^2_{\max}(G) \,\Big|\, f \text{ continuous}, \sum_{e\in E_v} r_e(v) \,\underline{\widetilde{f}}'_e(v) = 0 \ \forall\, v \in V \Big\}.$$

The transformation $f \mapsto \widetilde{f} = \{r_e^{1/2} f_e\}_e$ yields a unitary operator from $\mathsf{L}_2(G, r_e(s)\mathrm{d}s_e)$ onto the unweighted space $\mathsf{L}_2(G)$. If we assume in addition that $\sup_e \|r_e''\|_\infty < \infty$, then the operator D transforms into

$$(\widetilde{D}f)_e = -\widetilde{f}_e'' + Q_e \widetilde{f}_e,$$

with domain $\operatorname{dom} \widetilde{D}$ given by

$$\Big\{ \widetilde{f} \in \mathsf{H}^2_{\max}(G) \,\Big|\, \underline{\widetilde{f}}(v) = \widetilde{f}(v)\underline{p}(v),\ \frac{1}{|\underline{p}(v)|^2} \sum_{e\in E_v} p_e(v) \,\underline{\widetilde{f}}'_e(v) + L(v)\widetilde{f}(v) = 0 \Big\},$$

(see Proposition 5.5.3 and Remark 6.7.5). Here, $\underline{p}(v) = \{p_e(v)\}_{e\in E_v}$ and $p_e(v) = r_e(v)^{1/2}$. Moreover the potentials on the edges and vertices are given by

$$Q_e := q_e^2 + q_e', \qquad q_e := \frac{1}{2}(\log r_e)' \qquad \text{and} \qquad L(v) := \sum_{e\in E_v} q_e(v) p_e(v)^2.$$

Note that

$$Q_e = \Big(\frac{r_e'}{2r_e}\Big)^2 + \Big(\frac{r_e'}{2r_e}\Big)' \qquad \text{and} \qquad L(v) = \frac{1}{2}\sum_{e\in E_v} r_e'(v).$$

In particular, we obtain a *Schrödinger-type operator* with edge potentials Q_e and *vertex potentials* (or *delta-interactions*) of strengths $L(v)$. Recall that $\widetilde{D}$ is still a *non-negative* operator since D is a non-negative operator, although $L(v)$ and Q_e may achieve negative values.

Remark 7.1.7. In [P06] we showed that the manifold operator is close to the above weighted graph operator D with the weaker assumption of a lower bound on r_e only near the vertices. Therefore, we may also allow infinite edge neighbourhoods with *decaying* radius function r_e. Note that the Hilbert spaces $\mathsf{L}_2(I_e, r_e(s)\mathrm{d}s_e)$ and $\mathsf{L}_2(I_e)$ differ in general.

For example we can treat horn-like shapes as in [DS92]. If the spectrum of the Laplacian on the corresponding edge neighbourhood $X_{\varepsilon,e}$ is $[0,\infty)$, then the spectral convergence following from the quasi-partial isometry of the forms does not give new information. Nevertheless, one can show the convergence of resonances or embedded eigenvalues in this case (see Sect. 7.2). On the other hand, the spectrum on $X_{\varepsilon,e}$ is purely discrete, for example for radial functions decaying fast enough like $r_e(s) = \mathrm{e}^{-s^\beta}$, $\beta > 1$ (cf. [EH89, DS92]). An example of a horn-like end with infinitely many spectral gaps in the essential spectrum was constructed in [Lo01, Thm. 3]. In principle, these results can be recovered with our analysis.

7.2 Convergence of Resonances

In this section we want to show the convergence of resonances of non-compact graph-like manifolds. Resonances are defined as (complex) eigenvalues of a complexly dilated operator (see Definition 3.8.3). Our spaces here have a finite number of cylindrical ends, and therefore the spectrum of each of the operators is always $[0,\infty)$ (as a set). The method of complex dilation allows to 'reveal' eigenvalues embedded in the essential spectrum, as well as poles of a meromorphic continuation of the resolvent. For simplicity, both, the embedded eigenvalues and the poles, will be called *resonances*.

Before explaining some concrete examples, recall that we have proven a quite general result on the convergence of resonances in Sect. 4.9: Once, we are able to show that the quadratic forms on the interior spaces (e.g. on a graph and a shrinking manifold) are δ_ε-quasi unitarily equivalent for some $\delta_\varepsilon \to 0$ (and some other natural conditions), we may conclude the convergence of resonances of the associated non-compact spaces with half-lines and cylinders attached.

For the convergence of resonances, the case of a decoupling limit operator as in the slowly decaying or Dirichlet case is also of interest. Namely, we obtain the non-trivial information that for each discrete eigenvalue (e.g. of a decoupled component of the graph) there is a resonance converging to this eigenvalue.

Let us first define the corresponding boundary triples (for the general notion, see Sect. 3.4). Assume that G is a non-compact metric graph with *finitely many* exterior edges

$$E_{\mathrm{ext}} := \{\, e \in E \mid \ell_e = \infty \,\}.$$

For technical reasons, we use a cut into an *interior* and *exterior* part *away* from the vertices. In particular, we denote by G_{int} the subspace of G consisting of all interior edges and all vertices, and the segment $[0, 1]$ for each exterior edge $e \in E_{\mathrm{ext}}$ (i.e. we cut at $s = 1$ along each half-line). Consequently, the exterior part G_{ext} consists of $|E_{\mathrm{ext}}|$-many disjoint half-lines $[1, \infty)$. When associating the corresponding graph-like manifold, we do not consider the boundary points $1 \in I_e = [0, \infty)$, $e \in E_{\mathrm{ext}}$ as vertices (see Remark 2.2.17). Since the boundary $\partial G := G_{\mathrm{int}} \cap G_{\mathrm{ext}}$ can be labelled by E_{ext} we use

$$\mathscr{G} := \mathbb{C}^{E_{\mathrm{ext}}}$$

as associated boundary space. The boundary maps

$$\Gamma_\bullet \colon \mathscr{H}^1_\bullet \longrightarrow \mathscr{G} \qquad \text{and} \qquad \Gamma'_\bullet \colon \mathscr{W}^2_\bullet \longrightarrow \mathscr{G}$$

are defined by

$$(\Gamma_{\mathrm{int}} f)_e := f_e(1_-), \qquad (\Gamma_{\mathrm{ext}} f)_e := f_e(1_+), \tag{7.6a}$$

$$(\Gamma'_{\mathrm{int}} f)_e := f'_e(1_-), \qquad (\Gamma'_{\mathrm{ext}} f)_e := f'_e(1_+), \tag{7.6b}$$

where $1_\pm$ denotes the limit from the left/right. The spaces $\mathscr{H}^1_\bullet$ and $\mathscr{W}^2_\bullet$ will be given later on, but for the boundary triple, only the contribution of $f_e \in \mathsf{H}^1(\mathbb{R}_+)$ resp. $\mathsf{H}^2(\mathbb{R}_+)$ and the quadratic form near ∂G play a role. Both are the same in all our examples below.

Let X_ε be a graph-like manifold associated with the metric graph G. We use the corresponding split into an interior and exterior part $X_{\varepsilon,\mathrm{int}}$ and $X_{\varepsilon,\mathrm{ext}}$. Namely, $X_{\varepsilon,\mathrm{int}}$ consists of all vertex neighbourhoods $X_{\varepsilon,v}$ (resp. $\acute{X}_{\varepsilon,v}$ if we need the attached truncated cones), all interior edge neighbourhoods $X_{\varepsilon,e}$, $e \in E_{\mathrm{int}}$ and the exterior edge neighbourhood segment $[0,1] \times \varepsilon Y_e$, $e \in E_{\mathrm{ext}}$. Consequently, the exterior part $X_{\varepsilon,\mathrm{ext}}$ is the disjoint union of the half-infinite cylinders $[1, \infty) \times \varepsilon Y_e$, $e \in E_{\mathrm{ext}}$. Denote by $Y_\varepsilon := X_{\varepsilon,\mathrm{int}} \cap X_{\varepsilon,\mathrm{ext}}$ the common boundary. Note that Y_ε is isometric to the disjoint union of all transversal spaces $Y_{\varepsilon,e} = \varepsilon Y_e$. The boundary space on the manifold is

$$\mathscr{G}_\varepsilon := \mathsf{L}_2(Y_\varepsilon) = \bigoplus_{e \in E_{\mathrm{ext}}} \mathsf{L}_2(Y_{\varepsilon,e})$$

with boundary operators

$$\Gamma_{\varepsilon,\bullet} \colon \mathscr{H}^1_{\varepsilon,\bullet} \longrightarrow \mathscr{G} \qquad \text{and} \qquad \Gamma'_{\varepsilon,\bullet} \colon \mathscr{W}^2_{\varepsilon,\bullet} \longrightarrow \mathscr{G}$$

given by

$$(\Gamma_{\varepsilon,\mathrm{int}} u)_e := u_e(1_-, \cdot), \qquad (\Gamma_{\varepsilon,\mathrm{ext}} u)_e := u_e(1_+, \cdot), \tag{7.7a}$$

$$(\Gamma'_{\varepsilon,\mathrm{int}} u)_e := u'_e(1_-, \cdot), \qquad (\Gamma'_{\varepsilon,\mathrm{ext}} u)_e := u'_e(1_+, \cdot). \tag{7.7b}$$

Here, $u' = \partial_s u$ denotes the longitudinal derivative. Moreover,

$$\mathscr{H}^1_{\varepsilon,\bullet} := \mathsf{H}^1(X_{\varepsilon,\bullet}), \qquad \mathscr{W}^2_{\varepsilon,\bullet} := \mathsf{H}^2(X_{\varepsilon,\bullet}),$$

and the Sobolev-maximal operator is the Laplacian on $X_{\varepsilon,\bullet}$.

Remark 7.2.1. Note that we have shown in Proposition 3.8.8 that the definition of resonances is independent of where we make the cut into an interior and exterior part. For computational reasons it easier to split the graph at the initial *vertices* $\partial_- e$ of the exterior edges $e \in E_{\mathrm{ext}}$. The jump condition then becomes part of the vertex condition of the metric graph.

Let us now assume that the interior graph is compact, i.e. that

$$|E_{\mathrm{int}}| < \infty,$$

where $E_{\mathrm{int}} := E \setminus E_{\mathrm{ext}}$, although some of the results remain true for non-compact interior spaces.

7.2.1 Fast Decaying Vertex Volume

We start again with the fast decaying case. For the notation of the spaces we refer to Sect. 7.1.1. Let us first define the dilated coupled boundary triple together with the complexly dilated operators H^θ and H_ε^θ. On the interior part of the graph, we set

$$\mathscr{H}^1_{\mathrm{int}} := \mathsf{H}^1_p(G_\bullet) \qquad \text{and} \qquad \mathscr{W}^2_{\mathrm{int}} := \Big\{ f \in \mathsf{H}^2_p(G) \,\Big|\, \sum_{e \in E_v} p_e \underline{\widetilde{f}}'_e(v) = 0 \ \forall v \in V \Big\}.$$

Recall the notation

$$\mathsf{H}^k_p(G_{\mathrm{int}}) := \big\{ f \in \mathsf{H}^k_{\max}(G_{\mathrm{int}}) \,\big|\, \forall v \in V \ \exists f(v) \in \mathbb{C} \colon \underline{f}(v) = f(v) p(v) \big\} \qquad (7.8)$$

and

$$p(v) = \{p_e\}_{e \in E_v}, \qquad p_e = (\operatorname{vol} Y_e)^{1/2}, \qquad \underline{f}(v) = \{f_e(v)\}_{e \in E_v}.$$

On the exterior part, we simply have

$$\mathscr{H}^1_{\mathrm{ext}} := \mathsf{H}^1(G_{\mathrm{ext}}) \quad \text{and} \quad \mathscr{W}^2_{\mathrm{ext}} := \mathsf{H}^2(G_{\mathrm{ext}}), \quad \text{where} \quad \mathsf{H}^k(G_{\mathrm{ext}}) = \bigoplus_{e \in E_{\mathrm{ext}}} \mathsf{H}^k(\mathbb{R}_+).$$

The Sobolev-maximal operator $\check{\Delta}_\bullet$ acts as $(\check{\Delta}_\bullet f)_e = -f_e''$ for $f \in \mathscr{W}^2_\bullet$. We denote by $\mathfrak{d}_\bullet$ the quadratic form given by

$$\mathfrak{d}_\bullet(f) := \sum_{e \in E_\bullet} \|f_e'\|^2, \qquad \operatorname{dom} \mathfrak{d}_\bullet := \mathsf{H}^1_p(G_\bullet).$$

The transversal operators resp. quadratic forms on the cylindrical ends are

$$K := 0, \qquad \mathfrak{k} = 0, \qquad K_\varepsilon := \bigoplus_{e \in E_{\mathrm{ext}}} \Delta_{Y_{\varepsilon,e}}, \qquad \mathfrak{k}_\varepsilon(\psi) = \sum_{e \in E_{\mathrm{ext}}} \|\mathrm{d}_{Y_e} \psi_e\|^2_{Y_{\varepsilon,e}}$$

with $\operatorname{dom} \mathfrak{k}_\varepsilon = \bigoplus_{e \in E_{\mathrm{ext}}} \mathsf{H}^1(Y_{\varepsilon,e})$. For the notion of (elliptic) boundary triples and the coupled versions we refer to Sects. 3.4–3.6.

Proposition 7.2.2. *Assume that the interior metric graph is compact with finitely many attached half-lines (i.e. $|E| < \infty$). Then the following assertions are true:*

1. *The triples $(\Gamma_{\mathrm{int}}, \Gamma'_{\mathrm{int}}, \mathscr{G})$ and $(\Gamma_{\varepsilon,\mathrm{int}}, \Gamma'_{\varepsilon,\mathrm{int}}, \mathscr{G}_\varepsilon)$ are elliptic boundary triples associated with the quadratic forms $\mathfrak{d}_{\mathrm{int}}$ and $\mathfrak{d}_{\varepsilon,\mathrm{int}}$, respectively. The boundary triple $(\Gamma_{\mathrm{int}}, \Gamma'_{\mathrm{int}}, \mathscr{G})$ is bounded.*
2. *The triples $(\Gamma_{\mathrm{ext}}, -\Gamma'_{\mathrm{ext}}, \mathscr{G})$ and $(\Gamma_{\varepsilon,\mathrm{ext}}, -\Gamma'_{\varepsilon,\mathrm{ext}}, \mathscr{G}_\varepsilon)$ are elliptic boundary triples with transversal operators $K = 0$ and K_ε, respectively. In addition, the triple $(\Gamma_{\mathrm{ext}}, \Gamma'_{\mathrm{ext}}, \mathscr{G})$ is bounded. Finally, K_ε has purely discrete spectrum given by*
$$\sigma(K_\varepsilon) = \left\{ \frac{\lambda_k(Y_e)}{\varepsilon^2} \,\middle|\, k \in \mathbb{N},\ e \in E_{\mathrm{ext}} \right\},$$
where $\lambda_k(Y_e)$ is the k-th eigenvalue of Y_e. Moreover, the eigenvalue 0 is of multiplicity $|E_{\mathrm{ext}}|$.
3. *The coupled boundary triples $(\Gamma, \Gamma', \mathscr{G})$ and $(\Gamma_\varepsilon, \Gamma'_\varepsilon, \mathscr{G}_\varepsilon)$ are elliptic.*

Proof. The ellipticity of the two interior boundary triples follow from Proposition 2.2.18 for the metric graph and from Theorems 3.4.39 and 3.4.40 for the manifold. For the exterior boundary triples, the ellipticity follows from Theorem 3.5.6. The ellipticity of the coupled boundary triples is shown in Propositions 3.6.13 and 3.6.11. The assertion on the spectrum of K_ε is a consequence of the fact that the boundary $Y_\varepsilon = \dot{\bigcup}_{e \in E_{\mathrm{ext}}} Y_{\varepsilon,e}$ is compact as finite union of compact manifolds. Note that $\lambda_1(Y_e) = 0$ for all $e \in E_{\mathrm{ext}}$. □

For convenience, we recall the notion of a complexly dilated operator, i.e. the Neumann operator of the associated dilated coupled boundary triple (see Sects. 3.6–3.7). On the metric graph, the dilated coupled operator acts as
$$(H^\theta f)_e = -f_e'', \qquad (H^\theta f)_e = -\mathrm{e}^{-2\theta} f_e''$$
for interior resp. exterior edges. The vertex conditions, especially the jump conditions at the boundary $G_{\mathrm{int}} \cap G_{\mathrm{ext}}$, are encoded in the domain
$$\operatorname{dom} H^\theta := \left\{ f \in \mathscr{W}^2_{\mathrm{int}} \oplus \mathscr{W}^2_{\mathrm{ext}} \,\middle|\, \Gamma_{\mathrm{ext}} f = \mathrm{e}^{\theta/2} \Gamma_{\mathrm{int}} f,\ \Gamma'_{\mathrm{ext}} f = \mathrm{e}^{\theta/2} \Gamma'_{\mathrm{int}} f \right\}.$$
On the manifold, we have
$$(H^\theta_\varepsilon u)_{\mathrm{int}} = \Delta_{X_{\varepsilon,\mathrm{int}}} u, \qquad (H^\theta_\varepsilon u)_e = -\mathrm{e}^{-2\theta} u_e'' + (\mathrm{id} \otimes \Delta_{Y_{\varepsilon,e}}) u_e$$
on the interior part resp. on exterior edges. The domain contains the jump conditions at the common boundary $Y_\varepsilon = \overline{X}_{\varepsilon,\mathrm{int}} \cap \overline{X}_{\varepsilon,\mathrm{ext}}$, i.e.,

$$\operatorname{dom} H_\varepsilon^\theta := \{ u \in \mathsf{H}^2(X_{\varepsilon,\mathrm{int}}) \oplus \mathsf{H}^2(X_{\varepsilon,\mathrm{ext}}) \mid \Gamma_{\varepsilon,\mathrm{ext}} u = \mathrm{e}^{\theta/2} \Gamma_{\varepsilon,\mathrm{int}} u,\ \Gamma'_{\varepsilon,\mathrm{ext}} u = \mathrm{e}^{\theta/2} \Gamma'_{\varepsilon,\mathrm{int}} u \}.$$

Let us summarise the facts about the graph operator $\Delta_G = H^0$ and its dilated counterpart H^θ (cf. Theorems 3.7.14 and 3.8.2):

Theorem 7.2.3. *Assume that the interior metric graph is compact with finitely many attached half-lines (i.e. $|E| < \infty$). Then the following assertions are true:*

1. *The complexly dilated coupled operator H^θ has spectrum in the sector $\Sigma_\vartheta = \{ z \in \mathbb{C} \mid |\arg z| \le \vartheta \}$, the family $\{H^\theta\}_\theta$ is self-adjoint and depends analytically on $\theta \in S_\vartheta = \mathbb{R} + (-\vartheta/2, \vartheta/2)\mathrm{i}$.*
2. *The essential spectrum of H^θ is given by*

$$\sigma_{\mathrm{ess}}(H^\theta) = \mathrm{e}^{-2\theta}[0,\infty)$$

   with multiplicity $|E_{\mathrm{ext}}|$.
3. *The spectrum $\sigma(H^\theta)$ depends only on $\operatorname{Im}\theta$ and $\sigma(H^{\overline{\theta}}) = \overline{\sigma(H^\theta)}$ (complex conjugation).*
4. *The discrete spectrum $\sigma_{\mathrm{disc}}(H^\theta)$ is locally constant in θ, i.e. if $0 < \operatorname{Im}\theta_1 \le \operatorname{Im}\theta_2 < \vartheta/2$ then $\sigma_{\mathrm{disc}}(H^{\theta_1}) \subset \sigma_{\mathrm{disc}}(H^{\theta_2})$.*
5. *For $\theta \notin \mathbb{R}$ we have $\sigma(H^\theta) \cap [0,\infty) = \sigma_{\mathrm{p}}(\Delta_G)$, where $\sigma_{\mathrm{p}}(\Delta_G)$ denotes the set of eigenvalues of Δ_G, i.e. the eigenvalues of Δ_G embedded in the continuous spectrum are unveiled.*
6. *The singular continuous spectrum of H is empty.*

We have exactly the same statement for the manifold operators $\Delta_\varepsilon = H_\varepsilon^0$ and H_ε^θ, except that

$$\sigma_{\mathrm{ess}}(H_\varepsilon^\theta) = \mathrm{e}^{-2\theta}[0,\infty) \cup \bigcup_{k=2,3,\dots;\ e \in E_{\mathrm{ext}}} \frac{\lambda_k(Y_e)}{\varepsilon^2}\mathrm{e}^{-2\theta}[0,\infty).$$

In particular, only the first branch (of multiplicity $|E_{\mathrm{ext}}|$) does not tend to ∞ as $\varepsilon \to 0$. Recall that a *resonance* of Δ_G is a discrete eigenvalue of H^θ for $|\operatorname{Im}\theta|$ large enough (see Definition 3.8.3). Note that the definition does not depend on θ by (4) of the previous theorem. Similarly, we define a resonance of Δ_{X_ε}.

Let the *boundary identification operator* be defined by

$$I_\varepsilon : \mathscr{G} = \mathbb{C}^{E_{\mathrm{ext}}} \longrightarrow \mathscr{G}_\varepsilon, \qquad (I_\varepsilon F)_e := \varepsilon^{-m/2} F_e \mathbb{1}_e,$$

i.e. we extend the single value F_e to a constant normalised function on Y_e. The adjoint is given by

$$(I_\varepsilon^* \varphi)_e = \varepsilon^{m/2} (\operatorname{vol} Y_e)^{1/2} f_e \varphi.$$

In particular, we obtain a closeness result for the boundary triples $(\Gamma_{\mathrm{int}}, \Gamma'_{\mathrm{int}}, \mathscr{G})$ and $(\Gamma_{\varepsilon,\mathrm{int}}, \Gamma'_{\varepsilon,\mathrm{int}}, \mathscr{G}_\varepsilon)$, using the first order identification operators J_ε^1 and $J_\varepsilon'^1$ defined on each star graph by (6.32) and (6.34). We assume that the cut-off function χ_e used in the definition of $J_\varepsilon'^1$ on the exterior edge e has support in $[0, 1]$. In particular, we then have

$$(J_\varepsilon^1 f)_e(y) = \varepsilon^{-m/2}(\operatorname{vol} Y_e)^{-1/2} f_e(1) \qquad \text{and} \qquad (J_\varepsilon'^1 u)_e = \varepsilon^{m/2}(\operatorname{vol} Y_e)^{1/2} \underline{f}_e u(1)$$

for $e \in E_{\mathrm{ext}}$. For the notion of δ-closeness of two boundary maps Γ_{int} and $\Gamma_{\varepsilon,\mathrm{int}}$ we refer to Definition 4.8.1.

Proposition 7.2.4. *The boundary maps Γ_{int} and $\Gamma_{\varepsilon,\mathrm{int}}$ associated with the forms $\mathfrak{d}_{\mathrm{int}}$ and $\mathfrak{d}_{\varepsilon,\mathrm{int}}$ are 0-close with the identification operators J_ε^1, $J_\varepsilon'^1$ and the boundary identification operator I_ε. Moreover, I_ε is $\hat{\delta}_\varepsilon$-partial isometrically equivalent of order 1 w.r.t. the forms $\mathfrak{k} = 0$ and $\mathfrak{k}_\varepsilon$, where $\hat{\delta}_\varepsilon = \varepsilon(\min_{e \in E_{\mathrm{ext}}} \lambda_2(Y_e))^{-1/2}$. Finally,* $\operatorname{ran} I_\varepsilon \subset \operatorname{dom} K_\varepsilon$ *and* $K_\varepsilon I_\varepsilon = I_\varepsilon K (= 0)$.

Proof. The 0-closeness follows from the definitions, namely,

$$(\Gamma_{\varepsilon,\mathrm{int}} J_\varepsilon^1 f)_e = \varepsilon^{-m/2} f_e(0) \mathbb{1}_e = (I_\varepsilon \Gamma_{\mathrm{int}} f)_e,$$
$$(\Gamma_{\mathrm{int}} J_{\varepsilon,\mathrm{int}}'^1 u)_e = \varepsilon^{m/2}(\operatorname{vol} Y_e)^{1/2} \underline{f}_e u(1) = I_\varepsilon^* \Gamma_{\varepsilon,\mathrm{int}} u.$$

Moreover, $I_\varepsilon^* I_\varepsilon F = F$, and by applying Proposition 5.1.1 on each component we have

$$\|\varphi - I_\varepsilon I_\varepsilon^* \varphi\|_{Y_\varepsilon}^2 = \varepsilon^m \sum_e \|\varphi_e - \langle \mathbb{1}_e, \varphi_e \rangle \mathbb{1}_e\|_Y^2 \le \frac{\varepsilon^2}{\min_{e \in E_{\mathrm{ext}}} \lambda_2(Y_e)} \big(\mathfrak{k}_\varepsilon(\varphi) + \|\varphi\|_{Y_\varepsilon}^2\big)$$

for $\varphi \in \operatorname{dom} \mathfrak{k}_\varepsilon = \mathsf{H}^1(Y_\varepsilon)$. Clearly, $\operatorname{ran} I_\varepsilon \subset \operatorname{dom} K_\varepsilon$ since the range consists of locally constant functions only, and K_ε is the Laplacian. Finally, the intertwining property $K_\varepsilon I_\varepsilon = I_\varepsilon K = 0$ of I_ε is obvious. □

Theorem 7.2.5. *Let $\vartheta \in (0, \pi/2)$ and $\theta \in S_\vartheta$, i.e. $|\operatorname{Im} \theta| < \vartheta/2$. Assume that the interior metric graph G_{int} is compact and that $|E_{\mathrm{ext}}| < \infty$. Then H^θ and H_ε^θ are δ_ε-partial isometrically equivalent, where $\delta_\varepsilon = \mathrm{O}(\varepsilon^{(\alpha d - m)/2})$. The error depends only on ϑ, on* $\operatorname{Re} \theta$ *and on the constants in* (7.1).

Proof. We apply Theorem 4.9.8 and check the necessary assumptions (1)–(5).

The boundary space $\mathscr{G}$ is finite-dimensional and therefore elliptic. Moreover, the coupled dilated boundary triple on the manifold is elliptic, see Proposition 7.2.2 (3).

For Theorem 4.9.8 (1), note the following. The partial isometry of the interior forms was shown in Theorem 7.1.2, together with the error estimates depending only on the constants mentioned in (7.1).

For (2)–(3) note the following. The closeness of the boundary maps and the intertwining condition $K_\varepsilon I_\varepsilon = I_\varepsilon K$ are fulfilled by Proposition 7.2.4. The last item (4) is obvious since $K = 0$. □

As a consequence, we obtain from Theorem 4.9.12:

Corollary 7.2.6. *Let λ be a resonance of Δ_G (i.e. a discrete eigenvalue of H^θ) with multiplicity μ. Then there exist μ resonances $\lambda_j(\varepsilon)$ of Δ_{X_ε} (i.e. discrete eigenvalues of H^θ_ε), $j = 1, \dots, \mu$, not necessarily mutually distinct, such that*

$$\lambda_j(\varepsilon) \to \lambda \qquad \textit{as} \qquad \varepsilon \to 0.$$

If, in addition, $\mu = 1$, i.e. if λ is a simple eigenvalue with normalised eigenvector ψ, then for each $\varepsilon > 0$ small enough, there exist a normalised eigenvector ψ_ε of H^θ_ε such that

$$\|J_\varepsilon \psi - \psi_\varepsilon\| = \mathrm{O}(\varepsilon^{(\alpha d - m)/2}) \qquad \textit{and} \qquad \|J^*_\varepsilon \psi_\varepsilon - \psi\| = \mathrm{O}(\varepsilon^{(\alpha d - m)/2}).$$

Finally, the convergence depends only on ϑ, on $\operatorname{Re}\theta$, on constants in (7.1) and on H^θ (more precisely on $\|(H^\theta - z)^{-1}\|$ for $z \in \partial D$, where $D \subset \mathbb{C}$ is chosen such that $\overline{D} \cap \sigma(H^\theta) = \{\lambda\}$).

We showed indeed a stronger result, namely the quasi-isometric equivalence for the associated sesquilinear forms, see Theorem 4.9.7. Moreover, other convergence results like the convergence of the resolvents or the spectral projections of the dilated operators follow (see Sect. 4.5).

7.2.2 Slowly Decaying Vertex Volume

Let us now describe the case of slowly decaying vertex volumes. For the notation of the spaces we refer to Sect. 7.1.2. Recall that in the slowly decaying case, the limit operator on the graph *decouples* (see Theorem 7.1.3). In order to obtain a *non-decoupled* interior graph, let us associate the slowly decaying vertex neighbourhoods $X_{\varepsilon,v}$ with scaling exponent $0 < \alpha < (d-1)/d$ only to the *coupling* vertices $\partial_- e$, $e \in E_{\mathrm{ext}}$. On the inner vertices $\mathring{V} := V \setminus \{\, \partial_- e \mid e \in E_{\mathrm{ext}} \,\}$, we assume a fast decaying vertex neighbourhood (for simplicity with scaling ε, i.e. $\alpha = 1$).

We only describe the notion differing from the fast decaying case, namely the limit quadratic form and operator. On the graph, the spaces are given by

$$\mathscr{H}_{\mathrm{int}} := \mathsf{L}_2(G_{\mathrm{int}}) \oplus \mathbb{C}^{E_{\mathrm{ext}}}, \qquad \mathscr{H}^1_{\mathrm{int}} := \mathsf{H}^1_p(G_{\mathrm{int}}) \oplus \mathbb{C}^{E_{\mathrm{ext}}},$$

$$\mathscr{W}^2_{\mathrm{int}} := \Big\{\, f \in \mathsf{H}^2_p(G_{\mathrm{int}}) \;\Big|\; \sum_{e \in E_v} p_e \underline{f}'_e(v) = 0 \quad \forall v \in \mathring{V} \,\Big\} \oplus \mathbb{C}^{E_{\mathrm{ext}}}$$

(see (7.8) for the definition of $\mathsf{H}^k_p(G_{\mathrm{int}})$), where we have Dirichlet vertex conditions at the *coupling vertices* only, i.e.

$$p(v) = 0, \qquad v \in V_{\mathrm{ext}} := \{\, \partial_- e \mid e \in E_{\mathrm{ext}} \,\}.$$

Moreover, the boundary maps on the interior graph are the same as in (7.6), replacing the symbol f by $\hat{f}$. The quadratic form $\mathfrak{d}_{\mathrm{int}}$ on the graph is given by

$$\mathfrak{d}_{\mathrm{int}}(\hat{f}) := \sum_{e \in E_{\mathrm{int}}} \|f_e'\|^2, \qquad \operatorname{dom} \mathfrak{d}_{\mathrm{int}} := \mathscr{H}^1_{\mathrm{int}}.$$

The corresponding Laplacian has the form $\Delta_{\mathrm{int}} = \Delta^{\mathrm{D}}_{G_{\mathrm{int}}} \oplus 0$. Here, $\Delta^{\mathrm{D}}_{G_{\mathrm{int}}}$ is the metric graph Laplacian associated with the quantum graph $(G_{\mathrm{int}}, \bigoplus_{v \in V} \mathbb{C}p(v)$ (see Sect. 2.2.2), fulfilling Dirichlet conditions on the connecting vertices $v \in V_{\mathrm{ext}}$.

The boundary triple on the exterior graph part G_{ext} is defined as in Sect. 7.2.1. Note that the complexly dilated operator on the graph is in this case *decoupled*, i.e. H^θ acts as before, but the domain is a direct sum. In particular, the dilated operator is the direct sum of $\Delta^{\mathrm{D}}_{G_{\mathrm{int}}}$, 0 (on $\mathbb{C}^{E_{\mathrm{ext}}}$) and $|E_{\mathrm{ext}}|$ copies of $-\mathrm{e}^{-2\theta}\partial_{ss}$, the dilated Laplacian on $\mathbb{R}_+$ with Dirichlet boundary condition at 0.

The solely discrete eigenvalues of H^θ are the ones from the interior operator $\Delta^{\mathrm{D}}_{G_{\mathrm{int}}}$ with Dirichlet conditions on V_{ext}. Moreover, 0 is an eigenvalue of multiplicity $|E_{\mathrm{ext}}|$, which is *not* revealed as $\operatorname{Im}\theta$ increases.

Theorem 7.2.7. *Let $\vartheta \in (0, \pi/2)$ and $\theta \in S_\vartheta$, i.e. $|\operatorname{Im}\theta| < \vartheta/2$. Assume that the interior metric graph G_{int} is compact and that $|E_{\mathrm{ext}}| < \infty$. Then H^θ and H^θ_ε are δ_ε-partial isometrically equivalent, where $\delta_\varepsilon = \mathrm{O}(\varepsilon^{\min\{(m-\alpha d)/2, \alpha\}})$. The error depends only on ϑ, on* $\operatorname{Re}\theta$*, and on the constants in* (7.2).

The proof is very similar to the proof of Theorem 7.2.5, except that for Theorem 4.9.8 (1), we use Theorem 7.1.3 together with the error estimates, depending only on the constants mentioned in (7.2).

Corollary 7.2.8. *Let $\lambda \in (0, \infty)$ be an eigenvalue of $\Delta^{\mathrm{D}}_{G_{\mathrm{int}}}$ with multiplicity μ. Then there exist μ resonances $\lambda_j(\varepsilon)$ of Δ_{X_ε} (i.e. discrete eigenvalues of H^θ_ε), $j = 1, \dots, \mu$, not necessarily mutually distinct, such that*

$$\lambda_j(\varepsilon) \to \lambda \qquad \textit{as} \qquad \varepsilon \to 0.$$

If, in addition, $\mu = 1$, i.e. if λ is a simple eigenvalue with normalised eigenvector ψ, then there exist normalised eigenvectors ψ_ε of H^θ_ε such that

$$\|J_\varepsilon \psi - \psi_\varepsilon\| = \mathrm{O}(\varepsilon^{\min\{(m-\alpha d)/2, \alpha\}}) \quad \textit{and} \quad \|J^*_\varepsilon \psi_\varepsilon - \psi\| = \mathrm{O}(\varepsilon^{\min\{(m-\alpha d)/2, \alpha\}}).$$

Finally, the convergence depend only on ϑ, on $\operatorname{Re}\theta$ *and $\Delta^{\mathrm{D}}_{G_{\mathrm{int}}}$ and on the constants in* (7.2).

7.2.3 The Borderline Case

We now describe the borderline case $\alpha = (d-1)/d$. Namely we assume that to each vertex of the graph we associate a vertex neighbourhood $X_{\varepsilon,v}$ with volume of the same order as the transversal manifold as explained in Sects. 6.8 and 6.10. For the notation of the spaces we refer to Sect. 7.1.3. On the manifold, we use the same setting as before. On the graph, the spaces are given by

$$\mathscr{H}_{\mathrm{int}} := \mathsf{L}_2(G_{\mathrm{int}}) \oplus \mathbb{C}^V, \quad \mathscr{H}^1_{\mathrm{int}} := \{\, \hat{f} \in \mathsf{H}^1_p(G_{\mathrm{int}}) \oplus \mathbb{C}^V \mid f(v) = L(v)F(v)\ \forall v \in V \,\},$$

$$\mathscr{W}^2_{\mathrm{int}} := \{\, \hat{f} \in \mathsf{H}^2_p(G_{\mathrm{int}}) \oplus \mathbb{C}^V \mid f(v) = L(v)F(v)\ \forall v \in V \,\},$$

where $\hat{f} = (f, F)$ and $L(v) = (\operatorname{vol} X_v)^{-1/2}$. The quadratic form $\mathfrak{d}_{\mathrm{int}}$ on the graph is given by

$$\mathfrak{d}_{\mathrm{int}}(\hat{f}) := \sum_{e \in E_{\mathrm{int}}} \|f'_e\|^2, \qquad \operatorname{dom} \mathfrak{d}_{\mathrm{int}} := \mathscr{H}^1_{\mathrm{int}}.$$

Again, both dilated operators, on the graph and on the manifold, fulfil the assertions of Theorem 7.2.3. The main result in the borderline case is the following:

Theorem 7.2.9. *Let $\vartheta \in (0, \pi/2)$ and $\theta \in S_\vartheta$, i.e. $|\mathrm{Im}\,\theta| < \vartheta/2$. Assume that the interior metric graph G_{int} is compact, and that $|E_{\mathrm{ext}}| < \infty$. Then H^θ and H^θ_ε are δ_ε-partial isometrically equivalent, where $\delta_\varepsilon = \mathrm{O}(\varepsilon^\alpha)$. The error depends only on ϑ, on* Re θ, *and on the constants in* (7.3).

The proof is again similar to the proof of Theorem 7.2.5, using now Theorem 7.1.5.

Denote by Δ the operator associated with the form $\mathfrak{d}$ (cf. Sect. 7.1.3). Not that this operator is in general not decoupled.

Corollary 7.2.10. *Let λ be a resonance of Δ (i.e. a discrete eigenvalue of H^θ) with multiplicity μ. Then there exist μ resonances $\lambda_j(\varepsilon)$ of Δ_{X_ε} (i.e. discrete eigenvalues of H^θ_ε), $j = 1, \dots, \mu$, not necessarily mutually distinct, such that*

$$\lambda_j(\varepsilon) \to \lambda \qquad \text{as} \qquad \varepsilon \to 0.$$

If, in addition, $\mu = 1$, i.e. if λ is a simple eigenvalue with normalised eigenvector ψ, then there exist normalised eigenvectors ψ_ε of H^θ_ε such that

$$\|J_\varepsilon \psi - \psi_\varepsilon\| = \mathrm{O}(\varepsilon^\alpha) \qquad \text{and} \qquad \|J^*_\varepsilon \psi_\varepsilon - \psi\| = \mathrm{O}(\varepsilon^\alpha).$$

Finally, the convergence depend only on ϑ, on Re θ, *on the constants in* (7.3) *and on H^θ (more precisely on $\|(H^\theta - z)^{-1}\|$, $z \in \partial D$, $\overline{D} \cap \sigma(H^\theta) = \{\lambda\}$).*

7.2.4 The Dirichlet Decoupled Case

We also obtain a closeness result for operators with Dirichlet conditions on $\partial_0 X_\varepsilon$. For the notation we refer to Sect. 7.1.4. We recall that

$$\lambda_1 := \inf_{e \in E} \lambda_1^{\mathrm{D}}(Y_e)$$

denotes the infimum of all lowest transversal eigenmodes. We have already seen on a star graph in Sect. 6.11 that only those edges survive in the limit operator which have the lowest eigenmode equal to λ_1 (the "thick" edges), namely

$$E_0 := \{ e \in E \mid \lambda_1^{\mathrm{D}}(Y_e) = \lambda_1 \}.$$

i.e. E_0 is the set of edges for which the infimum is *achieved*. We denote by

$$E_{0,\mathrm{int}} := E_{\mathrm{int}} \cap E_0 \qquad \text{and} \qquad E_{0,\mathrm{ext}} := E_{\mathrm{ext}} \cap E_0$$

the corresponding thick interior ($\ell_e < \infty$) and exterior ($\ell_e = \infty$) edges. In order to obtain non-trivial results, we assume that $E_{0,\mathrm{int}}$ and $E_{0,\mathrm{ext}}$ are both non-empty. Let G_0 be the metric subgraph generated by all thick edges E_0. We use the decomposition into an interior and exterior part as before on the exterior edges at $s = 1$ and denote the corresponding spaces by $G_{0,\mathrm{int}}$ and $G_{0,\mathrm{ext}}$.

On the graph, the spaces are given by

$$\mathscr{H}_{\mathrm{int}} := \mathsf{L}_2(G_{0,\mathrm{int}}), \qquad \mathscr{H}^1_{\mathrm{int}} := \mathsf{H}^1_0(G_{0,\mathrm{int}}), \qquad \mathscr{W}^2_{\mathrm{int}} := \mathsf{H}^2_0(G_{0,\mathrm{int}}),$$

where the Sobolev spaces $\mathsf{H}^k_0(G_{0,\mathrm{int}}) = \bigoplus_{e \in E_{0,\mathrm{int}}} \{ f \in \mathsf{H}^k(I_e) \mid f\restriction_{\partial I_e} = 0 \}$ are completely decoupled. The quadratic form $\mathfrak{d}_{\mathrm{int}}$ on the graph here is given by

$$\mathfrak{d}_{\mathrm{int}}(f) := \sum_{e \in E_{0,\mathrm{int}}} \|f_e'\|^2, \qquad \operatorname{dom} \mathfrak{d}_{\mathrm{int}} := \mathscr{H}^1_{\mathrm{int}},$$

and also decoupled. The corresponding objects on the exterior part are defined as in the previous cases. For the manifold we set

$$\mathscr{H}_{\varepsilon,\bullet} := \mathsf{L}_2(X_{\varepsilon,\bullet}), \qquad \mathscr{H}^1_{\varepsilon,\bullet} := \mathsf{H}^1_{\partial_0 X_\varepsilon}(X_{\varepsilon,\bullet}), \qquad \mathscr{W}^2_{\varepsilon,\bullet} := \mathsf{H}^2_{\partial_0 X_\varepsilon}(X_{\varepsilon,\bullet}),$$

where functions in $\mathsf{H}^1_{\partial_0 X_\varepsilon}(X_{\varepsilon,\bullet})$ vanish on $\partial_0 X_\varepsilon = \partial X_\varepsilon \cap X_{\varepsilon,\bullet}$, i.e. the boundary part coming from the manifold boundary ∂X_ε. The forms and operators are given by

$$\mathfrak{d}_{\varepsilon,\bullet}(u) := \|du\|^2_{X_{\varepsilon,\bullet}} - \frac{\lambda_1}{\varepsilon^2}\|u\|^2_{X_{\varepsilon,\bullet}}, \qquad \operatorname{dom} \mathfrak{d}_{\varepsilon,\bullet} := \mathscr{H}^1_{\varepsilon,\bullet},$$

$$\Delta_{\varepsilon,\bullet} = \Delta^{\mathrm{D}}_{X_{\varepsilon,\bullet}} - \frac{\lambda_1}{\varepsilon^2}, \qquad \lambda_1 := \min_{e \in E} \lambda_1^{\mathrm{D}}(Y_e).$$

The boundary spaces for the graph and the manifold are given by

$$\mathscr{G} := \mathbb{C}^{E_{0,\mathrm{ext}}}, \qquad \mathscr{G}_\varepsilon := \bigoplus_{e \in E_{\mathrm{ext}}} \mathsf{L}_2(Y_{\varepsilon,e}).$$

The boundary maps on the graph and manifolds are defined as before by the evaluation at $s = 1$ on each thick exterior edge. Moreover, the *boundary identification operator* in the Dirichlet case is given by

$$I_\varepsilon : \mathscr{G} \longrightarrow \mathscr{G}_\varepsilon, \quad (I_\varepsilon F)_e := \begin{cases} F_e \varphi_{\varepsilon,e}, & \text{if } e \in E_{0,\mathrm{ext}}, \\ 0, & \text{otherwise}, \end{cases}$$

i.e. we choose the F_e-multiple of the first Dirichlet eigenfunction on the thick exterior edges. The adjoint is given by $(I^*\psi)_e = \varepsilon^{m/2}\langle \varphi_e, \psi_e \rangle$ for $e \in E_{0,\mathrm{ext}}$. The transversal operators here are

$$K = 0 \qquad \text{and} \qquad K_\varepsilon = \Delta^{\mathrm{D}}_{Y_\varepsilon} - \frac{\lambda_1}{\varepsilon^2} = \bigoplus_{e \in E_{\mathrm{ext}}} \Big(\Delta^{\mathrm{D}}_{Y_{\varepsilon,e}} - \frac{\lambda_1}{\varepsilon^2} \Big)$$

with associated quadratic forms $\mathfrak{k} = 0$ on $\mathscr{G}$ and $\widetilde{\mathfrak{k}}(\psi) = \|d\psi\|^2_{Y_\varepsilon} - \varepsilon^{-2}\lambda_1 \|\psi\|^2_{Y_\varepsilon}$. The spectrum of the associated operator K_ε has again purely discrete spectrum given by

$$\sigma(K_\varepsilon) = \Big\{ \frac{\lambda^{\mathrm{D}}_k(Y_e) - \lambda_1}{\varepsilon^2} \,\Big|\, k \in \mathbb{N},\ e \in E_{0,\mathrm{ext}} \Big\},$$

where $\lambda^{\mathrm{D}}_k(Y_e)$ is the k-th Dirichlet eigenvalue of Y_e. Moreover, the eigenvalue 0 is of multiplicity $|E_{0,\mathrm{ext}}|$.

As in Theorem 6.11.7, we obtain a closeness result for the boundary triples (the identification operators J^1_ε and J'^1_ε were defined in Sect. 6.11.2):

Proposition 7.2.11. *The boundary maps Γ_{int} and $\Gamma_{\varepsilon,\mathrm{int}}$ associated with the forms $\mathfrak{d}_{\mathrm{int}}$ and $\mathfrak{d}_{\varepsilon,\mathrm{int}}$ are 0-close with the identification operators J^1_ε, J'^1_ε and the boundary identification operator I_ε.*

Moreover, I_ε is $\hat{\delta}_\varepsilon$-partial isometrically equivalent of order 1 w.r.t. the forms $\mathfrak{k} = 0$ and $\mathfrak{k}_\varepsilon$, where $\hat{\delta}_\varepsilon = \varepsilon(\min_{e \in E_{0,\mathrm{ext}}} \lambda_{2,e} - \lambda_1)^{-1/2}$, where $\lambda_{2,e}$ was defined in (6.71b). Finally, $K_\varepsilon I_\varepsilon = I_\varepsilon K (= 0)$.

Both dilated operators, on the graph and on the manifold, fulfil the assertions of Theorem 7.2.3. As in the slowly decaying (Neumann) case, the complexly dilated operator on the graph is *decoupled*, i.e. H^θ is unitarily equivalent with the direct sum with respect to a decomposition into the metric graph generated by all interior edges and the exterior *entire* edges.

In particular, the solely discrete eigenvalues of H^θ are the ones from the interior operator $\Delta^{\mathrm{D}}_{G_{\mathrm{int}}}$ with Dirichlet conditions on V_{ext}. Moreover, 0 is an eigenvalue of multiplicity $|E_{\mathrm{ext}}|$, which is *not* revealed as $\operatorname{Im}\theta$ increases.

The main result in the Dirichlet decoupled case is the following:

Theorem 7.2.12. *Let $\vartheta \in (0, \pi/2)$ and $\theta \in S_\vartheta$, i.e. $|\operatorname{Im}\theta| < \vartheta/2$. Assume that the interior metric graph G_{int} is compact, and that $|E_{\mathrm{ext}}| < \infty$. Then H^θ and H^θ_ε are*

δ_ε-partial isometrically equivalent, where $\delta_\varepsilon = \mathrm{O}(\varepsilon^{1/2})$. The error depends only on ϑ, on Re θ, *and on the constants in* (7.5).

The proof is exactly the same as the one of Theorem 7.2.5, except that for (1), we use Theorem 7.1.6 together with the error estimates, depending only on the constants mentioned in (7.5).

Corollary 7.2.13. *Let $\lambda \in (0, \infty)$ be an eigenvalue of $\Delta^{\mathrm{D}}_{G0,\mathrm{int}}$ with multiplicity μ, i.e. $\lambda = (k\pi/\ell_e)^2$ for some $e \in E$, $k \in \mathbb{N}$. Then there exist μ resonances $\lambda_j(\varepsilon)$ of $\Delta^{\mathrm{D}}_{X_\varepsilon}$ (i.e. discrete eigenvalues of H^θ_ε), $j = 1, \dots, \mu$, not necessarily mutually distinct, such that*

$$\lambda_j(\varepsilon) \to \lambda \qquad \text{as} \qquad \varepsilon \to 0.$$

If, in addition, $\mu = 1$, i.e. if λ is a simple eigenvalue with normalised eigenvector ψ, then there exist normalised eigenvectors ψ_ε of H^θ_ε such that

$$\|J_\varepsilon\psi - \psi_\varepsilon\| = \mathrm{O}(\varepsilon^{1/2}) \quad \text{and} \quad \|J^*_\varepsilon\psi_\varepsilon - \psi\| = \mathrm{O}(\varepsilon^{1/2}).$$

Finally, the convergence depend only on ϑ, on Re θ *and on the constants in* (7.5).

Appendix A

A.1 Convergence of Set Sequences

In this section we collect some facts on the convergence of a sequence of closed sets A_n to A, typically the sets will be spectra of operators and therefore be subsets of $\mathbb{R}_+$ or $\mathbb{C}$. Nevertheless, we need also a weighted distance, so we formulate this chapter in terms of a complete metric space (M, d) and assume $A_n, A \subset M$. We denote the open and closed metric balls by

$$B_\eta(x) := \{ y \in M \mid d(x, y) < \eta \} \qquad \text{and} \qquad \overline{B}_\eta(x) := \{ y \in M \mid d(x, y) \le \eta \}.$$

Moreover, for a set $B \subset M$ and $a \in M$, we define the distance of a to B by

$$d(a, B) := \inf_{b \in B} d(a, b) \in [0, \infty]. \tag{A.1}$$

Note that if B is compact, then the minimum is achieved, and we can replace inf by min.

Definition A.1.1. We define the myemphmaximal outside distance of A to B by

$$d_+(A, B) := \sup_{a \in A} d(a, B) \in [0, \infty].$$

Moreover, the *maximal inside distance* of A to B is given by

$$d_-(A, B) := d_+(B, A).$$

Finally, the *Hausdorff distance* of A and B is defined by

$$d(A, B) := \max\{d_+(A, B), d_-(A, B)\}.$$

O. Post, *Spectral Analysis on Graph-Like Spaces*, Lecture Notes in Mathematics 2039,
DOI 10.1007/978-3-642-23840-6, © Springer-Verlag Berlin Heidelberg 2012

The proof of the following proposition is standard:

Proposition A.1.2. *We have* $d_\pm(A, B) = d_\pm(\overline{A}, \overline{B})$ *and* $d(A, B) = d(\overline{A}, \overline{B})$. *Moreover, the Hausdorff distance is indeed a metric on the space*

$$\mathscr{K}(M) := \{ K \subset M \mid K \text{ compact} \}.$$

If (M, d) *is complete, so is* $(\mathscr{K}(M), d)$.

Note that the maximal outside and inside distances are *not* symmetric: Assume that A, B are compact. Then $d_+(A, B) = 0$ is equivalent to $A \subset B$, i.e. the maximal outside distance of A to B does not see points in $B \setminus A$, and similarly for d_-.

Lemma A.1.3. *Let* $\eta \geq$. *We have the following equivalent characterisations for compact subsets* $A, B \subset M$:

$$d_+(A, B) \leq \eta \quad \Leftrightarrow \quad \forall a \in A\ \exists b \in B : d(a, b) \leq \eta,$$

$$d_-(A, B) \leq \eta \quad \Leftrightarrow \quad \forall b \in B\ \exists a \in A : d(a, b) \leq \eta.$$

Note that this characterisation allows to define maps $f_\eta : A \longrightarrow B$ and $g_\eta : B \longrightarrow A$ such that $d(a, f_\eta(a)) \leq \eta$ and $d(b, g_\eta(b)) \leq \eta$. This point of view is useful when comparing two different metric spaces A, B in the *Gromov-Hausdorff distance* (see e.g. [Ka02, Ka06]).

Definition A.1.4. Let $A_n, A \in \mathscr{K}(M)$ be compact subsets of M.

1. A_n *converges from outside* to A ($A_n \searrow A$) if $d_+(A_n, A) \to 0$.
2. A_n *converges from inside* to A ($A_n \nearrow A$) if $d_-(A_n, A) \to 0$.
3. A_n *converges* to A ($A_n \to A$) if $d(A_n, A) \to 0$.

Remark A.1.5. Again, we warn from taking the terminology too literally: if $A_n \subset A$, then $d_+(A_n, A) = 0$, i.e. a sequence A_n *inside* A converges from outside to A, and similarly for d_-. In particular, if $A_n \searrow A$ converges from outside, then the limit can *suddenly expand*. For example, if $A_n = A_0 \subsetneq A$, then $d_+(A_n, A) = 0$. The convergence from outside does not note what happens *inside* A. Similarly, if $A_n \nearrow A$ converges from inside, then the limit can *suddenly collapse*. For example, if $A_n = A_0 \supsetneq A$, then $d_-(A_n, A) = 0$. The convergence form inside does not note what happens *outside* A. Nevertheless, if $A_n \to A$ (i.e. A_n converges to A in Hausdorff distance), the limit A cannot either expand nor collapse.

Let us give an element-wise characterisation of the convergence from inside and outside. Recall that an assertion (A_n) holds *eventually* if there exists $n_0 \in \mathbb{N}$ such that (A_n) holds for all $n \geq n_0$. Similarly, (A_n) holds *infinitely often* if for all $n_0 \in \mathbb{N}$ there exists $n \geq n_0$ such that (A_n) holds.

Proposition A.1.6. *Assume that* M *is a compact metric space. Let* $A_n, A \in \mathscr{K}(M)$ *be closed subsets. Then the following conditions are equivalent:*

$$A_n \searrow A \quad \Leftrightarrow \quad \forall x \in M \setminus A\ \exists \eta > 0\colon B_\eta(x) \cap A_n = \emptyset \text{ eventually},$$

$$A_n \nearrow A \quad \Leftrightarrow \quad \forall a \in A\ \exists a_n A_n\colon d(a_n, a) \to 0.$$

Proof. Assume that $A_n \searrow A$ then $d(a_n, A) \to 0$. Let $x \notin A$. Since A^c is open, there exists $\eta > 0$ such that $B_{2\eta}(x) \cap A = \emptyset$. Now, if $a_n \in B_\eta(x) \cap A_n$ infinitely often, then

$$d(a, a_n) \geq d(a, x) - d(x, a_n) > 2\eta - \eta = \eta$$

for all $a \in A$, i.e. $d(a_n, A) \geq \eta$, contradicting $d(a_n, A) \to 0$.

For the opposite direction, assume that $A_n \searrow A$ is not true, i.e. there exists $\eta > 0$ and a sequence $a_n \in A_n$ such that $d(a_n, A) \geq \eta$ infinitely often. Since M is compact, we can extract a convergence sub-sequence $a_{n_k} \to x$. In particular, $d(x, A) \geq \eta$, thus $x \in M \setminus A$. From the assumption, we have $B_\eta(x) \cap A_n = \emptyset$ eventually, in contradiction to $a_{n_k} \to x$.

The second equivalence is simpler: The convergence $A_n \nearrow A$ is equivalent to $d(a, A_n) \to 0$ *uniformly* in $a \in A$. In particular, the pointwise convergence $d(a, A_n) \to 0$ for all $a \in A$ follows. For the opposite direction, we use again the compactness of M in order to obtain the *uniform* convergence. □

A.2 Estimates on Abstract Fibred Spaces

In this section, we develop an abstract framework in order to prove Sobolev trace theorems and related estimates on graphs and manifolds. In the abstract setting, we consider (weakly) differentiable functions on an interval with values in a Hilbert space. We allow the Hilbert spaces along the interval to vary, and define abstractly *(warped) products* and spaces close to (warped) products. All relevant examples are covered, e.g., the Sobolev trace estimate for a manifold with boundary or estimates on cone-like manifolds. Our approach here is related (at least formally) to [B89, GG91], as well as to the abstract approach to so-called "Dirac systems" in [BBC08]. There are also relations to "half-line boundary triples" introduced in Sect. 3.5.

A.2.1 *Vector-Valued Integrals*

Let us start with some facts about vector-valued integrals. More details can be found e.g. in [Y80, GG91]. Let (M, μ) be a measure space and $\mathscr{X}$ a Banach space. A function $u\colon M \longrightarrow \mathscr{X}$ is said to be *weakly measurable* if $\varphi \circ u\colon M \longrightarrow \mathbb{C}$, $s \mapsto \varphi(u(s))$, is measurable for all $\varphi \in \mathscr{X}^*$ in the dual space. A function $u\colon M \longrightarrow \mathscr{X}$ is said to be a *step function* if its range $\operatorname{ran} u = u(M)$ is finite, say, $u(M) = \{x_1, \dots, x_k\}$; and if $\mu(u^{-1}(x_i)) < \infty$ for all $i = 1, \dots, k$, provided $x_i \neq 0$. A function $u\colon M \longrightarrow \mathscr{X}$ is said to be *strongly measurable* if there is a sequence of step functions $u_n\colon M \longrightarrow \mathscr{X}$ (an *approximating sequence*) such that $\|u_n(s) - u(s)\| \to 0$ for μ-almost all $s \in S$. By Pettis' theorem, a weakly measurable function is strongly measurable provided $\mathscr{X}$ is separable (or at least if the closed subspace generated by $\operatorname{ran} u$ is separable). In our application below, $\mathscr{X}$ will be a separable Hilbert space. Moreover, we use the case $\mathscr{X} = \mathscr{L}(\mathscr{H})$ and $u(s) = -\varphi(s)(H-s)^{-1}$ for a closed operator H (see Theorems 3.3.6 and 3.3.20). Note that in the latter situation, the closed subspace generated by the range of u is an abelian sub-algebra of $\mathscr{L}(\mathscr{H})$ and therefore separable.

For strongly measurable functions, the vector-valued integral is defined in the obvious way. A function $u: M \longrightarrow \mathscr{X}$ is said to be *Bochner-integrable*, if u is strongly measurable (with approximating sequence u_n) and if $\int_M \|u(s) - u_n(s)\| \mathrm{d}\mu(s) \to 0$. It can be seen that the definition of the integral in the next theorem is independent of the approximating sequence (two approximating sequences can be combined into a single approximating sequence). By Bochner's theorem, a strongly measurable function is Bochner-integrable iff $s \to \|u(s)\|$ is integrable:

Theorem A.2.1 (Bochner). *Let (M, μ) be a measure space, and $\mathscr{X}$ a Banach space. Assume that $u: M \longrightarrow \mathscr{X}$ is strongly measurable. Then u is (Bochner-) integrable iff*

$$\int_M \|u(s)\| \mathrm{d}\mu(s) < \infty. \tag{A.2}$$

In particular, if the integrability condition (A.2) *if fulfilled, then the vector-valued integral*

$$\int_M u(s) \mathrm{d}\mu(s) := \lim_{n\to\infty} \sum_{x \in \operatorname{ran} u_n \setminus \{0\}} \mu(u_n^{-1}(x))x \in \mathscr{X}$$

exists in the norm-topology on $\mathscr{X}$, where u_n is an approximating sequence. Moreover,

$$\Big\| \int_M u(s) \mathrm{d}\mu(s) \Big\| \le \int_M \|u(s)\| \mathrm{d}\mu(s).$$

Let now $\mathscr{X} = \mathscr{H}_0$ be a (separable) Hilbert space. Then weakly measurable functions are strongly measurable. Moreover, we have a vector-valued generalisation of the Cauchy-Schwarz inequality, namely

$$\Big| \Big\langle \int_M u(s) \mathrm{d}\mu(s), \int_M v(s) \mathrm{d}\mu(s) \Big\rangle_{\mathscr{H}_0} \Big| \overset{\mathrm{CS}}{\le} \int_M \|u(s)\|_{\mathscr{H}_0} \|v(s)\|_{\mathscr{H}_0} \mathrm{d}\mu(s), \tag{A.3}$$

provided the RHS is finite, i.e. $u, v \in \mathsf{L}_2(M, \mathscr{H}_0)$. Here, $\mathsf{L}_2(M, \mathscr{H}_0)$ is the space of (equivalence classes of) measurable functions $u: M \longrightarrow \mathscr{H}_0$ such that

$$\|u\|^2_{\mathsf{L}_2(M,\mathscr{H}_0)} := \int_M \|u(s)\|^2_{\mathscr{H}_0} \mathrm{d}\mu(s) < \infty.$$

In particular, $\mathsf{L}_2(M, \mathscr{H}_0)$ is a Hilbert space with inner product

$$\langle u, v\rangle_{\mathsf{L}_2(M,\mathscr{H}_0)} := \int_M \langle u(s), v(s)\rangle_{\mathscr{H}_0} \mathrm{d}\mu(s) = \Big\langle \int_M u(s) \mathrm{d}\mu(s), \int_M u(s) \mathrm{d}\mu(s) \Big\rangle_{\mathscr{H}_0}.$$

Here, the latter equality is true since the Bochner integral commutes with (bounded) linear operators. Note that $\mathsf{L}_2(M, \mathscr{H}_0)$ is also unitarily equivalent with $\mathsf{L}_2(M) \otimes \mathscr{H}_0$ and with the direct integral $\int_M^{\oplus} \mathscr{H}_0 \mathrm{d}\mu(s)$ with constant fibre, see the next subsection.

Moreover, in the vector-valued case, we also have a "scalar-vector" version of the Cauchy-Schwarz inequality, namely

$$\Big\| \int_M f(s) u(s) \mathrm{d}\mu(s) \Big\|^2_{\mathscr{H}_0} \overset{\mathrm{CS}}{\leq} \int_M |f(s)|^2 \mathrm{d}\mu(s) \int_M \|u(s)\|^2_{\mathscr{H}_0} \mathrm{d}\mu(s) \tag{A.4}$$

for $f \in \mathsf{L}_2(M, \mathbb{C})$ and $u \in \mathsf{L}_2(M, \mathscr{H}_0)$.

A.2.2 Fibred Spaces Over an Interval

Assume that $I \subset \mathbb{R}$ is an interval and that $\mathscr{H}(s)$ is a family of separable Hilbert spaces based on the same vector space $\mathscr{H}_0$ (i.e. $\mathscr{H}(s) = \mathscr{H}_0$ as vector space). It is unimportant whether we choose I to be open or closed, since individual points are not seen by the Lebesque measure $\mathrm{d}s$. We assume that $\mathscr{H}(s)$ is measurable, i.e. that $s \mapsto \|u_0\|^2_{\mathscr{H}(s)}$ is measurable for all $u_0 \in \mathscr{H}_0$. We set

$$\mathscr{H}(I) = \int_I^{\oplus} \mathscr{H}(s) \mathrm{d}s$$

(for an abstract definition of the direct integral of Hilbert spaces $\int_I^{\oplus} \mathscr{H}(s)\mathrm{d}s$, we refer e.g. to [Di69] or [RS80, Sec. XIII.16]). Equivalently, we can define $\mathscr{H}(I)$ as the completion of the space of continuous functions $\mathsf{C}(I, \mathscr{H}_0)$ under the norm $\|\cdot\|_{\mathscr{H}(I)}$, where

$$\|u\|^2_{\mathscr{H}(I)} := \int_I \|u(s)\|^2_{\mathscr{H}(s)} \mathrm{d}s < \infty.$$

Similarly, we define a fibred Sobolev space $\mathscr{H}^1(I)$ as the completion of the space of functions $\mathsf{C}^1(I, \mathscr{H}_0)$ under the norm $\|\cdot\|_{\mathscr{H}^1(I)}$, where

$$\|u\|^2_{\mathscr{H}^1(I)} := \int_I \big(\|u(s)\|^2_{\mathscr{H}(s)} + \|u'(s)\|^2_{\mathscr{H}(s)} \big) \mathrm{d}s < \infty.$$

We denote by $\mathscr{H}^k(s_0, s_1)$ the corresponding space $\mathscr{H}^k(I)$ with $I = (s_0, s_1)$. Note that

$$I_1 \subset I_2 \qquad \text{implies} \qquad \|u\|_{\mathscr{H}^k(I_1)} \leq \|u\|_{\mathscr{H}^k(I_2)}. \tag{A.5}$$

Since the Hilbert spaces $\mathscr{H}(s)$ and $\mathscr{H}_0$ differ only in their inner product, there exists a positive operator-valued, measurable function $s \mapsto A(s)$ such that

$$\|u_0\|^2_{\mathscr{H}(s)} = \langle u_0, u_0 \rangle_{\mathscr{H}(s)} = \langle u_0, A(s) u_0 \rangle_{\mathscr{H}_0}$$

for all $u_0 \in \mathscr{H}_0$ and almost all $s \in I$ with respect to the Lebesque measure. Moreover, there exist measurable functions $\rho_\pm : I \longrightarrow (0, \infty)$ such that

$$\rho_-(s)\|u_0\|^2_{\mathscr{H}_0} \le \|u_0\|^2_{\mathscr{H}(s)} \le \rho_+(s)\|u_0\|^2_{\mathscr{H}_0}$$

for almost all $s \in I$ and all $u_0 \in \mathscr{H}_0$. For example, we can choose $\rho_-(s) := \inf \sigma(A(s))$ and $\rho_+(s) := \sup \sigma(A(s))$.

We will now make more assumptions on the functions $\rho_\pm$.

Definition A.2.2.

1. We say that the fibred space $\mathscr{H}(I)$ is an *almost warped product* if there is a function ρ called *distortion function* and constants $\kappa_\pm > 0$ such that

$$\kappa_-\rho(s) \le \rho_-(s) \qquad \text{and} \qquad \rho_+(s) \le \kappa_+\rho(s)$$

   i.e.

$$\kappa_-\rho(s)\|u(s)\|^2_{\mathscr{H}_0} \le \|u(s)\|^2_{\mathscr{H}(s)} \le \kappa_+\rho(s)\|u(s)\|^2_{\mathscr{H}_0} \tag{A.6a}$$

   for almost all $s \in I$ and all $u_0 \in \mathscr{H}_0$. We call $\rho_\infty := \kappa_+/\kappa_-$ the *global relative distortion.*
2. We say that $\mathscr{H}(I)$ is a *warped product* if it is an almost warped product with relative distortion $\rho_\infty = 1$, i.e. if $\rho_-(s) = \rho_+(s) =: \rho(s)$, or equivalently, $A(s) = \rho(s)\,\mathrm{id}_{\mathscr{H}_0}$, or, what is the same,

$$\|u_0\|^2_{\mathscr{H}(s)} = \rho(s)\|u_0\|^2_{\mathscr{H}_0} \tag{A.6b}$$

   for all $u_0 \in \mathscr{H}_0$. We call ρ (resp. $\rho_\pm$) the *(upper/lower) distortion function.*

The case when the distortion functions are constant deserves a special name:

Definition A.2.3.

1. We say that $\mathscr{H}(I)$ is an *almost product* if it is an almost warped product with *constant* functions $\rho_\pm$, i.e. $\rho_\pm(s) = \rho_{\pm,0}$ for almost all $s \in I$. In this case, $\rho_\infty = \rho_{+,0}/\rho_{-,0}$.
2. We say that the fibred space $\mathscr{H}(I)$ is a *product* if it is an almost product with relative distortion $\rho_\infty = 1$, i.e. $\|u_0\|^2_{\mathscr{H}(s)} = \|u_0\|^2_{\mathscr{H}_0}$. In this case,

$$\mathscr{H}(I) = \mathsf{L}_2(I, \rho_0\mathscr{H}_0) \cong \mathsf{L}_2(I) \otimes \rho_0\mathscr{H}_0$$

   where ρ_0 denotes the constant value of the distortion function ρ and $\rho_0\mathscr{H}_0$ is the Hilbert space $\mathscr{H}_0$ with norm defined by $\|u_0\|^2_{\rho_0\mathscr{H}_0} := \rho_0\|u_0\|^2_{\mathscr{H}_0}$.

On an (almost) product $\mathscr{H}(I)$, the identification operator from $\mathscr{H}(I)$ onto $\mathsf{L}_2(I) \otimes \rho_0\mathscr{H}_0$ resp. $\mathsf{L}_2(I) \otimes \rho_{\pm,0}\mathscr{H}_0$ defines a bijective isomorphism resp. an isometry. By suitably inserting the distortion functions $\rho_\pm$ into the inequality (A.4), we can extend the inequality to (almost) warped products, as we will do in the following lemma:

Lemma A.2.4. *Assume that $\mathscr{H}(I)$ is an almost warped product and that $s_0 < s_1 < s_2$ where $s_i \in \overline{I}$. If $u \in \mathscr{H}^1(I)$ with $u(s_2) = 0$, then*

$$\|u(s_1)\|^2_{\mathscr{H}(s_1)} \le \rho_\infty \eta_1(s_1, s_2) \|u'\|^2_{\mathscr{H}(s_1,s_2)}, \tag{A.7a}$$

$$\|u\|^2_{\mathscr{H}(s_0,s_1)} \le \rho_\infty \eta_2(s_0, s_1, s_2) \|u'\|^2_{\mathscr{H}(s_0,s_2)}, \tag{A.7b}$$

where

$$\eta_1(s_1, s_2) := \rho(s_1) \int_{s_1}^{s_2} \frac{1}{\rho(s)} \mathrm{d}s \qquad \textit{and} \qquad \eta_2(s_0, s_1, s_2) := \int_{s_0}^{s_1} \eta_1(t, s_2, \rho) \mathrm{d}t.$$

Proof. By a density argument, we can assume that u is of class C^1. Then we have

$$u(t) = -\int_t^{s_2} u'(s) \mathrm{d}s.$$

Using the Cauchy-Schwarz inequality for vector-valued integrals (A.4) we obtain

$$\begin{aligned} \|u(t)\|^2_{\mathscr{H}(t)} &\overset{\text{CS}}{\le} \int_t^{s_2} \frac{\kappa_+ \rho(t)}{\kappa_- \rho(s)} \mathrm{d}s \int_t^{s_2} \frac{\kappa_- \rho(s)}{\kappa_+ \rho(t)} \|u'(s)\|^2_{\mathscr{H}(t)} \mathrm{d}s \\ &\le \rho_\infty \int_t^{s_2} \frac{\rho(t)}{\rho(s)} \int_t^{s_2} \|u'(s)\|^2_{\mathscr{H}(s)} \mathrm{d}s \end{aligned}$$

using (A.6a) and the first estimate follows for $t = s_1$. The existence of the vector-valued integral $u(t)$ is guaranteed once the integrals on the RHS are finite (cf. Theorem A.2.1). The second estimate follows by integrating over t from s_0 to s_1 and using the monotonicity of the norms (A.5). □

Remark A.2.5. Note that the condition $u(s_2) = 0$ is well-defined for functions in $\mathscr{H}^1(I)$ by a similar argument as above.

If u does not vanish at s_2, we can argue as follows.

Proposition A.2.6. *Assume that $\mathscr{H}(I)$ is an almost warped product and that $u \in \mathscr{H}^1(s_0, s_2)$, then*

$$\|u(s_1)\|^2_{\mathscr{H}(s_1)} \le 2\rho_\infty \widetilde{\eta}_1(s_1, s_2, \chi) \|u'\|^2_{\mathscr{H}(s_1,s_2)} + 2\rho_\infty \widetilde{\eta}_1(s_1, s_2, \chi') \|u\|^2_{\mathscr{H}(s_1,s_2)},$$

$$\|u\|^2_{\mathscr{H}(s_0,s_1)} \le 2\rho_\infty \widetilde{\eta}_2(s_0, s_1, s_2, \chi) \|u'\|^2_{\mathscr{H}(s_0,s_2)} + 2\rho_\infty \widetilde{\eta}_2(s_0, s_1, s_2, \chi') \|u\|^2_{\mathscr{H}(s_0,s_2)},$$

where

$$\widetilde{\eta}_1(s_1, s_2, \chi) := \rho(s_1) \int_{s_1}^{s_2} \frac{|\chi(s)|^2}{\rho(s)} \mathrm{d}s, \quad \widetilde{\eta}_2(s_0, s_1, s_2, \chi) := \int_{s_0}^{s_1} \eta_1(t, s_2, \chi, \rho) \mathrm{d}t,$$

and where χ is a Lipschitz function such that $\chi(s) = 1$ for $s_0 \leq s \leq s_1$ and $\chi(s_2) = 0$.

Proof. Set $\widetilde{u} := \chi u$. Then $\widetilde{u}$ fulfils the assumptions of Lemma A.2.4. Moreover $u(t) = \widetilde{u}(t)$ for $t \in [s_0, s_1]$. We start with

$$\|u(t)\|^2_{\mathscr{H}(t)} \leq 2 \left\| \int_t^{s_2} \chi(s) u'(s) \mathrm{d}s \right\|^2_{\mathscr{H}(t)} + 2 \left\| \int_t^{s_2} \chi'(s) u(s) \mathrm{d}s \right\|^2_{\mathscr{H}(t)},$$

and argue then as in the proof of Lemma A.2.4. The last estimate follows as before by integration. □

Let us now fix the cut-off function $\chi = \chi_a$: Assume that $0 < a \leq s_2 - s_1$. Denote by χ_a the continuous, piecewise affine linear function given by

$$\chi_a(s) = \begin{cases} 1, & \text{for } s_0 \leq s \leq s_1, \\ 1 - a^{-1}(s - s_1), & \text{for } s_1 \leq s \leq s_1 + a, \\ 0, & \text{for } s_1 + a \leq s \leq s_2. \end{cases} \tag{A.8}$$

We obtain the following corollary. For the definitions of the functions η_1 and η_2 see Lemma A.2.4.

Corollary A.2.7. *Assume that $\mathscr{H}(I)$ is an almost warped product. If $u \in \mathscr{H}^1(s_0, s_2)$ then*

$$\|u(s_1)\|^2_{\mathscr{H}(s_1)} \leq 2\rho_\infty \eta_1(s_1, s_1 + a) \Big(\|u'\|^2_{\mathscr{H}(s_1, s_2)} + \frac{1}{a^2} \|u\|^2_{\mathscr{H}(s_1, s_2)} \Big) \tag{A.9a}$$

$$\leq \frac{2\rho_\infty \eta_1(s_1, s_1 + a)}{\min\{1, a^2\}} \|u\|^2_{\mathscr{H}^1(s_1, s_2)} \tag{A.9a'}$$

for $0 < a \leq s_2 - s_1$ and $a \leq 1$. Moreover,

$$\|u\|^2_{\mathscr{H}(s_0, s_1)} \leq 2\rho_\infty \eta_2(s_0, s_1, s_1 + a) \Big(\|u'\|^2_{\mathscr{H}(s_0, s_2)} + \frac{1}{a^2} \|u\|^2_{\mathscr{H}(s_0, s_2)} \Big) \tag{A.9b}$$

$$\leq \frac{2\rho_\infty \eta_2(s_0, s_1, s_1 + a)}{\min\{1, a^2\}} \|u\|^2_{\mathscr{H}^1(s_0, s_2)}. \tag{A.9b'}$$

Proof. The result follows from Proposition A.2.6 by simply estimating $|\chi_a(s)| \leq 1$ and $|\chi_a'(s)| = 1/a$ for $s_1 \leq s \leq s_1 + a$. □

If $\mathscr{H}(I)$ is an almost product, we have the following estimate:

Corollary A.2.8. *Assume that $\mathscr{H}(I)$ is an almost product. If $u \in \mathscr{H}^1(s_0, s_2)$ then*

$$\|u(s_1)\|^2_{\mathscr{H}(s_1)} \le \rho_\infty\Big(a\|u'\|^2_{\mathscr{H}(s_1,s_2)} + \frac{2}{a}\|u\|^2_{\mathscr{H}(s_1,s_2)}\Big) \tag{A.10}$$

$$\le \frac{2\rho_\infty}{\min\{1, (s_2 - s_1)\}}\|u\|^2_{\mathscr{H}^1(s_1,s_2)} \tag{A.10'}$$

for $0 < a \le s_2 - s_1$ and $a \le 1$.

Proof. Again, the result follows from Proposition A.2.6 by evaluating the integrals $\widetilde{\eta}_i$ for χ_a. □

Remark A.2.9. Note that the estimate (A.7a) is optimal in the following sense: Assume that $\mathscr{H}(I)$ is a warped product and that $u(s) = (\int_{s_1}^{s_2} \rho(s)\mathrm{d}s)^{-1}\varphi_0$ with $\|\varphi_0\|_{\mathscr{H}_0} = 1$, i.e. u is constant and $\|u\|_{\mathscr{H}(s_1,s_2)} = 1$. Moreover, assume that $\chi'(s) < 0$ for all $s_1 \le s \le s_2$. Then we have

$$1 = \big|\chi(s_2) - \chi(s_1)\big|^2 = \Big(\int_{s_1}^{s_2} |\chi'(s)|\mathrm{d}s\Big)^2 \overset{\text{CS}}{\le} \int_{s_1}^{s_2} \rho(s)\mathrm{d}s \int_{s_1}^{s_2} \frac{|\chi'(s)|^2}{\rho(s)}\mathrm{d}s.$$

If the distortion function is $\rho(s) = -\chi'(s) = |\chi'(s)|$, then we have equality in the Cauchy-Schwarz inequality, and in particular,

$$\|u(s_1)\|^2_{\mathscr{H}(s_1)} = \frac{\rho(s_1)}{\int_{s_1}^{s_2} \rho(s)\mathrm{d}s} = \rho(s_1)\int_{s_1}^{s_2} \frac{|\chi'(s)|^2}{\rho(s)}\mathrm{d}s = \widetilde{\eta}_1(s_1, s_2, \chi')\|u\|^2_{\mathscr{H}(s_1,s_2)}$$

by the above equality of Cauchy-Schwarz, the definition of $\widetilde{\eta}_1$ and the fact that u is normalised on $\mathscr{H}(s_1, s_2)$. Therefore, estimate (A.7a) is optimal. A similar remark holds for the other estimates.

Let us now fix a normalised vector $\varphi_0 \in \mathscr{H}_0$. Denote by $P_0 u_0 := \langle\varphi_0, u_0\rangle_{\mathscr{H}_0}\varphi_0$ the corresponding orthogonal projection onto the subspace $\mathbb{C}\varphi_0$. In our applications later on, φ_0 will be an eigenvector associated with the lowest eigenfunction of an operator K on $\mathscr{H}_0$. For a warped product, the projection is independent of s:

Lemma A.2.10. *Let $\mathscr{H}(I)$ be a warped product. Then $\varphi_0 \ne 0$ in $\mathscr{H}(s)$ for almost all $s \in I$, and $\varphi_s := \varphi_0/\|\varphi_0\|_{\mathscr{H}(s)} = \rho(s)^{-1/2}\varphi_0$ is normalised in $\mathscr{H}_0$. Moreover, P_0 is also the orthogonal projection onto φ_s in the Hilbert space $\mathscr{H}(s)$. In particular,*

$$\|P_0 u_0\|^2_{\mathscr{H}(s)} \overset{\text{CS}}{\le} \|u_0\|^2_{\mathscr{H}(s)}$$

for $u_0 \in \mathscr{H}(s) = \mathscr{H}_0$.

Proof. We have $\|\varphi_0\|^2_{\mathscr{H}(s)} = \rho(s)\|\varphi_0\|^2_{\mathscr{H}_0} = \rho(s) > 0$ for almost all $s \in I$; in particular, $\varphi_0 \ne 0 \in \mathscr{H}(s)$. Moreover, the orthogonal projection onto φ_s in $\mathscr{H}(s)$

is given by

$$\langle \varphi_s, u_0\rangle_{\mathscr{H}(s)}\varphi_s = \rho(s)^{-1}\langle \varphi_0, u_0\rangle_{\mathscr{H}(s)}\varphi_0 = \langle \varphi_0, u_0\rangle_{\mathscr{H}_0}\varphi_0 = P_0u_0.$$

The last estimate follows from Cauchy-Schwarz. □

Proposition A.2.11. *Let $\mathscr{H}(I)$ be a warped product. Then*

$$\begin{aligned}\rho(s_1)\|P_0u(s_2) - P_0u(s_1)\|^2_{\mathscr{H}_0} &= \rho(s_1)\big|\langle \varphi_0, u(s_2)\rangle_{\mathscr{H}_0} - \langle \varphi_0, u(s_1)\rangle_{\mathscr{H}_0}\big|^2\\ &\le \eta_1(s_1,s_2)\|u'\|^2_{\mathscr{H}(s_1,s_2)}\end{aligned}$$

for $u \in \mathscr{H}^1(I)$.

Proof. We have

$$\begin{aligned}\big\|\langle \varphi_0, u(s_2) - u(s_1)\rangle_{\mathscr{H}_0}\varphi_0\big\|^2_{\mathscr{H}_0} &\overset{\text{CS}}{\le} \|u(s_2) - u(s_1)\|^2_{\mathscr{H}_0}\\ &= \Big\|\Big\langle \varphi_0, \int_{s_1}^{s_2} u'(s)\mathrm{d}s\Big\rangle_{\mathscr{H}_0}\varphi_0\Big\|^2_{\mathscr{H}_0}\\ &\overset{\text{CS}}{\le} \int_{s_1}^{s_2}\frac{1}{\rho(s)}\mathrm{d}s \int_{s_1}^{s_2}\|u'(s)\|^2_{\mathscr{H}_0}\rho(s)\mathrm{d}s\end{aligned}$$

using Cauchy-Schwarz for vector-valued integrals (A.4) for the second inequality. The result follows by the definition of η_1 in Lemma A.2.4. □

A.2.3 Examples: Cones and Cylinders

We will give now some special cases coming from a warped product metric on a manifold as applied in Chap. 6. Denote by $X = I \times_s Y$ the warped product of $I = [0,\ell]$ and the compact Riemannian manifold Y with radius function $r\colon I \longrightarrow (0,\infty)$, i.e. X is equipped with the metric $g = \mathrm{d}s^2 + r(s)^2h$, where h is the metric on Y (see Definition 5.3.2). Then we have

$$\|u\|^2_{\mathsf{L}_2(X,g)} = \int_I \|u(s)\|^2_{\mathsf{L}_2(Y,h)} r(s)^m \mathrm{d}s.$$

In particular, $\mathsf{L}_2(X,g) = \mathscr{H}(I)$ is a warped product with distortion function $\rho(s) = r(s)^m$ and fibre Hilbert space $\mathscr{H}_0 = \mathsf{L}_2(Y,h)$.

For the next corollary, assume that r is constant, say $r = 1$.

Corollary A.2.12 (Product). *Assume that $I = [0,\ell]$ and $X = I \times Y$ with product metric $g = \mathrm{d}s^2 + h$. Then $\mathscr{H}(I)$ is a product (in the sense of Definition A.2.3). Moreover,*

$$\|u(0,\cdot)\|^2_{\mathsf{L}_2(Y)} \le a\|u'\|^2_{\mathsf{L}_2(X)} + \frac{2}{a}\|u\|^2_{\mathsf{L}_2(X)} \le \frac{2}{\min\{1,\ell\}}\|u\|^2_{\mathsf{H}^1(X)} \tag{A.11}$$

for all $0 < a \le \min\{1,\ell\}$ *and* $u \in \mathsf{H}^1(X)$.

Proof. The result follows from Corollary A.2.8. Note that $\mathsf{H}^1(X) \subset \mathscr{H}^1(I)$ since

$$\|u\|^2_{\mathscr{H}^1(I)} = \|u\|^2_{\mathsf{L}_2(X)} + \|u'\|^2_{\mathsf{L}_2(X)} \le \|u\|^2_{\mathsf{L}_2(X)} + \|u'\|^2_{\mathsf{L}_2(X)} + \|\mathrm{d}_Y u\|^2_{\mathsf{L}_2(X)} = \|u\|^2_{\mathsf{H}^1(X)}$$

for $u \in \mathsf{H}^1(X)$. □

We return to the abstract framework, and calculate the functions η_1, η_2 of Lemma A.2.4 for special choices of the distortion function $\rho = r^m$. Let us first define some universal functions occurring in the estimates later on. *Universal* refers here to the fact that the functions depend only on the parameter m.

Definition A.2.13. We set

$$p_m(\tau) := \frac{1}{1-\tau}\int_\tau^1 u^m \mathrm{d}u \qquad \text{and} \qquad \widetilde{p}_m(\tau) := \frac{1}{1-\tau}\int_\tau^1 p_m(t)\mathrm{d}t.$$

We have the following properties of the functions p_m and $\widetilde{p}_m$:

Lemma A.2.14.

1. *We have*

$$p_m(\tau) = \begin{cases} \dfrac{1}{m+1}\displaystyle\sum_{i=0}^m \tau^i, & \text{if } m \ge 0, \\ -\dfrac{\log\tau}{1-\tau}, & \text{if } m = -1, \end{cases}$$

and

$$\widetilde{p}_m(\tau) = \begin{cases} \dfrac{1}{m+1}\displaystyle\sum_{i=0}^m \frac{1}{i+1}\sum_{j=0}^i \tau^j, & \text{if } m \ge 0, \\ \dfrac{\operatorname{dilog}\tau + \tau(1-\log\tau) - 1}{1-\tau} & \text{if } m = -1, \end{cases}$$

where $\operatorname{dilog}\tau := \int_1^\tau \frac{\log t}{1-t}\mathrm{d}t$. *Moreover,* p_m *and* $\widetilde{p}_m$ *are polynomials of degree indicated by the subscript (for non-negative subscripts) with continuous extension* $p_m(1) = 1$ *and* $\widetilde{p}_m(1) = 1$. *In addition,* p_{-1} *and* $\widetilde{p}_{-1}$ *extend to an analytic functions also in* $\tau = 1$ *by setting* $p_{-1}(1) := 1$ *and* $\widetilde{p}_{-1}(1) := 1$.
2. *For* $0 < \tau \le 1$ *and* $m \ge 0$, *we have*

$$\frac{1}{m+1} \le p_m(\tau), \widetilde{p}_m(\tau) \le 1, \qquad -\log\tau \le p_{-1}(\tau) \le 1 - \log\tau.$$

Moreover, we have the asymptotic behaviour

$$p_m(\tau) \sim_0 \frac{1}{m+1}, \qquad p_{-1}(\tau) \sim_0 -\log\tau,$$

$$\widetilde{p}_m(\tau) \sim_0 \frac{1}{m+1}\sum_{i=0}^{m}\frac{1}{i+1}, \qquad \widetilde{p}_{-1}(\tau) \sim_0 \frac{\pi^2}{6}-1,$$

at $\tau = 0$*. Here,* $f(\tau) \sim_a g(\tau)$ *means that* $\lim_{\tau\to a} f(\tau)/g(\tau) = 1$*.*

3. *For* $1 \leq \tau$ *and* $m \geq 0$*, we have*

$$\frac{\tau^m}{m+1} \leq p_m(\tau) \leq \tau^m, \qquad \frac{\tau^m}{(m+1)^2} \leq \widetilde{p}_m(\tau) \leq \tau^m.$$

In addition, we have

$$p_m(\tau) \sim_\infty \frac{\tau^m}{m+1}, \qquad p_{-1}(\tau) \sim_\infty \frac{\log\tau}{\tau},$$

$$\widetilde{p}_m(\tau) \sim_\infty \frac{\tau^m}{(m+1)^2} \qquad \widetilde{p}_{-1}(\tau) \sim_\infty \log\tau.$$

Proof. The properties follow by direct calculation, e.g., for $\tau \neq 1$ and $m > -1$, we have

$$p_m(\tau) = \frac{1}{1-\tau}\int_1^{1/\tau}\frac{1}{t^{m+2}}\mathrm{d}t = \frac{1}{(m+1)}\cdot\frac{1-\tau^{m+1}}{1-\tau} = \frac{1}{m+1}\sum_{i=0}^{m}\tau^i$$

substituting $u = 1/t$. For $m = -1$ the result follows similarly. Moreover,

$$\begin{aligned}\widetilde{p}_m(\tau) = \frac{1}{1-\tau}\int_\tau^1 p_m(\tau)\tau\mathrm{d}\tau &= \frac{1}{m+1}\cdot\frac{1}{1-\tau}\sum_{i=0}^{m}\int_\tau^1\tau^i\mathrm{d}\tau \\ &= \frac{1}{m+1}\sum_{i=0}^{m}\frac{1}{i+1}\cdot\frac{1-\tau^{i+1}}{1-\tau} \\ &= \frac{1}{m+1}\sum_{i=0}^{m}\frac{1}{i+1}\sum_{j=0}^{i}\tau^j\end{aligned}$$

for $m > -1$ and similarly for $m = -1$. The other assertions can be checked easily. □

In the following lemma, it is convenient to use the differences $a = s_1 - s_0$ and $b = s_2 - s_1$:

Definition A.2.15. We set $s_0 = 0$, $s_1 = a$ and $s_2 = a + b$ for $a, b > 0$ and define

$$\eta_1(b) := \eta_1(a, a+b) \qquad \text{and} \qquad \eta_2(a,b) := \eta_2(0, a, a+b).$$

Moreover, we set

$$\operatorname{vol}(s_0, s_1) := \int_{s_0}^{s_1} \rho(s)\mathrm{d}s, \qquad \operatorname{vol}(a) := \operatorname{vol}(0, a).$$

Let us now fix the radius function r, and therefore the distortion function $\rho = r^m$: Note that if r is constant, then the resulting warped product manifold $X = I \times_r Y$ is a product, or a *cylinder*. If r is affine linear, then $X = I \times_r Y$ is a cone. Recall that we allowed weak regularity on r and therefore on the metric of the manifold (see Sect. 5.1). Therefore, we may consider the following (piecewise) affine linear function r, i.e. we consider two cones (if $r_{i-1} \neq r_i$) or cylinders (if $r_{i-1} = r_i$) attached together:

Lemma A.2.16.

1. ***(Almost) Product:*** *If* $\rho(s) = r_0^m$, *then*

$$\eta_1(b) = b,$$
$$\eta_2(a,b) = \int_{s_0}^{s_1} (s_2 - t)\mathrm{d}t = \frac{1}{2}\big((a+b)^2 - a^2\big) = a(b + a/2) \leq a(a+b),$$
$$\operatorname{vol}(a) = a r_0^m.$$

2. ***(Almost) warped product*** *If* $\rho(s) = r(s)^m$, *where* r *is the continuous affine linear function such that* $r(s_i) = r_i$, $i = 0, 1, 2$, *and* $\tau_i := r_{i-1}/r_i$ *are the relative radii, then*

$$\eta_1(b) = b p_{m-2}(\tau_2)\tau_2$$
$$ab p_{m-2}(\tau_2)\tau_2 p_m(\tau_1) \leq \eta_2(a,b) \leq a\big(a\widetilde{p}_{m-2}(\tau_1) + b p_{m-2}(\tau_2)\tau_2 p_m(\tau_1)\big),$$
$$\operatorname{vol}(a) = a r_0^m p_m(\tau_1^{-1}),$$

Note that η_1 and η_2 depend only on the ratio $\tau_1 = r_0/r_1$ and $\tau_2 = r_1/r_2$ an *not* on the values r_i itself.

Proof. The first result is an easy calculation. For the second, note that

$$r(s) := r_i + (s - s_i)\frac{r_j - r_i}{s_j - s_i}, \qquad 0 \leq i < j \leq 2, \quad s_i \leq s \leq s_j.$$

Moreover, for $r_1 \neq r_2$ and $m > 1$, we have

$$\eta_1(s_1, s_2) = \int_{s_1}^{s_2} \left(\frac{r(s)}{r_1}\right)^{-m} \mathrm{d}s = \frac{b\tau_2}{1 - \tau_2} \int_1^{1/\tau_2} \tau^{-m}\mathrm{d}\tau = b\tau_2 p_{m-2}(\tau_2)$$

using the substitution $\tau = r(s)/r_1$ and the fact that $\mathrm{d}\tau = (1-\tau_2)(b\tau_2)^{-1}\mathrm{d}s$. If $m = 1$, then

$$\eta_1(s_1, s_2) = -\frac{b\tau_2}{1-\tau_2}\log \tau_2 = bp_{-1}(\tau_2)\tau_2.$$

In addition, if $r_1 = r_2$, i.e. $\tau_2 = 1$, then $\eta_1(b) = b$ by the first part. But $p_m(1) = 1$ for $m = -1, 0, 1, \dots$, so that the formula for η_1 also holds in this case. In order to estimate η_2, we observe that

$$\eta_1(t, s_2) = \eta_1(t, s_1) + \frac{\rho(t)}{\rho(s_1)}\eta_1(s_1, s_2).$$

The integral over the first term can be estimated as

$$\begin{aligned}\int_{s_0}^{s_1} \eta_1(t, s_1)\mathrm{d}t &= \int_{s_0}^{s_1} (s_1 - t)p_{m-2}\Big(\frac{r(t)}{r_1}\Big)\frac{r(t)}{r_1}\mathrm{d}t \\ &\le a \int_{s_0}^{s_1} p_{m-2}\Big(\frac{r(t)}{r_1}\Big)\frac{r(t)}{r_1}\mathrm{d}t = \frac{a^2}{1-\tau_1}\int_{\tau_1}^{1} p_{m-2}(\tau)\tau\mathrm{d}\tau\end{aligned}$$

using the substitution $\tau = r(t)/r_1$ and $\mathrm{d}\tau = (1-\tau_1)a^{-1}\mathrm{d}t$.

The second term can be integrated by

$$\int_{s_0}^{s_1} \frac{\rho(t)}{\rho(s_1)}\mathrm{d}t = \int_{s_0}^{s_1}\Big(\frac{r(t)}{r_1}\Big)^m \mathrm{d}t = \frac{a}{1-\tau_1}\int_{\tau_1}^{1}\tau^m\mathrm{d}\tau = ap_m(\tau_1).$$

The calculation for $\mathrm{vol}(a)$ follows similarly. □

We will use the following special case in Sects. 6.3 and 6.8 on warped products:

Proposition A.2.17. *Assume that $I = [0, a+b]$, where $0 < a \le b$. Let ρ be the continuous, piecewise affine linear function with $\rho(0) = r_0$, $\rho(a) = r_1$ and $\rho(a+b) = r_1$, where $r_0 \le r_1$. Then $\mathscr{H}(I)$ is a warped product and,*

$$\|u\|^2_{\mathscr{H}(0,a)} \le 4ab\Big(\|u'\|^2_{\mathscr{H}(0,a+b)} + \frac{1}{a^2}\|u\|^2_{\mathscr{H}(0,a+b)}\Big) \quad \textit{and} \tag{A.12a}$$

$$\frac{a}{m+1}r_1^m \le \mathrm{vol}(a) \le ar_1^m \tag{A.12b}$$

for $u \in \mathscr{H}^1(I)$. Moreover, if P_0 is a projection in $\mathscr{H}_0$ with one-dimensional range, then

$$r_0^m\|P_0u(0)\|^2_{\mathscr{H}_0} \le 2ap_{m-2}\Big(\frac{r_0}{r_1}\Big)\frac{r_0}{r_1} + 2r_0^m\|P_0u(a)\|^2_{\mathscr{H}_0} \tag{A.12c}$$

for $u \in \mathscr{H}^1(I)$.

Proof. The first estimate follows from Corollary A.2.7 and the estimate

$$\eta_2(a,b) \le a\big(a\widetilde{p}_{m-2}(\tau_1) + bp_m(\tau_1)\big) \le a^2 + ab \le 2ab.$$

Here, we used Lemma A.2.16 with the setting $\tau_1 = r_0/r_1 \le 1$, $\tau_2 = 1$, $\widetilde{p}_{m-2}(\tau_1) \le 1$, $p_m(\tau_1) \le 1$ and $a \le b$. The second estimate estimate is a consequence of Lemma A.2.16 and

$$\frac{1}{(m+1)}\tau_1^{-1} \le p_m(\tau_1^{-1}) \le \tau_1^{-m}$$

using Lemma A.2.14. Finally, the last estimate can be seen from Proposition A.2.11 and $\eta_1(a) = ap_{m-2}(\tau_1)\tau_1$. □

Let us finish this section with some more examples illustrating the abstract setting. The possibly simplest example is given by $\mathscr{H}(s) = \mathbb{C}$. The estimate follows from Corollary A.2.8:

Corollary A.2.18 (Interval). *Let $I = [0,\ell]$. Assume that $\mathscr{H}(s) = \mathbb{C}$ and that the distortion function is constant, say $\rho(s) = 1$. Then $\mathscr{H}(I) = \mathsf{L}_2(I)$ is a product and*

$$|f(0)|^2 \le a\|f'\|^2_{\mathsf{L}_2(I)} + \frac{2}{a}\|f\|^2_{\mathsf{L}_2(I)} \le \frac{2}{\ell_-}\|f\|^2_{\mathsf{H}^1(I)}$$

for all $0 < a \le \ell_- = \min\{1,\ell\}$ and $f \in \mathsf{H}^1(I) = \mathscr{H}^1(I)$.

The following result is only needed in the special case $\mathscr{H}(s) = \mathbb{C}$:

Proposition A.2.19. *Assume that $I = [0,\ell]$, then*

$$\|f'\|^2_{\mathsf{L}_2(I)} \le \frac{1025}{\ell_-^2}\|f\|^2_{\mathsf{L}_2(I)} + 2\|f''\|^2_{\mathsf{L}_2(I)} \le \frac{1025}{\ell_-^2}\Big(\|f\|^2_{\mathsf{L}_2(I)} + \|f''\|^2_{\mathsf{L}_2(I)}\Big)$$

for $f \in \mathsf{H}^2(I)$.

Proof. Partial integration and Cauchy-Young's inequality yield

$$\|f'\|^2 \overset{\text{CY}}{\le} \frac{1}{2}\|f\|^2 + \frac{1}{2}\|f''\|^2 + \big|f(0)f'(0)\big| + \big|f(\ell)f'(\ell)\big|.$$

The boundary term at $s = 0$ can be estimated by

$$\begin{aligned}
\big|f(0)f'(0)\big| &\overset{\text{CY}}{\le} \frac{\eta}{2}|f'(0)|^2 + \frac{1}{2\eta}|f(0)|^2 \\
&\le \frac{\eta}{2}\Big(b'\|f''\|^2 + \frac{2}{b'}\|f'\|^2\Big) + \frac{1}{2\eta}\Big(b\|f'\|^2 + \frac{2}{b}\|f\|^2\Big) \\
&= \frac{1}{\eta b}\|f\|^2 + \frac{\eta b'}{2}\|f''\|^2 + \frac{1}{2}\Big(\frac{2\eta}{b'} + \frac{b}{\eta}\Big)\|f'\|^2
\end{aligned}$$

for $\eta >$ and $b, b' \in (0, \ell]$, applying Corollary A.2.18 to f 'and f'. A similar result holds for the boundary term at $s = \ell$, so that we end up with the inequality

$$\|f'\|^2 \leq \Big(\frac{1}{2} + \frac{2}{\eta b}\Big)\|f\|^2 + \Big(\frac{1}{2} + \eta b'\Big)\|f''\|^2 + \Big(\frac{2\eta}{b'} + \frac{b}{\eta}\Big)\|f'\|^2.$$

If we set $\eta := a/8$, $b := a/32$ and $b' := a$ for $0 < a \leq \ell$, then the coefficient of $\|f'\|^2$ on the RHS equals $1/2$. Bringing this term on the LHS and multiplying by 2 yields the desired estimate with $a = \ell_-$. Note that $1 + 4/(\eta b) = 1 + 1{,}024/a^2 \leq 1{,}025/a^2$ and $12\eta b' = 1 + a^2/4 \leq 2$ since $a \leq 1$. □

More generally than in Corollary A.2.18, we may assume that the measure on I is weighted, i.e. we replace the Lebesque measure ds by $w(s)$ds. Then from Corollary A.2.7 we conclude (with the estimate $\eta_1(a) \leq \omega a$):

Corollary A.2.20 (Interval with weights). *Assume that* $I = [0, \ell]$, *that* $w\colon (0, \ell) \longrightarrow (0, \infty)$ *is a measurable function, and that* $\mathscr{H}(s) = \mathbb{C}w(s)$. *Then* $\mathscr{H}(I) = \mathsf{L}_2(I, w\mathrm{d}s)$ *is a warped product with distortion function* $\rho = w$. *If in addition*

$$\omega := \frac{w(0)}{\inf w([0, \min\{1, \ell\}])} < \infty,$$

then

$$|f(0)|^2 \leq 2a\omega\|f'\|^2_{\mathsf{L}_2(I,w(s)\mathrm{d}s)} + \frac{2\omega}{a}\|f\|^2_{\mathsf{L}_2(I,w(s)\mathrm{d}s)} \leq \frac{2\omega}{\min\{1, a\}}\|f\|^2_{\mathsf{H}^1(I,w(s)\mathrm{d}s)}$$

for $0 < a \leq \min\{1, \ell\}$ *and* $f \in \mathsf{H}^1(I, w(s)\mathrm{d}s) = \mathscr{H}^1(I)$.

Let us finally give another example, which is needed for topological perturbations like removing balls from a manifold or adding handles to it. We will present such constructions in a subsequent work.

Example A.2.21 (Polar coordinates). Assume that (X, g) is a Riemannian manifold. Here, we apply the warped product structure of a metric g on a manifold X given in polar coordinates.

Denote by $\kappa_x(r)$ the maximal absolute value of the sectional curvature on $\overline{B}_x(r)$. Let $r_\kappa(x)$ be the maximal radius $r > 0$ such that $r \leq \pi/(2\sqrt{\kappa_x(r)})$ with the convention $1/0 = \infty$. We set

$$r_0(x) := \min\{\operatorname{inj\,rad} x, r_\kappa(x)\}. \tag{A.13}$$

For $x \in X$, we can parametrise $M_x := B(x, r_0(x))$ with *polar coordinates* $(s, y) \in I_x \times Y$, $I_x := (0, r_0(x))$, $Y = \mathbb{S}^m$. The metric in these coordinates is given by

$$g = \mathrm{d}s^2 + h_s,$$

where h_s is an s-depend metric on Y. We denote the standard metric on $\mathbb{S}^{d-1}$ by h. Recall that the warped product metric $g_0 = \mathrm{d}s^2 + s^2 h$ is the *flat* metric around x, i.e. $\mathbb{R}_+ \times_r \mathbb{S}^m \cong \mathbb{R}^{m+1}$, where $r(s) = s$ is the radius function (see Definition 5.3.2). From [Au82, Thm. 1.53] and the definition of $r_0(x)$ in (A.13), it follows that we have the estimate

$$\mathrm{d}s^2 + s^2\Big(\frac{2}{\pi}\Big)^2 h \le g = \mathrm{d}s^2 + h_s \le \mathrm{d}s^2 + s^2\Big(\sinh\Big(\frac{\pi}{2}\Big)\frac{2}{\pi}\Big)^2 h.$$

In particular, $\mathsf{L}_2(B_x)$ is an almost warped product with distortion function $\rho(s) = s^m$ and relative distortion $\rho_\infty = (\sinh(\pi/2))^m$.

References

[AC71] J. Aguilar and J. M. Combes, *A class of analytic perturbations for one-body Schrödinger Hamiltonians*, Comm. Math. Phys. **22** (1971), 269–279.

[AS00] M. Aizenman and J. H. Schenker, *The creation of spectral gaps by graph decoration*, Lett. Math. Phys. **53** (2000), 253–262.

[AADH94] S. Alama, M. Avellaneda, P. Deift, and R. Hempel, *On the existence of eigenvalues of a divergence-form operator $A + \lambda B$ in a gap of $\sigma(A)$*, Asymptotic Anal. **8** (1994), 311–344.

[ADH89] S. Alama, P. A. Deift, and R. Hempel, *Eigenvalue branches of the Schrödinger operator $H - \lambda W$ in a gap of $\sigma(H)$*, Commun. Math. Phys. **121** (1989), 291–321.

[ACF07] S. Albeverio, C. Cacciapuoti, and D. Finco, *Coupling in the singular limit of thin quantum waveguides*, J. Math. Phys. **48** (2007), 032103.

[AGH+05] S. Albeverio, F. Gesztesy, R. Høegh-Krohn, and H. Holden, *Solvable models in quantum mechanics*, second ed., AMS Chelsea Publishing, Providence, RI, 2005, With an appendix by Pavel Exner.

[AK00] S. Albeverio and P. Kurasov, *Singular perturbations of differential operators*, London Mathematical Society Lecture Note Series, vol. 271, Cambridge University Press, Cambridge, 2000, Solvable Schrödinger type operators.

[Al83] S. Alexander, *Superconductivity of networks. A percolation approach to the effects of disorder*, Phys. Rev. B (3) **27** (1983), 1541–1557.

[ALM04] C. Amovilli, F. E. Leys, and N. H. March, *Topology, connectivity, and electronic structure of C and B cages and the corresponding nanotubes*, J. Chem. Inf. Comput. Sci. **44** (2004), 122–135.

[ACh92] M. T. Anderson and J. Cheeger, *C^α-compactness for manifolds with Ricci curvature and injectivity radius bounded below*, J. Differential Geom. **35** (1992), 265–281.

[ACP09] C. Anné, G. Carron, and O. Post, *Gaps in the differential forms spectrum on cyclic coverings*, Math. Z. **262** (2009), 57–90.

[AC93] C. Anné and B. Colbois, *Opérateur de Hodge-Laplace sur des variétés compactes privées d'un nombre fini de boules*, J. Funct. Anal. **115** (1993), 190–211.

[AC95] ———, *Spectre du Laplacien agissant sur les p-formes différentielles et écrasement d'anses*, Math. Ann. **303** (1995), 545–573.

[A87] C. Anné, *Spectre du Laplacien et écrasement d'anses*, Ann. Sci. Éc. Norm. Super., IV. Sér. **20** (1987), 271–280.

[A90] ———, *Fonctions propres sur des variétés avec des anses fines, application à la multiplicité*, Comm. Partial Differential Equations **15** (1990), 1617–1630.

[A94] ———, *Laplaciens en interaction*, Manuscr. Math. **83** (1994), 59–74.

[AtE10] W. Arendt and A. F. M. ter Elst, *The Dirichlet-to-Neumann operator on rough domains*, Preprint `arXiv:1005.0875` (2010).

O. Post, *Spectral Analysis on Graph-Like Spaces*, Lecture Notes in Mathematics 2039,
DOI 10.1007/978-3-642-23840-6, © Springer-Verlag Berlin Heidelberg 2012

[Ar00] Y. Arlinskii, *Abstract boundary conditions for maximal sectorial extensions of sectorial operators*, Math. Nachr. **209** (2000), 5–36.

[Au82] T. Aubin, *Nonlinear analysis on manifolds. Monge-Ampère equations*, Grundlehren der Mathematischen Wissenschaften, vol. 252, Springer, New York, 1982.

[AHL⁺01] P. Auscher, S. Hofmann, M. Lacey, J. Lewis, A. McIntosh, and P. Tchamitchian, *The solution of Kato's conjectures*, C. R. Acad. Sci. Paris Sér. I Math. **332** (2001), 601–606.

[ABGM91] Y. Avishai, D. Bessis, B. G. Giraud, and G. Mantica, *Quantum bound states in open geometries*, Phys. Rev. B **44** (1991), 8028–8034.

[AEY94] J. E. Avron, P. Exner, and Y. Last, *Periodic Schrödinger operators with large gaps and Wannier-Stark ladders* , Phys. Rev. Lett. **72** (1994), 896–899.

[BBK07] A. Badanin, J. Brüning, and E. Korotyaev, *Schrödinger operators on armchair nanotubes. II*, Preprint `arXiv:0707.3900` (2007).

[BF06] M. Baker and X. Faber, *Metrized graphs, Laplacian operators, and electrical networks*, Quantum graphs and their applications, Contemp. Math., vol. 415, Amer. Math. Soc., Providence, RI, 2006, pp. 15–33.

[BR07] M. Baker and R. Rumely, *Harmonic analysis on metrized graphs*, Canad. J. Math. **59** (2007), 225–275.

[BBC08] W. Ballmann, J. Brüning, and G. Carron, *Regularity and index theory for Dirac-Schrödinger systems with Lipschitz coefficients*, J. Math. Pures Appl. (9) **89** (2008), 429–476.

[BC71] E. Balslev and J. M. Combes, *Spectral properties of many-body Schrödinger operators with dilatation-analytic interactions*, Comm. Math. Phys. **22** (1971), 280–294.

[BSS06] R. Band, T. Shapira, and U. Smilansky, *Nodal domains on isospectral quantum graphs: the resolution of isospectrality*, J. Phys. A **39** (2006), 13999–14014.

[BBK06] M. T. Barlow, R. F. Bass, and T. Kumagai, *Stability of parabolic Harnack inequalities on metric measure spaces*, J. Math. Soc. Japan **58** (2006), 485–519.

[BCh05] R. A. Bartnik and P. T. Chruściel, *Boundary value problems for Dirac-type equations*, J. Reine Angew. Math. **579** (2005), 13–73.

[Ba10] H. Baumgärtel, *Resonances of quantum mechanical scattering systems and Lax-Phillips scattering theory*, J. Math. Phys. **51** (2010), 113508, 20.

[Be73] J. T. Beale, *Scattering frequencies of resonators*, Commun. pure appl. Math. **26** (1973), 549–563.

[BL07] J. Behrndt and M. Langer, *Boundary value problems for elliptic partial differential operators on bounded domains*, J. Funct. Anal. **243** (2007), 536–565.

[BL10] ———, *Elliptic operators, Dirichlet-to-Neumann maps and quasi boundary triples*, Preprint (2010).

[BBG94] P. Bérard, G. Besson, and S. Gallot, *Embedding Riemannian manifolds by their heat kernel*, Geom. Funct. Anal. **4** (1994), 373–398.

[Bi53] M. Š. Birman, *On the theory of self-adjoint extensions of positive definite operators*, Doklady Akad. Nauk SSSR (N.S.) **91** (1953), 189–191.

[BCD89] P. Briet, J.-M. Combes, and P. Duclos, *Spectral stability under tunneling*, Comm. Math. Phys. **126** (1989), 133–156.

[BGW09] B. M. Brown, G. Grubb, and I. G. Wood, *M-functions for closed extensions of adjoint pairs of operators with applications to elliptic boundary problems*, Math. Nachr. **282** (2009), 314–347.

[BHM⁺09] B. M. Brown, J. Hinchcliffe, M. Marletta, S. Naboko, and I. Wood, *The abstract Titchmarsh-Weyl M-function for adjoint operator pairs and its relation to the spectrum*, Integral Equations Operator Theory **63** (2009), 297–320.

[BMNW08] B. M. Brown, M. Marletta, S. Naboko, and I. Wood, *Boundary triplets and M-functions for non-selfadjoint operators, with applications to elliptic PDEs and block operator matrices*, J. Lond. Math. Soc. (2) **77** (2008), 700–718.

[BHM94] R. M. Brown, P. D. Hislop, and A. Martinez, *Lower bounds on the interaction between cavities connected by a thin tube*, Duke Math. J. **73** (1994), 163–176.

[B89] J. Brüning, *On Schrödinger operators with discrete spectrum*, J. Funct. Anal. **85** (1989), 117–150.

[BEG03] J. Brüning, P. Exner, and V. Geyler, *Large gaps in point-coupled periodic systems of manifolds*, J. Phys. A **36** (2003), 4875–4890.

[BG03] J. Brüning and V. Geyler, *Scattering on compact manifolds with infinitely thin horns*, J. Math. Phys. **44** (2003), 371–405.

[BG05] ———, *Geometric scattering on compact Riemannian manifolds and spectral theory of automorphic functions*, Preprint (2005).

[BGL05] J. Brüning, V. Geyler, and I. Lobanov, *Spectral properties of Schrödinger operators on decorated graphs*, Mat. Zametki **77** (2005), 152–156.

[BGP07] J. Brüning, V. Geyler, and K. Pankrashkin, *Cantor and band spectra for periodic quantum graphs with magnetic fields*, Comm. Math. Phys. **269** (2007), 87–105.

[BGP08] ———, *Spectra of self-adjoint extensions and applications to solvable Schrödinger operators*, Rev. Math. Phys. **20** (2008), 1–70.

[BL92] J. Brüning and M. Lesch, *Hilbert complexes*, J. Funct. Anal. **108** (1992), 88–132.

[CE07] C. Cacciapuoti and P. Exner, *Nontrivial edge coupling from a Dirichlet network squeezing: the case of a bent waveguide*, J. Phys. A **40** (2007), L511–L523.

[Ca00] R. Carlson, *Nonclassical Sturm-Liouville problems and Schrödinger operators on radial trees*, Electron. J. Differential Equations (2000), No. 71, 24 pp. (electronic).

[CW07] D. I. Cartwright and W. Woess, *The spectrum of the averaging operator on a network (metric graph)*, Illinois J. Math. **51** (2007), 805–830.

[Ct97] C. Cattaneo, *The spectrum of the continuous Laplacian on a graph*, Monatsh. Math. **124** (1997), 215–235.

[Ch84] I. Chavel, *Eigenvalues in Riemannian geometry*, Academic Press, Orlando, 1984.

[CF78] I. Chavel and E. A. Feldman, *Spectra of domains in compact manifolds*, J. Funct. Anal. **30** (1978), 198–222.

[CF81] ———, *Spectra of manifolds with small handles*, Comment. Math. Helv. **56** (1981), 83–102.

[CLPZ02] G. A. Chechkin, D. Lukkassen, A. L. Piatnitski, and V. V. Zhikov, *On homogenization of networks and junctions*, Asymptot. Anal. **30** (2002), 61–80.

[Che76] S. Y. Cheng, *Eigenfunctions and nodal sets*, Comment. Math. Helv. **51** (1976), 43–55.

[CE04] T. Cheon and P. Exner, *An approximation to δ' couplings on graphs*, J. Phys. A **37** (2004), L329–L335.

[CET10] T. Cheon, P. Exner, and O. Turek, *Approximation of a general singular vertex coupling in quantum graphs*, Ann. Physics **325** (2010), 548–578.

[Chu97] F. Chung, *Spectral graph theory*, CBMS Regional Conference Series in Mathematics, vol. 92, Published for the Conference Board of the Mathematical Sciences, Washington, DC, 1997.

[CGY96] F. Chung, A. Grigor′yan, and S.-T. Yau, *Upper bounds for eigenvalues of the discrete and continuous Laplace operators*, Adv. Math. **117** (1996), 165–178.

[CGY00] ———, *Higher eigenvalues and isoperimetric inequalities on Riemannian manifolds and graphs*, Comm. Anal. Geom. **8** (2000), 969–1026.

[CC91] B. Colbois and G. Courtois, *Convergence de variétés et convergence du spectre du Laplacien*, Ann. Sci. École Norm. Sup. (4) **24** (1991), 507–518.

[CdV86] Y. Colin de Verdière, *Sur la multiplicité de la première valeur propre non nulle du laplacien*, Comment. Math. Helv. **61** (1986), 254–270.

[CdV87] ———, *Construction de laplaciens dont une partie finie du spectre est donnée*, Ann. Sci. École Norm. Sup. (4) **20** (1987), 599–615.

[CdV98] ———, *Spectres de graphes*, Cours Spécialisés [Specialized Courses], vol. 4, Société Mathématique de France, Paris, 1998.

[Co69] J.-M. Combes, *Relatively compact interactions in many particle systems*, Commun. Math. Phys. **12** (1969), 283–295.

[CDKS87] J.-M. Combes, P. Duclos, M. Klein, and R. Seiler, *The shape resonance*, Comm. Math. Phys. **110** (1987), 215–236.

[CT73] J.-M. Combes and L. Thomas, *Asymptotic behaviour of eigenfunctions for multiparticle Schrödinger operators*, Comm. Math. Phys. **34** (1973), 251–270.

[CMM94] E. Curtis, E. Mooers, and J. Morrow, *Finding the conductors in circular networks from boundary measurements*, RAIRO Modél. Math. Anal. Numér. **28** (1994), 781–814.

[CIM98] E. B. Curtis, D. Ingerman, and J. A. Morrow, *Circular planar graphs and resistor networks*, Linear Algebra Appl. **283** (1998), 115–150.

[CM91] E. B. Curtis and J. A. Morrow, *The Dirichlet to Neumann map for a resistor network*, SIAM J. Appl. Math. **51** (1991), 1011–1029.

[CM00] ———, *Inverse problems for electrical networks*, World Scientific, Singapore, 2000.

[DT82] B. E. J. Dahlberg and E. Trubowitz, *A remark on two-dimensional periodic potentials*, Comment. Math. Helv. **57** (1982), 130–134.

[Da08] D. Daners, *Domain perturbation for linear and semi-linear boundary value problems*, Handbook of differential equations: stationary partial differential equations. Vol. VI, Handb. Differ. Equ., Elsevier/North-Holland, Amsterdam, 2008, pp. 1–81.

[D95] E. B. Davies, *Spectral theory and differential operators*, Cambridge University Press, Cambridge, 1995.

[D97] ———, *Spectral enclosures and complex resonances for general self-adjoint operators*, LMS J. Comput. Math. **1** (1998), 42–74 (electronic).

[DS92] E. B. Davies and B. Simon, *Spectral properties of Neumann Laplacian of horns*, Geom. Funct. Anal. **2** (1992), 105–117.

[DEL10] E. B. Davies, P. Exner, and J. Lipovský, *Non-Weyl asymptotics for quantum graphs with general coupling conditions*, J. Phys. A **43** (2010), 474013, 16.

[DN00] B. Dekoninck and S. Nicaise, *The eigenvalue problem for networks of beams*, Linear Algebra Appl. **314** (2000), 165–189.

[DT04] G. F. Dell'Antonio and L. Tenuta, *Semiclassical analysis of constrained quantum systems*, J. Phys. A **37** (2004), 5605–5624.

[DT06] ———, *Quantum graphs as holonomic constraints*, J. Math. Phys. **47** (2006), 072102, 21.

[DeA07] G. Dell'Antonio, *Dynamics on quantum grpahs as constrained systems*, Rep. Math. Phys. **59** (2007), 267–279.

[DHMS00] V. Derkach, S. Hassi, M. Malamud, and H. de Snoo, *Generalized resolvents of symmetric operators and admissibility*, Methods Funct. Anal. Topology **6** (2000), 24–55.

[DHMS06] ———, *Boundary relations and their Weyl families*, Trans. Amer. Math. Soc. **358** (2006), 5351–5400 (electronic).

[DM91] V. Derkach and M. Malamud, *Generalized resolvents and the boundary value problems for Hermitian operators with gaps*, J. Funct. Anal. **95** (1991), 141–242.

[DM95] ———, *The extension theory of Hermitian operators and the moment problem*, J. Math. Sci. (New York) **73** (1995), 141–242.

[Di69] J. Dixmier, *Les algèbres d'opérateurs dans l'espace hilbertien (algèbres de von Neumann)*, Gauthier-Villars Éditeur, Paris, 1969.

[Do84] J. Dodziuk, *Difference equations, isoperimetric inequality and transience of certain random walks*, Trans. Amer. Math. Soc. **284** (1984), 787–794.

[EH89] W. D. Evans and D. J. Harris, *On the approximation numbers of Sobolev embeddings for irregular domains*, Quart. J. Math. Oxford Ser. (2) **40** (1989), 13–42.

[Ex95] P. Exner, *Lattice Kronig-Penney models*, Phys. Rev. Lett. **74** (1995), 3503–3506.

[Ex97] ———, *A duality between Schrödinger operators on graphs and certain Jacobi matrices*, Ann. Inst. H. Poincaré Phys. Théor. **66** (1997), 359–371.

[EG96] P. Exner and R. Gawlista, *Band spectra of rectangular graph superlattices*, Phys. Rev. B **53** (1996), 7275–7286.

[EKK+08] P. Exner, J. P. Keating, P. Kuchment, T. Sunada, and A. Teplyaev (eds.), *Analysis on graphs and its applications*, Proc. Symp. Pure Math., vol. 77, Providence, R.I., Amer. Math. Soc., 2008.

[EP05] P. Exner and O. Post, *Convergence of spectra of graph-like thin manifolds*, Journal of Geometry and Physics **54** (2005), 77–115.

[EP07] ———, *Convergence of resonances on thin branched quantum wave guides*, J. Math. Phys. **48** (2007), 092104 (43pp).

[EP09] ———, *Approximation of quantum graph vertex couplings by scaled Schrödinger operators on thin branched manifolds*, J. Phys. A **42** (2009), 415305 (22pp).

[EŠ89] P. Exner and P. Šeba, *Electrons in semiconductor microstructures: a challenge to operator theorists*, Schrödinger operators, standard and nonstandard (Dubna, 1988), World Sci. Publishing, Teaneck, NJ, 1989, pp. 78–100.

[EKW10] P. Exner, P. Kuchment, and B. Winn, *On the location of spectral edges in $\mathbb{Z}$-periodic media*, J. Phys. A **43** (2010), 474022, 8.

[FH05] K. Fissmer and U. Hamenstädt, *Spectral convergence of manifold pairs*, Comment. Math. Helv. **80** (2005), 725–754.

[Fre96] M. Freidlin, *Markov processes and differential equations: asymptotic problems*, Lectures in Mathematics ETH Zürich, Birkhäuser Verlag, Basel, 1996.

[FW93] M. I. Freidlin and A. D. Wentzell, *Diffusion processes on graphs and the averaging principle*, Ann. Probab. **21** (1993), 2215–2245.

[FT04a] J. Friedman and J.-P. Tillich, *Calculus on graphs*, Preprint `arXiv:cs.DM/0408028` (2004).

[FT04b] ———, *Wave equations for graphs and the edge-based Laplacian*, Pacific J. Math. **216** (2004), 229–266.

[Fri34] K. Friedrichs, *Spektraltheorie halbbeschränkter Operatoren und Anwendung auf die Spektralzerlegung von Differentialoperatoren*, Math. Ann. **109** (1934), 465–487.

[Fu87] K. Fukaya, *Collapsing of Riemannian manifolds and eigenvalues of Laplace operator*, Invent. Math. **87** (1987), 517–547.

[FKW07] S. Fulling, P. Kuchment, and J. H. Wilson, *Index theorems for quantum graphs*, J. Phys. A **40** (2007), 14165–14180.

[GO91] B. Gaveau and M. Okada, *Differential forms and heat diffusion on one-dimensional singular varieties*, Bull. Sci. Math. **115** (1991), 61–79.

[GnS06] S. Gnutzmann and U. Smilansky, *Quantum graphs: Applications to quantum chaos and universal spectral statistics*, Adv. in Phys. **55** (2006), 527–625.

[GG91] V. I. Gorbachuk and M. L. Gorbachuk, *Boundary value problems for operator differential equations*, Mathematics and its Applications (Soviet Series), vol. 48, Kluwer Academic Publishers Group, Dordrecht, 1991.

[GY83] S. Graffi and K. Yajima, *Exterior complex scaling and the AC-Stark effect in a Coulomb field*, Comm. Math. Phys. **89** (1983), 277–301.

[G08a] D. Grieser, *Spectra of graph neighborhoods and scattering*, Proc. Lond. Math. Soc. (3) **97** (2008), 718–752.

[G08b] ———, *Thin tubes in mathematical physics, global analysis and spectral geometry*, in [EKK$^+$08] (2008), 565–593.

[GJ09] D. Grieser and D. Jerison, *Asymptotics of eigenfunctions on plane domains*, Pacific J. Math. **240** (2009), 109–133.

[Gr03] A. Grigor′yan, *Heat kernels and function theory on metric measure spaces*, Heat kernels and analysis on manifolds, graphs, and metric spaces (Paris, 2002), Contemp. Math., vol. 338, Amer. Math. Soc., Providence, RI, 2003, pp. 143–172.

[Gr10] ———, *Heat kernels on metric measure spaces with regular volume growth*, Handbook of geometric analysis, No. 2, Adv. Lect. Math. (ALM), vol. 13, Int. Press, Somerville, MA, 2010, pp. 1–60.

[GSC99] A. Grigor′yan and L. Saloff-Coste, *Heat kernel on connected sums of Riemannian manifolds*, Math. Res. Lett. **6** (1999), 307–321.

[Grv85] P. Grisvard, *Elliptic problems in nonsmooth domains*, Monographs and Studies in Mathematics, vol. 24, Pitman (Advanced Publishing Program), Boston, MA, 1985.

[Gro81] M. Gromov, *Structures métriques pour les variétés riemanniennes*, Textes Mathématiques, vol. 1, CEDIC, Paris, 1981, Edited by J. Lafontaine and P. Pansu.

[Gro99] ______, *Metric structures for Riemannian and non-Riemannian spaces*, Progress in Mathematics, vol. 152, Birkhäuser Boston Inc., Boston, MA, 1999.

[Gru68] G. Grubb, *A characterization of the non-local boundary value problems associated with an elliptic operator*, Ann. Scuola Norm. Sup. Pisa (3) **22** (1968), 425–513.

[Gru08] ______, *Krein resolvent formulas for elliptic boundary problems in nonsmooth domains*, Rend. Semin. Mat. Univ. Politec. Torino **66** (2008), 271–297.

[Gru10a] ______, *Extension theory for elliptic partial differential operators with pseudodifferential methods*, Preprint `arXiv:1008.1081` (2010).

[Gru10b] ______, *Krein-like extensions and the lower boundedness problem for elliptic operators on exterior domains*, Preprint `arXiv:1002.4549` (2010).

[GSm01] B. Gutkin and U. Smilansky, *Can one hear the shape of a graph?*, J. Phys. A **34** (2001), 6061–6068.

[Ha00] M. Harmer, *Hermitian symplectic geometry and extension theory*, J. Phys. A **33** (2000), 9193–9203.

[HKSW07] J. M. Harrison, P. Kuchment, A. Sobolev, and B. Winn, *On occurrence of spectral edges for periodic operators inside the Brillouin zone*, J. Phys. A **40** (2007), 7597–7618.

[He96] E. Hebey, *Sobolev spaces on Riemannian manifolds*, Lecture Notes in Mathematics. Vol. 1635, Springer, Berlin, 1996.

[HM87] B. Helffer and A. Martinez, *Comparaison entre les diverses notions de résonances*, Helv. Phys. Acta **60** (1987), 992–1003.

[HSS91] R. Hempel, L. A. Seco, and B. Simon, *The essential spectrum of Neumann Laplacians on some bounded singular domains*, J. Funct. Anal. **102** (1991), 448–483.

[H06] R. Hempel, *On the lowest eigenvalue of the Laplacian with Neumann boundary condition at a small obstacle*, J. Comput. Appl. Math. **194** (2006), 54–74.

[HP03] R. Hempel and O. Post, *Spectral gaps for periodic elliptic operators with high contrast: an overview*, Progress in analysis, Vol. I, II (Berlin, 2001), World Sci. Publ., River Edge, NJ, 2003, pp. 577–587.

[HS99] Y. Higuchi and T. Shirai, *A remark on the spectrum of magnetic Laplacianon a graph*, Yokohama Math. J. **47** (1999), 129–141.

[HS04] ______, *Some spectral and geometric properties for infinite graphs*, Discrete geometric analysis, Contemp. Math., vol. 347, Amer. Math. Soc., Providence, RI, 2004, pp. 29–56.

[Hi92] P. D. Hislop, *Singular perturbations of Dirichlet and Neumann domains and resonances for obstacle scattering*, Astérisque (1992), 8, 197–216, Méthodes semi-classiques, Vol. 2 (Nantes, 1991).

[HiL01] P. D. Hislop and C. V. Lutzer, *Spectral asymptotics of the Dirichlet-to-Neumann map on multiply connected domains in* $\mathbb{R}^d$, Inverse Problems **17** (2001), 1717–1741.

[HiM91] P. D. Hislop and A. Martinez, *Scattering resonances of a Helmholtz resonator*, Indiana Univ. Math. J. **40** (1991), 767–788.

[HiS89] P. D. Hislop and I. M. Sigal, *Semiclassical theory of shape resonances in quantum mechanics*, Mem. Amer. Math. Soc. **78** (1989), 123.

[HiP09] P. D. Hislop and O. Post, *Anderson localization for radial tree-like quantum graphs*, Waves Random Complex Media **19** (2009), 216–261.

[Hu86] W. Hunziker, *Distortion analyticity and molecular resonance curves*, Ann. Inst. H. Poincaré Phys. Théor. **45** (1986), 339–358.

[JS10] P. Joly and A. Semin, *Study of propagation of acoustic waves in junction of thin slots*, Preprint (2010).

[Ju01] C. M. Judge, *Tracking eigenvalues to the frontier of moduli space. I. Convergence and spectral accumulation*, J. Funct. Anal. **184** (2001), 273–290.

[Ka66] M. Kac, *Can one hear the shape of a drum?* Am. Math. Mon. **73** (1966), 1–23.

[KKVW09] U. Kant, T. Klauss, J. Voigt, and M. Weber, *Dirichlet forms for singular one-dimensional operators and on graphs*, J. Evol. Equ. **9** (2009), 637–659.

[KP88] L. Karp and M. Pinsky, *First-order asymptotics of the principal eigenvalue of tubular neighborhoods*, Geometry of random motion (Ithaca, N.Y., 1987), Contemp. Math., vol. 73, Amer. Math. Soc., Providence, RI, 1988, pp. 105–119.

[Ka02] A. Kasue, *Convergence of Riemannian manifolds and Laplace operators. I*, Ann. Inst. Fourier (Grenoble) **52** (2002), 1219–1257.

[Ka06] ———, *Convergence of Riemannian manifolds and Laplace operators. II*, Potential Anal. **24** (2006), 137–194.

[KK94] A. Kasue and H. Kumura, *Spectral convergence of Riemannian manifolds*, Tohoku Math. J. (2) **46** (1994), 147–179.

[KK96] ———, *Spectral convergence of Riemannian manifolds. II*, Tohoku Math. J. (2) **48** (1996), 71–120.

[K66] T. Kato, *Perturbation theory for linear operators*, Springer-Verlag, Berlin, 1966.

[K67] ———, *Scattering theory with two Hilbert spaces*, J. Functional Analysis **1** (1967), 342–369.

[KL07] E. Korotyaev and I. Lobanov, *Schrödinger operators on zigzag nanotubes*, Ann. Henri Poincaré **8** (2007), 1151–1176.

[KPS07] V. Kostrykin, J. Potthoff, and R. Schrader, *Heat kernels on metric graphs and a trace formula*, Adventures in Mathematical Physics, Contemp. Math., vol. 447, Amer. Math. Soc., Providence, RI, 2007, pp. 175–198.

[KS99] V. Kostrykin and R. Schrader, *Kirchhoff's rule for quantum wires*, J. Phys. A **32** (1999), 595–630.

[KS03] ———, *Quantum wires with magnetic fluxes*, Comm. Math. Phys. **237** (2003), 161–179, Dedicated to Rudolf Haag.

[KS06] ———, *Laplacians on metric graphs: eigenvalues, resolvents and semigroups*, Quantum graphs and their applications, Contemp. Math., vol. 415, Amer. Math. Soc., Providence, RI, 2006, pp. 201–225.

[Ko00] S. Kosugi, *A semilinear elliptic equation in a thin network-shaped domain*, J. Math. Soc. Japan **52** (2000), 673–697.

[Ko02] ———, *Semilinear elliptic equations on thin network-shaped domains with variable thickness*, J. Differential Equations **183** (2002), 165–188.

[KSm97] T. Kottos and U. Smilansky, *Quantum chaos on graphs*, Phys. Rev. Lett. **79** (1997), 4794–4797.

[KSm99] ———, *Periodic orbit theory and spectral statistics for quantum graphs*, Ann. Physics **274** (1999), 76–124.

[KSm03] ———, *Quantum graphs: a simple model for chaotic scattering*, J. Phys. A **36** (2003), 3501–3524, Random matrix theory.

[KOZ94] S. M. Kozlov, O. A. Oleĭnik, and V. V. Zhikov, *Homogenization of differential operators and integral functionals*, Springer, Berlin, 1994.

[Kr47] M. Krein, *The theory of self-adjoint extensions of semi-bounded Hermitian transformations and its applications. I*, Rec. Math. [Mat. Sbornik] N.S. **20(62)** (1947), 431–495.

[Ku01] P. Kuchment, *The mathematics of photonic crystals*, Mathematical modeling in optical science, SIAM, Philadelphia, PA, 2001, pp. 207–272.

[Ku04] ———, *Quantum graphs: I. Some basic structures*, Waves Random Media **14** (2004), S107–S128.

[Ku05] ———, *Quantum graphs: II. Some spectral properties of quantum and combinatorial graphs*, J. Phys. A **38** (2005), 4887–4900.

[Ku08] ———, *Quantum graphs: an introduction and a brief survey*, in [EKK$^+$08] (2008), 291–312.

[KuP07] P. Kuchment and O. Post, *On the spectra of carbon nano-structures*, Comm. Math. Phys. **275** (2007), 805–826.

[KuZ01] P. Kuchment and H. Zeng, *Convergence of spectra of mesoscopic systems collapsing onto a graph*, J. Math. Anal. Appl. **258** (2001), 671–700.

[KuZ03] ______, *Asymptotics of spectra of Neumann Laplacians in thin domains*, Advances in differential equations and mathematical physics (Birmingham, AL, 2002), Contemp. Math., vol. 327, Amer. Math. Soc., Providence, RI, 2003, pp. 199–213.

[Ks08] P. Kurasov, *Graph Laplacians and topology*, Ark. Mat. **46** (2008), 95–111.

[KN05] P. Kurasov and M. Nowaczyk, *Inverse spectral problem for quantum graphs*, J. Phys. A **38** (2005), 4901–4915.

[KN06] ______, *Corrigendum to [KN05]*, J. Phys. A **39** (2006), 993.

[KZh98] S. Kusuoka and X. Y. Zhou, *Waves on fractal-like manifolds and effective energy propagation*, Probab. Theory Related Fields **110** (1998), 473–495.

[KSh03] K. Kuwae and T. Shioya, *Convergence of spectral structures: a functional analytic theory and its applications to spectral geometry*, Comm. Anal. Geom. **11** (2003), 599–673.

[LU01] M. Lassas and G. Uhlmann, *On determining a Riemannian manifold from the Dirichlet-to-Neumann map*, Ann. Sci. École Norm. Sup. (4) **34** (2001), 771–787.

[LPPV08] D. Lenz, N. Peyerimhoff, O. Post, and I. Veselić, *Continuity properties of the integrated density of states on manifolds*, Jpn. J. Math. **3** (2008), 121–161.

[LY86] P. Li and S.-T. Yau, *On the parabolic kernel of the Schrödinger operator*, Acta Math. **156** (1986), 153–201.

[LM68] J.-L. Lions and E. Magenes, *Problèmes aux limites non homogènes et applications. Vol. 1*, Travaux et Recherches Mathématiques, No. 17, Dunod, Paris, 1968.

[LP07] F. Lledó and O. Post, *Generating spectral gaps by geometry*, Prospects in Mathematical Physics, Young Researchers Symposium of the 14th International Congress on Mathematical Physics, Lisbon, July 2003, Contemporary Mathematics, vol. 437, 2007, pp. 159–169.

[LP08a] ______, *Eigenvalue bracketing for discrete and metric graphs*, J. Math. Anal. Appl. **348** (2008), 806–833.

[LP08b] ______, *Existence of spectral gaps, covering manifolds and residually finite groups*, Rev. Math. Phys. **20** (2008), 199–231.

[Lo01] J. Lott, *On the spectrum of a finite-volume negatively-curved manifold*, Amer. J. Math. **123** (2001), 185–205.

[Lü02] W. Lück, *L^2-invariants: theory and applications to geometry and K-theory*, Ergebnisse der Mathematik und ihrer Grenzgebiete., vol. 44, Springer, Berlin, 2002.

[Ma10] M. M. Malamud, *Spectral theory of elliptic operators in exterior domains*, Russ. J. Math. Phys. **17** (2010), 96–125.

[McI72] A. McIntosh, *On the comparability of $A^{1/2}$ and $A^{*1/2}$*, Proc. Amer. Math. Soc. **32** (1972), 430–434.

[Me01] T. A. Mel′nyk, *Hausdorff convergence and asymptotic estimates of the spectrum of a perturbed operator*, Z. Anal. Anwendungen **20** (2001), 941–957.

[Me03] ______, *A scheme for studying the spectrum of a family of perturbed operators and its application to spectral problems in thick junctions*, Nelīnīīnī Koliv. **6** (2003), 233–251.

[Me08] ______, *Homogenization of a boundary-value problem with a nonlinear boundary condition in a thick junction of type* $3:2:1$, Math. Methods Appl. Sci. **31** (2008), 1005–1027.

[MW89] B. Mohar and W. Woess, *A survey on spectra of infinite graphs*, Bull. London Math. Soc. **21** (1989), 209–234.

[MV06] S. Molchanov and B. Vainberg, *Transition from a network of thin fibers to the quantum graph: an explicitly solvable model*, Quantum graphs and their applications, Contemp. Math., vol. 415, Amer. Math. Soc., Providence, RI, 2006, pp. 227–239.

[MV07] ______, *Scattering solutions in networks of thin fibers: small diameter asymptotics*, Comm. Math. Phys. **273** (2007), 533–559.

[MV08] ______, *Laplace operator in networks of thin fibers: spectrum near the threshold*, Stochastic analysis in mathematical physics, World Sci. Publ., Hackensack, NJ, 2008, pp. 69–93.

[Mo94] U. Mosco, *Composite media and asymptotic Dirichlet forms*, J. Funct. Anal. **123** (1994), 368–421.

[MNP10] D. Mugnolo, R. Nittka, and O. Post, *Convergence of sectorial operators on varying Hilbert spaces*, Preprint arXiv:1007.3932 (2010).

[MM06] J. Müller and W. Müller, *Regularized determinants of Laplace-type operators, analytic surgery, and relative determinants*, Duke Math. J. **133** (2006), 259–312.

[Na88] A. I. Nachman, *Reconstructions from boundary measurements*, Ann. of Math. (2) **128** (1988), 531–576.

[Ni87] S. Nicaise, *Approche spectrale des problèmes de diffusion sur les réseaux*, Séminaire de Théorie du Potentiel, Paris, No. 8, Lecture Notes in Math., vol. 1235, Springer, Berlin, 1987, pp. 120–140.

[No07] M. Nowaczyk, *Inverse spectral problem for quantum graphs with rationally dependent edges*, Operator theory, analysis and mathematical physics, Oper. Theory Adv. Appl., vol. 174, Birkhäuser, Basel, 2007, pp. 105–116.

[OSY92] O. A. Oleĭnik, A. S. Shamaev, and G. A. Yosifian, *Mathematical problems in elasticity and homogenization*, Studies in Mathematics and its Applications, vol. 26, North-Holland Publishing Co., Amsterdam, 1992.

[Oz81] S. Ozawa, *Singular variation of domains and eigenvalues of the Laplacian*, Duke Math. J. **48** (1981), 767–778.

[Oz82] ———, *Spectra of domains with small spherical Neumann boundary*, Proc. Japan Acad. Ser. A Math. Sci. **58** (1982), 190–192.

[Pan06] K. Pankrashkin, *Spectra of Schrödinger operators on equilateral quantum graphs*, Lett. Math. Phys. **77** (2006), 139–154.

[Par08] L. Parnovski, *Bethe-Sommerfeld conjecture*, Ann. Henri Poincaré **9** (2008), 457–508.

[Pas04] S. E. Pastukhova, *On the convergence of hyperbolic semigroups in a variable Hilbert space*, Tr. Semin. im. I. G. Petrovskogo (2004), 215–249, 343.

[Pav02] B. Pavlov, *Resonance quantum switch: matching domains*, Surveys in analysis and operator theory (Canberra, 2001), Proc. Centre Math. Appl. Austral. Nat. Univ., vol. 40, Austral. Nat. Univ., Canberra, 2002, pp. 127–156.

[Pav07] ———, *A star-graph model via operator extension*, Math. Proc. Cambridge Philos. Soc. **142** (2007), 365–384.

[PKWA10] R. C. Penner, M. Knudsen, C. Wiuf, and J. E. Andersen, *Fatgraph models of proteins*, Comm. Pure Appl. Math. **63** (2010), 1249–1297.

[PWZ08] Y. Pinchover, G. Wolansky, and D. Zelig, *Spectral properties of Schrödinger operators on radial N-dimensional infinite trees*, Israel J. Math. **165** (2008), 281–328.

[Pc08] A. Posilicano, *Self-adjoint extensions of restrictions*, Oper. Matrices **2** (2008), 483–506.

[PcR09] A. Posilicano and L. Raimondi, *Krein's resolvent formula for self-adjoint extensions of symmetric second-order elliptic differential operators*, J. Phys. A **42** (2009), 015204 (11pp).

[P03a] O. Post, *Eigenvalues in spectral gaps of a perturbed periodic manifold*, Mathematische Nachrichten **261–262** (2003), 141–162.

[P03b] ———, *Periodic manifolds with spectral gaps*, J. Diff. Equations **187** (2003), 23–45.

[P05] ———, *Branched quantum wave guides with Dirichlet boundary conditions: the decoupling case*, Journal of Physics A: Mathematical and General **38** (2005), 4917–4931.

[P06] ———, *Spectral convergence of quasi-one-dimensional spaces*, Ann. Henri Poincaré **7** (2006), 933–973.

[P07] ———, *First order operators and boundary triples*, Russ. J. Math. Phys. **14** (2007), 482–492.

[P08] ———, *Equilateral quantum graphs and boundary triples*, in [EKK$^+$08] (2008), 469–490.

[P09a] ———, *First order approach and index theorems for discrete and metric graphs*, Ann. Henri Poincaré **10** (2009), 823–866.

[P09b] ______, *Spectral analysis of metric graphs and related spaces*, in "Limits of graphs in group theory", eds. G. Arzhantseva and A. Valette, Presses Polytechniques et Universitaires Romandes, 109–140 (2009), 109–140.

[RT75] J. Rauch and M. Taylor, *Potential and scattering theory on wildly perturbed domains*, J. Funct. Anal. **18** (1975), 27–59.

[RS80] M. Reed and B. Simon, *Methods of modern mathematical physics I–IV*, Academic Press, New York, 1980.

[Ros97] S. Rosenberg, *The Laplacian on a Riemannian manifold*, London Mathematical Society Student Texts. 31, Cambridge University Press, Cambridge, 1997.

[Rot84] J.-P. Roth, *Le spectre du laplacien sur un graphe*, Théorie du potentiel (Orsay, 1983), Lecture Notes in Math., vol. 1096, Springer, Berlin, 1984, pp. 521–539.

[RuS01a] J. Rubinstein and M. Schatzman, *Variational problems on multiply connected thin strips. I. Basic estimates and convergence of the Laplacian spectrum*, Arch. Ration. Mech. Anal. **160** (2001), 271–308.

[RuS01b] ______, *Variational problems on multiply connected thin strips. II. Convergence of the Ginzburg-Landau functional*, Arch. Ration. Mech. Anal. **160** (2001), 309–324.

[RSc53] K. Ruedenberg and C. W. Scherr, *Free–electron network model for conjugated systems, I. Theory*, J. Chem. Phys. **21** (1953), 1565–1581.

[Ry07] V. Ryzhov, *A general boundary value problem and its Weyl function*, Opuscula Math. **27** (2007), 305–331.

[Sa00] Y. Saito, *The limiting equation for Neumann Laplacians on shrinking domains.*, Electron. J. Differ. Equ. **31** (2000), 25 p.

[SP80] E. Sánchez-Palencia, *Nonhomogeneous media and vibration theory*, Lecture Notes in Physics, vol. 127, Springer, Berlin, 1980.

[SRW89] R. L. Schult, D. G. Ravenhall, and H. W. Wyld, *Quantum bound states in a classically unbound system of crossed wires*, Phys. Rev. B **39** (1989), 5476–5479.

[Sh00] T. Shirai, *The spectrum of infinite regular line graphs*, Trans. Amer. Math. Soc. **352** (2000), 115–132.

[Si72] B. Simon, *Quadratic form techniques and the Balslev-Combes theorem*, Comm. Math. Phys. **27** (1972), 1–9.

[Si79] ______, *The definition of molecular resonance curves by the method of exterior complex scaling*, Phys. Lett. A **71** (1979), 211–214.

[Si96] ______, *Operators with singular continuous spectrum. VI. Graph Laplacians and Laplace-Beltrami operators*, Proc. Amer. Math. Soc. **124** (1996), 1177–1182.

[Sk79] M. M. Skriganov, *Proof of the Bethe-Sommerfeld conjecture in dimension* 2, Dokl. Akad. Nauk SSSR **248** (1979), 39–42.

[Sk85] ______, *The spectrum band structure of the three-dimensional Schrödinger operator with periodic potential*, Invent. Math. **80** (1985), 107–121.

[Sm07] U. Smilansky, *Quantum chaos on discrete graphs*, J. Phys. A **40** (2007), F621–F630.

[SmS06] U. Smilansky and M. Solomyak, *The quantum graph as a limit of a network of physical wires*, Quantum graphs and their applications, Contemp. Math., vol. 415, Amer. Math. Soc., Providence, RI, 2006, pp. 283–291.

[So04] M. Solomyak, *On the spectrum of the Laplacian on regular metric trees*, Waves Random Media **14** (2004), S155–S171, Special section on quantum graphs.

[Su08] T. Sunada, *Discrete geometric analysis*, Analysison Graphs and its Applications (Providence, R.I.) (P. Exner, J. P. Keating, P. Kuchment, T. Sunada, and A. Teplayaev, eds.), Proc. Symp. Pure Math., vol. 77, Amer. Math. Soc., 2008, pp. 51–83.

[Ta96] M. E. Taylor, *Partial differential equations. Basic theory*, Springer-Verlag, New York, 1996.

[Tel83] N. Teleman, *The index of signature operators on Lipschitz manifolds*, Inst. Hautes Études Sci. Publ. Math. (1983), 39–78 (1984).

[Tep98] A. Teplyaev, *Spectral analysis on infinite Sierpiński gaskets*, J. Funct. Anal. **159** (1998), 537–567.

[TW09] S. Teufel and J. Wachsmuth, *Effective hamiltonians for constrained quantum systems*, Preprint `arXiv:0907.0351` (2009).

[Vi52] M. I. Vishik, *On general boundary problems for elliptic differential equations*, Trudy Moskov. Mat. Obšč. **1** (1952), 187–246 (English translation in Am. Math. Soc., Transl. (2) 24 (1963), 107–172).

[vB85] J. von Below, *A characteristic equation associated to an eigenvalue problem on C^2-networks*, Linear Algebra Appl. **71** (1985), 309–325.

[vN30] J. von Neumann, *Allgemeine Eigenwerttheorie Hermitescher Funktionaloperatoren*, Math. Ann. **102** (1930), 49–131.

[W84] J. Weidmann, *Stetige Abhängigkeit der Eigenwerte und Eigenfunktionen elliptischer Differentialoperatoren vom Gebiet*, Math. Scand. **54** (1984), 51–69.

[Y80] K. Yosida, *Functional analysis*, sixth ed., Grundlehren der Mathematischen Wissenschaften, vol. 123, Springer-Verlag, Berlin, 1980.

[Ze05] D. Zelig, *Properties of solutions of partial differential equations on human lung-shaped domains*, Ph.D. thesis, Ph-D thesis Technion, Haifa, 2005.

[Zh02] V. V. Zhikov, *Homogenization of elasticity problems on singular structures*, Izv. Ross. Akad. Nauk Ser. Mat. **66** (2002), 81–148.

[ZP07] V. V. Zhikov and S. E. Pastukhova, *On the Trotter-Kato theorem in a variable space*, Funktsional. Anal. i Prilozhen. **41** (2007), 22–29, 96.

[Zw99] M. Zworski, *Resonances in physics and geometry*, Notices Amer. Math. Soc. **46** (1999), 319–328.

Notation

We give an overview of some general notation used in this work. Other commonly used symbols are listed in the index.

General notation

- $\mathbb{N} = \{1, 2, 3, \dots\}$, $\mathbb{N}_0 = \{0, 1, 2, \dots\}$,
- $\mathbb{Z} = \{\dots, -2, -1, 0, 1, 2, \dots\}$,
- $\mathbb{R}_+ = [0, \infty)$, $\partial_+ I := \overline{I} \cap \overline{\mathbb{R}_+ \setminus I}$ denotes the topological boundary in the *relative* topology $\mathbb{R}_+$ ($I \subset \mathbb{R}_+$), e.g., $\partial_+ [0, \lambda] = \{\lambda\}$.
- We use the short hand notation

$$\partial_s f(s) = \frac{\partial f(s)}{\partial s}$$

  Moreover, $\mathrm{d}u$ denotes the exterior derivative of $u \in \mathsf{C}^\infty(M)$.

Sets and topology

- $|A|$ denotes the number of elements in the set A
- $C = A \mathbin{\dot\cup} B$ means that $C = A \cup B$ and that A, B are disjoint ($A \cap B = \emptyset$)

  Let A, B, C be subsets of a topological space X.

- $\overline{A}$ denotes the *closure* of A, $\mathring{A}$ denotes the *interior* of A.

O. Post, *Spectral Analysis on Graph-Like Spaces*, Lecture Notes in Mathematics 2039,
DOI 10.1007/978-3-642-23840-6, © Springer-Verlag Berlin Heidelberg 2012

Measure spaces

Let X be a measure space.

- $X = A \mathbin{\overline{\cup}} B$ ("X is the disjoint union of A and B up to measure 0") means that $X = A \cup B$ and that $A \cap B$ has measure 0.
- $X = \overline{\bigcup}_i A_i$ ("X is the disjoint union of all A_i up to measure 0") means that $X = \bigcup_i A_i$ and that $A_i \cap A_j$ has measure 0 for all $i \neq j$.
- $\operatorname{len}(I)$ denotes the Lebesque measure of I (length of the interval $I \subset \mathbb{R}$).

Hilbert spaces, operators and function spaces

All our Hilbert spaces are assumed to be separable, i.e. having a *countable* orthonormal basis. The inner product and all sesquilinear forms are anti-linear in their *first* argument.

- $a \overset{\text{CS}}{\leq} b$ means that we applied the Cauchy-Schwarz inequality.
- The Cauchy-Young reads as

$$2|\langle f, g\rangle| \overset{\text{CY}}{\leq} \eta \|f\|^2 + \frac{1}{\eta}\|g\|^2 \tag{1}$$

for any $\eta > 0$. In particular, if $\eta = 1$,

$$2|\langle f, g\rangle| \leq \|f\|^2 + \|g\|^2.$$

- $\mathscr{H}_1 \oplus \mathscr{H}_2$ denotes the *orthogonal sum* of the Hilbert spaces $\mathscr{H}_1$ and $\mathscr{H}_2$; elements of $\mathscr{H}_1 \oplus \mathscr{H}_2$ are written as $f = (f_1, f_2)$ or $f = f_1 \oplus f_2$. Similarly, elements of $\bigoplus_e \mathscr{H}_e$ written as $f = \{f_e\}_e$ or sometimes as $\bigoplus_e f_e$.
- $\mathscr{H}_1 \dot{+} \mathscr{H}_2$ denotes the *topological sum* of the spaces $\mathscr{H}_i \subset \mathscr{H}$, i.e., the direct (but not necessarily orthogonal) sum, and $\mathscr{H}_i$ is assumed to be closed in $\mathscr{H}$.
- If a linear operator A is not defined for all elements of $\mathscr{H}$, we write $\operatorname{dom} A$ for the *domain of* A. Similarly, for a quadratic form $\mathfrak{h}$, we write $\operatorname{dom} \mathfrak{h}$ for its domain. The corresponding sesquilinear form is then definded on $\mathfrak{h}\colon \operatorname{dom} \mathfrak{h} \times \operatorname{dom} \mathfrak{h} \longrightarrow \mathbb{C}$.
- For a linear operator $A\colon \mathscr{H} \longrightarrow \mathscr{G}$ we denote the *range* of A by $\operatorname{ran} A := \{ g \in \mathscr{G} \mid \exists h \in \operatorname{dom} A\colon Ah = g \}$ and the *kernel* of A by $\ker A := \{ h \in \operatorname{dom} A \mid Ah = 0 \}$.
- $\mathsf{C}_\mathsf{c}^\infty(I)$ denotes the space of smooth functions with compact support in $\mathring{I}$.
- Sobolev spaces $\mathsf{H}^k(I) := \{ f \in \mathsf{L}_2(I) \mid f^{(i)} \in \mathsf{L}_2(I) \quad \forall\, i = 1, \dots, k \}$ with norm $\|f\|^2_{\mathsf{H}^k(I)} := \sum_{i=0}^k \|f^{(i)}\|^2_{\mathsf{L}_2(I)}$, where $f^{(i)}$ is the i-th (weak) derivative, $\mathring{\mathsf{H}}^k(I)$ is the closure of $\mathsf{C}_\mathsf{c}^\infty(I)$ w.r.t. the norm $\|\cdot\|_{\mathsf{H}^k(I)}$.

Index

Symbols

O. Post, *Spectral Analysis on Graph-Like Spaces*, Lecture Notes in Mathematics 2039, DOI 10.1007/978-3-642-23840-6, © Springer-Verlag Berlin Heidelberg 2012

D

E

F

G

H

I

J

S

X

Y

Z

LECTURE NOTES IN MATHEMATICS

Springer

Edited by J.-M. Morel, B. Teissier; P.K. Maini

Editorial Policy (for the publication of monographs)

1. Lecture Notes aim to report new developments in all areas of mathematics and their applications - quickly, informally and at a high level. Mathematical texts analysing new developments in modelling and numerical simulation are welcome.
   Monograph manuscripts should be reasonably self-contained and rounded off. Thus they may, and often will, present not only results of the author but also related work by other people. They may be based on specialised lecture courses. Furthermore, the manuscripts should provide sufficient motivation, examples and applications. This clearly distinguishes Lecture Notes from journal articles or technical reports which normally are very concise. Articles intended for a journal but too long to be accepted by most journals, usually do not have this "lecture notes" character. For similar reasons it is unusual for doctoral theses to be accepted for the Lecture Notes series, though habilitation theses may be appropriate.

2. Manuscripts should be submitted either online at www.editorialmanager.com/lnm to Springer's mathematics editorial in Heidelberg, or to one of the series editors. In general, manuscripts will be sent out to 2 external referees for evaluation. If a decision cannot yet be reached on the basis of the first 2 reports, further referees may be contacted: The author will be informed of this. A final decision to publish can be made only on the basis of the complete manuscript, however a refereeing process leading to a preliminary decision can be based on a pre-final or incomplete manuscript. The strict minimum amount of material that will be considered should include a detailed outline describing the planned contents of each chapter, a bibliography and several sample chapters.
   Authors should be aware that incomplete or insufficiently close to final manuscripts almost always result in longer refereeing times and nevertheless unclear referees' recommendations, making further refereeing of a final draft necessary.
   Authors should also be aware that parallel submission of their manuscript to another publisher while under consideration for LNM will in general lead to immediate rejection.

3. Manuscripts should in general be submitted in English. Final manuscripts should contain at least 100 pages of mathematical text and should always include
   - a table of contents;
   - an informative introduction, with adequate motivation and perhaps some historical remarks: it should be accessible to a reader not intimately familiar with the topic treated;
   - a subject index: as a rule this is genuinely helpful for the reader.

   For evaluation purposes, manuscripts may be submitted in print or electronic form (print form is still preferred by most referees), in the latter case preferably as pdf- or zipped psfiles. Lecture Notes volumes are, as a rule, printed digitally from the authors' files. To ensure best results, authors are asked to use the LaTeX2e style files available from Springer's web-server at:

   ftp://ftp.springer.de/pub/tex/latex/svmonot1/ (for monographs) and
   ftp://ftp.springer.de/pub/tex/latex/svmultt1/ (for summer schools/tutorials).

Additional technical instructions, if necessary, are available on request from lnm@springer.com.

4. Careful preparation of the manuscripts will help keep production time short besides ensuring satisfactory appearance of the finished book in print and online. After acceptance of the manuscript authors will be asked to prepare the final LaTeX source files and also the corresponding dvi-, pdf- or zipped ps-file. The LaTeX source files are essential for producing the full-text online version of the book (see http://www.springerlink.com/openurl.asp?genre=journal&issn=0075-8434 for the existing online volumes of LNM). The actual production of a Lecture Notes volume takes approximately 12 weeks.

5. Authors receive a total of 50 free copies of their volume, but no royalties. They are entitled to a discount of 33.3 % on the price of Springer books purchased for their personal use, if ordering directly from Springer.

6. Commitment to publish is made by letter of intent rather than by signing a formal contract. Springer-Verlag secures the copyright for each volume. Authors are free to reuse material contained in their LNM volumes in later publications: a brief written (or e-mail) request for formal permission is sufficient.

Addresses:

Professor J.-M. Morel, CMLA,
École Normale Supérieure de Cachan,
61 Avenue du Président Wilson, 94235 Cachan Cedex, France
E-mail: morel@cmla.ens-cachan.fr

Professor B. Teissier, Institut Mathématique de Jussieu,
UMR 7586 du CNRS, Équipe "Géométrie et Dynamique",
175 rue du Chevaleret
75013 Paris, France
E-mail: teissier@math.jussieu.fr

For the "Mathematical Biosciences Subseries" of LNM:

Professor P. K. Maini, Center for Mathematical Biology,
Mathematical Institute, 24-29 St Giles,
Oxford OX1 3LP, UK
E-mail : maini@maths.ox.ac.uk

Springer, Mathematics Editorial, Tiergartenstr. 17,
69121 Heidelberg, Germany,
Tel.: +49 (6221) 4876-8259

Fax: +49 (6221) 4876-8259
E-mail: lnm@springer.com